ENVIRONMENTAL SCIENCE

Systems and Solutions

FIFTH EDITION

Michael L. McKinney
University of Tennessee, Knoxville

Robert M. Schoch
Boston University

Logan Yonavjak

With contributions from Stacy Zell

JONES & BARTLETT
LEARNING

World Headquarters
Jones & Bartlett Learning
5 Wall Street
Burlington, MA 01803
978-443-5000
info@jblearning.com
www.jblearning.com

Jones & Bartlett Learning books and products are available through most bookstores and online booksellers. To contact Jones & Bartlett Learning directly, call 800-832-0034, fax 978-443-8000, or visit our website, www.jblearning.com.

Substantial discounts on bulk quantities of Jones & Bartlett Learning publications are available to corporations, professional associations, and other qualified organizations. For details and specific discount information, contact the special sales department at Jones & Bartlett Learning via the above contact information or send an email to specialsales@jblearning.com.

Production Credits

Chief Executive Officer: Ty Field
President: James Homer
SVP, Editor-in-Chief: Michael Johnson
SVP, Chief Marketing Officer: Alison M. Pendergast
Publisher: Cathleen Sether
Senior Acquisitions Editor: Erin O'Connor
Editorial Assistant: Rachel Isaacs
Editorial Assistant: Michelle Bradbury
Production Editor: Leah Corrigan

Senior Marketing Manager: Andrea DeFronzo
V.P., Manufacturing and Inventory Control: Therese Connell
Composition: Circle Graphics, Inc.
Cover Design: Kristin E. Parker
Rights & Photo Research Associate: Lauren Miller
Cover Image: © NaughtyNut/ShutterStock, Inc.
Printing and Binding: Courier Companies
Cover Printing: Courier Companies

To order this product, use ISBN: 978-1-4496-6139-7

Library of Congress Cataloging-in-Publication Data

McKinney, Michael L.
  Environmental science : systems and solutions / Michael McKinney, Robert Schoch, Logan Yonavjak. — 5th ed.
     p. cm.
  Includes bibliographical references and index.
  ISBN 978-1-4496-2833-8 (pbk.) – ISBN 978-1-4496-6139-7 (pac) 1. Environmental sciences. 2. Pollution–Environmental aspects. 3. Environmentalism. I. Schoch, Robert M. II. Yonavjak, Logan. III. Title.
  GE105.M39 2012
  363.7—dc23
                          2012003662
6048

Printed in the United States of America
16 15 14 13 12   10 9 8 7 6 5 4 3 2 1

To my children, Jeannie, Michael, Holli, and Maddi
MLM

To my sons, Nicholas Schoch and Edward Schoch
RMS

To my parents, Liane Salgado and Don Yonavjak
LY

Satellite Image of North America

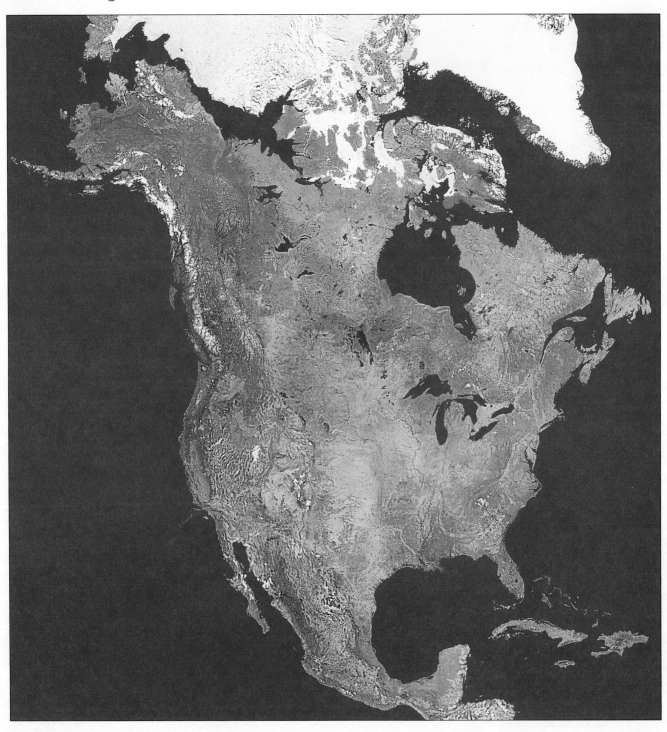

Map of North America

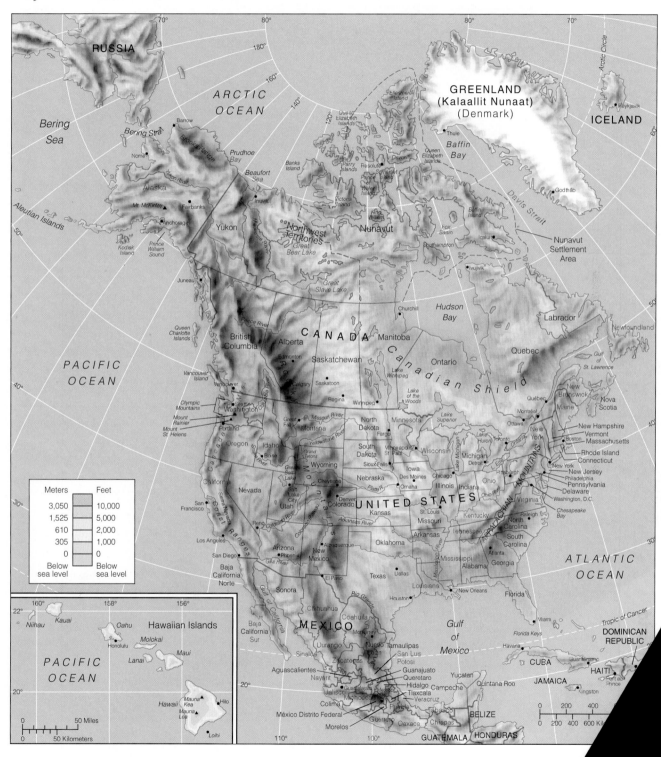

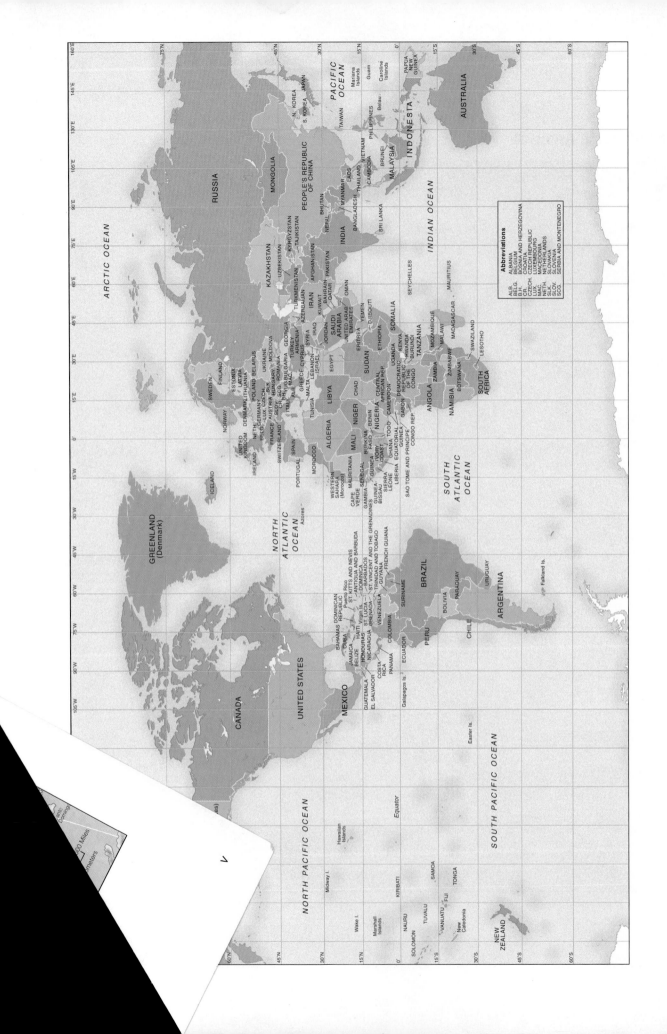

Physical Map of the World

ARCTIC OCEAN

ASIA

GOBI DESERT

HIMALAYA MTS.

HINDU KUSH

DECCAN PLATEAU

Indus R.

Ganges R.

Yangtze R.

URAL MTS.

Volga R.

Ural R.

SYRIAN DESERT

Nile R.

EUROPE

ALPS

AFRICA

SAHARA

KALAHARI DESERT

NAMIB DESERT

Cape of Good Hope

INDIAN OCEAN

PACIFIC OCEAN

AUSTRALIA

GREAT SANDY DESERT

ANTARCTICA

SOUTH ATLANTIC OCEAN

NORTH ATLANTIC OCEAN

NORTH AMERICA

CANADIAN SHIELD

APPALACHIAN MTS.

Mississippi R.

Rio Grande

ROCKY MOUNTAINS

ALASKA RANGE

Arctic Circle

SOUTH AMERICA

Amazon R.

ANDES MOUNTAINS

Cape Horn

SOUTH PACIFIC OCEAN

NORTH PACIFIC OCEAN

Tropic of Cancer

Equator

Tropic of Capricorn

Antarctic Circle

Ice cap
Tundra
Forest
Grassland
Desert
Mountains

165°E 180° 165°W 150°W 135°W 120°W 105°W 90°W 75°W 60°W 45°W 30°W 15°W 0° 15°E 30°E 45°E 60°E 75°E 90°E 105°E 130°E 145°E

75°N 60°N 45°N 30°N 15°N 0° 15°S 30°S 45°S 60°S

About the Authors

Michael L. McKinney is Director of the Environmental Studies Program at the University of Tennessee, Knoxville. He is also a Professor in the Geological Science Department and the Ecology & Evolutionary Biology Department. Since 1985, he has taught a variety of courses, focusing on environmental science and biodiversity issues at the undergraduate level.

Dr. McKinney has two master's degrees, one from the University of Colorado at Boulder and one from the University of Florida. He received his Ph.D. from Yale University in 1985. Since that time, he has published several books and dozens of technical articles. Most of his recent research has focused on conservation biology. Dr. McKinney has received several teaching awards and a prestigious University award for creative research. He is currently working on a book documenting the harmful impact of urban sprawl on native species.

In addition to his scholarly work, Dr. McKinney is very active in promoting environmental solutions where he lives, the Southern Appalachian bioregion. He is on the Board of Directors of the Foothills Land Conservancy, which is the major private land trust that creates wilderness preserves around the Smoky Mountain National Park. In 2001, Dr. McKinney received the Environmental Achievement award from the city's main newspaper, the *Knoxville News-Sentinel*, given to the individual who has done the most to promote a better environment. Dr. McKinney is also an active member of the Tennessee Citizens for Wilderness Planning, the East Tennessee Sierra Club (Harvey Broome Chapter), the Southern Alliance for Clean Energy, the Tennessee Clean Water Network, and Ijams Nature Center. He writes a bimonthly column called the "Suburban Ecologist" in the *Hellbender*, the environmental newspaper of East Tennessee.

Dr. McKinney lives in Knoxville, Tennessee, where he greatly enjoys hiking and promoting sustainable living.

Robert M. Schoch is on the faculty of the College of General Studies, Boston University, where he has specialized since 1984 in teaching undergraduate science, including environmental science, biology, physical science, geology, geography, and science and public policy courses. Dr. Schoch always includes a strong environmental component in any course he teaches. He is a recipient of his college's Peyton Richter Award for interdisciplinary teaching.

Dr. Schoch received his Ph.D. in geology and geophysics from Yale University in 1983 and is the author or coauthor of both technical and popular books, including *Phylogeny Reconstruction in Paleontology*, *Stratigraphy: Principles and Methods*, *Voices of the Rocks*, *Voyages of the Pyramid Builders*, and *Pyramid Quest*. Dr. Schoch is keenly interested in how environmental factors have helped shape ancient and modern civilizations. Along these lines, he has undertaken fieldwork in Egypt, Peru, Japan, and Bosnia. Understanding past environmental changes is important as we face future challenges.

Besides his academic and scholarly studies, Dr. Schoch is an active environmental advocate who stresses a pragmatic, hands-on approach. In this connection, he helped to found a local community land trust, and for many years sat on the board of directors, devoted to protecting land from harmful development. Likewise, Dr. Schoch takes an active part in "green" politics and for over a decade served as an elected member of the city council of Attleboro, Massachusetts.

Logan Yonavjak is a graduate of the University of North Carolina at Chapel Hill where she earned her bachelor's degree in Geography, with a Geographic Information Systems (GIS) concentration. Before attending UNC, Logan was Dr. Schoch's student at Boston University, and she traveled to Egypt and Peru as a member of expeditions related to Dr. Schoch's research.

Logan has long been committed to the conservation and preservation of the natural world. While still an undergraduate, she was involved with many activities on both of her campuses, including active membership in student government, working at a local non-profit, serving as an undergraduate representative on several sustainability-focused committees, and in 2006 helping to plan a statewide conference for North Carolina students to learn ways to reduce greenhouse gas emissions through renewable energy projects. Currently her primary focus is applying environmentally friendly and sustainable ideas, models, and techniques to city planning. She hopes to obtain a master's degree in City and Regional Planning in the next few years.

Logan currently lives in Chapel Hill, North Carolina, where she has resided most of her life.

Brief Contents

Contents

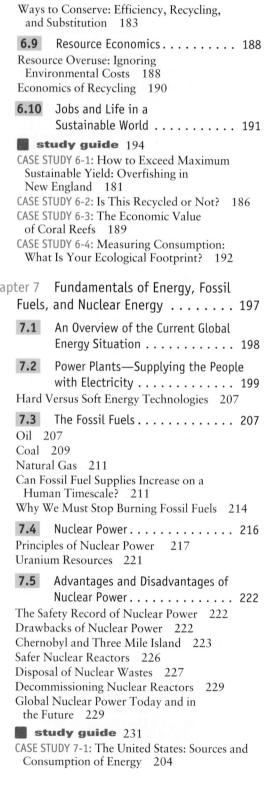

SECTION 4
Dealing with Environmental Degradation 408

SECTION 5
Social Solutions to Environmental
Concerns   576

Preface

The future which we hold in trust for our own children will be shaped by our fairness to other people's children.

—Marian Wright Edelman

Nothing is more honorable to any large mass of people assembled for the purpose of a fair discussion, than that kind and respectful attention that is yielded not only to your political friends, but to those who are opposed to you in politics.

—Stephen Douglas, from the Lincoln–Douglas debates

THE CRITICAL IMPORTANCE of environmental science and environmental studies cannot be disputed as virtually everyone is aware of the issues—be they global warming, the depletion of the ozone layer, the controversy over nuclear power, or the continuing problems of water pollution and solid waste disposal. *Environmental Science: Systems and Solutions, Fifth Edition,* offers the basic principles necessary to understand and address these multi-faceted and often very complex current environmental concerns.

We wrote this book to serve as a comprehensive overview and synthesis of environmental science. *Environmental Science: Systems and Solutions* provides the reader with the basic factual data necessary to understand current environmental issues. But to know the raw facts is not enough. A well-informed person must understand how various aspects of the natural environment interconnect with each other and with human society. We thus use a systems approach as a means of organizing complex information in a way that highlights connections for the reader. The systems approach allows the reader to take in the information without feeling overwhelmed, as often happens when large amounts of information are presented in a disorganized fashion. With a subject as diverse as environmental science, it is easy to get lost in the details. We have always kept the "big picture" in mind.

All too often environmental discussions become bogged down in partisan rhetoric or "gloom and doom" tactics. Our intention is not to preach but to inform. Accordingly, in approaching what is often an extremely controversial subject, we have adopted an objective and practical perspective that tries to highlight what is going right in dealing with modern environmental problems. Furthermore, we have consciously aimed at being both fair and balanced (presenting differing opinions and information) in our approach to many controversial issues.

A key concept among modern environmentalists is sustainability. In this book we have adopted the sustainability paradigm: We focus on sustainable technologies and economic systems and the ways that sustainable development can be implemented around the world. Our emphasis is on specific examples that can give concrete meaning to the concept: Sustainable technological and social solutions to environmental problems are discussed throughout the book. We hope to inspire the reader to move beyond simple awareness of current environmental problems to become an active promoter of sustainable solutions to those problems.

Organization and What's New in this Edition

Building on the framework of the four previous editions, we have rewritten the text to improve the discourse. Furthermore, recent disasters and noteworthy updates are reflected in this edition. We have updated case studies that cover topics relevant to the current environmental situation, including captive breeding, Hurricane Katrina, the Colorado River, sustainable agriculture practices, overpopulation concerns, pollution, and measuring ecological footprints. Additional changes include updated statistics throughout the text, revised and updated figures and tables, and more coverage of sustainability, climate change, fossil fuels, national parks, and water resources. We believe that all of these changes will make the book both more timely and more accessible to the reader.

The five sections of the book are:

Section 1, The Environment and People (Chapters 1 and 2), introduces the systems approach and gives an overview of environmental science in Chapter 1, while Chapter 2 focuses on the increasing impact that the growing human population has had on all natural systems.

Section 2, The Environment of Life on Planet Earth (Chapters 3 through 5), describes how natural systems work, including both biological systems and physical systems. Here we introduce such concepts as populations, communities, ecosystems, the distribution of life on Earth, biogeochemical cycles, weather patterns and climatic zones, the rock cycle and plate tectonics, and natural hazards.

Section 3, Resource Use and Management (Chapters 6 through 13), deals with issues surrounding the use of natural resources by human society. Chapter 6 introduces the broad principles of resource management. The following chapters address energy use, water use, mineral use, and the use of biological resources (including agriculture and soil resources). A major theme is that humans have been rapidly depleting many of these resources and that we must begin using them in a sustainable manner if we are to survive and flourish in the future.

Section 4, Dealing with Environmental Degradation (Chapters 14 through 18), concentrates on various forms of pollution and waste—the results of dumping large amounts of the by-products of human society into the environment. Chapter 14 introduces the principles of pollution control, toxicology, and risk, while subsequent chapters deal with such subjects as water pollution, air pollution, the destruction of the ozone layer, global climate change, municipal solid waste, and hazardous waste. Every chapter includes discussions of how we can limit or mitigate the effects of excessive pollution, especially by limiting the production of pollutants in the first place, as well as by increased efficiency, reuse, recycling, and substitutions.

Section 5, Social Solutions to Environmental Concerns (Chapters 19 and 20), includes discussions of economic, social, historical, and legal aspects of environmental issues. A major emphasis of the book is on solutions to current environmental concerns. Woven throughout the text are discussions and examples of environmentally friendly technological, legal, and economic solutions. We firmly believe that sustainable and realistic solutions must be implemented and that the root causes of the environmental problems we now face must be addressed. Such problems cannot be solved using science and technology alone; the human aspect must also be taken into account.

Using This Book for a Course in Environmental Science or Environmental Studies

We designed this book to be accessible to introductory non-major students, but it has enough depth and breadth to be used in a majors course. It can be adapted to either an environmental science course or an environmental studies course, and it can be used for either one or two semesters. Also, we designed the book so that the chapters need not necessarily be used in the order in which they appear. In particular, depending on the nature and emphasis of a specific course, an instructor may choose to use the chapters of Section 5 (Social Solutions to Environmental Concerns) at either the beginning or end of the course, or these or other chapters may be omitted entirely.

Assuming a standard 15 full weeks for a semester (usually about a week is lost due to holidays, exams, and the like), the chapters of this text might be assigned according to one of the following schedules:

For a comprehensive environmental science and environmental studies course:

Week 1: Chapters 1 & 2, An Overview of Environmental Science and Human Population Growth

Week 2: Chapter 3, An Overview of the Biosphere (including Populations, Communities, and Ecosystems) and Biogeochemical Cycles

Week 3: Chapters 4 & 5, The Distribution of Life on Earth and the Physical Environment of Earth

Week 4: Chapter 6, People and Natural Resources

Week 5: Chapter 7, Fossil Fuels and Nuclear Energy

Week 6: Chapter 8, Renewable (including Hydropower) and Alternative Energy Sources

Week 7: Chapters 9 & 10, Water and Mineral Resources

Week 8: Chapter 11, Conserving Biological Resources

Week 9: Chapter 12, Land Resources and Management

Week 10: Chapter 13, Food and Agriculture

Week 11: Chapters 14 & 15, Principles of Pollution Control and Water Pollution

Week 12: Chapter 16, Local and Regional Air Pollution

Week 13: Chapter 17, Destruction of the Ozone Layer and Global Climate Change

Week 14: Chapter 18, Municipal Solid Waste and Hazardous Waste

Week 15: Chapters 19 & 20, Economic, Historical, Social, and Legal Aspects of Current Environmental Concerns

For a basic environmental science course:

Week 1: Chapters 1 & 2, An Overview of Environmental Science and Human Population Growth

Week 2: Chapter 3, An Overview of the Biosphere (including Populations, Communities, and Ecosystems) and Biogeochemical Cycles

Week 3: Chapter 4, The Distribution of Life on Earth

Week 4: Chapter 5, The Workings of Planet Earth and Natural Hazards

Week 5: Chapter 6, People and Natural Resources

Week 6: Chapter 7, Fossil Fuels and Nuclear Energy

Week 7: Chapter 8, Renewable (including Hydropower) and Alternative Energy Sources

Week 8: Chapters 9 & 10, Water and Mineral Resources

Week 9: Chapter 11, Conserving Biological Resources

Week 10: Chapter 12, Land Resources and Management

Week 11: Chapter 13, Food and Agriculture

Week 12: Chapters 14 & 15, Principles of Pollution Control and Water Pollution

Week 13: Chapter 16, Local and Regional Air Pollution

Week 14: Chapter 17, Destruction of the Ozone Layer and Global Climate Change

Week 15: Chapter 18, Municipal Solid Waste and Hazardous Waste

For a general environmental studies course (emphasizing social and historical aspects):

Week 1: Chapter 1, An Overview of Environmental Science

Week 2: Chapter 17, Destruction of the Ozone Layer and Global Climate Change—Examples of the impacts humans are having on the environment

Week 3: Chapter 20, Historical, Cultural, and Legal Aspects of Current Environmental Concerns

Week 4: Chapter 2, Human Population Growth

Week 5: Chapter 6, People and Natural Resources

Week 6: Chapter 7, Fossil Fuels and Nuclear Energy

Week 7: Chapter 8, Renewable (including Hydropower) and Alternative Energy Sources

Week 9: Chapters 9 & 10, Water and Mineral Resources

Week 10: Chapters 11 & 12, Conserving Biological Resources and Land Resource Management

Week 11: Chapter 13, Food and Agriculture

Week 12: Chapters 14 & 15, Principles of Pollution Control and Water Pollution

Week 13: Chapter 16, Local and Regional Air Pollution

Week 14: Chapter 18, Municipal Solid Waste and Hazardous Waste

Week 15: Chapter 19, Environmental Economics

If this book is used for a two-semester course, some of the chapters should be used over a period longer than one week. In particular, we recommend that the following chapters be split as indicated and extended over two weeks:

Chapter 3, Populations, Communities, and Ecosystems/Biogeochemical Cycles

Chapter 4, Evolution and Biodiversity/Biomes on Earth

Chapter 5, General Physical Environment/Natural Hazards

Chapter 7, Fundamentals of Energy & Fossil Fuels/Nuclear Energy

Chapter 8, Renewable Energy/Energy Conservation

Chapter 13, Food/Soil Resources

Chapter 14, Pollution Control/Toxicology

Chapter 17, Destruction of the Ozone Layer/Global Climate Change

Chapter 18, Municipal Solid Waste/Hazardous Waste

Chapter 20, Historical and Social Perspectives/Environmental Law and Decision Making

If these chapters are used as suggested, then chapter or subchapter readings from the text will easily fit into a two-semester schedule (approximately 30 full weeks).

Pedagogical Features

Each chapter uses the same basic organizational format. Following an opening photograph and learning objectives, the chapter begins with an introduction that offers an overview of the subject matter of the chapter and places it in context.

We have written the text to be interesting and accessible to the average reader, and we have profusely illustrated it with diagrams, charts, tables, and photographs demonstrating basic concepts and key ideas. Throughout the text key terms denoting important concepts are in **boldface** type.

A Study Guide at the end of each chapter includes a bulleted summary, a list of the chapter's key terms, and several kinds of questions. Answers to the odd-numbered questions are on the text's website at http://environment.jbpub.com/mckinney/5e.

- Study Questions. These review questions generally test objective knowledge and require fairly short answers. Some require more analytical and critical thinking skills.
- What's the Evidence? Unique to this book, these innovative questions ask the reader to

review the authors' arguments and decide whether the authors have successfully supported their conclusions. The questions then challenge the reader to support their own position if they disagree.

- Calculations. Calculation questions, written at a pre-calculus level or lower, are provided for courses that have a quantitative component.
- Illustration and Table Review. These questions are designed to help readers strengthen their data interpretation skills.

This book includes several special features. On pages iv and v are maps of North America showing the physical geography and political boundaries of all the states and provinces of the United States, Mexico, and Canada. On pages vi and vii are physical and political maps of the world. These maps will serve as handy reference guides for the reader when various states, provinces, and countries are mentioned in the text. It is increasingly important that everyone be familiar with basic global political geography.

Section Opening Model. We use a global systems framework to reinforce the interconnectivity of natural and human systems and the importance of the input-throughput-output concept. The systems model, first shown in the Introduction (Figure Intro.-1), is repeated at the beginning of each later section to highlight the connections and the book's organization.

Case Studies. Thought-provoking case studies provide detailed examples of interesting environmental applications, experimental work, and controversies. Many of the Case Studies provide follow-up questions that ask the reader to examine the facts, the arguments, and the conclusions.

Fundamentals boxes. This set of boxes reviews quantitative information covering human population equations and statistics, energy and thermodynamics, and common measures of energy and power.

On-line appendices. The text's website includes selected major pieces of U.S. environmental legislation, selected pieces of international environmental legislation, and a list of selected environmental organizations and government agencies. We have placed them on-line rather than in the text so they can be more readily updated. These appendices should be useful to students as they peruse the text or decide to delve deeper into environmental issues.

The book concludes with English/Metric Conversion Tables, a glossary of key terms, and a detailed index.

Web Enhancement

Environmental Science: Systems and Solutions, Fifth Edition, is supported by an extensive website at http://environment.jbpub.com/mckinney/5e/. Students will find a variety of study aids and resources, all designed to explore in more depth the basic science, concepts, and controversies of environmental science.

The central learning component of the site is an interactive study guide with activities that will help students make full use of today's learning technology. Students will find practice quizzes, virtual flashcards of key terms, crossword puzzles, and a fully searchable glossary. The site also includes answers to the study questions from end of each chapter and web links to useful independent websites that offer additional coverage or alternative perspectives on material discussed in the text.

Online Appendices. To save on paper and allow for updating, we've placed the following appendices online. Links to websites offering information on careers in environmental science and environmental studies are also included in this area.

- Selected Major Pieces of U.S. Environmental Legislation
- Selected Pieces of International Environmental Legislation
- Selected Environmental Organizations

Ancillary Materials For the Instructor

To assist you in teaching this course and supplying your students with the best in teaching aids, Jones & Bartlett Learning, in conjunction with Stacy Zell of the University of Rio Grande in Rio Grande, Ohio, has prepared a complete ancillary package available to all adopters of the text. Additional information and review copies of any of the following items are available through your Jones & Bartlett Learning Sales Representative.

The Instructor's Manual, provided as a downloadable text file, includes complete chapter lecture outlines, learning objectives, discussions of common student misconceptions, and answers to the even-numbered study questions in the text.

The test bank is available as electronic text files. The test bank contains approximately 2,000 multiple-choice, true/false, fill-in-the-blank, matching, short-answer, analogy, and quantitative questions.

Instructor's Media CD

The Instructor's Media CD provides instructors who have adopted the text with the following traditional ancillary materials, all of which are cross-platform for Windows and Macintosh systems.

The PowerPoint Lecture Outline presentation package provides lecture notes, graphs, and images for each chapter of *Environmental Science*. Instructors with the Microsoft PowerPoint software can customize the outlines, art, and order of presentation.

The PowerPoint Image Bank provides a library of all of the art, tables, and photographs in the text to which Jones & Bartlett Learning holds the copyright or has permission to reproduce.

Michael L. McKinney
Robert M. Schoch
Logan Yonavjak

Acknowledgments

As authors, we are ultimately responsible for the content of this book, but dozens of people have provided help, encouragement, and advice. In particular, we are grateful for the advice of many teachers and practitioners of environmental science. Due to its depth and breadth, environmental science contains far more information than only three people can master, and we drew heavily on the expertise of people who have specialized in its many subfields. We therefore wish to express our deep appreciation to the reviewers of various editions of this book:

Clark E. Adams	Texas A & M University	James R. Karr	University of Washington
David A. Adams	North Carolina University	Michael G. King	College of the Redwoods
John W. Adams	University of Texas, San Antonio	Clifford B. Knight	East Carolina University
Michael Albert	University of Wisconsin, River Falls	Cindy M. Lee	Clemson University
		Jack Lutz	University of New Hampshire
Preston Aldrich	Benedictine University	Timothy F. Lyon	Ball State University
Sara E. Alexander	Baylor University	Theodore L. Maguder	St. Petersburg Junior College
Gary L. Anderson	Santa Rosa Junior College	Kenneth E. Mantai	SUNY, Fredonia
Richard D. Bates	Rancho Santiago College	Heidi Marcum	Baylor University
Deborah Beal	Illinois College	Priscilla Mattson	Middlesex Community College
Mark C. Belk	Brigham Young University		
Charles F. Bennett	University of California, Los Angeles	W. D. McBryde	Central Texas College
		Kevin McCartney	University of Maine, Presque Isle
William Berry	University of California, Berkeley	Richard L. Meyer	University of Kansas
Keith Bildstein	Winthrop College	Henry R. Mushinsky	University of South Florida
Jennifer Cole	Northeastern University	Muthena Naseri	Moorpark College
Gerald Collier	San Diego State University	Arnold L. O'Brien	University of Massachusetts, Lowell
Harold Cones	Christopher Newport University	Charles E. Olmsted	University of Northern Colorado
Carl F. Chuey	Youngstown State University		
Darren Divine	University of Nevada, Las Vegas	Ogochukwu Onyiri	Campbellsville University
		Nancy Ostiguy	California State University, Sacramento
Mary Lou Dolan	St. Mary-of-the-Woods College	Richard A. Paull	University of Wisconsin, Milwaukee
Lorraine Doucet	University of New Hampshire, Manchester	Adrienne Peacock	Douglas College
JodyLee Estrada Duek	Pima Community College, Desert Vista	Charles R. Peebles	Michigan State University
		R. H. Pemble	Moorhead State University
Nicholas P. Dunning	University of Cincinnati	Chris E. Petersen	College of DuPage
L. M. Ehrhart	University of Central Florida	Dennis M. Richter	University of Wisconsin, Whitewater
Gina Gupta	Clemson University		
Jason Haugland	Adams State College	Gordon C. Robinson	University of Manitoba
George W. Hinman	Washington State University	C. Lee Rockett	Bowling Green State University
Gary J. James	Orange Coast College		
Robert L. Janiskee	University of South Carolina	Paul Rowland	Northern Arizona University

Robert Sanford	University of Southern Maine	Jerry Towle	California State University, Fresno
David B. Scott	Dalhousie University	Lee B. Waian	Saddleback College
Joseph Shostell	Pennsylvania State University, Fayette	Linda Wallace	University of Oklahoma
Jon Stanley	Metropolitan State College of Denver	Joel Weintraub	California State University, Fullerton
Ray Sumner	Los Angeles Valley College	Frank Williams	Langara College
R. Bruce Sundrud	Harrisburg Area Community College	Richard J. Wright	Valencia Community College
		Bruce Wyman	McNeese State University
Peter G. Sutterlin	Wichita State University	Todd Yetter	University of the Cumberlands
Christopher Thoms	Knox College	Craig ZumBrunnen	University of Washington
S. Carl Tobin	Utah Valley State College		

We thank all of the wonderful people who made the revision and updating of this book possible; in particular, we must mention Joanne Revak at Circle Graphics and Leah Corrigan, Rachel Isaacs, Erin O'Connor, and Lauren Miller at Jones & Bartlett Learning.

We would also like to give a special thanks to Stacy Zell of the University of Rio Grande in Rio Grande, Ohio, who updated all of the statistics, art, and case studies in the fifth edition, and who offered invaluable assistance throughout the production of this book. Without her help, this edition would not have come to fruition.

Finally, we are deeply indebted to the patience and support of our families and friends while we worked on this book. In particular, RMS thanks his good friend and colleague Colette M. Dowell for her encouragement as he worked on this revision. LY thanks her mother, Liane Salgado, for instilling in her an appreciation of the environment from a young age.

MLM
RMS
LY

The Environment and People

"We stand now where two roads diverge. But . . . they are not equally fair. The road we have long been traveling is deceptively easy, a smooth superhighway on which we progress with great speed, but at its end lies disaster. The other fork of the road—the one less traveled by—offers our last, our only chance to reach a destination that assures the preservation of the earth."

Rachel Carson (1907–1964), environmentalist and author, *Silent Spring*

Introduction: A Guide to the Environment

This text provides an overview of the natural environment and the way people are increasingly affecting it. Through an understanding of environmental science, you will be in a position to make sound decisions that will affect the future of the entire planet. An important aspect of genuine understanding is the ability to relate seemingly diverse and disparate facts and phenomena to one another within a larger picture. A truism in environmental science is that everything is related to everything else. The environment is a finely tuned, grand, and extremely complex system in which if one component is modified or tampered with there will be ramifications in other, often unexpected, portions of the system. Therefore, it is particularly beneficial to use a **systems approach** in analyzing global environmental issues (discussed in more detail later in this chapter).

FIGURE INTRO.-1 shows how the text is subdivided into the five major sections (each containing a number of chapters). These sections reflect the input–output, or throughput, processes that can be used to describe the global environment, how the environmental system is modified, and solutions to environmental problems (such as the depletion of resources or pollution of nature) brought on by human activities.

- *Section 1: The Environment and People.* The natural environment and the people living on planet Earth form a complex, integrated system, as introduced here. The environmental science overview and discussion on human population growth also describe the explosive growth of the human population, the development of modern technological and industrial society, and some of the ramifications that these have had for the natural environment. Many environmentalists consider human population growth, or more precisely the impact of people on the environment, to be the single most important issue that the world currently faces. Studying Figure Intro.-1, you will see that conceptually

Population **Consumption**

FIGURE INTRO.-1 A system can be technically defined as a "set of components functioning together as a whole." A systems view allows us to isolate a part of the world and focus on those aspects that interact more closely than others. For example, a cell in the system we call a human body generally interacts much more closely with other cells in the body than with the outside world. By focusing only on the cells that function in digestion, we confine our view further to the digestive system. The key point here is that most systems are hierarchical: They are composed of smaller sets of systems made of smaller interacting parts.

1. The Environment and Humans (Chapters 1–2)

Lithosphere **Biosphere**

Hydrosphere **Atmosphere**

2. The Environment of Life on Planet Earth (Chapters 3–5)

Lithosphere **Biosphere**

Hydrosphere **Atmosphere**

3. Resource Use and Management (Chapters 6–13)

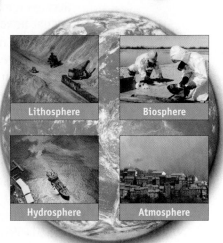

Lithosphere **Biosphere**

Hydrosphere **Atmosphere**

4. Dealing with Environmental Degradation (Chapters 14–18)

Lithosphere **Biosphere**

Hydrosphere **Atmosphere**

5. Environmental Issues: Social Aspects and Solutions (Chapters 19–20)

FIGURE INTRO.-2 Even parts of remote Antarctica, shown here, have become dumping grounds for human waste.

the raw elements of the natural environment pass through human society to produce the global environment in which we currently live. For all practical purposes, there is no pristine wilderness remaining anywhere on Earth (**FIGURE INTRO.-2**).

- *Section 2: The Environment of Life on Planet Earth.* Earth and life form an interconnected system based on energy flow and matter cycles. Study of these natural systems includes the disciplines of biology, chemistry, geology, climatology, and many other natural sciences. Learning the basic principles of these fields permits a better understanding of natural systems and how they can be used, protected, or disturbed. People use, modify, and are dependent on these natural systems.

- *Section 3: Resource Use and Management.* Simply by using resources, we run the risk of overusing and depleting them. Depletion of environmental resources is one of the two basic types of environmental disturbance. It includes depletion of water, minerals (including fossil fuels), soils, wild-

life and wildlife habitat, and many other resources. To avoid depletion problems, we must learn to manage our resources in a more sustainable manner. Issues surrounding depletion and resource scarcity are discussed in Section 3.

- *Section 4: Dealing with Environmental Degradation.* Discharging waste into the environment is the other basic type of environmental disturbance. It includes pollution of air, water, land, and biological communities, but there are ways to use resources productively without the levels of pollution and concomitant environmental damage that have been so widespread in the past. Environmentally friendly options to address problems of pollution are explored in Section 4.

- *Section 5: Social Solutions to Environmental Concerns.* Technological solutions are insufficient unless society is willing to use them. Social solutions are interwoven throughout this text, but these final chapters allow us to focus on specific social institutions and their legal and economic aspects. Ultimately, through a combination of scientific analysis, technological advances, and the continuing development of social institutions, humanity must lessen its detrimental impact on the environment if we are to avoid living on an impoverished planet in a very bleak postindustrial age.

The global environment of the future (represented in Section 5; refer to Figure Intro.-1) will not be the preindustrial environment of the past (represented in Section 2) because we (Section 1) have a tremendous impact on the environment (Sections 1, 3, and 4). Our actions today will have a profound effect on the environment of the future. By our deeds we will determine whether the environment of the future will be more or less hospitable to us and numerous other species.

Ancient earth and stonework in the Peruvian Andes. See Case Study 13-1.

Much of the world's land area has been disturbed by humans. Historically, profit-driven environmental decisions and unintentional consequences, such as industrial waste seepage into waterways, has left a negative impact on local biodiversity and the sustainability of related ecosystem services. In recent years, the increase in protected areas and community involvement aided by social media, as well as the current trend toward more sustainable business practices, have begun to reverse damage to the environment in some areas. Still, the loss of agricultural land to increased desertification, the mismanagement of critical resources such as water, and the persistent political unrest in many of the world's most critical areas present an ongoing call to action for global citizens committed to the process of preserving and protecting the array of ecosystem services that support life on earth.

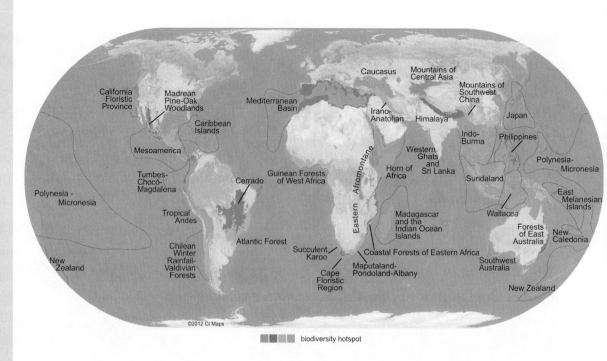

biodiversity hotspot

1

Environmental Science: An Overview

Chapter Objectives

After reading this chapter, you should be able to explain or describe the following:

- What is meant by environmental science
- How the environment works as a system
- The key traits of the environmental system
- The concepts of consumption and exponential growth
- A brief history of environmental movements
- The concept of a commons
- Throughput
- What is meant by true costs and why people avoid them

The critical importance of environmental science and environmental studies cannot be disputed. Virtually everyone is aware of environmental issues—global warming, the depletion of the ozone layer, the controversy over nuclear power, the global loss of species and biodiversity, the introduction of hormone-altering pollutants into the natural environment, the use of and controversy surrounding genetically engineered crops, or the continuing problem of solid waste disposal. No citizen of the Earth can afford to be ignorant of environmental issues.

Our planet has existed for more than 4 billion years, yet never before has one species dominated the Earth and other species so completely. Humanity now stands at a unique crossroads: because of human technological capabilities and impacts, the next few decades will witness drastic changes in the Earth and its inhabitants. It is our individual responsibility to try to influence the outcome of such changes so that the welfare of both people and the environment is best served.

Although this text presents many important facts, it also has three larger goals. One is to help you sort through the huge amount of environmental information available and focus on important "core" issues. Today, especially with the widespread availability of the Internet and the World Wide Web, little bits of information on almost any environmental issue are readily available; nevertheless, even as we become overwhelmed with factoids, the bigger picture, a larger perspective, is often lost. It often can be difficult to distinguish the truly important from the merely interesting or even the downright trivial. As the old saying indicates, it can be hard to see the forest because all of the trees block the view.

Seeing the big picture will help you to understand differing points of view of individuals or groups who have economic, political, or even personal stakes in the possible solutions. Environmental issues are complex. With every solution comes a different set of potential advantages and disadvantages. The second goal of this text then is to help you understand the problems, evaluate the solutions with all the stakeholders in mind, and make up your own mind.

The third goal of this text is to get you, the reader, seriously thinking about how you can put an understanding of environmental issues to work to effectively help your household, community, or nation make the fundamental changes needed to build a world that can sustain many generations of people with a decent standard of living, while minimizing the human impact on the natural environment. We encourage you to become actively involved in environmental concerns.

What Is Environmental Science?

Environmental science involves all fields of natural science as they bear on the physical and biological environment around us. Aspects of biology, geology, chemistry, physics, meteorology, and many other disciplines must be considered when studying environmental science. A major component of modern environmental science involves addressing current environmental problems brought about directly by human activity. Therefore, in addition to presenting the scientific concepts underlying environmental issues, we also discuss social aspects of environmental problems. Not only technology and scientific understanding, but also laws, ethics, economics, and other aspects of human behavior will play a key role in solving our current environmental dilemmas.

Avoiding "Information Overload"

One of the challenges we face today is how to cope with information overload. Newspapers, radio, television, and the Internet/ World Wide Web bombard us daily with data and statistics. Feeling overwhelmed, most people react by "tuning out."

Rather than quit trying to assimilate all this information, we might recall Henry David Thoreau's advice to "simplify! simplify!" Thoreau has become a hero to many environmentalists because he long ago predicted many of the problems that we now face. His advice to simplify was a response to what he saw as a tendency for civilization to become increasingly more complex and removed from the natural world. This, he said, led to anxieties and spiritual impoverishment despite material wealth. On a pragmatic level, it also leads to a weakened and damaged natural environment.

We can heed Thoreau's advice by mentally stepping back and keeping our priorities set on what we think is important. The systems approach to environmental science greatly heeds the process. In this way, we can focus on the information that we can use, instead of trying to learn it all (which no one can do). Ideally, we can strive to seek what might be called **environmental wisdom**. Wisdom is the ability to sort through all of the facts and information to make reasonable decisions and plan sensible long-term strategies. Wisdom is gained through education and practical experience, so environmental wisdom takes time to develop. Wisdom also means that we take a broad view in solving problems and weigh all kinds of information, social and economic as well as scientific and technical. Environmental science is often called **holistic**, meaning that it seeks connections among all aspects of a problem.

A lack of environmental wisdom is costly in many ways. It is costly to other species, to our quality of life, to future generations, and often to human happiness itself; however, currently the most easily measured costs are economic. A good example is illustrated in **FIGURE 1-1**, which shows that money spent on environmental problems does not always correlate directly to the actual magnitude of the problem. For instance, problems related to land preservation and restoration receive more money than those associated with global change (such as fixing the hole in the ozone layer or addressing clean air issues related to the increased incidence of asthma in children living in urban

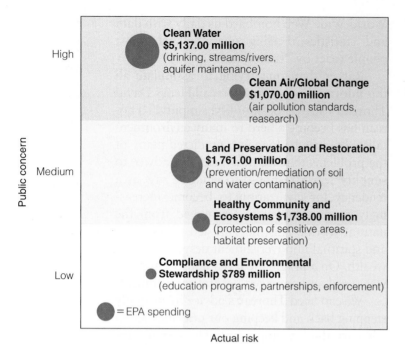

High

Medium

Low

Public concern

Clean Water
$5,137.00 million
(drinking, streams/rivers,
aquifer maintenance)

Clean Air/Global Change
$1,070.00 million
(air pollution standards,
reasearch)

Land Preservation and Restoration
$1,761.00 million
(prevention/remediation of soil
and water contamination)

Healthy Community and
Ecosystems $1,738.00 million
(protection of sensitive areas,
habitat preservation)

Compliance and Environmental
Stewardship $789 million
(education programs, partnerships, enforcement)

= EPA spending

Actual risk

FIGURE 1-1 The actual risk of a hazard compared to the amount of public concern about the risk. There is not always a direct correlation between the actual magnitude of a problem or its associated risks and the amount of money spent to mitigate the problem. As a result, large amounts of money (size of dot) are spent on issues with less direct risks, such as soil remediation projects, while at the same time problems such as air pollution and global climate change with more immediate consequences receive less money. (*Source:* U.S. Environmental Protection Agency and World Water.org.)

areas). Such spending inefficiencies occur because people often lack adequate information about the true risk of environmental problems.

Building a Sustainable World

Many environmentalists point out that past efforts have tended to focus on short-term emergency actions rather than long-term so-

FIGURE 1-2 Cleaning up toxic waste is an expensive after-the-fact approach. It is much cheaper to design methods that produce less waste.

lutions. Examples of this approach include cleanup of wastes and pollution after they are produced (**FIGURE 1-2**) and trying to save species only when they are nearly extinct. Besides being less effective, such piecemeal, late-acting remedial solutions are almost always the most expensive way to solve environmental problems.

In the last decade, the rising costs and inefficiencies of cleaning up pollution after it is produced have led to a search for better approaches to solving environmental problems. Generally, holistic approaches have been able to solve problems more cheaply and efficiently. By examining society and the environment as an interconnected system, we often can solve many problems at once. The **input** of materials and energy through society is commonly called **throughput**. Environmental resources (or inputs) are referred to as **sources** of throughput. Environmental reservoirs that receive throughput are called **sinks** and are the ultimate repository of societal **output**, which could become input and throughput again, as in recycled materials. As illustrated in **FIGURE 1-3**, environmental problems arise from (1) resource depletion and (2) pollution. Past efforts at pollution control were largely "end-of-pipe" solutions, cleaning up waste after it was produced. Reducing the flow of material through society can also control pollution. Such **input reduction**, which conserves resources and reduces pollution at the same time, is now widely accepted by environmental economists and others as a better solution to most environmental problems. Input reduction illustrates the kind of fundamental change needed to build a society that can be maintained for many years without degrading the environment. The works of chemist Michael Braungart and architect and designer William McDonough illustrate this last point. Together they have put to use nature's practice that nothing goes to waste and in the end waste equals food. Through their "cradle to cradle" certification process, factories and other businesses worldwide are proving that industrial design that promotes sustainability through reduced negative impact on the environment can occur through

Natural Resources

Environmental Problems:
Resource Depletion
or Saturation (pollution)

Input
High or low use

Output
Possibilities

Or

Throughput
Population   Consumption

Sustainability

SOURCES

SINKS

FIGURE 1-3 People use resources from the natural environment and deposit wastes back into the natural environment. Materials and energy that people draw from are known as "sources," and environmental reservoirs that receive the products of human society are known as "sinks." Environmental problems can arise from resource depletion and pollution of sinks. Controlling consumption and reducing the flow of material through society can alleviate many environmental problems.

creative problem-solving and ingenuity, and these solutions are profitable for the company and society that commit to them.

Figure 1-3 also identifies the two basic causes of most current environmental problems: human population pressures and consumption of material goods. Increases in population and consumption have led to increased resource depletion and pollution. Both need to be addressed if long-term solutions are to be achieved.

1.1 The Environment as a System: An Overview

The entire environment of planet Earth can be divided into four spheres (refer to Figure Intro.-1): the biosphere (living organisms of Earth), atmosphere (the gaseous envelope surrounding the planet), hydrosphere (the liquid water on the surface of Earth), and lithosphere (the stony or rocky matter composing the bulk of the surface of Earth). Matter cycles and energy flows through these spheres (**FIGURE 1-4**). A systems approach provides a convenient overview and manner of understanding how our environment works.

What Is a System?

A **system** can be defined technically as a "set of components functioning together as a whole." A systems view allows us to isolate a part of

the world and focus on those aspects that interact more closely than others. For example, a cell in the system we call a human body generally interacts much more closely with other cells in the body than with the outside world. By focusing only on the cells that function in digestion, we confine our view further to the digestive system. The key point here is that most systems are *hierarchical*: they are composed of smaller sets of systems made of smaller interacting parts.

Three Key Traits of the Environmental System

We can analyze the global environment in terms of three system traits: openness, integration, and complexity.

Openness refers to whether a system is isolated from other systems. An **open system** is not isolated in that it exchanges matter and/or energy with other systems. A **closed system** is isolated and exchanges nothing.

The law of entropy means that energy cannot be fully recycled; rather, "high-quality" energy is degraded to "lower quality" energy (such as waste heat). Therefore, any system that does not have a renewing supply of energy from outside eventually will cease to exist. Not surprisingly, the Earth is an open system in terms of energy. Figure 1-4 shows how energy flows from the sun and is often radiated back into space. In contrast, the

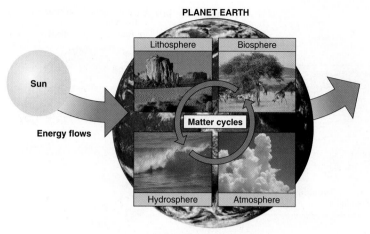

PLANET EARTH

Sun

Energy flows

Lithosphere

Biosphere

Matter cycles

Hydrosphere

Atmosphere

FIGURE 1-4 The Earth is open to energy, which flows into and out of it. However, the Earth is essentially closed to matter, which cycles over and over through the four spheres.

Earth, for all practical purposes, is a closed system in terms of matter. If we discount the relatively small amount of matter added from meteorites and other space debris, the Earth contains all of the matter it will ever have. Driven by energy from the sun, this matter cycles over and over among the four spheres, often moving back and forth among the gaseous, liquid, and solid states and participating in the metabolism of living things.

Integration refers to the strength of the interactions among the parts of the system. For instance, the human body is a highly integrated system whose cells are interdependent and in close communication. The loss of certain cells, such as those composing the heart or brain, can result in death of all of the other cells in the system (the whole organism) because the cells are so interdependent. At the other extreme are systems with very weak integration, such as the cells in a colony of single-celled organisms (such as the green algae *Volvox*). Removal of many cells (to a point) will have little effect on the remaining cells because they are less dependent on each other.

The degree of integration of the global environmental system is a matter of debate. At one extreme are scientists who argue that the global system is a superorganism: many complex pathways intimately interconnect the lithosphere, hydrosphere, atmosphere, and biosphere. According to this **Gaia hypothesis**, the Earth is similar to an organism, and its component parts are so integrated that they are like cells in a living body. However, many

scientists believe the global environment is less integrated than the Gaia hypothesis argues. This does not mean that the environment is "unconnected" or even as weakly connected as a colony of cells. As we describe in future chapters, many kinds of matter cycles and energy flows interconnect the spheres and cycle within the spheres as well. The true level of integration in the global system probably is somewhat less than a "superorganism" but considerably more than a loose collection of independent parts.

Complexity is often defined as how many kinds of parts a system has. This definition conforms to our intuition: a tiny insect seems more complex to us than a large rock because it has many more types of "parts." The insect has more complex molecules, as well as more different types of molecules, and these are used to construct cells and organs. This example also illustrates that complexity is often hierarchical, with smaller components being used to construct larger ones.

As you would expect, the environment is enormously complex. The four spheres, with their matter cycles and energy flows, have trillions and trillions of different components operating at many spatial and temporal scales. Organisms, soils, rainwater, air, and many other components interact in complicated ways. The individual spheres themselves are equally complex. Even with advanced computers, no one has been able to predict accurately and precisely the weather, or even climate, very far into the future because the atmosphere is so complex. Indeed, the many interactions make unpredictability a basic characteristic of complex systems (**CASE STUDY 1-1**). This inability to predict how the environment will respond to changing conditions (which is why it is so dangerous to tamper) is perhaps the major reason for so much controversy about and inaction with regard to environmental problems.

Major Obstacles: Delayed and Unpredictable Impacts. Social responses to environmental problems are greatly hindered by two of the key traits of the environmental system: its moderate integration and high complexity. Any system that is integrated, such as the environment, can transmit disturbances from one part of the system to another. Integration results from connectedness so that resource depletion or pollution of one part of the environment can have cascading, or domino, effects into other parts. For example, burning sulfur-rich coal affects the atmosphere as air pollution, but it also affects the hydrosphere when it falls as acid rain to acidify lakes. The biosphere is also affected because aquatic organisms in the lake can die from the more acidic lake water. The burning coal can even affect the lithosphere when the acid rain dissolves limestone and other alkaline rocks to form caves and sinkholes. This example shows how just one activity, burning coal, can affect all four spheres of the environment. Such wide-ranging cascading effects are anything but rare, as we see in later chapters. This connectedness of the environment means that virtually any action has a number of consequences, many of which are unforeseen and unintended. Such cascades are so important that the biologist Garrett Hardin has formulated the **first law of ecology:** "We can never do merely one thing." This is often called the "law of unintended consequences."

That the environment is only moderately integrated ("loosely connected") greatly hinders our ability to observe, predict, and ultimately correct the unintended consequences of our actions. The indirect connections and interactions in the environment create delays, or long lag times, before cascading effects become visible (**FIGURE 1-5**). Returning to the acid rain example, it may take many years of coal burning before the trees on mountaintops or fish populations in lakes are affected. In other cases, the impacts can be delayed for centuries, millennia, or even longer. Environmental impacts on a global scale usually take an especially long time to occur. Many decades and probably centuries will pass before the full effects of added atmospheric carbon dioxide on global temperature are observable. This lag time is one of the main reasons for the debate over global warming. To make matters more complicated, corrective measures may themselves have unanticipated consequences.

As we have seen, the complexity of the environment leads to unpredictability, which hinders social responses to environmental

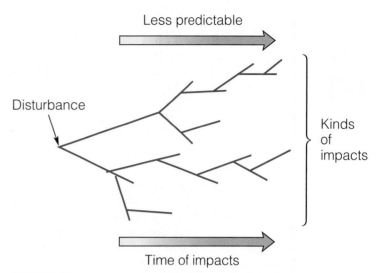

FIGURE 1-5 "We can never do merely one thing." Whenever we do something, we cause a cascade of impacts. These impacts become less predictable with time.

problems. As Figure 1-5 illustrates, the unintended cascades we cause not only take a long time, but also occur as unexpected, complicated chains of events. Take the example of global warming and disease. Strong evidence has shown for many years that many tropical diseases are spreading because of global warming. In 1993, an article was published in *The Lancet* concluding that a 1991 cholera outbreak in South America was related to localized warming of Pacific Ocean waters caused by global climate changes. The article argued that the warming had caused the rapid growth of plankton that harbors the cholera bacterium, leading to thousands of deaths (**FIGURE 1-6**). Since then, many other studies have linked global warming with the

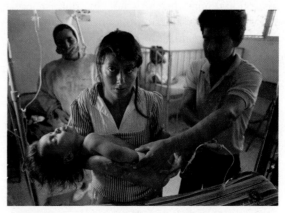

FIGURE 1-6 In a hospital in Ecuador, a mother holds her cholera-stricken son. The cholera outbreak occurred in Ecuador in June 1991.

1.1 The Environment as a System: An Overview

Systems and Chaos Theories: Ways to Study Complexity

http://environment.jbpub.com
/mckinney/5e/
for more information

Studying complex systems can be difficult because they have many parts that often interact in different ways. Over the last few decades, researchers have developed several methods of studying complexity. For example, systems theory and chaos theory try to produce general "laws" of complexity. Such laws would not only make complex systems more understandable, but they also would allow us to predict more accurately how these systems will behave. Think of how important such predictions could be for a complex system such as the stock market or for weather and climate forecasting!

None of these theories has been entirely successful in providing a complete understanding of complexity or producing accurate predictions of how any particular complex system will behave. Nevertheless, these theories have provided a better idea of how complex systems will generally behave under a given set of conditions. Consequently, we have a general understanding of how a lake ecosystem will respond to excess nutrients even though we cannot specify every event.

Systems theory (or general systems theory) was one of the first widely used attempts to find "laws" of complex systems. It traditionally has focused on how systems are regulated and become unregulated. Systems theory treats a complex system as a "black box" with inputs and outputs. Such a system is kept at equilibrium by negative feedback processes, defined as processes that counteract perturbations. An example is a thermostat that turns a furnace on to produce heat when a house is cold and turns on air conditioning when it is hot. In contrast, positive feedback processes amplify perturbations. For example, a cooling global climate can cause more snow to remain on the ground, which leads to more global cooling because the snow reflects sunlight back into space. This in turn causes yet more snow, and so on (a snowball effect).

Systems theory often has been criticized as too general or vague because by treating a system as a "black box," the theory omits many of the details of how the system operates. More recent efforts to study complex systems have focused on more mathematical, rigorous descriptions of them. One theory that has received much attention is chaos theory. A chaotic system is one in which the workings are sensitive to even the slightest change: this slight perturbation can become greatly amplified through positive feedback. The classic example is the weather, as E. Lorenz, who helped establish chaos theory in the early 1960s, first described. Lorenz created a set of equations that precisely described atmospheric conditions and showed how even tiny changes in one of the parameters could cause a massive alteration of the weather in a few days. This is often called the "butterfly effect," with the idea that the flapping of a butterfly's wings in South America could eventually affect the weather in North America (**FIGURE 1**). By creating tiny changes in atmospheric turbulence, which in turn creates cascading effects on larger airflows, the butterfly's wing flapping could have a major impact. (Of course, the chances that it actually will produce such a major impact are extremely low.)

Chaos theory's most important finding so far is that even simple systems, such as several atoms, often have chaotic properties. How can we make predictions when a minute unseen change, as in the butterfly effect, can have cascading effects? Many theorists think it will always be impossible to predict accurately precise future behaviors in complex systems, which have even more potential for chaos than do simple systems. Nevertheless, by applying chaos theory, patterns of regularity can be discerned and studied.

FIGURE 1 Can a butterfly in the rain forest affect the weather in North America?

spread of diseases. In a 2007 article in the *Online Journal of Issues in Nursing*, Brenda Afzal indicated that not only is insect-borne illness a challenge, but warming trends leading to migration of insects into new habitats and other factors such as daily air quality caused by intensified pollution also are becoming increasingly urgent issues for global health as Earth's climate continues to change. The World Health Organization and the National Institutes of Health indicate that these trends are not found only in developing nations: West Nile virus, which made its comeback in the United States in 1994, is now found in 44 states, and more alarming is the fact that the mosquitoes that carry dengue fever are found in counties within the United States where 173.5 million Americans live.

Along with safe food, water scarcity, and the impact of extreme weather on communication and transportation systems, U.S. intelligence officials list the spread of disease as one of their top four climate-change–related security concerns.

In addition to having many interactions, complex systems are unpredictable because some of the interactions exhibit positive feedback. **Positive feedback** occurs when part of a system responds to change in a way that magnifies the initial change. For example, evidence indicates that a slight increase in average global temperature can cause a further increase by melting some of the glaciers and snow that reflect sunlight back into space. Instead of reflecting light, more of the Earth's surface becomes available to absorb heat. Another example is poverty in developing countries that results from overpopulation; the poverty leads to high reproductive rates and thus still more overpopulation. In nontechnical terms, positive feedback is often referred to as a "snowball effect" or "vicious circle."

Society in the Environmental System

FIGURE 1-7 illustrates how modern society is embedded within the environment, being dependent on it for the materials and energy needed to maintain civilization. It also shows how industrialized society accelerates the cycling of

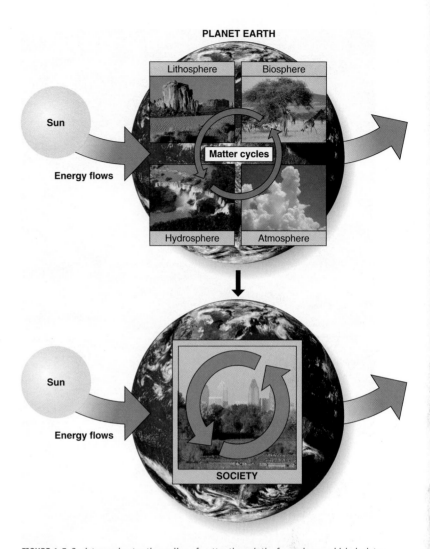

FIGURE 1-7 Society accelerates the cycling of matter through the four spheres, which depletes resources (matter inputs to society) and causes pollution (matter outputs by society). This acceleration is increasing because of increasing population (P) and consumption (C). Slowing population growth and reducing personal consumption will reduce this acceleration.

matter and the flow of energy through itself and the four spheres.

Fundamentally, all environmental problems involve either depletion of sources (consumption) or pollution of sinks (production of waste). Therefore, these two processes can help us measure the net environmental impact of society. Depletion occurs when the accelerated cycling and flow remove matter and energy faster than natural processes are renewing them. Conversely, pollution occurs when the accelerated cycling and flow are discharged into the environment, overwhelming the local natural purification processes (**FIGURE 1-8**).

1.1 The Environment as a System: An Overview   13

FIGURE 1-8 Pollution occurs when natural purification processes are overwhelmed, such as by large amounts of nutrients or poisons. Shown here is acid rain damage.

Problems of environmental depletion and pollution both exhibit the delayed and unpredictable impacts we have discussed. Much of the remainder of this text is concerned with these issues (Figure Intro.-1).

1.2 What Is Environmental Impact?

Environmental impact refers to the alteration of the natural environment by human activity. We have already identified two basic types of environmental impact: resource depletion and pollution (Figure 1-3). In other words, there is either too much input (resources)

and/or too much output (pollution). Input reduction seeks to slow both depletion and pollution, whereas output reduction slows only pollution.

Population and consumption are two main forces accelerating resource depletion and pollution. The following equation is a simple way to remember this:

$$\text{Impact} = \text{Population} \times \text{Consumption}$$
$$\text{or } I = P \times C$$

It is easy to see why the number of people affects impact (**FIGURE 1-9**). Consumption historically has tended to increase the effect of each person. For example, a baby born in the United States has many times the impact of a baby born in a developing country because over a lifetime the U.S. resident consumes many more resources, close to 20 times more, such as fossil fuels, and produces much more pollution. Although the United States accounts for only about 5% of the world's population, researchers have pointed out that the United States consumes nearly a quarter of all the energy used worldwide.

Current research indicates that population growth rates remain high but have begun to decrease slightly for most industrialized countries, while continuing to rise extremely rapidly in many developing nations. Today (2011), 95% of all population growth occurs in areas most challenged by poverty, civil unrest, malnutrition, and the lack of resources needed to disrupt the cycle. Despite differences in growth rate, consumption of nonrenewable

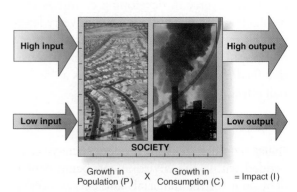

FIGURE 1-9 Personal consumption (C) has increased the throughput per person. This is multiplied by the number of people (population or P) to obtain the total resource depletion and pollution produced by society.

resources continues to increase at alarming rates. This worldwide trend has led to a rapid increase in environmental impact. This section briefly examines both of these factors.

Exponential Growth

Multiplicative processes cause **exponential growth** of any kind. It occurs in population growth because biological reproduction is inherently multiplicative: most organisms have the ability to produce more offspring than are necessary to simply replace the parental generation. If conditions allow, the population of each successive generation will be larger than the previous generation. Consider, for example, a pond where the algae cover starts from a single algal cell and doubles in size each day. For a long time, you would see nothing happening. However, after larger areas were covered, the algae would spread quickly: one day, the pond would be only half covered; the next day, it would be completely covered.

Exponential Growth of the Human Population

The current world population (2011) stands at approximately 7 billion people. For millions of years, relatively few people were on Earth at any given time. The total human population on Earth remained quite low until the development of agriculture. Even then, population growth did not become explosive until the 1900s, when it was aided by the global spread of industry and modern medicine. Currently, each year the world population experiences a net gain of approximately 77 million people, the equivalent of nearly one-third of the population of the United States. Since 1960, the world population has more than doubled. At the current world population growth rate (estimated to be approximately 1.2%), the world population will double again in less than 60 years. In the last 50 years, there has been more population growth than in the entire history of human existence. Even if population growth rates stabilize around the year 2050, as many scientists predict they might, world population would still be about 9 billion people, an increase of approximately 30% over the 2011 population. The stabilization of population growth rates is speculative, based on several factors, including increasing access to education and birth control for women in developing nations, increasing financial assistance and programs promoting more sustainable ways of life, and natural causes such as diseases assisting in the stabilization of the population. Because population growth is most dramatic in the most politically unstable regions of the world, there is disagreement about the accuracy of stability models. One fact remains, Earth's resources are being consumed at rates never before seen with consequences that are hard to predict.

How long will human population growth continue before it encounters environmental limitations? The answer depends on how many people the Earth can support, a question that is much debated, in part because the answer depends on how high a standard of living one assumes. However, many estimates predict that the Earth can adequately and sustainably support between 6 and 8 billion people at a reasonable standard of living. If this is true, population growth clearly must decline very soon if **overshoot** is to be avoided. Almost all of the United Nations, the World Bank, and other organizations' population projections indicate that world population probably will not stabilize until it reaches 9 billion, sometime in the middle to late 21st century.

Exponential Growth of Consumption

The second basic factor in our impact equation, $I = P \times C$, is consumption. Here, we use the term **consumption** in the economic sense to refer to the purchase and use of material goods, energy, and services in our daily lives. Consumption includes all of the things we do and buy to meet our wants and needs. Consumption has increased the overall environmental impact by increasing the impact per person:

$$\text{Overall impact} = \text{population} \times \text{consumption}$$
$$= \text{number of individuals}$$
$$\times (\text{impact/individual})$$

This impact per person occurs through an increase in both resource depletion and pollution. Because natural resources are used to

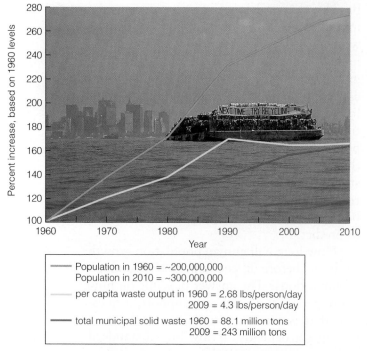

FIGURE 1-10 Trends in U.S. total solid waste production, output per person, and population. The background photo shows Greenpeace protesters on a New York garbage barge. (*Sources:* The chart is based on data from the U.S. Environmental Protection Agency and the U.S. Bureau of the Census.)

Chart axis label: Percent increase, based on 1960 levels

Legend:
- Population in 1960 = ~200,000,000
  Population in 2010 = ~300,000,000
- per capita waste output in 1960 = 2.68 lbs/person/day
  2009 = 4.3 lbs/person/day
- total municipal solid waste 1960 = 88.1 million tons
  2009 = 243 million tons

X-axis: Year (1960, 1970, 1980, 1990, 2000, 2010)

make the increasing amount of things we buy, consumption has tended to increase per capita resource use and pollution. **FIGURE 1-10** illustrates that solid-waste output has grown dramatically in the United States in the last few decades.

However, personal consumption in wealthy nations could be reduced easily with no reduction in quality of life. For example, many new sustainable or "green" technologies are being developed that can greatly reduce individual impact. More efficient technologies, such as fuel-efficient cars, use fewer resources and produce less pollution when an individual uses them. Alternative technologies can eliminate many impacts altogether. For example, replacing coal-burning machines with solar-powered machines conserves nonrenewable fossil fuels and eliminates many air pollutants that burning coal releases. Other current trends include encouraging pharmaceutical plants to recycle solvents instead of incinerating them and fast food restaurants to use completely compostable dining ware made from potato starch and the increased use of home energy audits to reduce waste and energy consumption.

Exponential Growth of Environmental Impact

The exponential increases of population and traditional industrial technology (which correlate with increasing consumption) have caused an exponential increase in environmental impact. The increase in total solid waste shown in Figure 1-10 is typical of the pattern seen in many other kinds of pollution. A society that produces such waste is likely to be consuming many resources that ultimately generate the waste. For example, global consumption of fossil fuel has risen exponentially because of the growing world population and the spread of technologies that use fossil fuels (increasing per capita use of fuels). As consumptive technology and population grow, more materials and energy move through society. This accelerates the depletion of environmental resources. In addition, the materials and energy that move through society must have somewhere to go when society is finished with them. Solid waste, air and water pollution, and other outputs ultimately end up in the environment.

In summary, there are two basic kinds of environmental impacts: depletion and pollution. Both have increased because growing consumption of resources increases pollution as throughput is accelerated. The main causes of increased throughput have been the growth of (1) population and (2) consumption.

1.3 A Brief History of Environmental Impact

Considering that our human species (*Homo sapiens*) has inhabited Earth for 100,000 or more years, the current massive global environmental impact of people is a very recent development. Indeed, our relationship with the environment has changed dramatically as we and our technology have evolved.

Conceptually, we can distinguish five basic developmental stages in the relationship between people and their environment (**FIGURE 1-11**). The economic activity people engage in using the technologies available to them at the time largely determines these stages. In turn, this activity affects how people have an impact on the environment. At any one time on Earth, different human populations may

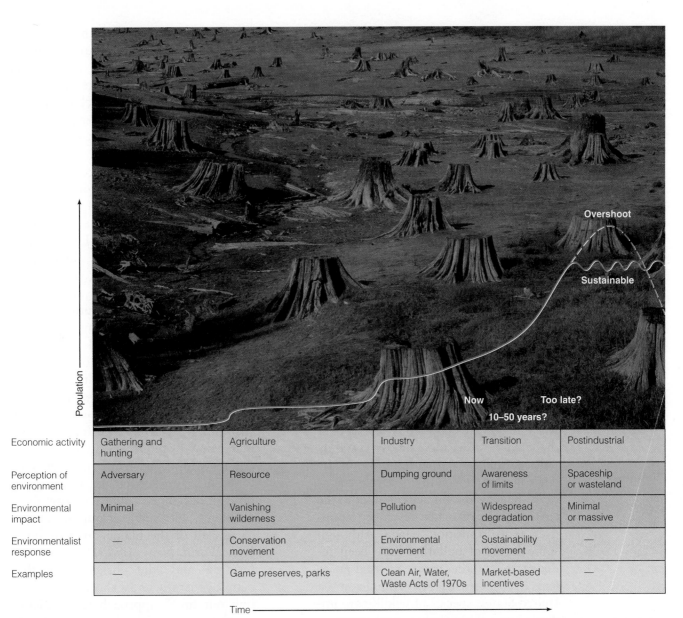

Economic activity	Gathering and hunting	Agriculture	Industry	Transition	Postindustrial
Perception of environment	Adversary	Resource	Dumping ground	Awareness of limits	Spaceship or wasteland
Environmental impact	Minimal	Vanishing wilderness	Pollution	Widespread degradation	Minimal or massive
Environmentalist response	—	Conservation movement	Environmental movement	Sustainability movement	—
Examples	—	Game preserves, parks	Clean Air, Water, Waste Acts of 1970s	Market-based incentives	—

Time ⟶

FIGURE 1-11 Population and personal consumption have increased the human impact on the environment in an exponential fashion. Environmentalists have responded with the conservation movement, the environmental movement, and the sustainability movement. Many people believe that we are in a transition stage and that people have no more than 50 years to prevent "overshoot" and attain sustainability.

be in different stages simultaneously, yet the general global trend has been to progress through these stages.

1. *Gathering and hunting.* Early humans were largely at the mercy of their environment, so they might have viewed it in adversarial terms. Weather, predators, food shortages, and disease were constant threats. However, during this stage, humans also had a familiarity with the environment that in general has been lost.

Population levels typically remained low, and any human-induced effects on the environment (such as purposefully setting forest fires) were relatively localized.

2. *Agriculture and conservationism.* The shift from hunting and gathering to cultivating food is one of the most profound milestones in human evolution. It allowed a great increase in population size and permitted people to settle in large towns and cities; however, agriculture also had a major impact on the environment. People

began to view land as a resource to be exploited as needed. As land was cleared and cultivated, the wilderness vanished.

3. *Industry and environmentalism.* As nations industrialize, population grows faster, and the environment is increasingly perceived as both a source of raw materials and a place to dispose of the concentrated waste by-products of industry. The result is a rapid increase in air and water pollution, as well as problems with solid and hazardous waste disposal. Toward the end of this stage, pollution becomes so widespread that antipollution social movements emerge. In the United States, these social movements began in the early 1960s and peaked in the 1970s. When people talk about classic "environmentalism," this antipollution movement may be what they mean.

4. *Transition and sustainability.* Today we live in a time of rapid change and technological advancement. The decisions we as a global community make over the next few decades almost certainly will determine the long-term fate of the environment for many future generations. For this reason, we can refer to the current stage of many developed countries as a "transition" stage. Although some forms of pollution have been reduced, many other environmental problems have increased. For example, in the United States, species of wildlife are imperiled as habitat is destroyed unnecessarily. Recently released reports indicate that the catastrophic events surrounding the April 2010 Deepwater Horizon rig explosion and oil well blowout in the Gulf of Mexico that killed 11 people were preventable and caused primarily by lack of communication, poor management, and profit-driven greed. Such reports underscore the urgency for sustainable industry practices grounded in policies, decisions, and practices that hold stewardship for the environment in higher regard. The actions we undertake in the first part of the 21st century will affect subsequent generations' sustainability in ways seen at no other time

in history. Among other challenges we face are groundwater contamination, including endocrine-altering by-products from pharmaceutical waste that interrupt normal adolescent development and the presence of many thousands of hazardous and radioactive waste sites that likely will not be cleaned up for centuries. Despite efforts in recycling and precycling (the reduction of packaging material by manufacturers), the amount of solid waste produced per person remains high. Globally, the Environmental Protection Agency has cited global warming, ozone depletion, and increasing species extinction as the greatest environmental threats to future generations. As developing nations grow, it is important for them to look beyond the often-inadequate example of developed nations' use of the environment. Current models for sustainable development include technological advances that are sustainable by design. Both the rapidly increasing populations in developing countries and the increasing demands for goods and services by the populations of developed countries cause these problems. No single group or nation carries all of the blame. In their book *New World, New Mind* (1989), Robert Ornstein and Paul Ehrlich argue that people generally had a short-term, "putting out fires" approach to problems; another term might be a "band-aid" approach. Only recently, with the sustainability movement, has there been widespread interest in addressing the systematic causes that underlie these problems. **Sustainability** means meeting the needs of today without reducing the quality of life for future generations. This includes not reducing the quality of the future environment. Sustainability can be achieved in many ways, such as through the use of sustainable ("green") technologies that use renewable resources (for instance, solar power) and recycling or upcycling (converting waste materials or useless products into usable materials or better products) of material components. These technologies allow a **sustainable economy** that produces wealth and provides jobs for many human generations without degrading the environment. An example of the practical application of sustainable design is looking at materials as natural components of an ecosystem. Nature recycles and does not waste. Waste in nature ends up as the basic components of food, achieved by natural decomposition of biological by-products. Industry that commits to material reuse, renewable energy use, water stewardship, and social responsibility can qualify for the "Cradle to Cradle" recognition, which means that the manufacturer commits to safe materials that can be reused or composted and accepts the responsibility of business-as-stewards in the use of environmental materials. Not only does this model make sense for future generations, but it also is profitable for business owners.

5. *Postindustrial stage—sustainability or overshoot?* The next (future) stage of society has been called the "postindustrial" stage. Just what form postindustrial society will take is open to speculation and debate, but two possible alternatives have been proposed. One is a sustainable future, where the human population stabilizes and technology becomes less environmentally harmful. The second alternative is overshoot, where population climbs so high and technology is so harmful that the environment is degraded to the point that relatively few people can be supported, at least with a decent standard of living. Which alternative comes into being could depend on the collective actions of humanity during this first half of the 21st century.

1.4 The Environment as a Commons

In 1968, the biologist Garrett Hardin wrote the famous essay "The Tragedy of the Commons." He argued that property that many people hold in common will be destroyed or at least overused until it deteriorates. He gave the example of a pasture where each herdsman in the village can keep his cattle. The herdsman who overgrazes the most will also benefit the most. Each cow that a herdsman adds will benefit the owner, but the community as a whole will bear the cost of overgrazing. Because the benefit of adding another cow goes to the individual and the cost of overgrazing goes to the community, the "rational" choice of each individual is to add cows. Thus, the commons rewards behaviors that lead to deterioration, such as overgrazing, and punishes individuals who show restraint. Those who add fewer cows will simply obtain fewer benefits while the commons itself deteriorates because of the individuals who continue to add cows. This problem with common property was known long before Hardin's eloquent essay. For instance, the ancient Greek philosopher Aristotle noted, "What is common to the greatest number has the least care bestowed upon it."

Hardin's pasture exemplifies the problems that arise when any part of our natural environment is treated as common property. However, privatization can lead to private exploitation of resources as well, and the key is well-managed property, whether private or public. Public space is important but must be taken care of. Unless there is some kind of regulation, overexploitation will likely occur via both input and output impacts.

1. Commonly held resources, such as the pasture, will become depleted through excessive consumption.
2. Commonly held environmental sinks will be overwhelmed by pollution.

Many local, regional, and global environmental problems illustrate this view of the commons as a source or a sink:

	Problem
Atmosphere as global common sink	Global warming, ozone loss
Atmosphere as regional common sink	Acid rain
Atmosphere as local common sink	Urban smog
Ocean as global common sink	Ocean pollution
Ocean as global common resource	Many fish species overfished
Rain forest as common resources	Global warming promoted by deforestation, biodiversity reduced by deforestation

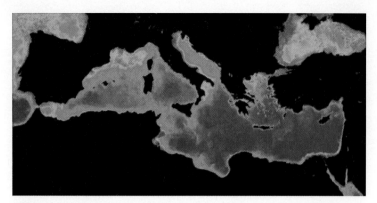

FIGURE 1-12 A computer-enhanced image of pollution in the Mediterranean Sea. Red, yellow, and orange areas are concentrations of plankton growth promoted by discharge of raw sewage. (Blue indicates clear water.) Many coastal cities lack sewage treatment.

Many nations share some of these commons. The international nature of many environmental problems adds greatly to the complexity of solving them because international agreements are required. Consider the Mediterranean Sea. A confined shallow ocean basin surrounded by many nations, this sea is one of the most over fished and polluted large bodies of water on Earth (**FIGURE 1-12**).

1.5 Saving the Commons

If we are to save the environmental commons, the throughput of matter and energy through all societies must be reduced. This will slow both depletion and pollution. Increasing population (P) and consumption (C) have driven increased impact (I), so their growth must be reduced.

We are in the transition stage (as previously discussed). Environmental problems have become so widespread that they demand large-scale regional and global solutions involving many aspects of society. Furthermore, countries such as China and India are in the process of rapid industrialization that consumes a tremendous amount of Earth's resources. It is important that these economies learn to work in a sustainable manner instead of using their resources in a wasteful and inefficient manner, as do the United States and other countries. Likewise, the United States must improve. Beginning in the early 1980s, a sustainability movement

emerged to try to deal with these problems. Unlike the conservation and antipollution movements of the past, which emphasized specific problems rather than a global approach, the sustainability movement seeks long-term coexistence with the environment. As mentioned, sustainability means meeting the needs of today without reducing the quality of life for future generations.

The sustainability movement uses three approaches not attempted by previous environmental movements. First, it focuses explicitly on trying to reduce society's use of all resources; therefore, emphasis is on input reduction, as opposed to end-of-pipe solutions. *Waste is viewed as a symptom, not a cause, of the environmental crisis.*

Second, the sustainability movement is more holistic. It realizes the necessity of addressing the social, and especially economic, causes of environmental degradation. This has led to an increasing appreciation of the role of poverty and other economic factors that encourage people to deplete resources and pollute. Market-based solutions are becoming more popular, and less emphasis is placed on the legal solutions used in the past. For example, many experts now agree that it is often more effective and cheaper for society to tax coal, gasoline, and other polluting substances than to pass laws specifying how much pollution may be emitted. For instance, higher gasoline prices encourage people to drive less or buy fuel-efficient cars.

Third, the sustainability movement has encouraged the growth of thousands of local community action groups, as opposed to the national groups that dominated the conservationism and environmentalism periods. Many of the national and international environmental organizations arose well before 1980. For instance, the Sierra Club was founded in 1892, the National Audubon Society in 1905, the National Wildlife Federation in 1936, and the World Wildlife Fund in 1961. Since the exposure of the toxic dump at Love Canal, New York (1978), and the Three Mile Island nuclear accident in Pennsylvania (1979), residents of communities have become increasingly active in addressing local environmental problems. Vocal debates over incinerators, landfills, land development, mountain top removal, water contamination, and many other environmental issues now often dominate the local news (**FIGURE 1-13**). Such local participation is often called **grassroots activism**. Recently, grassroots groups with common interests have begun to network by establishing regional and national newsletters and computer nets. By the mid-1990s, many of the national organizations had begun to experience a decline in membership. Although some people suggest that this decline means environmental interest is waning in the United States, others argue that it simply reflects the transfer of environmental allegiance from national to grassroots organizations.

All three characteristics of the sustainability movement historically arose from the desire to find better ways to solve widespread environmental problems. Grassroots activism is often the best way to deal with local issues. The rise of input reduction and economic approaches reflects the need to reduce social costs. Although litigation and lobbying are still used, the sustainability movement also uses direct action and lifestyle changes. In keeping with its more holistic approach, many members of the sustainability movement are also **ecocentric** ("environment centered") instead of **anthropocentric** ("human centered"). In other words, they seek to preserve nature for reasons beyond simply improving the quality of human life. Nature is viewed as having a high "intrinsic" value that is not related to human needs. Nature is worth protecting for its own sake. As the sustainability movement spreads, addressing the needs of the 21st century, it will continue to develop many new strategies and approaches to further the goal of sustaining the environment for the generations that follow.

Reducing Consumption: Defusing the Bomb of the North

Rapid population growth is often described as a "time bomb" that will greatly degrade the environment in coming years. Although population growth is especially rapid in the Southern Hemisphere, the industrialized nations of the Northern Hemisphere are also contributing to environmental degradation through their habit of excessive and wasteful

FIGURE 1-13 Citizen protest is an effective and growing way to promote local environmental sustainability.

resource consumption developed during a time of abundant resources. Consumption may be viewed as the environmental "time bomb" of the North. The solution is to reduce consumption or C in the impact equation $I = P \times C$.

There are three basic ways to reduce consumption. None of these ways needs to involve painful self-sacrifice or a lowered quality of life. In contrast, reducing consumption can improve the quality of life in many ways by improving human health and the environment.

- *Reduce material needs.* Psychological studies have statistically documented what many religions have long taught: increasing a person's material wealth will rarely produce an increase in happiness or contentment. The "treadmill of wealth" says that the more we accumulate, the more we want. When Americans are surveyed, most respond that they have far too much "stuff."
- *Use less technology to meet our needs.* Long before Thoreau, many people observed that some machines might have detrimental effects on people. Riding bicycles to work (instead of driving cars), buying products with fewer artificial chemicals and less superfluous packaging, and using fewer unnecessary convenience appliances are but a few ways that people have reduced their reliance on technology.

- *Use sustainable technology to meet our needs.* Another way to reduce technology's impact is to use technologies that are much more "environmentally friendly" than the fossil fuel-based, industrial technologies of the past. **Sustainable technology** permits people to meet their needs with minimum impact on the environment. It produces a sustainable economy that provides jobs for many generations without degrading the environment.

There are many kinds of sustainable technologies, ranging from direct solar and wind power to recycling. Although advances in pollution cleanup technologies are often heralded, true sustainability results from input reduction: slowing resource depletion also slows pollution. The three basic ways that sustainable technologies achieve input reduction, or conservation, are (1) efficiency improvements, (2) reuse and recycle, and (3) substitution.

Efficiency improvements reduce the flow of throughput by decreasing the per capita resource use. The United States, more than almost any other country, uses technologies developed during times of abundant resources. As a result, it generates much waste, providing enormous opportunities to save many resources by relatively simple changes in existing technologies. The amount of waste is so great that, in many cases, the large amount of resources saved will more than compensate for the cost of investment to make the change.

Let us use the example of energy, which drives all economies. It is estimated that changes in U.S. technology, ranging from high-mileage cars to superefficient heating and cooling systems to new lighting technologies, could reduce overall energy consumption by as much as 80%. For example, the incandescent light bulb converts 95% of the electricity it uses to waste heat instead of light. Compact fluorescent bulbs (CFLs) generate 70% less heat and use between 50% and 80% less energy to produce light. According to the Environmental Protection Agency, if every household in the United States replaced incandescent lights with compact fluorescent lights, the society would benefit by a reduction of greenhouse gas equivalent to removing 800,000 cars from the road. Recent discussion about the use of compact fluorescent bulbs often mentions that the bulbs contain mercury. Although it is true that the bulbs contain a small amount of mercury (5 grams; a little less than the weight of a modern quarter), The Union of Concerned Scientists points out that the bulbs can be recycled and that the energy savings must take into account that "the average coal-fired power plant emits only 3.2 milligrams of mercury for each CFL running six hours per day for five years, but emits nearly 15 milligrams of mercury for an incandescent bulb running the same amount of time" (www.ucsusa.org/publications/greentips/let-there-be-fluorescent.html December 13, 2011). Even more efficient are light-emitting diode (LED) lights that are becoming more readily available as costs decrease, making them more attractive to consumers. LEDs can last as long as 50,000 hours, more than eight times as long as CFLs, and also contain no mercury, making this evolution of technology a promising answer to home energy efficiency challenges.

In addition to reducing environmental damage, efficiency improvements have two immediate economic advantages. First, as noted, the improvement usually pays for itself. The cost of a technological change that increases efficiency is usually recovered within a few years and sometimes almost immediately. For instance, a study by a major beverage company found that each reduction of one thousandth of an inch (0.001) reduces aluminum costs by $1 million. (Of course, there is a limit to how thin aluminum cans can be and remain structurally sound.) In addition, increased efficiency tends to produce many more jobs than do wasteful technologies. The traditional energy industries, such as oil, coal, and nuclear power, are much less labor intensive than the energy-conservation industry that designs and maintains many kinds of energy-conservation equipment.

Many other resources in the United States could be used much more efficiently. Here are just a couple of examples:

1. *Wood.* The United States converts only about 50% of raw timber (compared with 70% in Japan) directly into furniture and other refined timber products.

CHAPTER 1 Environmental Science: An Overview

2. *Water.* The United States uses more than twice as much water per person as most other nations on Earth. The average U.S. farmer typically could cut water use by more than half by adopting water-conservation measures, such as micro-irrigation that pipes water directly to crops instead of using evaporation-prone irrigation ditches. In the home, installing a low-flow toilet can save a family of four more than 1,350 gallons of water per month.

Reuse and recycling are the second best ways to accomplish input reduction. **Reuse** refers to using the same resource over and over in the same form. An example would be soda bottles that are returned, sanitized, and refilled. **Recycling** refers to using the same resource over and over but in modified form. For instance, the soda bottles could be melted to produce new soda bottles. Wastewater, paper, plastics, and many other resources can be recycled. In general, reuse is less costly than recycling because the resource is not modified. Both measures are often less costly than extracting "virgin" resources, such as aluminum ore or cutting trees for paper, because they usually consume less energy and fewer natural resources than does making products from virgin materials. For example, recycling aluminum cans saves as much as 95% of the energy cost of cans made from newly mined aluminum. Recycling and reuse are less costly in both economic and environmental terms than is using natural raw materials.

Substitution of one resource for another can benefit the environment in a number of ways. A renewable resource can be substituted for a nonrenewable one, or a less-polluting resource can substitute for a highly polluting resource. Often the newly substituted resource provides both benefits. Substituting renewable, cleaner alternative fuels, such as solar and wind energy, for fossil fuels is an example. Another example is making products from paper instead of plastics, which are generally made from fossil fuels and last longer in the environment (although there are now biodegradable plastics, and plastics produced from plants have been developed recently). Like conservation, reuse, and recycling,

substitution often yields economic benefits. Studies routinely show that substitution of solar and wind power technologies for the fossil fuel energy now consumed in the United States would produce three to five times as many jobs as now exist in fossil fuel industries.

Some people are surprised to learn that many sustainable technologies were invented centuries ago. Wind power and water power are examples. The solar cell was invented in the 1950s. These technologies have failed to become widespread largely for social and economic reasons, not technical ones. When nonrenewable resources are cheap, they will be wasted because people have no incentive to increase efficiency, recycle/reuse, and substitute renewable resources. The current interest in solar energy has reached record highs, with industry journals indicating a 26% increase in jobs in the solar industry throughout 2011. This is particularly impressive when considering the current average unemployment rate (2011) is as great as 10% in some states. Combining other renewable energy industry data results in trends predicting approximately 2.5 million people by 2025 will have jobs related to renewable energy. Cost of installation is becoming more attractive to consumers, as are incentive plans and infrastructure improvements. For instance, more than 42 states and the District of Columbia now have net metering rules that allow an individual owner of a solar energy system to sell excess electricity back to the grid. In other nations such as Germany and Japan where governmental and private incentive plans are in place, growth in renewable energy soared in recent years.

1.6 Promoting Sustainability by Paying the True Costs of Nature's Benefits

The economic forces that promote population growth and overconsumption have arisen because the goods and services that the environment provides, such as water purification, erosion control, and carbon sequestration, have been undervalued in the past. These and other benefits collectively are known as **ecosystem services.** Essentially, the true costs of using the environment as both a source and a sink have not been incorporated into global economic activity. This trend, according to the Millennium Ecosystem Assessment (a 2005 United Nations-sponsored report designed to assess the state of the world's ecosystems), has led to a degradation of more than 60% of the ecosystem services that were evaluated degraded over the last 50 years. **FIGURE 1-14A** shows the situation when resources are cheap and sinks are free and treated as "commons." There is much throughput, which leads to high rates of resource depletion and pollution. It also contributes greatly to the rapid population growth of developing nations. Poverty is strongly correlated with high population growth rates, and people in resource-rich tropical developing nations are often underpaid for their resources (as compared with prices in many developed nations).

Market prices that do not reflect all the true costs of a product or service are called **market failure** (or "externality") by economists. Most ecosystem services are either undervalued or considered free in our current economic system. Many environmental problems ultimately can be traced to market failures. A few examples are listed:

1. The cost of gasoline does not include the cost of smog and other urban air pollutants, contamination of groundwater by leaking underground oil tanks, oil spills, and many other well-known impacts of gasoline use.
2. Electricity from nuclear energy does not include the cost of disposing of nuclear waste; electricity produced by coal burning omits the cost of most air pollution.
3. Water used by many U.S. farmers does not reflect the fact that in many regions the groundwater is being depleted much faster than it is being replenished by rainfall.

Many economists suggest society can correct market failures by imposing **green fees** that adjust the costs of products and services to include environmental costs (**FIGURE 1-14B**). Fees that increase the price of a resource are particularly effective because they promote conservation of the resource and also reduce pollution by reducing throughput. For example, a "gas tax" or "carbon tax" covering all fossil fuels would encourage reduced and more efficient use of these fuels. Charging for garbage by the bag motivates people to reduce the waste they produce.

Another emerging solution to address market failures is known collectively as payments for ecosystem services (PES), which involve offering incentives to landowners in exchange for managing their land to provide an ecological service, such as carbon sequestration and storage, water quality, endangered species habitat, and wetlands. PES programs are contracts between the consumers of ecosystem services and the suppliers of these services. The beneficiaries of the ecosystem services are willing to pay a price that is lower than their welfare gain because of these services, whereas the providers of the service are willing to accept a payment that is greater than the cost of providing the

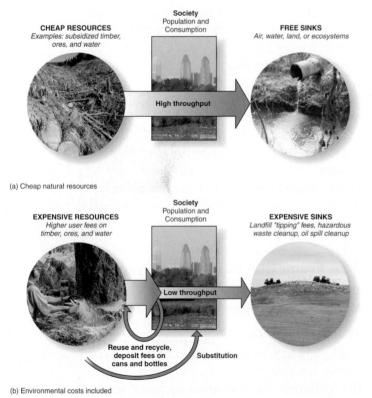

(a) Cheap natural resources

(b) Environmental costs included

FIGURE 1-14 (a) Excluding environmental costs by allowing cheap resources and free sinks promotes high throughput (and thus much depletion and pollution). (b) Including environmental costs by imposing user fees and deposit fees promotes an increase in efficiency, recycling, and all other forms of input reduction.

service. Some examples of these programs include:

- The Conservation Reserve Program: A U.S. federal government program in which landowners are paid to retire portions of land that are considered "environmentally sensitive." Land-owners agree to plant cover crops to help improve water quality, control soil erosion, and enhance habitat for wildlife.

- Quito, Ecuador, established a water fund in the early 2000s to restore up-stream forests and better protect na-tional parks to reduce sedimentation of the city's water supply. The fund's principal was raised by the city's water utility (via a water user levy), a local brewer, a bottler, and a hydroelectric company. The principal was invested in stocks and other financial instru-ments and allowed to grow before interest earnings were used to finance forest restoration projects. An inde-pendent governing body selects the conservation projects. By late 2010, more than 2 million trees had been planted and more than 5,000 acres of land have been restored.

Although the United States is the world's largest resource consumer and polluter (by most measures), it unfortunately lags far be-hind most of the industrialized world in the use of green fees and other economic incentives to conserve resources and reduce pollution. The general trend has been the continuation of cheap resources and free sinks. In fact, in many cases, the United States actively discour-ages conservation and encourages pollution by subsidies that reward harmful activities (as opposed to green taxes that discourage them).

FIGURE 1-15 shows an example of great im-portance, the historically relatively cheap cost of oil (affecting retail prices of gasoline at the pump) in the United States between the early 1980s and the beginning of the 21st cen-tury. The cheap costs led to a reversal of the previous trend toward more fuel-efficient cars. However, with the terrorist attacks of September 11, 2001, and the subsequent wars in Afghanistan and Iraq, along with natural disasters such as Hurricane Katrina, the 2010 BP oil spill in the Gulf of Mexico, and the un-rest in Libya and the Middle East more gen-erally, the price of oil surged. Many observers think that a "greener" economy is inevitable, not only because of the increasing cost of oil, but also because of consumer demand

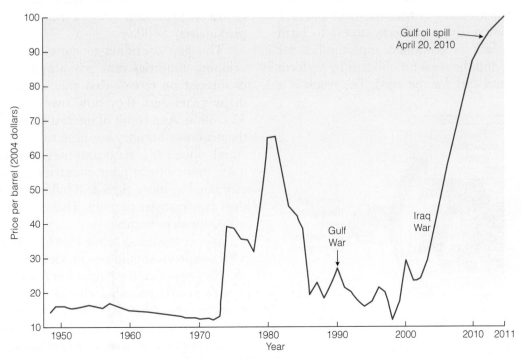

FIGURE 1-15 Price of oil per barrel. (*Source:* U.S. Department of Energy.)

1.6 Promoting Sustainability by Paying the True Costs of Nature's Benefits

for environmental quality and the economic benefits of sustainable technologies, such as increased efficiency, reduced resource imports, and more jobs. There are signs that U.S. business is responding to "green" pressures. More companies are advertising that their products are made from reused or recycled materials or are "environmentally friendly." The rapid increase in gasoline costs in 2005 led to a dramatic decline in sales of large vehicles, such as sport utility vehicles (SUVs), so that automobile makers became much more interested in producing smaller and more fuel-efficient vehicles, such as those with hybrid engines. This trend has been accompanied by increased discussion of green fees and the removal of government subsidies on timber, grazing lands, ore deposits, and oil companies.

Reducing Population: Defusing the Bomb of the South

More than 90% of world population growth occurs in developing nations. Most developing nations are in the Southern Hemisphere; thus the cause of throughput is especially important there. To defuse this so-called **population bomb,** we must address its central causes, which often are economic. **FIGURE 1-16** illustrates how population growth is often driven by poverty. Poor people tend to have more offspring for a variety of reasons, including lack of education about and easy access to birth control, lack of economic opportunities for women, and the need for children to perform chores and care for the aged. The result is a

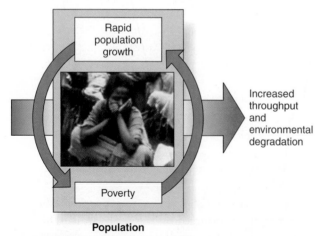

Population

FIGURE 1-16 Poverty and population growth feed on each other, and both produce environmental degradation.

"vicious circle" in which population growth leads to poverty because increasing numbers of people reduce the standard of living. Impoverished people tend not to focus on environmental concerns but must turn their attention to simply surviving from one day to the next.

Can this "poverty–population–environment cycle" of human misery and environmental destruction be broken? Many people still subscribe to the **fallacy of enlightenment,** or the idea that education will solve the problem. Education is very important, but often it is not enough; realistic solutions that remove the root causes of the problem must be found. Attempting to protect the environment by legal means, such as creating game reserves, is often not effective either because desperate people will break laws. Decades of experience show that the economic causes of this cycle must be addressed. Eliminating poverty will not only reduce many causes of population growth, but it also will reduce environmental degradation.

Reducing Population Growth by Paying True Costs

The gap in wealth between the developed and developing nations has widened at an increasing rate. Between 1961 and 1999, the gross economic product per capita increased $15,000 for industrial countries. Africa's gross economic product per capita increased by approximately $600.

This gap continues to grow because developing countries now pay a huge amount of interest on money that was lent to them many years ago; they now owe more than $2 trillion. As a result of interest on this debt, the net flow of money has been from South to North since 1982. Rich countries now receive more money from poor countries (variously estimated at more than $50 billion per year) than they transfer to them. This is sometimes called the **debt bomb**.

Some economists argue that foreign aid to poor countries should not be viewed as charity. They note that these poor but resource-rich nations would not need charity if they were paid the true value for their resources. They maintain that if people in the developed Northern Hemisphere wish to save rain forests, they should pay for the environmental goods and services the forests produce. Although it is often difficult to estimate the true environmental costs and values in most cases, nearly all estimates indicate that developing nations are greatly underpaid for their resources. As payments for ecosystem service markets develop (discussed in the section on promoting sustainability by paying the true costs of nature's benefits), the true costs of environmental goods and services will be better valued in the economic system. Programs such as the United Nations Collaborative Program on Reducing Emissions from Deforestation and Forest Degradation in Developing Countries (REDD) are structured to help alleviate the debt bomb issue.

Appropriate payment for ecosystem services could help eliminate poverty, probably much more effectively than foreign aid donations. These payments would break the vicious poverty–population growth cycle and reduce population growth. A key necessity is that this increased wealth be used to buy sustainable technologies that focus on efficiency, recycling/reuse, and renewable resources.

1.7 The Role of the Individual

The environmentalist Wendell Berry said that the roots of all environmental problems ultimately lie in the values of the individuals who comprise society. As the writer Paul Hawken put it, "The environment is not being degraded by corporate presidents; it is being degraded by popular demand." By this, he means that companies produce only things that people buy, and they cause depletion and pollution only as long as society permits such behavior to be profitable.

Values on the Here and Now:
Why We Avoid True Costs

Why do individuals have values that lead to environmental degradation by "popular demand?" Many writers have argued that most large-scale problems arise because individuals are not good at dealing with problems beyond their own immediate situation. We tend to focus on the "here and now."

Discounting the future results from focusing on the "now": Environmental costs of our actions on future generations are not fully paid,

mostly because they are not fully valued in the current economic system. For instance, cheap gasoline and minerals lead to rapid depletion of these nonrenewable resources, making them unavailable to future generations. These resources are cheap because their current cost does not include the future effects of their use, such as global warming or other pollution hazards. **Discounting by distance** results from focusing on the "here": environmental costs of our actions on people living in another area are not fully paid. For instance, cheap ivory, imported animals, and tropical timber lead to rapid depletion of those tropical resources, thereby degrading the environment for the people who live there. These resources are so cheap because their cost does not incorporate their full value to their local environment.

A Solution: Values Beyond the Self and
Recognizing the True Value of
Ecosystem Services

It can be argued that the single greatest obstacle to building a sustainable society is this tendency of the human mind to focus on the here and now. It limits our ability to make the systemic social changes needed to solve regional and global problems. This is why David Ehrenfeld, in *The Arrogance of Humanism* (1981), warns that history shows that people have never successfully managed anything for very long. How can we expect to solve global environmental problems with such limitations?

FIGURE 1-17 illustrates how people can extend their sphere of concern, beginning with close relatives and progressing through other social groupings to include all people. Ultimately, all living things can be included, as in the philosophy of "**deep ecology.**" If we are to place a higher value on the environment of the future and in other parts of the world, such extended spheres of concern are needed. If many people adopt this view, society will be more willing to pay green taxes, vote for politicians who favor sustainable lifestyles, and in general, promote solutions that require looking beyond the here and now.

Most sociologists believe that the trend has been in the opposite direction until now. The growth of modern civilization has promoted individualism that has progressively shrunk the

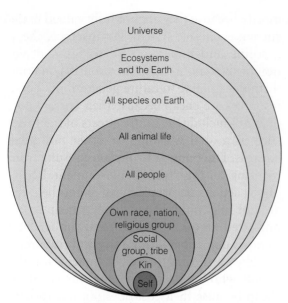

FIGURE 1-17 An ethical sequence in which the individual concerns extend outward beyond the self to progressively more inclusive levels. (*Source:* Adapted from R. Noss, Essay: Issues of Scale, in P.L. Fideler and S. Jain, eds., Conservation Biology. Chapman & Hall, 1992, p. 240.

sphere of concern. Large cities have fostered a loss of community and the fragmentation of social groups, including even families. Can this trend be reversed? No one knows, but some encouraging signs have appeared. For example, the grassroots activism of the current sustainability movement is based on community concerns. There are also many examples of exceptional unselfishness in individuals (**CASE STUDY 1-2**).

Payments for ecosystem service markets, whereby the true cost of an environmental service or benefit is incorporated into the costs of decision-making, can help people realize more directly the value of environmental services. Other initiatives, such as fair trade labeling, organic food, and green products, are also making it easier for people to become more environmentally literate and make choices that result in less environmental harm.

1.8 Toward a Sustainable World

Although no one knows if sustainability will be attained, there is certainly no shortage of debate and speculation. Historically, two schools of thought have existed on the urgency of overpopulation and resource limitation, and they tend to be polarized. At one extreme are **cornucopians**, who argue that human ingenuity has always overcome environmental limitations. They suggest, for example, that inventions

such as genetic engineering could make agriculture more productive so that the planet will be able to support many more people at a high standard of living. At the other extreme are **cassandras** (also known as neo-Malthusians), who argue that people have always altered the environment and managed things poorly; they insist that exponential growth of populations and consumptive technology will finally degrade the environment so much that it will lead to overshoot (see Figure 1-11). Cornucopian and cassandran views have been common, in various forms, throughout history.

As is often the case, extreme views can be counterproductive. Too much optimism, such as a cornucopian view, can lull people into inactivity because they think that the future will take care of itself. Too much pessimism, such as a cassandran view, can cause despair, which leads to inaction because people think that the future is bleak no matter what they do. Thus, many writers suggest a more moderate view that acknowledges that urgent environmental problems exist. These problems need to be addressed within the next few years, or at least decades, or they may lead to large-scale environmental degradation and overshoot. This moderate view relies on the **precautionary principle**, which says that in the face of uncertainty the best course of action is to assume that a potential problem is real and should be addressed. In other words, we are "better safe than sorry." The worst that can happen with this approach is that society will become more efficient, less wasteful, and less polluting, even if environmental problems prove to be not as bad as some people argue. On the other hand, the results can be disastrous if the precautionary principle is not used and problems are indeed as urgent as some people believe.

Assuming that a person wishes to rely on the precautionary principle, which specific environmental problems should be of most concern? A good starting place is the following priority list based on material published by the U.S. Environmental Protection Agency and other sources (**TABLE 1-1**):

Highest Priority: Species extinction, ecological disruption and habitat destruction, stratospheric ozone depletion, and global climate change.

What Are Your Values? Would You Decline $600,000 to Save a Rain Forest?

Although many of us say that we are "pro environment," we never really know how strong our values are until we are asked to make personal sacrifices for them. For example, in this chapter we have asserted that the environment of future generations is often discounted or devalued. Most of us would like to change that situation, but just how much are we willing to sacrifice?

Consider the case of Miguel Sanchez, a subsistence farmer in Costa Rica. He was offered $600,000 for a piece of land that he and his family have tended for many years. This land consists of a spectacular old-growth rain forest and a beautiful black sand beach in an area where land is rapidly being developed. A hotel developer offered the money. Sanchez declined, knowing that this was more money than he, his children, and all of his grandchildren would ever accumulate. Sanchez explained his decision: "I have no desire to live anywhere else. Money can be evil. People will ask me for money."

Miguel knows that cleared forest means less rain, less water in the rivers, less escape from the blazing tropical sun; however, he is reluctant to talk about his decision. Hundreds of other farmers and fishermen in his area feel the same way and are fighting development. (*Source:* Adapted from A. Carothers, "Letters from Costa Rica," *E Magazine,* September 1993.)

Questions

1. How does Miguel's sacrifice compare with sacrifices that you have made to promote long-term environmental health? What would you do if you were in Miguel's position?
2. Comment on this statement from writer Andre Carothers: America has "a skewed moral universe, where people regularly swoon over each other's negligible acts of charity." The writer's point is that we make a big deal about recycling cans, buying organic foods, and giving tiny donations to environmental funds. Is this a fair criticism or unjustified cynicism?

Middle Priority: Problems associated with pesticides and herbicides, lake and river pollution, acid rain, and airborne toxics. *Lowest Priority:* Oil spills, groundwater pollution and depletion, airborne radiation, waste dumping, thermal pollution to water bodies, and acid runoff.

The highest priority problems are global and are very long term in their scope. Lower priority problems tend to be geographically more localized and can (in theory) be solved within shorter time spans.

Visualizing a Sustainable World

Recall that sustainability means meeting the needs of today without degrading the environment for future generations. Table 1-1 summarizes many of the changes that are necessary to create a sustainable society. Most of these changes are discussed later in the text, but for now, some major points can be noted. First, many aspects of our society are affected: scientific paradigms, role of the human, and values toward nature, land, and people. Second, many of our social institutions are affected: religion, education, political systems, and economic systems. Third, technology and agriculture are affected. Fourth, many of these changes ultimately are based on increasing the importance of our communities, from encouraging community values among individuals to creating decentralized, community-based economies and political systems.

The need for such pervasive changes is reflected in academic disciplines such as the social sciences and humanities, which traditionally have studied society as an entity distinct from the natural environment or at most only moderately influenced by the natural

	Industrial Age	**Sustainability or Ecological Age**
TABLE 1-1	**Transition From the Industrial to the Ecological Age***	
Scientific paradigms	Mechanistic Earth as inert matter Determinism Atomism	Organismic Gaia: Earth as a superorganism Indeterminancy, probability Holism/systems theory
Role of the human	Conquest of nature Individual versus world Resource management	Living as part of nature Extended sense of self Ecological stewardship
Values in relationship to nature	Nature as resource Exploit or conserve Anthropocentric/humanist Nature has instrumental value	Preserve biodiversity Protect ecosystem integrity Biocentric/ecocentric Nature has intrinsic value
Relation to land	Land use; farming, herding Competing for territory Owning "real estate"	Land ethic: think like mountain Dwelling in place Reinhabiting the bioregion
Human/social values	Sexism, patriarchy Racism, ethnocentrism Hierarchies of class and caste	Ecofeminism, partnership Respect and value differences Social ecology, egalitarianism
Theology and religion	Nature as background Nature as demonic/frightening Transcendent divinity Monotheism and atheism	Animism: everything lives Nature as sacred Immanent divinity Pantheism and panentheism
Education and research	Specialized disciplines	Multidisciplinary, integrative
Political systems	Nation-state sovereignty Centralized national authority Cultural homogeneity National security focus Militarism	Multinational federations Decentralized bioregions Pluralistic societies Humans and environment focus Commitment to nonviolence
Economic systems	Multinational corporations Competition Limitless progress "Economic development" No accounting of nature	Community-based economies Cooperation Limits to growth Steady state, sustainability Economics based on ecology
Technology	Addiction to fossil fuels Profit-driven technologies Waste overload Exploitation/consumerism	Reliance on renewables Appropriate technologies Recycling, reusing Protect and restore ecosystems
Agriculture	Monoculture farming Agribusiness, factory farms Chemical fertilizers and pesticides Vulnerable high-yield hybrids	Polyculture and permaculture Community and family farms Biological pest control Preservation of genetic diversity

*Some of these may be considered speculative attempts to understand the personal and humanistic aspects of sustainable living.
Source: Adapted from R. Metzner, "The Emerging Ecological World View," in M. Tucker and J. Grim, eds. World Views and Ecology. Cranbury, NJ: Associated University Presses, 1993, pp. 170–171.

world. For example, economics once (and still largely does) focused almost exclusively on the flow of goods through society, ignoring the costs of depletion and pollution that the flows impose on the environment. Similarly, other disciplines such as ecology studied the environment as a largely separate entity. The new field of environmental economics is a more holistic discipline that includes both society and the environment. It incorporates externalities into the overall profit or loss of the product. It studies not just the flow of goods within society but also the effects on the environment and makes

an effort to value the true costs of ecosystem services.

Other socially oriented disciplines are also rapidly developing an interest in studies that incorporate the environment. Findings from such fields are included throughout this text and are emphasized in the section on social solutions to environmental concerns.

Supporting Services

Supporting services are those that are necessary for the production of all other ecosystem services. They differ from provisioning, regulating, and cultural services in that their impacts on people are often indirect or occur over a long time, whereas changes in the other categories have relatively direct and short-term impacts on people. (Some services, such as erosion regulation, can be categorized as both a supporting and a regulating service, depending on the time scale and immediacy of their impact on people).

- *Soil formation.* Because many provisioning services depend on soil fertility, the rate of soil formation influences human well-being in many ways.
- *Photosynthesis.* Photosynthesis produces oxygen necessary for most living organisms.
- *Primary production.* The assimilation or accumulation of energy and nutrients by organisms.
- *Nutrient cycling.* Approximately 20 nutrients essential for life, including nitrogen and phosphorus, cycle through ecosystems and are maintained at different concentrations in different parts of ecosystems.
- *Water cycling.* Water cycles through ecosystems and is essential for living organisms.

Source: http://www.greenfacts.org/en/ecosys tems/toolboxes/box2-1-services.htm

■ study guide

SUMMARY

- Environmental science uses a holistic approach to realize the interconnected aspects of environmental problems and apply solutions.
- Human–environmental interactions evolve through five basic stages: gathering and hunting, agriculture and conservationism, industry and environmentalism, transition and sustainability, and the postindustrial stage.
- We are in the transition phase, which has produced the sustainability movement.
- The sustainability movement
  - focuses on reducing societal consumption of resources.
  - seeks to solve holistically the structural social causes of environmental problems such as poverty.
  - relies on grassroots activism.
- The postindustrial stage will likely witness either sustainability or catastrophic overshoot of human civilization.
- Environmental impact, the alteration of the natural environment by human activity, involves either resource depletion or pollution.
- Impact is promoted by population and consumption: $I = P \times C$.
- The world's population continues to increase by about 77 million people a year.
- Traditionally, resource-intensive industrial technology and consumption have increased exponentially.
- Impact is often measured as throughput—the movement of materials and energy through society.

KEY TERMS

anthropocentric	ecosystem services	input
cassandras	efficiency improvements	input reduction
closed system	environmental science	integration
complexity	environmental wisdom	market failure
consumption	exponential growth	open system
cornucopians	fallacy of enlightenment	openness
debt bomb	first law of ecology	output
deep ecology	Gaia hypothesis	overshoot
discounting by distance	grassroots activism	population bomb
discounting the future	green fees	positive feedback
ecocentric	holistic	precautionary principle

recycling
reuse
sinks
sources

substitution
sustainability
sustainable economy
sustainable technology

system
systems approach
throughput

STUDY QUESTIONS

1. What are the major goals of this text?
2. Define environmental science. What relevance does it have to people?
3. Name several costs of a lack of environmental wisdom.
4. What is often the most expensive way to solve environmental problems?
5. What is the throughput model and the systems approach as applied to environmental science?
6. Why is input reduction a good solution to many environmental problems?
7. Name the two basic causes of environmental problems.
8. What are the key traits of the environmental system?
9. State and explain the "first law of ecology."
10. What is environmental impact? Identify two basic types of environmental impact and explain their causes.
11. Why was the cultivation of food such an important development in human technological evolution?
12. What are the five basic developmental stages in the relationship between people and the environment? Briefly describe each.
13. Define sustainable technology and a sustainable economy. Define "ecosystem services" and name a few examples and which category they fall under according to the Millennium Ecosystem Assessment. Read over Case Study 1-2 and think about the ecosystem services that landowners provide to you and your family.
14. Explain the concept of "The Tragedy of the Commons" and how it applies to environmental issues.
15. What are some basic ways that have been suggested to reduce consumption? Does a reduction in consumption need to entail a decreased quality of life?
16. What is a green fee? What does "market failure" refer to?
17. What is "discounting the future?"
18. What is the "precautionary principle," and how can it be applied to environmental issues?

WHAT'S THE EVIDENCE?

1. The authors contend that many people suffer from information overload and thus "tune out." Do you really believe that this is true? Isn't it better to have more information rather than less? Cite examples that either support or contradict the authors' assertion relative to information overload and a lack of environmental wisdom on the part of some people.
2. The authors suggest in the context of the history of environmental impact that the United States and other developed nations are in a transition stage and that the decisions we make during the next few decades (in your lifetime) will determine the long-term fate of the environment. Do you believe this assertion? Is it simply alarmist rhetoric on the part of the authors? Cite specific and general examples to support your position.

ILLUSTRATION AND TABLE REVIEW

1. Study Figure 1-10. Which of the three statistics shown (total solid waste, per capita solid waste output, population) has increased most quickly during the last few decades? Why? How are these factors related to one another?
2. Using Figure 1-15, determine the approximate price per barrel of oil in 1980.
3. Using Figure 1-15, determine the approximate price per barrel of oil in 2000. How does this compare with the price 20 years earlier?

http://environment.jbpub.com/mckinney/5e/

http://environment.jbpub.com
/mckinney/5e/

Connect to this text's website at http://environment.jbpub.com/mckinney/5e/. This site features eLearning, an online review area that provides quizzes, chapter outlines, and other tools to help you study for your class. You can also follow useful links for more in-depth information.

In Calcutta, India, pollution remains a significant local problem as it enters the Hoogly River. The 260 km (160 mile) long river is connected to the Ganges at its point of origin and flows along the west Bengal region until it enters the Bay of Bengal. The navigable river provides water for irrigation, industry, and human and livestock consumption. Believed by many to be holy, it is also the site of ceremonial baths. Local and international groups, such as The Rivers of the World Foundation, are currently conducting cleanup efforts along this important river, as well as at many other river sites around the world through the work of student volunteers and local community partnerships. See: http://rowfoundation.org.

Human Population Growth

2

E arth is now a very crowded planet. Currently (2011), approximately 7 billion people inhabit Earth. This is just a little over a decade since the planet's human population reached 6 billion in 1999. More than 15,148 people are added each hour, more than 363,554 people a day. The global population increased by approximately 76.4 million persons in 2010 (132.7 million babies were born, and 56.2 million individuals died).

Human overpopulation is one of the central issues in environmental science. As noted in the overview of environmental science, high population levels increase both major types of environmental problems: (1) resource use and (2) pollution and waste. For example, our species alone uses, either directly or by diverting it from other uses, an estimated 40% of the world's terrestrial green plant production. The other 60% is divided among the remaining 5 to 50 million terrestrial species (each species being composed of untold numbers of individuals) with which we share the globe.

Demography is the study of the size, growth, density, distribution, and other characteristics of human populations. No longer can population, or overpopulation, be viewed simply as a cause of environmental destruction; rather, it must be viewed as a major symptom of underlying social, economic, and environmental issues. Poverty, high fertility, and environmental degradation go hand in hand, each reinforcing the other. In this chapter, we review the early development of human society, trace the

Chapter Objectives

After reading this chapter, you should be able to explain or describe the following:

- The basic trends in world population growth over the last 10,000 years
- The chief factors affecting population growth on a global and national level
- The broad distribution of humans over the surface of Earth
- The concept of carrying capacity
- Typical age structures (population age profiles) of industrial and nonindustrial countries
- Infant mortality rates and total fertility rates
- Potential consequences of overpopulation
- Recent population trends in the United States and Canada
- The concept of the demographic transition model
- Methods that can be used to reduce population growth

33

initially slow and subsequently **exponential growth** of the human population, and begin to explore the ways in which human population pressure has progressively modified the environment. We also briefly discuss what is being done to address the current "population crisis."

2.1 World Population Changes Over Time

Natural animal populations often exhibit an "S-shaped" growth curve—slow growth ("**lag phase**"), then rapid growth ("**exponential phase**"), then slowing as limits are reached, and finally a leveling (**FIGURE 2-1**). The same curve can be applied to human populations. On a global scale, the human population is still in the rapid growth phase.

Starting Slow: The "Lag Phase"

About 1 million years ago, the total population of our ancestors was perhaps only 125,000. From this small size, the population slowly increased (**FIGURE 2-2A**). By about 8,000 B.C., it numbered between 5 and 10 million. For comparison, the population of Arizona is just over 6 million, and the population of Michigan is just over 10 million. The combined population of Ontario, Manitoba, and Saskatchewan is just over 15 million. This slow increase was due to (1) the invention and development of tools that enabled food and other necessities to be procured more efficiently and as a result allowed more people to be supported in the same area and (2) migration to new areas, as from Europe to the Americas. Although we focus on chipped stone tools because they are most readily preserved in the archaeological record, many organic tools (e.g., woven nets, baskets, clothing, tents, and other shelters) were invented. Also "discovered" and used during this period was fire—to heat, cook food, and scare away predatory animals.

Before 10,000 years ago, people had spread through most of the world, reaching all of the mainland land masses except Antarctica. The islands of the Caribbean, Polynesia, Madagascar, and New Zealand were colonized later. This simple spread of people to new areas allowed the world population to increase. The extent to which early people affected and permanently modified their environments remains in question. Some researchers believe that through overhunting and competition for environmental resources, the Upper Paleolithic hunters of Europe, Africa, Asia, and the Americas were responsible for global patterns of extinction among large mammals, such as the woolly mammoth and the woolly rhinoceros, during the later Pleistocene (around 100,000 to 10,000 years ago). If this is accurate, human modification of the environment is nothing new; many habitats that we think of as "natural" may in fact be the product of human intervention over tens of thousands of years. What is very important to understand is that the extent, degree, and rate at which people are modifying their environment in the 21st century are unprecedented; the changes are happening much faster than we or the rest of nature can adapt to the new conditions.

The Agricultural (Neolithic) Revolution: Beginning Exponential Growth

The advent of domestication and agriculture about 10,000 years ago led to a sharp rise in the human **birth rate** for reasons discussed later here. Although often referred to as the **agricultural (Neolithic) revolution**, this was actually a gradual process that extended over 8 or more millennia and occurred independently at different rates in different parts of the world. Full domestication of many plants and animals seems to have been established in the Old World by 8,000 to 5,000 B.C. After agriculture and domestication were entrenched, the delicate near balance between births and deaths, which had apparently held

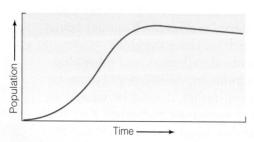

FIGURE 2-1 A typical organismal population growth curve.

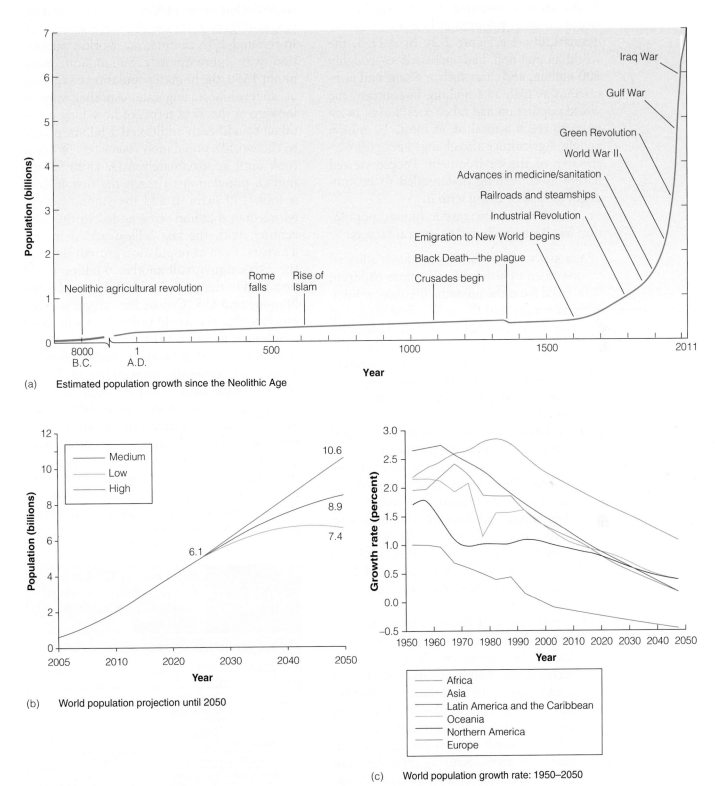

(a) Estimated population growth since the Neolithic Age

(b) World population projection until 2050

(c) World population growth rate: 1950–2050

FIGURE 2-2 World human population changes over time. (a) Estimated population growth in billions since the Neolithic age. (b) World population projections until 2050 (medium-, high-, and low-fertility variants) in billions. The time scales for the two graphs are different. The first graph is in 100-year increments and appears to show a vertical and unending increase in population. The second graph is in 10-year increments and shows the population beginning to level out at less than 10 billion in approximately 50 years (using the medium or low variant). (c) World population growth rate, 1950 to 2005, with projections to 2050. *Sources:* (a) U.S. Census Bureau, International Programs Center, 2001. (b) United Nations Department of Economic and Social Affairs, Population Division, World Population Prospects: The 2004 Revision, Highlights 2005. (c) U.S. Census Bureau, International Data Base, 2005.

for hundreds of millennia, was broken, and the world's human population began to increase dramatically (see Figure 2-2). By A.D. 1, the world population had increased to roughly 300 million, and cities such as Rome had populations as high as 1 million. In contrast, the world population had taken a million or more years to reach a total of, at most, 10 million people. Agriculture also changed people's perspective of the environment. People viewed nature as something that needed to be subdued, instead of living with it.

The dramatic increase in human population has been attributed to several factors:

1. Settlement on farms may have allowed women to bear and raise more children; freed from the nomadic lifestyle, women no longer had to carry young offspring for great distances. Previously, females may have chosen to raise fewer children, perhaps using natural contraceptives and practicing primitive forms of abortion or infanticide to get rid of unwanted offspring.

2. Children are more useful in agricultural communities than in hunter and gatherer cultures and thus are more highly valued. This would have increased the incentive to have more children.

3. Agriculture and domestication may have made softer foods available, which allowed mothers to wean their children earlier. As a result, women could bear additional children over the course of their lifetimes.

4. Agriculture and domestication, by their very nature, allowed and promoted higher densities of people. Indeed, with farming, one family or group of persons could raise more food than they personally needed. This surplus led directly to the rise of cities and civilization because it allowed people to develop and concentrate on manufacturing, trading, and other specializations. In a classic example of positive feedback, this in turn led to rapid advances in technology, art, and other innovations.

After late ancient times, the world's population slowly but steadily increased, except for a slight decline in the 14th century because of the plague (Figure 2-2a; see also Case Study 2-2). In the mid-17th century, the world's population was approximately 500 million. Since about 1650, the human population has grown at an ever-increasing rate. Another way of looking at this is in terms of how long it has taken to add each additional 1 billion people to the world's population (TABLE 2-1). While it took until approximately A.D. 1800 for the human population to reach the first billion, it took 130 years to add the second billion. More than 4 billion were added in the 20th century, with the last billion added in just 12 years. Even as population growth slows in the 21st century, still another 4 billion could be added to the globe by 2100. Recent United Nations and U.S. Census Bureau projections estimate that the world population will reach 7 billion in the year 2011 and possibly 9.2 to 9.3 billion in 2050 (using a medium-fertility projection, generally considered as "most likely"; FIGURE 2-2B). However, such estimates are based on many uncertain variables, such as how quickly modern **birth control** methods will spread and **fertility rates** drop, especially in developing countries. Even assuming that continued advances are made in decreasing global fertility rates and increasing the use of contraceptives worldwide, the global population may not stabilize until it reaches

TABLE 2-1	How Long Has It Taken to Add an Additional 1 Billion People to Earth's Human Population?
World Population reached	
1 billion in approximately 1800	
2 billion in 1930 (130 years later)	
3 billion in 1960 (30 years later)	
4 billion in 1974 (14 years later)	
5 billion in 1987 (13 years later)	
6 billion in 1999 (12 years later)	
7 billion in 2013 (14 years later) 2012 (13 years later)	
8 billion in 2027 (14 years later)	
9 billion in 2045 (18 years later) 2044 (17 years later)	
2100 10 Billion people	

Based on Census 2010 data projections. census.gov International Database
Sources: United Nations Population Division. *World Population Prospects: The 2004 Revision, February 2005,* and U.S. Census Bureau, International Data Base, 2005.

approximately 10 billion toward the end of the 21st century!

Carrying Capacity

Many natural populations of organisms increase very rapidly when invading a new area, but then population growth slows as the upper limit of the number of individuals that the area can support is approached. Eventually, an equilibrium population size is reached. This pattern can be conceptualized in terms of the **carrying capacity** for the particular population of organisms in a given area. The carrying capacity can be thought of as the number of individuals of a certain population that can be supported in a certain area for a prolonged period of time given the resources of that area. If the population's needs overwhelm the resources, the population can crash.

Applying these generalizations to human world population growth, we can ask the following: What is the carrying capacity of Earth relative to people? This question has yet to be satisfactorily answered. Some believe that people have already exceeded the world's carrying capacity (see Case Study 2-1 for information about an early proponent of this idea). How could this be? These are very complex issues to deal with, and there are many views from either side; however, they must be discussed if people want to continue into the future and leave subsequent generations with a sustainable, manageable, and stable set of resources. When a species lives within its carrying capacity, it does not degrade the resources on which it depends. While degrading our resources (e.g., through topsoil erosion and the pumping of underground water for irrigation faster than it is renewed), we produce enough food globally to feed our population of 7 billion, but an estimated 600 to 800 million of those people are undernourished (estimates vary depending on the definition of "undernourished"). Given current global food production, it has been suggested that our 7 billion could be fed adequately if everyone kept to a vegetarian diet.

On the other hand, some persons believe that we have not even begun to approach the physical limits on human population size (such as the cornucopians mentioned in our overview of environmental science). With appropriate technology and distribution systems for foodstuffs and other necessities, some argue that Earth might be able to support 40 to 50 billion people (although for how long is another issue). The human carrying capacity also depends on the standard of living that is acceptable for the average human. To a certain extent, we can conceptually increase the world's carrying capacity by lowering the standards at which we are willing to exist.

One can argue that the world's human carrying capacity has systematically changed over time. Although physical and biological factors, such as the withdrawal of the ice and general global warming at the end of the Ice Age about 10,000 years ago and the evolution or introduction of new plant or animal species, may have allowed human densities to increase, the prime factor increasing the human carrying capacity has probably been our ability to grow our own food and our increasingly advanced technology and culture. The development of stone tools, the harnessing of fire, and especially agriculture and domestication allowed the human population to increase from the hundreds of thousands to the hundreds of millions. Essentially, people increased the carrying capacity of Earth for themselves.

Consider the last several hundred years. Specific developments that have increased the human population since the mid-17th century include advances in agriculture, such as growing legumes, which replace nitrogen in the soils. This allowed fields to be cultivated continuously, rather than having to lie fallow every third year; the result was increased food production that could support a growing population. Also important was the development of modern theories of disease (the "germ theory") and sanitation, which decreased infant mortality and increased life expectancy. Vaccines for diseases were developed. Food preservation and storage were improved. Of course, the **Industrial Revolution** allowed the cheap and efficient mass production of necessary commodities. Since the mid-19th century, we have seen continued improvements in mass production of foods and goods, increasing advances in medical fields, and the so-called **Green Revolution**, which was the

Thomas Malthus, the Original Population Pessimist

The English political economist Thomas Robert Malthus (1766–1834) is generally credited with being the first modern pessimistic thinker concerning population growth rates and the overpopulation problem. Indeed, the term **neo-Malthusian** is often used to refer to those who believe that the modern rapid increase in human population is detrimental. Neo-Malthusians generally believe that we will run out of resources and seriously damage or destroy our environment unless we can control our population.

In his *An Essay on the Principle of Population, As It Affects the Future Improvement of Society* (1798, revised and enlarged in 1803), Malthus suggested that although the size of a population increases geometrically or exponentially, the means that support the population tend to increase only arithmetically. Increasing populations invariably outstrip their resource bases (**FIGURE 1**). As this occurs, the poor get poorer and more desperate, leading to misery and vice. Essentially, people eventually will overshoot their carrying capacity. If people do not intervene of their own accord, such as through "moral restraint" (restraint in reproduction), the population increase ultimately will be checked by natural means, such as widespread famine, disease, and possible warfare. Although Malthus did not necessarily advocate direct birth control, he did suggest that early marriages should be avoided and self-restraint should be cultivated. Most neo-Malthusians are strong proponents of accessible birth control and family planning services.

Since his *Essay* first appeared, Malthus's ideas have been widely discussed—both admired and heavily criticized—and have greatly influenced subsequent generations. The effects of his writings have been profound in economic, political, social, and biological circles. It was from reading a version of Malthus's *Essay* that Charles Darwin hit on the idea of natural selection and "survival of the fittest" as the major mechanism underlying biological evolution. Scientists today still draw influence from Malthus and find the work valuable in understanding today's world population challenges. One example of this is the book *Beyond Malthus: Nineteen Dimensions of the Population Challenge*, published in 1999, which includes the ongoing research of Lester Brown and the work of researchers at the Earth Policy Institute. Much of the work at the Earth Policy Institute documents ongoing environmental challenges and strives to make recommendations for increased problem-solving; some of those ideas are fleshed out in the 2009 Lester Brown book *Plan 4.0: Mobilizing to Save Civilization.*

FIGURE 1 A "Malthusian view" of an overcrowded London of the future is shown in this 1851 George Cruikshank etching.

industrialization of agriculture of the mid-20th century that enabled greatly increased crop yields (see our discussion of food and soil resources for more information). All of these developments have not only increased the world's theoretical human carrying capacity, but also have promoted the actual growth of the world population.

However, in the last few decades, major environmental changes, such as the buildup of greenhouse gases and the destruction of the ozone layer, have been taking place on a global level. Global warming also threatens to decrease the amount of land on earth because of sea level rise. Some observers have interpreted these changes as indications that we have finally reached, and perhaps begun to exceed, Earth's carrying capacity for people.

Growth Rate and Doubling Time of the World's Human Population

Although the human population **growth rate** generally decreased during the last quarter century, over the broad course of human history, estimates show that the doubling time for the human population progressively shrank. Between 10,000 years ago and A.D. 1650, the **annual growth rate** was less than 0.1% a year, for a **doubling time** of more than 1,000 years. However, between 1650 and approximately 1800, the human population doubled again, reaching 1 billion at the start of the 19th century. This was a doubling time of only 150 years, or an average annual growth rate of just over 0.46%. In the next 130 years or so, the population doubled again, with an average annual growth rate approaching 0.54%. In the late 1960s and into the 1970s, the annual growth rate hovered around 2.0%, so the doubling time was about 35 years. An annual growth rate of 2.0% theoretically may seem small, but in reality it is enormous—especially in comparison with the annual growth rate during the vast majority of human history and prehistory. The human species is adding more and more individuals at a faster and faster pace.

Between 1970 and 2005, the annual rate of growth of the world's population began to decrease, from about 2.0% to approximately 1.26%. According to some U.N. projections,

using a medium-fertility projection, the annual growth rate could continue to decline until it reaches about 0.34% in 2045 to 2050. In the last half of the 21st century, the growth rate may continue to decline until it reaches zero or possibly even a negative number (i.e., the population would shrink) in future centuries. This steady decline in the annual growth rate of the world population is quite significant—but any growth rate above zero means that the already huge global population is continuing to increase in size. If there are 9.1 billion people on Earth in 2050 and the growth rate at that time is 0.34%, that will mean that nearly 32 million people will be added to the world's population in 2050 (**FIGURE 2-3**). However, this is fewer than the number of people currently added to the world on an annual basis. An annual growth rate of 0.34% corresponds to a doubling time for the population of about 200 years.

Today (2011), the world population stands at 7 billion, and opinions differ as to where it will stand in the future. In 1988, the United Nations predicted a global human population of 6.25 billion in the year 2000 (an overestimation); however, that was when the growth rate was 1.7%, and the global average fertility (children per woman) was approximately 3.6. It is possible the world's population finally may be coming under control. Another billion people were added to Earth in the 12 years from 1987 to 1999; however, the growth rate continues to drop dramatically, and the end may be in sight. Today, the global average fertility rate is 2.49, and many countries, especially in Europe, have below-replacement fertility levels. Their populations are decreasing, rather than increasing (see further discussion later here). After the global population is stabilized at around 10 billion, it could even begin to fall during the 22nd century as populations age and fertility rates continue to decline.

In our overview of environmental science, we discussed the conceptual equation $I = P \times C$ (impact = population × consumption). In the future, it is hoped the population (P) aspect of this equation will stabilize, and society can focus solely on environmentally friendly technologies that can serve to feed,

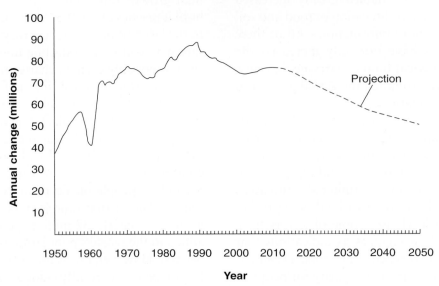

Annual World Population Change: 1950–2050

Projection

FIGURE 2-3 Annual increases in world population, 1950 to 2005, with projections to 2050. *Source:* U.S. Census Bureau, International Data Base, 2010.

clothe, shelter, and otherwise provide for all the people on Earth while also lessening the negative human impact on the planet. When we (MLM and RMS) began writing the first edition of this text in the late 1980s, we were much less optimistic about the long-term future of people on Earth than we are now.

2.2 Distribution of the Earth's Human Population

The current human population is distributed somewhat unevenly over the Earth (**FIGURE 2-4; TABLE 2-2**)—even more so in terms of access to and use of resources. A select few tend to be

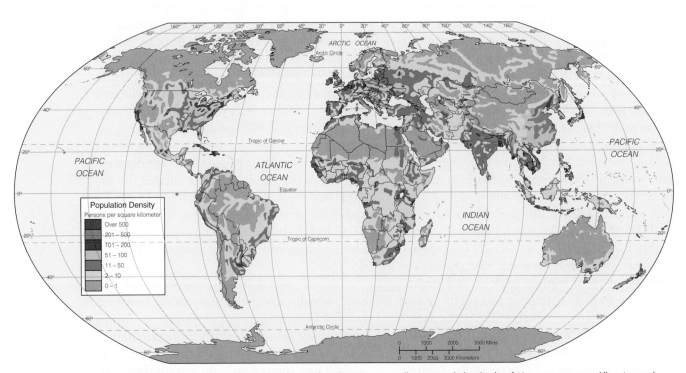

FIGURE 2-4 Map of world population densities. One square kilometer equals approximately 0.386 square mile, so a population density of 10 persons per square kilometer equals approximately 26 persons per square mile.

TABLE 2-2	Population Distribution for World and Major Areas, 1750 to 2050 (Millions)							
Major Area	**1750**	**1800**	**1850**	**1900**	**1950**	**2005**	**2050**	**(Medium Projection)**
World	791	1,000	1,262	1,650	2,521	6,465	9,149	9,076
Africa	106	109	111	133	221	906	1,998	1,937
Asia	502	649	809	947	1,402	3,905	5,231	5,217
Europe	163	208	276	408	547	728	691	653
Latin America and the Caribbean	16	25	38	74	167	561	729	783
Northern America	2	7	26	82	172	331	448	438
Oceania	2	2	2	6	13	33	51	48

Sources: United Nations, 1973. The Determinants and Consequences of Population Trends, Vol. 1. New York: United Nations. United Nations. *World Population Prospects: The 2000 Revision, February 2001.* New York: United Nations. United Nations Department of Economic and Social Affairs, Population Division. *World Population Prospects: The 2004 Revision, Highlights, 2005.* http://www.census.gov/population/international/

moderately to very well off, whereas many lead mediocre, marginal, or substandard existences. Likewise, such basic statistics as infant mortality, crude birth and death rates, and longevity vary widely from country to country. As **FIGURES 2-5** and **2-6** and Table 2-2 demonstrate, past and predicted future changes in population size differ among the continents and between the **more developed countries (MDCs,** basically the countries of North America, Europe, Japan, Australia, New Zealand, and the former Soviet Union)

and the **less developed countries (LDCs).** The population growth curve for Europe is essentially flat, but it is rising sharply for Asia and Africa.

Region	Population Increase per year	Current Population Growth Rate
Asia	49 million	1.05%
Africa	18 million	2.20%
Latin America and the Caribbean	7 million	0.99%

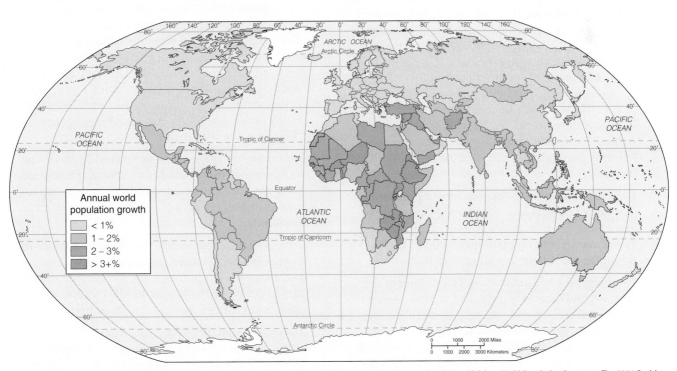

FIGURE 2-5 Expected annual population growth rates around the world in 2005 to 2010. *Source:* United Nations Population Division, *World Population Prospects: The 2004 Revision, February 2005.*

The basic equation describing the growth of a population (the change in the size of a population, represented by ΔN, that takes place over some interval of time, represented by Δt) is $\Delta N/\Delta t = rN$, where N is the population size, t is time, and r is the intrinsic rate of increase of the population (birth and recruitment rate minus death and emigration rate). If we are discussing the global human population as a whole, there is no recruitment or emigration; the intrinsic rate of increase is simply the birth rate minus the death rate.

In discussing human populations, the following basic statistics are commonly used:

Crude birth rate: The number of births per year per 1,000 members of a population. The crude birth rate is commonly determined by dividing the total number of births in the given population by the mid-year population size (which gives the number of births per individual per year) and then multiplying by 1,000.

Crude death rate: The number of deaths per year per 1,000 members of a population. The crude death rate is calculated in a manner that is comparable to that of the crude birth rate.

Rate of natural increase (essentially r): The crude birth rate minus the crude death rate. The rate of natural increase can be expressed in terms of number of additional people per 1,000 members of the population at mid-year. A rate of natural increase of 20 per 1,000 (5.02) means that the population is increasing overall by 20 individuals per 1,000 members of the population each year. A population of 10,000 at the midpoint of one year would grow to a population of 10,200 by the midpoint of the following year. When the rate of natural increase is a negative number, the population is decreasing in size over time, rather than increasing.

Percent annual growth (or change) of a population: The rate of natural increase expressed as a percentage of the given population. For example, if the rate of natural increase is 20 per 1,000 (20/1,000 = 0.02), the percentage annual growth is 2%.

Doubling time of a population: The amount of time that a population of a given size at time zero, increasing at a fixed rate, will take to double in size. Essentially, the annual growth rate of a population is equivalent to the compounding of interest on money in the bank; that is, as dollars of interest added to the account will themselves earn interest, so too will persons added to a population give birth to more people. To demonstrate, if the annual growth rate is 1.0%, the doubling time will be 70 years (rather than 100 years as it would be if "interest" were not compounded). Likewise, the doubling times of populations with growth rates of 2%, 3%, or 4% per year are 35, 24, and 17 years, respectively. The doubling time of a population can be calculated roughly by dividing the percentage annual growth into 70. For example, a population with an annual growth of 2% will double in size in about 35 years (70/2.0 = 35).

General fertility rate: The total number of births in a population in any given year as a function of the total number of women in their reproductive years, variously defined as from age 15 to age 44 or 49 years. Often, the general fertility rate is expressed in terms of number of births per 1,000 women in their reproductive years; it is calculated in a manner similar to that used to calculate the crude birth and death rates, except that the number of births for a given year is divided only by the mid-year number of reproductive-aged women in the population.

Total fertility rate (TFR): Basically, the number of children a woman of a given population will have, on average, during her childbearing years. The replacement level generally is considered to be approximately 2.1 children per woman. (Women must replace not only themselves, but also their male partners. All other factors being equal, more males tend to be born than females—about 105.5 males for every 100 females—and not all children make it to their reproductive years.)

Infant mortality rate: The number of babies dying before their first birthday, given that they are born alive. Babies dying shortly after birth are included in both the crude birth rate and the crude death rate.

Life expectancy at birth: The average number of years a typical newborn can expect to live. Life expectancy can change over time; for instance, in certain populations the biggest hurdle to the newborn may be survival of the first 5 years of life.

Carrying capacity: The equation $\Delta N/\Delta t = rN$ maps an exponential J-shaped curve of population growth. This equation predicts astronomical population sizes after relatively short periods of time, and it does not take into account any factors that may limit the growth of a population. The equation for the J-shaped curve can be modified as follows:

$$\Delta N/\Delta t = rN\,[(K - N)/K]$$

where K is the equilibrium limit of the population, or the carrying capacity. This gives an S-shaped curve. The population size increases but eventually levels as factors that limit population growth take effect—such as lack of food, space to live, and so forth.

At the present time (2010 statistics), Asia has a population growth rate of 1.05%. Africa is growing at double that rate (2.20%). Latin American and the Caribbean are growing at a much lower rate, which is just less than 1%. Europe has the slowest growth rate at 0.03%. Eastern Europe actually has a declining population with a growth rate of −0.03%. As the 21st century begins, only 10 countries accounted for 60% of the world's population growth (TABLE 2-3). Of these, China and India accounted for about a third of the world's population growth. The United States continues to be the third most populated country and is projected to stay so through 2050, with an estimated population of 439 million (partly due to immigration), after India (2050 estimated population of 1.656 billion) and China (2050 estimated population of 1.303 billion). However, by 2050 China may be experiencing negative population growth

(−0.33%). The United Nations has predicted that the more developed regions of the world (including Japan and all of Europe) will reach a collective population peak of about 1.2 billion in 2050 and will decline after that. During the same time, the population of the less developed regions will continue to increase in size (with the fastest growth in Africa), reaching a projected 7.9 billion in 2050.

2.3 Age Structures

When comparing populations between different countries and regions, one must consider the **age structures** or the population age profiles. The age structure of a particular population is essentially a frozen profile of the population at any one instant; the age structure of a population often will change over time.

There are two common patterns in age structures today. A typical Western industrial-

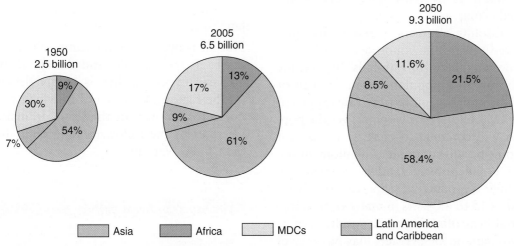

FIGURE 2-6 The shifting balance of the world's population by region, 1950 to 2050. No growth in the population of the present MDCs is predicted between 2000 and 2050, yet the size of the world's population will continue to increase. Therefore, people of those industrial countries will account for a much smaller percentage of the world's population in 2050. In contrast, the populations of Asia and Africa will increase sharply, with the highest growth rates occurring in Africa. In 2050, the number of Asians could be nearly the number of all people on Earth in 1998. *Source:* United Nations Population Division, *World Population Prospects: The 2004 Revision, February 2005.*

TABLE 2-3	Top 10 Contributors to World Population Growth, Mid 2005 (Net Annual Additions in Thousands)		
		Population	
Rank	Country or Area	2001	2010
1.	China	1.275 billion	1.338 billion
2.	India	1.008 billion	1.188 billion
3.	United States	283 million	310 million
4.	Indonesia	212 million	242 million
5.	Brazil	170 million	201 million
6.	Pakistan	141 million	184 million
7.	Bangladesh	129 million	156 million
8.	Nigeria	113 million	152 million
9.	Russia	142 million	139 million
10.	Japan	127 million	127 million

Source: U.S. Census Bureau, International Data Base, www.census.gov

ized country, which is an example of an MDC, has a relatively "flat," "vertical," or uniform age structure profile (**FIGURE 2-7A**); that is, in each age category from about age 0 to 30 years or older, there are approximately the same number of people (or even slightly more people in older categories until about age 40; of course, as people get much older than 50, their numbers drop dramatically as they die of both natural causes and complicating factors associated with age such as degenerative disease and an increased incidence of heart attacks and strokes.). In comparison, a typical LDC (**FIGURE 2-7B**) has an age profile that is strongly skewed toward the younger categories, indicating that there are many more young people than old people.

In absolute numbers, a population that is skewed toward the young will continue to grow even as the birth rate falls. In fact, the fertility rate of a population can drop to or below the replacement fertility rate (approximately 2.1 children per woman), yet the population will continue to increase in size for some time because of **population momentum**. As the population ages, more women will reach their reproductive years (generously regarded as 15 to 49 years in many U.N. statistics) and bear offspring. As a result, even as each woman, on the average, may have fewer children, many more women will be having children. In contrast, a flat age structure profile combined with replacement fertility will result in stable population numbers. As indi-

viduals age and move from one age category to the next and ultimately to death, they will be replaced by equal numbers of individuals being born.

FIGURE 2-8 shows the age structure profiles for the world as a whole in 2010 and the year 2050 (estimated projection). Currently, there are an estimated 1.7 billion people on Earth of prime reproductive age (15 to 29) and another 4.1 billion people younger than 15 who will enter their reproductive phase in the next generation. This is a primary reason the world's population will continue to grow in the 21st century even as fertility rates decline.

Life Expectancies and Infant Mortality Rates

Not all peoples have the same chance of living to a ripe old age, and the age to which you can expect to live is highly dependent on the region of the world in which you are born. To demonstrate this point dramatically, we refer to the following data compiled by the United Nations.

Life expectancies at birth in years (2010)	
Total world	68.9
Western Europe	81
North America	80.1
Latin America and the Caribbean	74.5
Asia	70.3
Sub-Saharan Africa	53.5

Fig. 8

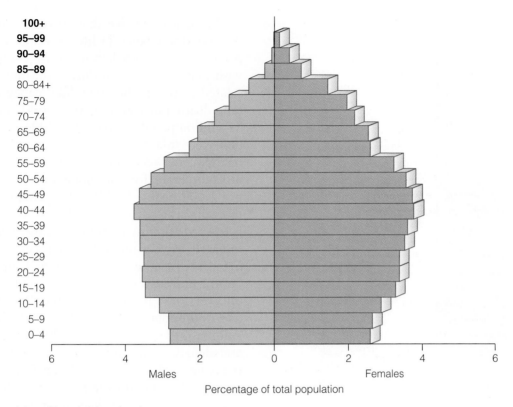

(a)   More developed regions

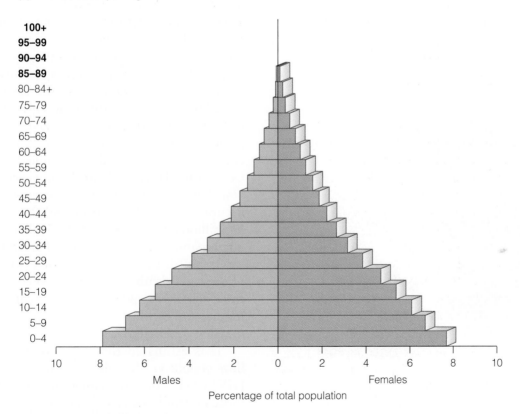

(b)   Least developed countries

FIGURE 2-7 Examples of age structure profiles in two dimensions: (a) more developed regions, 2002 and 2050 (projected) and (b) less developed regions, 2002 and 2050 (projected). An age structure profile for a particular population records the relative numbers of people in different age categories, in these cases by 5-year intervals (0 to 4 years, 5 to 9 years, 10 to 14 years, 15 to 19 years, and so forth). *Source:* United Nations Population Division, *World Population Prospects: The 2004 Revision, February 2005.*

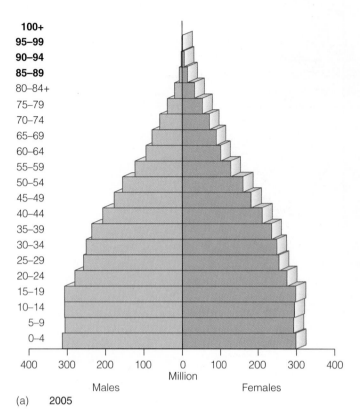

(a)   2005

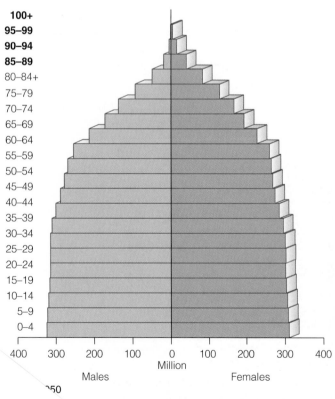

550

structure profiles of the world as a whole in (a) 2005 and (b) 2050
Source: United Nations Population Division, *World Population
on, February 2005.*

45

We can easily see that there is a nearly 30-year discrepancy in **life expectancy** for a person from sub-Saharan Africa compared with that of a Western European. Intimately related to this is the large discrepancy in infant death rates around the world. Of the 57 million people who died in 2004, 6.9 million were infants who died before their first birthday. Between 2000 and 2005, the **infant mortality** rate for the globe was 57 deaths per 1,000 births. In 2009, the global infant mortality rate dropped to 44.13. The trend is toward continual gradual decline. Current research from a number of sources, including UNICEF (the United Nations Children's Fund), indicates that globally 22,000 babies die every day. To put this number in perspective, the population of College Park, Maryland, is just a little over 24,000. Pneumonia and dehydration from diarrhea are the top two causes of death for children 5 years and younger. In 2009, of the 8.1 million children younger than 5 who died, 6.6 million were living in South Asia or Sub-Saharan Africa. The lack of clean drinking water and unsanitary conditions are the major contributing factors. One factor that recently has had a major effect on life expectancies in many sub-Saharan African countries is the spread of HIV/AIDS, as well as the resurgence of various other parasitic and infectious (communicable) diseases, including polio and whooping cough. According to data compiled by the Centers for Disease Control and Prevention, worldwide in 2005 (the latest year for which accurate statistics are available), more than 14 million deaths of a total of 57 million were attributable to infectious and parasitic diseases. Six diseases are responsible for 90% of the total deaths attributable to communicable diseases experienced each year: pneumonia, tuberculosis, diarrhea, HIV/AIDS, measles, and malaria.

It is estimated that more than 40 million people worldwide were infected with HIV/AIDS as of 2005. Ninety-four percent of those infections occurred in developing countries. AIDS has killed an estimated 25 million people since the epidemic began (World Health Organization, 2005). In 2005, 3 million people died of AIDS-related diseases—

more than half a million were children. In 2009, approximately 260,000 children died of HIV of a total of an estimated 1.8 million deaths of HIV-related illness. There were approximately 5 million new HIV infections in just 2005 and 2.6 million new cases in 2010. Health officials indicate that the level of new cases is stabilizing worldwide, but in some regions HIV is still rapidly spreading. For instance, in central Europe and Asia, needle-related spread of HIV is on the rise.

Overall, as of 2005, nearly 26 million sub-Saharan Africans were infected with HIV, but some countries are harder hit than others. By comparison, in 2010 sub-Saharan HIV infection is down 20%, attributable in part to the antiretroviral therapy more commonly available there after 2004. In Botswana, 90% of the people infected with HIV received antiretroviral therapy by 2009. In addition, increased education and access to protection has contributed to the decline. Despite decline of the new infection rate, the number of people living with HIV in sub-Saharan Africa is still the highest in the world (22.5 million or 68% of total people living with AIDS). In 2009, more than 70% of all deaths attributable to AIDS occurred in sub-Saharan Africa. In 2005, the country the most devastated by HIV/AIDS was Botswana, where HIV infected approximately one of every three adults. As a result of this epidemic, the life expectancy of Botswanans dropped from 61 years in 1995 to an estimated 44 years in 2000 and a mere 34 years in 2005. Furthermore, the annual population growth rate of the country has steadily declined from 3.5% in 1985 to 1.6% in 2000. By 2025, Botswana's growth rate is expected to be 0.8% or less. Currently, the small nation of Swaziland has the world's highest incidence of HIV, at 26%. Another country very hard hit by HIV/AIDS is Zimbabwe, where estimates place the proportion of adults infected as up to a third. As in Botswana, both the life expectancy and annual population growth rate of Zimbabwe have been declining. Without AIDS, Zimbabwe could have expected an annual growth of 3.5% between 2000 and 2005; it actually experienced a negative growth rate in 2005 (−0.5%) as a direct result of the HIV epidemic. The growth rate is expected to reach 1.7 by the year 2025. In all, 35 sub-Saharan countries are particularly hard hit by HIV/AIDS—a "hard hit" country is one containing either more than 1 million infected individuals or an adult prevalence of infection greater than 2%. However, it is not just in Africa that HIV/AIDS is a major problem. China is experiencing a growing epidemic as well. Other parts of Asia (Cambodia, India, Myanmar, and Thailand) have been particularly hard hit, and in Latin America and the Caribbean area, the countries of Brazil and Haiti have major segments of their populations that are infected. In Asia, the increase is attributed primarily to needle use and sex workers. There were 300,000 AIDS-related deaths in 2009 in Asia, which was an increase from 2001, when AIDS-related deaths totaled 200,000. In Eastern Europe, the cases of HIV are also rising, primarily in the young adult demographic. The Russian Federation and Ukraine have 90% of the cases in Eastern Europe. In North America and Western Europe, the number of people living with HIV continues to increase (2.3 million in 2009, which is up 30% from 2001). The Middle East and North Africa have also seen an increase in new HIV cases, from 36,000 in 2001 to 75,000 in 2009. AIDS is causing decreases in life expectancies and a slowing of population growth in many of the countries where its incidence has not yet leveled. In the hardest hit areas, such as sub-Saharan Africa, overall population growth will continue to increase even as localized areas experience extreme loss of life.

Total Fertility Rates

The TFR is an important statistic when looking at human populations around the world. The TFR is basically the number of children a woman of a given population will have, on average, during her childbearing years. **Replacement level** TFR is generally considered to be approximately 2.1 children per woman. (Women must replace not only themselves, but their male partners as well. All other factors being equal, more males tend to be born than females—about 105.5 male births for every 100 females born—and not all children make it to their reproductive years.)

Based on U.N. data, during the period of 1995 to 2000, the TFR was at or below the replacement level (2.1) in 64 countries, or 44% of the world's population. This includes most of the countries of the more developed regions along with a number of developing countries, particularly in Eastern Asia (China, e.g., has a TFR of 1.1). These data are indicative of a dramatic decline in TFRs around the globe. For instance, during the period of 1965 to 1970, not a single major country had a TFR below replacement level; China had a TFR of 6.0. The global TFR was approximately 5.9, whereas today it is estimated at 2.4 and expected to decrease to 2.0 by 2050.

Of course, the flip side of these data is that approximately 56% of the planet's population is still characterized by above-replacement fertility. Furthermore, just because a country has achieved a TFR below replacement level does not mean that the size of the population will decline in the near future. Data from 2009 indicate that practically all developed nations have reached a 2.1 TFR and that 28% of developing nations have also reached a TFR of 2.1%. The developing nations with a TFR of 2.1% also are home to 25% of the world's population. Six of the 10 countries with the highest fertility still have TFRs in excess of 6. Overall, the worldwide trend shows a steady decrease in worldwide TFRs. A large proportion of young persons in a population, as well as **immigration**, can cause an increase even when the TFR is below 2.1. For instance, the population of China, with a TFR of 1.7, is still expected to grow from 1.27 billion in 2000 to an estimated 1.46 billion in 2050 (based on medium variant U.N. projections), an addition of more than 187 million people. However, China's estimates have improved. Projections in 1998 predicted that China would have an

TABLE 2-4	Median Age by Major Area, 1950, 2000, and 2050 (Medium Variant Projections)		
	Median Age (Years)		
	1950	**2000**	**2050**
World total	23.6	26.5	38.4
More developed regions	28.6	37.4	45.6
Less developed regions	21.4	24.3	37.2
Africa	19	18.4	28.5
Asia	22	26.2	40.2
Europe	29.2	37.7	46.6
Latin America and the Caribbean	20.1	24.4	47.7
Northern America	29.8	35.6	42.1
Oceania	27.9	30.9	39.1

Source: United Nations Population Division. *World Population Prospects: The 2004 Revision,* February 2005.

additional 220 million people by 2050. The population of the world as a whole will continue to increase until the global TFR is at or below replacement level and the age structure distribution of the planet is no longer skewed toward the young side.

Aging of the World Population

With the decline of fertility rates, the age structure of the world's population is changing. The distinct trend is toward older median ages and higher proportions of older people (60 years and older) relative to children younger than 15 years (**TABLES 2-4** and **2-5**). In general, the less developed regions are lagging behind the more developed regions in these trends. In approximately 2050, the less developed regions will be at the stage the more developed regions were in about 2000. This, along with discrepancies in TFRs between more and less developed regions, is a primary reason most of the world's population growth during the 21st century will take place in the less developed regions.

TABLE 2-5	Percentages of the Population Less Younger Than 15 Years Old and 60 and Older					
	Younger Than 15 Years			**60 Years and Older**		
	1950	**2000**	**2050**	**1950**	**2000**	**2050**
World	34%	29.8%	19.6%	8%	10%	21.9%
More developed regions	27%	18.2%	15.4%	12%	19.5%	32.6%
Less developed regions	38%	32.7%	20.3%	6%	7.6%	20.2%

Source: Based on U.S. Census Bureau, International Programs Center, 2001.

2.4 The Consequences of Overpopulation

Rapid population growth and overpopulation have many far-reaching effects ecologically, economically, and societally. The increasing population is putting a greater and greater burden on the Earth's natural resource base and environment. As Paul Ehrlich pointed out:

> One can think of our species as having inherited from Earth a one-time bonanza of nonrenewable resources. These include fossil fuels, high-grade ores, deep agricultural soils, abundant groundwater, and the plethora of plants, animals, and microorganisms. These accumulate on time-scales ranging from millennia (soils) to hundreds of millions of years (ores) but are being consumed and dispersed on time-scales of centuries (fuels, ores) or even decades (water, soils, species).*

Most people readily acknowledge that fossil fuels, such as oil and coal, are nonrenewable; however, many fail to realize that, from the human perspective, soils and much fresh water that is pumped from underground aquifers are also nonrenewable resources (see our discussions of water resources, land resources and management, and food and soil resources). Topsoils are being eroded at a tremendous rate, and in many regions, the water table is being drawn down to alarmingly low levels. Another example is phosphorus, an extremely important element that is mined from nonrenewable rock deposits primarily to make modern fertilizers. A 1971 study concluded that known supplies of phosphorus would be exhausted by 2100 and that without phosphate fertilizers, the Earth can support only 1 to 2 billion people. Phosphate mines in Florida (2005) supply 75% of all farmers in the United States and 25% of all the phosphate in the world. The best reserves are used up, and Florida residents are against possible southern expansion of the mines mainly because of two factors. One issue is that a mine was abandoned in 2001 leaving a $160 million

*P. Ehrlich. Populations of people and other living things. In H. J. de Blij, editor. Earth '88: Changing Geographic Perspectives. Washington, D.C.: National Geographic Society, 1988, pp. 302–315.

cleanup bill and more than 3.78 billion liters (more than 1 billion gallons) of contaminated wastewater that has on occasion leaked into area tributaries. A second issue is that phosphogypsum (a by-product of processing phosphate ore into fertilizers) is slightly radioactive, and there is not much public confidence in allowing use of additional lands closer to drinking water reserves when mining companies have not had the public trust. Mining phosphates in other countries would increase agricultural costs. So the debate about what to do and the race to find ingenious ways of extracting the phosphorus in poorer ore remain current challenges for which there are no immediate solutions.

Such ecological damage is not solely a function of more people. Given the discrepancies in affluence and consumption among the Earth's peoples, not everyone has an equal impact on the environment (remember the equation $I = P \times C$ from our overview of environmental science). Persons in rich, industrialized countries (the MDCs) typically cause much more ecosystem damage per capita than do persons in poor, nonindustrialized countries (LDCs). Based on such considerations, Paul Ehrlich has suggested that one new American baby and 250 new Bangladeshi babies pose an equal threat to the environment. The United States contains less than 5% of the world's population but imports more goods than it exports. The United States produces about one quarter of the world's pollution and consumes approximately 25% of the world's energy resources. U.S. citizens use approximately 12 times as much energy per capita as the average citizen in a developing country. On the other hand, the United States also produces 24% of the world's gross domestic product.

Rapid population growth and overpopulation lead to increased urbanization, increased unemployment, and in many cases, spreading poverty. Projections indicate that 59% of the world's population will live in urban areas by 2030, and 69% by 2050. In the developing world, many cities are growing at phenomenal rates and to phenomenal sizes (TABLE 2-6), but much of the growth is in the form of slums, shanty towns, and squatter settlements that lack such necessities as adequate housing, safe drinking water, and proper sanitation systems;

TABLE 2-6	Selected Examples of Rapid Population Growth in Cities Located in Developing Countries			
	Population in Millions			
			Projection for	
City	1950	1995	2015	2025
Mumbai (Bombay), India	2.9	18.0	21.7	25.81
Lagos, Nigeria	0.28	13.4	14.1	15.8
Dhaka, Bangladesh	0.41	12.3	16.6	20.9
Sao Paulo, Brazil	2.4	17.7	21.3	21.6
Karachi, Pakistan	1.0	11.7	14.8	18.7

Source: United Nations Population Division. *World Urbanization Prospects: The 1999 Revision.*
http://esa.un.org/unpd/wup/index.htm

however, overpopulation does not always cause the extreme poverty. Economic short-sightedness, government mismanagement, or deliberately harmful political policies can also be the cause of suffering.

Population growth leads invariably to competition for limited resources, often culminating in outright armed conflict. This may take the form of increasing local crime and violence, particularly among various ethnic and racial groups, some of whom may have migrated voluntarily or been displaced forcefully because of increasing population pressures. Sometimes the conflict escalates to actual warfare, primarily domestic civil wars. There have been more than 20 million war-related deaths in the world since 1945, and more than 75% of such deaths came in domestic civil wars. The majority of these deaths have occurred in highly populated countries. Bangladesh, with one of the highest population densities in the world, accounts for at least 1.5 million of these deaths.

2.5 The Population of the United States and Canada

The 2006 population of the United States was approximately 300 million, or 4.6% of the world's population at that time, making it the third most populous country on the planet (based on U.N. statistics). Although the TFR in the United States is about 2.02 (2.1 is generally accepted as replacement-level fertility),

the population is still growing by approximately 2.7 million every year, in part because of immigration. It is projected to reach approximately 403 million in 2050. However, the rate of growth is slowing. By 2050, the rate is projected to level at 0.36%. For comparison, the current growth rate is 0.90%.

Another way to look at the growth is by percentage of increase. The world's population increased by approximately 13.2% between 1990 and 2000. For comparison, the U.S. population increases per decade have been as high as 18.5% (**TABLE 2-7**).

The continued growth of the U.S. population in the late 20th century was attributable primarily to two factors: (1) The age structure of the population was such that a large number of women (many of whom were of the baby boom generation, discussed later) were in or entering their reproductive years and bearing children; consequently, the birth rate was appreciably higher than the death rate. (2) Large numbers of immigrants, legal (about 600,000 a year) and illegal (numbers unknown), continued to enter the country. The proportion of the U.S. population growth attributable to immigration rose by more than 20% between World War II and the end of the century. Often foreign-born women in the United States had higher fertility rates than did American-born women. As the 20th century closed, it was estimated that immigration accounted for more than 25% of the growth of the U.S. population.

The fact that the U.S. population is still growing faster than populations in most of

TABLE 2-7	United States Population Change From 1940 to 2010	
Census Year	**Population**	**Percentage Change From Preceding Census**
2010	308,700,000	9.7
2000	281,421,906	13.2
1990	248,709,873	9.8
1980	226,542,199	11.4
1970	203,302,031	13.4
1960	179,323,175	18.5
1950	151,325,798	14.5
1940	132,164,569	7.3

Source: U.S. Census Bureau. Selected Historical Census Data—1790 to 1990, Census 2000 Ranking Tables for States: Population in 2000 and Population Change from 1990 to 2000.
www.census.gov/prod/cen2010/briefs/c2010br-01.pdf

the other industrialized nations is worrisome, not only for the citizens of the United States but also for the Earth. The United States, it can be argued, consumes much more than its fair share of world resources and produces an equivalent amount of pollution and waste; this inordinate consumption will only get worse as the U.S. population increases.

Even within the borders of the United States, the increasing population pressure undoubtedly will carry serious consequences. The infrastructure and social services will be further stressed. Dumpsites for waste, both hazardous and nonhazardous, are in short supply. Freshwater supplies are already pressed; periodical water shortages have been reported around the country, and these will only increase. The agricultural capacity of the United States is decreasing, even as its population is increasing, because of the conversion of an estimated 1.2 million hectares (about 3 million acres) of agricultural land each year to other uses. This current rate of loss is roughly equivalent to losing 0.8 hectares (about 2 acres) of agricultural land every minute.

Another concern is that the age structure of the United States underwent major changes in the 20th century and continues to shift. After World War II (during the years of 1946 to 1964), birth rates soared in the United States, giving rise to the baby boom generation. In the middle and late 1960s, birth rates underwent a dramatic decrease. What this means is that there is a large segment of the U.S. population that demands appropriate resources relative to their age. In the 1950s, the baby boomers needed to be educated, and schools were built. Later, colleges and universities expanded to meet the needs of the baby boomers. In the 1970s, baby boomers entering the work force drove unemployment rates down and fueled the economy with increased consumption of goods and services. In the early 21st century, baby boomers are starting to retire, and some fear this may put a strain on certain aspects of the American economy, including Social Security. In the late 1990s, the ratio of working to retired stood at about 5 to 1, but by 2020, it may be only 2.5 to 1. The exact consequences of this demographic shift are uncertain.

Although Canada is much less densely populated than the United States, in the past it has had an even higher population growth rate than the United States; however, Canada's birth rate is now projected to decline more quickly (TFR for 1995 to 2005 was about 1.6), and the population has been aging; the Canadian population also underwent a 1950s baby boom. In 1950, the Canadian population stood at 13.74 million, in 1960 at 17.9 million, in 1990 at 26.64 million, and in 2000 at 30.75 million. In 2005, the population was estimated at 32 million. It is projected to reach 35 million in 2015, 39 million around 2025, and perhaps 44 million about 2050. Geographically, Canada is a huge country, with an area slightly greater than the United States (including Alaska and Hawaii). Yet large portions of Canada, such as the Arctic regions, are inhospitable to dense human habitation. Consequently, more than three quarters of all Canadians are urban dwellers. The majority of Canadians live within about 322 kilometers (200 miles) of the U.S. border, many being concentrated in the metropolitan areas of Toronto, Montreal, and Vancouver.

2.6 Solving the World's Population Problem

Industrialization, Economic Development, and the Demographic Transition

How do we solve the world's population problem? Many people have subscribed to the following proposition, as succinctly stated by Dr. Nathan Keyfitz:

> *The one thing that we know for sure about population is that people in the developed condition do not have many children; whatever problems the rich countries face, rapid population growth is not one of them. Hence, all that is needed is for the poor countries to develop and they too will be spared the troubles arising from rapid population growth.*[†]

[†]N. Keyfitz. Population growth can prevent the development that would slow population growth. In: J. T. Mathews, editor. Preserving the Global Environment: The Challenge of Shared Leadership (pp. 39–77). New York: W. W. Norton, 1991, pp. 39–40.

This is the theory of **demographic transition**— essentially, that as a nation undergoes technological and economic development, its population growth rate will decrease. According to this theory, early in its history a nation will be characterized by both high birth rates and high death rates, and population growth will be relatively slow. With initial economic and technological development, the death rate will drop, while the birth rate remains high. Usually, this trend is ascribed to better nutrition, increased and advanced medical technology, more adequate shelter, and the like. However, cultural norms are modified more slowly, and parents continue to produce large families, even though they do not "need" to because a higher percentage of their offspring survives and reaches adulthood. In addition, with initial economic and technological development, children will be in less demand as a labor force and will no longer be an economic asset. Still, because children continue to be produced at high rates, the nation will undergo rapid population growth, and the size of the population will increase dramatically.

Finally, industrial, technological, and economic development will mature and peak, and the culture will change to meet the new conditions. Parents will no longer produce as many children as did previous generations. Why this is the case is still under investigation. As nations industrialize, many of the attributes and nuances of the indigenous culture are progressively lost. In many Westernized societies, children are no longer needed as a form of "social security" to support the parents in their old age. High-technology material goods are in demand, the dietary intake tends to include more meat, and having a large number of children is de-emphasized (**FIGURE 2-9**). In terms of a cost–benefit analysis, having more children is an economic burden. Children in industrialized countries want more material goods and are more of an economic strain on the family. Until they are adults, children produce little, if anything. In one way or another, consumer goods appear to substitute for children. At any rate, values and cultural norms change, and the birth rate falls until it approximates the death rate; during this time, the nation's population growth

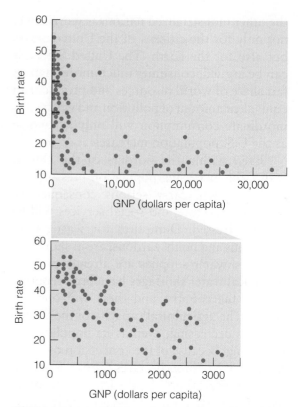

FIGURE 2-9 Increasing GNP (gross national product) per capita correlates with decreased birth rates (crude birth rates per 1,000 members of the population), based on a study of 107 countries. *Source:* H. R. Pulliam, N. M. Haddad. Human Population Growth and the Carrying Capacity Concept. Bulletin of the Ecological Society of America, September 1994:141–157.

rate slows, perhaps even reaching zero or a negative number.

Problems with the Demographic Transition Model

The demographic transition model seems to describe well the population histories of some technologically advanced nations, especially those of Europe. However, does this model hold the key to the world's population problem? This is now being seriously questioned. For instance, countries in Africa, Asia, and Latin America have been heavily affected by Westernization. Their death rates have been lowered, but their birth rates remain high. In many of these countries, children are viewed as a financial benefit rather than a drain. Consequently, their populations are growing at staggering rates. These nations are partly, but not fully, industrialized. With partial industrialization, their environments and resources are being destroyed, but they do not have the money to industrialize further. The longer a nation remains stuck halfway through the demographic transition, with high birth rates and low death rates, the larger its population will become. The larger the population becomes, the poorer the nation will be, the more poverty will manifest itself, and the harder it will be to end the cycle. A partial developmental program seems to be much worse than no developmental program; partial development has a very strong destabilizing effect. As Jodi Jacobson has pointed out, "Slower economic growth in developing countries plagued by debt, dwindling exports, and environmental degradation means that governments can no longer rely on socioeconomic gains to help reduce births."[‡]

The demographic transition model has also been criticized on another related issue. The industrialized, highly developed nations of the north (such as the United States and the European nations) reached their current level of development and affluence not only by degrading their own immediate environments, but also by exploiting the territory of the developing nations in the Southern Hemisphere. For example, for every hectare (or 2.5 acres) of land farmed in the United Kingdom, another 2 hectares (or 5 acres) of land are farmed elsewhere in the world in support of the United Kingdom. Perhaps the classic example is the Netherlands, with a very dense population of about 390 people per square kilometer (or 1,018 people per square mile); in contrast, the United States has a density of about 29 per square kilometer (77 people per square mile). The Netherlands can support such a dense population only by importing massive amounts of foodstuffs and other resources. Essentially, the argument is that it would be impossible for the entire world to consist of only developed countries; in order for a would-be developing country to become a fully developed country, it needs suitable developing countries to exploit.

It is also questionable whether the correlation between development and industrialization and lower fertility seen in Europe and the United States is a causal relationship. Not only are fertility rates lower in the industrialized nations, but so are rates of malnutrition, infant mortality, and illiteracy. This evidence suggests increasing levels of education, nutrition, and infant survivorship, rather than

industrialization and development per se, primarily cause fertility declines. A case in point would appear to be Costa Rica, which cut its fertility rate by 53% between 1960 and 1985 without major industrial development or a major effort in family planning. Instead, Costa Rica promoted education and health issues and waged a war on poverty.

Another factor that must be considered is that as the growth rate of a particular population declines, the age structure profile of the population will change. For example, the typical LDC currently has an age structure profile that is skewed toward the younger age groups. However, as population growth rates decline and birth rates approximate death rates, the population will "mature" and have many more persons in the older age categories than previously. Such a shift in age structure can seriously affect the culture and society. More adults in their prime will need jobs, and older people, who were once relatively rare, will demand a larger proportion of the resources with fewer younger people to support them.

Contraceptives, Abortion, and Reproductive Rights

An obvious way to address the population problem is to continue to expand family planning services, the distribution of contraceptives (birth control devices; see **FIGURE 2-10**), and the availability of safe, legal abortions while promoting the "**reproductive rights**" of women (essentially the right to be able to choose not to bear children) throughout the world. Indeed, much progress has been made in these areas, but it is also a highly sensitive, emotionally charged subject (**FIGURE 2-11**). Contraception, abortion, and reproductive rights in general go against the beliefs of some major religions and cultural values. Some people in developing countries view family planning as genocide because the developed, industrialized nations are encouraging the undeveloped and developing nations to curtail their populations. Many African governments traditionally have opposed family planning in the belief that curtailing population growth would hurt them economically; without people, leaders of these countries felt they could not reach their full economic potential. This view is changing, however.

[‡]J. Jacobson. Planning the global family. In L. R. Brown, et al. State of the World. (pp. 150–169). New York: W. W. Norton, 1988, p. 151.

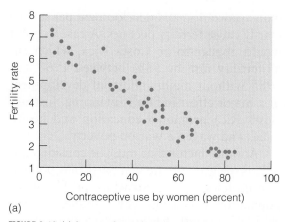

(a)

(b)

FIGURE 2-10 (a) Contraceptive use by women strongly correlates with decreased fertility rates (based on a study of 50 countries). (b) Women attending a family planning class in India. *Source:* (a) H. R. Pulliam, N. M. Haddad. Human Population Growth and the Carrying Capacity Concept. Bulletin of the Ecological Society of America, September 1994:141–157.

Many people in developed countries such as the United States and Canada often take for granted easy access to contraception; however, access to and use of various methods of contraception vary widely between developed and developing regions (**FIGURE 2-12**). Of all regions, Africa is the farthest behind in the use of contraception; more than three quarters of African couples do not use any form of contraception. In contrast, the rate of nonuse in developed countries is less than one third. Population Action International estimates that more than 100 million women in the developing world have an "unmet need" for contraception and family planning services. These women have indicated that they do not want to have children, at least not at the present, yet they are not using any form of contraception. In some cases, this nonuse of contraception may be because of deeply held religious or cultural practices, but in many cases, it is simply due to a lack of information about and accessibility to methods of contraception. Because of nonexistent or inadequate preventative measures to stop conceptions, approximately 50 million abortions are performed annually around the world. Many of these abortions occur in developing countries under unsafe (and sometimes illegal)

FIGURE 2-11 Birth control, family planning, reproductive rights, and especially abortion are highly sensitive, emotionally charged issues for many people. Even in the United States, where abortion is legal, many find it morally repugnant and wish to make it illegal.

CHAPTER 2 Human Population Growth

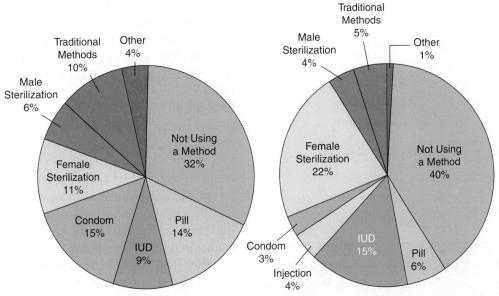

More Developed Countries

Less Developed Countries

FIGURE 2-12 Contraceptive use in more developed and less developed countries at the turn of the 20th to 21st century. *Source:* Population Reference Bureau, 2005.

conditions, resulting in an estimated 50,000 to 100,000 deaths of the women involved each year. Numerous studies from around the world have demonstrated that, in general, greater access to more effective contraception methods, such as sterilization, pills, condoms, and intrauterine devices (all of which are typically much more effective than using methods that rely on withdrawal before ejaculation or "rhythm" or calendar methods) is positively correlated with declining numbers of abortions.

Female Education and Status

Empirically, it has been found that one of the best ways to decrease the growth rate of a particular population is to increase the average educational level and societal status of women in that population (**FIGURE 2-13**). Female education apparently is successful in decreasing fertility rates for several reasons.

- Women typically apply their knowledge to improving the home situation of their families—for instance, by assuring better nutrition, health care, sanitation, and such. Rates of infant and child mortality decline as a result, causing the population to become more receptive to the notion of smaller families.

- Educated women are also more likely to be willing and able to use contraceptive methods effectively.
- Through education, women can gain status and prestige in ways other than through bearing numerous offspring.
- Educated women can choose to pursue career opportunities that do not involve staying home and having children.

Educating the women of the world is relatively inexpensive. Estimates are that increasing the education level of all women in developing nations to at least equality with that of men would cost only about $6.5 billion per year, less than is spent on lawn care annually in the United States and much less than is spent on video games each year in industrialized nations.

Unfortunately, the status of women is relatively low in many non-Westernized societies, including Islamic societies, traditional Latin American, and traditional African societies. For example, in many traditional African societies, women carry a large economic and labor burden but have almost no legal independence or rights. In these societies, a groom buys his wife from the bride's family and pays a high price in the expectation that she will be fertile and produce many offspring that can contribute to the family income. As long as a woman's social

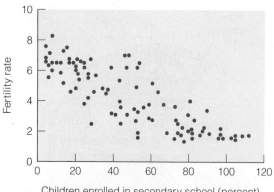

Fertility rate vs Children enrolled in secondary school (percent)

(a)

(b)

FIGURE 2-13 (a) Increased average education levels correlate with decreased fertility rates, based on a study of 105 countries. This correlation holds at all levels, among both males and females, and at all ages. However, particularly important for decreasing the growth rate of a population is the education of women. Note that UNESCO (United Nations Education, Scientific and Cultural Organization) defines the school enrollment ratio as total enrollment, regardless of age, divided by the population of the age group that typically corresponds to a specific education level (if enough younger or older children or adults are enrolled in school, the ratio can be greater than 100%). (b) Women attending class in Afghanistan. *Source:* (a) H. R. Pulliam, N. M. Haddad. Human population growth and the carrying capacity concept. Bulletin of the Ecological Society of America, September 1994:141–157.

and economic status is tied to the number of children she bears, little headway can be expected in the area of decreasing fertility rates.

Another sign of the generally low status of women in developing countries is the low priority placed on women's health. It has been estimated that roughly one-half million women of reproductive age die in developing counties each year of complications of pregnancy, childbirth, and abortion. A Worldwatch Institute report estimated that 60% of these deaths could be avoided by spending less than $2.00 a year on each woman in those countries.

Of course, educating women and improving their status will irrevocably alter these traditional societies. Some have questioned whether it is "right" for Westerners to interfere in the natural development and order of non-European cultures.

Economic Incentives, Disincentives, and Government Regulation of Childbearing

Private organizations and national governments have attempted to control population growth through voluntary programs, such as increasing the accessibility of modern birth control without mandating its use. However, as the overpopulation problem has intensified, many governments have decided they must regulate childbearing.

To encourage their citizens to limit voluntarily the number of offspring, governments have offered a variety of economic incentives for using contraceptives, being sterilized, or limiting the number of children in a family (**FIGURE 2-14**).

For example, in parts of Taiwan during the 1970s, families that had two children or fewer received annual payments, and in Thailand, some families who used contraceptives could rent a team of water buffalo at a reduced rate. South Korea and Pakistan allow income tax deductions for families with two children or fewer. The government of Singapore has withheld employment benefits, housing subsidies, and preferred school admissions from families with more than three children. In an attempt to reduce its population, China implemented a one-child policy in 1979. Exactly how far the Chinese government has gone to enforce this policy has been

FIGURE 2-14 The People's Republic of China is currently the most populous nation on Earth, although by the middle of the 21st century, this distinction will be held by India. Here, a roadside sculpture in Beijing promotes China's one-child policy.

difficult to verify. In the extreme, sterilizations and abortions allegedly have been performed without the full consent of the individual concerned. China has made exceptions to allow very poor couples to have a second child if the first is a girl. Rural Chinese prefer boys to help support their agrarian way of life.

Many people question whether such policies are appropriate. For one thing, bribes and payoffs offered may be most attractive to the poor, and some people may not fully understand what sterilization entails. In addition, the case has been strongly argued that childbearing is and should remain a private family matter. However, given the nature of the world today, it can also be argued that childbearing is no longer a "private" matter; the number of children a couple bears affects more than that couple and their immediate family. In the 1970s, the United States was a leader in family planning investments in the United States and in developing countries, but these policies have been lax in recent years as more conservative government officials have taken office. Most governments of developing countries now acknowledge the need for family planning, and according to the United Nations, very few countries (e.g., Cambodia, Iraq, Laos, and Saudi Arabia) actively restrict access to family planning services.

Population Control Around the World

Government efforts to curb birth rates have met with notable successes, as well as some failures. Great strides have been made in contraceptive use and availability; 9% of the world had access to contraceptives during the period from 1960 to 1965, but by the end of the 20th century, this figure had increased to an estimated 53% (married women between 15 and 49 years) and slowly continues to rise. Yet this is not enough. The United Nations estimates that if we are to achieve a stable world population, 75% of all couples worldwide will need to use contraception. Contraceptive use is also very unevenly distributed (see Figure 2-12). This is part of the reason more than 95% of global population growth is projected to take place in the developing nations of Africa, Asia, and Latin America during the next 35 years.

In the 1990s and early 2000s, the typical completed family in the developing nations, excluding China, had three to five children (and in many families, especially in East Asia, three or less). Although this number represents a drop from the typical six children of the early 1960s, a U.N. Population Fund report observed that cutting the fertility rate from four to two children most likely will prove more difficult than reducing it from six to four. Will contraceptive use continue to increase and fertility rates and average completed family sizes continue to decrease at the rates predicted? It is difficult to know; however, it is just this cut in fertility rate and completed family size—from about four children to two—that is essential if we are eventually to stabilize the world's population. The world has made great strides along these lines in the past few decades. Fertility rates in more than 60 countries are already at or below replacement levels. Let us hope that things portend well for the future.

Disease and Global Population: The Plague and AIDS

VISIT

http://environment.jbpub.com
/mckinney/5e/
for more information

In the mid-14th century, Europe, as well as much of Asia and Africa, was devastated by an outbreak of plague, which contemporaries called the Black Death (**FIGURE 1**). The disease actually took three forms: bubonic plague, pneumonic plague, and septicemic plague (all caused by the bacterium variously known as *Pasteurella pestis* or *Yersinia pestis*). It is estimated that at least 25 million and perhaps as many as 75 million of Europe's population of 100 million died between 1347 and 1351. As the plague raged, the social structure of Europe was destroyed. Contemporary accounts report that government and law enforcement, religious ceremonies, and medical practice disappeared in areas where the plague was worst. In an ecological sense, the plague can be viewed as a classic case of a density-dependent mechanism that served to limit the population. Six and a half centuries later, it should serve as a warning to us as we continue to overcrowd our planet.

Medieval Europeans had no idea what caused the plague or how to control it. It is now known to be caused by a bacterium that can be carried by rodents, such as rats and squirrels, and is transmitted from rodents to people by fleas. For tens of thousands, or even millions, of years, populations of *Y. pestis* have been living in the guts of fleas that feed on rats and infect them with the plague. After the rat dies, the fleas seek another host, carrying with them the plague bacilli. Eventually, the plague-carrying flea also dies, but often not before it has infected other mammalian hosts and indirectly infected other fleas that feed on the same host.

Even today bubonic plague is not well understood, and isolated cases and small outbreaks continue to occur among people. We should remember that even with our advanced medical knowledge and technology, we could conceivably find ourselves facing an unknown or poorly understood, but rampant and devastating, disease. The current AIDS crisis is a roughly analogous situation (**FIGURE 2**). According to the latest data from USAID, the governmental body that assists in the development of struggling nations, globally from 1980 through 2009, an estimated 60 million have been infected with HIV (the cause of AIDS), and over 25 million have succumbed to the disease. At the current rate of infection over 7,000 new cases of HIV occur around the world each day, making AIDS the fourth leading cause of death. To date, more than 16 million children have lost one or both of their parents because of the epidemic. In modern times, disease outbreaks cannot be isolated in the same way that they were in the time of the Black Death. Increased globalization puts people all over the world in contact with one another. Certainly, the ability to transmit information, increased infrastructure, and so forth

FIGURE 1 Victims of the Black Death appear in this 14th-century French fresco. A physician lances a plague-caused bubo (a swollen and inflamed lymph node) on a woman's neck; on the left is another victim with an enlarged bubo under his arm. *Source:* The Granger Collection, New York.

play a leading role that was not present during the time of the plague. This can be both a positive and negative thing. On the one hand, increased inter-communication can result in increased awareness, resource aid, and responses for disease outbreaks from people and institutions all over the world. On the other hand, it can make the disease all the more difficult to isolate and control.

The United Nations has taken a leading role in combating AIDS, especially in African countries, where rates of infection can reach 30%; typically, epidemics curb themselves at that level. Africa still remains the leading epicenter of the global crisis, according to an update released by the Joint U.N. Programme on HIV/AIDS (UNAIDS). This report confirmed that Africa accounts for almost 80% of the 3 million annual fatalities worldwide and more than 60% of the 5 million new infections. Many of these figures are exacerbated by lack of proper funding, inadequate access to health care, poor sanitary conditions in impoverished nations, overcrowding, increases in dangerous drug use, and unsafe sex practices in both heterosexual and homosexual individuals.

The U.N. Millennium goal is to halt and reverse the epidemic by 2015 through HIV pre-vention, care, treatment, and impact alleviation programs. The response from many leading coun-tries has been very positive, and funding for AIDS programs in developing countries, especially from the United States, has increased dramatically—from U.S. $2 billion in 2001 to an estimated U.S. $8 billion in 2008. However, resources still fall short of what is needed to effectively turn back the epidemic. Despite encouraging signs, the report also outlines serious challenges that need urgent attention to achieve the intended goal of reversing the epidemic. Access to HIV treatment and preven-tion services remains low. In 2009, there was an estimated 7.7 billion dollar gap between monetary support needed and monetary support received to address worldwide AIDS services. Globally, only one in five persons has access to prevention ser-vices, and in 2003 targeted prevention services reached only 16% of sex workers, 11% of men who have sex with men, 20% of street children, and less than 5% of the world's 13 million inject-ing drug users. Although 2010 estimates vary, worldwide over 6.5 million people are in need of HIV treatment and are not getting it for a vari-ety of reasons including governmental barriers, access to clinics, and education about treatment. On a more positive note, according to the Secre-tary General's report, progress has been made on several fronts since 2001. Worldwide, the number of people receiving counseling and testing services

VISIT

http://environment.jbpub.com
/mckinney/5e/
for more information

FIGURE 2 The AIDS Memorial Quilt on display in Washington, D.C. There are more than 46,000 panels with approximately 84,000 names, representing approximately 17.5% of the AIDS deaths in the United States.

Disease and Global Population: The Plague and AIDS

doubled in the 4-year period of 2001–2005. In 2009, the number of women accessing services to prevent mother-to-child HIV transmission in middle to low income countries was up to 53%. Internationally, education programs targeted at youth continue to increase. In 2010, the Secretary General noted that the use of antiviral treatments increased 10-fold within the years of 2005 to 2008 in several areas, but cautioned that at the current funding level it will most likely not be possible to halt the spread of AIDS by 2015 as previously predicted by international organizations. While there are many promising areas of improvement in the treatment of AIDS, many challenges remain.

AIDS takes a toll in human lives but also has a devastating effect on local economies and social structures. AIDS reduces the number of healthy workers in their prime and increases the number of dependent people, including the sick, the young, and the old. Scarce resources are diverted to caring for AIDS victims, and as AIDS continues to spread, the productivity of the workforce continues to decrease. Children in particular are being hit hard; teachers are being lost to AIDS, and many children must leave school to help support the family after a parent contracts AIDS. The full effect of the AIDS epidemic will not be seen for another generation or more.

■ study guide

SUMMARY

- The Earth's human population has grown explosively in the last few centuries, particularly during the last few decades, reaching a total (2009 data) of 6.9 billion people.
- For hundreds of thousands or millions of years, our ancestors subsisted as gatherers and hunters at very low population levels.
- Major increases in human population came with the agricultural revolution (about 10,000 years ago) and the Industrial Revolution (in the last 250 years).
- Current population growth is not distributed evenly over the globe; the greatest increases are occurring in the developing countries
- In some Western industrialized regions, such as Europe, the population has stabilized or is even declining.
- The United States has a high population growth rate compared with other developed countries, but part of the increase in population is attributable to immigration.
- Population is one of the major factors in the $I = P \times C$ equation (impact equals population times consumption) and is a major concern of environmentalists.

- Many different solutions have been suggested as ways of dealing with overpopulation.
- Increased industrialization and economic development will force a demographic transition, halting population growth in developing countries.
- Increasing medical care, family planning services, and the use of contraceptives are all ways to improve women's reproductive rights.
- Studies indicate that increasing the educational level and social status of women in particular; positively correlate to an increase in a woman's educational level and the decrease in fertility rates.
- Some governments have used voluntary (such as economic incentives and disincentives) and mandatory (allegations of forced contraceptive use, sterilizations, and abortions) means to regulate population growth.
- The subject of human population is a complex, emotionally charged issue. The potential ramifications of increasing human numbers are a continuing theme that will recur in every chapter of this text.

KEY TERMS

age structure

agricultural (Neolithic)
  revolution

annual growth rate

birth control

birth rate

carrying capacity

demographic transition

demography

doubling time

exponential growth

exponential phase

fertility rates

Green Revolution

growth rate

immigration

Industrial Revolution

infant mortality

lag phase

less developed countries (LDCs)

life expectancy

more developed countries (MDCs)

neo-Malthusian

population momentum

replacement level

reproductive rights

total fertility rate (TFR)

STUDY QUESTIONS

1. Briefly summarize the history of human population growth over the last million years.

2. Explain how the agricultural revolution spurred human population growth.

3. List some factors involved in the exponential growth of the human population over the last 4 centuries. What was the Industrial Revolution? What was the Green Revolution?

4. What parts of the world currently have high rates of population growth? Where are there low rates of population growth?

5. What are some of the potential consequences of global overpopulation?

6. Describe several alternative scenarios for global population growth through the year 2100.

7. Discuss the thesis that there is no such thing as human "overpopulation."

8. How have the growth rate and doubling time of the world's population changed over the millennia?

9. Compare and contrast the age structure profiles of populations in developing versus industrialized countries.

10. Explain the concepts of live expectancy, infant mortality rates, and TFRs. What types of factors affect each of these in different parts of the world?

11. Briefly discuss how the equation $I = P \times C$ relates to the problem of global human overpopulation.

12. Describe the concept of the demographic transition model. What criticisms has this model received?

13. How does the education of women bear on the issue of global population?

14. Describe some of the controversies surrounding various methods of birth control and abortion.

15. Discuss the appropriateness of voluntary versus mandatory government regulation of childbearing.

WHAT'S THE EVIDENCE?

1. Throughout this chapter, the authors present many statistics concerning populations on a global and national scale over time. How reliable do you think these statistics are? Can we be sure of how many people inhabited Earth in, for instance, 1800? How precisely can we know the number of people that are alive today? As a project, research population statistics from various sources (e.g., the U.S. Bureau of the Census, the Population Division of the United Nations, and the World Resources Institute), and compare the various numbers and the data that are used to support them.

2. The authors suggest that the theory of the demographic transition—that increasing technological and economic development inevitably leads to a decrease in population growth rate—is not always true. What evidence is marshaled against the demographic transition model? Are you convinced that the demographic transition model, long held to be true by many scholars of demographic theory, is fundamentally flawed?

CALCULATIONS

1. If the population of Nigeria was 126.93 million at the midpoint of 1995 and had increased to 130.77 million by the midpoint of 1996, what was Nigeria's percentage annual growth over this period? At this annual growth rate, approximately how long will Nigeria's population take to double? (For information on population equations and statistics, see the Fundamentals Box.)
2. If the population of El Salvador was 5.77 million at the midpoint of 1995 and had increased to 5.88 million by the midpoint of 1996, what was El Salvador's percentage annual growth over this period? At this annual growth rate, approximately how long will El Salvador's population take to double?
3. Visit the U.S. government census website and write down the U.S. population and the world population. Return to the website in 24 hours and record the U.S. population and world population. Calculate how many people were added to the U.S. and world populations in the 24-hour period.

ILLUSTRATION AND TABLE REVIEW

1. Using the graph in Figure 2-2a, an estimate of the global human population in 1000 A.D. and 1900 A.D. By what amount did the population change over this period? What percentage change is this?
2. Using the projections in Figure 2-2b, extrapolate each curve to the year 2150. Based on your extrapolations, what might the global world population be in 2150? Is there any validity to such an estimate based on extrapolating a projection? On what assumptions is such an extrapolation based?
3. Based on Figure 2-7, what percentage of the population of the less developed regions was 9 years old or younger in 2002? What percentage of the population of the more developed regions was 9 years old or younger in 2002?
4. Using the figures in Table 2-3 and assuming a total world population of 6.15 billion in 2001, what percentage of the human population lived in Pakistan in 2001? In Nigeria? In China? In India? In the United States?
5. Of the cities listed in Table 2-6, which had the fastest rate of growth between 1950 and 1995? By what percentage did Karachi grow between 1950 and 1995? By what percentage did Lagos grow between 1950 and 1995?

http://environment.jbpub.com/mckinney/5e/

http://environment.jbpub.com/mckinney/5e/

Connect to this text's website at http://environment.jbpub.com/mckinney/5e/. This site features eLearning, an online review area that provides quizzes, chapter outlines, and other tools to help you study for your class. You can also follow useful links for more in-depth information.

With Earth's population just over 7 billion people, living sustainably has never been more important. The United Nations projects that another 2.5 billion people will be added to the world's population by 2050. When will Earth reach its carrying capacity, or has it already? While research may differ about this important question, there is no doubt that the choices consumers make, the products manufacturers offer, and the policies that govern resource management make an impact on the health of the planet, for better or for worse. Human consumption and population growth are two of the largest challenges facing our global community. The viability of future generations rests in the collective choices of the global community; cooperation across traditional geographic boundaries is essential.

2.6 Solving the World's Population Problem

The Environment of Life on Planet Earth

"Tug on anything at all and you'll find it connected to everything else in the universe."

John Muir, Founder of the Sierra Club

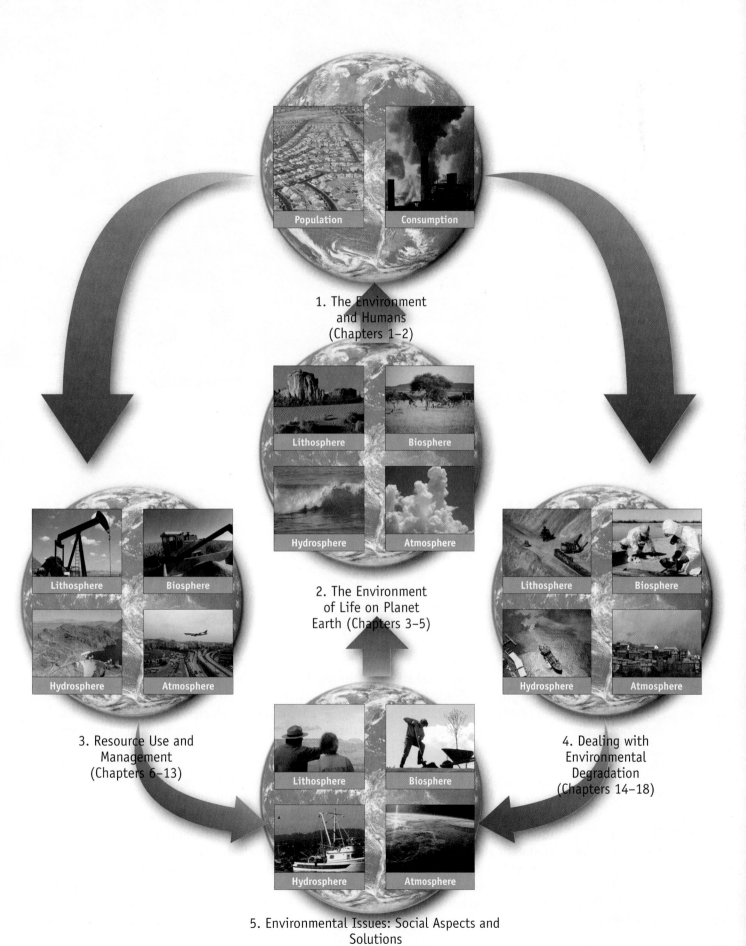

1. The Environment
and Humans
(Chapters 1–2)

Population

Consumption

Lithosphere

Biosphere

Hydrosphere

Atmosphere

2. The Environment
of Life on Planet
Earth (Chapters 3–5)

Lithosphere

Biosphere

Hydrosphere

Atmosphere

3. Resource Use and
Management
(Chapters 6–13)

Lithosphere

Biosphere

Hydrosphere

Atmosphere

4. Dealing with
Environmental
Degradation
(Chapters 14–18)

Lithosphere

Biosphere

Hydrosphere

Atmosphere

5. Environmental Issues: Social Aspects and
Solutions
(Chapters 19–20)

Of the 97% of water covering the earth, only 3% is fresh water. Of that, much is locked in deep underground aquifers, leaving only 1% readily available to be cycled through the biogeochemical cycles that support life. This becomes apparent in areas of the world with frequent droughts. Over hunted in the early 1900s, elephant worldwide populations, such as pictured sharing waterhole in Namibia, remain critically endangered. With human development encroaching on critical habitats throughout the world and the persistence of poachers remaining one of the biggest threats to endangered species, the work of conservation organizations remains critical. Although we live in a time of incredible biodiversity, securing critical resources, such as water and land security, for these endangered animals remains an international challenge with increased urgency. See: http://www.desertelephant .org/about-ehra.html

3

The Biosphere: Populations, Communities, Ecosystems, and Biogeochemical Cycles

Chapter Objectives

After reading this chapter, you should be able to explain or describe the following:

- The four spheres of the environment and how they are interrelated
- The major aspects of the physical environment
- The basic science of ecology
- The basic principles of population dynamics
- Human impact on natural populations
- Biosphere interactions (communities and ecosystems)
- Ecosystem productivity
- Matter cycling
- The major biogeochemical cycles and their features
- Major energy flows on Earth

The environment is everything that surrounds you, including the air, the land, the oceans, and all living things. For convenience, the natural environment can be subdivided into two parts: (1) the physical environment, which includes nonliving things, and (2) the biological environment, which includes all life forms. The physical environment can be further subdivided into the three basic states of physical matter: solid, liquid, and gas. This division creates four "spheres" that compose the natural environment. The three physical spheres are called the **lithosphere** ("*lithos*" = rock), **hydrosphere** ("*hydro*" = water), and **atmosphere** ("*atmos*" = vapor). The biological environment is called the **biosphere** ("*bios*" = life).

FIGURE 3-1 shows how the four spheres form the outermost layers of the planet Earth. A very thick molten layer of rock belonging to the mantle and a heavy metallic core underlies them. Three major points about Figure 3-1 should be stressed. First, these four spheres make the Earth a dynamic planet. The lithosphere creates rocks as hot magma, generated by the "internal heat engine" of radioactive minerals. Sometimes this magma erupts onto the surface in the form of volcanoes. The atmosphere erodes the rocks with chemically reactive gases. The hydrosphere also erodes rocks as well as

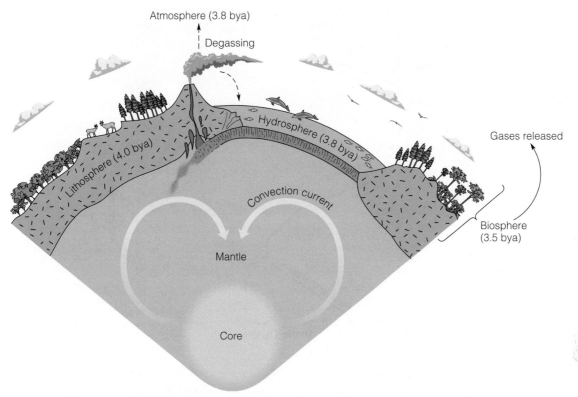

FIGURE 3-1 Spheres of the environment. The dates given in billions of years ago (bya) are approximate times when each of the major spheres of the environment originated: lithosphere 4.0 bya, atmosphere and hydrosphere 3.8 bya, and biosphere 3.5 bya. Note that the Earth is a dynamic planet and that all of the spheres have evolved and changed since its origin.

transporting them as sediments that are often ultimately deposited in ocean basins. The moving lithosphere then carries these sediments to great depth, where they are reheated and remelted into magma, beginning the cycle anew. The life of the biosphere inhabits parts of the other spheres, relying on the cycling of chemicals through all of the spheres. Life also affects the other spheres by altering the cycling of chemicals, such as oxygen.

Second, the figure illustrates that the spheres closely interact with one another. Matter is transported both within and between the spheres in various kinds of cycles. For instance, gases are expelled from the lithosphere (by volcanoes) and the biosphere (by plants and animals) into the atmosphere. Similarly, water vapor moves from the oceans to the atmosphere and back again.

Third, these dynamic and interactive spheres have evolved through time. Figure 3-1 depicts the environment as it exists today, but the Earth is about 4.6 billion years old. The current lithosphere, originally formed from a molten layer on the hot, young Earth,

probably originated some 4 billion years ago. The oceans and atmosphere date to at least 3.8 billion years ago, and the earliest known fossils (single-celled microbes) representing the biosphere are at least 3.5 billion years old.

3.1 Water, Air, and Energy—Three Major Aspects of the Physical Environment

The Earth's surface is almost entirely covered with water—not just the seas and oceans, but numerous lakes, ponds, swamps, and even morning dew on the "dry" land. It is enveloped in an atmosphere composed predominantly of nitrogen, or N_2 (78.1%), and oxygen, or O_2 (20.9%). It is also bathed in radiation—energy in the form of heat (infrared radiation) and light—from the sun. All of these components work together to make the surface conditions of our planet very active and constantly changing. In addition, these components underpin the supporting ecosystem services, which include nutrient cycling, primary production

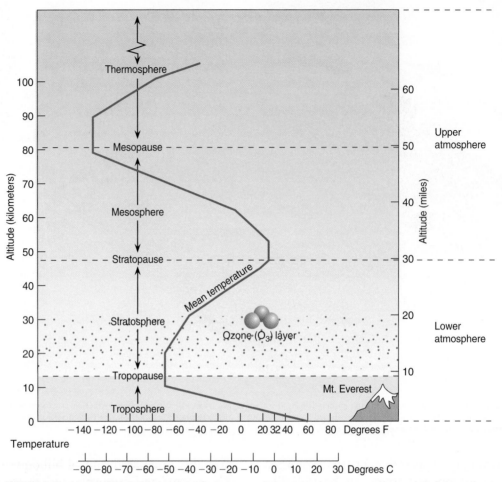

FIGURE 3-2 The structure of the Earth's atmosphere. The tropopause, stratopause, and mesopause are the boundaries between the troposphere and stratosphere, stratosphere and mesosphere, and mesosphere and thermosphere, respectively.

(photosynthesis), and water cycling, that maintain the other ecosystem services.

The Earth's atmospheric envelope is divided into layers, distinguished mainly by the changing temperature gradient encountered as one moves up into the atmosphere (**FIGURE 3-2**): the troposphere, stratosphere, mesosphere, and thermosphere. The changing temperature gradient is the result of the interplay of the energy received directly from the sun, the gaseous constitution and density of the atmosphere at different levels, and the distance the energy (heat) reflected and emitted from the Earth's surface reaches into the atmosphere.

In terms of energy flow, the Earth's surface is an open system: it constantly receives energy from the sun, and it continuously loses most of this energy back into space. Approximately 30% of the incoming solar energy reaching the upper atmosphere is im-

mediately reflected back to space, and another 19% is absorbed by water vapor in the atmosphere; only about 51% reaches Earth's surface. Very little (less than 1%) of the sun's energy reaching Earth directly supports living organisms; that is the amount that goes into **photosynthesis**. Most of the energy reaching Earth goes into evaporating surface waters, where it re-enters the atmosphere and is eventually dissipated into space as heat (electromagnetic radiation in the infrared range).

The movement of water about the surface of Earth, driven by energy from the Sun, constitutes the **hydrologic cycle** (**FIGURE 3-3**). More than 97% of the water on or near the Earth's surface is found in the seas and oceans; the remaining water is bound up as ice (primarily in the polar ice caps), stored temporarily as groundwater, found in freshwater lakes and streams, or takes the form of atmospheric

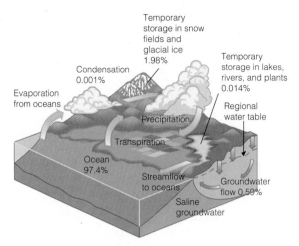

FIGURE 3-3 The hydrologic cycle.

water vapor. Only about 0.0015% is available to people in the form of accessible freshwater. The shifting of water over the globe through the hydrologic cycle redistributes heat and generally makes for more equitable climates. The hydrologic cycle also plays a prominent role in the weathering and decomposition of rocks. Together with other atmospheric movements (influenced by many factors, including the revolution of the Earth around the sun to give the seasons and the movement of predominant winds over the surface of Earth; a review of the dynamic Earth and natural hazards provides more information), the hydrologic cycle is also responsible for **weather,** or the short-term, daily fluctuations in the atmospheric temperature, precipitation, and winds. **Climate** is the long-term average of these same variables over time ranges of a year to thousands or millions of years.

3.2 Biosphere Interactions: Populations

The biosphere is hierarchical: organisms, composed of atoms, molecules, and cells, are grouped into populations. Populations form communities, which then form ecosystems. Finally, ecosystems, when considered together, form the biosphere, which subsumes all life on Earth. **FIGURE 3-4** illustrates the interconnectedness of life; each level is composed of the units in the level below. Each level is dynamic, with many interactions occurring among its units. For instance, the organisms that compose

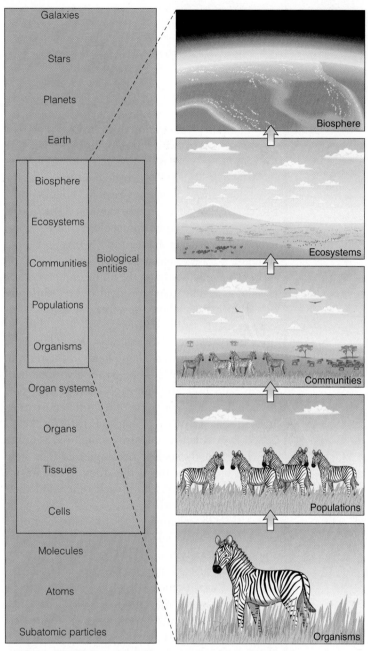

FIGURE 3-4 The cosmos can be depicted as a hierarchy from subatomic particles up through galaxies. Life, or biological entities, is at an intermediate level, ranging from cells through the biosphere.

populations interact in many ways, as do the populations that compose communities.

Ecology is the study of how organisms interact with each other and their physical environment. Similarly, biology is the study of life. Because interactions occur at many levels in the hierarchy of life, ecology is very complex and often yields generalities and predictions that are less precise than we might wish. Partly because of this complexity, ecology has come

What Is Ecology?

The term "eco-" is derived from the Greek word *oikos*, which means "home." Because *logia* means "study of" in Latin, ecology is the study of home. The word economics has the same derivation: *nomos* is Greek for "managing"; therefore, economics means managing the house. Ironically, these two words, ecology and economics, which derive from the same word, are often thought to represent opposing interests. Even more ironically, studying the home and managing the home seem to have resulted in very different sets of priorities about how the home (i.e., our environment) should be treated.

Henry David Thoreau was apparently the first to use the word ecology; he used the term in 1858 in a letter but did not give it a specific definition (and possibly Thoreau's ecology was simply a misspelling for some other intended word). Instead, the German biologist Ernst Haeckel is generally credited with introducing, in 1866, the word (originally spelled oecologie) as it is now used in biology to mean the study of organisms and their interactions with each other and their physical environment.

Although scientists still use this technical definition, ecology has come to mean many other things, especially to nonscientists. As environmental awareness has risen over the last few decades and environmental issues have come to be discussed in social rather than scientific contexts, ecology has taken on new meanings. Many people who express concern about the environment call themselves ecologists and are interested in ecology. This label has even been extended to political ideologies that reflect these concerns. For example, the political scientist William Ophuls has said

that ecology is a profoundly conservative doctrine in its social implications.

The confusion arises when ecology is expanded beyond its restricted meaning as a branch of biology that studies natural environments (ecosystems) to refer to the social and political ideas of people who are actively concerned with preserving those natural environments. For reasons of clarity, we might do better to use environmentalist as a general term for someone who actively wants to preserve the natural environment, while reserving ecologist for scientists who study it. (Of course, a person can be, and often is, both an ecologist and an environmentalist.) Ophuls should say that environmentalism is a conservative doctrine because the study of something is not a doctrine at all: a doctrine is a system of beliefs.

In addition to being expanded to mean social and political environmental action, ecology has been generalized as an academic term. Many students, perhaps even yourself, have enrolled in ecology courses expecting to learn about water pollution, solar power, and many other aspects of environmental problems. They are often surprised to find that the course focuses on the study of natural communities, unaltered by humans. Although the study of natural laws governing such communities is basic to understanding how people affect them, many students desire a broader perspective that includes pollution and other ways that we modify the natural world. Environmental science, such as the course you are enrolled in now, has arisen in recent years to fulfill this need. Because environmental science is such a broad area of study, it is taught in a variety of traditional science departments, including biology, geology, chemistry, and geography.

to have a somewhat different meaning in popular usage (see **CASE STUDY 3-1**, What Is Ecology?). The science of ecology provides us with a general understanding of how interactions in the hierarchy of life occur. Interactions among coexisting populations are among the most fundamental, so let us begin there.

Population Dynamics

A **population** is a group of individuals of the same species living in the same area. There is no limit on the size of the area; a population may be all the bass in a lake or all the black bears in Alaska. Whatever the species or area, all populations undergo three distinct phases during their existence: (1) growth, (2) stability, and (3) decline (**FIGURE 3-5**).

Growth occurs when available resources exceed the number of individuals able to exploit them, such as when a population is first introduced into an area. Individuals tend to reproduce rapidly, and death rates are relatively low because of the relatively abundant resources. Stability occurs when population growth levels, as the environment becomes saturated with individuals of that population; however, a population "crash" usually precedes stability because the rapidly growing population abruptly exceeds the resources, causing an "overshoot" (Figure 3-5). Even stable populations fluctuate, sometimes widely, so stability does not mean that population abundance is unchanging. Usually, stability is the longest phase by far. Decline refers to the inevitable decrease in abundance that leads to extinction of all populations in the long term. Each of these three phases is affected by many factors; the study of how these factors interact is called population dynamics.

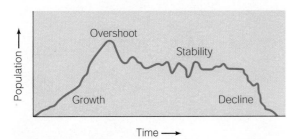

FIGURE 3-5 Initially, a population will sometimes undergo exponential growth until it exceeds the capacity of the environment to support it, causing overshoot. This phase is sometimes followed by a period of relative stability and eventually by decline.

Growth of Populations

Nearly all populations have one key trait in common: they will grow exponentially if left unchecked. This exponential growth occurs because individuals usually produce many more than one offspring in their lifetime. Furthermore, each of those offspring can, in turn, have many more. As a result, most populations have the potential to increase very rapidly. The potential for increase in a given population is called the **intrinsic rate of increase**, symbolized by r.

The exact intrinsic rate varies among populations depending on several factors, many of them genetically determined. Two of the most basic are the birth rate and death rate:

Intrinsic rate of increase = birth rate − death rate

This equation shows that the net rate of increase in a population is governed not only by the rate at which the population can multiply, but also by the rate at which individuals die. For example, bacterial populations can have birth rates of millions per day, whereas the time between births for elephants can be 3 to 4 years or longer, and orangutans once every 8 years on average. However, bacteria also die by the millions every day, so the net intrinsic rate is dramatically lower than if it were determined by births alone.

Population Stability: Regulation of Growth

Common sense tells us that nothing grows forever, and this is certainly true of populations. Growth is limited by the complex interaction of many factors, including other species.

The simplest way to examine abundance limits is to consider a species of microbe in a glass container (but the same analysis can be applied to any biological population, such as deer on an island). When a few individuals are first admitted to the container, they will likely experience rapid population growth for the reasons we have just discussed. However, at some point, the limited space and food in the container will cause the multiplicative growth phase to enter a slower growth phase as net reproduction slows. Growth continues to slow until the population size stabilizes—that is, fluctuating around some average size. This average abundance where the population levels is the **carrying capacity**, or the maximum population size that an environment can sustain for a long time. As noted, the carrying capacity is usually briefly exceeded by an overshoot phase before the population levels in the stability phase.

Four Basic Abundance Controls

What causes population size to level and stabilize? Clearly, the ultimate cause is to be found in the environment: factors external to the population must be limiting the rate at which individuals can reproduce and survive. Ecologists typically divide these environmental factors into two basic categories: the (1) physical environment and (2) biological environment. Within the biological environment, we can identify three subcategories of biological interactions that can limit population growth:

1. Physical environment
   Physical limitations
2. Biological environment
   Competition
   Predation
   Symbiosis

In the discussion on the **environment and people**, we noted that a system can be defined as a "set of components functioning together as a whole." The abundance controls on populations that are discussed here are in fact natural systems that act to together to control the sizes of populations.

Limitations of the physical environment include any number of constraints, such as water supply, space availability, or soil and light in the case of plants. The population's **habitat**, which is the place where it lives, determines all constraints of the physical environment. It takes only one aspect of the physical environment to limit population growth. For instance the Los Angeles area, with its idyllic soil, light, and temperature but scarce water, could not support its enormous human population without the water being piped in from the north and east. This observation has come to be called the **law of the minimum**: growth is limited by the resource in the shortest supply.

Competition occurs when organisms require the same limited resource. Competition can take many forms, but two of the most important are exploitative and interference competition. Exploitative, or "scramble," competition occurs when both competing populations have equal access to the resource: the population that exploits the resource the fastest is the "winner." Interference competition occurs when one of the competitors prevents the other from gaining access. For instance, coyotes and red foxes share the same habitat and diet. Coyotes attack red foxes but usually do not eat them. The threat of a coyote attack is enough to force the foxes to abandon their natural habitat (**FIGURE 3-6**). Arguably, humans are currently the most competitive species on the planet, in both exploitative and interference terms. In the Los Angeles example, the early developers of the city "interfered" with the Owens Valley ranchers and fruit growers by appropriating virtually all of their water.

(a)

(b)

FIGURE 3-6 Red foxes are often on the losing end of interference competition from (a) coyotes, and (b) the red foxes can be forced into a habitat they do not adapt to easily.

A basic determinant of any competition is the organism's **niche**, which is often defined as the organism's "occupation" or how it lives. The niche can be thought of as how the organism fits into the local "system," such as the local habitat, ecosystem, or biome. The niche includes what the organism eats, how it eats, and what eats the organism. The degree to which the niches of competing species overlap indicates the extent of similar resources required and therefore the strength of the competition between the species.

Competitive exclusion occurs when niche overlap is very great and competition is so intense that one species eliminates the second species from an area. Although competitive exclusion has received much theoretical attention, some ecologists think it is relatively rare in nature. In contrast, species often compete for only one or a few resources, but each species can also use resources that the others will not. For example, foxes and owls in a forest may compete for mice, but each will also eat prey that the other usually does not pursue. Such cases of minor niche overlap do not result in complete competitive exclusion. Instead, each competitor merely limits the abundance of the other. When the single-celled eukaryote *Paramecium aurelia* is raised alone, its population size is nearly twice as large as when it must share resources with a second species of pond water organism.

An important result of competitive exclusion is that if one of two competing species is removed, the remaining species may increase in number. The population of *P. aurelia* rebounds after the competitor is removed. This is often called **ecological release** because the species is "released" from one of the factors limiting its abundance. In this case, some ecologists would call it "competitive release," specifying release from competition. **Predation** occurs when organisms consume other living organisms. Predators can be divided into (1) carnivores, which prey on animals; (2) herbivores, which prey on plants; and (3) omnivores, which prey on both. As with competition, the extent to which predation limits abundance of prey varies considerably. In the most extreme cases, a predator drives the prey species to extinction. Much more commonly, predators simply limit abundance and do not drive the prey to extinction. In the case of many herbivores, they eat some leaves or fruit, but they don't kill the plant.

Why don't predators keep reproducing until they have eaten all of the prey? At least three major reasons have been identified.

1. Prey species have often evolved protective traits, such as camouflage, poisons, spines, large size, and so forth.
2. Prey species often flee to refuges such as burrows and treetops where predators cannot reach them. Experiments have shown that an insect or microbe predator population in a container will consume the prey population into extinction unless the prey can escape to a refuge.
3. Predators tend to engage in *prey switching*. When one prey species becomes scarce, predators switch to another prey, thereby allowing the first prey species to rebound in numbers.

Ecological release is also seen in predation control of abundance. When predator populations are greatly reduced by disease or other causes or when prey populations move to an area without major predators (such as islands), the prey species may exhibit "predatory release" and expand in the same way populations expand when experiencing competitive release. Today, people are a major cause of predator release both by killing off predators, such as mountain lions, and by introducing species into areas without major predators.

Symbiosis (sym = "together"; biosis = "life") encompasses many kinds of interactions, sometimes including competition and predation. As **TABLE 3-1** shows, these interactions can be classified according to whether they benefit or inhibit one or both populations that are interacting. **Mutualism** benefits both species. An example of mutualism is the coexistence of algae (known as zooanthellae) within the tissue of the tiny animals that build coral reefs. The algae are provided with a protected living space while the coral animals are supplied with nutrients from the algae's photosynthesis. Similarly, a lichen is

TABLE 3-1	Symbolic Classification of Symbioses*	
Form of Symbiosis	**Species A**	**Species B**
Mutualism	1	1
Predation and parasitism	1	2
Commensalism	1	0
Competition	2	2
Amensalism	0	2
*1 = benefit, 2 = detriment, 0 = no effect.		

FIGURE 3-8 Commensal organisms, such as these barnacles, derive benefits from the turtle without harming it.

actually a fungus and algae growing together symbiotically. Mutualism is a factor regulating the abundance of certain types of algae because their numbers are influenced by the abundance of coral or fungal hosts.

Parasitism is similar to predation in that one species benefits while harming the other species (**FIGURE 3-7**). The main difference is that parasites act more slowly than predators and do not always kill the prey (host). **Commensalism** occurs when one species benefits and the other is not affected. For example, Spanish moss hangs from trees for support but causes the trees no great harm or benefit. Barnacles attach to crab shells or other organisms in a similar way, deriving food from the surrounding water (**FIGURE 3-8**). **Amensalism** occurs when one population inhibits another while being unaffected itself. Harm to one species is simply an incidental by-product of the actions of another. For example, when elephants crash through vegetation, they often have a

detrimental effect on it, while gaining relatively few benefits (of course, the elephants get where they want to go).

The Real World: Complex Interaction of Abundance Controls

Our brief description of the various controls on abundance (physical fluctuations, competition, predation, and symbioses) has greatly oversimplified the concept by discussing each control separately. In reality, not just a single control but several controls, often acting simultaneously and forming a system, govern most natural populations. For example, when two populations of beetles interact in the laboratory, competition leads to a high abundance of one species and a very low abundance of another. When a third species, a parasite, is introduced, the formerly abundant species becomes the rarer one because it is more susceptible to the parasite (**FIGURE 3-9**).

Now consider that natural communities are often composed of hundreds to many thousands of species. Any single population will be affected by changes in the population abundance of its many competitors, predators, and symbionts such as commensals. At the same time, changes in physical conditions can influence any of the populations, with each responding differently. Thus, the abundances of all species are nearly always fluctuating in response to physical conditions

FIGURE 3-7 The wasp eggs on this tomato hornworm are parasitic; when the larvae hatch, they will eat the caterpillar.

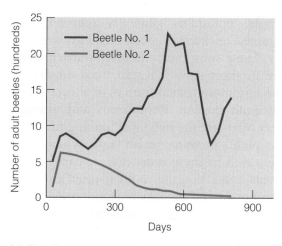

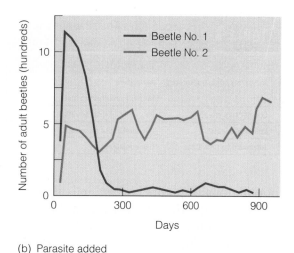

(a) Parasite absent

(b) Parasite added

FIGURE 3-9 (a) With two competing species of beetles, the superior competitor can drive the other to a very low abundance. (b) The addition of a parasite can drive the formerly abundant species to rarity and allow the formerly rare species to increase. *Source:* Modified from T. Park. Ecological Monographs 18, 1948:265–308.

and other species. In addition, this complex interaction of abundance controls in natural ecosystems means that small changes in one control, such as the abundance of a prey species, can have a cascading domino effect throughout the ecosystem.

Population Decline

Although even stable populations rise and fall in abundance during fluctuations, given enough time, eventually most populations decline to zero and become extinct. **Extinction** is the elimination of all individuals in a group. In a local extinction, all individuals in a population are lost, but a new population can be re-established from other populations of that species living elsewhere. Species extinction occurs when all populations of the species become extinct.

The inevitability of population decline is clear when we consider that more than 99% of all species that have ever existed are now extinct. When examining past extinctions with current trends, there are several key points to keep in mind. The great majority of these became extinct before humans, so decline and extinction are natural processes. Ultimately, environmental change causes decline and extinction: one or more of the abundance controls becomes altered, leading to an abrupt or gradual decline in abundance. For instance, many extinctions in the fossil record resulted

largely from physical changes such as cooling climates or asteroid impacts. Similarly, the fossils reveal cases where competition from new groups apparently caused decline, such as when placental mammals invaded South America and replaced most marsupials.

Human Impact on Population Growth and Decline

Increasing human populations and traditional industrial technology have caused progressively greater disruption in natural populations. Pollution, agriculture, and many other kinds of human alterations of the environment have destabilized populations by affecting the various abundance controls. Depending on the population, people may cause (1) population growth by removing previous limitations or (2) population decline by imposing new limitations.

Human Impact on Population Growth

Populations grow when there is an excess of resources relative to the number of individuals available to exploit them. People commonly contribute to population growth in the four ways listed in **TABLE 3-2**; each involves removing a control on abundance.

An increase of available resources can be planned or unplanned. Agriculture and animal domestication are obvious examples of how people can deliberately increase the

TABLE 3-2	Four Ways That People Cause Population Growth in Nature
	Examples
Increase available resources	Agriculture
	Nutrient pollution in lakes
Competitive release	Poisoning of insect pests
Predator release	Overhunting of large carnivores
Introduce to new areas	Game releases

populations of plants and animals that they favor. We do this by providing far more food and other resources than those organisms would find in their natural state. The effect of people on cat numbers provides a striking example. In England alone, domestic cats occur in such numbers that more than 300,000 cats must be euthanized each year. Pet overpopulation is a huge concern in the United States. Domesticated cats, besides consuming resources, collectively kill around 4.4 million songbirds *per day*, as well as other wildlife. Yet before people began to domesticate them a few thousand years ago (to catch mice in grain storage areas), the small, wild ancestors of our pet cats were somewhat rare and probably were limited to the Middle East and Europe, a relatively small area compared with their current range. Similarly, corn (maize), potatoes, and many other domesticated crops had much smaller ancestral ranges.

Unplanned increases in available resources by people also have major effects. For example, pollutants generally represent unplanned releases of substances into water or air. These substances are often nutrients that are in short supply. Recall from the law of the minimum that resources in shortest supply limit growth. Phosphorus and nitrogen very often are the two most **limiting nutrients** for plants in water or on land. When fertilizers rich in these nutrients are carried into rivers and lakes, the result is often runaway plant growth. This enrichment of nutrients in waters is called **eutrophication**. Eutrophication is a classic example of "too much of a good thing" because the decomposition of the accumulating plants uses up increasing amounts of oxygen, causing fish and other organisms to suffocate (**FIGURE 3-10**).

Competitive release is common when people try to eliminate a population of one species, allowing its competitors to increase in numbers. The most economically important examples occur when farmers try to eradicate pests from crops with pesticides. Because poison tolerance varies among species, some pests will not be killed by the poisons and will actually increase in numbers when their competitors are gone. This is called a "secondary pest outbreak." A dramatic example occurred in Central America, where cotton crops were sprayed in 1950 to kill boll weevils. The effort was highly successful until 1955, when populations of cotton aphids and cotton bollworms soared. When a new pesticide was used to remove them, five other secondary pests emerged. Such experiences have led to new methods of pest control (a review of food and soil resources and principles of pollution control, toxicology, and risk provides more information).

Predator release is common where people hunt, trap, or otherwise reduce populations of predators, allowing the prey species' population to increase. For example, large mammalian predators such as wolves and panthers have long been the target of ranchers and farmers because they prey on domesticated animals. The result has been a rapid increase in the predators' natural prey. Deer have shown an especially spectacular rise. Indeed, most experts estimate that more deer are now living in the United States than were here be-

FIGURE 3-10 An overabundance of phosphorous and nitrogen, often from fertilizers, can cause eutrophication in a lake.

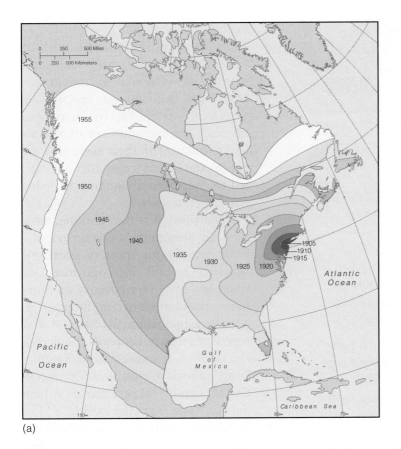

(a)

(b)

FIGURE 3-11 (a) This map shows the rapid migration of the European starling across North America after its introduction in New York. (b) The European starling is now one of the most common birds in North America. *Source:* (a) Adapted from C. B. Cox, P. Moore. Biogeography. Cambridge, MA: Blackwell, 1993, p. 62.

fore Europeans arrived. Unfortunately, the excess populations often lead to overgrazing and death by starvation. For this reason, people's controlled game hunting of certain animals is a justifiable activity, at least based on ecological criteria. By prudently culling deer and other prey, people are essentially carrying out the role of predators that no longer exist.

The introduction of nonnative (also called "exotic") species into new areas is, along with such issues as habitat destruction and global climate change (a review of global air pollution provides more information), one of the most severe alterations of nature carried out by people so far. Few people realize the enormous scale on which we have, either accidentally or purposely, transferred organisms from one area to another. For example, more than 1,500 insect species and 24 families of fish have been introduced successfully into North America. The extent of this introduction is even more impressive considering that most of the populations die out when initially introduced. An analysis of ballast water in a

tanker revealed that it contained the live larvae of 367 species of marine organisms carried from Japan to the Oregon coast. Such ballast waters are discharged routinely. After an organism is introduced and established, expansion can be quite rapid. The European starling covered much of North America in just a few decades after it was introduced in New York for a Shakespearean play in the early 1900s (**FIGURE 3-11**).

Human Impact on Population Decline

People are causing the decline and extinction of species at thousands of times the natural rate in nearly all parts of the biosphere. Given that biological diversity underpins critical ecosystem functioning, the loss of species has a profound impact on the provisioning of critical ecosystem services. A review of conserving biological resources provides information on the causes of extinction, but here we note that people alter the environment in four basic ways to cause population decline (**TABLE 3-3**). All are directly related to

TABLE 3-3	Four Ways That People Cause Population Decline and Extinction*	
	Examples	
Change physical environment:		
1. Habitat disruption	Draining a swamp, toxic pollution	
Change biological environment:		
2. Introduce new species	New predator	
3. Overkill	Big-game hunting	
4. Secondary extinctions	Loss of food species	

*The actions are listed in approximate order of importance in causing extinctions. In other words, habitat disruption causes the most extinctions today; secondary extinctions cause the least.

the abundance controls. The first affects the physical environment and the other three ways affect the biological environment.

- Habitat disruption occurs when people disturb the physical environment in which a population lives. This disturbance can range from minor, such as mild chemical changes from air pollution, to major, such as total destruction of a forest. California's red-legged frog, made famous in Mark Twain's short story "The Celebrated Jumping Frog of Calaveras County" was once common in most of the state. This large frog can now be found only in isolated, undisturbed pockets along California's central coast and in a few sites in the Sierra Nevada (**FIGURE 3-12**).

FIGURE 3-12 California's red-legged frog.

- People can introduce species that are competitors, predators, or symbionts (including diseases and parasites) in the native biological system (see **CASE STUDY 3-2**). Island species are especially susceptible to introduced species. The introduction of rabbits to Australia, with their voracious appetite and rapid reproduction rate, led to decreased abundance of many native marsupials.
- Human overkill is the shooting, trapping, or poisoning of certain populations, usually for sport or economic reasons. Overkill since our ancestors' times has been very effective in eliminating populations of large animals (especially large mammals) because they reproduce slowly. Leopards, elephants, rhinoceroses, pandas, and many other large animals comprise a disproportionate number of threatened and endangered species in the world.
- Secondary extinctions occur when a population is lost due to the extinction of another population on which it depends, such as a food species.

It is not necessary for these environmental changes to reduce a population to zero in order to cause extinction. Even if many individuals survive, the population may never recover if it becomes too small.

Population Range

In addition to varying through time, abundance may vary geographically. Populations tend to have a maximum abundance near the center of their geographic range, which is the total area occupied by the population (**FIGURE 3-13**). This *central maximum* occurs where the physical and biological factors that control abundance are the most favorable. As one moves away from this central optimum into the zone of physiological stress, abundance generally begins to decline. This decline is usually gradual because both physical and biological limiting factors tend to follow a gradient. Physical conditions may change gradually (e.g., latitudinal temperature) or abruptly (such as where the water in a lake meets the shoreline), as can

CHAPTER 3 The Biosphere: Populations, Communities, Ecosystems, and Biogeochemical Cycles

Captive Breeding

VISIT

http://environment.jbpub.com
/mckinney/5e/
for more information

When the hope for survival of a species is no longer possible in the wild, conservationists turn to captive breeding. Captive breeding, sometimes called *ex situ* conservation, is seen as a last resort conservation method, but it has been effective for some species.

Most captive propagation programs in zoos focus on large birds and mammals, partly because they face greater extinction threats than do most small animals, but also because these are more successful in attracting customers. Their popularity has led to some problems because the mammals receive an unrealistic amount of attention compared with other organisms that make up just as important a part of dynamic ecosystems. Zoos are very costly to maintain and can usually realistically sustain only small portions of a population. In addition, large mammals have performed particularly poorly in captive breeding programs.

When one thinks of captive breeding, the marquee mammals, such as pandas, usually come to mind, but a lot of captive breeding is done with plants. Not surprisingly, it is much easier to breed and sustain populations of plants than animals; they are easier to care for, easier to breed, and easier to relocate, and you do not have to muck out any pens.

Whether it is the last surviving plant of its species or a disappearing animal, the goal of most captive breeding programs is eventual reintroduction into the wild. The results from most of these attempts are a mixed bag, as are most environmental endeavors. A habitat may be intact enough to allow the species to reintroduce successfully, but if that habitat is not well protected or the animal itself is not protected, the species will quickly decline again. Consider the California condor: in the first few years the U.S. Fish and Wildlife Service began reintroducing captive-bred condors to the wild, five condors died because of contact with power poles or lines.

In other cases, it is difficult to convince local citizens that reintroduction of "vermin" is a good idea. Gray wolves once numbered in the millions but were extirpated from the western United States by the 1930s. The Fish and Wildlife Service wanted to reintroduce the wolves into Montana, Idaho, and Wyoming. Because of the wolves' threat to livestock, ranchers were against the reintroductions. A nonprofit organization that promised to reimburse the cost of the killed livestock alleviated this, but some ranchers believed it was within their right to kill any protected wolf they caught attacking their livestock. The issue of reintroduction remains contentious and complex.

The problem with captive breeding programs is that they often focus on populations with very few individuals. Many of these individuals do not breed well in captivity, and often there are not enough members to make it a truly diverse population. Obviously, not every species can be bred in captivity. The lack of knowledge, funding, and other resources makes it impossible to breed the thousands of fungus, plant, shellfish, insect, fish, amphibian, reptile, bird, mammal, and other species that are threatened in the wild. A better method of conservation would be to practice the precautionary principle: conserving habitat and species before species need to be bred in captivity. Prevention always makes the most sense.

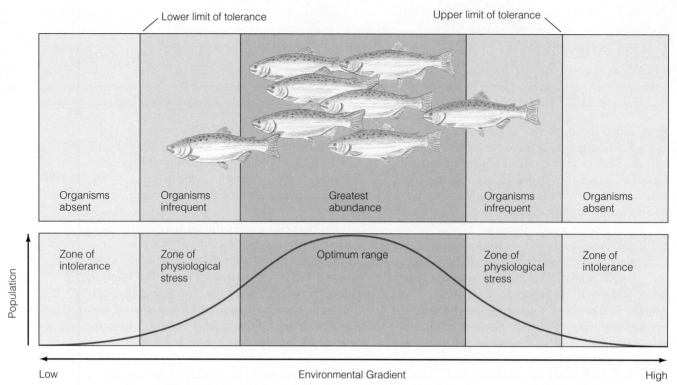

FIGURE 3-13 Organisms tolerate a range of conditions but thrive in an optimum range.

biological limits, such as the abundance of competitors and predators. The zone of physiological stress grades into or meets the zone of intolerance, where the population is absent. Some limiting factor has become so great that the species can no longer survive. Around deep-sea thermal vents, an extreme example, the zones of physiological stress and intolerance may be very narrow indeed. Water coming out of the vent may be as much as 400°C, yet the water temperature only 2 to 3 cm (an inch or so) from the vent can be 2°C, nearly freezing! Organisms living near the vent use the chemical nutrients pouring out of the vent but may not be able to tolerate the high temperatures of the vent itself. Other organisms, called extremophiles, have adapted to life in the thermal vents and would perish outside this environment.

Abundance is almost never uniformly distributed throughout a population's geographic range because the environment is rarely uniform enough to follow perfect gradients. Instead, many irregularities occur; ecologists call this the patchiness of the environment. In populations that inhabit a very patchy environment, individuals are commonly clumped

together with gaps in between. The size of the geographic range of populations also varies among species. **Endemic species** are localized and may have just one population that inhabits only a small area. This pattern is especially common in tropical organisms and in organisms that are highly specialized to live on resources with limited distributions or that have narrow environmental tolerances (**FIGURE 3-14**).

Human Impact on Population Ranges

People have both decreased and expanded the geographic ranges of populations. On the one hand, range decrease such as habitat destruction has commonly correlated with declining populations. If the habitat disappears, the population's geographic range can be reduced to zero. This is a major reason tropical extinctions are particularly destructive; with the high number of endemic populations, it does not take much habitat destruction to reduce the geographic range to zero. On the other hand, we saw how people have often expanded geographic range by domesticating and introducing wild species into new areas (see Figure 3-11).

FIGURE 3-14 The giant panda, which generally relies on bamboo shoots and roots as a major part of its diet, is an example of a specialized species. Such species are easily pushed to extinction.

3.3 Biosphere Interactions: Communities and Ecosystems

Now that we understand how population dynamics work, we can carry this concept into an examination of communities and ecosystems. A **community** consists of all populations that inhabit a certain area. The size of this area can range from very small, such as a puddle of water, to large regions encompassing many thousands of square kilometers (or hundreds of square miles). An **ecosystem** is the community plus its physical environment. It makes sense, then, that ecosystems can also differ in their size.

Because ecosystems include both organisms and their physical environment, the study of ecosystems often tends to focus on the movement of physical components, such as the flow of energy and the cycling of matter through the "system." This approach is often called the "functional" view. In contrast, be-

cause communities consist only of organisms, one can focus on describing how organisms are distributed in communities through time and space: this approach is called the "structural" view. We turn first to the community, or structural, view.

Community Structure

Discovering the various spatial patterns by which species are distributed has been one of the major accomplishments of ecology over the last century. Two of the most important patterns are the open structure of communities and the relative rarity of most species in communities.

Open Structure of Communities

Most communities are open. That is, open communities have populations of different species distributed with varied abundance peaks and range boundaries. For example, populations of the many plants comprising a typical forest community have highly varied ranges. Tolerance to moisture is a major determinant: some species do best in wetter areas, whereas others thrive in a wide range of moisture. As a result, population ranges very often overlap to various degrees depending on the similarity of the populations' tolerances to moisture. Because changes in the physical environment are usually more gradual, population abundance between communities also changes gradually. Open communities are more complicated to study because it is difficult to characterize, describe, and even name a community when it changes gradually in nearly every geographic direction.

Areas where drastic changes occur in the physical environment are the major exception to the generally open structure of communities. In such areas, boundaries between communities can be sharp because population tolerances are abruptly exceeded. A beach where land and sea come together is an example. A closed community forms a discrete unit with well-defined boundaries, called **ecotones** (see **FIGURE 3-15** for a diagrammatic comparison of closed and open communities).

A major implication of open communities is that communities are not tightly integrated assemblages of organisms that can be destroyed in an all-or-nothing fashion. Some

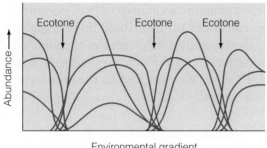

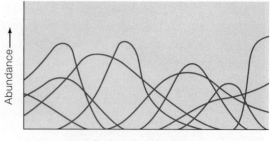

(a) Closed communities

(b) Open communities

FIGURE 3-15 Each curve represents the abundance of a single species. (a) In closed communities, the species boundaries tend to coincide. (b) In open communities, the species boundaries are more randomly distributed. *Source:* R. E. Ricklefs. Ecology. New York: W. H. Freeman, 1990, p. 659. Copyright 1990 by W. H. Freeman and Company.

ecologists once argued that communities were closed, highly integrated units forming a "superorganism." They maintained that if we destroyed just one or a few populations, the whole superorganism would die, just as an organism dies if a key organ is removed. However, the fact that species come and go in communities means that communities are generally not as integrated as the superorganismic concept suggests. This point is significant because many nonscientists still accept the superorganismic concept of biological communities.

Most Species Are Rare in Communities

A second basic population pattern is the relatively low abundance of most populations in communities. As we have seen, populations tend to have their peak abundance near the center of their range, where optimum conditions prevail. This high-population abundance of only a few species is found in nearly all natural communities from the deep ocean to high mountain slopes and among many kinds of organisms. In the previous example of the forest community, optimum moisture level seems to be a primary determinant of abundance. However, at any given point on the gradient, a large percentage of the individuals belong to just a few tree species (such as beech, maple, oak, and pine). Even at their maximum abundance, any other species present are usually represented by only a few individuals.

What causes this abundance dominance by a few species? This question is currently being debated among ecologists, but most agree that the general cause is related to resource partitioning. In any environment, a limited amount of resources is available. Because of their evolutionary history, individuals of only a few species are best able to exploit a large part of the available resources. These species were the first to evolve the ability to obtain and eat a common food in the community. Other species must partition the remaining resources and are therefore less abundant. However, a species that is very rare in one community may be very abundant in a nearby community if conditions are different enough that the species' particular adaptations are more effective in exploiting the resources there.

Community Diversity

Diversity refers to how many kinds of organisms occur in a community and is often expressed in terms of species richness, or the number of species in a community. Many factors influence diversity, and the importance of any single factor varies with the particular community. Recently, researchers have found an example of species diversity that illustrates this concept in dramatic fashion. The researchers discovered that the diversity of soil bacteria living in the arid sands of a desert far outstrips the number of soil bacteria living in a tropical community. It appears that the primary factor controlling the diversity of soil bacteria is soil pH. Tropical forest soil is more acidic, so it is home to fewer bacterial species than the more neutral desert soils.

The **latitudinal diversity gradient** describes how species richness in most terrestrial groups steadily decreases going away from the equator. In other words, richer communities are generally found in tropical areas (exceptions

CHAPTER 3 The Biosphere: Populations, Communities, Ecosystems, and Biogeochemical Cycles

include the soil bacteria described previously). For instance, 1 hectare (2.5 acres) of tropical forest typically contains from 40 to 100 tree species. In contrast, a typical temperate zone forest has about 10 to 30 tree species, whereas a taiga forest in northern Canada has only 1 to 5 species. Furthermore, the number of insect species living on those trees increases with the kinds of trees (resources) available to exploit; consequently, tropical forests also have many more kinds of insects and other animal species per hectare. As a result of this richness, habitat destruction in tropical countries generally leads to many more extinctions per acre than does destruction elsewhere.

Ecologists generally agree that this gradient is largely due to four interrelated factors: (1) environmental stability, (2) community age, (3) length of growing season, and (4) supply of nutrients. Greater environmental stability in equatorial areas means that communities are exposed to less environmental change on a daily, seasonal, and even 100-year basis. This stability allows more kinds of species to thrive because high disturbance or stress generally reduces diversity. Equatorial communities are older because they have been less disturbed by advancing ice sheets and other climatic changes over the long span of geologic time. Thus, the process of evolution has been going on long enough to create new species (**FIGURE 3-16**). The longer growing season in equatorial areas leads to more photosynthesis and plant growth, which forms the food base for these ecosystems. As we will see, higher plant productivity supports a greater diversity of organisms that depend on the plants.

Another important diversity trend is the **depth diversity gradient** found in aquatic communities. This gradient shows how species richness increases with water depth, to about 2,000 m (6,560 ft) deep, and then begins to decline. This gradient is due to (1) environmental stability and (2) nutrients. Environmental stability increases as one moves away from the higher water energies of the beach and shoreline. Stability allows more species to thrive. Similarly, as one moves offshore, the amount of nutrients from land runoff begins to diminish. As a result, although deep water

FIGURE 3-16 The Verreaux's sifaka is a large lemur endemic to Madagascar. The species (*Propithecus verreauxi*) is split into four subspecies, each of which has a distinctive appearance and is found in isolated and separate ranges on the island.

is very stable, it contains insufficient nutrients to permit the high productivity seen in shallower waters. Marine life depends especially on land runoff to supply limiting nutrients such as phosphorus.

To summarize, four major factors may increase diversity in a community: *increasing environmental stability, age, growing season,* and *nutrients*. Stability provides an accommodating environment for diversity to proliferate. Age provides the time, and the last two factors provide the energy and nutrients to supply many types of organisms. In general, the more of these factors a community has, the more species rich it will be.

Community Change Through Time

Biological communities change through time. This is not surprising because the physical environment that ultimately supports life is always changing. Whether we perceive change as "fast" or "slow" depends almost entirely on the time scale we are using. Two time scales are particularly useful for examining change in communities: ecological time and geological time. Ecological time focuses on community events that occur over tens to hundreds of years. These events are most relevant to our own human time scale. Geological time focuses on community events that are longer, on the order of thousands of years or more, such as the long-term development or evolution of communities and the contained species.

Community Succession

Community succession is the sequential replacement of species in a community by immigration of new species and the local extinction of old ones. A disturbance that creates unoccupied habitats for colonizing species initiates community succession. These colonizers usually have a hardy nature and are adapted for widespread dispersal and rapid growth, characteristics that enable them to become the first species to appear and thrive.

This initial community of colonizing species is called the **pioneer community**. Eventually, other species migrate into the community. These new species are usually poorer dispersers and grow more slowly than the colonizers, but they are more efficient specialists and better competitors and therefore begin to replace the colonizers. This process continues, as still newer species migrate in, until the **climax community** is reached. For example, in the southeastern United States, the climax community species in many areas are hardwoods such as oak trees. The climax community continues to change, although at a much slower pace.

Succession has been most fully documented in forest communities (**FIGURE 3-17**). Pioneering plant species include lichens, mosses, and

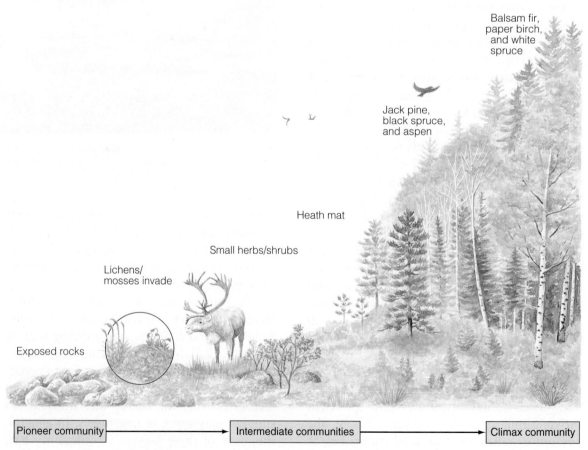

Balsam fir, paper birch, and white spruce

Jack pine, black spruce, and aspen

Heath mat

Small herbs/shrubs

Lichens/mosses invade

Exposed rocks

| Pioneer community | → | Intermediate communities | → | Climax community |

FIGURE 3-17 Ecological succession occurs when biological communities become established in a sequence, from pioneer through climax communities.

CHAPTER 3 The Biosphere: Populations, Communities, Ecosystems, and Biogeochemical Cycles

herbs, which give way to shrubs, small trees, and finally large trees in the climax community. Different animals, such as various bird and mammal species, also appear in sequence. The animal sequence is determined largely by the appearance of the plants upon which they rely.

Succession is characterized by a number of trends (**TABLE 3-4**). One of the most basic is changes in productivity. Pioneering plants tend to be smaller and exhibit rapid growth that maximizes productivity. In later stages, as more specialized, slower-growing species begin to migrate in, productivity declines. As these later species immigrate, diversity increases because the more specialized species more finely subdivide the resources. In addition, the later species generally are characterized by larger sizes and longer life cycles. Together, these trends result in more biomass in later stages because living tissue accumulates. **Biomass** is the total weight of living tissue in a community. Finally, later stages tend to have populations controlled mainly by biological or density-dependent controls, such as competition and predation. In contrast, early stages mainly show physical or density-independent

controls, such as physical disturbance. Some communities, such as an estuary or a beach, are generally in a constant state of physical disturbance, so early-successional populations become permanent dwellers (**FIGURE 3-18**).

Succession occurs because each community stage, from pioneer to climax, prepares the way for the stage that follows. Each preceding community alters soil conditions, nutrient availability, temperature, and many other environmental traits. For example, the pioneering stage of a forest stabilizes the soil of a bare patch of land, begins to accumulate nutrients in the soil, attracts pollinating insects, retains water, and provides ground shade, among many other processes that make the environment more livable for later stages. The process of "preparation" is ironic in the sense that species in each community often bring about their own demise. Yet these early

FIGURE 3-18 The environmental conditions along a rocky shoreline and the physical assault of waves can be so extreme that less-resilient succession species never replace the tough colonizers, such as the barnacles, starfish, seaweeds, and sea anemones.

TABLE 3-4	Trends in Ecological Succession	
	Stage in Ecosystem Development	
Attribute	**Early**	**Late**
Biomass	Small	Large
Productivity	High	Low
Food chains	Short	Long, complex
Species diversity	Low	High
Niche specialization	Broad	Narrow
Feeding relationships	General	Specialized
Size of individuals	Smaller	Larger
Life cycles	Short, simple	Long, complex
Population control mechanisms	Physical	Biological
Fluctuations	More pronounced	Less pronounced
Mineral cycles	Open	More or less closed
Stability	Low	High
Potential yield to humans	High	Low

Source: Adapted from R. Smith, Elements of Ecology and Field Biology. New York: Harper & Row, 1977, Table 8-9: Trends in Ecological Succession.

species are adapted by evolution to be colonists and can nearly always migrate to other, newly disturbed areas that permit them to persist. Nearly all natural landscapes (including sea-bottom areas) consist of a mosaic of undisturbed patches intermixed with patches that are disturbed to varying degrees. Ecologists think that this "patchiness" of the natural environment is crucial for maintaining diversity because it allows species from different stages of succession to exist simultaneously.

Human Disturbance of Communities

Community structure is determined by species distributions; whenever species distributions change, the structure is altered. People can change natural species distributions in many ways, from introducing new predators into pristine ecosystems to outright annihilation of large areas of habitat; however, the basic effect of nearly all human activity is community simplification: the reduction of overall species diversity (number of species).

In many cases, people simplify communities on purpose. The farmer's agricultural and the suburbanite's horticultural communities of plants, insects, and other animals are common examples. In such cases, we seek to grow only certain species, creating a much lower diversity than normally is found in that area. The extreme case is called **monoculture**, meaning that only one particular species is grown. An example is a wheat field (**FIGURE 3-19**). Monocultures and other forms of extreme community simplification are very susceptible to diseases and other forms of destruction, such as the blight that caused the Irish Potato Famine between 1845 and 1852. It is interesting to note that most of the plant species we cultivate for food and pleasure are species from pioneering communities. For example, corn (maize), wheat, and many other plants are grasses that originally were adapted to colonizing disturbed areas. People favor them as foods because they are fast-growing, rapidly reproducing organisms.

In other cases, people inadvertently simplify communities. Construction, road building, pollution, and many other aspects of development act as disturbances that simplify communities. It is important to note that such stressed communities are simplified not only by having fewer species, but also by having some species that are superabundant. Although most species cannot tolerate the stressful conditions, some find the new conditions beneficial. For instance, some organisms thrive in highly polluted waters and even use the pollutants as food. Even in these inadvertent disturbances, we often favor early-successional species. Whether we are building roads, farms, cities, or lawns, one of our first actions is to bulldoze or otherwise remove the climax community. Because colonizing disturbed environments is what early-successional species are adapted to do, they have tended to thrive as we have expanded. Indeed, the term *weed* is virtually synonymous with early-successional species, which also include "weedy" animals, such as some rodents and many insects.

Ecosystems and Community Function

Although communities vary in structure, certain basic processes, or functions, unite them all. The most basic processes are (1) energy flow and (2) matter cycling. All organisms must eat (take in energy and matter) to stay alive, causing energy and matter to move through the community. All energy and matter ultimately come from and return to the

FIGURE 3-19 Farmers who plant vast expanses of one plant, such as the wheat shown here, are more at risk of widespread crop damage caused by plant-specific pests and diseases.

physical environment. Thus, we must observe the ecosystem (community plus physical environment) to understand the complete process.

Energy flows and matter cycles through all four of the environmental spheres. The movement of energy and matter through ecosystems represents movement through one of those spheres, the biosphere, back into and through the other three physical spheres.

Energy Flow Through Ecosystems

Energy flows are routed through feeding relationships in the ecosystem. An organism's niche ("occupation") in the ecosystem is closely associated with feeding. The **food web** and the biomass pyramid represent energy flow through any ecosystem. The food web describes the complex interrelationships by which organisms consume other organisms. The food web in **FIGURE 3-20** illustrates how even aquatic and land organisms prey on each other.

The **biomass pyramid** provides a more basic understanding of energy flow (**FIGURE 3-21**). Biomass is the weight of living matter. The first trophic (feeding) level consists of the producers in the ecosystem, which produce the food that all other organisms use. Usually, the producers are plants, producing the food by photosynthesis; in a few deep-sea ecosystems, organisms produce food by chemosynthesis based on heat energy and compounds from underwater hydrothermal vents instead of the sun's energy.

All levels above this first level contain consumers. First-order, or *primary*, consumers

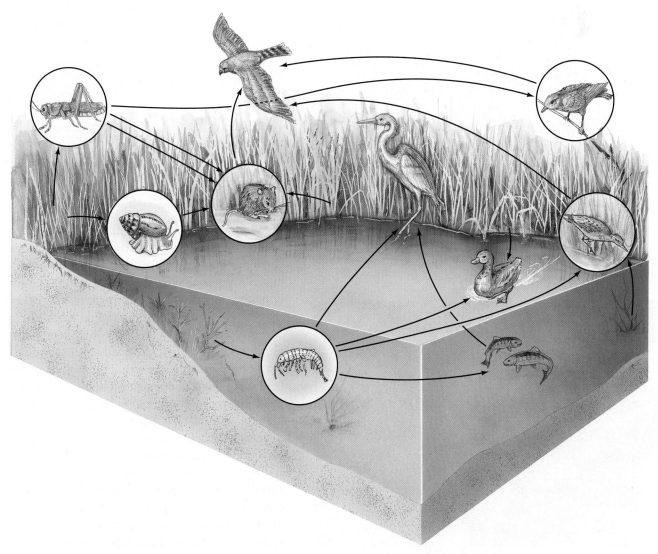

FIGURE 3-20 A food web showing the interaction between aquatic and land organisms. The arrows indicate the direction of energy flow.

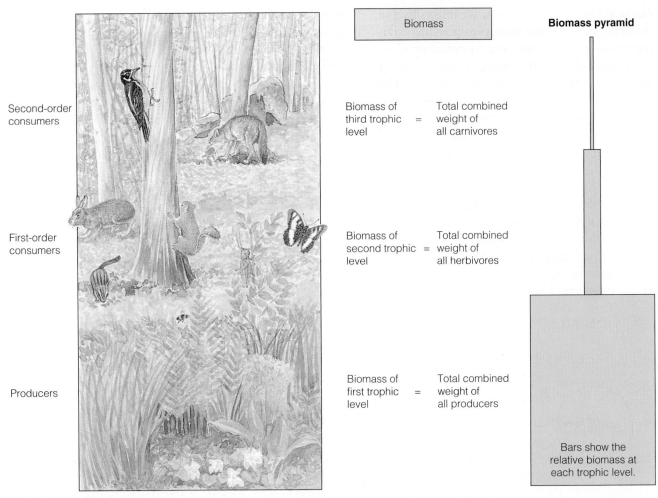

Second-order consumers

First-order consumers

Producers

Biomass

| Biomass of third trophic level | = | Total combined weight of all carnivores |

| Biomass of second trophic level | = | Total combined weight of all herbivores |

| Biomass of first trophic level | = | Total combined weight of all producers |

Biomass pyramid

Bars show the relative biomass at each trophic level.

FIGURE 3-21 A biomass pyramid. In most land food webs, biomass decreases from one feeding (trophic) level to the next highest.

FIGURE 3-22 The gray wolf is a predator high on the biomass pyramid.

are generally herbivores that directly consume the producers, deriving energy from the chemical energy stored in the producers' bodies. As a marine example, first-order consumers include the crustaceans and other organisms that eat the phytoplankton. In a forest ecosystem, first-order consumers include deer and other plant eaters. Above the first-order consumers are the second-order (or *secondary*) consumers, which feed on the first-order consumers. In a marine ecosystem, second-order consumers may consist of fish, lobsters, and other species. In a forest ecosystem, second-order consumers include wolves, panthers, and other meat eaters (carnivores) that eat the deer and other first-order consumers (**FIGURE 3-22**). Third-, fourth-, and even higher-order consumers can occur in some ecosystems. Decomposers are a special type of consumer. Decomposers, such as

CHAPTER 3 The Biosphere: Populations, Communities, Ecosystems, and Biogeochemical Cycles

many bacteria, consume the tissue of dead organisms from all levels of the food pyramid. Although they are inconspicuous, decomposers are extremely important in energy flow; in virtually all ecosystems, they consume the largest part of the energy flow.

Why does the biomass pyramid form? Biomass declines with each higher trophic level because progressively less food is available. Much of the food an animal consumes is not passed on to the animal that eats it. Instead, much of the food is (1) lost as undigested waste or (2) "burned up" by the animal's metabolism to produce heat (**FIGURE 3-23**). For example, a deer excretes about 25% of its ingested calories as undigested waste. Of the 75% that is digested, most is lost as metabolic waste products (such as urine) and, especially, body heat generated from movement and other kinds of maintenance. Of all the calories the deer eats, fewer than 20% are converted into the deer's body tissue, which can be eaten by wolves or other animals that feed on the deer. Insects and other cold-blooded organisms can convert ingested calories into tissue much more efficiently than can mammals because they have slower metabolisms. Even so, these organisms convert less than 50% of ingested calories into tissue. The result is a "leakage" of energy between each feeding level. This inefficiency is why feeding relationships form a pyramid. In general, about 80% to 95% of the energy is lost in the transfer between each level, depending on the organisms involved. Because so little energy is left, very few ecosystems have food pyramids with more than five levels. This is also why large carnivores are rare: they are the organisms at the top.

Ecosystem Productivity

The amount of food generated by producers at the base of the food pyramid varies greatly among ecosystems. Productivity is the rate at which biomass is produced in a community.

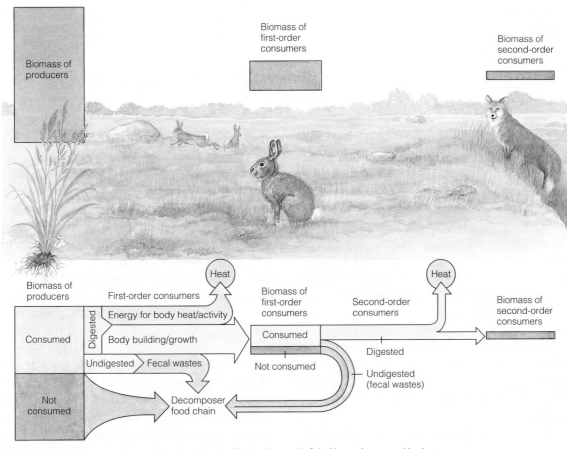

FIGURE 3-23 Flow of biomass and energy through a food pyramid. Note that much of the biomass is consumed by decomposers.

TABLE 3-5	Ecosystems and Productivity: Net Primary Productivity, Per Unit Area Per Year (g/m² or t/km² per year)*			
Ecosystem Type	Area (10⁶ km²)†	Normal Range	Mean	World Net Primary Production (10⁹ t per year)‡
Tropical rain forest	17.0	1,000–3,500	2,200	37.4
Tropical seasonal forest	7.5	1,000–2,500	1,600	12.0
Temperate evergreen forest	5.0	600–2,500	1,300	6.5
Temperate deciduous forest	7.0	600–2,500	1,200	8.4
Boreal northern forest	12.0	400–2,000	800	9.6
Woodland and shrubland	8.5	250–1,200	700	6.0
Savanna	15.0	200–2,000	900	13.5
Temperate grassland	9.0	200–1,500	600	5.4
Tundra and alpine	8.0	10–400	140	1.1
Desert and semidesert shrub	18.0	10–250	90	1.6
Extreme desert, rock, sand, and ice	24.0	0–10	3	0.07
Cultivated land	14.0	100–3,500	650	9.1
Swamp and marsh	2.0	800–3,500	2,000	4.0
Lake and stream	2.0	100–1,500	250	0.5
Total continental	149		773	115
Open ocean	33	2.02–400	125	41.5
Upwelling zones	0.4	400–1,000	500	0.2
Continental shelf	26.6	200–600	360	9.6
Reefs	0.6	500–4,000	2,500	1.6
Estuaries	1.4	200–3,500	1,500	2.1
Total marine	361		152	55.0
Full total	510		333	170

*t/km² = g/m² = metric tons/km² = approximately 2.85 tons per square mile.
†10⁶ km² = approximately 386,000 square miles.
‡10⁹ t = 1 billion metric tons = approximately 1.102 billion tons.
Source: Begon M, Harper J, Townsend C. Ecology, 2d ed. Cambridge, MA: Blackwell, 1990. Reprinted by permission of Blackwell Science, Inc.

Net primary productivity (NPP) is the rate at which producer, usually plant, biomass is created. Among the most productive terrestrial ecosystems are tropical forests and swamps (TABLE 3-5), which produce plant biomass (NPP) at many times the rate of deserts. Temperate communities such as grasslands and temperate forests have intermediate productivities. The main reason for this pattern is that productivity on land increases where the growing season is longer. As FIGURE 3-24 illustrates, productivity tends to increase toward the equator where winters are milder and shorter. Deserts are the exception to this trend because the lack of water limits growth, even though the growing season is long.

In terms of productivity per unit area, among the most productive aquatic ecosystems are estuaries and reefs, which may be as much as 10 times more productive than certain other freshwater or marine ecosystems despite their small areas (Table 3-5). By this measure, the open ocean is the least productive by far. This point is crucial because 90% of the open ocean, which covers more than half of the Earth's surface, is essentially a "marine desert" in terms of productivity per unit area. However, taken as a very large whole, the open ocean does make a significant contribution to the world's net primary productivity (Table 3-5).

Nutrient availability, rather than length of growing season, tends to be the main limiting factor in marine ecosystems. The open ocean is relatively "starved" for some nutrients, especially phosphorus, because the source of the nutrients is runoff from land.

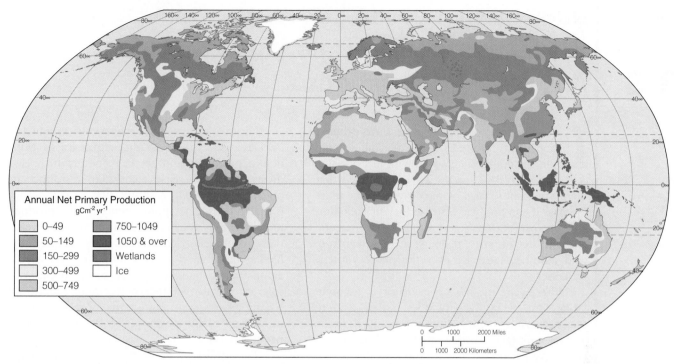

FIGURE 3-24 Annual net primary productivity (NPP) on the Earth's land surface. In photosynthesis, plants extract carbon from the atmosphere and use the carbon to form biomass. Therefore, one way to measure NPP is in terms of the amount of carbon converted into biomass per unit area per year. In this figure, NPP is measured using the units grams of carbon per square meter per year (g C/m²/yr; 1 gram C per m² per year is approximately 0.029 ounce C per yd² per year). Note that productivity is high in tropical forests and very low in arid regions. *Source:* Adapted from J. M. Melillo, et al. Nature, 1993:237. With permission from Nature. Copyright 1993 Macmillan Magazines Limited.

However, zones of upwelling, the upward transport of water to the nearshore surface from the ocean depths, can be highly productive. The upwelling currents often carry many nutrients that have settled and been swept up from the ocean bottom. The upwelling zones west of Peru, which support a productive fishing industry, are an example.

Net secondary productivity (NSP) is the rate at which consumer and decomposer biomass is produced. In other words, NSP includes all biomass except plants. A general rule of ecology is that primary net productivity and secondary net productivity are correlated: communities that have high primary productivity almost always have high secondary productivity. If the base of the food pyramid is producing much biomass, the organisms that consume and decompose plants will usually produce more biomass, too.

Human Disturbance of Energy Flow and Productivity

The extent to which people have altered ecosystem energy flow is demonstrated by a startling statistic: nearly 40% of the potential terrestrial NPP and about 2% of the oceanic ecosystem NPP are directly used, diverted, or lost (such as when forests are paved over to construct shopping malls) because of the activities of humans. This means that nearly half of the energy potentially converted by land plants is largely not available to species in natural ecosystems to trickle upward into the food pyramids. If we add the NPP of the aquatic food pyramid to that of the land pyramid, people redirect an estimated 25% of global NPP.

Another way people disturb productivity is by causing extinctions. Case Study 3-3 discusses evidence that people not only "usurp" natural productivity, but also reduce what ecosystems can produce.

Matter Cycling through Ecosystems

The second basic ecosystem function, matter cycling, occurs because, unlike energy, matter is not always converted into less useful forms when used. Dozens of elements are cycled through ecosystems in biogeochemical cycles,

CASE STUDY 3-3

Does Extinction Reduce Ecosystem Productivity? How Experimental Ecology Answers Key Questions

VISIT

http://environment.jbpub.com
/mckinney/5e/
for more information

If an ecosystem loses plant species, is its primary productivity reduced? After all, the ground could be completely covered with just one species. However, some ecologists have argued that the more plant species in an ecosystem, the more biomass it can produce because they provide buffers against seasonal and other environmental changes. For instance, if one species suffers from cold, another can take over the photosynthetic processes.

After years of debate, evidence is accumulating that plant diversity does indeed tend to increase primary productivity. John H. Lawton and his colleagues at the Imperial College in England have performed a series of experiments that measured productivity of ecosystems under environmentally controlled laboratory conditions. As **FIGURE 1** shows, their general finding was that plant productivity remained relatively high during the initial loss of species in very species-rich communities. However, as the number of species continued to decline, productivity began to decrease until species-poor communities, with one to five plant species, showed significantly lower productivity. Much more work is needed to verify this finding, but it shows how experimental methods can answer crucial environmental questions that otherwise become embroiled in fruitless debates.

Another series of experiments conducted by David C. Tilman of the University of Minnesota produced another key finding: Increased diversity also increases ecosystem resistance to disturbance. **FIGURE 2** shows that in Minnesota grasslands, species-poor communities produce much less relative biomass during drought years than do species-rich communities. Apparently, having more species helps buffer the ecosystem against disturbances because some species can tolerate the disturbance better than others. By having more species, an ecosystem is more likely to have at least some species that can continue to produce biomass.

The studies by Lawton and Tilman illustrate how ecological experiments can answer crucial questions about how extinction is affecting the environment. Species loss not only diminishes our world aesthetically and economically, but it also apparently impairs (1) ecosystem functioning such as biomass production and (2) ecosystem resistance to disturbance. A species-poor ecosystem has lower productivity during normal years, and this reduced productivity is even more drastically lowered during times of stress.

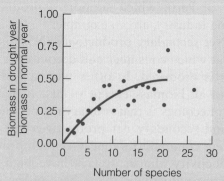

FIGURE 2 Communities with more species produce relatively more biomass in drought years than do communities with fewer species. *Source:* Redrawn from J. Lockwood, S. Pimm. Do species matter? Current Biology 1994;4:456. Reprinted with permission from Current Science.

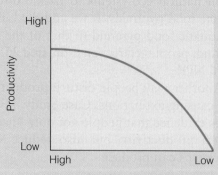

FIGURE 1 Declines in the number of species ultimately lead to declines in productivity.

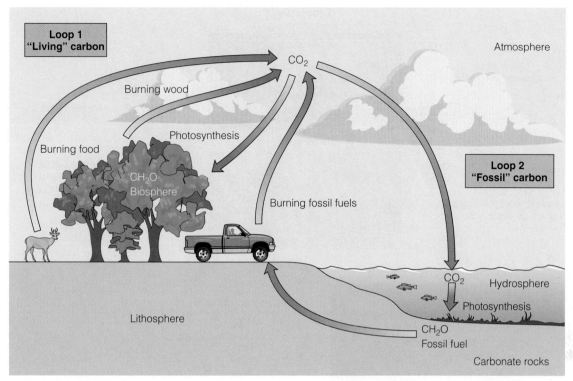

FIGURE 3-25 The carbon cycle. Loop 1 is "living" carbon that is still actively circulating among living organisms and their environment. Loop 2 is "fossil" carbon; it consists of carbon that is bound in molecules such as coal deposits that are deeply buried until released by burning or some other process. Combustion (and respiration) and photosynthesis ultimately cause carbon to move through both cycles.

which carry the elements through living tissue and the physical environment such as water, air, and rocks. An example is the carbon cycle (**FIGURE 3-25**), which has a major influence on global climate—carbon dioxide is a potent "greenhouse gas" (a review of global air pollution provides more information). Most of the elements that cycle through ecosystems are trace elements, used in small amounts by organisms. However, living things use carbon, hydrogen, oxygen, nitrogen, sulfur, and phosphorus in large amounts. Because organisms both metabolize and store these elements, ecosystems exert great control over how fast elements cycle. Some elements cycle in a matter of days, whereas others may be buried for millions of years. Carbon may spend millions of years underground stored in fossil fuels such as coal and oil or as limestone.

Ecosystems are generally very efficient in cycling matter, in that most matter is cycled over and over within the ecosystem itself (**FIGURE 3-26A**). For example, the carbon atoms in a plant will be incorporated into a deer. These, in turn, will be incorporated into the tissue of a wolf that eats the deer. When the wolf dies, decomposers will incorporate the same carbon atoms. All of these changes take place within the ecosystem. Nevertheless, a small amount of matter will be lost from the ecosystem over time. Leaching from rainfall will carry carbon from decaying organic matter, leaves, and so forth. In undisturbed ecosystems, this output loss is roughly balanced by an equal input gain of similar materials. For instance, carbon enters the ecosystem via weathering of rocks and is carried into the ecosystem by rainwater. In undisturbed natural ecosystems, both the input and the output are small relative to the amount of matter "locked up" and recycled within the biomass of the ecosystem itself.

Both the rate and efficiency of matter cycling vary between ecosystems. The cycling of matter is generally faster in tropical ecosystems, such as tropical rain forests and coral reefs, because biochemical reaction rates tend to increase with temperature. Matter cycling is also especially efficient in tropical ecosystems, where high rainfall will leach

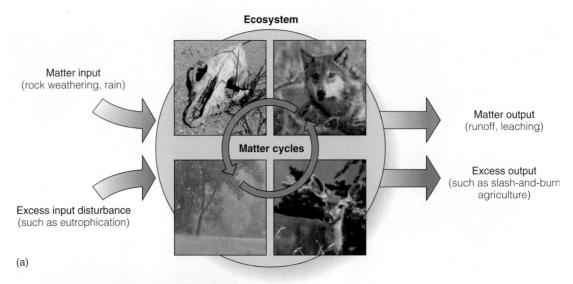

Ecosystem

Matter input
(rock weathering, rain)

Matter cycles

Matter output
(runoff, leaching)

Excess output
(such as slash-and-burn
agriculture)

Excess input disturbance
(such as eutrophication)

(a)

(b)

FIGURE 3-26 (a) A healthy ecosystem cycles most of its matter over and over through the food web. Excess input and excess output are symptoms of an unhealthy ecosystem. An example of excess input, where too much matter enters into the system, is fertilizer or nutrient excess. An example of excess output is the rapid loss of matter caused by slash-and-burn agriculture. (b) Slash-and-burn farming of the rain forest releases nutrients normally stored in plant biomass.

elements from the soil unless plants incorporate them quickly and efficiently into their tissue. Similarly, coral reefs thrive mainly in nutrient-poor tropical waters so the elements in the nutrients must be used quickly and recycled very efficiently into the tissues of the marine life.

Human Disturbance of Matter Cycling

Matter cycling in the ecosystem is disturbed when people alter the balance between the input and output of matter by creating (1) excess output or (2) excess input (**FIGURE 3-26B**).

Excess output occurs when people suddenly release the large quantity of matter retained in the biomass of the ecosystem. For instance, in **slash-and-burn agriculture**, trees are cut down and burned. The burning releases the nutrients into the soil for agriculture. Unfortunately, the nutrients are quickly leached out of the soil in areas where rainfall is heavy, such as in the tropics, where slash-and-burn techniques are common. This massive output of matter from the ecosystem is not fully replaced by input for many hundreds or perhaps thousands of years. During this time, the area can sustain only a relatively barren ecosystem with a fraction of its former diversity. In the meantime, farmers must move on and burn another area of tropical forest to produce arable land. This practice is contributing to massive tropical deforestation worldwide and the release of billions of tons of carbon into the atmosphere annually. Another example of excess output is the massive burning of fossil fuels. These emissions are contributing to global climate change (a review of global air pollution provides more information).

Disturbance by excess input commonly occurs when runoff from agricultural activity carries large amounts of fertilizer, organic waste, and other nutrients into natural ecosystems. This also destroys diversity because the excess nutrients cause eutrophication, leading to unrestrained growth of some organisms, such as algae in a lake. When the algae die, the decay of their now-abundant bodies by the action of bacteria uses up so much oxygen and produces so much carbon dioxide the water becomes anoxic. The low (or completely zero) concentrations of dissolved oxygen kill the fish and many other organisms. Excessive runoff of fertilizers from farms upriver can also contribute to large coastal dead zones. One of these dead zones has been observed where the Mississippi River empties into the Gulf of Mexico and another in the Chesapeake Bay.

3.4 Biogeochemical Cycles: An Introduction

When observing the cycles of matter such as water, rocks, nutrients, and other substances within and among the spheres, scientists often find it useful to focus on the cycles of chemical elements that compose those substances. This approach can be used to simplify our models of environmental cycles because just a few basic elements participate in many of the most important cycles on Earth. These cycles of chemical elements through the atmosphere, lithosphere, hydrosphere, and biosphere are called **biogeochemical cycles**.

Among the most important biogeochemical cycles are the six that transport the six elements most important to life: carbon, hy-

TABLE 3-6	Atomic Composition by Weight of Three Representative Organisms		
Element	Human (%)	Alfalfa (%)	Bacterium (%)
Oxygen	62.81	77.90	73.68
Carbon	19.37	11.34	12.14
Hydrogen	9.31	8.72	9.94
Nitrogen	5.14	0.83	3.04
Phosphorus	0.63	0.71	0.60
Sulfur	0.64	0.10	0.32
Total	97.90	99.60	99.72

drogen, oxygen, nitrogen, phosphorus, and sulfur. Of the approximately 90 elements that occur naturally on Earth, these six comprise the majority of atoms in the tissue of all living things. As **TABLE 3-6** shows, oxygen alone accounts for more than 62% of the weight of the human body and more than 77% of the weight of the alfalfa plant. Carbon and oxygen together account for more than 80% of the weight of a human.

TABLE 3-7 shows the relative abundances of the most common elements in the Earth's crust. Oxygen is the most abundant, just as it is most common in the human body. However, the second most common human element, carbon, is hundreds of times rarer in the crust. Instead, silicon, which is virtually absent from the human body, is quite abundant in the crust. You can see other major discrepancies by comparing Tables 3-6 and 3-7. These

TABLE 3-7	The Relative Abundance by Weight of Some Chemical Elements in the Earth's Crust
Element (Chemical Symbol)	Relative Abundance (%)
Oxygen (O)	46.6
Silicon (Si)	27.7
Aluminum (Al)	8.1
Iron (Fe)	5.0
Calcium (Ca)	3.6
Sodium (Na)	2.8
Potassium (K)	2.6
Magnesium (Mg)	2.1
Phosphorus (P)	0.07
Carbon (C)	0.03
Nitrogen (N)	Trace

discrepancies illustrate how life is chemically distinct from its environment. Without biogeochemical cycles to transport and store temporary concentrations of matter for food and other uses, life could not survive.

Each of the many biogeochemical cycles has different pathways of transport and temporary storage reservoirs. The carbon cycle in Figure 3-25 is a typical biogeochemical cycle. Like many cycles, the **carbon cycle** appears at first glance to be quite complex with many pathways (arrows), but closer inspection shows that these pathways are based on just two processes: withdrawal from and addition to the atmosphere.

1. Withdrawal of carbon is largely driven by photosynthesis, whereby plants take carbon out of the atmosphere where it resides as carbon dioxide. The CO_2 is combined with water (H_2O) to form biochemical molecules such as sugars [$(CH_2O)n$] and oxygen. Photosynthesis is conveniently written as follows:

$$CO_2 + H_2O + energy \rightarrow (CH_2O)n + O_2$$

This reaction is called photosynthesis because it requires energy from the sun (*photo* = light; *synthesis* = combine).

Figure 3-25 also shows two "loops" of photosynthesis. Loop 1 illustrates the pathway of "living" carbon in the ongoing photosynthesis of modern plants. Loop 2 shows how "fossil" carbon forms. Fossil carbon is stored carbon that has been temporarily withdrawn from use by living organisms. Carbon is stored in the lithosphere, when plants, such as tiny plankton in marine and fresh waters, die, sink to the bottom, and are buried. After millions of years of burial, these dead plants and the carbon in them can become fossil fuels such as petroleum. Clams and other ocean shellfish withdraw carbon for use in constructing their shells. When the shellfish die, their shells contribute to the enormous amounts of limestone. The large majority of the Earth's carbon now resides in the oceans. As **FIGURE 3-27** illustrates, the ocean stores much more carbon than is found in the other three sinks—the atmosphere, lithosphere (geological), and living organisms on land (terrestrial biosphere).

2. The addition of carbon to the atmosphere often occurs from combustion. Combustion or "burning" is essentially

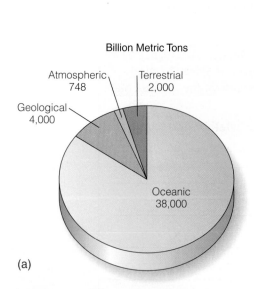

Billion Metric Tons

(a)

(b)

FIGURE 3-27 (a) Major reservoirs of the carbon cycle, in billion tons of carbon. *Source:* Adapted from W. M. Post, et al. The global carbon cycle. American Scientist, 1990;78:315. (b) The oceans are the largest reservoir by far. Limestone, such as the white cliffs of Dover, is formed from shells of clams and other marine life.

CHAPTER 3 The Biosphere: Populations, Communities, Ecosystems, and Biogeochemical Cycles

the reverse of photosynthesis; oxygen (O_2) is combined with plant matter [$(CH_2O)n$] to release CO_2 and H_2O:

$$(CH_2O)n + O_2 \rightarrow CO_2 + H_2O + energy$$

Setting fire to either living matter (loop 1) or fossil fuel (loop 2) releases carbon dioxide. We also carry out combustion when we digest food: our bodies take the oxygen we inhale and use it to break down the biochemical molecules we eat. We then use the energy given off to move around, grow, and maintain our bodies. The carbon dioxide produced is exhaled into the atmosphere (loop 1). This process of biological combustion is called **respiration.**

Another important biogeochemical cycle is the **nitrogen cycle** (**FIGURE 3-28**). Nitrogen is of fundamental importance to all organisms. It is a necessary component of DNA (the genetic material), RNA, amino acids (of which proteins are formed), and many other organic molecules. Nitrogen is also quite abundant, or so it would seem, because the atmosphere is 78% nitrogen gas (N_2) by volume. However, organisms cannot use this atmospheric nitrogen; it must be turned into a more chemically reactive form, such as ammonium (NH_4^+) or nitrate (NO_3^-). Most of the usable nitrogen is converted from atmospheric N_2 by certain **nitrogen-fixing** bacteria in the soil, in root nodules of legumes, and by certain other types of plants. These bacteria produce ammonia (NH_3). Most soils are slightly acidic, and the ammonia can gain a hydrogen ion (H+) to become NH_4^+, which plants can use. Ammonia may also evaporate into the atmosphere,

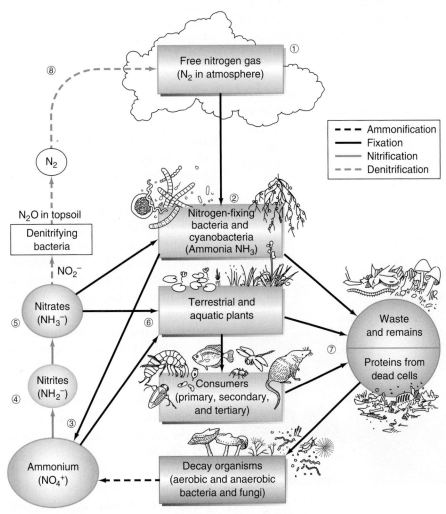

FIGURE 3-28 The nitrogen cycle. The circled numbers indicate the pathway.

where it can form NH_4^+ that is returned in rainfall.

In the soil, aerobic bacteria can also transform NH_4^+ to nitrite (NO_2^-) and then nitrate (NO_3^-) in the process known as nitrification. Various plants can also use the nitrate. Certain bacteria can convert nitrate to N_2, which is released to the atmosphere (the process is known as **denitrification**). Still other bacteria, as well as various fungi, decompose dead organic matter and in the process turn nitrogen bound up in organic molecules back into ammonium (this is known as **ammonification**), which other organisms can reassimilate. In fact, decomposition and reassimilation of nitrogen account for the majority of nitrogen cycled through most natural ecosystems. By eating plants or other animals, animals gain necessary nitrogen. In modern times, nitrogen fixation is also carried out artificially during the manufacturing of fertilizers, and this has significantly increased the amount of nitrogen chemically available to organisms.

Biogeochemical Cycles: Major Features

The biogeochemical cycles as a group can be analyzed in terms of a number of important features. These features include the cycles' pathways, their *rates of cycling*, and the *degree* to which they are being disturbed by human activities.

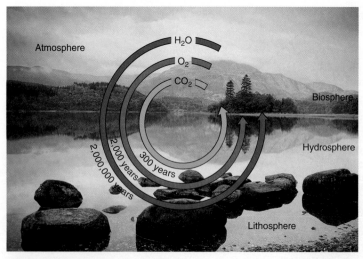

FIGURE 3-29 Recycling rates of water, oxygen, and carbon dioxide through the four spheres.
Source: L. Laporte. Encounter with the Earth. San Francisco: Canfield, 1975, p. 22. Modified by permission of Leo F. Laporte.

A Variety of Pathways

Each biogeochemical cycle has many different pathways: many chemical and physical processes help to cycle each atom. For instance, the carbon cycle transports carbon through any of the four spheres. From the atmosphere, the carbon dioxide (CO_2 molecule) dissolves in water in the hydrosphere, where plankton use the carbon to grow, thereby moving the carbon into the biosphere. If the plankton are buried and converted to fossil fuel such as oil, the carbon becomes part of the lithosphere. When the oil is burned, the carbon is released back into the atmosphere, where it may recycle through a different set of pathways. The next time, a tropical tree, instead of plankton, may absorb the CO_2 gas. (In the case of carbon, more is being released than absorbed, which is contributing to global climate change—see a review of global air pollution for more information.) Of course, each element has a different set of potential biogeochemical pathways. For example, phosphorus often combines with different atoms than carbon does to form different molecules and undergo different chemical reactions.

Variable Rates of Cycling

Biogeochemical cycles vary in their rate of cycling. **FIGURE 3-29** shows the average amount of time water, oxygen, and carbon dioxide molecules take to make a complete cycle through the four spheres. Carbon takes only hundreds of years to cycle, whereas water takes about 2 million years. These times are only approximate because specific molecules may cycle much more rapidly or slowly, depending on the pathway they follow. A small number of carbon atoms (fewer than 1 in 10,000) in the active, living loop, for instance, may be diverted and stored as oil deposits for more than 200 million years and cycle through the much-slower fossil loop.

Why do substances cycle at such different rates? Two major determinants are (1) the chemical reactivity of the substance and (2) whether it has a gaseous phase (occurs in the atmosphere) somewhere in the cycle. The high chemical reactivity of carbon causes it to participate in many chemical pathways and is a major reason it cycles so quickly. In addition, carbon is abundant as the gas carbon dioxide. Because gas molecules move much more quickly than more tightly bonded molecules in liquids or solids, the existence of a gaseous phase allows the substance to be transported more rapidly.

Although oxygen and water cycle more slowly than carbon dioxide, they cycle at relatively fast rates compared with many other substances. Like carbon dioxide, oxygen and water are chemically reactive and have a major gaseous phase. For instance, the average water molecule has a **residence time** of 10 days in the atmosphere, during which time it may move thousands of kilometers (thousands of miles)

before falling back to the surface of Earth as a liquid. To find a substance that has an extremely slow cycling time, we should look for one that has no gaseous phase and is also relatively unreactive in natural systems. Phosphorus is a good example because it not only has a very slow cycling rate but is one of the six most important elements of life. Because of its chemical and physical properties, phosphorus does not form a gas and does not readily combine with other substances. Its main mode of transport is water, which moves much more slowly than air, and even in water, phosphorus is relatively insoluble. As **FIGURE 3-30** illustrates, large amounts of phosphorus become

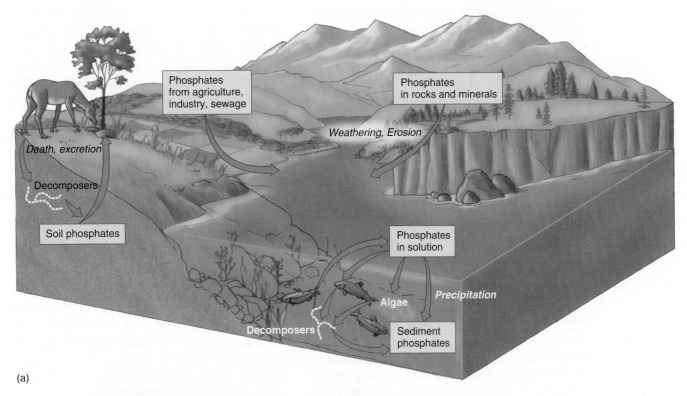

(a)

(b)

FIGURE 3-30 (a) The global phosphorus cycle. The amount that flows on Earth is much smaller than the amount stored in rocks and sediments. (b) Phosphate is mined in Florida and other areas where ancient ocean waters deposited phosphorus-rich sediments.

"locked up" in storage for long periods of time as sediments in the deep ocean and Earth's crust. Only relatively slow and rare events, such as upwelling ocean currents from the deep sea or weathering of phosphorus-rich rocks, recycle the phosphorus.

Instead of the few hundreds to few millions of years typical of cycles with a gaseous phase, phosphorus requires many tens of millions of years to complete its biogeochemical cycle. This slow cycling rate drastically reduces the availability of this critical nutrient, which has profound effects for the biosphere. Phosphorus is usually the nutrient in shortest supply in most ecosystems and is considered the limiting nutrient. You will recall the sudden availability of limiting nutrients in natural systems causes rapid growth.

The Effects of Human Activity

Biogeochemical cycles are crucial to all life but are being greatly disturbed by human activity. As human population and technology rapidly increase, huge quantities of the Earth's materials are extracted and redistributed through all of the spheres. The net result has been the disturbance of nearly all biogeochemical cycles. The most common type of disturbance is acceleration of the cycles: materials are being rapidly mined and otherwise extracted from storage reservoirs (sources) and, after use, are rapidly deposited back into the environment (sinks). This increased rate of cycling from source to sink leads to the two basic environmental problems: depletion and pollution (see discussion on the **environment and people**). Indeed, a basic definition of pollution is a temporary concentration of a chemical above levels that normally occur in its biogeochemical cycle.

No one really knows just how drastic or dangerous this acceleration of natural cycles will ultimately prove to be. However, no one doubts that major consequences will occur, and because biogeochemical cycles are global in nature, many of these consequences will occur on a global scale. Carbon provides a prominent example of how we are actively disturbing a major cycle, with potentially drastic consequences. The burning of fossil fuels has released increasing amounts of

carbon dioxide into the atmosphere. Global release of carbon has increased exponentially from about 1 billion metric tons (1.1 billion tons) per year in 1940 to an estimated 6.3 billion metric tons (6.9 billion tons) per year at the end of the century. It is estimated that plant life and the oceans absorb about half of the 6.3 billion metric tons per year, but the remainder is accumulating in the atmosphere in the form of carbon dioxide.

This accumulation of atmospheric carbon dioxide has many potential global consequences. The most publicized is global warming (see a review of global air pollution for more information). Carbon dioxide is a "greenhouse gas," meaning that it increases the ability of the atmosphere to trap heat. How much carbon dioxide can be added to the atmosphere before significant global warming will occur is much debated. As another example, many studies have shown that the rate of plant growth will generally increase from increasing photosynthesis.

Although the consequences of disturbing the carbon cycle may be especially dramatic—rising temperatures could cause dramatic changes in weather and climate, shift rainfall patterns, melt glaciers, and raise sea levels—people are increasingly disturbing all of the biogeochemical cycles. We cover these and other disturbances in detail in later chapters.

Energy Flows

The **first law of thermodynamics** says that energy cannot be created or destroyed but can be transformed. The **second law of thermodynamics** says that when energy is transformed from one kind to another, it is degraded, meaning that the energy becomes less capable of doing useful work. For example, typically less than half of the chemical energy in gasoline is converted to the energy of motion in a car. Similarly, photosynthesis uses solar energy to create food from carbon and other atoms. Food represents chemical energy, which is stored in the bonds between atoms; photosynthesis is a transformation from solar to chemical energy. As **FIGURE 3-31** illustrates, this transformation is far from 100% efficient. Most of the incoming energy is "lost" as heat. Heat is considered to be low-quality energy and is capable of doing less work than high-quality energy.

Entropy refers to the amount of low-quality energy in a system. If entropy is very high, matter will tend to disorganize to simpler states. The second law of thermodynamics is sometimes called the law of entropy because all energy transformations will increase the entropy of a system unless new high-quality energy, such as sunlight, enters the system to replenish it. Some energy transformations are more efficient than others. By carefully refining our technology, people have managed to achieve much greater efficiencies. For example, most solar (photovoltaic) cells convert 15% to 30% of the sunlight's energy to electricity. In 2009, Spectrolab in California invented a photovoltaic cell that converts slightly more than 40%. Other examples of energy conversion are the use of nuclear energy to generate electrical energy and the conversion of gasoline (chemical energy) into mechanical energy ("energy of motion"). However, in each of these and all other energy transformations, engineers long ago accepted that no matter how advanced our technology, some loss of usable energy will always occur, if only a small percentage.

Because of this loss, energy cannot be recycled like matter. Ultimately, all of the energy in a system will become relatively useless (transformed to heat) unless new energy flows into the system. We say that matter cycles, but energy must flow from one source to another. On Earth, the greatest majority of new energy flows from the sun. This energy originates with nuclear reactions at the sun's core and travels as light energy across 150 million kilometers (93 million miles) to strike the Earth. Upon striking the Earth, the energy is transformed in many ways, depending on where it strikes. Collectively, the various flow pathways of all energy on Earth are called the Earth's **energy budget**. More than 80% of the incoming light is either directly reflected back into space or absorbed and reradiated back into space as heat. The remaining energy powers the hydrologic cycle by evaporation, generates wind, powers photosynthesis, and in general drives many of the cycles within and between the spheres that we discussed

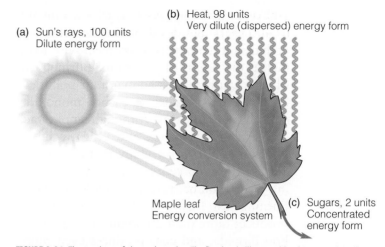

(a) Sun's rays, 100 units
Dilute energy form

(b) Heat, 98 units
Very dilute (dispersed) energy form

Maple leaf
Energy conversion system

(c) Sugars, 2 units
Concentrated energy form

FIGURE 3-31 The two laws of thermodynamics. The first law is illustrated by the conversion of sun energy (a) to food (chemical) energy (c). The second law dictates that heat loss (b) during conversion causes the amount of usable food (chemical) energy to be less than the sun energy. In this case, it is much less. *Source:* Adapted from E. P. Odum. Ecology and Our Endangered Life Support Systems. Sunderland, MA: Sinauer, 1989, p. 70.

earlier. Photosynthesis, for all of its importance in sustaining most life on Earth, including humans, uses a tiny fraction, just 0.06%, of the solar radiation received by Earth, and given the inefficiency of photosynthesis, much of this is wasted.

Not all of the Earth's energy comes from the sun. A small fraction comes from two other sources. One is the moon's gravitational pull, which causes tides in the ocean. This tidal energy is thousands of times less than the amount of energy provided by the sun. The second source of energy is the Earth's own internal **geothermal energy**, which is generated by radioactive minerals deep within the Earth. The heat diffuses outward to the Earth's surface and melts rocks to drive the tectonic cycle of the lithosphere and the emission of volcanic gases into the atmosphere. It also produces the heat that creates deep-sea vents in the ocean floor, where hot magma provides nutrients and energy to rich marine communities.

Human Use of Energy Flows

We are using ever-greater amounts of the energy flows on Earth. Modern civilization is built on fossil fuels. Because they take so long to form, fossil fuels are called **non-renewable resources**. If rates of use continue to rise, most estimates indicate that the world supply of oil will be used up within the next

100 years and world coal supplies within 300 to 400 years. Besides depletion, another and perhaps more immediate problem with fossil fuels is the release of pollutants. Burning fossil fuels causes almost all major air pollution, including acid rain, smog, carbon monoxide, and greenhouse gases. Although smokestack devices, fuel cleansing, and other technical solutions can control many of these pollutants, there is no economical way of removing the carbon because so much is produced. Burning just 1 gallon (3.8 liters) of gasoline produces more than 20 pounds (9 kg) of carbon dioxide, and coal produces even more. Fossil fuels also cause many other pollution problems, such as seepage from storage tanks into groundwater and ocean spills.

Instead of using "fossilized" solar energy, it would be less damaging to use the solar energy flow as it strikes the Earth today. Because the sun will last about 5 billion more years, solar energy will not soon be depleted, and the potential supply is huge. In just 1 month, the Earth intercepts more energy from the Sun than is contained in all the fossil fuels on the planet. Solar energy (which includes indirectly wind power, hydropower, and biomass energy) along with tidal and geothermal energy, is a form of **renewable energy**.

▪ study guide
SUMMARY

- Earth's natural environment can be divided into the lithosphere, hydrosphere, atmosphere, and biosphere.
- Water, air, and energy are three major aspects of the physical environment.
- Ecology studies how organisms interact with each other and the physical environment.
- Populations of organisms typically go through three distinct phases: growth, stability, and decline.
- Human populations can have an impact on and cause disruption of natural populations.
- Populations interact with one another and their physical environment in communities and ecosystems.
- Communities can change through the sequential replacement of certain species by others in a process known as community succession.

- Energy flow through an ecosystem can be studied using food webs and biomass pyramids.
- Productivity is the rate at which a community produces biomass.
- Energy flows and matter cycles through ecosystems.
- Biogeochemical cycles are cycles of elements and substances (e.g., carbon and nitrogen) that move through the atmosphere, lithosphere, hydrosphere, and biosphere.
- According to the laws of thermodynamics, energy cannot be created or destroyed, but it is degraded (less capable of doing work) when transformed from one form of energy to another.

KEY TERMS

amensalism	climate	diversity
ammonification	climax community	ecological release
atmosphere	commensalism	ecology
biogeochemical cycles	community	ecosystem
biomass	community succession	ecotones
biomass pyramid	competition	endemic species
biosphere	competitive exclusion	energy budget
carbon cycle	denitrification	entropy
carrying capacity	depth diversity gradient	environment and people

eutrophication
extinction
first law of thermodynamics
food web
geothermal energy
habitat
hydrologic cycle
hydrosphere
intrinsic rate of increase
latitudinal diversity gradient
law of the minimum

limiting nutrients
lithosphere
monoculture
mutualism
net primary productivity (NPP)
net secondary productivity (NSP)
niche
nitrogen cycle
nitrogen fixing
nonrenewable resources
parasitism

photosynthesis
pioneer community
population
predation
renewable energy
residence time
respiration
second law of thermodynamics
slash-and-burn agriculture
symbiosis
weather

STUDY QUESTIONS

1. Briefly describe the four spheres that compose the natural environment. What are a few ways that people have affected each?

2. Describe the composition of Earth's atmosphere, including the major constituent gases and the layers of the atmospheric envelope.

3. Briefly describe the energy flow on the surface of Earth. What is the hydrologic cycle?

4. Define ecology. Must an ecologist necessarily be an environmentalist?

5. What are the typical phases that a population of organisms will pass through over time?

6. List the basic categories of abundance controls that regulate the sizes of populations in nature. Give an example of how people have, perhaps inadvertently, affected the size of a specific population of organisms.

7. What is the "law of the minimum?"

8. Discuss the difference between closed and open structure communities.

9. How can a community change over time? What are some of the basic trends in community succession?

10. How do people affect the biosphere? Include the following in your discussion: population growth, population decline, population ranges, communities, energy flow and productivity, and matter cycling.

11. Distinguish between food webs and biomass pyramids. Why are both important to consider when discussing ecosystems? How does each relate to energy flow through an ecosystem?

12. How does net primary productivity generally correlate with latitude?

13. Name the six elements most important to life.

14. Review the carbon cycle, including the two loops of the carbon cycle. How does the carbon cycle illustrate the major features of a biogeochemical cycle?

15. Where is the majority of carbon stored?

16. Why do different substances cycle at different rates?

17. Why do large amounts of phosphorus become locked up?

18. What is the limiting nutrient in many ecosystems?

19. Describe an example of a biogeochemical cycle that people have disturbed.

20. How is energy different from matter? How is it similar?

21. Why cannot energy be recycled like matter?

22. What are the two laws of thermodynamics? What is entropy?

WHAT'S THE EVIDENCE?

1. The authors assert that people, in many cases inadvertently, have had major effects on some populations of organisms. Do the authors present enough evidence to convince you that this is the case? In some cases, might a natural population have increased or decreased even if people had not altered its environment? Are human-induced effects necessarily detrimental?

2. The authors state that people redirect an estimated 25% of global net primary productivity. What exactly does this mean? Does this sound reasonable to you? That is, is it believable that people could redirect such a large percentage of NPP? Or would you expect it to be larger? Cite some examples of how people redirect NPP.

CALCULATIONS

1. If plants are only 2% efficient in using sunlight that they capture, and they capture less than 0.1% of all sunlight striking the Earth, what is the total percentage of sunlight striking the Earth that the plants actually use?

2. If a population doubles each month and begins with 10 individuals, what will the size of the population be after 1 year?

ILLUSTRATION AND TABLE REVIEW

1. Based on Table 3-5, which ecosystem type covers the largest area of Earth? What percentage is this of the total area of Earth?

2. Based on Table 3-5, which ecosystem type is the most productive? What percentage of the total area of Earth does it cover?

http://environment.jbpub.com/mckinney/5e/

http://environment.jbpub.com /mckinney/5e/

Connect to this text's website at http://environment.jbpub.com/mckinney/5e/. This site features eLearning, an online review area that provides quizzes, chapter outlines, and other tools to help you study for your class. You can also follow useful links for more in-depth information.

The Distribution of Life on Earth

4

This is a special time in Earth's long history. The continents are widely separated. The global climate is relatively cool compared with what it has been for most of Earth's history, and life has had time to adapt to its physical environment without interruption from a cataclysmic event. The Earth holds an unusually high number of species.

In this chapter, we examine how species originated and discuss estimates of the number of species on Earth today. We see that life arose rather early in Earth's history and has diversified into many species through the process of evolution via natural selection. Indeed, there are now so many species that biologists have scarcely begun the task of classifying and describing them all. Sadly, as species continue to disappear as a result of direct and indirect human activities, some species clearly will be gone before they are found and described. No one really knows how fast species are disappearing, but scientists put the numbers between 1,000 and 10,000 times what the normal or "natural" rate would be if people were not affecting the planet. A lack of professional biologists who are trained to study insects and other groups that are so abundant hinders the efforts to find and describe them. The information biologists can provide is very important in debates over how fast species are going extinct and what percentage of Earth's species are disappearing.

Chapter Objectives

After reading this chapter, you should be able to explain or describe the following:

- How life may have originated
- How life evolved and diversified by natural selection
- How to measure biodiversity
- The seven major land biomes of the world
- The two major aquatic biomes of the world

We also see how species are distributed on Earth. Species are neither randomly nor evenly distributed because their distribution is closely related to the physical conditions to which they are adapted. Rainfall and temperature are especially crucial in determining which suites of species can survive in any given region. Some species are narrowly adapted to very limited conditions in one area, whereas other species, such as the wolf, were once found across many continents. Human impacts, such as on species distributions, are not randomly or evenly distributed. Some regions, such as deserts, are more sensitive than others to our disturbance.

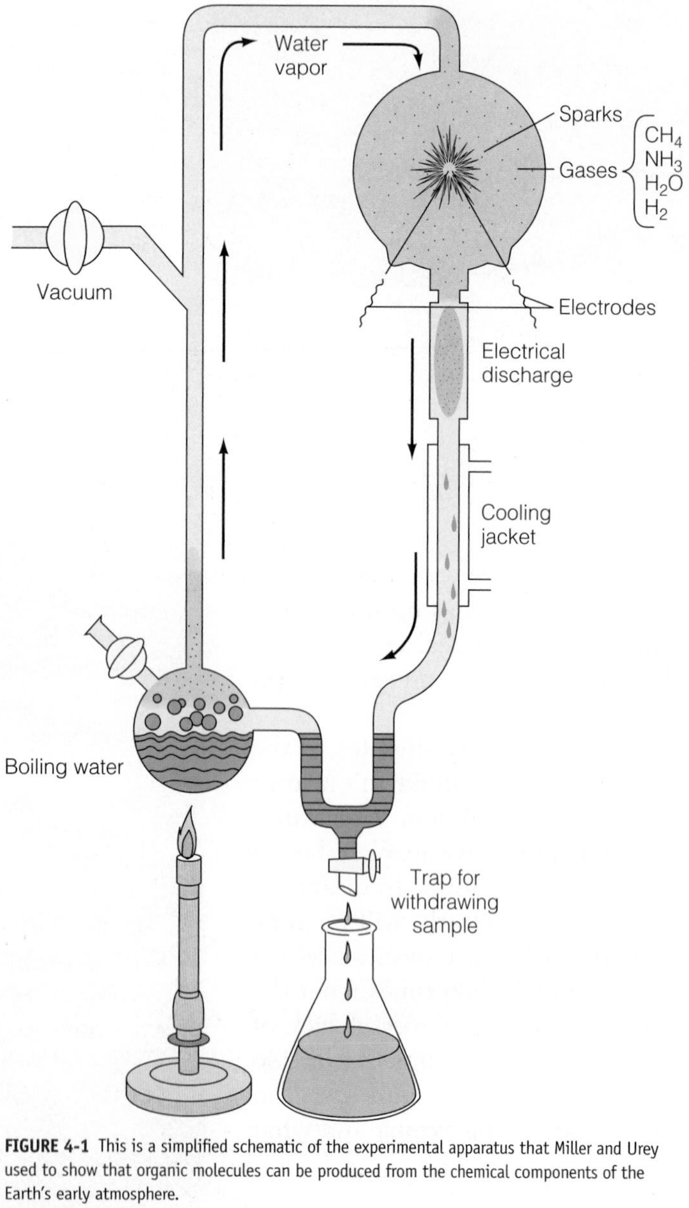

FIGURE 4-1 This is a simplified schematic of the experimental apparatus that Miller and Urey used to show that organic molecules can be produced from the chemical components of the Earth's early atmosphere.

Other regions, such as grasslands, with rich soils for crops, are affected because they are resource rich and we can rapidly exploit them.

4.1 Evolution of the Biosphere

Primitive fossilized bacteria are at least 3.5 billion years old. This indicates that life arose not long after the molten Earth first became cool enough to support life. This very early appearance of life implies that natural processes readily produce life under appropriate conditions. Beginning in the early 1950s with the work of Stanley Miller and Harold Urey, scientists have shown that the complex molecules possessed by all living things are readily produced under laboratory conditions that duplicate early environments on Earth. As **FIGURE 4-1** shows, the early atmosphere is thought to have been composed of ammonia (NH_3), methane (CH_4), water vapor (H_2O), and other gases. When these are subjected to electricity, which simulates lightning and sunlight, chemical reactions occur that produce **amino acids**. Amino acids are complex molecules that are the building blocks of proteins. Proteins make up enzymes and many other components of life, such as muscles, hair, and skin.

Of course, protein molecules alone are not living things. Organisms are composed of molecules that are organized in very complex ways. The basic organizational unit of life is the cell. Remarkably, in the late 1950s, Sidney Fox found that heated amino acids can form cell-like structures sometimes called **protocells**. These structures are not true cells but have many cell-like properties, such as being semipermeable to certain materials.

Producing amino acids and protocells in the laboratory is far from creating life in the laboratory. Even the simplest bacteria are considerably more complex than these protein and protocellular building blocks. Nevertheless, the readiness with which these first steps toward life occur, combined with the fossil record, support the idea that life readily arose through natural processes.

Evolution Through Natural Selection

After life originated, it began to diversify into different kinds of organisms through biological evolution. As Charles Darwin documented in 1859, biological evolution occurs because of **natural selection** of individual variation:

1. Nearly all populations exhibit variation among individuals.
2. Individuals with advantageous traits will tend to have more fertile offspring.
3. Advantageous traits will therefore become widespread in populations.

Variation in neck length in giraffes is an example. We know from fossils that early giraffes had relatively short necks. However, a few individuals had genes that produced slightly longer necks. In some localities, these giraffes had a feeding advantage because they could browse on leaves in taller trees. Consequently, in populations living where tall trees were common, longer-necked giraffes tended to have more offspring and passed on genes for longer neck growth so that longer necks became more common.

If this process occurs with many traits over a long period of time, the population will eventually become very different from other populations and will create a new species. **FIGURE 4-2** illustrates how isolation of populations promotes speciation. The different populations are exposed to different environments, which favor different traits, causing the populations to diverge through time. When do two different populations become two different species? Biologists generally define a **species** as a group of individuals that can interbreed to produce fertile offspring (this definition applies only to sexual organisms; the issue of defining nonsexual species is a matter of debate in biological circles). Therefore, a new species is formed when members of a diverging population can no longer successfully mate with populations of the ancestral species. Closely related species, which have often diverged relatively recently, are grouped together within the same genus. Similar genera are then grouped together within the same family. This method of classifying species according to hierarchically nested categories is a form of **taxonomy**. Applying this system to humans, we have the following:

Kingdom Animalia
Phylum Chordata
Class Mammalia
Order Primates
Family Hominidae
Genus *Homo*
Species *H. sapiens**

Diversification of the Biosphere

Where does individual variation come from? This question is crucial because without the initial variation in the population, natural selection would have nothing to act on. This question troubled Darwin because he was not aware of the work of Gregor Mendel, who is credited with discovering the laws of inherited variation in 1865. Indeed, initially Mendel's work was overlooked, and scientists rediscovered it only in the early 1900s. Mendel discovered that **genes**, which are the basic units of heredity, pass on traits. According to the Human Genome Project, humans have an estimated 30,000 to 40,000 genes that determine our traits. Variation in a population, or gene pool, occurs because individuals possess different sets of genes that produce different traits. How do different sets of genes arise? Reproduction is one way. Genes are shuffled when sperm and egg cells are fused, causing offspring from the same parents to have different genes. A second cause of genetic variation is **mutation**, which is a spontaneous change in a gene. Genes are composed of **DNA** molecules, and mutations occur when DNA molecules are altered, such as during DNA replication. Mutation is the ultimate source of all genetic variation.

Patterns of Diversification

Evolution through natural selection has produced an increasingly diverse biosphere, with the total number of species becoming greater through time. Initially, evolution was relatively slow. The right side of **FIGURE 4-3** summarizes the evolution of life on Earth. For about 2 billion years after the first appearance of fossils, relatively few species of simple single-celled organisms, such as various bacteria and Cyanobacteria (formerly

---

*A species name must always be associated with a generic name or abbreviation (in this case *H.* for *Homo*). Humans are the species *Homo sapiens*; it is incorrect to call humans simply *sapiens*.

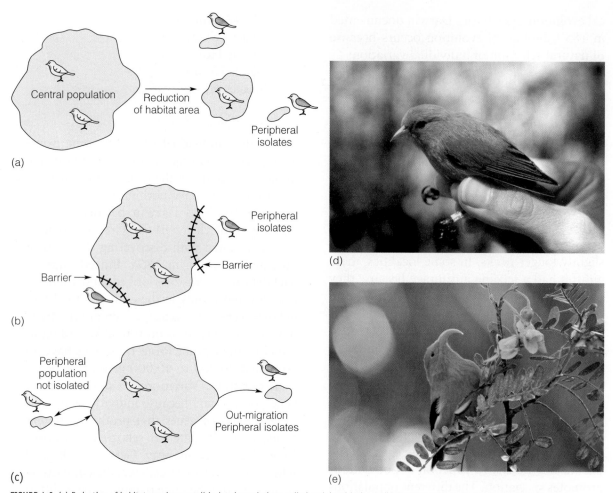

FIGURE 4-2 (a) Reduction of habitat may leave small isolated populations, called peripheral isolates. (b) Barriers may form to isolate small populations. (c) Out-migration to an isolated area may form small separated populations. In all cases (a–c), the small populations can evolve into new species. Isolated by thousands of miles of ocean, the Hawaiian honeycreeper has evolved into many forms, including species with short (d) and long (e) beaks.

known as blue-green algae), appear in the fossil record. A major change occurred about 1.5 billion years ago when more complex cells, called **eukaryotes,** evolved. These cells had a true nucleus, chromosomes, and specialized cellular organelles such as mitochondria. Apparently, such complex cells were a primary impetus of increasing rates of evolution. Multicellular organisms, including sponge-like and jellyfish-like creatures, appeared in the oceans by at least 1 billion years ago. These probably evolved from colonies of single-celled eukaryotes, such as protozoa, that became progressively more specialized and integrated. About 570 million years ago, the fossil record shows a rapid diversification, sometimes called the **explosion of life** or the **Cambrian explosion,** when most of the major groups of animals first appear. As shown in

Figure 4-3, this explosion (which occurred during the Cambrian Period of Earth history) corresponds to the time when modern oxygen levels were attained in the atmosphere. This permitted the evolution of more complex animals, which have a greater metabolic need for oxygen.

After the explosion of life, living things diversified into new environments. As shown in Figure 4-3, life colonized the land (lithosphere) and the air (atmosphere) during the Paleozoic Era, which began with the explosion of life and ended with a global mass extinction. The Mesozoic Era, sometimes called the age of dinosaurs, was the second major era, and it also ended with a mass extinction. The third and last era is the Cenozoic Era, sometimes called the age of mammals. Although catastrophes, especially from aster-

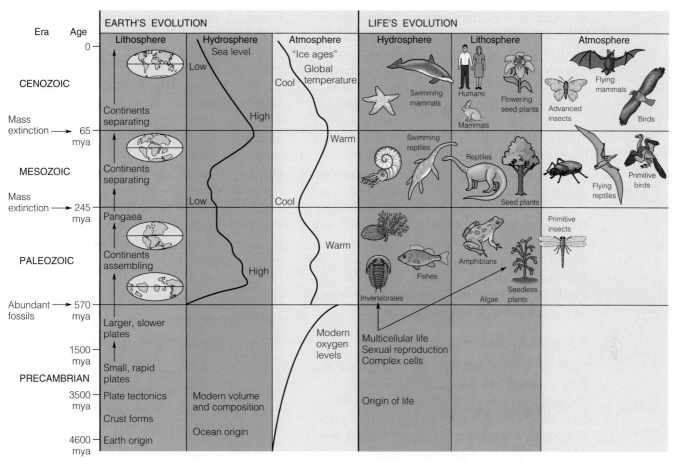

FIGURE 4-3 The left side of this diagram summarizes the changes in the Earth's continents, oceans, and atmosphere, whereas the right side summarizes the evolution of life on Earth.

oid and comet impacts and global climate change, have temporarily caused species numbers to decrease through mass extinctions at very rare intervals, the overall trend throughout these eras is toward increasing numbers of species. Life has adapted to new environments and found new ways of doing things through mutation and natural selection. We now live in a biosphere that is one of the most diverse in life's long history.

4.2 What Is Biodiversity?

Biological diversity, or **biodiversity**, has many definitions, but they all involve the variety of living things in a given area. For instance, the U.S. Office of Technology Assessment defines biodiversity as "the variety and variability among living organisms and the ecological complexes in which they occur." One reason for the lack of a precise definition is that life

is hierarchical. For example, genes occur in cells. Cells occur in organisms, and organisms occur in ecosystems (a review of the biosphere provides more information). At what level do we measure the diversity of life in an area? The number of genes? Organisms? Ecosystems? All can be used.

Measuring Biodiversity

Although one can measure biodiversity by counting the variety of genes or ecosystems in an area, the most common method is to count species (**CASE STUDY 4-1**). *Species diversity* is usually measured as **species richness**, which is the number of species that occur in an area. Using species richness is largely a matter of convenience: it is easier to tabulate the number of species in an area than to count genes (the genetic diversity) or ecosystems. Fortunately, species diversity is generally a good indicator of genetic diversity and ecosystem diversity as well. Nevertheless, even species richness omits important information about biodiversity, such as the abundance of each species. Species evenness is the distribution of individuals among species in a community.

Biodiversity can be measured at all geographic scales, from local to regional to global. **FIGURE 4-4A** illustrates low diversity at both the local and regional scales. Regions with low diversity often have fewer species at the local level as well. As **FIGURE 4-4B** illustrates, regions with high diversity tend to be composed of local biological communities with high diversity. For instance, tropical rain forests and tropical coral reefs tend to have very high species richness at both the local and regional level compared with areas of similar size in temperate zones.

Global Biodiversity: How Many Species?

No one knows how many species live on Earth (**CASE STUDY 4-2**). **Taxonomists**, the biologists who classify and describe organisms, have described approximately 1.8 to 2.0 million species. Of these, 56% are insects, and 14% are plants; vertebrates such as birds, mammals, and fishes are just 3%. Only about 15% of described species occur in the oceans.

However, these 2 million species are a highly biased sample and may not reflect the true species richness of the biosphere. For one thing, vertebrates are much more widely studied and, therefore much better known than nonvertebrates: only one or two new species of birds are described each year, whereas hundreds of new invertebrates are found. A recent study reported more than 4,000 bacteria species in a single gram of Norwegian soil. Most of these species were previously unknown.

Another problem is that most biologists have been concentrated in North America and Europe, but the tropics have the most species. Similarly, most biologists study life on land, but the oceans cover 70–75% of the Earth and contain unknown numbers of species. For instance, after a December 2004 tsunami that hit Southeast Asia, many rarely seen species were found washed up on the shores. It is likely that taxonomists have considerably underestimated the number of species existing in the tropics and the oceans.

How Estimates Are Made

Describing all species on Earth could take centuries, so biologists have devised a number of ways to develop immediate biodiversity estimates of species numbers from limited information. The following are just three examples of the many methods of estimating global diversity from small samples:

1. *Rain forest insect samples.* Terry Erwin of the Smithsonian Institution has become well known for his studies of the very diverse tropical insect communities. Because most species are insects and most insects are tropical, these studies are important for global estimates (**FIGURE 4-5**). Erwin used insecticides to kill and collect all of the insects in the canopy (upper branches) of a tropical tree. In Panama, for example, 1,200 beetle species were found in the canopy of a single tree. Because about 40% of all insects are beetles, we can estimate that perhaps 3,000 insect species occur there; however, only about two-thirds of the species occur in the canopy (the rest occur on roots, bark, and other places). The total number of insects on this one tree is thus estimated at 4,500 species! Because many of these insects occur only on one tree species and because there are about 50,000 species of tropical trees, it seems that many tropical insect species must exist. Erwin estimates the insect species on Earth at more than 30 million.

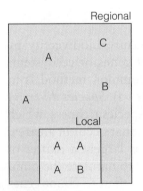

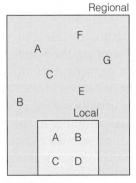

(a) Low diversity   (b) High diversity

FIGURE 4-4 Letters (A, B, C, and so forth) represent species living in an area: (a) low diversity at both the local and regional scales; (b) high diversity at both scales.

How Does Biodiversity Affect the Provisioning of Ecosystem Services?

VISIT

http://environment.jbpub.com
/mckinney/5e/
for more information

Over the past 50 years, humans have changed ecosystems more rapidly and extensively than in any comparable period of time in human history, largely to meet rapidly growing demands for food, fresh water, timber, fiber, and fuel. This has resulted in a substantial and largely irreversible loss in the diversity of life on earth. The first major finding of the Millennium Ecosystem Assessment (MA, 2005a) clearly links the rapid and widespread loss of biodiversity on Earth to the growing intensity of many human pressures on biodiversity. According to the MA's Biodiversity Synthesis (MA, 2005b), the most important direct drivers of biodiversity loss and ecosystem service changes are habitat change (such as land use changes or physical modification of rivers), climate change, invasive alien species, overexploitation, and pollution. Thus, biodiversity loss is linked to "the degradation of many ecosystem services [and] could grow significantly worse during the first half of this century [. . .]" (MA, 2005b).

The Millennium Ecosystem Assessment indicates that human activities over the last 50 years have affected ecosystems across the globe more quickly and profoundly than any other time. In fact, we are now in the midst of one of the major extinction events on Earth. The report states: "Appropriating major parts of the energy flow through the food web and altering the fabric of the land cover to favor the species of greatest value have increased the rate of species extinction 100 to 1,000 times the rate before human dominance on Earth." Current trends, such as population growth and climate change, affect ecosystems and the biodiversity that underpins these ecosystems negatively; yet, without major intervention, the threat to global biodiversity will worsen.

In addition to the moral reasons to preserve biodiversity for its own sake, biodiversity provides a plethora of ecosystem services that are critical to not only overall human well-being but also our very survival on earth. Biodiversity can affect ecosystem services both directly and indirectly. Directly, humans derive most of their essential food and fibers from animals and plants. More indirectly, biodiversity can affect the provision of ecosystem services through its influence on ecosystem processes that are essential to Earth's life support systems.

"By affecting the magnitude, pace, and temporal continuity by which *energy* and materials are circulated through ecosystems, biodiversity influences the provision of regulating ecosystem services, such as *pollination* and seed dispersal of useful plants, regulation of climatic conditions suitable to humans and the animals and plants they consider important, the control of agricultural pests and diseases, and the regulation of human health. Also, by affecting nutrient and water cycling, and soil formation and fertility, biodiversity indirectly supports the production of food, fiber, potable water, shelter, and medicines." (Diaz, 2010).

Based on available evidence, it is difficult, if almost impossible, to conclude how many species are needed to preserve different ecosystem services. However, moral reasons and the precautionary principle suggest that in all ecosystems as many existing species as possible should be preserved. In general, because most ecosystem services are provided at the local scale, if we are to preserve the regulating and supporting services that ecosystems provide to humans, conservation efforts need to focus on preserving or restoring their biotic integrity (whether inherently species-poor or species-rich), rather than on simply maximizing the number of species present (Diaz, 2010).

Regarding the decrease of ecosystem services as a result of rapid biodiversity loss, not all communities will be similarly affected. In fact, it is more likely that people who rely most directly on ecosystem services, such as subsistence farmers and fishers, the rural poor, and traditional societies, face the most serious and immediate risks. In large part this results from these individuals and communities relying most directly on the benefits provided by the biodiversity of natural ecosystems in terms of food security, sustained access to medicinal goods, construction materials, and fuel. Furthermore, these communities and individuals are the most likely to suffer from storms and floods that may increase as a result of ecosystem degradation. Finally, less privileged socioeconomic sectors are also less able to substitute purchased

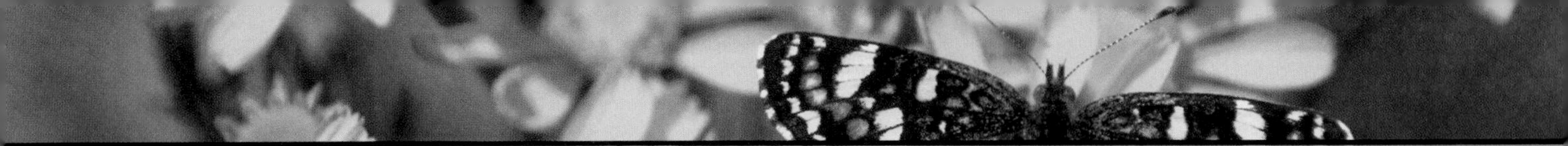

How Does Biodiversity Affect the Provisioning of Ecosystem Services?

goods and services and have less purchasing power and often less political power. Thus, the loss of biodiversity-dependent ecosystem services is likely to accentuate inequality and marginalization of the most vulnerable sectors of society.

However, during the past few decades, more people have become aware of the impact on biodiversity and the threats of those impacts on critical ecosystem services. Many countries have biodiversity protection programs of varying degrees of effectiveness, and several international treaties and agreements coordinate measures to slow or halt the loss of biodiversity. For example, the United Nations Environment Programme has played a major role in the establishment of major multilateral environmental agreements, including the Convention on Biodiversity. This convention came into force on December 29, 1993, during the 1992 earth summit in Rio de Janeiro. The United Nations declared 2010 as the International Year of Biodiversity to help draw attention to the importance of global biodiversity conservation efforts.

References

Millennium Ecosystem Assessment (MA). 2005a. *Ecosystems and human well-being: Synthesis*. Island Press, Washington, DC.

Millennium Ecosystem Assessment (MA). 2005b. *Ecosystems and human well-being: Biodiversity synthesis*. World Resources Institute, Washington, DC.

Diaz S. 2010. Biodiversity and ecosystem services. Online at: http://www.eoearth.org/article/Bio-diversity_and_eco system_services?topic=49575

For more information on international efforts: http://www.unep.org/iyb/unepwork.asp

2. *Ecological ratios.* Another method is to use well-studied groups to predict the diversity of less studied groups that are associated with them. For example, in Europe, there are about six fungus species for each plant species. Plant species have been relatively well described, and it is estimated that 270,000 plant species exist worldwide. Thus, there may be as many as 1.6 million fungus species if the 6:1 ratio is applicable throughout the world. Only 69,000 fungus species have been described.

3. *Species area curves.* The species area curve has been very influential since the 1960s. Whenever the number of species is counted in a gradually enlarged area of sampling, the result is a curve as in **FIGURE 4-6.** The number of species rises rapidly at first, but it slows as the area of sampling increases because the same species are encountered again and again. Repeated surveys in the tropics and temperate areas have shown that small areas of rain forest contain many more species, often more than 100 times more species, than comparable areas of temperature forest. By using the known shape of the species area curve, one can predict how many more species will be found in larger unsampled areas of tropical and temperate regions.

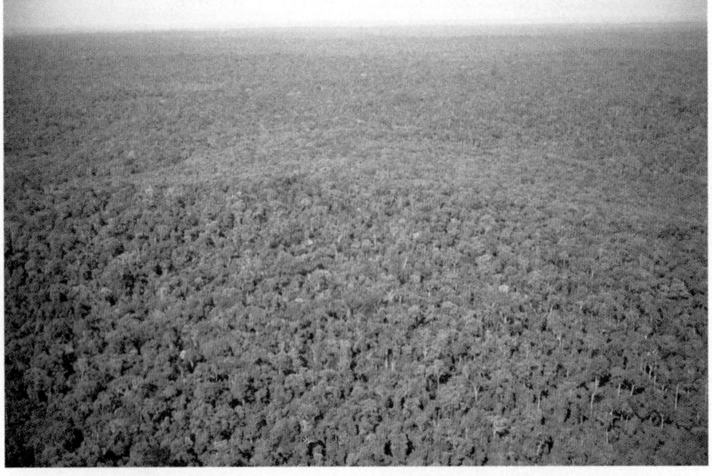

FIGURE 4-5 Each of the tropical trees in this expanse can be home to hundreds of insect species.

Hellbenders and Alligators:
The Asia-U.S. Connection

VISIT

http://environment.jbpub.com
/mckinney/5e/
for more information

The geographic distribution of species is partly dependent on their biological adaptations, such as how well a species can tolerate cold winters. However, in many cases, the history of a group also plays an important role in its current distribution. A good example of this is the presence of some plants and animals in eastern North America.

As early as 1750, the famous biologist Linnaeus noticed a curious similarity between many species in eastern North America and species in Eastern Asia. A century later, Charles Darwin discussed this relationship as well. For example, a giant salamander called the hellbender, found only in the southern Appalachian region, belongs to a salamander family with all of its other species in Asia (**FIGURE 1**). Similarly, there are only two species of alligators in the world: the American alligator, which is found throughout the southeastern United States, and the Chinese alligator. Primitive fishes also show this pattern. The paddlefish is an archaic fish with a long snout able to detect magnetic fields of its prey in murky waters. The only species of this family outside of the United States occurs in the Yangtze River system of China.

An even more striking relationship is seen among plants: the broadleaf deciduous forests of eastern Asia and eastern North America have a great resemblance to one another. Most notable are similarities among primitive flowering plants, descended from plant families that appeared more than 70 million years ago when dinosaurs were still alive. These include members of the magnolia family, which has many species in both eastern Asia and eastern North America. In some cases, the species themselves are very similar, such as the tulip poplar that occurs in both regions. Even some herbs have striking similarities. The medicinal herb ginseng has but two closely related species, with one occurring in eastern North America and the other in eastern Asia.

What is the cause of this biological connection between Asia and eastern North America? Many people have suggested the North American plants and animals may have colonized North America in the same way as the American Indians. These early arrivals migrated across the Bering Sea land bridge from Asia through Alaska several thousand years ago; however, the very ancient geological age of nearly all of these groups suggest that they became separated much longer than a few thousand years ago. Paleobiologists have found that about 40 million years ago there was a land-bridge connection across the North Atlantic. It ranged across northern Asia through northern Europe into Greenland and northern Canada. The climate of the Earth was much warmer at this time, so it permitted the migration of species adapted to warm climates. By 30 million years ago, the Earth had cooled to the point where such migration was no longer possible. These groups have been isolated from Asia ever since.

This example illustrates how understanding the history of life helps us value its preservation. We see that many of the species that constitute the eastern North American forests are part of a very ancient ecosystem. These species belong to groups with a very long evolutionary heritage, and they contain very ancient genetic diversity.

Questions

1. Name three species that might be called "living fossils" because they have a very long evolutionary heritage.
2. Botanists have long noted that ornamental plants from eastern Asia are among the easiest plants for landscapers to cultivate in eastern North America. Why is this?

FIGURE 1 Hellbender salamander.

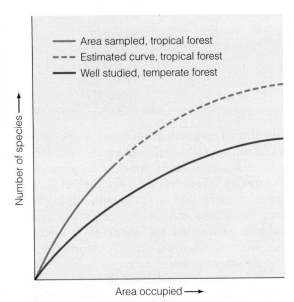

FIGURE 4-6 A species area curve plots the number of species found in increasingly larger areas. The temperate forests are well studied compared to tropical forests. Only small areas have been examined in most tropical forests so data are derived from extrapolated projections as indicated by the dashed line.

Biodiversity Today: A Rare Wealth in Time

All methods of estimating global biodiversity involve a certain amount of extrapolation; so understandably, there is still wide disagreement over how many species exist. Estimates range from as low as 3 million species to as high as 100 million or more. However, biologists generally agree that fewer species live in the ocean than on land. Estimates for the ocean range from 1 million to 10 million species. Despite having fewer species, however, the oceans have more fundamental biodiversity: 32 phyla are found in the oceans compared with just 12 phyla on land. A phylum, such as echinoderms or mollusks, is a much more distinct taxonomic unit of biodiversity than a species.

Whatever the exact number of species today, the fossil record indicates that we live during a special time of Earth's history. Detailed compilations of fossil data show that the number of families in the oceans has generally increased through time (**FIGURE 4-7**). Evolution has produced new families that have added to overall biodiversity. Estimates based on species produce similar results. We apparently live at a time when global biodiversity is near or at its all-time peak. As we discussed in our review of the biosphere, the individual organisms composing this diver-

sity are organized into populations, and all of the populations of a certain geographic area form a community. In the remainder of this chapter, we briefly describe some of the broad types of natural communities that occur on Earth today (**CASE STUDY 4-3**).

4.3 Biomes and Communities

A **biome** is a large-scale category that includes many communities of a similar nature. Many thousands of communities exist on the Earth, but rather than describe each in detail, we examine the basic categories into which communities can be grouped.

The most basic distinction is between terrestrial (land)- and aquatic (water)-dwelling communities. There are seven major types of terrestrial biomes and two major types of aquatic biomes:

1. *Terrestrial.* Tundra, grassland, savanna, desert, taiga, temperate forest, tropical forest (including tropical rain forests)
2. *Aquatic.* Marine, freshwater

Both terrestrial and aquatic biomes (and thus the communities within them) are largely determined by climate, especially temperature. Climate is so important because it affects many aspects of the physical environment: rainfall, air and water temperature, soil conditions, and so forth. However, secondary factors such as local nutrient availability are also important. In all cases, biomes illustrate the key point that species often will adapt to physical conditions in similar ways, no matter what their evolutionary heritage. A desert biome in the western United States looks superficially similar to a desert biome in North Africa even though the plants have different ancestries.

Terrestrial Biomes

Of the seven basic land biome types commonly distinguished (tropical forests, savanna, grasslands, deserts, temperate forests, taiga, and tundra—each of these is described and illustrated), the tundra and desert biomes represent adaptations to the extreme conditions of very low temperature and low water, respectively. Not surprisingly, communities in these

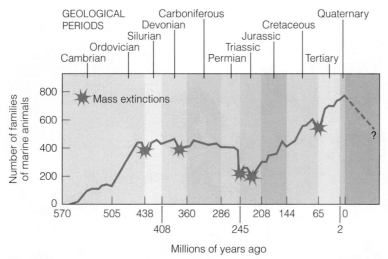

FIGURE 4-7 The number of families in the ocean has generally increased until the present. Five mass extinctions have temporarily lowered biodiversity. Will a sixth mass extinction occur in the near future? *Source:* Modified from D. Raup, J. Sepkoski. Mass extinctions in the marine fossil record. Science 1982;215:1501–1503.

biomes tend to have the least number of species because organisms have difficulty adapting to the extreme physical conditions. In contrast, the tropical rain forests tend to be richest in species, in part because the tropics have the most moderate overall conditions.

FIGURE 4-8 illustrates the approximate distribution of some of the major land biomes by precipitation, altitude, and latitude. This figure demonstrates the importance of temperature, which decreases with both increasing altitude,

and increasing latitude; similar changes result in both cases. In addition, note the importance of precipitation, ranging from the very moist equatorial regions to the desert biome. A more detailed global view in **FIGURE 4-9** shows some of the true complexities of the latitudinal pattern. For example, tropical forests are not always neatly confined to equatorial areas.

We now describe seven major terrestrial biomes (listed in approximate order going from the equator to the poles).

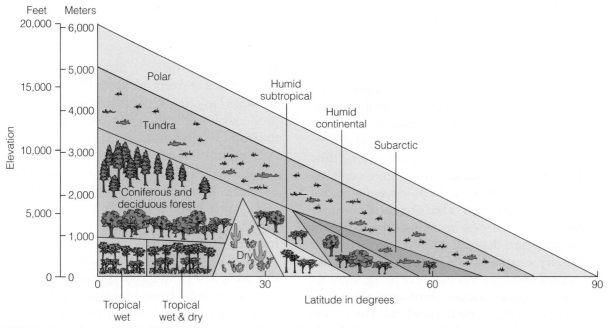

FIGURE 4-8 Vegetation changes with altitude in the same way as with latitude because both changes are associated with decreasing temperatures.

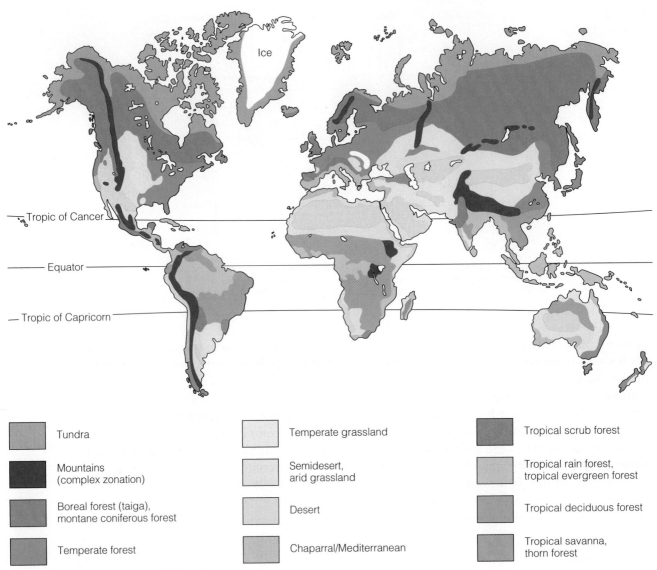

Ice

Tropic of Cancer

Equator

Tropic of Capricorn

■	Tundra	■	Temperate grassland	■	Tropical scrub forest
■	Mountains (complex zonation)	■	Semidesert, arid grassland	■	Tropical rain forest, tropical evergreen forest
■	Boreal forest (taiga), montane coniferous forest	■	Desert	■	Tropical deciduous forest
■	Temperate forest	■	Chaparral/Mediterranean	■	Tropical savanna, thorn forest

FIGURE 4-9 The major terrestrial biomes of the world.

Tropical Forests

Although rain forests are the best known kind of tropical forest, there are actually several types. Tropical seasonal forests occur in dry areas of the tropics, where pronounced dry periods cause the trees to be deciduous (drop their leaves). Tropical seasonal forests are common in parts of Southeast Asia, especially India, where they are often called "monsoon forests" because the rain-bearing monsoon winds carry the moisture that restarts forest growth after the dry season. In contrast, tropical rain forests are evergreen forests where plants show no major seasonal changes. Another type of tropical forest is the tropical woodland. This occurs in parts of the tropics where rainfall is relatively

rare throughout the year, such as areas of Central America and central Africa. Because of the moisture limitations, photosynthesis and plant productivity are only about one-third that of the tropical rain forest.

Tropical forests are found on most major continents, mainly in equatorial regions (Figure 4-9). These represent the most complex and diverse biome, containing perhaps more than 50% of the world's species while occupying only 7% of the land area. Unfortunately, people have deforested approximately half of this area; and consequently, many rain forest species are on the brink of extinction. The high biodiversity of tropical rain forests is partly due to the relatively constant temperatures and high water availability: daily and seasonal changes (fluctuations) are usually less than a total of 5°C (9°F). Rainfall is very heavy, ranging from 200 to 450 centimeters per year (80 to 175 inches). In comparison, Seattle, Washington, receives an average of 100 centimeters (37 inches) of precipitation annually. The average temperature of the tropical rain forest biome is about 25°C (77°F to 80°F). In comparison, New Orleans, Louisiana, has an average temperature of just 20°C (68°F).

This lack of pronounced seasonal temperatures, including longer days and more direct sunlight, provides more opportunity for plant growth and productivity in this biome. The result is an ecosystem of very high species diversity and structural complexity (**FIGURE 4-10**). This structural complexity is seen in the many layers of trees forming a multistoried canopy. The canopy blocks out most of the sunlight so that, in contrast to popular conception, the forest floor is relatively dark and does not support dense vegetation. Plants living at ground level must have special adaptations for coping with the lack of sunlight. For example, many plants have very broad leaves (such as the common "elephant ear" plant) to maximize surface area and very dark green pigmentation (chlorophyll) to maximize light absorption. Another common adaptation is found among epiphytes (plants that grow on the branches of trees) and vines, which grow up tree trunks to capture light in gaps high in the canopy.

The high species diversity of tropical rain forests has also led to the misconception that they must have very fertile soils. In fact, the soils here are quite poor (low in phosphorus, nitrogen, and other nutrients) because they are depleted by the intense competition among the abundant fast-growing plants for nutrients. The hot, humid climate makes tropical rain forests an ideal environment for bacteria, fungi, termites, and other soil organisms, which quickly decompose matter on the forest floor. Most of the free nutrients remaining in the soil are leached away by the high rainfall. The result is that nearly all nutrients are tied up in the biomass of the plants and animals living in the rain forest. The only way to release these nutrients for farming is to cut and burn the vegetation by "slash-and-burn" agriculture. Unlike the temperate forest biome (discussed later), tropical forests on different continents vary widely in the genera and families of plants that compose them. This is probably a result of the relative isolation and rapid evolution of species in tropical ecosystems. In addition, most tropical forest plants (especially trees) belong to botanical families that are rarely, if ever, found in temperate latitudes. This gives the rain forest its exotic appearance to visitors from North America and other temperate latitudes.

In most ecosystems, there is a strong correlation between the diversity of plant species and the diversity of animal species; so it is not surprising that the diverse tropical plant life supports a proportionately rich diversity of animal life. It has been estimated, for example, that more than 90% of earth's insect species are found in the rain forests. This high diversity is true for many other animal groups, such as mammals and especially birds, which are much more diverse in the tropics than other biomes (**FIGURE 4-11**).

FIGURE 4-10 The complex (multilayered) structure of a tropical rain forest.

FIGURE 4-11 Macaws range from southern Mexico down into central South America. This is a blue and gold macaw.

Savannas

Savannas occur in tropical and subtropical areas that are not wet enough to support rain forests. Probably the best known of these are the savannas of Africa and South America, but they also occur in tropical regions of Austra-

FIGURE 4-12 Savanna in Kenya.

lia and Asia (Figure 4-9). These regions are warm and have prolonged dry seasons with annual rainfall averaging 90 to 150 cm (35 to 60 inches). Such climatic conditions support vegetation of open grasslands with scattered shrubs and trees (see **FIGURE 4-12**). The grasses grow rapidly during the rainy season and dry up during the dry season while the few trees survive by growing deep root systems to access groundwater supplies. In areas of increasing rainfall, woodland begins to replace the savanna as tree density increases. Some savanna-like environments with dense scrubby evergreen vegetation are called **chaparral**. Examples occur in California and the Mediterranean region (Figure 4-9).

Many people are familiar with the big-game animals of the African savanna, including such large herbivores as wildebeests, zebra, rhinoceroses, and buffalo. The most notorious predators of these creatures are the prides of lions that roam the savanna. Savannas also support many species of insects, including ants, beetles, grasshoppers, and termites. The termites may be familiar because of the huge mounds they build. These mounds are important to the savanna ecosystem because they are made of digested plant material. Their construction accelerates the breakdown of these nutrients, thereby enhancing soil quality. The deep and complex passageways in these mounds help carry badly needed rainwater into the soil, instead of the water being carried away or evaporating.

During the dry season, savannas periodically experience large grass fires. The vegetation has evolved ways to be fire resistant, such as rapid regeneration from unburned roots. People have often taken advantage of this process by purposely setting fires to promote new growth that is favorable for grazing animals.

Deserts

Deserts occur where rainfall is very scarce, less than 25 centimeters (10 inches) per year. In warm regions, the problem of scarce water is aggravated by high evaporation and water loss. As a result, deserts have among the lowest rates of plant growth and productivity of any area on earth (see our review of the biosphere). An important consequence is that desert ecosystems will take a longer time to recover from human disturbances, often centuries, than nearly all other ecosystems. The most extreme deserts, including the Sahara of Africa and Gobi of Mongolia, average less than 10 cm (4 inches) of rain per year, sometimes going for years with no rain at all (**FIGURE 4-13**). Some areas of the Atacama Desert in Chile reportedly have never seen recorded rainfall and only occasional cloud cover. As a result, extreme deserts support very little life of any kind. Most extreme deserts occur in a belt that is about 20 to 30 degrees latitude north and south of the equator (see Figure 4-9). This is because rising hot equatorial air produces heavy tropical rains so that when the cooler air sinks to the Earth at about 20 to 30 degrees north and south latitude, it is quite arid.

Semideserts, such as those of the southwestern United States, are less extreme and receive between 10 and 25 cm (4 to 10 inches) of rainfall per year. These extra few inches of rain can support a fairly diverse array of species, usually typified by widely spaced, thorny "desert scrub" vegetation. Desert plants, often called **xerophytes**, have many interesting adaptations that allow them to survive the scarce and unpredictable water supply (**FIGURE 4-14**). Cacti are the most familiar example to many people; water-storing plants such as these are called succulents. Other adaptations can include the following:

- Wide spacing of plants allows maximum moisture per plant.
- Hard exteriors maximize water retention.
- Thorns reduce the consumption of the plants by animals seeking to extract the stored water.
- Deciduous plants shed leaves during dry periods to conserve water.
- Nighttime gas exchange (transpiration) is favored.
- Rapid growth occurs during brief periods of rainfall.
- Very long tap roots, especially in trees, can reach into deep underground water supplies.
- Nearly all perennial plant species can go dormant (a time of reduced photosynthesis) during periods of very high temperature.

FIGURE 4-13 The lack of rain means almost nothing can grow in this part of the Sahara desert in Morocco.

Annual plants are common in deserts. In these plants, the adult dies and the species survives as seeds that germinate when enough rainfall occurs. Small trees, such as the Joshua tree and mesquite, can occur where water supplies are adequate, but large trees such as the desert cottonwood occur only at oases (springs) or dry riverbeds where deep groundwater supplies are present.

Animal adaptations are often similar to those of plants. Nearly all animals have mechanisms that increase the efficiency by which water is retained. The ability of the camel to go for weeks without drinking is legendary. Desert animals tend to be small, with some exceptions, such as the camel, mule deer, and

FIGURE 4-14 A variety of cacti and other semidesert plants.

kangaroo. Reptiles are generally more abundant than warm-blooded animals. Reptiles are ectothermic; their behavior regulates their body temperatures. By cooling their body temperatures, they can reduce their energy needs to wait out the dry seasons or low food sources. Also very common is the habit of restricting activity to nighttime, with daytime spent in deep burrows as a means of escaping the heat. Desert rattlesnakes and small rodents illustrate this. Many animals, such as the desert toad, take this one step farther and enter a dormant stage for long periods, becoming revived during the brief periods of rain. It is during this period that reproduction usually occurs.

Temperate Grasslands

Temperate regions with scarce rainfall, about 25 to 75 cm (10 to 30 inches) per year, tend to have grasses and other herbaceous forms as the most prominent plants (**FIGURE 4-15**). Examples include the Great Plains of North America, the pampas of South America, and the steppes of Russia (see Figure 4-9). Tall-grass prairies are grasslands, with plants often standing more than 1 meter tall, of the moister areas. Short-grass prairies consist of clumps of short grasses, such as bunchgrass, that grow in drier regions. The organic matter added annually when the grasses die forms the rich soils of grasslands. The grass roots and the relative lack of rain hold the soils in place. Because of the aridity, grasslands regularly experience fires, especially

during the dry season. Trees, although not common, occur locally along stream banks and steep slopes where shade from the sun provides more moisture and protection from fires.

People have drastically altered much of this biome because economically this is our most important biome. Temperate grasslands provide the richest land in the world for grain crops such as wheat and grazing for sheep, cattle, and other food animals. For example, it is estimated that farming and grazing have drastically modified more than 95% of the original Great Plains biome in the last 200 years. Although many species of grass and the insects supported in the great prairie were lost before they could be counted accurately or their complex relationships understood, modern prairie grass species diversity have been reduced from approximately more than 200 distinct species to about 20 species. In recent years, individual landowners have taken an interest in reintroducing a wider diversity of prairie grasses, but overgrazing still threatens the once highly impressive biodiversity of the Great Plains.

Burrowing animals such as prairie dogs populate grasslands in their natural state, and native herbivorous mammals, for example bison in North America, extensively graze the area. Although grasses have evolved to tolerate significant amounts of grazing, the introduction of nonnative grazers such as cattle has led to overgrazing because, unlike native grazers, they tend to eat the plant faster than it grows. It may not always be obvious that nonnative grazers are more destructive than native species. For instance, why did not the millions of bison that once roamed the American Plains cause just as much impact as some herded cattle do now? In fact, bison could cause localized, temporary damage in the sense of short-term overgrazing, but the bison were migratory and moved on to other areas, allowing the plants to grow back. In addition, bison under natural conditions could be more selective than domesticated cattle in what they ate, focusing, for instance, on just certain tender or tasty parts of a plant. Some bison would also selectively feed on a variety of leaves, twigs, and bark. The overall effect was that bison could live sustainably in their natural environment.

FIGURE 4-15 A rancher and his son look out over their herd of cattle on temperate grasslands.

Another problem is the invasion of nonnative grasses, including cheatgrass in the United States, which replace the native grasses after such disturbances as fire and overgrazing. Because of this extensive loss of native grassland plants, substantial efforts have been made toward the ecological restoration of the prairies in North America.

Temperate Forests

Temperate regions with adequate rainfall, 75 to 150 cm (30 to 60 inches) per year, support forests of broad-leaved deciduous trees that show colorful seasonal changes before dropping their leaves (**FIGURE 4-16A**). The seasonal leaf dropping is an adaptation to the very cold winters when water becomes frozen and photosynthesis is difficult or impossible. Examples are found in Europe, parts of Asia, and the eastern United States (Figure 4-9). These regions share many closely related tree species, including the oaks, chestnuts, beeches, and birches, indicating a shared common evolutionary heritage dating geologic periods when the continents were interconnected (see our review of the dynamic Earth and natural hazards). Within the United States, oak and hickory dominate in the Southeast, and beech and maple dominate in the Northeast; however, substantial numbers of conifers (evergreen cone-bearing trees such as pines) occur in many of these forests, mainly as early colonizers of disturbed areas or on poorer soils to which deciduous trees are not well adapted.

These forests lack the spectacular diversity of tropical forests but are still more diverse than coniferous forests. Compared with the tropical rain forests to the south, temperate forests have a well-developed understory vegetation of many shrubs, herbs, ferns,

(a)   (b)

FIGURE 4-16 (a) Colorful fall foliage in an eastern U.S. deciduous forest. (b) Cougar.

mosses, and small trees because so much light reaches the ground when the tall trees seasonally lose their leaves. In addition, unlike the rain forest, where much animal life is found in the trees, a diverse array of species dwells on the ground. Large herbivores include elk and deer, and carnivores include wolves, cougars, and foxes (**FIGURE 4-16B**).

In the eastern United States, this biome has been drastically modified throughout nearly all of its range. Europeans cleared these forests for farming during the early years of colonization. In some areas, especially the northeastern United States, where the land is no longer farmed, ecological succession is beginning to return the land to its former state. Young temperate forests are now common.

Taiga

Taiga, also called coniferous or boreal forests, occur in a broad belt in northern North America and Asia (Figure 4-9). Very few taigas occur in the Southern Hemisphere because ocean waters mainly occupy these latitudes. Diversity and plant productivity are relatively low because of the stress imposed by long, cold winters with little precipitation. Trees are dominated by conifers (evergreens), which are tolerant of dry, cold conditions (**FIGURE 4-17**). Prominent types of evergreens include spruce, firs, and pines whose needles conserve water and withstand freezing better than do leaves.

During the long summer days, plants grow rapidly. Marshes, ponds, and lakes are common, often forming in depressions carved out by the glaciers present in the area just a few thousand years ago.

Common animals of the taiga include herbivores, such as moose, elk, deer, and the snowshoe hare, and carnivores, such as timber wolves, brown bears, lynx, and wolverines. The animals have adapted to the cold in various ways, including exceptionally thick fur and fat storage abilities that help sustain them through the long winters.

Compared with other biomes, the taiga has been less modified by agriculture. One reason is that cleared taiga makes poor cropland. In addition to the cold climate and short growing season, the thin soils are very acidic (pH of 4.5 to 5.0) compared with the soils of grasslands or temperate forests (pH of 6 to 7). The microbial decay of pine needles and other coniferous ground litter produce the acidic soil; however, clear-cutting for timber drastically modifies extensive tracts of the lower-latitude taiga.

Tundra

Arctic tundra occurs in Canada and Eurasia's high northern latitudes, between the taiga and shores of the polar seas (Figure 4-9). As with taiga, little tundra occurs in the Southern Hemisphere because ocean waters occupy these latitudes. Tundra is the low-lying vegetation that occupies an extensive treeless plain whose topsoil is frozen all year except for about 2 months during summer. Below this is permafrost soil, which is frozen all year long. This inhospitable region is not only very cold but also very dry. It receives less than 25 cm (10 inches) per year, or about the same as the desert biome. During the winter, the sun barely rises above the horizon and only for a few hours a day. Few species are able to adapt to such harsh conditions, and so the tundra has relatively low plant species diversity. Not surprisingly, this region also has relatively low plant productivity because of the limited opportunity for photosynthesis.

During the brief summer thaw, the tundra becomes an almost impassable bog, pocketed with countless lakes and streams (**FIGURE 4-18**).

FIGURE 4-17 Taiga.

Lichens (algae and fungi symbionts), grasses, mosses, and small shrubs are dominant plants. With a growing season of less than 8 weeks, these species must grow and reproduce rapidly. During this period, the tundra teems with life as animals seek to capitalize on these rich but short-lived plant resources. Birds arrive from the south, and animals, including caribou (reindeer), migrate into the area for feeding (Figure 4-18b). The air is filled with billions of insects. More permanent residents include the arctic hare, arctic fox, and snowy owl. Some small rodents such as lemmings survive the winters underground, although they often show drastic population fluctuations. Like the taiga, the harsh nature of this biome has kept it from being modified or occupied by people to any degree.

Alpine tundra occurs in lower latitudes where the altitude is sufficiently high (Figure 4-18). Above the tree line on the slopes of mountains, plant and animal diversity is low. The soil can only support plants similar to those found in the arctic tundra: lichens, grasses, mosses, and shrubs that hug the ground to retain heat and withstand the high winds. The dominant animals are rodents. Mountain goats, elk, sheep, and bears are seasonal residents. Compared with the boggy soil of the arctic tundra, alpine tundra soil is well drained.

Aquatic Biomes

Living conditions in water tend to be much less harsh than on land. The limiting factors in aquatic biomes are different in several important ways. Recall that temperature is a limiting factor on land, but water has many fewer temperature fluctuations. In aquatic environments, the water provides the added benefit of more buoyancy for organisms as support against gravity. The amount of light available in aquatic biomes is a limiting factor in ways that are not experienced on land. Water is a powerful solvent and also readily carries many substances in suspension; these substances, which range from toxins to nutrients, influence life locally. All of these differences occur because water, as a liquid medium, is much denser than the gaseous air. Most importantly, perhaps, getting enough water is not a problem. Living in water eliminates the danger of

(a)

(b)

FIGURE 4-18 (a) The Arctic tundra. (b) Caribou.

drying out that most land organisms must face every day. The fact that water in the frozen state floats and expands also is a noteworthy point since this property allows organisms to live in water under the ice even when temperatures are below freezing, and also makes possible life in mud at the bottom of a lake, since most large bodies of water never freeze all the way to the bottom. Not surprisingly, life originated in the shallow oceans and took many millions of years to adapt to land.

Aquatic communities do not divide into distinctive latitudinal biomes like those found on land. Water readily transports heat, and this property enables warm currents, such as the Gulf Stream, to warm large areas even near the poles. This prevents a simple latitudinal gradient from forming. It is the water's salinity that is particularly important in determining what can live in a particular environment. Therefore, ecologists often designate only two aquatic biomes: the marine biome and the freshwater biome (**FIGURE 4-19**).

Marine Biome

Marine water differs from freshwater in containing more dissolved minerals (salts) of various kinds. On average, marine water has about 3.5% salt, mainly sodium chloride, but with many hundreds of other materials as

well. The marine biome is the largest biome by far (more than 70% of the Earth's surface) but can be divided relatively easily into the **benthic** zone (on the bottom) and the **pelagic** zone (in the water column). Because water depth and the amount of light affect the distribution of organisms, biologists use them to subdivide these zones into smaller zones.

The pelagic zone includes the relatively shallow water zone over the continental shelves and the deep ocean zone. Pelagic creatures include planktonic organisms (floaters), such as diatoms, and nektonic organisms (swimmers), such as fish, turtles, whales, and squids (**FIGURE 4-20**). Bottom-dwelling organisms can be considered benthic regardless of whether they are attached or not. Benthic creatures include burrowers such as worms, crawlers such as snails, and stationary filter feeders such as barnacles (**FIGURE 4-21**). Benthic communities (Figure 4-19) are further subdivided by depth: littoral (shore), continental shelf, and abyssal (deep sea). Littoral species have adaptations to avoid drying out. For example, barnacles can tightly seal their hard exterior to retain water when the tide goes out. Many other crustaceans dig deeply into the wet sand. The type of shoreline—sandy, muddy, or rocky—subdivides the littoral communities. Deep-water benthic species

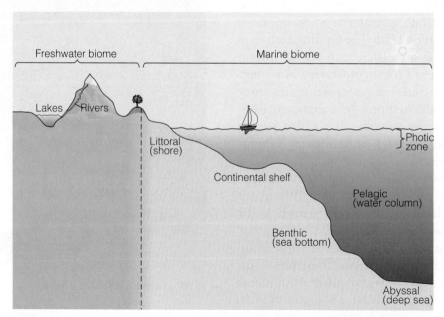

FIGURE 4-19 The two aquatic biomes are freshwater and marine. The marine biome can be divided into benthic and pelagic zones.

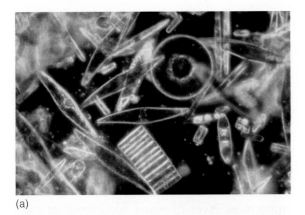

(a)

(b)

FIGURE 4-20 (a) Diatoms are microscopic phytoplankton responsible for much of the ocean's primary production. (b) French angelfish.

(a)

(b)

FIGURE 4-21 (a) This bloodworm lives in salt marsh sediments. Some of these worms are free living, whereas others are tube builders. (b) Spider crabs around a deep sea vent.

are adapted for withstanding the crushing water pressures found in the deep sea.

The **photic zone** is the upper part of the biome where light penetrates. The depth of this zone depends on the clarity of the water—as shallow as 5 m or as deep as 100 m in tropical waters. The photic zone is the main zone of photosynthesis and therefore is crucial to life in the biome. The photosynthesizing plants, which form the base of the marine food pyramid, are mainly tiny planktonic organisms including diatoms.

Although the marine biome covers nearly three-fourths of the Earth's surface, most of the open ocean supports relatively little life. Much of the ocean is too deep for light to penetrate and drive the photosynthetic food base. The deep ocean is poorly oxygenated, which also restricts the ability of species to live there. The creatures in the deep ocean are largely dependent on the dead organic material that drifts down from above for food. The most pro-

ductive areas of the ocean are near the shore, where upwelling brings the nutrients from the sea bottom into the photic zone and where nutrients in the ocean flow in from land.

Estuaries are transitional ecosystems between the ocean and freshwater biome. Estuaries, such as the Chesapeake Bay of Maryland and Virginia, occur where rivers and streams flow into the ocean (**FIGURE 4-22**). Five major rivers and over 100 smaller tributaries from six states flow into the Chesapeake Bay. This produces wide fluctuations in the salinity of these waters to which organisms must adapt. On the other hand, the flowing rivers provide a rich supply of nutrients that are rarely rivaled anywhere else on Earth. An estuary's shallow water means tides and waves keep large amounts of nutrients suspended in the water along with oxygen, and light is available for increased photosynthetic activity. As a result, estuaries can support large populations of organisms. Typical animals that

FIGURE 4-22 A satellite image of Chesapeake Bay.

live in estuaries include oysters, which form reefs that support many other species, such as sponges and barnacles that attach to the reef. The reef traps nutrients, and this attracts many fish and other species. A single oyster reef has been found to support more than 300 species of organisms.

Freshwater Biome

Liquid freshwater covers only 2% of the Earth. The freshwater biome includes both running waters, such as rivers and streams, and standing waters, such as lakes and ponds. In general, rivers and lakes are larger and more permanent than streams and ponds. The faster motion of running waters tends to keep them more highly oxygenated and easier to clean up after being polluted. A river or stream is an open ecosystem in that it derives most of its organic material from upstream or runoff from the land. Many species feed on this detritus. Of special ecological importance is the **riparian zone**. This is the area along the river or stream bank, consisting of cattails, reeds, willows, river birches, and other plants that tolerate high moisture. These plants are very important buffers that

reduce flooding impacts and water pollution by absorbing water. They are also critical habitat for many animal species including numerous highly endangered wetland species. Riparian plants are also important for aquatic life: they help protect the water from direct sunlight and reduce heat stress and provide many nutrients for aquatic life in the form of leaves and other detritus.

The slower motion of water in lakes and ponds leads to stratification of the water: the uppermost layer of water has plenty of oxygen while the oxygen decreases with depth. The uppermost layer is also much warmer during the summer and cooler during the winter than the lower layers. There is a sharp boundary, the **thermocline**, between the warm surface water and the colder deeper waters. Summertime swimmers often notice this abrupt transition to cold water when diving. Most of the mixing between the uppermost and deeper layers occurs during seasonal changes known as spring and fall overturn. As with rivers and streams, lakes also have a transitional zone with land, sometimes called the **shore zone**. It functions in the same way as the riparian zone, reducing flooding and heat stress and providing nutrients and critical habitat for many wetland and aquatic species.

As in the ocean, freshwater has two basic life zones: the bottom-dwelling benthic organisms and the swimming pelagic creatures (**FIGURE 4-23**). Familiar benthic organisms include freshwater mussels that filter nutrients from the water and a variety of insects and other crawling or burrowing invertebrates such as

FIGURE 4-23 Beavers are highly dependent on water and can be considered pelagic mammals.

FIGURE 4-24 Red-eared sliders prefer freshwater habitat with slow-moving water, mud bottoms, and a nice log or other handy place for basking in the sun.

crayfish. Fish, of course, are the most familiar swimming organisms, ranging from catfish that prefer bottom waters to perch and trout that prefer shallower depths. Amphibians are abundant in many freshwaters because they require water to reproduce. Other vertebrates include turtles and a few water-adapted mammals, such as otters, beavers, and muskrats. Algae are usually an especially important type of plant in freshwater ecosystems, although water lilies, various "grasses," and other plants occur as well.

Marshes and swamps are classified as **wetlands**. The ground in these ecosystems are waterlogged or submerged. Because the anaerobic conditions slow organic decomposition, wetlands accumulate rich layers of organic material. Reeds and grasses dominate marshes, whereas trees dominate swamps. Wetlands support a high diversity of organisms and carry out important functions for the biosphere despite the small land area they occupy on Earth. For example, wetlands help filter out pollutants, replenish groundwater supplies, and provide natural flood control. (Saltwater marshes help protect against storm surges.) Wetlands are home to a staggering number of bird, amphibian, insect, and reptile species (**FIGURE 4-24**).

CASE STUDY 4-3

Conservation and the Geography of People

We see in this chapter how species are not randomly distributed on Earth. As you might suspect, people are not randomly distributed either. Far more people live in some areas, such as near coastlines, than in other areas, such as deserts. Do you suppose that large concentrations of people tend to occupy the same areas that are favorable for high species diversity and density? If true, this has enormous implications for conservation.

This hypothesis seems likely at first glance. We have seen, for example, that biomes such as the desert and tundra have a low productivity. This not only lowers the species diversity and density of these areas, but it also makes them less favorable places for agriculture. In addition, many people prefer to avoid the cold climate of the tundra or the heat of the desert; however, this is only a hypothesis based on the initial evidence presented in this chapter. What does more detailed evidence indicate?

An article that Andrew Balmford and colleagues (2001) published in the journal *Science* analyzed a map of sub-Saharan Africa at a very high degree of resolution. They compared the number of bird, mammal, snake, and amphibian species in each square on a grid with the human population density of that square. They found that, at least for most of Africa, the scenario seems, unfortunately for most nonhuman species, to be true. People do tend to occupy the same areas that are richest in species diversity. This is not entirely because of plant productivity. Other reasons are that people prefer to live in areas of fertile soils, at low elevations, and near water bodies. All of these geographic factors also tend to support relatively high numbers of species.

Another important aspect of species geography is how it relates to the nature reserves that are set aside to protect species. Are these nature preserves geographically selected so that they protect areas of high species diversity? An article by J. Michael Scott and colleagues (2001) published in the journal *Ecological Applications* examined the distribution of more than 2,500 protected areas in the lower 48 states of the United States. They found that less than 6% of the total area is protected and that the reserves were disproportionately located in areas of higher elevations and poor soils. This pattern occurs because land that is set aside for preserves historically has been mountainous land that was not productive for agriculture and was of little **economic value**. As you might suspect, this is bad for the preservation of biodiversity because many areas of high species diversity, such as wetland and riparian habitats, occur only at low elevations. A more effective reserve system must add land that is more representative of all habitats, even if it requires setting aside land that is desirable for agriculture and other human uses.

References

1. Balmford A, Moore J, Brooks T, Burgess N, Hansen L, Williams P, Rahbek C. Conservation conflicts across Africa. *Science* 2001;291:2616–2619.

2. Scott JM, Davis FW, McGhie G, Wright R, Groves C, Estes J. Nature reserves: Do they capture the full range of America's biological diversity? *Ecological Applications* 2001;11:999–1007.

Questions

1. Name two state parks in your state. Do you know anything about their history or why they were located where they are?
2. California has the highest population of any state and also the most native plant species. What reasons can you give for its high rank for both population and biodiversity?

SUMMARY

- Primitive bacteria arose at least 3.5 billion years ago.
- New species originate through natural selection, which passes on beneficial genes.
- Biodiversity is often measured as number of species.
- Rain forest sampling, ecological ratios, and species area curves estimate biodiversity.
- Biomes are suites of communities that share many basic adaptations.

- Climate, especially temperature and rainfall, largely determine biomes.
- The seven terrestrial biomes are tropical forests, savanna, grasslands, deserts, temperate forests, taiga, and tundra.
- The two aquatic biomes are marine and freshwater.

KEY TERMS

amino acids
benthic
biodiversity
biome
Cambrian explosion
chaparral
DNA
economic value
estuaries

eukaryotes
explosion of life
genes
mutation
natural selection
pelagic
photic zone
protocells
riparian zone

shore zone
species
species area curve
species richness
taxonomists
taxonomy
thermocline
wetlands
xerophytes

STUDY QUESTIONS

1. What are amino acids and what is the significance of the amino acid experiments of Miller and Urey and later Fox?
2. Describe the process of evolution by natural selection. How has the physical environment affected the history of life?
3. If a population consisted of a small number of individuals, what effect might natural selection have on its survival?
4. Describe how Mendel's discovery helped in the understanding of variation.
5. Explain two basic causes of variation.
6. Approximately how many genes do humans have? Why is this number significant?
7. What factors may have caused the "explosion of life"?
8. Define biodiversity. Is there only one way to define this concept? How can biodiversity be measured or estimated?

9. Explain why tropical rain forests have such high biodiversity.
10. Do tropical rain forests have fertile soils? Why or why not?
11. Describe four adaptations that plants can have for survival in the desert.
12. How are the two fundamental aquatic biomes recognized?
13. Describe why riparian zones are so important.
14. What is the sharp boundary between warm and cold lake waters?
15. Explain why much of the ocean can be considered a "biological desert."
16. Name the major biomes of Earth. Which biome or biomes do you live in? Would you consider it one of the more favorable biomes for human habitation? Why or why not?

WHAT'S THE EVIDENCE?

1. The authors state that new species may arise as a result of natural selection of individual variation and the isolation of populations. What evidence is given to support this? Can you think of evidence for other causes of evolution? If so, list them.

2. The authors state that the number of species on Earth could be anywhere from 5 to 100 million species. What is the evidence given to support this estimate?

CALCULATIONS

1. If taxonomists have described approximately 1.8 million different species and 56% of these are insects, how many of the described species are insects?

2. If taxonomists have described approximately 1.8 million species but there are actually 30 million different species on Earth, what percentage of the species have taxonomists described?

ILLUSTRATION REVIEW

1. Using Figure 4-8, explain how there can be a polar climate at the equator.

2. In which biomes would you expect to find the type of diversity shown in Figure 4-4a?

http://environment.jbpub.com/mckinney/5e/

http://environment.jbpub.com
/mckinney/5e/

Connect to this text's website at http://environment.jbpub.com/mckinney/5e/. This site features eLearning, an online review area that provides quizzes, chapter outlines, and other tools to help you study for your class. You can also follow useful links for more in-depth information.

The Dynamic Earth and Natural Hazards

5

Our physical Earth is still a young and evolving planet. We, as small creatures living on the surface of the globe, are still subject to the mercies of nature in many respects, even as we affect the global ecological balance. We must never forget that we cannot totally isolate ourselves from our physical surroundings on this planet. In August 1999, a large dust storm moving over Phoenix, Arizona, caused 90-minute airport delays and wind gusts of nearly 50 miles per hour. Although dust storms are prevalent in the desert southwest, their magnitude and frequency have intensified over the last decade.

The Earth has evolved from a relatively homogeneous collection of particles (cosmic dust) to form a well-differentiated, complex, dynamic, recycling machine. Our planet may seem eternal and unchanging; nevertheless, it was very different more than 4 billion years ago, and it will continue to evolve and change. From a human perspective, such changes are slow and gradual, taking hundreds of millions of years. We must have a clear understanding of the delicate, and in many ways fortuitous, development of the present state of the Earth to comprehend fully just how fragile our environment really is. Furthermore, despite our sometime hubris, to this day humanity is virtually helpless in comparison to some of the major forces of nature. Certain phenomena that can be tragic to us, such as the eruption of a volcano that sends plumes of ash into the atmosphere, with resultant worldwide weather changes that destroy crops and bring about mass starvation and epidemics (and indeed, this has happened, for instance, in the early 19th century after the 1815 eruption of the volcano Tambora in Indonesia), are in reality just the normal workings of planet Earth. In fact, "natural hazards" are an integral part of the natural world. We, as small creatures living on the surface of the globe, are still subject to the mercies of nature in many respects even as we tamper with the global ecological balance. We must never forget that we cannot totally isolate ourselves from our physical surroundings on Earth.

Chapter Objectives

After reading this chapter, you should be able to explain or describe the following:

- The basic internal structure of Earth
- Plate tectonics
- The composition of matter
- The rock cycle
- Climate, weather, and large-scale atmospheric and seasonal cycles
- The origin of the solar system and Earth
- The basic types of natural hazards

131

5.1 Workings of Planet Earth Today

Earth and Its Neighboring Planets

Earth is the third planet from the center of the solar system (moving away from the sun, the planets are as follows: Mercury, Venus, Earth, Mars, Jupiter, Saturn, Uranus, Neptune, and Pluto [although Pluto is not technically a planet, according to recently changed definitions, but rather a large object that belongs to the Kuiper Belt of space debris beyond Neptune]), but it is very different from the other planets. Unlike its two immediate neighbors, Venus and Mars, which both have atmospheres that are 95% to 97% carbon dioxide, Earth has an atmosphere composed of about 78% nitrogen, 21% oxygen, and only a trace of carbon dioxide (approximately 0.03% to 0.04%). As a result, much of the sunlight and heat reaching the Earth is scattered back into space, and Earth has a combined average surface air and water temperature of approximately 12°C or 53.6°F. At this temperature, water can exist in a liquid state, making life possible on Earth. As far as we are aware, no other planet has anything resembling life (although it has been suggested that structures in a meteorite that originated from Mars may represent microscopic organisms, and it is certainly conceivable, even highly probable, that life could exist elsewhere in the solar system and universe).

The gases surrounding the Earth (**FIGURE 5-1**) and the liquids (primarily water) on its surface are continually swirling and moving, causing the degradation, erosion, and destruction of topographic highs such as mountains. Why then does the Earth's surface have any relief at all? The reason is that the Earth has an "internal engine."

Present-Day Structure of the Earth

To understand natural processes on the surface of the Earth, we must have a clear understanding of the structure of the Earth. The internal makeup of the Earth accounts for volcanoes and earthquakes. It even affects the composition of the atmosphere, oceans, and ultimately the nature of life on the Earth. The Earth's interior is seething and churning. Rocks slide and move plastically around one another. The interior is very hot, and at breaks and cracks, molten rock erupts at the surface

FIGURE 5-1 The Earth seen from the moon. The Earth is a "living planet" with an active atmosphere and an internal heat engine driving many geological processes. In contrast, the moon is a "dead planet" (our moon is larger than Pluto and nearly as large as Mercury) with virtually no atmosphere or active geological processes on its surface. As a result, the moon's surface is pockmarked with craters that have not eroded away.

as volcanoes. In other areas, ancient rock is dragged back toward the center of Earth, only to be reheated, melted, and rejuvenated. Most of the Earth's surface is relatively young rock. In comparison, some of Earth's close neighbors—the Moon, Mercury, and Mars—are relatively inactive. The Moon, for instance, shows little or no evidence of plate tectonic activity, and its surface is covered with very ancient rocks, generally dating back more than 3 billion years. Mars has large volcanoes but apparently shows no evidence of true plate tectonic activity except perhaps very early in its history. Of the planets close to Earth, only Venus shows some definitive evidence of limited crustal movements comparable to the plate tectonics seen on Earth; however, most of that planet's geological activity seems to be in the form of active volcanoes situated over stationary "hot spots" where molten rock wells up from the interior as a result of thermal convection cells (discussed later).

If we could cut a slice through the Earth, we would see that it is formed of concentric rings of differing constitutions—similar to the layers of an onion (**FIGURE 5-2**). From the outside, the first major layer is the Earth's gaseous envelope or atmosphere. It extends in rarefied form for hundreds of miles above the Earth's surface, although almost the total mass is contained within the bottom 20 kilometers (12 to 13 miles). The Earth's surface is covered with seawater or dry land (including land that contains freshwater lakes or ice patches) in a ratio of approximately 7 to 3. If all water were removed from the surface, we would observe two basic terranes: ocean basins, which average about 5 kilometers (3 miles) depth below sea level and are floored with basaltic-type rocks, and continents and continental islands, which on average rise a couple of hundred meters above sea level and are founded on granitic-type rocks. Actual ocean basins cover only about two thirds of Earth's surface; in some places, the oceans lap onto the shallow continental margins.

The thin **oceanic crust** is mainly basaltic rock averaging only about 5 to 10 kilometers (3 to 6 miles) thick. It is composed of about 50% (by weight) silicon dioxide (perhaps most familiar as the mineral quartz) and contains higher amounts of iron, magnesium, and calcium than are generally seen in continental crust. The **continental crust**, mainly granitic rock, ranging from about 20 kilometers to 70 kilometers (12 to 43 miles) thick, is about two-thirds silicon dioxide.

Below the crust is the **mantle**, which represents nearly 83% of Earth's volume, with a thickness of about 2,900 kilometers (1,800 miles). It is composed of about 45% silicon dioxide and 38% magnesium oxide. The relatively rigid upper layer of the mantle with its attached crust is known as the **lithosphere**. The lithosphere is the major unit of movement in plate tectonics (discussed later here). Below the lithosphere lies another layer of the mantle, the relatively weak and soft **asthenosphere** (literally, weak or glassy sphere; see Figure 5-2).

Beneath the mantle is the Earth's **core**. The outer core consists of a liquid iron–nickel (mostly iron) alloy. Movements within the outer core probably are responsible for generating the Earth's magnetic field. The inner core is composed of a solid iron–nickel alloy (see **CASE STUDY 5-1**).

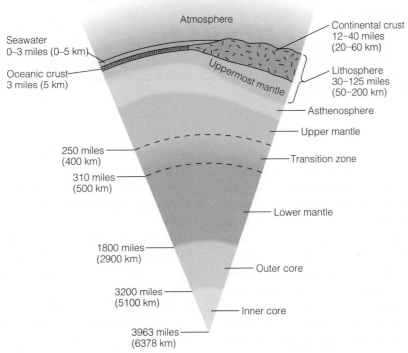

FIGURE 5-2 Schematic section through the Earth.

Evolution and Plate Tectonics

VISIT

http://environment.jbpub.com
/mckinney/5e/
for more information

Before the adoption of the idea of plate tectonics, it had not been known that the positions and shapes of the continents had changed throughout Earth's history. There had been many cases where the distributions of plants, animals, and specific geologic formations had been quite difficult to explain. For instance, why were primitive marsupials found in South America and Australia but nowhere else? How had they gotten from one place to the other?

Scientists came up with some bizarre convoluted theories of mysterious land bridges rising across ocean basins and then sinking beneath the oceans again. Plate tectonics and paleontology demonstrated that marsupials arose in what is now Antarctica, during the early Eocene (which lasted from 56 to 34 million years ago), when Australia, Antarctica, and South America formed one continent known as Gondwana. After spreading across all three land masses, they became isolated when Antarctica became polar, glaciated, and separated from the other continents.

Why was the stratigraphy of the Appalachians so similar to the stratigraphy of the mountains in Scandinavia, Ireland, and Scotland, and why did fossil records also indicate striking similarities? Further research explained that they had once been part of the same Caledonian mountain range, five times their current size before hundreds of millions of years of erosion softened them to their current size, and they were separated by the opening of the Atlantic Ocean, beginning about 200 million years ago.

Wallace's Line is a distinct imaginary boundary that was noticed by Alfred Russell Wallace who codiscovered the theory of natural selection with Charles Darwin in the late 19th century. The line demarcated a distinct, sudden, and inexplicable discontinuity between Borneo and Sulawesi in Indonesia and all the other islands to the east and west of the line. Animals on the west were similar to the fauna of Asia and those to the east strongly resembled the fauna of Australia.

Islands that were very far apart actually had similar species, if they were on the same side of Wallace's line. Islands like Lombok and Bali were close together (about 20 miles apart), but their biotas had almost nothing in common. Plate tectonics showed that Wallace's Line was the line between two continental plates. The two sides of Wallace's Line were two land masses that only recently became neighbors. During times of lower global sea level, land bridges had connected the various islands on the western side, allowing for species to migrate between islands, while no land bridges had existed between the two sides of the line until relatively recently.

Francis Bacon noted in 1620, that the continental margins of Africa and South America seemed to "fit together." Once it was determined in the middle decades of the 20th century that the continents moved from place to place, this obvious pattern made sense and also served to explain a lot of geographic mysteries in plant, animal, and geologic distribution that exists both in the fossil record and on the continents today.

Plate Tectonics

If you look carefully at a map and imagine that the continents are pieces of a jigsaw puzzle, you will notice that the east coasts of North and South America appear to fit into the west coasts of Europe and Africa. This simple observation and a wealth of other data led to the hypothesis of drifting continents, which the German scientist Alfred Wegener proposed in 1912. Today, virtually all earth scientists accept a variation of the theory of continental drift, known as **plate tectonics**. Indeed, plate tectonics forms the unifying theory for most of the geological structures observed on the surface of the planet. It is the active process of plate tectonics that not only moves continents, but also raises mountains, creates new sea floor, destroys and recycles old sea floor, and causes volcanoes to erupt and earthquakes to occur. Plate tectonics plays a major role in natural biogeochemical cycles on Earth—cycles that human activities are now disrupting (a review of the biosphere provides more information).

Today, the Earth's lithosphere is divided into about eight major tectonic plates and numerous smaller ones (**FIGURE 5-3**). These

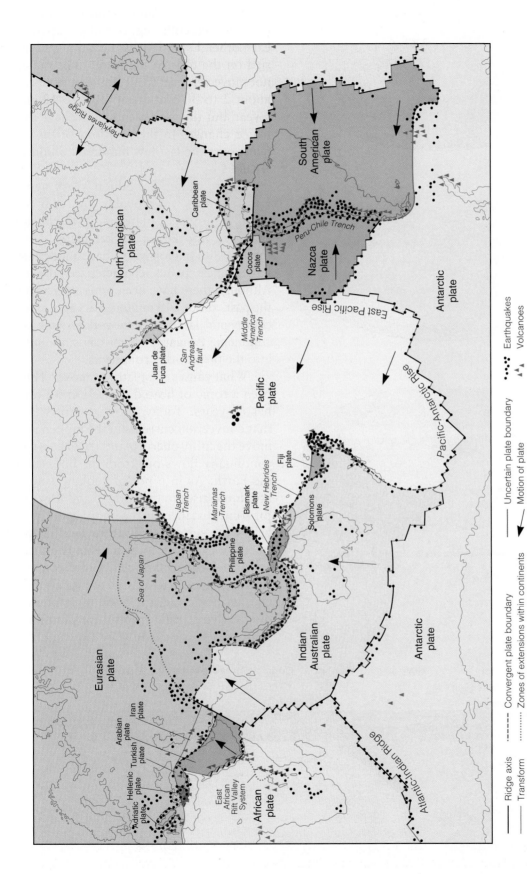

FIGURE 5-3 The Earth's tectonic plates. The plates are in continuous motion relative to one another. Notice the relationship between plate boundaries, intense earthquakes, and major volcanic eruptions.

(a)

(b)

(c)

FIGURE 5-4 Continental movements over the last 240 million years. (a) The Earth about 240 million years ago. (b) The Earth about 70 million years ago. (c) The modern Earth.

plates are in continuous motion, sliding past one another, colliding, or pulling apart from each other. Essentially, the continents are carried on the tops of the plates. The plates do not move very fast from a human perspective, about 2 to 30 centimeters (0.8 to 12 inches) a year, but this is fast enough to have caused major changes in the positions of the continents over the last few hundred million years (**FIGURE 5-4**).

Before about 2.5 billion years ago, the Earth's surface was probably covered by small, rapidly moving "platelets," which slowly coalesced into larger, thicker plates. Only in the last 800 million years have modern-style, large, thick plates characterized the crust. Although there have been other movements in the past, about 240 million years ago, all of the continental land mass formed a single supercontinent, **Pangaea**, which has since split up into the present-day continents.

What causes the plates to move? This has been a topic of heated discussion for a number of years. Today, most scientists believe that convection currents in the molten mantle move the lithospheric plates (**FIGURE 5-5**). Heat continually flows out from the hot center of the Earth. As in a pot of cooking soup, the hot liquid rises to the surface and flows there for some distance; as the liquid cools, it sinks below the surface, only to be heated up and start the cycle once again. This is known as a **thermal convection cell**. Giant **convection cells** are believed to exist in the liquid mantle, and as they cycle, the flowing mantle drags the overlying, rigid lithosphere plates along.

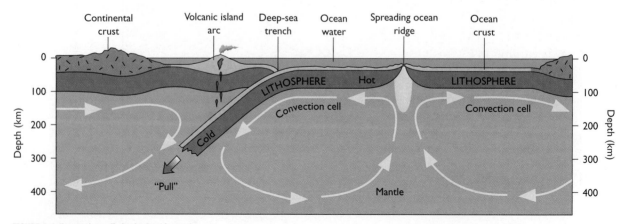

FIGURE 5-5 Convection cells in the Earth's mantle.

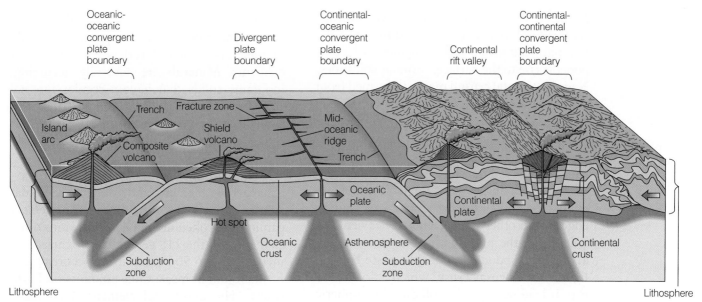

Oceanic-oceanic convergent plate boundary

Divergent plate boundary

Continental-oceanic convergent plate boundary

Continental rift valley

Continental-continental convergent plate boundary

Trench · Fracture zone
Island arc
Shield volcano
Composite volcano
Mid-oceanic ridge
Trench
Oceanic plate
Continental plate

Hot spot
Oceanic crust
Asthenosphere
Continental crust

Subduction zone
Subduction zone

Lithosphere · Lithosphere

FIGURE 5-6 Principal types of plate boundaries.

As the Earth cools over time, the rates of motion of the plates are decreasing, and they are continuing to enlarge and thicken.

At divergent plate boundaries, places where two lithospheric plates are moving away from each other, a gap or void is left in the solid crust (**FIGURE 5-6**). Initially, the rock in this area may start to collapse, forming a rift valley, but as the plates move apart, hot molten rock from the mantle wells up to fill the void and form new oceanic crust. This process is often referred to as sea-floor spreading. The classic example of a divergent plate boundary is the Mid-Atlantic Ridge, which runs roughly north–south through the Atlantic Ocean.

Along deep-sea rift zones, hot water (about 350°C, or 660°F) emanates from hydrothermal vents. These vents often support unique ecological communities that are based on geothermal energy rather than the solar energy most ecosystems require (**FIGURE 5-7**). Bacteria use the energy found in hydrogen sulfide (H_2S) and similar chemicals emanating from the vents to produce organic molecules from inorganic molecules. Crabs, clams, tube worms, and fish then feed on the bacteria. These vents dramatically illustrate the interrelationships between geological processes and living organisms.

In some places, convergent plate boundaries occur where two lithospheric plates converge or collide (see Figure 5-6). If the colliding edges of the two plates both bear oceanic

crust, either plate may be forced, or subducted, under the other plate. A subduction zone is formed under the leading edge of the plate that remains on top. On the surface, where the

FIGURE 5-7 A deep-sea hydrothermal vent supports a unique biotic community, which is based on geothermal energy rather than the solar energy most systems require. Bacteria use the energy found in hydrogen sulfide and similar chemicals to produce organic molecules. Crabs, clams, giant tube worms, and fishes then feed on the bacteria.

plates are down-warped, a deep trench forms. As the subducted plate moves down into the hot mantle, it melts, and the lighter rock components rise to the surface, forming volcanoes. New continental crustal material, formed from the recycled oceanic crust, collects behind the trench. The northern and western edges of the Pacific Ocean (known as the "ring of fire" for their numerous volcanoes) are lined with convergent plate boundaries (see Figure 5-3).

If the leading edge of one colliding plate contains continental crust and the leading edge of the other plate contains heavier oceanic crust, the plate with oceanic crust will always be subducted under the plate bearing lighter continental crust, which "floats" on top. If leading edges of colliding lithospheric plates both contain continental crust, neither can be subducted because they are both relatively light. Instead, the continents crash into one another, crumple, and deform, often raising imposing mountain ranges. The mighty Himalayas are the result of two continental land masses crashing into one another. Eventually, the plates lock, and relative motion between them stops.

Finally, two plates may slide past one another; their edges grind and slip against each other along what are commonly termed transform faults (also known as strike-slip faults). This is the least dramatic type of relative plate motion in that it usually does not involve either volcanoes or mountain building. Earthquakes are common along transform faults, as residents of California well know, where the Pacific plate slides north past the North American plate. The surficial rocks do not flow past one another slowly and smoothly, but tend to "hang up." Pressure builds and then is suddenly released—the rock "jumps," and an **earthquake** occurs.

The Composition of Matter on the Earth

Rocks are by far the most common substances on the Earth's surface. Indeed, everything is rock or made primarily of materials that were once components of rocks. A clear understanding of rocks and the rock cycle is necessary in order to understand fully the cycling of matter through the environment (a review of the biosphere provides more information).

The Structure of Rocks

Rocks themselves are composed primarily of minerals, which in turn are composed of elements. **Minerals** are naturally occurring, inorganic solids that have a regular internal structure and composition—they are said to be crystalline. Some well-known minerals include quartz, diamond, garnet, and pyrite. In all, several thousand different minerals are known.

The **elements** such as iron, hydrogen, oxygen, carbon, mercury, and gold are the fundamental substances of our world. An element cannot be broken down chemically into other elements. Of the 118 generally recognized elements, close to 90 are naturally occurring elements found in measurable amounts on Earth today. The additional elements have been synthesized artificially under extreme laboratory conditions. The artificially synthesized elements are highly unstable (radioactive) and sometimes exist for only a fraction of a second.

Atoms are the smallest units of an element that retain the physical and chemical properties of the element. One or more atoms of the same type can combine to make a molecule. When molecules of different types combine, they form **compounds**. The chemical abbreviation (formula) for water (H_2O) tells us that there are two atoms of hydrogen and one atom of oxygen in the compound that makes water. A collection of similar molecules, like those that constitute a glass of water, is often called a substance. When combined together, substances can take on new properties. Atoms themselves are divisible into even smaller particles. In the center of each atom is a nucleus composed of protons and neutrons (**FIGURE 5-8**). Protons carry positive charges, and neutrons are electrically neutral. Most of an atom's mass is contained in its nucleus. The number of protons defines the element. No two elements have the same number of protons; when you look at a periodic table of the elements you will see that the elements are numbered. The number is the number of protons in the nucleus. For example, all atoms of hydrogen have only one proton in their nucleus, whereas all atoms of carbon have six protons, and those of gold have 79 protons in their nucleus. The number of protons in the nucleus is referred to as the atom's atomic number and is characteristic of

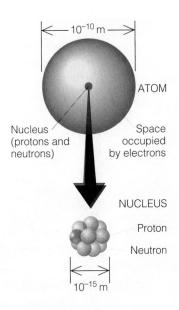

FIGURE 5-8 Basic structure of an atom.

Properties of Atomic Particles

Particle	Charge	Mass	Location	Symbol
Proton	Positive	1.673×10^{-24}g	Nucleus	p
Neutron	Neutral	1.675×10^{-24}g	Nucleus	n
Electron	Negative	9.110×10^{-28}g	Orbitals	e

the atoms of a particular element. The number of protons and neutrons in the nucleus is the atom's atomic mass number. Atoms of the same element can have varying numbers of neutrons in their nucleus. In nature, one can find carbon atoms with atomic mass numbers of 12, 13, or 14. These variants are known as isotopes of the element carbon. Some isotopes are unstable and undergo radioactive decay by emitting particles spontaneously so as to change into a more stable form of atom (a review of the fundamentals of energy, fossil fuels, and nuclear energy provides more information).

"Orbiting" (they are not literally orbiting in an everyday sense) around the protons and neutrons of the atom's nucleus are electrons (Figure 5-8), small negatively charged particles that electrically balance the protons in an electrically neutral atom. If a particular atom has more electrons than protons, it will be a negatively charged ion. If it has fewer electrons than protons, it will be a positively charged ion.

Single atoms are much too small for us to perceive in isolation. Substances that we are familiar with are composed of tremendous numbers of atoms, which are almost always bonded together in various arrangements. Depending on how the atoms are arranged, we refer to states of matter composed of atoms as solids, liquids, or gases.

Rocks: Their Origin and the Rock Cycle

Rocks are classified into three basic categories: igneous, sedimentary, and metamorphic.

Igneous rocks formed, or crystallized, from hot molten rock known as magma ("igneous" = fire rock). The lava that flows out of a volcano and hardens into rock is a well-known type of igneous rock. Igneous rocks that formed from lava flows or volcanic ash ejected from a volcano are known as extrusive igneous rocks. However, most igneous rocks form deep underground as liquid magma slowly cools and crystallizes; such rocks are known as intrusive igneous rocks. Typical igneous rocks include basalt and granite. Igneous rocks are the most common rocks, forming the bulk of the Earth's crust and its entire rocky interior (where it is not molten). However, on the surface of the Earth, a blanket of sediments or sedimentary rocks often covers igneous rocks.

Over time, the rocks weather, broken down by both mechanical agents (running water, ice, wind, or the roots of trees) and chemical agents (acids and solvents found in nature, such as the carbonic acid of rain water), and disintegrate into fragments. These fragments are known as sediments; they generally consist of either small pieces of rock or single mineral grains and come in various sizes, ranging from gravel, to sand, to silt, to mud. Gravity and running water, wind, or ice flows (glaciers) physically transport the sediments from higher to lower elevations until they are deposited in a resting place. As piles of sediments accumulate, they become compacted and cemented together, or even crystallized. Minerals may precipitate between the sediment grains, binding them tightly together to form a **sedimentary rock** (**FIGURE 5-9**). Sedimentary rocks cover approximately 75% of Earth's surface. Among the most common sedimentary rocks are sandstones and shales.

FIGURE 5-9 The Grand Canyon exposes many different types of sedimentary rocks, including limestone, slate, shale, and sandstone.

FIGURE 5-10 The travertine terraces of Mammoth Hot Springs in Yellowstone National Park are formed when carbon dioxide combines with hot groundwater to form a carbonic acid solution. This solution then dissolves limestone as it moves up to the surface hot springs. When it reaches open air, the calcium carbonate can no longer remain in solution and precipitates as travertine.

Not all sedimentary rocks are formed from the weathered remains of other rocks. For example, many limestones are formed from sediment that consists entirely of the old calcium carbonate shells or skeletons of sea organisms. Fossils, the remains of ancient organisms, are almost always found encapsulated in sedimentary rocks such as limestones, sandstones, and shales. Peats, coals, oils, and other fossil fuels, formed of the fragments of many plants and other organisms, are components of some sedimentary rock layers (and in such cases, a "rock" such as coal may be composed of nonmineral material). Other types of sedimentary rocks are formed from the precipitation, or "sedimentation," of substances out of an aqueous solution, perhaps as it evaporates (**FIGURE 5-10**). For instance, large deposits of rock salt (halite) often form in this way.

Like igneous rocks, sedimentary rocks weather and break down. The sediments formed from sedimentary rocks can then be recycled to form other sedimentary rocks. The remaining class is **metamorphic rocks** ("metamorphic" = change of form). Metamorphic rocks are made when rock—igneous, sedimentary, or in some cases another metamorphic rock—is subjected to great temperature (but not hot enough to completely melt the rock) or pressure or both. The rock undergoes some combination of mineralogical changes (certain minerals are turned into other minerals) and/or textural changes (mineral grains may grow in size and/or reorient). In some cases, hot fluids released at the edge of a molten rock body that penetrates preexisting rock may produce metamorphic rocks. Some common metamorphic rocks include slate (produced from the metamorphism of shale or mudstone) and marble (produced from limestone).

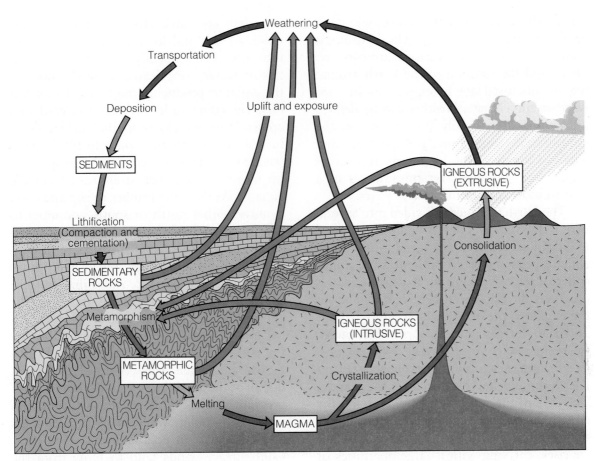

FIGURE 5-11 The rock cycle.

Geologists often refer to the **rock cycle** (**FIGURE 5-11**). Effectively, all rock/earth material on the Earth's surface begins as igneous rock. Initial crystallization may take place deep underground or at a rift zone where two lithospheric plates diverge or by extrusion from a volcano. This new rock may weather directly into sediments or become metamorphosed. Ultimately, it may be carried to a subduction zone, where it is remelted, rises toward the surface, and eventually is added to the continental crust. From here it may be metamorphosed or remelted again, or it may be weathered at the surface, with the fragments eventually forming sedimentary rocks. The sedimentary rock may be weathered, broken into fragments, and made into other sedimentary rocks, or more sediments may be piled on top of it until it is buried deep below the surface. With increasing temperatures and pressures, the sediments may be metamorphosed and eventually find themselves in a setting where they actually melt. If so, they will form genuine igneous rocks on cooling and crystallization, thus continuing the rock cycle.

By human standards, the rock cycle is very slow, taking on the order of millions to hundreds of millions of years. Furthermore, it is not a complete or closed cycle. After continental (granitic type) material is formed, it remains continental material, even if it cycles from igneous to sedimentary, to metamorphic, back to igneous, and so on. As described in the section on plate tectonics, continental material is too light to be subducted back into the mantle; thus, it is not recycled in that sense. More continental crustal material is in existence today than ever before in Earth's history.

The Surface of the Earth: Climate and Weather

In our review of the biosphere, we discussed the Earth's atmosphere, energy flow from the sun to the Earth, and the hydrologic cycle driven by the sun's energy. The **hydrologic cycle**, combined with various atmospheric cycles and the rotation of the Earth around the sun (discussed later here), is responsible for weather and climate. **Weather** can be defined as the short-term daily fluctuations in the atmospheric temperature, precipitation, and winds. **Climate** is the long-term average of these same variables over time ranges of a year to thousands or millions of years.

The Earth's surface is divided into several major climatic belts that roughly correspond to latitudinal changes (north and south); these climate belts are based on large-scale **atmospheric cycles**. Working out from the equator, one finds the rain forest belt (centered about the equator), the desert belts (centered about 20 to 30 degrees latitude north and south), the temperate regions, and the polar regions. Convection cells form in the atmosphere similar to those believed to occur in the mantle. The rays of the sun fall most directly on the equator, making the equatorial regions hotter than the rest of the Earth's surface. The hot air rises, carrying with it large quantities of water vapor (warm air can hold more mois-

ture than cool air). However, the rising air quickly cools and loses its capacity to hold moisture. Consequently, enormous quantities of water are released in the vicinity of the equator, producing the tropical rain forests. The rising air leaves a relative void, and cooler, drier air closer to the surface flows in from the north and south to fill the void (**FIGURE 5-12**). This new cool air heats up, picks up moisture, and then itself rises from the equator. As the air continues to rise and cool, it moves either north or south, continues to lose moisture, and eventually begins to sink as dry air in the region of 25 to 35 degrees latitude north and south, giving rise to the subtropical high-pressure zones that produce some of the world's great deserts. After the cool air approaches the surface, it flows south or north toward the equator. On reaching the equator, the air again heats and absorbs moisture, and the cycle continues. However, not all of the air sinks at the subtropical high-pressure zones. Some of the air continues toward the poles, where it eventually sinks and takes part in polar atmospheric cycles. Cold surface winds tend to blow from the poles toward the equator.

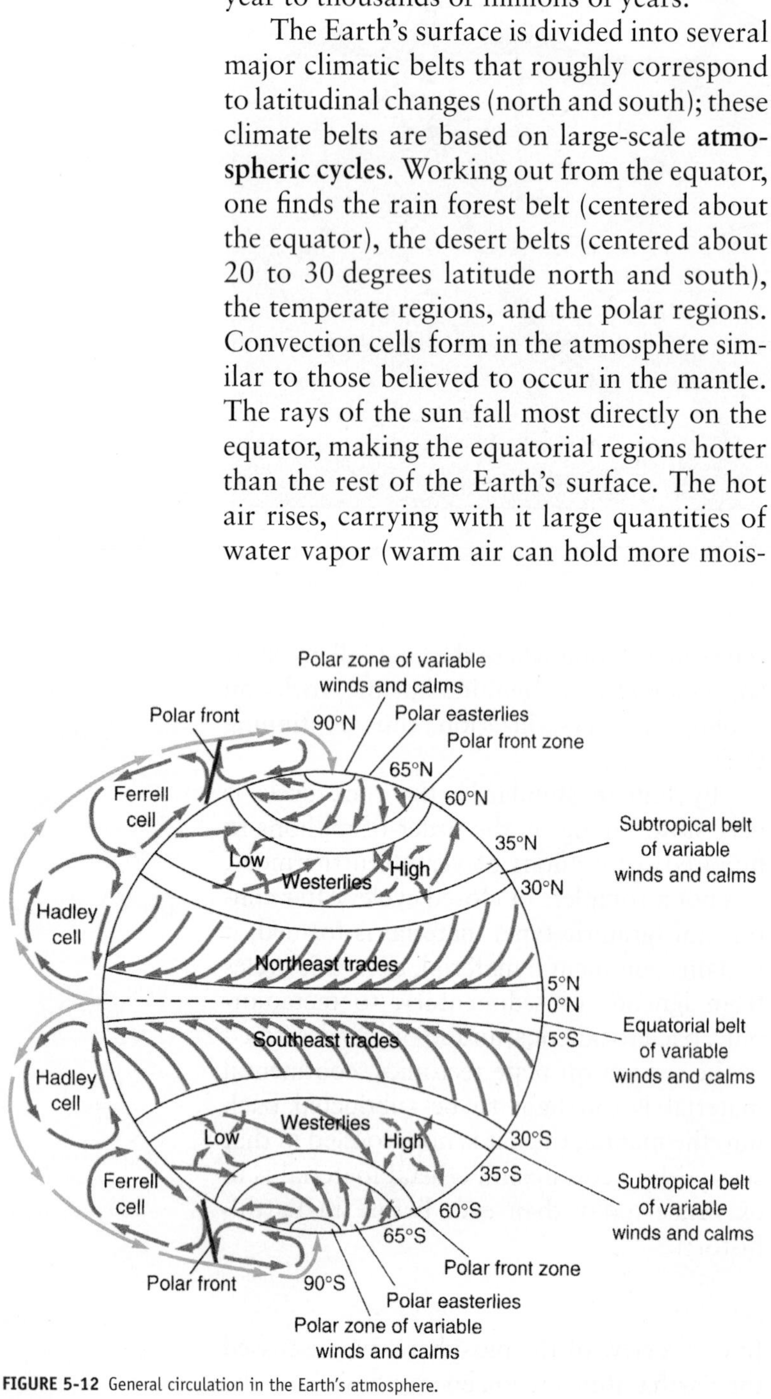

Of course, the atmospheric cycle is not quite that simple. Two effects in particular modify it. First, the Earth spins on its axis, causing the air in the convection cells to cycle at an angle, rather than directly north and south. This is known as the **Coriolis effect**. To understand the Coriolis effect, imagine yourself standing at the spinning center of a merry-go-round (which represents the geographical North Pole). The merry-go-round (Earth) is rotating toward the east. If you throw a ball to someone at the outside edge of the merry-go-round, it will appear to curve away from where you aimed it (to the west)—the ball flies straight, but the Earth turns under it as it travels through the air (**FIGURE 5-13**). This effect produces the trade winds that once carried sailing vessels across the oceans. Second, the Earth's topography— the positions of its mountains, valleys, lakes, and oceans—modifies both weather and climate. The oceans that cover much of the surface contribute greatly to the overall climate (**FIGURE 5-14**). Moving air masses create ocean

FIGURE 5-12 General circulation in the Earth's atmosphere.

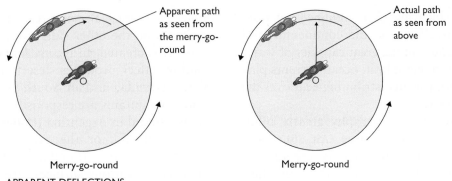

Apparent path
as seen from
the merry-go-
round

Actual path
as seen from
above

Merry-go-round

Merry-go-round

APPARENT DEFLECTIONS

FIGURE 5-13 A simplified demonstration of the Coriolis effect.

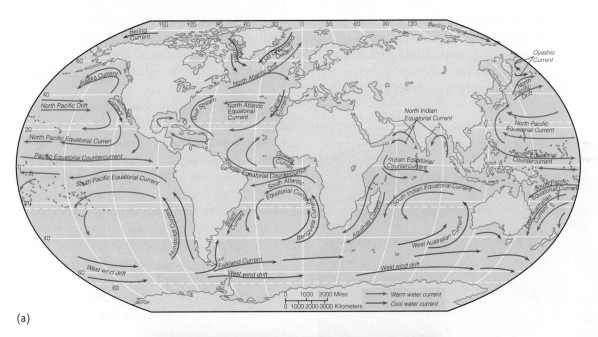

(a)

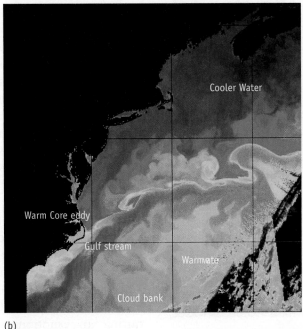

(b)

FIGURE 5-14 (a) Global surface ocean currents. (b) "Heat" photo of the Gulf Stream flowing northward past the U.S. coastline.

currents, which are then modified by both the Coriolis effect and the continental land masses. Because of the great capacity of water to absorb and retain heat, ocean currents play an important role in distributing heat over the Earth's surface.

On land, the topography greatly influences weather patterns (see **CASE STUDY 5-2**). For example, mountains may create rain shadows, causing deserts to form on their leeward sides. When a warm moist air mass encounters a mountain, the air begins to sweep up the mountain's windward side. In so doing, the air rises, cools, and loses much of its moisture as precipitation. Once across the mountain, the now dry air flows down the other side, becoming warmer as it reaches lower elevations. Because the air has left its moisture on the other side of the mountain, a desert is created. In North America, a rain shadow effect produces deserts east of the Sierra Nevada, and in South America, the Andes Mountains are responsible for the dry deserts found in Argentina (**FIGURE 5-15**).

The **albedo**, or the proportion of the incoming solar radiation that is reflected back to space by various surfaces, also influences local and global temperature and climate. Large, vegetation-free deserts and blankets of ice or snow have a large reflective capacity—a high albedo. Oceans and other large bodies of water, which have the ability to absorb heat, as well as areas covered with thick vegetation, have a low albedo.

Dust in the atmosphere can also have a high albedo and reflect significant amounts of radiation including heat back into space. The dust and ash that the volcano Tambora emitted in 1815 had worldwide ramifications for the weather, including a "year without a summer" for New England in 1816. The impact of a huge meteorite could throw up blankets of dust and particulate matter that would have the same effect. Conversely, other additions to the atmosphere can lead to higher temperatures. Various greenhouse gases, most notably carbon dioxide, when added to the atmosphere (either artificially or naturally) create a blanket that allows radiation to enter but blocks the escape of heat, thus leading to global warming (a review of global air pollution provides more information).

There are five gases normally released during a volcanic eruption; all pose hazards except the water vapor. Sulfur dioxide can affect local weather after an eruption by affecting the ability of the sun's rays to reach localized area. Sulfur contributes to acid rain. Chlorine destroys the protective ozone layer, while fluorine in high quantities can kill aquatic animals and livestock that eat contaminated vegetation when rain carrying fluorine falls. It is important to keep in mind that unlike human-created disasters, global volcanic eruptions are part of Earth's natural cycle. Carbon dioxide is also released during an eruption. On average, humans emit 110 billions tons of carbon dioxide per

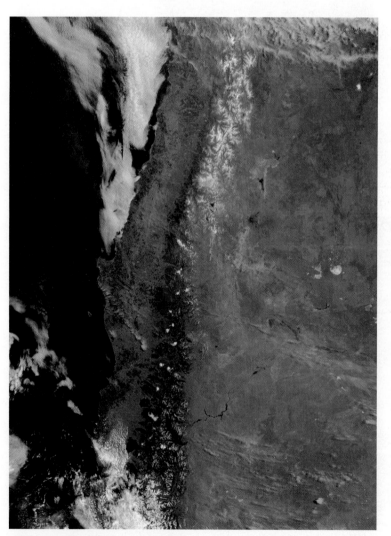

FIGURE 5-15 This satellite image of Chile (left) and Argentina (right) illustrates the "rain-shadow effect" caused by the Andes Mountains.

Vernal equinox: Sun is at the equator.

Winter solstice, late December: Sun is farthest south.

23-½°

Summer solstice: Sun is farthest north.

Autumnal equinox: Sun is at the equator.

FIGURE 5-16 The tilt of the Earth toward or away from the sun causes the changing seasons (seasons of the Northern Hemisphere labeled here).

year, whereas all the volcanoes in 1 year emit approximately 10 billion tons. Despite the intensity of nature's fury, humans still contribute to excessive carbon emissions more than ten times more than all volcanic eruptions each year.

Rotation, Orbits, and Seasons

In addition to its short-term, day-to-day fluctuations, weather is seasonal, changing on an annual basis. These cyclical variations in the weather are due to changes in the orientation of the Earth's axis. As you know, the Earth rotates on its axis once a day as it orbits around the sun once a year. The Earth's axis of rotation is not perpendicular to the plane of its orbit but rather is inclined at an angle of about 23.5 degrees. Consequently, as the Earth follows its orbit, different parts of the surface are exposed to more direct rays from the sun (**FIGURE 5-16**). In the Northern Hemisphere during the summer, the axis is tilted toward the sun, and the sunlight has less of the Earth's atmosphere to go through. The rays are more intense. During the winter months, when the axis is tilted away from the sun, the sunlight reaches the surface at an oblique angle. The rays travel through more atmosphere and are less intense. The seasons create fluctuating surface temperatures that stay within a moderate range. Without the seasons, large amounts of heat might be permanently trapped at the equator, and huge regions of permanently frozen wasteland would exist at moderate and high latitudes.

5.2 Origin and Physical Development of the Earth

The Earth is not a stagnant, unchanging planet. Since its origin around 4.6 billion years ago, our planet has been progressively developing and evolving. Atmospheric composition, climate and weather, continental and ocean positions, surface topography, sea level heights, and many other factors have varied in the past.

Therefore, if we are to evaluate the modern environment in a holistic context (taking into account the functional relationships between the parts and the whole), we need to take a historical perspective. For instance, to evaluate the current concentration of greenhouse gases and the predictions of global warming, we need to consider the changing composition of the atmosphere through geologic time and the oscillations in climate that have occurred over the past few million years (a review of global air pollution provides more information). Similarly, we must compare current extinction rates of species with extinction rates over the last 500 million years (a review of conservation of biological resources provides more information).

Origin of the Solar System and Earth

The current universe is believed to be approximately 13.7 billion years, according to the most recent studies. However, our little bit of the universe—our sun and solar system—is very much younger. The sun and solar system originated about 5 billion years ago when a gas and dust cloud in our region of the galaxy began to collapse and coalesce. The material rotated more and more swiftly around its center, flattening the cloud into a disc shape.

In the middle of the disc, the proto-sun condensed, and at various distances from the center, eddies concentrated particles in clusters that formed planetesimals (small planet-like bodies). Through gravitational attraction and accretion, these bodies eventually became our familiar planets. By about 4.6 or 4.5 billion years ago, the sun ignited, giving off tremendous heat and light. The influence of the sun may have produced the basic arrangement of the planets we observe today, with the innermost planets being the "rocky" planets (Mercury, Venus, Earth and its moon, and Mars), which can withstand the higher temperatures closer to the sun, and the major outer planets (Jupiter, Saturn, Uranus, and Neptune) being frozen, gassy, or icy planets. Pluto appears to be a rocky planet-like object.

On the Earth, a period of intense heating caused the various components to melt and separate into layers of heavier and lighter components. The dense iron–nickel components sank to the core, whereas the lighter silicates rose and formed the mantle and crust. In all likelihood, the main source for the heat was the radioactivity of elements trapped within the planet.

Origin of the Oceans

By about 4 billion years ago, the Earth probably had an ocean. We can assign an approximate date because among the oldest known rocks, formed some 3.8 billion years ago, are sedimentary rocks that show evidence of water weathering and erosion. However, where did the water come from? Various hypotheses have been proposed. Many scientists believe that during the early accretion of the Earth, volcanoes released water (initially in the form of water vapor) and large amounts of other gases. Icy comets that collided with primitive Earth may also have deposited a significant amount of water.

Over time, the oceans have accumulated salts as weathered material from the rocks of the continents has washed into them. The oceans contain ions, or salts, of such elements as sodium, magnesium, and chlorine, along with smaller amounts of calcium, silica, potassium, and many other elements, and various dissolved gases, including oxygen, carbon dioxide, and sulfur dioxide.

Origin of the Atmosphere

The earliest atmosphere of which we have any direct evidence, that of about 3.5 to 4 billion years ago, was very different than the atmosphere of today. It almost certainly originated from the Earth's interior as the planet underwent melting and segregation into a core, mantle, and crust. This early atmosphere probably contained large amounts of nitrogen gas, appreciable amounts of carbon dioxide, some methane and ammonia, and virtually no molecular oxygen.

So where did the current atmosphere's oxygen come from? Two basic processes are known to produce oxygen gas in some abundance in the natural world: (1) the breakdown of water into hydrogen gas and oxygen gas by ultraviolet radiation and (2) plant photosynthesis, the process of using the energy of sunlight to combine carbon dioxide and water into carbohydrates, giving off free oxygen gas as a by-product. The latter process is much more efficient at producing oxygen.

Many scientists believe that the high oxygen level of our current atmosphere is largely the result of photosynthetic organisms dumping this once-lethal toxic by-product into the environment over the last 3 billion years. As the levels of oxygen in the atmosphere increased, organisms evolved the ability to use oxygen in an efficient form of metabolism known as aerobic respiration; ultimately, these changes allowed the evolution of very complex, multicellular life forms. These organisms have evolved over the course of Earth's history into millions of different types of unique organisms, including people.

El Niño and Integrated Earth Systems

VISIT

http://environment.jbpub.com
/mckinney/5e/
for more information

The El Niño Southern Oscillation (ENSO) is a periodic weather phenomenon that occurs approximately every 3 to 7 years (as measured during the last century) over the tropical Pacific Ocean and influences rainfall and temperatures around the globe. An analysis of El Niño demonstrates just how integrated and finely balanced various global systems can be, with changes in one region having worldwide repercussions. This should be kept in mind as we consider potential global climate change (a review of global air pollution provides more information) and other possible human-produced assaults on the environment. Indeed, some researchers question whether some weather phenomena that in recent years have been attributed to global warming are in reality part of the El Niño phenomenon or whether El Niño events may have increased in intensity and frequency during the last 100 years (and especially during the last few decades), perhaps driven by increasing global warming.

Under normal conditions, the western portion of the tropical Pacific Ocean, east of Australia, has the highest surface water temperatures (refer to **FIGURE 1A**) and is a relatively low-pressure area (meaning the air is heating and rising). In contrast, over the eastern portion of the tropical Pacific, air pressures are higher. Air is descending to the surface of the ocean, and the air flows from the high pressure to low pressure zones, creating westward blowing winds (the trade winds). The winds blowing against the surface of the ocean water from east to west set up surface currents in the same direction and actually cause warm water to "pile up" along the edge of the western tropical Pacific basin, and the movement of the warm water to the west reinforces the pressure gradient, which in turn reinforces the strength of the trade winds. As a result of the water piling up, the ocean's surface can actually be about 20 inches (a half meter) higher in Indonesia than Ecuador. This piling up of warm water in the western Pacific also causes relatively cold and deep water to well up along the west coast of South America, bringing with it rich nutrients important to the fish and other ocean life of the region. (In the oceans, deep waters tend to be relatively rich in nutrients because it is in the shallow, warm, upper layers that much life lives, depleting the local nutrients, and as organisms die, some of their remains tend to descend into the deep waters and dissolve, adding nutrients to the cold depths. In places of deep-water upwelling, these nutrients are brought to the surface and support an abundance of life.)

However, during El Niño episodes, the relative air pressure differences over the western and eastern regions of the tropical Pacific decrease or even reverse, with higher pressures and descending air in the west and lower pressures and ascending air in the east (see **FIGURE 1B**). This is known as the "Southern Oscillation" in air pressures

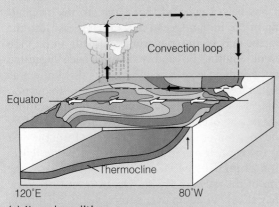

(a) Normal conditions

FIGURE 1 (a) Schematic diagram of normal (non-El Niño) conditions in the Pacific. North and South America are shown schematically on the right, and a portion of Australia is shown schematically on the left. Warmest surface waters are shown in red and coolest surface waters in light blue. Thick white arrows show the direction of surface ocean currents. The thermocline marks the transition between relatively warm surface waters and cooler deep waters. (b) El Niño conditions in the Pacific. Notice the reversal of direction of surface currents, the change in surface water temperatures, the lowering of the thermocline along the western coast of South America (meaning that deep, nutrient-rich waters are not coming to the surface there as they do during non-El Niño conditions), and the change in atmospheric convection patterns. *Source:* National Oceanic and Atmospheric Administration.

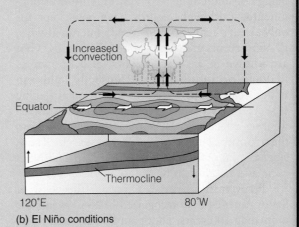

(b) El Niño conditions

El Niño and Integrated Earth Systems

and causes the warm water that is normally piled up on the western end of the tropical Pacific to "slosh" or flow back toward the east, and if the pressure is high enough in the west and low enough in the east, winds and surface ocean currents may actually reverse directions. In addition, the upwelling of cold, deep waters along the west coast of South America is decreased.

An El Niño episode not only affects the tropical Pacific, but also has ramifications for weather patterns worldwide, in part by modifying and displacing the jet streams (high altitude wind currents). Although they can vary greatly depending on the magnitude and duration of a typical El Niño event (which may last a number of months or more than a year from beginning to end), some of the effects of an El Niño include droughts in the western Pacific, such as in the region of Australia, Indonesia, and southeast Asia; warm weather in parts of Alaska, northwestern and northeastern Canada, and India; cooler weather in the southeastern United States; and increased rainfall (with possible flooding) in Peru as well as parts of the southeastern coast of South America, the southern United States (especially the southeast), and parts of Europe and central Asia. Outbreaks of infectious diseases have been correlated with extreme weather conditions caused by El Niño events, as have major fires in Australia, Indonesia, and elsewhere.

During the last 2 decades, El Niño events have occurred during 1982/1983, 1986/1987, 1991/1992, 1993, 1994, 1997/1998, 2002/2003, and 2006. Of these, 1982/1983 and 1997/1998 were particularly strong events, with 1997/1998 being considered the strongest El Niño ever documented (although records have been kept for only about a century). Sometimes, but not always, El Niño events are preceded or followed by what are known as La Niña events (also known as El Viejo or anti-El Niño). Unusually cold and widespread temperatures in the tropical Pacific characterize La Niña events, and the worldwide weather effects of a La Niña can be more or less the opposite of an El Niño. Many researchers consider El Niño to be the warm phase of the ENSO cycle and La Niña to be the cold phase.

Peruvian fishermen working off the coast of South America have recognized El Niño events for more than 100 years. The Spanish name itself refers to "The Little One," "Little Boy," or "Christ child" because the phenomenon often begins around the season of Christmas and the New Year. (Originally "El Niño" referred to warm ocean currents that appeared around Christmastime every year.) La Niña refers to "Little Girl" (as opposed to "Little Boy") and "El Viejo," an alternative name for the cold phase of ENSO, means "The Old One."

No one really knows what causes the initiation of an El Niño event. It appears that in many cases, relatively short-lived winds might blow against the normal trade winds, pushing warm water to the east. This eastward movement of warm water could lower the air pressure in the central and eastern tropical Pacific (because the warm water will heat the air above it and cause it to rise), which in turn will weaken the westward moving trade winds that depend on a strong air pressure gradient from east to west, and in a self-perpetuating cycle, the warm water could continue to flow east (remember that the warm water normally is piled to the west and held there by the trade winds) even as the trade winds continue to weaken; ultimately, a full-fledged El Niño might occur.

As far as scientists are able to determine, essentially stochastic, or random, processes may cause the initiating wind described in the last paragraph. It has been suggested that large storms originating in the Indian Ocean and moving east into the Pacific may be the triggering factor for an El Niño event. Another possibility that has been suggested is the influence of Rossby waves (also known as planetary waves), which are very low-frequency, long, slow-moving waves found in both the atmosphere and oceans. Rossby waves are generated by the rotation of Earth and travel only from the east to the west. It has been speculated that the reflection of the Rossby waves in the ocean off of the rim of the western Pacific Ocean basin somehow may slowly build up over several years and initiate a chain of events leading to an El Niño. It is also quite conceivable that not all El Niño events are caused by the same triggering factors.

The El Niño phenomenon is just one example that illustrates how complex and deeply integrated the various systems of Earth are and also points to the fact that our understanding of the workings of Earth is far from complete. We should keep this in mind as we ponder the possible outcomes of human-induced environmental change, whether on a local level or a global level.

5.3 Natural Hazards

People have always had to deal with "unpredictable" **natural hazards**—earthquakes, volcanic eruptions, floods, avalanches, droughts, fires, tornadoes, hurricanes, and so forth (**TABLE 5-1**). For most of human history, such phenomena were beyond human control: they were neither caused by people nor was there much people could do to predict their occurrences or mitigate (make less severe) their consequences.

Although many acts of nature still cannot be controlled, we have learned to better predict their occurrences and mitigate their effects. Furthermore, we have come to understand that some types of phenomena have a large **anthropogenic** (produced or caused by people) component. Human deforestation in hilly regions may cause, or at least exacerbate, flash floods and avalanches. "Desertification," the changing of fertile land into deserts, is also largely a result of unsustainable agricultural practices combined with droughts. Human interference with the atmosphere, such as the emission of high levels of greenhouse gases, may cause or magnify abnormal droughts, storms, and other unusual weather phenomena. Earthquakes, at least on a local level, may be caused by the pumping of fluid wastes into rocks lying deep below the surface.

As the human population increases, the damage that naturally occurring, periodic "disasters" impart is magnified. A coastal area may be hit by a major storm once every century. A river may swell over its "normal" banks covering the floodplain only once every few centuries, or a naturally set forest fire may periodically burn off dead underbrush and litter on the forest floor. Viewed on a temporal scale of millennia (thousands of years), such events form an integral part of the natural ecosystem cycle; from a natural, holistic perspective, they are not disasters at all. However, when people inhabit a coastal area or a river's floodplain and depend on a forest for lumber products or for scenic beauty, these events are perceived as terrible disasters—and so they are from a human perspective. However, they are disasters that given a basic knowledge of the natural world, should have been easily anticipated. In this section, we briefly introduce some of the basic types of natural hazards that civilization must face as we continue to live in the natural environment.

Earthquakes and Volcanoes

Earthquakes and volcanic eruptions are geological phenomena that people may be able to predict (although not always) but can virtually never control. With few exceptions, earthquakes and volcanoes are caused by processes that take place deep within the crust and mantle. At least minor to moderate earthquakes usually accompany volcanic eruptions, but very large earthquakes can occur in the absence of volcanic activity. As we have discussed, earthquakes and volcanic activity are associated with the boundaries of the lithospheric plates (refer to Figure 5-3) and can be explained in terms of moving and/or subducting plates. However, earthquakes (and less commonly volcanoes) can also occur in the middle of plates; indeed, no place on the surface of Earth is immune to earthquake activity. Approximately 80% of all earthquakes and volcanoes occur along a geographic area known as the "ring of fire" (see **FIGURE 5-17**).

Earthquakes

Earthquakes are essentially shock waves that originate when large masses of rocks suddenly move relative to each other below the Earth's surface. For instance, along a plate boundary where two lithospheric plates are sliding past each other, the plates may "hang up," allowing strain (frictional drag) to accumulate until it is finally relieved by rock movement—causing

TABLE 5-1	Estimated Deaths Caused by Major Natural Hazards, 1960 to 2000		
Hazard Type	**Estimated Total Deaths**	**Examples of Major Events**	**Deaths**
Tropical cyclones	850,000	East Pakistan (Bangladesh) 1970	500,000
Earthquakes	650,000	Tangshan, China 1976	250,000
Floods	60,000	Vietnam 1964	8,000
Avalanches, mudslides	50,000	Peru 1970	25,000
Volcanic eruptions	36,000	Colombia 1985	23,000

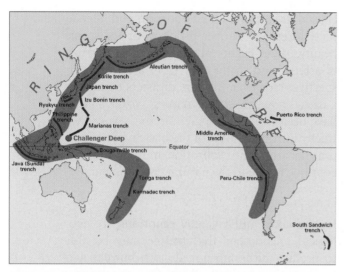

FIGURE 5-17 Ring of Fire. Visit http://earthquake.usgs.gov/earthquakes/ to view the earthquake activity happening every day anywhere in the world. Reproduced from U.S. Geological Survey.

an earthquake. Earthquakes can originate as much as 700 kilometers (450 miles) below the surface. When the sudden rock movement occurs, various types of shock, or seismic, waves are transmitted through the Earth, causing the shaking or trembling felt by people on the surface. An earthquake's vibrations can travel through the Earth and be detected around the world. Seismic waves from the largest earthquakes (greater than magnitude 8.3 on the Richter scale) can reverberate inside the Earth and make the planet "ring."

Earthquakes are a constant threat to human life and well-being in some areas. Perhaps the most devastating set of earthquakes occurred near Shensi, China, in 1556, killing an esti-

mated 830,000 people. A dozen earthquakes that caused the death of 100,000 people or more have been recorded. Estimates are that, on average, earthquakes kill at least 10,000 people a year and cause about $500 million in property damage. Japan, a nation always at risk for earthquakes, suffered devastating consequences in 1995 when a quake hit the Kobe area, killing more than 5,000 people, injuring another 25,000, and initially leaving more than 300,000 homeless. More than 17,000 people were killed in the Izmit, Turkey, earthquake in 1999. In October 2005, a quake struck Pakistan, killing more than 80,000 people and leaving more than 2 million people homeless (**FIGURE 5-18**). Estimates vary greatly, but an estimated 92,000 to 316,000 people died as a result of the 2010 magnitude 7.0 Haiti earthquake.

Japan, a nation always at risk for earthquakes, suffered devastating consequences in 1995 when a quake hit the Kobe area, killing more than 5,000 people, injuring another 25,000, and initially leaving more than 300,000 homeless.

At that time, however, no one could have predicted the devastation that would result in 2011 from the "Great East Japan Earthquake" and tsunami, which occurred on March 11, 2011, at 2:45 pm Japanese local time. This 9.0 magnitude earthquake shook Japan during the school day at the height of afternoon activity. With the epicenter located 81 miles (130 km) off the east coast of Sendai, Japan, a massive tsunami formed. Within about 30 minutes the initial tsunami wave made its devastating landfall. The tsunami measured approximately 40 m (130 ft) at its highest. In the days that followed, there were more than 1,000 aftershocks recorded after the main event, with at least 60 measuring a magnitude of over 6.0. Although official death toll figures of from 11,000 to over 20,000 differ based on whether those still missing are figured in with the deceased, the widespread disaster left more than 400,000 displaced, making it the largest natural disaster in Japan in at least 100 years. It was also the strongest earthquake in the history of Japan and one of the top five strongest earthquakes since worldwide standardized record keeping began in 1900. To read more about this event and the resulting nuclear accident, see **CASE STUDY 5-3**, Compounded Natural Disasters.

FIGURE 5-18 A heap of rubble is all that remains of a 19-story housing complex after a magnitude 7.6 earthquake struck Islamabad, Pakistan, in October 2005.

CHAPTER 5 The Dynamic Earth and Natural Hazards

Compounded Natural Disasters: The Great East Japan Earthquake, Tsunami, and Nuclear Power Disaster of 2011

VISIT

http://environment.jbpub.com /mckinney/5e/ for more information

In the hours of the afternoon on March 11, 2011, following the immediate damage of the magnitude 9.0 earthquake and subsequent tsunami, Japanese citizens were further challenged by news that the tsunami had significantly damaged six nuclear reactors at the Fukushima Daiichi nuclear power plant. This event resulted in radiation exposure to workers, the need for local evacuations, and evidence of low level nuclear radiation in water and food in areas nearly 100 miles around the plant. Increased radiation levels, still below the level governmental officials deemed safe, were documented south of Tokyo weeks later and then again months after the event. Japanese officials stated that the backup power systems at the plant were not properly protected and that the plant overall lacked safety features found in newer plants.

When the backup power systems failed to provide continuous cooling to the uranium fuel, the fuel overheated causing a series of explosions. Initially, more than 200,000 people were evacuated as a result of the nuclear power plant accident. Eight months later, more than 160,000 people still cannot return to their homes because of the intensity of radiation. As of this writing (January 2012), radiation is still being detected in food such as beef, vegetables, and fish in varying distances from the accident, making it harder for residents to obtain reliable food sources if they do chose to return. When the event first happened, the Japanese government was criticized by international media as being slow to report damage and hesitant to engage the international community in problem-solving regarding its nuclear reaction situation. At the same time, international media who were granted access to the area often reported speculative data, making it difficult for families to know the true extent of the damage, especially around the nuclear power plant. Where social media was working, hand-held devices streamed data in live-time reports from amateur reporters attempting to communicate the gravity of the situation.

At this writing (January 2012), there have been more than 3,000 cases of radiation exposure to workers at the plant complex, but less than 40 officially reported deaths resulting from incidents of exposure or injury to workers who were required to stay on site attempting to cool the reactors and contain the damage since the first explosion

there just after the tsunami. According to official Japanese reports, most of the deaths were injury related and not due to immediate nuclear exposure. A total of 700 workers remained engaged in plant operations aimed at culling further damage immediately after the event. Although efforts were taken to reduce exposure, cumulative effects may not be observed for years, especially in the 50 workers who remained after radiation levels soared and all but the most essential employees were released. One month after the event, the level of the event was raised to a 7 on The International Nuclear Events Scale (INES), which is the highest level, putting it at the top of nuclear disasters, which includes Chernobyl.

It was not until December 2011 that Japanese officials completely gained control of the damaged nuclear power reactors. The cost of the accident is estimated at close to $10 billion, with more costs accruing as decisions are still needed regarding building a new town for displaced residents or decontaminating the present area, which would require at least 100 million metric tons of soil to be removed and stored safely somewhere. This figure does not include long-term costs to damaged ecosystems and the services they provide.

Underscoring the fact that large scale natural disasters have far reaching effects, across the Pacific Ocean, many countries issued evacuations along the coasts because of the predicted tsunami waves. In Hawaii, coastal resorts were affected with sand and high waves entering ground floor rooms. In Santa Cruz harbor, docks and boats received minor damage. The sustained energy of the wave coming in continuously for several minutes, not the height of the wave, was the cause of the minor damage that occurred on the western U.S. coastline. Florida officials over 7,000 miles from the epicenter noted that groundwater tables rose and fell a few inches within 30 minutes of the quake. The cloud of nuclear radiation moved across the ocean to North America and then Europe, though deemed a minor threat by most scientists. At this writing (January 2012), the Japanese government is currently assessing its nuclear power program and has taken several nuclear reactors off line.

As global citizens, there is no community in the world immune to natural disaster challenges,

Compounded Natural Disasters: The Great East Japan Earthquake, Tsunami, and Nuclear Power Disaster of 2011

the shared threat of accidental nuclear exposure, or to the long-term environmental consequences that occur after such events. Can energy disasters like the Japanese Fukushima Daiichi event be avoided with better disaster scenario planning and more rigorous enforcement and regulations? Residents in Japan are voicing their concern about public safety nuclear power by organizing and petitioning the government to shut down outdated facilities. Social media is assisting mothers who are organizing to raise concerns that radiation levels are not being reported accurately, some having had radiation levels in breast milk documented by their physicians. Compounding the effects of the natural disaster in Japan is the lack of safety updates that made the nuclear disaster so hard to bring under control; infuriating to Japanese citizen groups is the fact that the plant's safety issues were known previously. The cost of cleanup and real cost of environmental damage far exceeds what safety improvements might have cost. In recent years, events such as the ongoing nuclear accident and natural disaster cleanup in Japan bring worldwide attention the connection and vulnerabilities of countries to natural disaster

and infrastructure management that often compounds the negative effects of natural events.

With the dawn of social media and hand-held portable devices, often information received by individuals living the experience locally is more reliable than official reports and can serve to bring immediate attention to an emerging situation or ongoing problem. If there is anything positive to come from such incidents, it is the realization that events in one part of the world accentuate our global connections. In the race to provide energy for a rapidly increasing human population, safety and environmental protection will most probably remain a citizen priority as events like this one remind us that all forms of life that share this fragile planet are connected through climate patterns, shared resources, environmental values, and an increasingly complex global economy. In this challenge there is the potential for innovative solutions and future technologies that can minimize future compounding affects of catastrophes through broadening international pressure and increasing citizen involvement in policies that protect and preserve ecosystems and the services they provide. With a more paramount role in the international discourse weighing long term pros and cons in the process of future energy development,

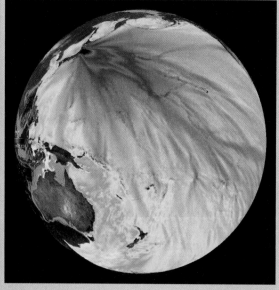

(a)

(b)

FIGURE 1 (a) This is the wave energy graphic from satellite imagery from: http://sos.noaa.gov/datasets/Ocean/japan_quake_tsunami.html. (b) Smoke pours out of unit three; unit four is to the right. This is from March 18, 2011. http://photos.oregonlive.com/photo-essay/2011/03/japan_earthquake_and_tsunami_a_3.html

(c)

(d)

VISIT

http://environment.jbpub.com
/mckinney/5e/
for more information

FIGURE 1 (c) Sendai 1 week later, under water. http://photos.oregonlive.com/photo-essay/2011/03/japan_earthquake_and_tsunami_a_3.html. (d) http://neic.usgs.gov/neis/bulletin/neic_b0001b9f_l.html.

global citizens are well positioned now more than ever before to have access to the information needed to shape policies that will govern decisions in the future and for generations to come.

References

http://sos.noaa.gov/datasets/Ocean/japan_quake_tsunami.html
  This site also has a QuickTime video with merged data from before and after the earthquake showing the building wave in the form of satellite data.

http://www.nytimes.com/interactive/2011/03/13/world/asia/satellite-photos-japan-before-and-after-tsunami.html
  Series of before and after photos good for website:

http://www.reuters.com/article/2011/03/19/us-japan-quake-chernobyl-sarcophagus-idUSTRE72I2UZ20110319

http://pubs.usgs.gov/gip/dynamic/fire.html

http://earthquake.usgs.gov/earthquakes/

http://www.chernobyl-international.com/information-sheets.html#sequence

http://www.voanews.com/english/news/Japan-Death-Toll-Rises-Above-7000-118292399.html

http://news.yahoo.com/s/ap/as_japan_earthquake

http://www.nei.org/newsandevents/information-on-the-japanese-earthquake-and-reactors-in-that-region/

http://www.nytimes.com/2011/03/16/world/asia/16workers.html?pagewanted=all

http://topics.nytimes.com/top/news/business/energy-environment/atomic-energy/index.html

www.usgs.gov

http://www.nature.com/news/specials/japanquake/index.html#lc

http://fukushimaupdates.wordpress.com/2011/07/20/tepco-status-update-of-fukushima-radiation-exposed-workers/#more-1811

The intensity, magnitude, or strength of an earthquake is commonly measured either on the Richter scale (in North America) or the Mercalli scale (in Europe). The Richter scale, devised by Charles Richter in 1935, is based on the amplitude of the seismic waves recorded by seismographs (instruments that record the motions of the Earth's surface) coming from a particular earthquake. The **Richter scale** measures the energy that an earthquake releases, or in crude terms, how much "shaking" of the ground occurs during an earthquake. The Richter scale is a logarithmic scale so that every unit corresponds to a 10-fold increase in the amplitude of the seismic waves; a 7.0 earthquake is characterized by seismic waves with an amplitude of 100 times that seen in a 5.0 earthquake. Theoretically, the Richter scale has no upper limit, but some of the largest recorded earthquakes (such as the Lisbon earthquake of 1755 immortalized in Voltaire's *Candide*) have been ranked at about 8.9 to 9.0 (**TABLE 5-2**).

TABLE 5-2	Some of the Largest Earthquakes by Magnitude on the Richter Scale		
Magnitude	**Year**	**Location**	
9.5	1960	Chile	
9.2	1964	Prince William Sound, Alaska	
9.0	1755	Lisbon, Portugal	
9.0	1952	Kamchatka	
9.0	2004	Sumatra, Indonesia	
9.0	**2011**	**Near the East Coast of Honshu, Japan**	
8.9	1977	Sumba, Indonesia	
8.8	1906	Coast of Ecuador	
8.8	**2010**	**Chile**	
8.7	2005	Sumatra, Indonesia	
8.6	1906	Andes (Colombia)	
8.6	1950	North Assam, India	
8.5	1933	Japanese trench	
8.4	1911	Tienshan, China	
8.2	1923	Tokyo, Japan	
8.2	1976	Tangshan, China	
8.2	1977	Argentina	
8.2	1994	Bolivia	
8.1	1979	Indonesia	
8.1	1985	Mexico City, Mexico	
8.0	**2007**	**Peru**	
7.8	1906	San Francisco, United States	
7.0	**2010**	**Haiti**	

Source: United States Geological Survey, Earthquake Hazards Program, World Earthquake Information.

The **Mercalli scale** (**TABLE 5-3**) is more qualitative, based primarily on observations of the effects caused by an earthquake close to its origin. This scale classifies earthquakes into a dozen basic categories, from instrumental (earthquakes so small that they can be detected only on seismographs) to catastrophic (earthquakes where local destruction is virtually total). The Mercalli scale is easily related to the Richter scale in an approximate way (Table 5-3).

Recently, much research has gone into devising ways to predict earthquakes in both the long term and the short term. Some work has attempted to discover long-term earthquake cycles, based on the movements of the lithospheric plates and also perhaps correlated with such phenomena as meteoritic impacts, variations in the Earth's geomagnetic field, sunspot activity, and variations in the length of the day. Such work is very difficult to pursue, and no method has been able to predict long-term earthquake events with any degree of accuracy.

As for the short term, researchers have identified several precursors that often signal the probability of a major earthquake in earthquake-prone areas. The most commonly used precursors are unusual land deformation, seismic activity, geomagnetic and geoelectric activity, groundwater fluctuations, and natural phenomena such as animal behavior and unusual weather conditions (see **CASE STUDY 5-4**).

Although we usually do not think of earthquakes as being caused by human activity, some earthquakes are indeed anthropogenically induced (caused by people). Nuclear blasts, conventional blasting, mining activities, fluid injection and extraction from rocks deep underground, and the building of dams and reservoirs have induced earthquakes. A now classic example of human-induced faulting and seismic activity occurred near Denver in the early 1960s. Nerve-gas waste was disposed of by pumping it down a well deep enough to be below groundwater supplies. Pumping the waste down the well at high pressures triggered a series of earthquakes. Since then, it has been verified experimentally in oil fields that pumping fluids into the ground (and in some cases pumping out the oil) can induce seismic activity.

TABLE 5-3   The Mercalli Scale of Earthquake Intensity

Scale	Intensity	Description of Effect	Maximum Acceleration (mm sec^{22})	Corresponding Richter Scale
I	Instrumental	Not felt except by a very few under especially favorable circumstances.	< 10	
II	Feeble	Felt only by a few persons at rest, especially on upper floors of buildings. Delicately suspended objects may swing.	< 25	
III	Slight	Felt quite noticeably indoors, especially on upper floors of buildings, but many people do not recognize it as an earthquake. Standing automobiles may rock slightly. Vibration like a passing truck.	< 50	< 4.2
IV	Moderate	During the day, felt indoors by many, outdoors by few. At night, some awakened. Dishes, windows, doors disturbed; walls make cracking sound. Sensation like heavy truck striking building. Standing automobiles rock noticeably.	< 100	
V	Slightly strong	Felt by nearly everyone, many awakened. Some dishes, windows, etc., broken; a few instances of cracked plaster; unstable objects overturned. Disturbances of trees, poles, and other tall objects sometimes noticed. Pendulum clocks may stop.	< 250	< 4.8
VI	Strong	Felt by all; many frightened and run outdoors. Some heavy furniture moved; a few instances of fallen plaster or damaged chimneys. Damage slight.	< 500	< 5.4
VII	Very strong	Everybody runs outdoors. Damage negligible in buildings of good design and construction; slight to moderate in well-built ordinary structures; considerable in poorly built or badly designed structures; some chimneys broken. Noticed by persons driving automobiles.	< 1,000	< 6.1
VIII	Destructive	Damage slight in specially designed structures; considerable in ordinary substantial buildings, with partial collapse; great in poorly built structures. Panel walls thrown out of frame structures. Fall of chimneys, factory stacks, columns, monuments, and walls. Heavy furniture overturned. Sand and mud ejected in small amounts. Changes in well water. Persons driving automobiles disturbed.	< 2,500	
IX	Ruinous	Damage considerable in specially designed structures; well-designed frame structures thrown out of plumb; great in substantial buildings, with partial collapse. Buildings shifted off foundations. Ground cracked conspicuously. Underground pipes broken.	< 5,000	< 6.9
X	Disastrous	Some well-built wooden structures destroyed; most masonry and frame structures destroyed with foundations destroyed; ground badly cracked. Rails bent. Landslides considerable from river banks and steep slopes. Shifted sand and mud. Water splashed (slopped) over river banks.	< 7,500	< 7.3
XI	Very Disastrous	Few, if any (masonry) structures remain standing. Bridges destroyed. Broad fissures in ground. Underground pipelines completely out of service. Earth slumps and land slips in soft ground. Rails bent greatly.	< 9,800	< 8.1
XII	Catastrophic	Damage total. Practically all works of construction are damaged greatly or destroyed. Waves seen on ground surface. Objects are thrown upward into the air.	> 9,800	> 8.1

Source: United States Geological Survey.

The most important cause of human-induced earthquakes seems to be the construction of large dams and reservoirs. In the case of at least six major dams around the world, including Hoover Dam on the Colorado River, the impounding of water in the dams' reservoirs has apparently induced earthquakes of a magnitude greater than 5 on the Richter scale (moderately strong earthquakes, capable of minor damage). More than 1,000 earthquakes of various magnitudes have been felt since Hoover Dam was constructed in 1935; before 1935, the area was not known for earthquake activity. Worldwide, dozens of dams have been associated with seismic phenomena. It is thought that water in the reservoirs may penetrate the underlying bedrock and cause rock slippage that generates earthquakes. The huge mass of water in the reservoir also exerts tremendous pressures on the underlying rocks, and this can cause down-warping or "sinks" and subsidence of the land's surface.

Because people have inadvertently caused earthquakes, some people suggest that in the future we might be able to intervene to

Predicting Earthquakes

Predicting earthquakes is a tricky business, perhaps more art than science. If predictions were correct and people could be evacuated from an area before an earthquake, the death toll could be kept at a minimum (although substantial financial losses might still be unavoidable). Currently, no method exists that is absolutely reliable, but seismologists are using a combination of observations that could suggest the likelihood of an event.

The careful periodic surveying of known benchmarks on the Earth's surface can lead to the detection of movement within the crust. Benchmarks may have moved relative to each other, suggesting possible faulting. In some areas where an active fault zone has been identified, accurate surveys across the fault zone can detect very slight movements (on the order of millimeters) that may be precursors of larger earthquake activity to come. In known earthquake-prone areas, instruments called tiltmeters can be used to monitor the surface of the Earth; if the surface starts tilting rapidly, an earthquake is usually imminent.

Daily monitoring of seismic activity is also useful in earthquake prone areas. Changes in the background seismic activity, especially increases in activity (sometimes referred to as foreshocks), can herald a major earthquake. Magnetic fields and electric currents flowing through the rocks locally characterize the Earth's surface. Anomalous geomagnetic and geoelectric activity may be detectable months, days, or hours before an earthquake strikes.

Some of the most interesting and useful predictors are also rather low-tech and poorly explained from a scientific perspective (indeed, they almost verge on superstition): namely, the behaviors of various animals in the days and hours before an area experiences an earthquake. The Japanese have observed that catfish exhibit unusual behavior—high activity and even jumping out of the water (**FIGURE 1**)—before an earthquake. All kinds of domestic animals become restless and exhibit odd behaviors before an earthquake: Dogs bark. Pigs become very aggressive. Horses refuse to go into their stables, and so forth. Wild animals also show unusual behavior: Burrowing animals leave the ground. Rats run around randomly, and worms crawl from the ground to the surface in large numbers. In the San Francisco area, the behavior of zoo and marine animals is monitored daily as part of the local earthquake prediction system.

It is possible that certain animals are sensitive to vibrations (or ultrasound) generated by small earthquakes undetectable to people that precede a major earthquake. Perhaps some animals are sensitive to the smell of methane that may leak from the ground before an earthquake. Another possibility is that electrostatic particles may be coming from the ground; furry and feathered animals, such as mammals and birds, generally are very sensitive to electrostatic charges.

Reports of unusual weather before major earthquakes have often been dismissed as nonsense. However, some researchers have suggested that degassing of methane and other gases from below the Earth's surface may precede a major earthquake and could account for unusual weather phenomena, such as strange mists, glowing skies, and flashes of light. The release of electrostatic particles before an earthquake, might help explain unusual weather phenomena.

So far, the Chinese have had the most success. They have been able to predict a number of major earthquakes in time to evacuate cities before they were hit. A technique they have used very successfully in China is to monitor groundwater levels in wells. Fluctuations in normal water levels may precede an earthquake by hours to more than a week.

FIGURE 1 The Japanese have used catfish behavior as a predictor of earthquakes, and according to Japanese legend, earthquakes are caused by the movement of a giant catfish. This antique illustration shows people attempting to subdue the giant catfish that causes earthquakes.

control natural earthquake activity. Perhaps, in an active earthquake zone, we could selectively induce a series of small earthquakes by either injecting or extracting fluids from the rocks. The small, relatively harmless earthquakes might relieve the strain on the rocks and thus allow us to avoid a single large, very destructive earthquake.

Volcanoes

Volcanoes are basically spots in the Earth's crust where hot, molten rock (magma) wells up to the surface. Active volcanoes are found almost exclusively in three geologic settings:

1. At convergent plate margins where one lithospheric plate subducts under another, melting the rock, which rises to the surface as volcanoes (e.g., in Indonesia and elsewhere along the Pacific rim)
2. At divergent plate margins where magma wells up to the surface as two lithospheric plates pull apart from each other, forming a rift (as in the middle of the Atlantic Ocean)
3. Over mantle hot spots, areas that lie over a hot mantle plume that breaks through the crust and spews molten rock onto the Earth's surface (as in the Hawaiian Islands)

Volcanoes can be classified according to whether they extrude predominantly basaltic or andesitic magma. Volcanoes at divergent plate boundaries and hot spots tend to produce basaltic magma that is relatively silica poor and originates from the mantle. Andesitic magma contains a higher percentage of silica and generally is formed from the remelting, differentiation, and recrystallization of previously existing crustal or mantle material. Thus andesitic volcanoes are commonly found in subduction zones where rock is remelted.

Basaltic magmas are hotter and much less viscous than andesitic magmas, which tend to contain a much higher percentage of gases (often predominantly water vapor). Consequently, whereas lava may flow smoothly out of the crater of a basaltic volcano, andesitic volcanoes tend to be much more explosive, shooting out steam and other gases, rock fragments of various sizes, and volcanic ash (**FIGURE 5-19**).

FIGURE 5-19 Volcanoes, evidence that the Earth is still geologically young and active, inspire fear and awe in people. Illustrated here is the 1980 eruption of Mount St. Helens, Washington.

Volcanic eruptions can affect people on both a local and a global scale. On a local level, volcanic eruptions can destroy local towns and cities. In Italy in A.D. 79, ash falls and mudflows caused by the eruption of Mount Vesuvius buried the towns of Pompeii and Herculaneum. The eruption of Mount Pelée on Martinique in 1902 destroyed the city of St. Pierre, killing some 30,000 people in a span of 2 minutes (**FIGURE 5-20**). Large volcanic eruptions can have worldwide effects by spewing dust, ash, and gases (including material that can form acid rain) into the atmosphere, affecting global weather patterns. For example, the 1883 eruption of Krakatoa, a volcano in the Sunda Straits

FIGURE 5-20 Seen here are the remains of St. Pierre, a once thriving city of 30,000 on the island of Martinique, that was destroyed by the May 8, 1902, eruption of Mount Pelée.

between Sumatra and Java, threw 18 cubic kilometers (4 cubic miles) of rock, ash, and other debris into the atmosphere to heights of 80 kilometers (50 miles). The materials initially reduced the amount of incoming solar radiation reaching the Earth's surface by an estimated 13%. Even 2 years later, the amount of incoming solar radiation over France was still 10% below normal. A volcano's potential to alter dramatically regional and global weather patterns is an important consideration to take into account as one debates whether recorded patterns of apparent climate change during the last century are partly or wholly induced by human activities.

Even with modern technology and knowledge, predicting future volcanic activity is difficult. We know where to expect active volcanoes relative to lithospheric plates, and we can usually determine whether a volcano is active or dormant (based on the time since its last eruption); however, we cannot predict exactly the timing and intensity of future eruptions. Researchers use many of the same techniques used for predicting earthquakes. Seismic activity, land deformation, and geomagnetic and geoelectric parameters can be monitored. Any anomalous behavior, such as a change in seismic activity or the sides of a volcano starting to bulge, can herald an imminent eruption; however, estimating the strength and type of eruption that will occur is not easy. The analysis of gases given off by volcanoes holds some promise for predicting volcanic activity, at least in the short term, but sampling gases being vented from an active volcano that may be near eruption can be very difficult and dangerous (**FIGURE 5-21**).

Land Instability

A very widespread form of natural geologic hazard, usually brought on by human ignorance of the principles of geology and soil mechanics, is the collapse of soil or weathered rock material. Landslides, rockfalls, and avalanches may bury roads, homes, and other buildings. Soils may fail to support buildings, thus causing them to collapse. Such phenomena may occur simply because people build in areas that are geologically unstable and thus unsuitable. Human activity, such as clear cut-

FIGURE 5-21 Scientists retrieve a sample from an opening into a lava tube.

ting a mountainside or pumping out groundwater, may induce land instability. Nearly 1,000 people died on February 17, 2006, when heavy rains triggered a landslide into the Philippines village of Guinsaugon. Of course, natural phenomena such as earthquakes and volcanic eruptions may also induce landslides, rockfalls, avalanches, or surface subsidence.

Weather Hazards

Hurricanes, typhoons, tornadoes, droughts, floods, heat waves, wind storms, dust storms, and other "irregular" weather patterns can wreak havoc on human settlements and animal habitats. Here we briefly mention some major types of storm hazards. Because of possible global warming brought on by the greenhouse effect (a review of global air pollution provides more information), some researchers expect a dramatic increase in irregular weather patterns, particularly the number of violent storms, in the relatively near future (see further discussion in our review of global air pollution).

Tropical Cyclones

Tropical **cyclones** are intense storms that develop over warm tropical seas (**FIGURE 5-22**). When they occur in North America, tropical cyclones generally are referred to as hurricanes; when they occur in Southeast Asia, they are known as typhoons. Only storms in the Indian Ocean and around Australia are commonly or colloquially called cyclones. Approximately 100 to 120 tropical cyclones develop worldwide every year, and many of them pose hazards to coastal areas. Some can travel far inland as well. Every 5 years or so a hurricane along the east coast of North America crosses the Appalachian Mountains and enters the Great Lakes region.

In terms of deaths, tropical cyclones are among the worst natural hazards that people currently face. Tropical cyclones in the United States have killed about 15,000 citizens during the past 125 years. In fact, the greatest loss of life in the United States from a natural disaster occurred when an unnamed hurricane unexpectedly slammed into Galveston, Texas, on September 8, 1900. More than 8,000 people died from this tragedy.

Hurricane Katrina (2005) picked up tremendous momentum off the Gulf Coast and hit just east of New Orleans, Louisiana, about 70% of which lies below sea level. Katrina hit with 140-mph winds and caused a 20-foot storm surge. Economically, the city was devastated, and more than 1 million people were forced to evacuate the city. It is estimated that the economic damage alone was $300 billion or more, the most expensive natural disaster in U.S. history (see **CASE STUDY 5-5**). Hurricanes are a natural phenomena, but climatologists are debating what role global climate change may have had on the increasing intensity of these storm systems over the past decade.

Between 1960 and 2005, approximately 1 million people died as a result of tropical cyclones compared with approximately 750,000 people killed by earthquakes. However, of the people who died in tropical cyclones, 500,000 lost their lives in a single event—when a cyclone hit Bangladesh (then East Pakistan) in 1970. Bangladesh was hit by another tropical cyclone that caused 100,000 deaths in 1985 and by yet another that killed 140,000 people in 1991. The death rates were so high because Bangladesh is situated in a low-lying deltaic area that, given the regularity of storms, is virtually unsuitable for human habitation, and yet the country was, and still is, immensely overcrowded.

People can do very little about tropical cyclones or most natural climatic hazards except to take preventive measures long before the disasters hit. For instance, areas prone to tropical cyclones should not be heavily settled, but this is not a realistic expectation for poverty-stricken areas where people have

FIGURE 5-22 Tropical cyclones, more commonly known as hurricanes and typhoons, are among the most destructive of natural hazards. This house was destroyed by Hurricane Katrina.

few options. Early predictions of impending storms, community-awareness campaigns, warning systems and enhanced communications, and cyclone shelters can help the public take proper precautions, such as evacuating low-lying areas that will be hardest hit. However, as the United States saw with Hurricanes Katrina (August 2005) and Rita (September 2005), evacuating hundreds of thousands to millions of people at once is nearly impossible.

Tornadoes

Tornadoes are a classic American phenomenon; about 80% of all tornadoes occur in the United States, mostly in the Great Plains (about 600 a year in this area). The typical tornado consists of a rapidly rotating vortex of air that forms a funnel (**FIGURE 5-23**); tornadoes may be associated with hurricanes but not necessarily. A tornado may be relatively harmless as long as the funnel does not touch the surface of the Earth. However, when they do touch down, tornadoes are among the most intense and destructive phenomena found in nature. As a result of winds of 400 to 500 kilometers/hour (250 to 300 miles/hour) or greater and changes in air pressure found within the funnel, tornadoes can lift objects weighing hundreds of tons. A tornado that moves over water can suck in millions of tons of water, perhaps temporarily draining a

FIGURE 5-23 With a vortex of rotating air and wind speeds as high as 300 miles an hour (500 kilometers/hour), a single tornado can cause incredible destruction in a small area. The tornado shown here is in its early stage of formation near Union City, Oklahoma.

river or lake and even carrying the fishes or organisms in the lake to a new area.

Despite their destructive force, tornadoes that touch the surface of the Earth are fairly localized phenomena. Their contact with the ground generally lasts only half an hour or less; at most, the ground path of a single tornado might be 1 kilometer (1,100 yd) wide and a few tens of kilometers (10 or 20 miles) long.

Tornado data from April and May 2011 broke many U.S. records. In April alone, 753 tornados broke out, resulting in 364 deaths. April 2011 was the most active month on

FIGURE 5-24 A dust storm approaching a Kansas town in 1935.

record, including the long track EF-5 April 27 tornado in Alabama, which was the largest single tornado ever recorded and killed 78 people. From April 25 to 28, more than 350 tornados in 21 states caused unprecedented destruction. In May, the devastating Joplin County, Missouri, tornado was the single most deadly tornado since standardized record keeping began in 1950 and the fifth most deadly tornado in U.S. history. The 2011 tornado season resulted in 549 deaths and billions of dollars of damage that will take years to repair.

Although most people associate tornadoes with trouble, tornadoes serve a number of important geological and ecological functions. They transport large quantities of clay and silt particles from one area to another. They also serve as an important biotic dispersal mechanism; tornadoes have been proposed as the primary mechanism by which various species are dispersed rapidly over fairly large areas. Seeds, pollen, microorganisms, insects, and even such animals as fishes, frogs, toads, and turtles can be transported alive for several hundred kilometers (a few hundred miles) in a tornado (and literally "rain" fish after the energy of the tornado dissipates). Tornadoes may explain how aquatic species could migrate between unconnected lakes in parts of North America.

Dust Storms

Tornadoes are not the only weather phenomenon associated with strong winds that pose a hazard for people. Much more important, in terms of their frequency and worldwide extent, are dust storms. In 1901, a dust storm carried approximately 150 million metric tons (165 million tons) of dust from the Sahara in Africa to western Europe and the Ural Mountains. In 1928, a dust storm removed an estimated 15 million metric tons (16.5 million tons) of soil from Ukraine and deposited it in Romania and Poland. The dust storms that occurred in the midwestern United States during the 1930s are legendary (**FIGURE 5-24**). The resulting loss of topsoil severely damaged more than 1 million hectares (2.5 million acres) of agricultural land. This occurrence was greatly exacerbated because of desertification, which also played a role in the historic 2011 Arizona dust storms. In August 1999, a large dust storm moving over Phoenix, Arizona, caused

Hurricane Katrina: A Natural Disaster Exacerbated by Human Actions

Hurricane Katrina, which struck the Gulf Coast in late August of 2005, was a "wake-up call" for much of the American public, and many important lessons about natural disasters were learned from it. This hurricane was the costliest in U.S. history, with an estimated cost exceeding $300 billion in terms of property damage and economic losses to manufacturing, tourism, oil, and many other industries. Much of the damage occurred in New Orleans, where 80% of the city was flooded, with some parts under 6 meters (20 feet) of water. As a result, many thousands of buildings were lost or damaged (**FIGURE 1**). Hundreds of lives were lost.

One lesson that was learned from this tragedy was the importance of requiring citizens along the coast to evacuate. In the past, many cities and agencies have been fairly lax about requiring evacuations, and the evacuations themselves often have been poorly planned. Another important lesson was the importance of a well-organized plan for disaster relief after a natural hazard occurs. The Federal Emergency Management Agency (FEMA) was widely criticized for its relatively poor response to the disaster, but the city of New Orleans and the state of Louisiana also received much criticism for poor management of the crisis and the recovery period afterward.

Another key lesson of Katrina was the need to emphasize the major role people play in increasing the damages incurred by natural disasters. Human activities very often inadvertently increase the damage, often dramatically. The New Orleans area is a tragic example of people compromising the natural defenses of their environment. To begin, much of New Orleans is below sea level, and many human activities have actually accelerated the sinking of the city to ever-lower depths. For example, pumping of groundwater from under the city has contributed to the sinking. Ironically, the construction of levees around the city to protect it has increased the subsidence because the levees, channels, and flood-control facilities interfered with the normal process whereby the Mississippi River deposits sediment and builds up the river delta where it meets the Gulf of Mexico. Interrupting this process has caused the land to dry out and sink. In addition, people have destroyed most of the cypress trees, mangroves, and other vegetation at the mouth of the river. Out on the delta and barrier islands, the lack of sediment has led to the erosion and loss of the marshes and barrier islands that served as natural storm protection system. Historically, these features helped blunt storm surges before they could strike the populated inland areas. Allowing the river to renew the delta and barrier islands will help guard against future flood surges. Investing in revegetation efforts for mangroves and other native vegetation along the coast will help prevent more severe flooding in the future.

VISIT

http://environment.jbpub.com
/mckinney/5e/
for more information

FIGURE 1 The storm surge from Hurricane Katrina flooded 80% of New Orleans.

90-minute airport delays and wind gust of nearly 50 miles per hour. (See chapter 5 cover photo.) Although dust storms are prevalent in the desert southwest, their magnitude and frequency have intensified during the last decade.

In 2005, scientists reported on dust storms from China's Gobi Desert. Human activity, such as overgrazing, contributed to an increase in dust storms late in the 20th century. Historically, northwestern China experienced a dust storm every 31 years on average. However, after 1990, the average changed to one storm a year.

In recent years, satellite tracking of dust storms and dust movements around the planet has demonstrated just how pervasive dust movements are on Earth. By one estimate, 11.8 million metric tons a year (13 million tons) are blown or drift (sometimes at altitudes of more than 10,000 ft) from Africa to the Northeastern Amazon Basin in Brazil (**FIGURE 5-25**). In 2001, a huge Asian dust cloud blew east across the Pacific Ocean and across the entire continental United States, and some remnants of it made their way to the mid-Atlantic Ocean. Some of the dust below your feet or on your windowsill could literally be from anywhere in the world. The dust that is transported around the world serves a vital function—it brings fresh nutrients from one part of the globe to another. Iron and various minerals in desert dust that lands in the oceans provide essential nutrients on which the marine organisms depend. According to some research, the canopies of Central and South American rain forests receive valuable

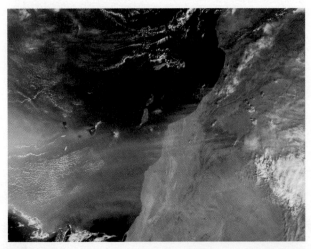

FIGURE 5-25 A satellite image of sand and dust blowing out to sea from the Sahara.

nutrients from the dust that is transported across the Atlantic from the Sahara Desert. It is not only inorganic or nonliving dust (finally ground minerals and nonliving organic matter) that can be transported around the globe, but also airborne microbes and pollens. Dust that global atmospheric winds and currents carry demonstrates once again that our planet is ultimately a single, integrated system.

Floods

A **flood**, quite simply, is a high flow of water that overruns its normal confinement area and covers land that is usually dry. Natural floods are a normal part of the physical environment, but because they often seem "unusual" from a human perspective—occurring irregularly over periods of decades, centuries, or millennia—they may take us by surprise. Cumulatively, floods are among the most destructive of all natural hazards. In China, flooding along the Yangtze and Yellow Rivers has taken the lives of millions of people during the 150 years. Many were drowned or died of flood-associated diseases or flood-caused starvation when crops were destroyed. In the United States, floods have caused billions of dollars in damage and killed hundreds of people (**FIGURE 5-26**). The human losses are greatly exacerbated by the fact that riverfront properties are often considered prime real estate, many towns and farms are located in natural floodplains, and large cities are often situated along rivers. The record-breaking Mississippi flood of 2011 affected seven states directly, resulting in billions of dollars of damage. The combination of unusually high rainfall, spring melting, and ongoing land-use issues along the Mississippi exacerbated the extent of this historic flood.

Other than storm surges, where the sea level rises locally relative to a low coastal area (such as during a hurricane), two basic types of floods can be distinguished: riverine floods and flash floods. Riverine floods occur along major streams or rivers when particularly heavy rainfalls or the rapid melting of snow causes a large amount of water to flow through the **drainage basin**. For example, the worst Canadian flood event to date occurred in the Saguenay River valley of Québec in 1996, when 290 millimeters (11.5 inches) of rain fell in less than 36 hours. The flood caused an estimated

FIGURE 5-26 Clean up and salvage begin after flooding in the Midwest during June 1994.

$1 billion in damages and at least 10 deaths, and more than 50 towns were inundated. Riverine floods may also occur when dams, either natural or human made, break because of the strain imposed by high water levels and flood conditions or other causes. Flash floods generally occur in drainage basins containing small, shallow, or ephemeral (periodically dry) streams when heavy rainfall overloads the system with water.

In the natural environment, rivers and streams are generally short-term phenomena. Rivers continually change their course and flood their banks; the flooding serves to transport and redistribute sediments and nutrients that the water carried. The annual floods of the Nile, bringing fresh nutrients to restore the fertility of the land, were the lifeblood of the ancient Egyptian civilization. Thus, flooding can be viewed as a small part of the larger hydrologic and rock cycles.

However, when people develop a drainage basin, they tend to build permanent structures as if the rivers and streams are permanent, sedentary features of the landscape. All too often people do not properly consider the natural periodic occurrence of floods. Consequently, when the occasional major flood does occur (perhaps only once every 100 or 1,000 years), the toll in human life and property loss can be immense. To make matters worse, many human activities inadvertently promote flooding. Farming, overgrazing, deforestation, paving large expanses, and other aspects of development limit the ability of water to seep into the ground (infiltration). The result is more runoff—more water will travel overland when it rains—which can promote flooding. Mining, construction, and other human activities can cause stream channels to become filled with sediment, hindering their ability to carry water quickly, and this too promotes flooding.

Even human activities that are specifically directed at controlling or avoiding flood conditions can actually promote flooding. Dams built to control flood surges may burst and cause even more flooding. Even a properly maintained dam will eventually collect silt, ending its usefulness. Water reservoirs behind dams may flood large tracts of forest or agricultural land, and as these artificial pools sit stagnant, they may become a breeding ground for disease vectors and pests (a review of renewable and alternative energy sources provides more information). Dams also disrupt stream flow, causing areas downstream to undergo abnormal erosion due to sediment starvation (the sediments once carried by the stream are trapped behind the dam). Of course, artificial dams can severely disrupt some forms of aquatic wildlife.

Channelization, the artificial straightening of a stream or river to increase its capacity to carry large amounts of water quickly downstream, has been used in some areas as a flood-control measure. Ironically, channelization can further worsen flooding conditions downstream by carrying even more water at a faster rate. Likewise, for thousands of years, people have built dikes, levees, and flood walls to keep a stream in its channel and avoid flooding. However, when the levees or dikes are breached, the resulting flood can be devastating. All of the evidence points to the conclusion that artificial containment structures can be only temporary measures at best; ultimately, they are bound to fail. In 1927, a breach in an Army Corps of Engineers' levee allowed the Mississippi River to flood nearly 1 million hectares (more than 2.3 million acres) of land despite the efforts of 5,000 workers to repair the levee. In China, for hundreds of years, levees and dikes have artificially contained the Yellow River; periodically, these structures break, flooding as much as 26,000 square kilometers (10,000 square miles) of land at a time. Aquatic life often evolves to take advantage of different microhabitats along a stream or riverbank, and channelization can damage these microhabitats.

Today, rather than emphasizing engineering works, such as the construction of dams and levees to control flooding, many authorities are advocating stricter zoning laws and better floodplain management policies. Ultimately, development may need to be strictly limited in flood-prone areas.

Droughts

Droughts—periods of abnormally low rainfall over an extended period of time in a particular area—can have devastating effects. Agricultural productivity may drop dramatically, and even drinkable water for people may be in short supply. Cycles of droughts and abnormally high rainfall are a typical aspect of the climate in the long run. However, in the short run, people may settle in an area because of abnormally good (perhaps moister than usual) weather conditions. The favorable weather may last for decades, and then the area is hit hard when drier weather conditions set in again. Changing human land-use patterns, especially the destruction of tropical forests and human-induced desertification, may affect circulation patterns globally, increasing local droughts. In addition, it is possible that global warming because of the human-induced greenhouse effect may shift precipitation patterns, causing once well-watered areas to become drier, and visa versa.

A number of places around the world have experienced major droughts in recent years, in particular, Ethiopia, Sudan, and especially the Sahel region in (Chad, Niger, Mali, Burkina Faso, Mauritania, Senegal, and Gambia) south of the Sahara in Africa (**FIGURE 5-27**). This region is instructive, for it is a demonstration of a positive feedback loop discussed in our overview of environmental science. The declining rainfall in the area has led to reduced plant growth, which reduces the amount of **evapotranspiration** (water given off from the soil by evaporation and water given off from plants by transpiration), which decreases the local moisture content of the atmosphere, which further decreases the amount of rainfall locally. Over time, the soil dries out and heats up. The lack of vegetation allows the wind to carry more dust into the lower atmosphere, which can add to the heating of the air higher in the troposphere, contributing to atmospheric instability and reducing the amount of dew that forms at night. All of these factors enhance and accelerate drought conditions in the Sahel region.

Furthermore, human overpopulation has exacerbated the situation. Shrubs and trees are cut down and burned as fuel or fed to animals. Virtually all of the arable land (land fit to be cultivated or tilled) is under cultivation, often using inappropriate Western plowing, which leads to progressive destruction of the soil structure and erosion of the topsoil. A hard crust remains on the surface of the land, preventing water from infiltrating the soil on the rare occasions when rain occurs. Eventually, the ground becomes **desert pavement**,

FIGURE 5-27 The Sahel region of Africa has experienced extreme droughts in recent years, yet the people continue to attempt to farm the land in an effort to survive.

an asphalt-like phenomena that occurs when the soil is packed down very tightly like pavement. Together, the drought and the human activities are causing severe desertification in the once-semiarid Sahel region.

Fires

Fires are a natural part of many ecosystems; particularly severe natural fires may be the result of drought conditions and lightning (**FIGURE 5-28**). Since prehistoric times, people have set fires—some of which invariably get out of control. Forest fires, grassland fires, and bushfires are particularly feared in the United States and Australia, but they pose a threat to forests around the world (**TABLE 5-4**). By far the greatest known fire incident was the Great Siberian fire of 1915. Huge amounts of smoke were injected into the atmosphere, blocking incoming solar radiation and suppressing ground temperatures.

For the past century, a basic principle of U.S. wilderness management has been to fight fires, whether they are set by people or occur naturally. Timber interests saw fighting fires as preserving a resource; people interested in recreation or preservation saw the prevention of fires as a way to protect the wilderness.

By the 1940s, some foresters had begun to question the wisdom of suppressing all natural forest fires. Slowly, they came to realize that fires play a critical role in nature. Fires promote the decomposition of some forest litter (dead leaves, branches, and so on) and are an essential component of the biogeochemical nutrient cycle. Periodic fires increase the biotic diversity (the range of different organisms) of an area because they maintain more open habitats necessary for some species, especially certain birds and larger mammals.

FIGURE 5-28 Elk take refuge in the East Fork of Bitterroot River during an August 2000 fire in Montana. Alaskan Type I Incident Management Team.

Certain plants need periodic fires as part of their life cycle; for example, the seeds of the giant sequoia (California redwood) and the jack pine (*Pinus banksiana*) and longleaf pine (*Pinus palustris*) must be exposed to the heat of a fire before they will germinate.

Coastal Hazards

Coastal regions along the oceans and larger lakes are among the most sought-after places for human occupation; in the United States and around the world, more than half of the population lives on or near such coasts. Yet geologically, coastal areas can be very unstable and prone to change. Natural erosional processes are ongoing along many coasts, and human habitation may accelerate the process, for instance, by destroying wild plants that help stabilize soils and sands. In the United States, approximately 25% of the coastline experiences severe erosion (**FIGURE 5-29**). The shifting sand dunes and unconsolidated sediments found on many coasts provide a poor foundation for homes or other buildings.

TABLE 5-4	Examples of Major Fires in the Past Century		
Date	Name/Place	Hectares (Acres) Destroyed	Impact
1915	Great Siberian fire	100 million (~247 million)	
1871	Wisconsin and Michigan	1.7 million (4.2 million)	Killed 2,200 people
1825	New Brunswick and Maine	~1.2 million (3 million)	Killed 160 people
1922	Haileybury and Northern Ontario	518,000 (approximately 1.3 million)	Killed 43
February 1983	The Ash Wednesday fires of South Australia and Victoria (southern-Australia)	500,000+ (1.2 million)	Destroyed nearly 2 dozen towns. Killed 76 people, more than 300,000 sheep, and 18,000 cattle. Injured 3,500 people.

FIGURE 5-29 The unconsolidated sediments found along many coasts can spell disaster for the local homeowner when a major storm strikes. This house has started to tip after a Northeaster storm hit the Outer Banks of North Carolina.

Wave action and moderate storms can cause a shore to recede (in some cases at a rate of 20 meters [65 ft] a year), leaving structures dangling on the edges of cliffs or eventually collapsing into the water.

Storm surges—sudden local rises in sea level caused by low atmospheric pressure and hurricanes or typhoons blowing water toward shore—pose particular threats to coastal regions. Likewise, tsunamis can wreak havoc on coastal areas.

Tsunamis (Japanese for "harbor wave") are huge waves, sometimes caused by large undersea earthquakes (usually registering higher than 6.5 on the Richter scale), volcanoes, or landslides. Fortunately, tsunamis are relatively rare; however, they can be devastating. The waves can range in height from less than 1 meter to nearly 60 meters (200 ft).

In 1692, an earthquake destroyed Port Royal, Jamaica, and the resulting tsunami threw harbored ships inland over two-story buildings. The Lisbon earthquake of 1755 sent a wave across the Atlantic Ocean that temporarily raised the sea by 3 to 4 meters (10 to 13 feet) in Barbados.

On December 26, 2004, a huge undersea earthquake occurred in the Indian Ocean. The quake had a magnitude of 9.0 on the Richter scale and triggered a series of deadly tsunamis that killed more than 270,000 people (more than 168,000 in Indonesia alone), making it *the deadliest tsunami in recorded history*. The tsunami killed people over an area ranging from the immediate vicinity of the quake in Indonesia, Thailand, and the northwestern coast of Malaysia to thousands of kilometers away in Bangladesh, India, Sri Lanka, the Maldives, and even as far as Somalia, Kenya, and Tanzania in eastern Africa. The disaster produced a worldwide effort to help victims of the tragedy, with hundreds of millions of dollars being raised for disaster relief (**FIGURE 5-30**).

(a)

(b)

FIGURE 5-30 (a) The 2004 Indonesian tsunami smashes into a beachfront resort in Thailand. © J.T. and Caroline Malatesta via Birmingham News/AP Photos. (b) A mosque is the only structure left standing in this coastal village near Aceh, Sumatra, Indonesia.

SUMMARY

- Major components of the Earth's internal structure include the crust, lithosphere, asthenosphere, mantle, and core.
- The lithosphere (crust and upper mantle) is divided into plates that move relative to one another.
- Plate tectonics, the unifying theory of geology, can account for the distribution of major features (such as mountain ranges, earthquake zones, and volcanoes) on the surface of Earth.
- Rocks are formed primarily of minerals, which are in turn formed of atoms.

- Different types of rocks—igneous, metamorphic, and sedimentary—can be transformed from one to another during the rock cycle.
- Weather is short-term perturbations in the atmospheric/hydrologic cycles, and climate is average weather over many years.
- The Earth's axis of rotation and orbit accounts for seasonal changes in weather.
- The Earth originated about 4.6 billion years ago and has been undergoing changes ever since.
- Natural hazards from a human perspective include earthquakes, volcanic eruptions, floods, land instability, cyclones and other storms, droughts, and fires.

KEY TERMS

anthropogenic
albedo
asthenosphere
atmospheric cycles
atoms
climate
compound
continental crust
convection cells
core
Coriolis effect
cyclones
desert pavement

drainage basin
droughts
earthquakes
elements
evapotranspiration
flood
hydrologic cycle
igneous rocks
lithosphere
mantle
Mercalli scale
metamorphic rocks
minerals

natural hazards
oceanic crust
Pangaea
plate tectonics
Richter scale
rock cycle
sedimentary rock
thermal convection cells
tornadoes
tsunamis
volcanoes
weather

STUDY QUESTIONS

1. How does planet Earth compare with its nearest neighbors in the solar system?
2. If one could cut a slice through the Earth, what would be encountered?
3. Describe modern plate tectonic theory.
4. What is the composition of matter on Earth?
5. What are the differences and similarities among sedimentary, igneous, and metamorphic rocks? Where does each form and occur?
6. What is the rock cycle?
7. What is the hydrologic cycle?
8. What are some major factors contributing to large-scale atmospheric cycles and climatic belts?
9. What happens when warm, moisture-laden air rises?
10. Explain the causes of the seasons.
11. Using sketch diagrams as necessary, describe the major atmospheric circulation patterns on Earth.

12. How and when did our solar system and Earth originate?
13. What are some of the most dangerous natural hazards?
14. What approaches have scientists used in their attempts to predict future earthquake activity?
15. Describe some of the more important weather hazards. What, if anything, can people do to prepare for such disasters?
16. After a major flood, should people whose homes and businesses were destroyed by the flood be allowed to rebuild in the floodplain? If so, under what conditions and who should pay for the rebuilding (private citizens, insurance companies, or the federal government)? If you feel people should not be allowed to rebuild, why not? Justify your answer.
17. Why did the American Dust Bowl occur? What can be done to prevent such disasters in the future?

CALCULATIONS

1. When an earthquake occurs in the Earth's crust, caused by a rupture of the rock along a fault, seismic (shock) waves are released. Various types of seismic waves are generally recognized, including primary P-waves (compressional waves) and secondary S-waves (shear waves). P-waves travel faster than S-waves; in a certain context, P-waves may travel at a speed of 4.8 miles per second and S-waves at a speed of 2.7 miles per second. Given this information, the time, TS (in seconds), that S-waves require to travel a certain distance (D) in miles from the epicenter (the point on the Earth's surface at, or immediately above, the rupturing fault that caused the earthquake; the epicenter is generally most damaged by the shock waves) to a seismograph can be expressed as follows:

$$TS = D/2.7$$

Likewise, the time P-waves require to arrive can be expressed as

$$TP = D/4.8$$

The difference in arrival time between the S-waves and the P-waves is

$$TS \div TP = (D/2.7) \div (D/4.8)$$

or

$$TS \div TP = (2.1D)/13$$

Therefore, the distance (in miles) from the earthquake's epicenter to the seismograph is

$$D = 13(TS \div TP)/2.1$$

Using this equation, if the difference in arrival time between S-waves and P-waves for a certain earthquake is 6 minutes 23 seconds at a particular seismograph station, how far away is the epicenter of the earthquake from the seismograph station?

2. Using the equation in Question 1, if the difference in arrival time between S-waves and P-waves for a certain earthquake is 4 minutes 16 seconds at a particular seismograph station, how far away is the epicenter of the earthquake from the seismograph station?

ILLUSTRATION AND TABLE REVIEW

1. Study Table 5-3, The Mercalli Scale of Earthquake Intensity. Have you ever experienced an earthquake? If so, where would you judge that it fell on the Mercalli Scale? If not, either talk to someone firsthand about an earthquake experience and try to place it on the Mercalli Scale, or research an earthquake and place it on the Mercalli Scale.

2. Referring to Figure 5-12, at which latitudes do high-pressure zones occur? At which latitudes do low-pressure zones occur?

http://environment.jbpub.com/mckinney/5e/

http://environment.jbpub.com /mckinney/5e/

Connect to this text's website at http://environment.jbpub.com/mckinney/5e/. This site features eLearning, an online review area that provides quizzes, chapter outlines, and other tools to help you study for your class. You can also follow useful links for more in-depth information.

Christchurch cathedral, Christchurch New Zealand, after a 2011 earthquake.

Resource Use and Management

"Once we see our place, our part of the world, as surrounding us, we have already made a profound division between it and ourselves. We have given up the understanding . . . that we and our country create one another, depend on one another, are literally part of one another; that our land passes in and out of our bodies just as our bodies pass in and out of our land . . . it is for this reason that none of our basic problems is ever solved."

Wendell Berry, poet, essayist, and social commentator

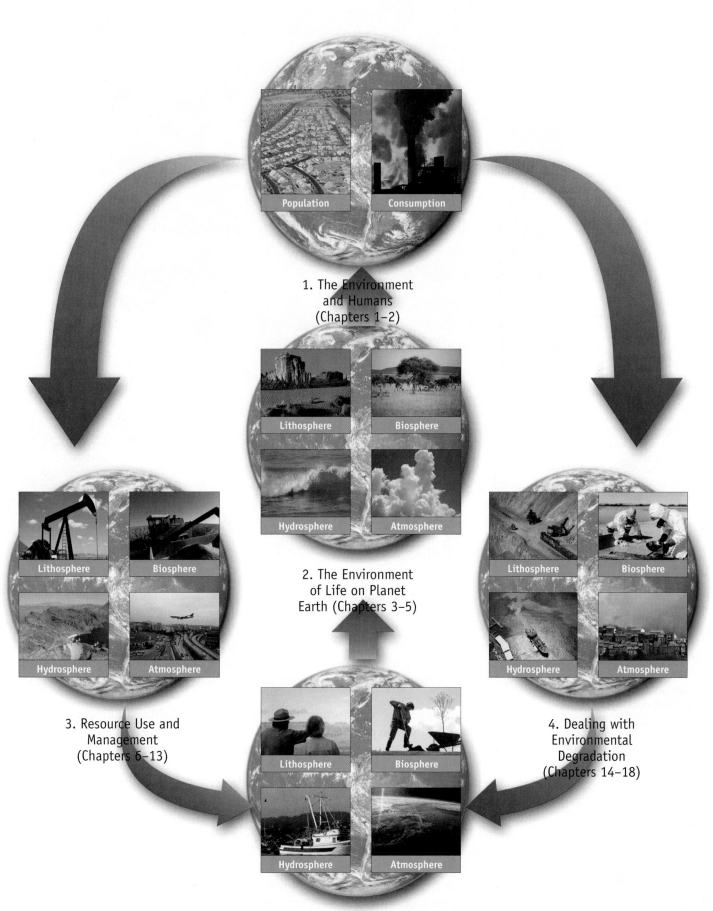

1. The Environment and Humans (Chapters 1–2)

Population

Consumption

2. The Environment of Life on Planet Earth (Chapters 3–5)

Lithosphere

Biosphere

Hydrosphere

Atmosphere

3. Resource Use and Management (Chapters 6–13)

Lithosphere

Biosphere

Hydrosphere

Atmosphere

4. Dealing with Environmental Degradation (Chapters 14–18)

Lithosphere

Biosphere

Hydrosphere

Atmosphere

5. Environmental Issues: Social Aspects and Solutions (Chapters 19–20)

Lithosphere

Biosphere

Hydrosphere

Atmosphere

6 People and Natural Resources

Chapter Objectives

After reading this chapter, you should be able to explain or describe the following:

- How and why resources are managed
- Three basic options for managing resources
- The five basic values of resources
- The importance of paying true environmental costs
- What causes the bubble pattern of resource depletion
- Problems with net yield and maximum sustainable yield
- How efficiency, recycling, and substitution help conserve resources

In this chapter, we take an overview of the nature of resources and how to better manage them. As we illustrate in the diagram for this section 3, natural resources from all four spheres may suffer from depletion. Depletion occurs when a resource is used faster than natural processes can replace it. For example, an oil deposit that took millions of years to form may be extracted and burned in just a few years, or a species that took many thousands of years to evolve may be driven to extinction in a few years.

The ideal goal of much resource management is sustainable resource use, which seeks to reduce or slow the rate of resource exploitation to the point where the resource can be replaced by nature over a reasonable time frame. Tree harvesting and ocean fishing are examples. In other cases, if the resource is in immediate threat of disappearing, sustainable management must go farther and actually "preserve" the resource by stopping its current exploitation. For instance, endangered species and endangered ecosystems often must be preserved immediately, or they will disappear rapidly.

Recall from our overview of environmental science that stopping or slowing the rate of resource use not only slows depletion but also usually has the added benefit of reducing pollution. A sustainable society would have "throughput" reduced to the point where natural processes renew exploited resources (inputs) and natural processes safely absorb pollution (outputs). Historically, resource management has not achieved this goal of sustainable use because of social, economic, and political pressures that emphasize rapid exploitation of resources.

6.1 Kinds of Resources

A **resource** is a source of raw materials that society uses. These materials include all types of matter and energy that are used to build and run society. Minerals, trees, soil, water, coal, and all other naturally occurring materials are resources. **Reserves** are the subset of resources that have been located and can be profitably extracted at the current market price. Raw materials that have been located but cannot be profitably extracted at the present time are simply called resources, as are those raw materials that have not been discovered.

Renewable resources can be replaced within a few human generations, but an entire ecosystem (if it should be destroyed) generally cannot. Examples of renewable resources include timber, food, and many alternative fuels, such as solar power, biomass, and hydropower. **Nonrenewable resources** cannot be replaced within a few human generations. Examples include fossil fuels, such as oil and coal, and ore deposits of metals. The phrase a "few human generations" is necessary because some resources are replaceable on very long, geologic time scales. Oil, coal, soils, and some metallic mineral deposits may form again over thousands to hundreds of millions of years. However, these rates of renewal are so many thousands of times slower than the rates of use that, for all intents, they are nonrenewable. In contrast, solar energy is actually supplied faster than we can use it, but the problem is harnessing it.

It is important to note that the definition of renewability is sometimes blurred. For example, very old groundwater in deserts may take centuries or even many thousands of years to replace, whereas groundwater in rainy tropical areas may be replaced in a few days. Thus, deep groundwater in deserts, sometimes called "fossil groundwater," is essentially a nonrenewable resource. Although nonrenewable resources cannot be replaced through natural processes on a human time scale, some (such as certain metals) can be recycled many times.

6.2 People Managing Resources

To some people, the concept of resource management reflects human arrogance. They argue that viewing the natural environment as a "resource" is a very narrow anthropocentric (human centered) approach to nature. In addition, some question whether we are able to manage resources effectively. Many debates in environmental ethics revolve around whether humans have a right to "tamper" with nature and, if so, how much tampering is justified. Ethics aside, the assumption that humans are able, as a practical reality, to manage nature effectively is not demonstrated by most of past human history. David Ehrenfeld (in his book *The Arrogance of Humanism*) has argued that humans have never effectively managed most resources for very long.

Despite these valid concerns, the need for resource management is inescapable. As human populations and energy-intensive technologies continue to grow, pressures to exploit the environment inevitably will increase. Proper management can help minimize and mitigate environmental damage. For instance, careful planning of water use could spare water for native ecosystems that would have been used for agriculture. Furthermore, management can help undo past damage. Elimination of alien (introduced) species is a common management strategy for some biological communities. Although resource management is not an attractive concept in some ways, it is preferable to the alternative, which is uncontrolled resource exploitation. Currently, global society faces many challenges inherited from antiquated methods of resource management. Re-examining the way we view our resources, our relationship with the Earth, and the legacy we want to leave for future generations is essential to the development of informed long-term and sustainable solutions.

6.3 What Is Resource Management?

Increasing resource use tends to increase environmental impacts. For example, mining tends to degrade the land more than tourism, but increasing resource use also tends to provide high short-term economic benefits. The history of the United States demonstrates that the economy has generally rewarded entrepreneurs who most rapidly exploited natural resources. This process seemed justified to most people because modern society traditionally has ignored most environmental costs, called **environmental externalities**. For instance, **benefit–cost analysis** is a method of comparing the benefits of an activity to its cost. When the benefits (calculated in a dollar amount) are greater than the costs (calculated in a dollar amount), there is said to be a net benefit to society. If most environmental costs are ignored, any short-term economic benefits of resource use will seem worthwhile: When environmental costs are very low (artificially), the benefit will be greater than the costs. For example, clear-cutting a virgin forest could yield profits (benefits) in the short term.

A more realistic way to analyze resource use is to include the long-term economic benefits of not using them. When this is done, less resource use often translates to greater economic benefits. For instance, the total economic value of a rain forest usually is greater if the forest is used over a long time span for multiple ecosystem benefits, including tourism, water purification, pharmaceuticals, native foods, and other uses, than if it is cut down for a one-time, short-term gain in lumber that leaves the forest unusable for decades or centuries. The total value of the rain forest is enhanced even more if long-term environmental benefits are included, such as the value of the forest to future generations.

6.4 Sustaining Resource Use: Preservation, Conservation, and Restoration

Careful use of natural resources can lead to overall sustainable long-term economic benefits and reduced environmental costs. Such use and management seek to minimize adverse impacts where possible. There are three basic options that resource management can apply to minimize resource use: preservation, conservation, and restoration.

Preservation generally refers to nonuse. A "preserve," national park, or wilderness area is an ecosystem that is set aside and (in theory at least) protected in its pristine, natural state. **Conservation** (input reduction) recognizes that some resource use by humans is unavoidable, so attempts try to minimize the unsustainable use of a natural resource. Use can be minimized through efficiency improvements, recycling or reuse, and substitution of other resources. Finally, **restoration** seeks to return a degraded resource to its original, or close to original, state. For example, attempts are being made to redirect the Kissimmee River of Florida into the path it originally followed before people altered it, thereby attempting to restore its natural hydrology regime. The rapidly growing field of restoration ecology is attempting to return many ecosystems, such as tall-grass prairies and wetlands, to their original state.

A Brief History of Preservation, Conservation, and Restoration in the United States

When the national parks and national forests were being established in the early 1900s under President Theodore Roosevelt, there was a lively debate over how much public land should be allotted to preservation and how much to conservation. National parks, where most forms of resource use except tourism are prohibited, are examples of preservation (**FIGURE 6-1**). The debate continues, with many environmental groups calling for setting aside more land as wilderness areas to be regulated by government agencies. Some groups, especially the **Nature Conservancy**, buy such land and set it aside as private preserves. In contrast, national forests (and most other federal lands) permit timber cutting, mining, grazing, and other uses. The promoters of conservation won the debate, and most federal land has permitted these resource uses. In theory, such uses represent "conservation" because the resources were supposed to be closely managed in a way that minimizes damage to the land. However, this was rarely realized in practice, and many federal and state public lands suffered extensive damage from overuse. More recently, better management has led to significantly improved rangeland and forests, especially in the United States and Canada. To date, the U.S. Fisheries and

Wildlife Agency protects more than 60 million hectares (150 million acres) of land for varied use. The Nature Conservancy oversees the use of 48 million hectares (119 million) acres of protected land nationally.

Recall from our overview of environmental science that the abuse of public lands is often called the **tragedy of the commons**. In this sense, a "commons" refers to public lands that people own and use. It is important to understand that holding property in common is not the root of the problem. The idea of publicly held property is a good one in theory, but in practice, it rarely works out to everyone's benefit. History shows that such public property is almost invariably abused when there are no strict controls or regulations implemented for protection. Thus, the *tragedy* stems from the lack of constraints or rules governing the use of the commons. For example, early American settlers found that farm animals much more frequently overgrazed shared lands because no single landowner had a vested interest in protecting the land. In the case of federal and state lands, they are overused and abused because few people view them as "their" responsibility, and they do not see themselves as the ones responsible for depleting the resource.

Restoration, the newest type of resource management, has become much more common since the 1990s and 2000s. Wetlands, rangelands, forests, fisheries, lakes, soils, and many other resources and environments are being restored by a growing number of restoration specialists. However, restoration can be very expensive; applying the precautionary principle and working toward preservation and conservation early on in the process are more cost-effective.

Looking back on the historical impact of mismanaged resources gives us a strong case for applying the **precautionary principle** ("better safe than sorry"), to current environmental challenges so that future generations do not have to suffer needless tragedy as those now and in the past have. Take for instance the collapse of the cod fisheries in Canada in the 1990s. The end result was the fact that 44,000 people lost their livelihood. Loss of the fisheries cost the Canadians billions of dollars. The cost of putting in place sustainable practices for the

FIGURE 6-1 The U.S. national park system represents an attempt to preserve nature. This geyser is one of many unique features in Yellowstone, the country's first national park.

cod industry would have prevented the cost of the government paying people for the loss of their livelihoods and the subsequent negative impact on support services as people no longer had income to spend locally.

The over-fishing of menhaden (*Brevoortia tyrannus*) at the mouth of the lower Chesapeake Bay on the east coast of the United States provides a valuable example of the complexity of applying the precautionary principle in action. Menhaden, a small nutrient-rich fish, is a keystone species absolutely vital to the health of not only the Chesapeake Bay estuarine ecosystem, but also to the larger marine environment. The striped bass (*Morone saxatilis*) has a documented life span of more than 30 years. Of the four worldwide breeding grounds for species of striped bass, the Chesapeake Bay is the most successful. As small young fish, the striped bass grow strong on a diet rich in menhaden; they migrate out to open ocean waters and return annually up the bay to breed again. A full-grown striped bass is an impressive fish and can weigh in at between 9 and 22.5 kilograms (20 and 50 pounds). The striped bass is in turn food for larger species of marine fish, but their early vitality is dependent largely on menhaden.

Although scientists, policy makers, sports fishermen, common citizens, environmentalists, and journalists, among others, have repeatedly documented the negative impact of over-fishing menhaden, especially by the commercial fishery at the lower end of the bay in Virginia, menhaden are still in decline. Despite the loud and active precautionary tale spelled out for policy makers, Virginia is the only East Coast state that allows menhaden to be aggressively harvested, mainly for livestock feed, pet food, and heart-healthy supplements.

Once a resource has been depleted and restoration of the environment is necessary there is often a wide range of stakeholders involved with varying data and methodology. The process of restoration can also be particularly contentious when groups or agencies disagree over the method or the extent of the restoration. Groups will block action because they fear the cause-and-effect relationships have not been fully investigated. In other words, nothing is done because of concerns that the solution might turn out to be worse than the problem. In this case, some argue that a variation of the precautionary principle in which some method of restoration is better than no restoration, should apply.

Because so many environments are highly degraded from past human abuses, restoration undoubtedly will become increasingly common. Restoration is most effective at the landscape level. For instance, it does little good to restore a lake acidified from mining runoff to its normal chemistry unless the surrounding land is treated to help reduce acidic runoff. Similarly, reintroducing wolves into a small forest is unlikely to be successful unless the natural ecosystem for the entire region is prepared to support them. A small forest is not large enough for a self-sustaining wolf population (more information can be obtained in a review of conserving biological resources).

6.5 Who Cares? The Many Values of Natural Resources

To this point, we have talked about the value of resources in economic terms. However, as **FIGURE 6-2** shows, people can place at least

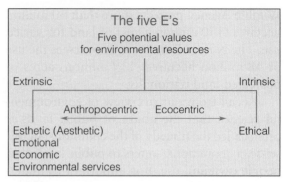

FIGURE 6-2 The five E's representing five potential values of environmental resources.

five values, sometimes called the **five Es**, on natural resources. One of the Es, **ethical value**, is what philosophers call an **intrinsic value**. This is the value of a thing unto itself, regardless of its value to humans. Does a mountain have as much right to exist as you do? Does a worm? If you say "yes," then these things have intrinsic value; however, can a resource, by definition something that has human utility value, still have intrinsic value? Intrinsic values are **ecocentric** (environment centered).

The other four Es are what philosophers call **extrinsic values** (Figure 6-2). These are values that are external to a resource's own right to exist, referring instead to the resource's ability to provide something that is of value to humans (managing ecosystems for the benefits or "ecosystem services" they provide to humans is one example of an extrinsic value approach). Such values are **anthropocentric** (human centered). Extrinsic values are more practical than intrinsic values and therefore tend to be more widely discussed in political and economic debates on resource management. **Esthetic** (aesthetic) **value** is the value of a resource in making the world more beautiful, more appealing to the senses, and generally more pleasant. The value you place on a mountain hike in the cool morning air might be an example. Some people place no value on this and would pay nothing for it. Others find it indispensable. **Emotional values** include the value of a resource beyond sensory enjoyment. For example, some people develop very strong emotional bonds to certain natural areas or certain plant or animal species. This is sometimes called a "sense of place." Many psychologists consider natural habitats to be important for mental health, especially in children.

Economic values traditionally are directly involved with tangible products that can be bought or sold: food, timber, energy, and so forth. As we have discussed, society needs to focus more on long-term economic values, which actually provide more income over the long run. The value of resources for tourism, native fruits, or other sustainable products is ultimately (over the long term) much greater than the value of their destructive uses. **Environmental service values** are the values of resources in providing tangible (such as timber and nontimber forest products) and intangible services that allow humans (and other life) to exist on Earth. For instance, plants help purify air and produce oxygen, and plant roots and soil microbes purify water; ultimately, all food relies on a variety of environmental services.

Some people place all five values on all environmental resources. How many values would you place on a forest? How many on a beach? Many people would place only economic values on such resources. Logging, mining, and other types of harvesting, some of which may destroy the resources (sustainable forestry may not destroy the resource, whereas other types or harvesting may), are called **direct values**. Most environmental problems arise when resources are appreciated for only their direct value and without concern for using such resources sustainably. Placing only direct values on natural resources often artificially "discounts" their true value to society and to future generations. For example, four of the five Es—esthetic, emotional, environmental services, and ethical values—generally are not viewed as direct values. They represent what economists call **indirect values**, meaning that they are valued in ways that do not involve direct mining, harvesting, or other destruction of the resources. If resource prices incorporated both indirect values and long-term direct values (environmental externalities), the prices would reflect the resources' true environmental cost.

Although the direct value of resources provides immediate financial rewards, harvesting the direct value often destroys (1) the long-term economic value and (2) many or all of the indirect values. If society included all of the indirect values of existing natural resources in the pricing of consumer products to reflect their **true environmental costs**, or externalities, it would encourage less destructive harvesting of natural resources. A major problem is that indirect values, such as ethical and emotional values, are often subjective and difficult to calculate (a review of environmental economics provides more information). Including the true environmental costs in the prices of products would encourage the market place to preserve and conserve resources. More sustainable uses of resources, such as recycling, organic farming, and ecotourism, would become more profitable businesses. As long as only direct (short-term) values are considered, overuse and exploitation will be encouraged and rewarded.

6.6 Patterns of Resource Depletion

There are two basic inputs from the environment: matter and energy. Matter constantly cycles through society and the environment, whereas energy primarily has a one-way flow. Because of this difference, matter and energy are depleted differently.

How Matter Resources Are Depleted

Being dispersed depletes matter resources. Ore deposits are unusually concentrated deposits of minerals that are normally found in more dilute form in the Earth's crust. When we mine and process the ore into metals to build cars and other refined products, the atoms may be eventually dispersed (such as when gears wear down grinding against one another) or lost to further human use when we dispose of the products in landfills and elsewhere. Of course, the materials in a landfill, which can be thought of as an **artificial ore**, may later be mined for metal content. Similarly, rapid erosion depletes soil not because the nutrients and minerals in the soil are destroyed, but because the soil is dispersed, ultimately into the oceans.

These examples are nonrenewable matter; when dispersed, molecules of metals and soils will stay dispersed unless we spend much energy and money to reconcentrate them. In the case of renewable matter resources, dispersal still occurs, such as when we build houses from timber, but we can regenerate the timber relatively quickly. Renewable matter resources are often biological resources that can be regrown.

How Energy Resources Are Depleted

Energy has a one-way flow through society because it is transformed to an unusable form, "waste heat," when we use it. Energy resources are therefore depleted when they are transformed this way. This is a key difference from matter, which can often be recollected and reconcentrated if society has enough cheap energy. In contrast, after energy is transformed, it is lost forever; waste heat can never be reconcentrated. If we burn oil or coal to release their chemical energy to drive an engine, that energy can never be reused.

However, we do have a major source of renewable energy: the sun. This source of renewable energy could potentially keep society running for many millions of years. Examples of the sun's energy include direct solar power, biomass, hydropower, and wind (a discussion of renewable and alternative energy sources provides more information8).

Bubble Pattern of Depletion

Unsustainable use of many resources exhibits a bubble pattern of depletion. The best-known example is the so-called **Hubbert's bubble** of oil depletion. This is named after M. King Hubbert, who accurately predicted the bubble pattern of oil depletion in the United States. Since the 1950s when his predictions were first made, they have proven to be strikingly accurate. U.S. oil production peaked around 1970 and has been declining since, as the richest reserves are steadily depleted (**FIGURE 6-3**). The United States produced 11% of the world's petroleum in 2009 and consumed 22%.

The bubble pattern has two main causes: exponential exploitation and exponential depletion. Because both use and exhaustion are exponential, they tend to form a mirror image. The left, or exploitation, side of the bubble is exponential because resources are exploited very quickly once society discovers their utility; however, because all resources on Earth are finite, limits to growth eventually occur, and demand begins to exceed supply. During this time, society usually intensifies its efforts to obtain more of the resource through further exploration and increased technological applications. However, these efforts usually soon encounter what economists call the **law of diminishing returns**, meaning that increasing efforts to extract the resource produce progressively smaller amounts. The result is shown in the right, or depletion, side of the bubble. Production declines exponentially because the most easily extracted concentrations of the resource become exhausted.

As supplies of the resource decline, prices rise, sometimes leading to unemployment and other unpleasant changes. Historically, society has responded to the increase in resource prices by switching to another resource. For example, in the 1800s, England switched from wood to coal as an energy source when the forests were decimated. Then, in the 1900s, the English economy switched to oil as its primary energy source because it became cheap and abundant. This pattern is common, whereby societies have tended to switch from one unsustainable resource to another. The way to break this "cycle of unsustainable use" is to switch to sustainable uses.

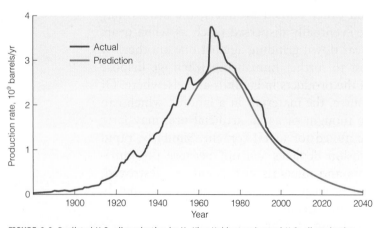

FIGURE 6-3 Predicted U.S. oil production by M. King Hubbert and actual U.S. oil production through 2005. Both curves peak around 1970. (*Source*: U.S. Department of Energy.)

6.7 Problems with Past Resource Management

Society can respond to a diminishing supply of a resource in two ways: (1) intensify efforts to extract more of the resource or (2) reduce the need for the resource, including finding alternatives. The first response typically has been and remains much more common. The usual result was a bubble pattern as depletion accelerated when supplies ran low. The mining of lower-grade, high-volume materials also led to pollution problems. Two influential concepts were advanced to justify these past intensification efforts to extract more of a diminishing resource.

Net Yield of Nonrenewable Resources

Intensified extraction of nonrenewable resources is based on the concept of **net yield**, which holds that a resource can continue to be extracted as long as the resources used in extraction do not exceed the resources gained. Because energy prices have been relatively low during the last century, miners have simply switched to lower grade deposits when high-grade deposits were depleted (**FIGURE 6-4**). For example, at the current market price of copper, copper ore that is as low as 0.5% copper content in the rock can profitably be mined. Switching to lower grade deposits greatly increases the available supplies because nearly all nonrenewable resources are characterized by an inverse quality curve:

higher grade deposits of coal, oil, minerals, and other resources are much rarer than lower grade deposits.

The pursuit of net yield has incurred great environmental costs by accelerating both depletion and pollution. Depletion accelerates because increasingly lower grade ores must be mined and more energy and other resources must be expended to extract the ore. Pollution accelerates because increasing amounts of waste rock are produced when lower grade ores are mined (**FIGURE 6-5**). The Goldstrike mine of Nevada, for example, the largest gold mine in the United States, moves 295,000 metric tons of rock and waste to produce just more than 45 kg (100 pounds) of gold. In 2004, globally 1,600 million metric tons of copper ore was mined to produce about 14.5 million metric tons of copper. In 2009, U.S. domestic copper production was 1.2 million metric tons, with approximately 35% of that copper coming from re-melted and recovered sources. Even with the increased efforts to reuse scrap and waste copper, some geologists worry that eventually humans might run out of ores to mine. Many environmentalists now think the bigger issue is mining-related environmental

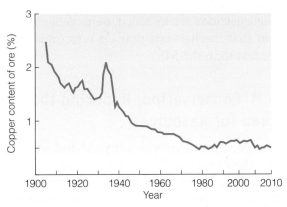

FIGURE 6-4 Since 1905, the quality of copper ore mined in the United States has declined from 2.5% to approximately 0.5%. (*Source:* U.S. Bureau of Mines.)

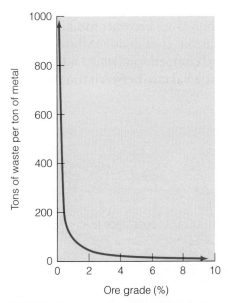

FIGURE 6-5 For all ores, the amount of waste rock and other waste materials increases exponentially as ore grade decreases. (*Source:* Adapted from D. H. Meadows, D. L. Meadows, and J. Randers. Beyond the Limits. [Post Mills, Vt.: Chelsea Green, 1992], p. 87.)

degradation. Mining may have to be curtailed because of the pollution generated by metal extraction, rather than because the ores are actually depleted.

Maximum Sustainable Yield of Renewable Resources

Another very influential concept in resource management has been the **maximum sustainable yield** (**MSY**), which holds that the optimal way to exploit a renewable resource is to harvest as much as possible to the point where the harvest rate equals the renewal rate. As an example, a person could withdraw as much groundwater as needed to the point where the withdrawal rate equaled the recharge rate from rainfall. Taking less could be considered underutilization, and taking more would lead to depletion.

MSY has been especially influential in the management of renewable biological resources, such as commercially important fish and wildlife. The basic principle was that fish, game, and other populations could be harvested to the point where the population's ability to reproduce itself was impaired. For instance, individuals could have trouble finding mates if the population was over-harvested. MSY actively encourages limited harvesting because it holds that if population abundance becomes too high, additional population growth is inhibited by crowding and competition (**FIGURE 6-6**). Thus, MSY aims at a balance between too much and too little harvesting to keep the population at some intermediate abundance.

Although it is widely practiced by state and federal government agencies regulating wildlife, forests, and fishing, MSY has come under heavy criticism by ecologists and others for both theoretical and practical reasons. An important theoretical shortcoming is that MSY does not take large environmental fluctuations into account: a bad winter or some other natural catastrophe could reduce the population unpredictably. More important perhaps are the many practical problems with MSY. Calculating the point at which population growth begins to slow from competition is very difficult. Indeed, in many cases, such as marine fish, merely estimating the population size can be difficult. Consequently, estimating MSY with certainty is virtually impossible, although more sophisticated methods, such as refined statistical sampling models for estimating population size, are improving the calculations. Still, the decline in commercial fishing in many areas indicates that humans are harvesting far more than the MSY (see **CASE STUDY 6-1**).

Despite these and other problems, many agencies continue to rely on MSY in some variation or another. A variation on the MSY concept known as the **optimum sustainable yield** (**OSY**) is gaining popularity with some resource managers. OSY states that the optimal harvestable rate for a renewable resource must consider many factors, not just the maximum yield. How will harvesting affect other species in the ecosystem? How will it affect other human uses of the ecosystem, such as recreation? When such additional benefit–cost questions are included, managers usually find that the harvests that OSY recommends are less than the MSY.

6.8 Conservation: Reducing the Need for Resources

The basic problem with net yield and MSY is that their main goal has been to maximize resource use. Their focus was on short-term economic gain, but long-term economic and other gains (such as esthetic, emotional, and environmental services—for instance, the service

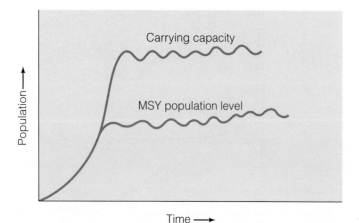

FIGURE 6-6 MSY occurs at population levels around one half of the carrying capacity.

How to Exceed Maximum Sustainable Yield: Overfishing in New England

VISIT

http://environment.jbpub.com
/mckinney/5e/
for more information

In 1981, the United States declared a 322-kilometer (200-mile) boundary around its shores, banning most foreign trawlers that had fished U.S. waters. The ban was an attempt to save the beleaguered New England fishing industry. The measure worked until Americans themselves began overfishing. Encouraged in part by a federal loan program, they began building more boats, equipping them with powerful fish-finding electronics. They lobbied to remove quotas on catches. For a while, they caught a lot more fish, but the trawler catch in New England peaked in 1983 and has since generally continued to fall (**FIGURE 1**).

FIGURE 1 New England fishermen are struggling to make a living because of the decline of many species and government restrictions.

The story is the same throughout the world. Stocks of flounder and haddock are near record lows. The cod population is down. Bluefin tuna and swordfish have been depleted. There have been booms and busts before, but scientists say that this time it is different. The fleets are so big and the technology so advanced that fish no longer have anywhere to hide. The global fishing industry has more than doubled in the last 3 decades. Dozens of species have been fished to commercial extinction, and the rate of growth in marine harvests has plummeted to nearly zero. In 2002, the United Nations Food and Agriculture Organization warned that about one quarter of the world's marine resources were either overexploited or already depleted. About half of the marine fishing grounds was classified as "fully exploited," meaning that increased fish production from these regions is unattainable. The Millennium Ecosystem Assessment (2005) reported that in all regions of the world, overfishing has depleted finfish, crustaceans, and mollusks, which are major contributors to local economies, food supplies, and ecosystem vitality. Unrest related to the current situation creates tensions in communities losing their subsistence. In Asia and Africa, increasing conflicts related to this alarming trend have been documented.

In addition, the composition of global catches has changed. Fishers respond to reduced population size by catching huge numbers of juvenile fish. Many of the fish have not had a chance to spawn, which accelerates the long-term decline. There is also a trend toward catching fish lower on the "food chain." As fishers deplete large, long-lived predatory species, such as cod, tuna, shark, and snapper, they move down to the next level—to species that tend to be smaller, shorter lived, and less valuable. As a result, fishers worldwide now fill their nets with plankton-eating species, such as mackerel, jacks, and sardines, and invertebrates, such as squid, oysters, mussels, and shrimp. Initially, this transition brings new bounty: At higher trophic levels, fishes are larger, but there are fewer of them. At lower levels, the species are smaller but more plentiful. Relieved of predatory pressures and competition for food, the smaller species are free to fill in the empty niche that their predators once occupied. This is one reason global fish

How to Exceed Maximum Sustainable Yield: Overfishing in New England

VISIT

http://environment.jbpub.com
/mckinney/5e/
for more information

catches have not declined even more than they have in the past 25 years, despite severe overfishing. At this rate, researchers such as Boris Worm, a noted marine research ecologist, predict that global fisheries will collapse by 2048.

Some people in the fishing industry say they cannot back off. They have to keep the boats going nonstop to make enough to survive. However, others and conservationists insist that current measures, such as minimum net mesh sizes and occasional closures of overfished waters, simply are not enough. Strict catch quotas, trip limits, even moratoria on new boats are needed, or more people will be forced out of the business in the long run. Legislation has been introduced to reduce the size of the New England fleet by buying out vessels with money from a tax on the diesel fuel the fishing boats use. Some have suggested that the government should subsidize the fleets by meeting mortgages for those who do not fish a certain number of days, and severe restrictions are being placed on fishing in some areas, such as the once-fertile Georges Bank ecosystem off the coast of Massachusetts. Some way of rationally allocating fishery resources is needed to guarantee that the industry remains intact.

This is a classic example of the "tragedy of the commons." Many individuals and nations use the oceans. Thus, many nations must agree on laws that govern access to ocean resources, and perhaps more difficult, these laws must be enforced in

ways that prevent overfishing. The U.N. Food and Agriculture Organization works internationally to promote a sustainable code of conduct for commercial fisheries and also addresses other important issues, such as promoting practices to reduce the 20 million tons of annual by-product discards.

One solution is to set aside marine "reserves"—strategically located areas where fishing is prohibited to rebuild fish populations and restore the marine ecosystem. However, 95% of the world's fishers are from developing countries and work in traditional fisheries where implementation of no-fishing zones has met with tremendous resistance among people economically dependent on fishing.

The successful establishment of marine protection programs requires that the immediate and long-term economic benefits for the fisheries be secured through incentives. Co-management of marine resources by governments, private fisheries, and communities should encourage and reward fisheries in their effort to manage marine resources sustainably.

Questions

1. How has the composition of catches changed through time as overfishing occurred?
2. How does overfishing relate to the tragedy of the commons? What are some possible solutions?

that a forest provides in producing oxygen and removing carbon dioxide from the atmosphere) are greater if the emphasis is shifted to reducing resource consumption. In addition, instead of maximizing resource use, the focus should be on accomplishing more with the resources that are used.

Thus, conservation or input reduction by lowering resource use will be a fundamental part of a sustainable society. Recall from our discussion of environmental science that conservation (1) slows depletion of resources, (2) reduces pollution by slowing the flow of matter and energy (throughput) through soci-

ety, and (3) saves money. For example, burning less coal by increasing a power plant's efficiency not only saves coal but also produces less acid rain and other pollution. Taking a precautionary approach, it is also often cheaper to design power plants to burn less coal than to pay for all the pollution control devices needed to trap the air pollution in the smokestack and dispose of the trapped pollutants. Until recently, supplies of most resources were so abundant, especially in the United States, that little attention was paid to input reduction; as a result, depletion rates were high, and scrubbers and other forms of output, or "end-of-pipe," reduction controlled pollution.

Ways to Conserve: Efficiency, Recycling, and Substitution

FIGURE 6-7 shows three basic ways to conserve or reduce the need for a resource: *efficiency improvements* (input reduction), *reuse/recycling,* and *substitution.*

Efficiency improvement generally is most effective and economically cheapest because many technologies and activities are wasteful and inefficient. However, different resources will require different methods of input reduction. It is very difficult, for example, to find affordable substitutes for water in many of its uses, such as agriculture and drinking, so increased efficiency and reuse/recycling of wastewater are common.

Efficiency Improvements

Efficiency improvements occur when the same task is accomplished with fewer resources. An example would be lighter, more fuel-efficient cars to conserve fuel and building materials. Average fuel efficiency on new vehicles sold in the United States has risen 7% from 23.1 miles per gallon in 1980 to 24.7 mpg in 2004. Recent hybrids (2010) have achieved fuel efficiencies greater than 40 mpg.

Such cars perform the same tasks as less fuel-efficient cars but use fewer resources. As another example, about two thirds of the water used in irrigation is lost to evaporation. Using "microirrigation," in which water is carried by pipes and sprayed through small holes, reduces water loss to less than 20%.

Japan in particular and the European countries are leaders in devising technologies that improve efficiency. The **energy intensity index,** which equals energy consumption/gross national product (GNP), measures how much energy is used to produce the same unit of wealth. The index for the United States in the late 1990s was about 0.015 petajoules consumed per $1 million GNP (approximately 14.5 billion BTUs per $1 million GNP). In contrast, the index for Japan was about 0.006 petajoules per $1 million GNP (5.7 billion BTUs per $1 million GNP). Japan uses less than half the energy to produce a given amount

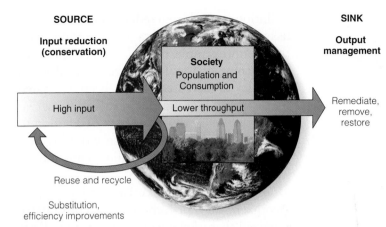

FIGURE 6-7 Input reduction (= conservation) can be achieved by efficiency improvements, reuse/recycling, and substitution. All three methods reduce the flow of materials and energy (throughput) into and out of society, slowing both depletion and pollution.

of wealth, from manufacturing cars to heating hotel rooms. Material efficiency indexes, which indicate usage of other resources, show a similar pattern of waste for the United States. For example, Japanese technologies convert about 70% of raw timber into finished wood products, whereas in the United States, only about 50% is converted.

The relative inefficiency of resource use in the United States is a direct result of its historical endowment of abundant resources. The spacious tracts of land still present in parts of the United States create a misinformed notion among some people that resources and land are infinite. Costs for energy and other resources were historically low, so efficiency was not encouraged. However, ultimately inefficiency is costly to any economy. According to many experts, energy inefficiencies account for at least 50% of U.S. energy use. With such inefficiency, money invested in energy-saving technologies would more than pay for itself by reduced energy costs. The Worldwatch Institute estimates that U.S. electricity could be reduced by 70% through efficiency gains at no net cost because energy savings would equal the investment needed to make the changes. Increased efficiency leading to conservation of minerals and other resources would achieve similar savings.

Besides the economic savings, environmental savings from efficiency increases are also enormous. Less depletion saves more resources for future generations. Less resources extracted means less degradation of the land.

Pollution control from reduced processing and usage can also be significant. To continue with the energy example, the pollutants released by the burning of fossil fuels are reduced when efficiency is increased. It has been estimated that national investments in energy conservation would not only save from $10 billion to $110 billion per year, but also would reduce greenhouse gas (global warming) emissions as much as 40%.

Researchers at Duke University and the Georgia Institute of Technology recently (2010) reported that implementing just nine policies across commercial, residential, and commercial sites could generate 380,000 new jobs, save consumers $41 billion on energy bills, and conserve 32.55 billion liters (8.6 billion gallons) of water for the 10-year trial period in the District of Columbia and 16 southern states. These researchers estimate that for every dollar invested in this program over an extended 20 years, $2.25 will be generated as profit. The plan includes implementing new appliance standards, incentives for weatherization, upgrading utility plant efficiency, retrofitting older homes and factories with newer energy-saving equipment, and streamlining processes that generate power. Their model includes how individual policies overlap and work together to create a powerful and profitable example of attainable energy conservation.

Reuse and Recycling

Reuse occurs when the same resource is used again in the same form, such as refilling soda bottles. Recycling is similar, but the resource is not reused in the same form. For instance, soda bottles may be remelted to make new bottles or other glass containers. Like improved efficiency, reuse/recycling reduces depletion of resources and pollution from resource extraction and use (Figure 6-7). Reuse/recycling is useful in reducing solid waste, which is important because the number of landfills have been steadily decreasing from nearly 8,000 in 1988 to 1,654 in 2005. The area of a present day landfill can expand quite impressively. Take, for instance, the Rumpke landfill in northern Ohio. It is one of the largest in the nation,

currently occupying 93 hectares (230 acres) of a site that is more than 160 hectares (400 acres) large. At its highest point, it sits 318.5 meters (1,045 feet) above sea level (as a point of comparison, the highest point in the state, Campbell Hill sits at 472 meters [1,549 feet] above sea level). The largest methane recovery plant in the world converts 425 thousand cubic meters (15 million cubic feet) of methane gas a day into natural gas at the site. This translates into enough energy to supply 25,000 residents with power. Still, the problem of how to reduce waste entering the landfills remains a societal challenge. Despite that the Environmental Protection Agency (EPA) reports that 32% of municipal waste is recycled today as compared with just 10% in 1980, 54% of all waste still ends up in landfills. The majority of the waste is paper, followed by miscellaneous discards that do not fit neatly into categories of waste that are easy to identify and commercially viable to sort and recover, such as bottles, plastics, and metal. Yard trimmings and food waste are examples of other waste that can be composted instead of discarded. Overall, 245.7 million tons of municipal solid waste was collected in 2005. Each person in the United States generates approximately 4.5 pounds of municipal solid waste each day. The EPA documented that more than 54% of all discarded waste represents items related to cardboard and paper. Therefore, **precycling**, which is the manufacturers' reduction of packaging material, is becoming more important. For example, concentrated foods can often be packaged in smaller containers. Precycling is not recycling but is conservation by increased efficiency; the same task is accomplished, but fewer resources are used. Despite such efforts, the amount of solid waste in the United States continues to grow (a review of municipal solid waste and hazardous waste provide additional information).

The technological and economic aspects of recycling can be very complex. A simplified recycling scheme begins with **virgin resources**, which are the original natural resources being extracted. Both extraction, such as mining, and processing during product manufacture usually create pollution. The **recycling loop** begins before the purchased product is discarded; the discard is reprocessed into the same or perhaps another product. Unlike the reuse loop, the reprocessed product does not have to be the same as the original. For example, although paper is often recycled into more paper, car tires are often recycled into road asphalt, shoes, playground structures, and many other products. As long as the discard is being used in place of a virgin resource, recycling occurs. Unfortunately, some manufacturers claim their products are recycled even though the products do not meet this criterion of being used instead of a virgin resource (see **CASE STUDY 6-2**, Is This Recycled or Not?). Markets need to be developed for recycled products; it is not enough to simply recycle and then purchase materials that are not recycled.

The recycling loop is closed when someone buys a product containing recycled material (**FIGURE 6-8**). This slows depletion of virgin resources and reduces pollution in two basic ways. Most obviously, it reduces the amount of solid waste that would have been discarded into landfills, incinerators, and other means of disposal. In addition, recycling reduces the pollution that would have been produced

from the extraction of the virgin resources. Unfortunately, the recycling loop often is not closed because although people are willing to sort and return recyclable waste, there often is little consumer demand to purchase the recycled product. The reasons for this usually are economic: when products made from virgin resources are artificially cheaper, the loop goes unclosed. If the costs of products made from virgin resources were increased to reflect their true cost to the environment, the price differential would disappear, and the loop could be closed.

Substitution

Substitution occurs when one resource is used instead of another. Substitution can also help reduce both depletion and pollution problems. It helps with depletion because when one resource is being depleted, a more common substitute can be used at a cheaper price. For instance, aluminum, a very common metal in the Earth's crust, sometimes can be substituted for the much rarer and more expensive copper in making alloys, equipment, and other uses. Substitution helps with pollution problems when the extraction, processing, or disposal of the substituted resources produces less pollution. For example, many plastics last for 50 to 100 years in the environment before they significantly decompose. In addition, plastics are made of the nonrenewable resource petroleum, whereas trees are renewable. Such considerations have led to the substitution of paper for plastic in many items, such as drinking cups and food containers.

Although it can be useful in reducing resource depletion, substitution is often less desirable than efficiency improvements or reuse/recycling. Instead of reducing overall resource depletion, substitution often simply switches depletion from one resource to another. This can be satisfactory if the new resource is renewable, as in paper, or very abundant, as with glass made from sand. In addition, substitution often does not reduce pollution, solid waste, or other output problems. For instance, the use of paper products offers many environmental advantages over plastic, but it may do little to solve landfill space problems.

FIGURE 6-8 Pens and many other products can be made from recycled tires.

Is This Recycled or Not?

Since Earth Day 1990, many products touted as "environmentally friendly" have entered the market. Sometimes called green marketing or "greenwashing," promoting products as green can be very successful because of increasing concern about the environment. Unfortunately, green marketing has been plagued by misunderstandings and even false claims. A major reason is the lack of clear definitions. For example, what exactly does it mean when a product is advertised as being biodegradable or recycled (**FIGURE 1**)?

The definition of biodegradability came to widespread public attention when "biodegradable" trash bags were found to degrade only under laboratory conditions and not under those found in landfills. As a result, the company had to remove this claim, and the term biodegradable is now reserved for products that decompose under natural conditions. Another problem is that decomposition must involve true chemical destruction of the product into harmless components and not just reduction into smaller particles, such as smaller plastic pieces. When disposed of improperly, these plastics often end up in bodies of water. Not truly biodegradable, most plastics that end up in these toxic ocean deposits photodegrade into smaller and smaller pieces, releasing synthetic chemicals that are also harmful to life. Currently there are 417 dead zones in ocean waters worldwide, including one in the Baltic Sea measuring 70,000 square kilometers (27,000 square miles). There are also various "garbage patches" in the oceans, most prominently the Great Pacific Garbage Patch (also known as the Pacific Trash Vortex), an area in the central portion of the North Pacific Ocean possibly as large as twice the size of Texas, where plastics, chemical sludge, and various types of debris has collected. Because of the chemical makeup of plastics, they attract other harmful chemicals swirling in these dead zones making ingestion of the plastics not only problematic because they are indigestible, but also because of the deadly chemical cocktail the sea creature ingests. Mammals, birds, and turtles are the most likely creatures to ingest the plastics as they mistake the plastic for crustaceans and fish.

This accidental ingestion of plastics is proving deadly for many animals, perhaps most notably the albatross. Of the 21 species of albatross, 19 are threatened and 2 are near threatened. These magnificent birds occupy huge ocean ranges, can circle the globe, and display amazing dancing behavior during courtship. With a 3.3-meter (11-foot) wing span, the giant albatross has the largest wing span of any bird alive today. Human activity is directly linked to the declining numbers of all species of albatross. As recently documented, the Laysan albatross chicks studied on Midway Island, about 1,600 kilometers (1,000 miles) from any major city, are dying from the ingestion of toxic plastic. Of the birds examined, 97% had ingested plastics. Parent albatrosses fly hundreds of kilometers/miles for food for their chicks and return to their nests on remote island areas. The plastic flotsam on the top of the water attracts the albatross as fish or crustaceans would and they bring it back to the nest. The chicks that do not die unfortunately learn to look for this material when gathering food for their own chicks. More than 100,000 sea mammals and turtles die each year from eating plastics. For the Laysan albatross, the numbers are staggering; more than 40% of chicks born on Midway Island in the Pacific Ocean die before they ever get a chance to take flight over open waters. Plastic pellets from manufacturing and waste plastics from consumers enter the oceans through storm drains and rivers.

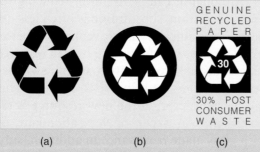

(a)   (b)   (c)

FIGURE 1 Three types of recycling symbols commonly used in the United States. (a) This symbol simply means that the object is potentially recyclable, not that it has been or will be recycled. (b) This symbol indicates that a product contains recycled material, but it does not indicate how much recycled material is in the product (it could be only a very small amount). (c) This symbol states explicitly the percentage of recycled content found in the product.

However, even if we define biodegradable as chemical decomposition in natural conditions, there is still plenty of room for vagueness. What are natural conditions? What if decomposition takes 5 years? Ten years? Biodegradable does not mean that all harmful byproducts are neutralized or that in the process of breaking down more toxic compounds are produced, as is the case with the albatross example. There is considerable variation in how different plastics break down. An average plastic bottle takes 450 years to break down, whereas a plastic bag might take 30 years.

The definition of a "recycled" product is perhaps even more difficult to specify. Many kinds of products are recycled in many ways, and the remanufactured products are often only partly composed of recycled material. Yet buying recycled material is essential to close the recycling loop. Paper recycling is a key illustration of the difficulties because it is so important: paper is the largest component of municipal solid waste, composing over 50% of landfill volume, but the United States has no national legal standards defining recycled paper. As a result, some paper that is sold as recycled actually contains only a tiny fraction of recycled paper. Furthermore, even this recycled paper can be preconsumer, meaning that it is wastepaper generated by paper mills, such as cuttings, scraps, and flawed batches. Such preconsumer waste often is not as desirable as recycling postconsumer waste, which is discarded by consumers; recycling postconsumer waste is more helpful to the environment because that is the source of so much landfill waste.

The lack of clear meaning for product environmental claims is harmful to nearly everyone. Consumers suffer when they pay extra for a product that is not what they think; companies that really try to produce environmentally friendly products suffer when they are undersold by companies that market more harmful products as environmentally friendly, and the environment suffers when consumers use harmful products when they think they are not.

The solution to such problems is to create clear definitions for environmental marketing claims. Those who falsely make such claims can then be legally prosecuted, or at least conscientious con-

sumers can avoid the products. Currently, state and local laws and nonbinding agreements between consumer groups and companies have established some standards. Unfortunately, these vary from place to place. Therefore, support is growing for a single set of federal guidelines that would apply all over the United States. For example, the EPA currently has a paper guideline specifying that recycled paper should contain at least 50% recycled fiber, but companies are not required to use this definition, which also omits how much of the recycled fiber must be postconsumer. In 1992, the EPA invited the U.S. Office of Consumer Affairs and the Federal Trade Commission to join in developing voluntary guidelines for the use of environmental terms on product labels.

Until enforceable federal regulations are established, consumers and private organizations must educate themselves about what is being marketed. Two private nonprofit organizations, Green Cross Certification Company and the Green Seal, test products and affix "seal of approval" logos if the product passes the tests. Similar organizations exist in other countries, such as the German Blue Angel program, Canada's Environmental Choice, and Japan's Eco-mark. Unfortunately, certification by the Green Cross and Green Seal is largely voluntary: many companies never submit products for certification. In addition, the United States has an Energy Star label that certifies and approves various green products; begun in 1992, various computer companies were the first to join. As of 2009, more than 17,000 public and private partners were part of the program, with items ranging over 60 different product categories.

Consumers should also be aware that many recycling logos on boxes and cans do not mean very much. For example, consider the familiar logo with arrows in a triangle that appears on many food and paper products. The arrows alone indicate that the product is recyclable; arrows within a circle mean "made from recycled materials." However, this logo alone does not indicate the amount of recycled material, how easily recycled the product is, or other key information. For example, many products are recyclable but are never

VISIT

http://environment.jbpub.com
/mckinney/5e/
for more information

Is This Recycled or Not?

recycled because the manufacturer does not repurchase its own recycled material. The recycling loop remains "unclosed."

If you want to be a green consumer, here are a few basic rules: avoid needless shopping (conservation is always the "greenest" activity you can do). Avoid vague environmental claims such as "environmentally safe." Where possible, try to buy products certified by the Green Cross or Green Seal. Seek out minimal packaging and reusable containers, and look for specific information that a company may provide such as "made

from 30% postconsumer recycled fiber" (this is preferable to a product made of, for example, 10% postconsumer recycled fiber).

Questions

1. What does "biodegradable" really mean? Do you think that paper is more biodegradable than plastic? Why?
2. What are some basic rules for being a green consumer? How often do you really try to follow these rules when shopping?

Indeed, paper is already the largest component of city garbage and sometimes takes up as much as 50% or more of landfill space. Current applications of vegetable starch in the creation of plastic-like utensils, plates, and takeout containers represent an exciting application of human ingenuity to solve environmental challenges, preserve resources for future generations, and reduce waste.

6.9 Resource Economics

All environmental problems are closely intertwined with economic causes. This is especially true of resources because their extraction and use are directly determined by their profitability.

Resource Overuse: Ignoring Environmental Costs

A basic reason resources are overused is that they are too cheap. The price paid for metals, timber, petroleum, and many other virgin natural resources does not reflect their true environmental costs. These true costs would include esthetic, emotional, environmental services, ethical values, and long-term economic values. The price paid by consumers for natural resources now usually omits many of these. For

instance, the price of metals rarely incorporates many of the environmental costs of mining: costs of water pollution on nearby aquatic ecosystems, esthetic loss from a cratered landscape, and so forth. In our review of environmental science, we saw how many environmental economists suggest that green fees, such as taxes, can incorporate environmental costs into natural resource use. Mining companies, manufacturers, and consumers all respond to higher prices by increasing efficiency, reusing/recycling more, and substituting more. It is difficult, under the current economic system, to evaluate how much an ecosystem, such as a coral reef (see **CASE STUDY 6-3**), contributes in numerical value to the economy; however, there are measures that have been developed to try and quantify these services that the environment provides. Another way to incorporate long-term and indirect values is to use basic market principles: consumers can demand sustainable products such as those made with recycled materials. When the profit made from such sustainable activities exceeds the profit made from the destruction of resources, sustainable activities will be rewarded. Currently, this is rarely the case, and unsustainable activities are more profitable.

The Economic Value of Coral Reefs

VISIT

http://environment.jbpub.com
/mckinney/5e/
for more information

Even though coral reefs occupy less than one percent of the Earth's marine environment, they are home to more than a quarter of all known marine fish species and tens of thousands of other species found nowhere else on earth. Reefs, in addition to tremendous biodiversity, provide numerous benefits or "ecosystem services" to people. Critical economic and social benefits associated with healthy coral reefs include high fishery yields, high tourism-related incomes, protection from coastal erosion, and good nutrition for coastal communities. Reefs also serve as a buffer against incoming storm surges, protecting local communities. Because of their tremendous biodiversity, reefs are also being actively explored for bioactive compounds for pharmaceuticals. In fact, a few high-value products have already been discovered.

However, as reefs continue to be degraded by human activities, such as overfishing, agriculture, and urban development, their ecosystem and economic benefits will continue to decline as well. Degradation of reefs results in a loss of fishing livelihoods, protein deficiencies and the increased potential for malnutrition, loss of tourism revenue, increased coastal erosion, and the need for investment to stabilize the shoreline. This destruction is occurring because the ecosystem services are not often valued in economic terms.

In 2005, the World Resources Institute began working in St. Lucia, Tobago, and Belize to more accurately measure the value of coral reefs in the Caribbean and to identify incentives for decision-makers to reduce threats to coral reefs. The Economic Valuation project seeks to measure the value of reefs in three areas: tourism, fisheries, and shoreline protection.

WRI commenced a study that looks at only three out of the many culturally and economically valuable services provided by these ecosystems in Belize, although there are many other goods and services that coastal and marine ecosystems provide to countries in the Caribbean. Even within this narrowed scope, this study finds that the country's coastal resources are extremely valuable. Belize's coral reefs and mangrove-lined coasts provide critical protection against erosion and wave-induced damages from tropical storms; they have supported artisanal fishing communities for generations; and they stand at the center of vibrant tourism industry, drawing snorkelers, divers, and sport fishermen from all over the world. The World Resources Institute (WRI), in collaboration with WWF Central America, assessed the economic contribution of these services at the national level and within individual Marine Protected Areas in Belize.

Coral reef- and mangrove-associated tourism contributed an estimated $150 million to $196 million to the national economy in 2007 (12% to 15% of GDP). Fishing is an important cultural tradition, as well as a safety net and livelihood for many coastal Belizeans. Annual economic benefits from reef and mangrove dependent fisheries is estimated at between $14 and $16 million. Reefs and mangroves also protect coastal properties from erosion and wave-induced damage, providing an estimated $231 to $347 million in avoided damages per year. By comparison, Belize's GDP in 2007 was $1.3 billion. These estimates capture only three of the many services provided by coral reefs and mangroves, and should not be considered the "total" value of these resources. These numbers should be regarded as a lower bound estimate.

Despite their importance, these benefits are frequently overlooked or underappreciated in coastal investment and policy decisions. Unchecked coastal development, overfishing, and pressures from tourism threaten the country's reefs, with the additional threats of warming seas, fiercer storms, and other climate-related changes looming on the horizon. Fish populations, including commercially valuable sport-fishing species and colorful reef fish, will diminish if they lose the mangrove forests they rely upon as critical nursery habitats. Coastal properties will become increasingly vulnerable to storms and erosion, and reef-related tourism will suffer as reefs and mangroves decline. Belize's government, NGOs, and private sector have begun to recognize the importance of coastal ecosystems to the economy. Nevertheless, the amount currently invested in protecting Belize's coral reefs and mangroves is small when compared to the contribution of these resources to the national economy.

Source: Adapted from: http://pdf.wri.org/coastal_capital_belize_brochure.pdf

Economics of Recycling

Recycling is an excellent example of the key role that economics plays in promoting resource conservation. Although about 80% of U.S. household trash can be recycled as compost, aluminum, paper, and many other components, only about 32% of municipal solid waste was being recycled in 2005 because the economic incentives were lacking. Returning used items for collection is not enough. The recycling loop is closed only if collected items are used again to make a product that is repurchased. The loop will often go unclosed if products made from recycled materials cost more than products made only from virgin resources. Consumers will often choose to buy the cheaper product, and the waste people thought they returned for recycling will be taken to landfills because no one will buy it.

Widespread recycling is unlikely to occur until it is economically feasible, even though the number of U.S. communities requiring recycling has risen dramatically and nearly half of all U.S. communities now have recycling programs. However, as many communities have learned, simply passing laws requiring recycling is not enough. Seattle, Washington, which resolved to recycle 60% of its garbage by 1998, is a good example. By 1993, Seattle had the highest rate of recycling of any major U.S. city, with 42% of its garbage being recycled and 90% of all single-family homes participating. By charging an extra fee for each barrel of garbage left for pickup, the city encouraged people to sort their waste and deposit it in recycling containers. Unfortunately, the city has had difficulty closing the loop on some materials, such as glass and paper, which were sometimes sent to landfills. Although the average price of used newspapers dropped from $20 to $5 per ton (0.907 metric tons) in 1991, it still was often cheaper to make paper from trees, so no recycler would buy the newspaper. However, more recently the price of paper has risen, and the demand for recycled paper has increased, especially since the federal government now requires all of its offices to use recycled paper. As becoming green is increasingly more important to cities across the nation, many strive to set the bar higher. In 2009, the city of San Fran-

cisco reached a national record for any city, recycling 72% of all collected waste, which includes waste from businesses and construction sites. Through enforcement of recycling mandates and an infrastructure to process recycling and composting of a wide variety of materials, San Francisco is striving for a "zero waste" goal by 2020.

Why is recycled material sometimes more expensive than virgin material? A major reason is the cost of energy, including the energy of human labor. The cost of **embodied energy**, or the energy used in producing a product, is often the main factor determining the product's retail cost. In the case of recycled paper, the machines used to make paper are actually smaller, so the price of running the machines is passed along to the consumer. To date, with increased competition and demand for recycled paper, recycled paper is often less expensive than virgin paper, and when it is more expensive, such as in high-end copier paper, it is about 2–3% more, with total paper budgets for 100% recycled paper differing by not more than 10%. The energy consumed in recycling some materials is much less than the energy consumed in processing virgin material. For example, mining and processing aluminum and many other metals are very energy intensive. Because recycling aluminum uses as much as 95% less energy, there is a strong cost incentive to reprocess it and close the loop. In everyday terms, recycling just one aluminum can saves enough energy to run a television set for 3 hours. The aluminum recycling industry is one of the fastest growing industries in the United States; it earns more than $1 billion annually and recycles an astounding 62 billion aluminum cans (enough to circle around the earth 171 times) per year, equal to about 350 cans per U.S. citizen. In the last 50 years, the aluminum industry has cut its energy use from 12 to 7 kilowatt hours needed to make one pound of aluminum. The aluminum industry in the United States is a $39.1 billion industry and growing. Combined with other recycling operations, studies show that recycling creates four jobs for every one job created by waste management and removal alone.

This example shows how both the environment and jobs can benefit from recycling. However, recycling other materials, such as newspaper and glass, saves considerably less energy than recycling aluminum. When the costs of human energy, such as collecting and sorting the newspaper and glass, are added, the slim profit margin may disappear, and the loop may sometimes be unclosed, as in the Seattle example.

The problem of closing the recycling loop has many possible solutions. One option is to appeal to consumers to purchase recycled products, even if they cost more. For instance, some students buy recycled notebook paper even if it is more expensive because they want to encourage recycling. To encourage such buying on a wide scale, society can levy higher "green taxes" on virgin resources to boost the price of products made from them. If the taxes are high enough, the recycled product will be cheaper, and the loop will be closed through consumer demand. Finally, human ingenuity and innovation may devise new recycling processes that lower costs or new products that appeal to consumers. For example, the growing problem of recycling used tires has led to dozens of products that incorporate them, from sandals and tennis shoes to asphalt. When products that people want are created, there is an economic incentive to close the loop. Many businesses are finding new uses for plastics, such as stuffing for ski jackets and even "lumber" to build benches, fences, and playground equipment.

6.10 Jobs and Life in a Sustainable World

A sustainable world meets today's needs without reducing the quality of life of future generations. Modern resource exploitation is not even close to being sustainable. **FIGURE 6-9** illustrates how the consumption of virgin raw materials such as minerals has increased much faster in the United States than the population. Since 1900, materials consumption increased about six times faster than population growth. The average American today consumes six times more resources than the average Ameri-

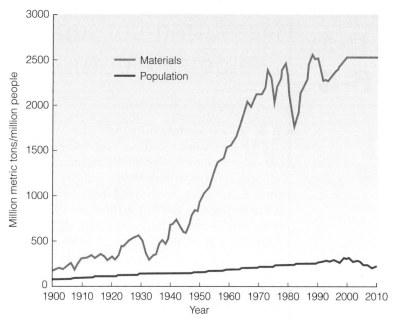

FIGURE 6-9 Consumption of materials in the United States has grown much faster than the population, according to the latest estimates for which accurate data are available. Note that 1 metric ton = approximately 1.102 tons. (*Source:* U.S. Bureau of Mines.)

can in 1900. Americans today have a very large **ecological footprint**. The term refers to how much of the planet is used to support one person's lifestyle (see **CASE STUDY 6-4**).

FIGURE 6-10 shows that many jobs in the U.S. high-throughput society are concentrated at both ends of the flow of materials through society—in resource extraction and waste cleanup. This leads to unsustainability because resources are being used at a vastly

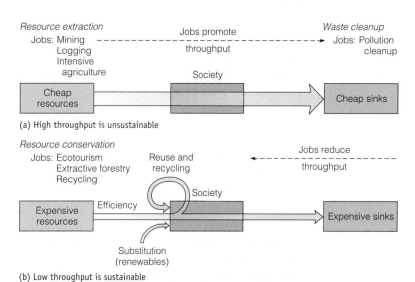

(a) High throughput is unsustainable

(b) Low throughput is sustainable

FIGURE 6-10 (a) Cheap resources and environmental sinks promote an unsustainable, high-throughput economy with jobs concentrated in resource extraction and waste cleanup. (b) Expensive resources and environmental sinks promote a low-throughput, sustainable economy with jobs concentrated in resource conservation (input reduction) activities.

Measuring Consumption: What Is Your Ecological Footprint?

We all depend on the natural environment for resources such as food, energy, the absorption of waste products and other life-support services. Sustainability requires that nature's productivity not be used more quickly than it can be renewed and that waste is not discharged more quickly than nature can absorb it. The ecological footprint that Mathis Wackernagel and William Rees developed is an increasingly popular way of measuring the sustainability of our lifestyles.

The ecological footprint measures how much land is required for people to support their lifestyle. For example, how much land is required to grow the crops for the food you eat and cover the housing you occupy, the roads you use, and the waste you produce? As you might expect, the footprint varies among nations and generally increases with rising wealth and consumption. Scientists usually report these data in global hectares. One global hectare is a measurement of the average biocapacity of all hectare measurements of any biologically productive region on the planet. A hectare is equal to 2.47 acres. For the purposes of our discussion, global hectares (gha) are used.

Research from the Global Footprint network and the European Environmental Agency indicates that Earth's biologically productive area is just over 11 billion hectares. In 2007, the human ecological footprint reached 2.7 gha per person, which is 24% more than the 1.8 gha per person footprint that scientists believe is the tipping point beyond which the Earth's natural systems might not be able to keep up with the resource depletion created by humans. At our current state of impact, humans are depleting the Earth's resources much faster than Earth can restore and balance sustainability. All Earth's inhabitants are not equally responsible for the detrimental effects currently documented. If you divide all the nations of the world into three main groups based on income data, you would find that the average person in a high-income country has a total ecological footprint of 6.4 gha; whereas, an average citizen of a middle-income country has an ecological footprint of 1.9 gha; and an average person in a low-income country has an ecological footprint of 0.8 gha. Although populations in poorer countries tend to be larger, resource consumption per person is much less. The average U.S. citizen has a total ecological foot print of 8.0 gha, whereas a Canadian citizen has a corresponding total ecological footprint of 7.1 gha. Elsewhere on the globe, the average European has a total ecological footprint of 4.7 gha; a person living on the continent of Africa has a total ecological footprint of 1.1 gha; someone living in Latin America, the Middle East, or Central Asia has a total ecological footprint much less at around 2 gha. In fact, if everyone on earth lived like the average U.S. citizen, it would require at least three Earths to provide all of the material and energy that currently used by the U.S. citizen. Estimates show that the ecological footprint of today's global consumption in food, forestry products, and fossil fuels alone might already exceed global carrying capacity by roughly 20–30%. The differences in consumption are staggering when you consider

higher rate than natural processes are replacing them. Figure 6-10 also shows what an economy could look like in a low-throughput, sustainable world. Virgin resources would be more costly, reflecting their true long-term value and promoting highly efficient use of resources, reuse/recycling, and substitution of more abundant and more renewable resources. Jobs would be concentrated at the input end of the throughput flow and would involve sustainable resource management activities, such as ecotourism, recycling, and the renewable energy industry (**FIGURE 6-11**). These and similar activities reduce throughput and slow nonrenewable resource depletion and pollution. In contrast, in a high-throughput society today, jobs actively promote nonrenewable resource depletion and pollution.

In many cases, sustainable activities produce more jobs than unsustainable ones. Extracting resources often produces few jobs. The U.S. Recycling Economic Information Study pub-

VISIT

http://environment.jbpub.com
/mckinney/5e/
for more information

a recent U.N. report that indicated the wealthiest one fifth of the world's population consumes 45% of all meat and fish, whereas the poorest one fifth consume less than 5%. The wealthiest one fifth of the world's population also consumes 58% of the total energy and 84% of all the paper produced worldwide, and owns 87% of all vehicles. The poorest one fifth consumes less than 4% of all energy produced, uses about 1.1% of all the world's supply of paper, and owns less than 1% of all vehicles.

Of course, even within nations, personal life-style choices have a strong influence on a person's ecological footprint. Do you drive a large car? How far do you commute? For example, compare how much ecologically productive land is necessary to commute by bicycle, bus, or car. Most of the car's land is required to absorb CO_2. Most of the biker's land is required to provide the extra food for quenching the biker's hunger. Obviously, the biker has a much smaller footprint for travel needs. The same is true for many other choices, such as housing, food, energy, and water consumption. For example, someone who takes longer showers is increasing his or her ecological footprint compared with someone who does not.

In general, as personal consumption increases, as it has been doing in wealthy nations during the past decades, so does one's ecological footprint. In recent years, movement toward better recycling methods, energy conservation, and an increase in human innovation, such as cradle-to-grave design methods, all serve to curtail the current out-of-balance use of global resources. One definite challenge globally is for developing nations to refrain from repeating prior mistakes as they strive for secure footholds in a global economy. As people join forces to curtail overconsumption, it stands to reason that the efforts to curb resource depletion need to have a larger impact on the nations most responsible while at the same time delivering the technology and resources needed to assist developing nations in the decisions they are making that will influence consumption and ecological footprints of the generations ahead.

Questions

1. What, exactly, does the ecological footprint measure? How can it be reduced?
2. Must wealth always produce a larger ecological footprint? Explain.

lished in 2001 indicated that the United States has more than 56,000 recycling and reuse centers that employ more than 1 million people, with an annual $236 billion in revenue. Approximately 2.7% of the gross domestic product (GDP) is related directly to the recycling and reuse industry. One of every three green jobs is in recycling reuse and remanufacturing, which includes upcycling (taking industrial "waste," such as materials scraps, and turning them into something more valuable, such as like rugs). Ecotourism and its support services, such as motels, guides, restaurants, and travel venues, are examples of sustainable practice that will grow an economy over subsequent generations. In recent Cradle to Cradle design models, large-ticket items such as cars would

FIGURE 6-11 "Solar farms," such as this one in California, can potentially provide many high-tech careers.

TABLE 6-1	Some Suggestions for Creating a Sustainable Society

Environmentally harmful (unsustainable) activities can be solved in many ways. All of these solutions use efficiency, recycling, and substitution promoted by economic incentives to reduce throughput.

Harmful Activity	Example	Solutions
Extraction	Mining high-grade ore and then moving on to a new site because the land is artificially cheap, while ignoring lower grade ore on the already-disrupted site	Higher land prices (through elimination of subsidies and addition of full costs of environmental disruption) would increase incentives to use more efficient extraction technologies, reducing the area of land disrupted.
Manufacture	Making paper from 90–100% virgin wood fiber	Most paperboard, paper packaging, and office paper can be made with less than 50% virgin input with no loss of quality, potentially saving millions of trees each year.
Product design	Designing discount products—from umbrellas to televisions to houses—that compete for low retail prices but do not last	Design emphasizing durability and repairability would reduce the number of times the consumer has to replace the product and would thus reduce materials consumption.
Community development	Planning communities in which residences are far from workplaces and services	Planning that puts people closer to what they need and with efficient use of already-developed land would reduce the use of cars and thus the need for materials-intensive construction projects such as roads and bridges.
Direct consumption	Stressing immediate convenience of consumption and disposal as the ultimate good, without considering the prospects for sustainable consumption	Making changes in our consumption patterns to promote a culture of conservation: Copying on both sides of the page, using canvas shopping bags, reading books from the library instead of buying new copies, and taking public transportation could ultimately save both money and materials.

Source: Modified from J. E. Young, A. Sachs. Worldwatch paper 121. The Next Efficiency Revolution. Washington, D.C., Worldwatch Institute, 1994. Copyright 1994. Used with permission of Worldwatch Institute.

be made from zero-waste materials such that as a car gets older or a person's needs change (like trading in a sports car for a minivan), the consumer could trade the car in and each piece of the car could be re-assembled or composted. Having a factory that produces a wide range of vehicles from interchangeable parts would save money, resources, and provide more stable employment. In Europe, this idea is being put into practice in the "End of Life Vehicle Directive" supported by Cradle to Cradle innovations.

TABLE 6-1 lists many other specific opportunities for jobs to make society more sustainable.

■ study guide

SUMMARY

- Resources are raw materials that society uses.
- The need for resource management is inescapable.
- Resources should be managed for both living and future generations.

- Preservation, conservation, and restoration manage resources for future generations.
- The "five Es" (esthetic, emotional, economic, environmental services, ethical) describe five key values of resources.

- Dispersion depletes matter resources.
- Energy resources are depleted when energy is changed to a less usable form.
- As a resource is depleted, we should conserve, not intensify, its use.
- Net yield and maximum sustainable yield (MSY) represent intensification of resource use.

- Three ways to conserve resources are raise efficiency, recycle/reuse, and substitute.
- Conservation typically is discouraged because natural resources are undervalued.
- Sustainable industries not only conserve resources but often produce more jobs.

KEY TERMS

anthropocentric
artificial ore
benefit–cost analysis
conservation
direct values
ecocentric
ecological footprint
economic values
embodied energy
emotional values
energy intensity index
environmental service values
environmental externalities
esthetic value
ethical value
extrinsic values
five Es
Hubbert's bubble
indirect values

intrinsic value
law of diminishing returns
maximum sustainable yield (MSY)
Nature Conservancy
net yield
nonrenewable resource
optimum sustainable yield (OSY)
precautionary principle
precycling
preservation
recycling loop
renewable resource
reserve
resource
restoration
tragedy of the commons
true environmental costs
virgin resources

STUDY QUESTIONS

1. How are "matter" resources depleted?
2. How are cost–benefit analyses sometimes used to analyze environmental problems?
3. What does conservation seek to minimize?
4. What is the "tragedy of the commons"?
5. List the "five Es."
6. Distinguish between extrinsic and intrinsic values. Give examples of both.
7. Distinguish between indirect and direct values. Give examples of both.
8. What are "true environmental costs?"
9. Distinguish between reserves, resources, renewable resources, and nonrenewable resources.
10. How is depletion of matter resources different from depletion of energy resources? Give examples.
11. How has the past pursuit of net yield incurred environmental costs?

12. What are the two main causes of the bubble pattern?
13. Distinguish between OSY and MSY.
14. MSY seeks to keep the harvest rate equal to what? How is MSY related to carrying capacity?
15. What are the criticisms of MSY?
16. List and briefly describe some basic ways to conserve and reduce the need for resources.
17. What is the energy intensity index? What does it mean? What is the historical reason the United States has a much higher index than Japan?
18. Give an example of a commonly recycled material in the United States. What materials are rarely recycled?
19. What happens to throughput in a sustainable world? What kinds of jobs are promoted?

WHAT'S THE EVIDENCE?

1. The authors assert that many resources, when used unsustainably, exhibit a bubble pattern of depletion because of exponential exploitation and exponential depletion. Cite some examples of the bubble pattern. Can you think of a resource that is currently being exploited or could potentially be exploited that might not fit the bubble pattern? If a resource does not fit the bubble pattern, does that necessarily mean that is it being used sustainably?

2. The authors suggest that in many cases, long-term economic and other gains (e.g., environmental services and esthetic values) can be increased by reducing resource use, at least in the short term. Can you cite examples either supporting or countering the authors' suggestion along these lines? Is this just a matter of how one defines "long-term economic and other gains" (perhaps one does not put any value on esthetic aspects of nature)?

3. The authors state that environmental sustainability often produces more jobs than unsustainability? What evidence is given to justify this? Can you think of evidence that contradicts this?

CALCULATIONS

1. Assuming an energy intensity index of 14.5 billion BTUs per $1 million GNP for the United States, how would this index change if the U.S. gross national product remained constant while energy consumption was reduced by 50%? How would it change if U.S. energy consumption increased by 10% while the gross national product remained constant?

2. Packaging accounts for 33% of U.S. municipal solid waste. In 2003, the average American discarded approximately 2.05 kg (4.43 pounds) of trash per day. How many kilograms of this 2 kg was packaging? How much of the 2 kg would be packaging if packaging were reduced to 20% of municipal solid waste?

ILLUSTRATION AND TABLE REVIEW

1. Based on Figure 6-3, what was the peak U.S. oil production (around 1970)? How much did this differ from Hubbert's predictions of the 1950s? For the decade from 1980 to 1990, were Hubbert's predictions more accurate than for the previous decade?

2. Based on Figure 6-9, what was the U.S. annual consumption of materials in 1950? In 1960? By what percentage did materials consumption increase between 1950 and 1960?

http://environment.jbpub.com/mckinney/5e/

http://environment.jbpub.com /mckinney/5e/

Connect to this text's website at http://environment.jbpub.com/mckinney/5e/. This site features eLearning, an online review area that provides quizzes, chapter outlines, and other tools to help you study for your class. You can also follow useful links for more in-depth information.

Traditionally, energy problems have been addressed simply by harnessing more energy—increasing the usable energy supply. This strategy involves both increasing the rate at which current technologies can produce energy and developing new energy-producing technologies. Another way to address the world's energy concerns is from the demand side. Attacking the demand side of energy use need not lead to a reduction in living standards or a change in lifestyle for the worse. Investing in an infrastructure that supports increased use of railways for mass transit is a cost-effective way for governments to reduce fossil fuel consumption by commuters and leisure travelers.

Fundamentals of Energy, Fossil Fuels, and Nuclear Energy

7

Chapter Objectives

After reading this chapter, you should be able to explain or describe the following:
- A brief overview of global energy use
- An overview of U.S. energy sources and consumption
- Electrical power plants and their functioning
- The concepts of hard and soft energy technologies
- The relative importance of the various fossil fuels (petroleum, coal, and natural gas)
- Pros and cons of using fossil fuels
- The principles, advantages, and disadvantages of nuclear power

Human society is dependent on a continuous flow of energy; indeed, all living organisms require **energy** (for a discussion of what energy is, see **FUNDAMENTALS 7-1**). Without energy, life could not exist. One of the prime concerns of any nation is to ensure that its citizens have ready access to the energy they need, whether that energy takes the form of food, heat for a home, power to drive machinery, electricity to run appliances, or gasoline to fuel an automobile. As technology has advanced, the amount of energy that humans use has increased dramatically. Energy consumption per capita in a technologically developed society today is at least 100 times greater than it was when humans first evolved and about 6 times greater than it was less than 200 years ago. Of course, more people are living on the Earth today than at any one time in the past, so total energy use has increased even more dramatically than per capita energy use: it has been estimated that today the world as a whole uses 70 times as much energy as it did in 1865.

Energy is necessary for industrial development; thus, most of the world's energy flow historically has been through industrialized countries. Recent data from governmental agencies indicate that a person in the United States uses 20 times more energy than a person in Australia (see **CASE STUDY 7-1**). A person in Australia uses 4 times the amount of energy as someone living in Ecuador, and someone in China uses more than 10 times that of a South African. Certainly, this massive use of energy has brought material prosperity to the industrialized world but at a substantial cost. The more energy people use, the greater their detrimental impact on the environment. One could argue that virtually all our environmental problems can be traced to energy use (or overuse). (Remember the equation $I = P \times C$ [Impact equals Population times Consumption] introduced in this text's overview of environmental science.) The use of **fossil fuels** (oil, **coal**, and natural gas) is the major source of air pollution, acid rain, and greenhouse gases. Mining for coal or pumping oil can destroy extensive tracts of land. Dams necessary for **hydroelectric power** production can wreak havoc on natural ecosystems and spread disease. Mining and processing of uranium to fuel nuclear power plants is fraught with hazards. Finally, conventional fossil fuel sources—on which we rely most heavily for personal and industrial purposes—are nonrenewable and are in the long run a limited resource.

Thus, humans' energy consumption raises two basic concerns: (1) Where will the energy come from in the future as we deplete the most convenient, high-quality supplies of fossil fuels on which much of modern industrial society is based? (2) How can we avoid the environmental degradation so often associated with the levels of energy consumption that characterize modern civilization? We can take two basic approaches to these energy problems: we can address the supply side or the demand side. We can view energy as a commodity that we produce and then attempt to maximize production of useful energy (such as refined oil or usable electricity); furthermore, some people suggest that we use more energy to solve any environmental problems that energy use may initially cause. Alternatively, we can consider energy to be a valuable resource and attempt to minimize its consumption, such as by increasing the efficiency of our energy-consuming devices, and thus minimize any detrimental environmental effects that energy use may entail.

7.1 An Overview of the Current Global Energy Situation

Worldwide consumption of energy has increased dramatically since the middle of the 20th century (**FIGURE 7-1**), and it continues to increase, although at different rates in different parts of the world (see **Figures 7-1a** and **7-1b**). Worldwide, total energy use is projected to grow from 412 quadrillion British thermal units (quads; see **Fundamentals 7-2**) in 2002 to 553 quads in 2015 and 645 quads in 2025. This includes the consumption of **petroleum** products (liquid petroleum products generally, often referred to as "oil"), coal, hydroelectric power (the major component of "renewables" on **Figure 7-1d**), nuclear power, geothermal power, wind power, and so on. In addition to commercial energy consumption, many people consume traditional fuels, such as fuelwood, charcoal, and animal and plant wastes. According to the United Nations in 2011, an estimated 2 billion people did not have access to modern energy services, and more than 3 billion people relied on traditional biomass for cooking in 2006. However, such traditional fuels are estimated to supply less than 10% of the energy supplied by the combined sources listed previously. The largest per capita consumptions of traditional fuels are in South America and Africa. In developing nations, traditional fuel use can lead to severe environmental degradation. One of the world's poorest countries, the island nation of Haiti, has just 1% of its forest remaining because Haitians have little else to use as fuel.

Locally and regionally, the energy mix may differ significantly from that of the world as a whole, and even on a global scale, there have been changes in the mix in recent years. As an example, until 1998, more coal was used on a global scale than natural gas, but since then, the use of the two energy sources has been about even worldwide (see **Figure 7-1d**). In many developing countries, coal remains dominant. In the Far East, several times as much coal is used as natural gas (in terms of energy equivalents). China, in particular, burns more coal than oil, but that is changing. China's industrialization and economic growth are increasing so rapidly that the country is now one of the largest importers

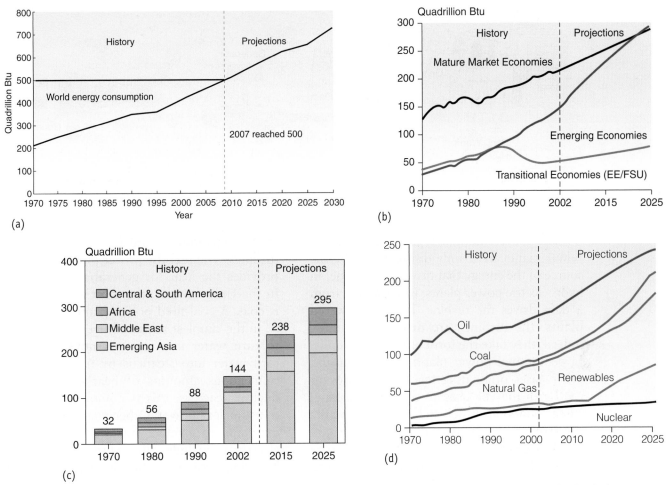

FIGURE 7-1 (a) World energy consumption 1970 through 2003, with projections to 2030. (b) World energy consumption by region, 1970 through 2002, with projections to 2025. In this graph, "Mature Market Economies" refers basically to North America, Western Europe, Japan, Australia, and New Zealand; "EE/FSU" is Eastern Europe and the former Soviet Union. (c) Energy use in the emerging economies by region, 1970 through 2002, with projections to 2025. (d) World energy consumption by fuel type, 1970 through 2002, with projections to 2025. (*Sources:* Adapted from United States Department of Energy, Energy Information Administration, International Energy Annual, 2002, 2003 (May–July 2005), 2005 and System for the Analysis of Global Energy Markets, 2005 and 2006.)

of oil (fourth largest by some estimates). The competition for the world's resources is intensifying as more countries industrialize.

7.2 Power Plants—Supplying the People with Electricity

In modern, industrialized societies such as the United States, much of the fossil fuel energy consumption goes to power vehicles and machinery directly or to heat buildings. Another large portion of the energy sector produces the **electricity** distributed to consumers for running electrical appliances, lighting, heating (to some extent), air conditioning, and so forth. Although electricity provides only about 12% of the world's energy, it is often viewed as indispensable to more developed

countries. Electricity is most often generated at electrical **power plants,** usually operated by private or semipublic utility companies, and then distributed to consumers.

Commercial power plants can take many forms, some of which are described in more detail later in this chapter and in the discussion of renewable and alternative energy sources. Using a turbine to drive a **generator** transforms mechanical energy into electrical energy. A **turbine** is simply a machine, something like an old-fashioned water wheel, that can convert the lateral motion of a flowing fluid (such as water, steam, or some other gas) into the rotational or turning motion of a shaft or axle (**FIGURE 7-2**). This turning motion can then be used to turn the generator that transforms mechanical energy into electricity.

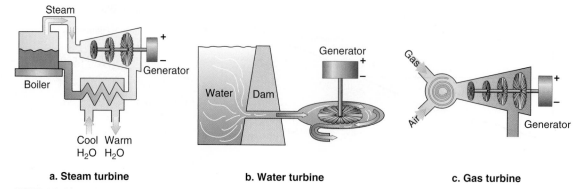

a. Steam turbine **b. Water turbine** **c. Gas turbine**

FIGURE 7-2 Using steam, water, or some other gas to spin a turbine, which in turn drives a generator, generates electricity.

The more common types of commercial power plants are mainly distinguished by the source of the energy that drives the turbine. In hydroelectric power plants, water falling from a dam drives the turbine. In wind-powered plants, the turbines are driven by wind—indeed, they take the form of huge windmills. In geothermal power plants, hot steam and other gases originating from the interior of the Earth may be used to drive the turbine. Most commonly, however, large power plants use artificially created steam. Water is heated to temperatures in excess of 300°C (approximately 600°F), and the resulting pressure of the expanding steam is used to turn the turbines. The fuel the power plant consumes provides the heat for generating the steam. In the United States, the most common fuel is coal. A coal-fired power plant (**FIGURE 7-3**) is, in the simplest sense, a huge furnace used to turn water into steam. Another way to turn water into steam is by harnessing the tremendous amounts of heat given off by nuclear reactors; this is commonly referred to as nuclear power. Other methods of heating water to produce steam, and ultimately electricity, include burning wood or other biomass, burning oil, burning natural gas, or

FIGURE 7-3 A coal-fired power plant, Shawville, Pennsylvania.

CHAPTER 7 Fundamentals of Energy, Fossil Fuels, and Nuclear Energy

Energy, Work, and Power

What is energy? To a physicist, *energy* is simply the ability to do work. But what is work? **Work (W)** is defined as a force (*F*) applied to a material object times the distance (*d*) that the material object is moved. This is expressed in the equation

W = Fd (Work equals Force times Distance)

A commonly used unit of force is the newton (named after Sir Isaac Newton), which is defined as the force that will accelerate a mass of 1 kg 1 m per second per second (in a vacuum with no frictional resistance). Distance can be expressed in meters. A force of 1 newton times 1 meter is the unit of work known as the **joule (J)**. Energy has the same units as work; thus, the energy that makes it possible to do a joule of work is a joule of energy.

Energy, the capacity to do work, can take many different forms and be stored in many different ways. Mechanical energy involves objects, their motion, and position. One form of mechanical energy is known as potential energy—energy stored by virtue of the position of an object. A brick raised over your head before you drop it has potential energy, as does the wound spring of a watch or the water behind a hydroelectric dam built across a river. When you drop the brick or the water is released from the dam, the potential energy is turned into kinetic energy, or the energy of motion. The water dropping from the dam can be used to turn the blades of a turbine. When the turbine is connected to a generator, it will convert the mechanical energy to electrical energy. Electrical energy involves the forces of charged particles, such as electrons, acting on one another.

Energy can take the form of heat, which is the random motion of atoms and molecules in a substance. As the brick you drop hits the ground, or the water of a waterfall hits bottom, the brick and the ground, or water, will heat up—the macroscopic kinetic energy will be converted to energy at an atomic and molecular level. You can demonstrate this convincingly by pounding hard on a nail with a steel hammer and then immediately feeling the heat radiating from the nail's head.

Often energy of any form is expressed in units based upon heat energy. One **calorie** (also known as a gram-calorie, and commonly abbreviated cal) is the quantity of energy that when converted completely to heat will warm 1 g of water 1 degree Celsius (or centigrade, commonly abbreviated C). A kilocalorie (or kilogram-calorie, commonly abbreviated kcal) is equal to 1,000 calories; sometimes the term Calorie (with a capital C) represents a kilocalorie, as in the rating of foods by energy con-

tent. One kilocalorie is equal to 4,184 joules of energy. Another commonly used energy unit is the British thermal unit (**BTU**). One BTU is the amount of energy that when converted to heat will raise the temperature of one pound of water 1 degree Fahrenheit (F). One kilocalorie is equal to 3.968 BTU.

When discussing energy consumption or production on a national or global scale, very large-scale units are used. Some of the more common units are listed in Fundamentals 7-2. Two frequently used units are the "**quad,**" or quadrillion BTU (Q), and the **petajoule** (1×10^{15} joules, abbreviated PJ). One quad is equivalent to approximately 1,054 petajoules. Another common unit is the metric ton of oil equivalent (toe), equal to 41.868 gigajoules. Thus, 1,000 metric tons of oil equivalent (1,000 toe) is equal to 41,868 gigajoules or 0.041868 petajoules, and a million metric tons of oil equivalent (1,000,000 toe) is equal to 41.868 petajoules. As an example, in 1999, the United States consumed nearly 38 quads worth of petroleum, or the equivalent of about 40,000 petajoules of oil (this represents about 950 million metric tons of oil equivalent, or 6.5 to 7 billion barrels of oil—the numbers are staggering). Data from recent calculations estimating the total amount of oil used worldwide since the late 1800s is at least 135 billion barrels, with some estimates putting the figure closer to 1 trillion.

Other forms that energy takes include the energy stored in the chemical bonds of substances (such as the energy in the food you eat or the gasoline that fills the tank of an automobile), the energy of light (electromagnetic radiation), and the energy that binds the particles composing the nucleus of an atom. Energy is the capacity to do work, but this is not the complete story. How fast work is done—how fast energy is converted from one form to another—is the concept of power. **Power** is defined as work (requiring energy) divided by the time period over which the work is done:

Power = Work/Time (Power equals
Work divided by Time)

A common unit of power is the **watt,** defined as one joule of work (or energy) per second. A kilowatt (kW) is equal to 1,000 watts. A **megawatt (MW)** is equal to 1,000 kilowatts, or 1,000,000 watts. A **terawatt (TW)** is equal to 1 billion (10^9) kilowatts, or 10^{12} watts. The unit horsepower is equal to three-quarters of a kilowatt, or 750 watts. Thus, a 1-horsepower electric motor can convert 750 joules of electrical energy into mechanical energy per second. If we multiply a unit of power by time, we arrive at another unit for energy. For example, a 1-kilowatt motor running for 1 hour will convert 1,000 joules of electrical energy per second into mechanical energy; because there are 3,600 seconds in an hour, it will convert 3,600,000 joules of energy in a kilowatt-hour (kWh). In other words, 1 kilowatt-hour is equal to 3,600,000 joules, or approximately 860.4 kilocalories (because 1 kilocalorie is equal to 4,184 joules).

The Laws of Thermodynamics

To study energy and energy conversions, one must know the laws of **thermodynamics** (the term thermodynamic means "heat movement"). The first law of thermodynamics is the law of the conservation of energy. Energy can be neither created nor destroyed—it is simply changed from one form to another (of course, it has been demonstrated that matter can be converted into energy, and vice versa, in nuclear reactions). Within an isolated (closed) system, the total quantity of energy will always be the same. However, not all energy is equivalent from a human perspective. One must take into consideration not only the quantity of energy, but also its form or quality. As energy is transformed from one type to another—for instance, from mechanical to electrical energy—the transformation will not be 100% efficient (some of the energy will be lost as waste heat). This leads to the second law of thermodynamics, which states that heat or energy cannot be transformed into work with 100% efficiency, and heat will always flow spontaneously from an object of higher temperature to an object of lower temperature. Another way of considering the second law of thermodynamics is to note that highly organized energy (such as electricity or mechanical energy) always ultimately degenerates into disorganized energy (such as heat). **Entropy**, the concept of the amount of disorder or randomness in a system, always increases. For example, a liter of gasoline represents a relatively ordered, low-entropy system. The large, complex organic molecules in the gasoline store much high-quality energy. When this gasoline is burned in a car, some of the stored energy is converted into high-quality, usable mechanical energy; however, much of the original energy is transformed into low-quality, high-entropy heat, and the gasoline is reordered into simple combustion products such as carbon dioxide, carbon monoxide, and water—randomness and entropy are increased in the system.

According to the laws of thermodynamics, entropy in the universe as a whole is continually increasing. In certain situations, entropy may decrease locally, but only at the expense of increasing entropy somewhere else. Living systems create and maintain order where there was previously disorder, but only by using vast quantities of energy. Plants convert sunlight—a form of energy—into complex biological molecules that contain high-quality energy. These organic molecules are the high-quality energy reservoir on which virtually all other life depends. People, in their quest for energy, are currently highly dependent on the high-quality energy stored in the fossil fuels (oil, coal, natural gas) formed from the organic molecules of countless organisms over millions of years.

As described by the second law of thermodynamics, no energy transformations are 100% efficient. However, from a pragmatic point of view, just how efficient (or inefficient) are our machines that transform energy from one form to another?

Efficiency is the useful work that is performed relative to the total energy input of a system. Thus, an engine that is 25% efficient will convert one-quarter of the energy contained in its fuel to mechanical work. In any engine, some energy is lost as heat because of friction. Furthermore, a power plant or any other heat engine usually achieves the most efficient transformations of energy at a slower rate than people desire. We prefer to have our energy converted quickly so that we can drive faster or heat more houses within a specific timeframe, even if this takes a heavy toll on the efficiency of the energy exchange, meaning that we use considerably more fuel.

using the focused energy of the sun to heat water to high temperatures.

An internal combustion engine or turbine can also be connected to a generator. An internal combustion turbine is essentially an engine that burns oil or natural gas, rather than heating water first. The engine uses the hot gases from the burned fossil fuel directly to turn the turbine. Conventional steam-generating power plants, whether coal burning or nuclear fired, typically are large and expensive. They cost billions of dollars each and take many years or even decades to build and put online. Once built, these large plants are relatively efficient in converting fuel into electrical energy. Conversely, standard internal combustion turbines are smaller, less expensive to purchase, and can be placed into operation more quickly. The drawback is that they are much less fuel-efficient than larger plants, and the fuel they burn—oil or natural gas—usually is more expensive than the fuel (such as coal or uranium) of larger, steam-based power plants. However, newly developed gas turbines and combined-cycle plants are greatly increasing realized efficiencies.

Utilities that produce electrical power must have a considerably larger generating capacity than they actually need at most times. In fact, to ensure reliable production, the local grid of power plants usually must be able to generate approximately 20% more electricity than is normally used at any one time. This extra generating capacity is known as the reserve capacity. Electricity is not used at an equal rate throughout the day or the year. Depending on the area, more electricity may be used during the day when businesses are operating and in hot climates when air conditioners are running at maximum capacity. The amount of electricity needed at the time of highest demand is known as the **peak load**. With current technology, it is very difficult to store electricity, and it is also not possible to turn large electrical power plants off when temporarily not needed. As a result, utilities must be able to satisfy peak load requirements, even if this means generating more electricity than is needed at other times of the day or year.

Another reason that utilities need to maintain a substantial reserve capacity is that at any one time more than 10% of the generating capacity may be shut down for repairs, routine maintenance, or changing the fuel (in nuclear plants). When a plant shuts down temporarily, power is supplied to its customers from the reserve capacity of other plants owned by the utility and connected to the same power grid (the set of power lines that connect power plants to their customers and to each other). Different electrical utility companies may purchase power from each other if their power lines interconnect. However, even with interconnecting lines, transporting electricity over long distances is inefficient and expensive because much of the electrical energy is lost in transit.

When the capacity of a utility's power plants is exceeded, **brownouts** and **blackouts** can occur. In a brownout, consumer demand is slightly more than what the utility can generate. The utility can still provide everyone electricity, but not at 100% of what homes and businesses need. The voltage the consumers receive is inadequate; the lights often dim, and machines do not work properly. In fact, machines and appliances can be ruined if underpowered for too long. Blackouts are often associated with major breakdowns, either of a power plant or in the grid that distributes power. In a classic blackout, a region the utility serves is left without electricity for an extended period of time. Localized blackouts can occur if demand significantly exceeds the utility's capacity to generate electricity. In such an instance, the utility may have to institute a rolling blackout—the utility stops and later restores electrical service to a sequence of neighborhoods in turn, each losing electricity for an hour or two during the day. Such a rolling blackout forcibly reduces the consumption of electricity. Needless to say, brownouts and blackouts are very disruptive to modern societies dependent on electricity.

For example, in August 2003, 21 power plants across the northeast United States and Canada shut down in a cascade effect within the span of just 3 minutes. New York, New Jersey, Vermont, Michigan, Ohio, Pennsylvania, Connecticut, Massachusetts, and the Canadian province of Ontario lost power. Cities affected included New York; Cleveland, Ohio; Detroit, Michigan; and Toronto and Ottawa, Canada. Power was lost for 4 days in some parts of the United States, and approximately 50 million people were affected by the outage. The outage was attributed to overgrown trees short-circuiting power lines, and ultimately, the power companies not recognizing the inadequacies and vulnerabilities of their antiquated power grid system (**FIGURE 7-4**).

FIGURE 7-4 Thousands of stranded commuters walk across the Brooklyn Bridge during a massive power outage in New York on August 14, 2003.

The United States: Sources and Consumption of Energy

Even today, much of the world, especially in developing countries, depends on the traditional forms of energy used for centuries, but modern industrialized Western society seems inconceivable without the intensive energy use to which North Americans and Europeans have grown accustomed. As an example of energy sources and consumption in a modern industrialized nation, let us take a look at the United States.

The United States is both the leading producer and consumer of primary energy in the world. Currently, the United States produces approximately 17% to 18% of the world's energy, but it consumes approximately 22% to 23% of the world's energy; thus, the United States consumes over 25% more energy than it produces. The next two largest producers and consumers of energy are China (producing about 11% to 12% of the world total and consuming about the same amount as it produces), and the Russian Federation (producing 9% to 10% of the world's energy, but accounting for only 6% to 7% of world consumption). Saudi Arabia is the world's fourth leading energy producer (accounting for approximately 5% of world production), yet its energy consumption is just 1% of the world total. Japan produces slightly more than 1% of the world's energy, yet accounts for more than 5% of world consumption.

Energy production and consumption in the United States have generally increased during the last 40 plus years (**FIGURE 1**), but consumption has grown faster than production. This means that energy imports (principally oil and other petroleum products and natural gas) have substantially increased over the decades.

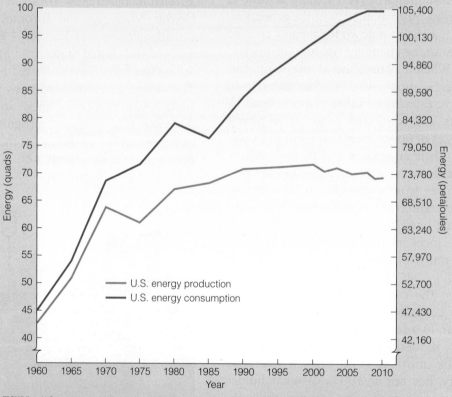

FIGURE 1 U.S. energy production and consumption, 1960–2010. In 1960, the United States produced almost as much energy as it consumed, but since then, the trend has been for consumption to outpace production. Today, more than 25% of the energy consumed in the United States is imported. (*Source:* Based on data from the Energy Information Administration, U.S. Department of Energy, as published in The World Almanac and Book of Facts, 2005. New York: World Almanac Books, 2005, p. 169.)

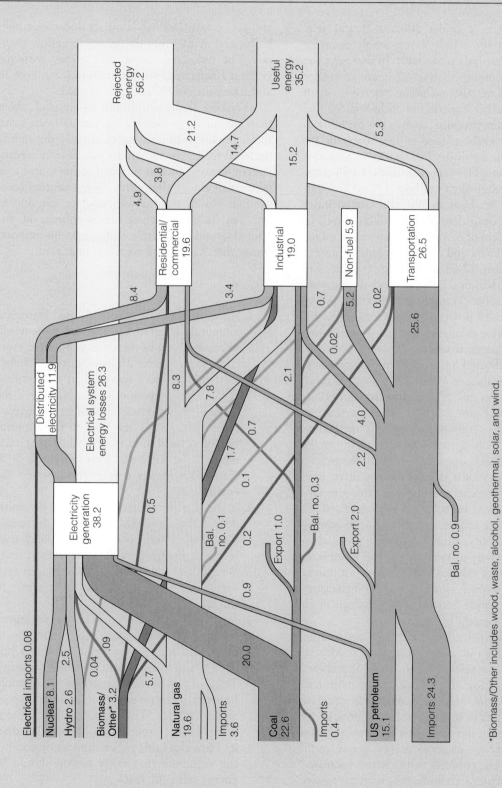

Electrical imports 0.08

Nuclear 8.1

Hydro 2.6 — 2.5

Biomass/ Other* 3.2 — 0.04 .09

Natural gas 19.6

Imports 3.6

Coal 22.6

Imports 0.4

US petroleum 15.1

Imports 24.3

Electricity generation 38.2

Distributed electricity 11.9

Electrical system energy losses 26.3

5.7

0.5

Bal. no. 0.1

0.2

Export 1.0

Bal. no. 0.3

Export 2.0

Bal. no. 0.9

0.9

20.0

1.7

0.1

0.7

2.2

4.0

25.6

8.4

3.4

8.3

7.8

2.1

0.7

0.02

5.2

0.02

Residential/ commercial 19.6

Industrial 19.0

Non-fuel 5.9

Transportation 26.5

Rejected energy 56.2

4.9

3.8

21.2

14.7

15.2

5.3

Useful energy 35.2

*Biomass/Other includes wood, waste, alcohol, geothermal, solar, and wind.

FIGURE 2 U.S. energy sources and consumption, 2002 (in quads). The numbers do not necessarily add up exactly because of rounding and estimations. Because of many uncertainties and discrepancies in statistical compilations, it is very difficult to carry out a precise and thorough analysis of the energy sources and consumption of a large country such as the United States. However, this analysis is representative of current source and consumption patterns. One quad equals approximately 1,054 petajoules. (*Source:* Lawrence Livermore National Laboratory and U.S. Department of Energy.)

The United States: Sources and Consumption of Energy

For example, consider 2002, a typical year and the latest for which a reasonably accurate and complete analysis is available. In that year, the United States consumed some 97 quads of energy and energy materials, excluding relatively small amounts of energy exports but including nonfuel uses of petroleum products, such as oils for lubrication and hydrocarbons used as raw materials for the petrochemical industry (see **FIGURE 2**). A quad is one quadrillion BTUs (approximately 1,054 petajoules) or the equivalent of about 171 to 172 million barrels of oil. Thus, in 2002, the United States consumed the equivalent of 16.6 billion barrels of oil (the equivalent of more than 55 barrels of oil for every man, woman, and child in the United States).

Of the nearly 97 quads of energy consumed in 2002, 26.3 quads, or 27%, were lost during electrical transmission. Transmitting electrical power through standard power lines is very inefficient. More than twice as much electrical power is lost during transmission from large, centralized power plants as is delivered in useful form to consumers. Looking at the data in Figure 2, we see that more than 25% of our energy goes toward transportation. Inspecting Figure 2 further, we notice that our modes of transportation are relatively inefficient. We waste about four times as much energy as we actually use. This is due in large part to the notoriously inefficient motor vehicles that we drive; with their low gas mileages, compared with the efficiencies that could be achieved even with today's technology, these vehicles waste most of the energy value of the gasoline. Most environmentalists advocate a major effort to increase the fuel efficiencies of cars and other vehicles. After energy lost during transmission through power lines, the inefficiency of our motor vehicle fleet is the largest single drain on the nation's energy budget.

Twenty-five percent of the energy used to heat, cool, and light our commercial and residential buildings, run our televisions and small appliances, and so forth is lost. Better insulation, more efficient refrigerators, and such could help decrease the energy losses in this sector. In addition, as we discuss further in the discussion of renewable and alternative energy sources, energy needs could be considerably reduced through the increased use of old and new technologies, such as passive and active solar heating, compact fluorescent light bulbs, and many other energy conservation methods. The industrial sector uses energy with the greatest efficiency in the United States, but even so about one-fifth of the energy is wasted.

Looking at the supply side of Figure 2, more than 80 quads of our energy comes from fossil fuels; thus, the United States is truly dependent on fossil fuels. The remainder comes from hydroelectric power, nuclear power, biomass, and to a very small extent alternative energy sources (geothermal power, wind power, solar power, and so on). For its single largest energy source, oil, the United States is heavily dependent on imports (more than half the oil we use is imported).

Questions

1. Some people think that the United States has a "moral obligation" to the rest of the world to cut down on its energy consumption, especially because the United States consumes substantially more energy than it produces. What is your opinion on this matter? If the United States suddenly slashed its energy consumption by 30% and stopped the importation of all fossil fuels, how would global trade relations be affected? How would a 30% cut in U.S. energy consumption benefit the environment?

2. If you were a cabinet member and the president charged you with promoting the reduction of U.S. energy consumption, which user sector in Figure 2 would you address first? Where do you think the most gains in energy savings can be made in the short term? In the long term?

3. Japan is the world's fourth largest consumer of primary energy, using more than 5% of the world supply. Yet Japan produces just over 1% of the world's primary energy. Thus, in global energy matters, Japan could be seen as the inverse of a country such as Saudi Arabia. Why does the world need both "Japans" and "Saudi Arabias," or does it? How does this either support or counter the argument that every nation should be energy self-sufficient?

Hard Versus Soft Energy Technologies

Physicist and energy consultant Amory B. Lovins has characterized large, modern electricity-generating power plants as **"hard" technologies**. Hard energy technologies depend on large-scale plants that are complex, expensive, and centralized. The energy they generate, such as electricity, must be distributed to consumers through an elaborate system of hardware, such as power lines. Large coal- or oil-burning electrical power plants, major hydroelectric plants, and nuclear power plants are all classic examples of hard technologies. Millions, if not billions, of dollars are needed to establish a hard technology and its accompanying infrastructure. Because of the substantial investments required, it can be very difficult to modify or abandon a hard technology once it is in place.

As Lovins points out, however, the end uses of the expensively produced energy by a hard technology may be mundane applications, for instance, heating water or open space in buildings. Rather than burning oil or coal at a power plant to generate electricity and then transmitting the electricity to the consumer who uses it to just generate heat, it is much more efficient to burn the oil or coal on site to generate heat directly. Better yet, passive solar systems might be used to heat (and even cool) a building.

The alternative to hard energy technologies are **"soft" technologies**, such as using passive solar energy, burning wood or coal in a stove, using an on-site windmill or small water wheel, and even increasing energy efficiencies and promoting conservation measures. Soft technologies are small-scale, local (thus they do not require elaborate distribution systems), and usually much more environmentally friendly than hard technologies. Soft energy technologies are often relatively inexpensive to install and operate, and they deliver energy that is of the appropriate nature—such as electricity from photovoltaic cells to run computers and other electronic devices and heat for space or water heating. Examples of both hard and soft technologies are discussed in this chapter and in the discussion of renewable and alternative energy sources.

7.3 The Fossil Fuels

The fossil fuels are currently the primary sources of commercial energy used worldwide for a number of reasons. Historically, fossil fuels have been available in a plentiful supply that was easy to mine or otherwise obtain. Fossil fuels provide a form of concentrated energy that is easily transported and can be put to many different uses. Historically, generating electricity using fossil fuels (specifically, natural gas and coal) typically has been much less expensive than using competing technologies, such as nuclear power plants, solar thermal, or wind power.

Oil

Oil and its products (liquid petroleum products) are the most widely used forms of commercial energy for the reasons mentioned previously. In particular, oil can be put to a variety of uses—from heating buildings and running machinery to (in modified form) fueling vehicles and laying down roads (made of tar and asphalt). Indeed, worldwide, more than half of all oil is used for transportation. This assortment of different organic compounds also forms the raw materials to manufacture plastics, synthetic rubber, fertilizers, medicines, detergents, and many products indispensable to the industrialized world.

Oil is a nonrenewable resource, however, and the majority of the remaining reserves are found in the Middle East. By most estimates, two-thirds of crude oil reserves are located in the Middle East, and a quarter of global oil reserves may be in Saudi Arabia. Iraq contains about 10% of global oil reserves, and Kuwait almost as much. The Persian Gulf War of 1990–1991 was effectively a battle over oil supplies, and it is quite possible that hostilities of the future could revolve around limited supplies of petroleum (some people view the 2003 U.S. invasion of Iraq as driven, at least in part, by concern over continued access to oil supplies). Some proponents who are in favor of the continued use of fossil fuels, such as oil, contend that current known reserves are a poor measure of how much oil actually still exists. They note that estimates of proven oil reserves, as well as coal reserves and natural gas reserves, have often been revised upward as more oil is found. This is true to a certain extent, but it is also true that in general over the past 2 decades, the new sources of fossil fuels that have been discovered have been smaller and of lower quality than those of past discoveries. In addition, because of their inaccessibility, some proven reserves are much more expensive (in terms of dollars and energy) than others. For instance, it can cost 5 to 10 times as much to produce a barrel of oil from Alaska or the North Sea as it does to produce a barrel of oil in the Middle East.

Even if we can stretch our use of fossil fuels by a few decades, either by finding new sources (such as undiscovered oil fields) or by using lower quality sources (such as **oil shale**, which is generally a lime-rich mudstone that does not actually contain oil, but primarily the hydrocarbon precursors known as kerogens, which can be processed into oil-like substances, or forcing the remaining oil out of "depleted" wells), the fact remains that fossil fuels are a finite, nonrenewable resource. They will run out eventually. Looking into the future, we see that fossil fuel resources may become so valuable as a source of the hydrocarbons used in manufacturing plastics, fertilizers, and so forth that the price will rise to the point where they will no longer be economically feasible as fuels. Furthermore, even before our last supplies of oil, coal, and gas are depleted, we may be forced to switch to non–carbon-emitting, nonfossil fuels to lessen deleterious effects of greenhouse gas accumulation, acid precipitation, and ground-level air pollution (see our discussions of local and regional air pollution and global air pollution).

Both the sources of fossil fuels (the absolute amount of fossil fuels that occur on Earth) and the sinks for fossil fuels (the ability of the Earth to successfully absorb the waste products of the burning of fossil fuels) are already limiting their use. Currently, the burning of oil and related liquid fuels releases about 9.5 billion metric tons of carbon dioxide per year into the atmosphere, more than 30% of all carbon emissions. In addition, burning oil releases such atmospheric pollutants as nitrogen oxides, sulfur dioxide, and various hydrocarbons.

Besides its contribution to atmospheric pollution, a particular concern with oil is that its use routinely results in huge amounts of raw oil being dumped into the environment (**FIGURE 7-5**). Some of this happens accidentally, as during an oil spill (a tanker may collide with shallow rocks, or a pipeline may burst), and sometimes oil is released intentionally as an act of terror or war. For instance, during the Persian Gulf War in 1990–1991, Iraq deliberately released an estimated 11 million barrels of oil from tankers off the coast of Kuwait. The *Exxon Valdez* disaster off the Alaskan coast in 1989 (March 24), which dumped a quarter million barrels of oil into Prince William Sound, was well publicized, but it was certainly not the largest recent spill. Later in the same year, on December 19, 1989, an explosion on the Iranian supertanker *Kharg-5* caused 450,000 barrels of crude oil to be released into the Atlantic 400 miles north of the Canary Islands; a 100-square-mile slick formed. In 1996, the supertanker *Sea Express* ran aground at the port of Milford Haven, Wales, spilling approximately half a million barrels of oil. The April 20, 2010, BP

FIGURE 7-5 An oil spill response worker takes three seabirds to be cleaned after the Exxon Valdez accident in Prince William Sound, off the coast of Alaska.

CHAPTER 7 Fundamentals of Energy, Fossil Fuels, and Nuclear Energy

Deepwater Horizon oil rig explosion and subsequent oil spill leaked almost 5 million barrels of oil into the Gulf of Mexico, making it the largest spill in United States history. The largest oil spill worldwide was in 1991, during the Persian Gulf War, when valves were deliberately opened on desert oil fields, sending approximately 6 to 8 million barrels of oil into the Persian Gulf. Since 1996, the amount of oil spilled from tankers has declined because of better hull design and navigational equipment. Although such spills are usually avoidable, much more crude oil is released intentionally during normal operations; the oil-carrying tanks of many ships are routinely flooded with seawater to clean them out, and the contaminated water is pumped back into the sea. An estimated 3 to 6 million metric tons, or roughly 22 to 44 million barrels, of oil are released into the world's oceans every year. Even more of the oil found in the ocean comes from rivers that carry untreated domestic and industrial hydrocarbon wastes to sea. However, natural seepage also accounts for an uncertain portion of the oil in the ocean.

According to U.S. Energy Information Administration data, world oil consumption was 63,114 thousand barrels a day in 1980; it dipped to 58,558 thousand barrels a day in 1983, and then steadily rose to 85,936 thousand barrels a day in 2007; in 2008 (the latest available data at the time of writing in mid-2011), world oil consumption was 85,462 thousand barrels a day. Today, oil provides about 40% of the world's energy and 96% of its transportation energy. Since the beginning of the "oil age," the world may possibly have consumed more than 875 billion barrels. Another 1,000 billion barrels of proved and probable reserves remain to be recovered.

Between now and 2020, world oil consumption is predicted to increase by about 60%. Transportation will be the fastest growing oil-consuming sector. By 2025, the number of cars will increase to more than 1.25 billion, from approximately 700 million today. Global consumption of gasoline could double.

The reason for the general decline and stabilizing of oil consumption between 1979 and 1990 seems to have stemmed from the energy crisis of the 1970s and the growing realization of the environmental damage that oil use incurred. More emphasis was placed on energy efficiency and the use of alternative energy sources. Most recently, however, oil production has again been rising because of several factors. Various developing countries are encouraging the exploration and development of their oil reserves, in part to meet the increasing demand for oil in these countries and in part to raise much-needed revenues. Central Asia in particular is experiencing an oil boom, and advanced technologies are allowing the location and exploitation of oil reserves that previously were economically inaccessible.

Currently, there is no imminent shortage of oil. Some analysts believe that oil prices and production will be driven by demand well into the future and that the theoretical maximum level of production will never be reached simply because the demand will not be there. It is hoped the world will break its heavy oil dependence in this next century. Still, until that time comes, oil will continue to supply approximately 40% of the globe's commercial energy (see CASE STUDY 7-2).

Coal

Like oil, coal is a fossil fuel, but in solid rather than liquid form. It is much more abundant than oil, and significant coal deposits are much more evenly distributed throughout the world than are major oil fields. Given these factors, some analysts predict that worldwide use of coal will continue to increase in the decades to come. In the United States, a primary method of coal extraction is mountaintop removal mining, a surface mining technique, authorized by the Surface Mining Control and Reclamation Act of 1977, whereby explosives are used to remove large areas of mountaintops in order to access underlying coal seams. In the United States, mountaintop removal is concentrated in central Appalachia, an area that includes southern West Virginia, eastern Kentucky, southwest Virginia, and eastern Tennessee. This area produces 33% of all U.S. coal, 40% of which comes from surface mining.

At current consumption rates, the world's supply of coal could probably last for about 200 years or more. Of course, this would be considerably shortened if coal consumption increased dramatically, perhaps as a replacement for oil after that resource is exhausted. A major concern over the burning of coal is that coal emits 25% more carbon dioxide into the atmosphere than an equivalent amount of oil and 80% more carbon dioxide than an equivalent amount of natural gas, promoting the greenhouse effect. In addition, coal burning may release varying amounts of sulfur dioxide, depending on the type of coal, and some nitrogen dioxide, major contributors to acid rain. New techniques are being developed to ensure the more complete burning of coal with less pollution. For example, in **fluidized bed combustion**, very small coal particles are burned at very high temperatures in the presence of limestone particles (the limestone helps capture sulfur and other pollutants) while air is blown through them.

Currently, coal is used directly in only a handful of major contexts, such as to produce electricity in coal-fired power plants, to heat large plants and buildings, and in the production of iron and steel. In the more developed countries, coal use on a smaller scale has been generally phased out—few people burn coal directly to heat their homes, for instance—because it is bulky and cumbersome to handle and transport on a small scale, dirty, and polluting. To overcome some of these disadvantages and increase the use of coal, various organizations are working on coal

gasification and liquefaction techniques that can produce natural gas, alcohol, and oil replacements from solid coal. The main drawback with such technologies is that at the moment they are not economically competitive with natural gas and petroleum products. Methane or gasoline produced from coal can cost over 50% more than the equivalent substances refined from natural gas or petroleum taken from the ground. In addition, although transforming coal into liquid and gaseous hydrocarbons may help mitigate pollution and the greenhouse effect, these new products are certainly not pollution free. At best, such techniques can serve only as short-term, stop-gap measures to fend off the fossil fuel crisis.

Since the middle of the 20th century, global consumption of coal has more than doubled, climbing from 1,074 million metric tons of oil equivalent per year in 1950 to 2,578 million metric tons of oil equivalent in 2003. Analysts are divided when it comes to predicting the future of worldwide coal use. On the one hand, it is an abundant, widely distributed fuel that both developed and developing countries use. United States coal consumption accounts for approximately a quarter of the world total. Many developing countries are dependent on coal, especially if they cannot afford oil, natural gas, or other sources of energy (**FIGURE 7-6**). China, with a population of more than 1.3 billion, extracts about 60% of its total energy requirements from coal. India is also a leading user of coal. If the developing world continues to expand its use of coal, worldwide consumption could be driven higher.

On the other hand, some analysts predict that worldwide coal use has already reached a plateau or will plateau in the next decade or two. General improvements in energy efficiency, as have been instituted in Russia and other republics of the former Soviet Union, may be eliminating the need for major increases in coal use. Increasingly, governments are instituting and enforcing environmental regulations that discourage the use of coal. When it comes to meeting environmental standards, coal is a poor match for other forms of energy. In addition, there is a trend toward eliminating the subsidies for coal production and use that historically have been

FIGURE 7-6 The cumulative effect of so many people in Asia cooking over coal fires like this man in India creates local health problems (respiratory ailments) and regional pollution problems.

provided by many governments, including Germany and the United Kingdom. However, coal workers often lose their jobs as a result, and major strikes by miners (e.g., in England in 1992) have slowed progress toward eliminating coal subsidies. Only time will tell whether worldwide coal use will increase or decrease.

Natural Gas

A fossil fuel that takes a gaseous form, **natural gas** is currently a very commonly used fuel. Natural gas is versatile and relatively abundant, so its use is growing rapidly. It also burns cleaner than other fossil fuels (producing less carbon dioxide and other pollutants than coal or oil).

Nevertheless, there is some fear that economies based on natural gas will not be able to last much longer than economies that depend on oil. Based on current consumption rates, the world's known natural gas reserves will be depleted by the middle or end of the 21st century. However, it is now widely believed that natural gas resources are much more abundant than once assumed. In the past, natural gas was often found as a by-product of the search for oil, and in some cases, the natural gas was simply released or burned off the top of oil pockets. Today, natural gas reserves are being discovered in areas not associated with oil deposits, and various research groups have suggested that our lack of understanding of natural gas geology may have led to overly conservative estimates of the extent of natural gas resources. It is now known that, whereas oil is primarily limited to relatively shallow deposits, methane (natural gas) can be found at much more extreme depths. At such depths, large amounts of natural gas may be highly compressed into small volumes. New exploration and drilling techniques are making it easier to locate and tap such natural gas resources but can also have a more dramatic impact on the local and global environment.

Natural gas extraction is sometimes performed through a process known as hydraulic fracturing ("fracking"), which results in the creation of fractures in rocks. This process is often done from a well bore drilled into reservoir rock formations in order to increase the rate and ultimate recovery of oil and natural gas. Some environmental and human health concerns possibly associated with hydraulic fracturing include the contamination of ground water, risks to air quality, the migration of gases and hydraulic fracturing chemicals to the surface, and the potential mishandling of waste. For example, in April 2010, the state of Pennsylvania banned Cabot Oil & Gas Corporation from further drilling in the state until it agrees to plug wells believed to be the source of contamination of the drinking water of 14 homes in Dimock Township. The investigation was initiated after a water well exploded on New Year's day in 2009.

Natural gas has also been discovered in association with coal seams. Another advantage of natural gas is that deposits are somewhat more evenly distributed around the globe than are major petroleum fields. Although the majority of proven reserves of oil occur in the Middle East, significant proven reserves of natural gas are found in the former Soviet Union and the Middle East, and smaller reserves are in North America, Africa, Latin America, and Western Europe.

Worldwide production (and likewise consumption) of natural gas has grown by an order of magnitude since the middle of the 20th century—from 171 million metric tons of oil equivalent in 1950 to 2,332 million metric tons of oil equivalent a year in 2003. Some analysts suggest that natural gas use may nearly double during the next 2 to 3 decades and then continue to be used at a level of about 3,700 to 4,000 million metric tons of oil equivalent a year for several decades. Some predict natural gas ultimately will be the fossil fuel to usher in the final transition to a solar–hydrogen and alternative energy global economy (see the discussion of renewable and alternative energy sources in this text).

Can Fossil Fuel Supplies Increase on a Human Timescale?

Of course, the simple answer to this question is no. As we have seen, the geological processes involved in the formation of fossil fuel resources are imperceptibly slow from a human perspective. Yet a superficial reading of certain statistics may suggest that our

The Hidden Costs of Foreign Oil Dependence

A secure supply of energy is critical for any nation, for without energy the country runs the risk of collapse. When a nation is heavily dependent on foreign oil (or foreign natural gas, coal, or uranium), it may need to spend huge sums of money and even enter into armed conflict to guard its energy supplies. This occurred in the Persian Gulf. In January 1991, the United States attacked Iraqi forces that had taken over the country of Kuwait, a small oil-rich nation on the northwestern corner of the Gulf, and threatened to disrupt the free flow of oil. Six weeks later, the Gulf War was over, but beyond the loss of life, the war's cost was approximately $40 billion, and considerable environmental damage had been done to the area. Even before the Gulf War, the hidden costs of Gulf oil meant that it was far from a bargain have estimated that before January 1991 the costs of the routine U.S. military maneuvers that took place in the Gulf to guard the oil added approximately $25 to the real cost of a barrel of oil. Between 1974 and 1989, the price of oil on the world market fluctuated between $12 and $48, so this hidden cost effectively doubled the cost of oil. Of course, this extra $25 per barrel was not added to the price of the oil but was paid by American taxpayers as part of the country's outlays for defense. Some analysts estimate that, when all of the costs are totaled (military training, equipment, transportation of troops

and supplies, maintaining a ready state of alert, and so forth), the United States effectively spends $50 billion or more a year protecting oil supplies in the Middle East. Of course, if the recent military operations in Iraq (beginning with the U.S. invasion in 2004) are seen, at least in part, as involving oil, the cost is even higher.

In addition to the cost in lives and money spent to wage the war and then rebuild the damaged infrastructure of Kuwait and Iraq, the war extracted a very high price in terms of environmental damage. During the war, Iraq resorted to ecological sabotage. Many Kuwaiti oil wells were deliberately damaged and set on fire (**FIGURE 1**), whereas oil reservoirs were allowed to leak onto the land and water. Millions of barrels of Kuwaiti oil leaked into the Gulf, forming an oil slick tens of miles long that damaged coasts and wildlife. Approximately 700 wells were set on fire, burning about 3 million barrels of oil a day. Even after the war ended in late February 1991, many of these wells continued to burn—the last one was not capped until November 1991. As they burned, they released millions of tons of smoke, carbon dioxide, carbon monoxide, sulfur dioxide, and other pollutants into the atmosphere. Millions of tons of acid rain came down in the Gulf region. During the summer of 1991, the smoke and air pollution caused

fossil fuel supplies have actually increased over time. For example, in 1970, the ratio of oil reserves to production was 31 years, but by 1989, the ratio had increased to 41 years. Even more dramatically, over the same period, the ratio of reserves to production for natural gas increased from 38 to 60 years! In the year 2000, more than a decade later, the ratio of oil reserves to production was 38 years, and for natural gas, it was still 60 years.

Superficially, it may appear that there was more oil and natural gas available in 2000 than 30 years earlier. Of course, this is

not the case—there were about 600 billion fewer barrels of oil (about 88 billion metric tons of oil equivalent) and 2,000 trillion fewer cubic feet (51 billion metric tons of oil equivalent) of natural gas in the world (the amount consumed between 1970 and 2000). What changed was that more oil deposits and natural gas deposits were discovered between 1970 and 2000, thus increasing the known reserves of these fossil fuels. Some might argue that the successful history of oil and gas exploration between 1970 and 2000 demonstrates that there are still plenty of deposits

FIGURE 1 The Bergan oil field fire in Kuwait is reflected in a pool of oil. The Iraqis deliberately set the fire during the Persian Gulf War in 1991.

the temperatures in Kuwait to be 10°F to 27°F (5.6°C to 15°C) lower than normal. Smoke and acid rain from the burning oil were detected more than 1,600 kilometers (more than 1,000 miles) away. Another hundred or so oil wells spilled oil over the Kuwaiti desert—reportedly, the lakes of oil were 2 meters (6 ft) deep in some areas. Scarce freshwater supplies throughout the region were contaminated.

Although the burning oil wells were dramatic and horrible, in hindsight, many experts suggest that the worst long-term damage to Iraq and Kuwait occurred when the vegetation and soils of the fragile desert ecosystems were subjected to massive bombings, trenches dug by bulldozers, and the military maneuvers of soldiers and motorized armored vehicles. As a result, an estimated 25% of the land surface of Kuwait has been damaged. In both countries, wind erosion has increased significantly, causing an increase in sandstorms and an intensification of desertification processes. In an area already heavily dependent on imports of food, the loss of what little soils there were can only worsen the situation.

References

P. R. Ehrlich and A. H. Ehrlich (1991. *Healing the Planet: Strategies for Resolving the Environmental Crisis*. Reading, MA: Addison-Wesley.)

VISIT

http://environment.jbpub.com /mckinney/5e/ for more information

to be discovered, but this too is a false argument. There is an absolute amount of fossil fuels still buried in the crust of the Earth. In addition, extraction requires tremendous resources and has an impact on the environment (see **CASE STUDY 7-3**). With exploration and discovery, some of this fixed amount is shifted from the category of "undiscovered reserves" to "known reserves." Every time we do this, the amount of undiscovered reserves is decreased. By definition, we can never know precisely the quantity of unknown reserves that remain (for they are "unknown"), but we

can estimate their quantity by various means. One way to gauge the extent of undiscovered reserves is by how easy it is to find them—the more abundant the undiscovered reserves, the easier they are to locate.

In fact, in many parts of the world in the last 30 years and since about 1940 in the continental United States, it has become progressively more difficult, with progressively more sophisticated instrumentation and deeper drilling rigs in more obscure places, to locate previously undiscovered reserves of oil. This clearly indicates that we are quickly locating the last major fields of undiscovered reserves. It would be unwise to anticipate that the rate of past oil discoveries will continue into the future.

Why We Must Stop Burning Fossil Fuels

The burning of coal, and to a lesser extent the burning of petroleum products and natural gas, has created environmental havoc on a global scale and has led to global warming, acid rain, air pollution, damage to the land's surface as a result of mining and drilling activities (especially harmful is strip-mining for coal), water pollution, and so forth (**FIGURE 7-7**). Whether we are in danger of running out of fossil fuels or not, most people who have seriously considered the matter agree that we must greatly curb our use of these fuels. One consideration is the degree to which the United States and many other industrialized societies are dependent on foreign oil; dependence on foreign oil

can have drastic negative consequences if such supplies are disrupted.

The fossil fuels are composed principally of carbon and hydrogen derived from once-living organisms. When a fossil fuel is burned, the carbon in the fuel combines with oxygen (O_2) from the air to form carbon dioxide (CO_2), and the hydrogen combines with atmospheric oxygen to form water (H_2O). During these reactions, energy in the form of heat is released, and the released heat also causes the reaction to proceed further.

A high-grade coal is almost pure carbon tainted by small quantities of other elements (such as sulfur and various metals). As the coal is burned, it gives off mostly carbon dioxide, plus small amounts of sulfur dioxide and other pollutants. Carbon dioxide is the dominant greenhouse gas. Natural gas, in contrast, contains a large amount of hydrogen (the formula for methane is CH_4), so when it is burned,

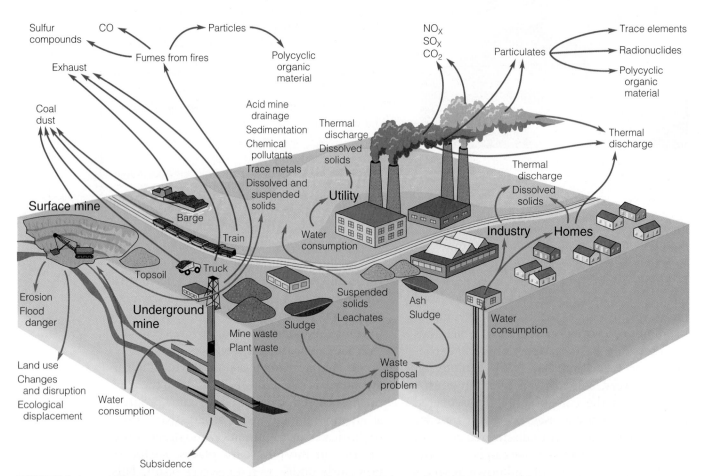

FIGURE 7-7 Environmental impacts of the coal fuel cycle. The oil and natural gas cycles cause similar impacts. (*Source:* Adapted from C. Hall, C. Cleveland, and R. Kaufman. Energy and resource quality. Niwot, CO: University Press of Colorado, 1986, p. 366.)

less carbon dioxide is emitted per unit of heat emitted. Some of the gas that methane gives off takes the form of water vapor, which does not appear to pose any environmental problems. For these reasons, natural gas is considered to burn much more cleanly than coal; nevertheless, even natural gas combustion produces significant amounts of carbon dioxide. In addition, unburned natural gas, if released into the atmosphere, is itself a powerful greenhouse gas. See this text's discussion of global air pollution for additional information about greenhouse gases and their effects.

Fossil fuel procurement is also visibly damaging to the environment. Oil and natural gas prospecting and drilling can involve the construction of access roads, the movement of heavy equipment, the erection of large drilling rigs or platforms (both on land and at sea), and so forth—all of which degrade the natural environment. Oil spills, either at wells or during transportation, have gained notoriety, but perhaps more damaging are "routine" leaks and emissions of petroleum compounds from wells, pipes, and tankers. Coal mining takes a heavy toll on the environment and is dangerous to the miners (**FIGURE 7-8**). **Strip mining** may involve the removal of as much as 30 meters (100 ft) of soil to get to the coal, and in the United States alone, it is estimated that more than 500,000 hectares (1.3 million acres) have been strip mined. For mountaintop removal mining, a type of strip mining, the U.S. Environmental Protection Agency (EPA) estimates that by the end of 2010, 1.4 million acres of Appalachian forests were disturbed or cleared by mountaintop removal, an area larger than Delaware. Many of the older strips have not yet been reclaimed. Another 26,000 hectares (65,000 acres) continue to be strip mined annually. With mountaintop removal mining operations, water quality declines as a result of dumping mining "spoil" or refuse from the land clearing and mining operations in existing streams. According to the U.S. EPA, since 1992, nearly 3,200 kilometers (2,000 miles) of Appalachian streams have been buried by mining refuse. In addition, 9 of every 10 streams downstream from surface mining operations have been found

FIGURE 7-8 Without any attempt at reclamation, a strip mine is very damaging to the land. The rows of rubble are called tailings, and nothing will grow on them for years to come.

to be polluted. Underground coal mines are slightly better. They can result in land subsidence, whereas water seeping into them can form acids (from the sulfur and other compounds that commonly occur in coal) that then leak out and contaminate land, streams, and lakes. Underground coal fires are often virtually impossible to extinguish and may smolder for decades; hundreds of such uncontrolled fires are reported in the United States each year. Coal dust can cause explosions. Coal washing, in which the usable coal is separated from impurities, results in unsightly and dangerous waste banks that can catch fire.

Very few professions in the world are as dangerous and physically demanding as coal mining. Black lung disease plagues many miners, and miners are regularly killed in accidents. The average coal miner dies several years prematurely as compared with the rest of the population at a similar socioeconomic level. China has the world's deadliest mines; more than 5,000 miners are killed every year.

7.4 Nuclear Power

Nuclear power supplies roughly 6.5% of the world's commercial energy overall but accounts for a much larger percentage of the electrical generating capacity in certain industrial countries. More than 430 commercial nuclear reactors are currently in operation in some 30 countries, and approximately 25 nuclear plants are under active construction. The United States, with just more than 100 nuclear power plants, generates 19.9% of its electricity from nuclear power, whereas many other countries depend on nuclear power for more than 25% of their electricity (**TABLE 7-1**). The leading user of nuclear power in this respect is France, which generates 78% of its electricity from nuclear power. Japan continues to invest in nuclear power, and India and China, among other countries, are developing their nuclear capacity. In contrast, Austria is nuclear free, and it, along with the Philippines, does not plan to develop nuclear power. Because of public pressure, Sweden has been dismantling its nuclear power plants.

The United States first developed nuclear power during World War II, but initially, research focused on military applications. The U.S. Atomic Energy Act of 1954 paved the way for private industry to develop nuclear energy, and by 1983, 80 nuclear power plants were operating in the United States. However, as nuclear plants proliferated in the 1960s and 1970s, the public became increasingly concerned about their safety. It was also discovered that nuclear power was not as inexpensive and efficient as had been predicted. In fact, in the United States, electricity that nuclear power plants generate currently is more expensive than electricity generated by any other major technology. Nuclear power plants are expensive to build; mining and refining **uranium** ore are considerable undertakings; accidents and malfunctions incur significant monetary losses, and the ultimate **decommissioning** of plants will require considerable sums of money (which ultimately could increase the cost of nuclear power to even higher levels).

Since 1978, there have been no new orders for nuclear power plants in the United States. Many other developed countries have overtaken the United States in the research, development, and implementation of nuclear power capability. Still, with more than 100 operating nuclear power plants, the United States has the largest total monetary investment in nuclear power. Some observers worry that the United States is no longer a leader in advanced peacetime nuclear technology. Others think that the country is still too dependent on nuclear power.

TABLE 7-1		Some of the Leading Countries in the Use of Nuclear Power	
Country	Number of Reactors	Electricity from Nuclear Generators (Percentage of Total)	Nuclear Energy Supplied, 2004 (Billions of Kilowatt-Hours)
Belgium	7	55.0	44.9
Bulgaria	4	41.6	15.6
Czech Republic	6	31.2	26.3
France	59	78.1	426.8
Germany	18	32.1	158.4
Hungary	4	33.8	11.2
Japan	54	29.3	273.8
Korea, S.	19	38.0	124.0
Lithuania	1	72.1	13.9
Slovakia	6	55.2	15.6
Slovenia	1	38.9	5.2
Sweden	11	51.8	75.0
Switzerland	5	40.0	25.4
Ukraine	15	51.1	81.8
United States	104	19.95	788.6

Source: International Atomic Energy Agency. Based on data from The World Almanac and Book of Facts, 2006. New York: World Almanac Books, 2006, p. 141.

Nuclear energy potentially encompasses two different but related types of reactions: fission and fusion. During nuclear **fission**, a radioactive isotope of a heavy element, such as a uranium or **plutonium atom**, is split into **daughter products** and simultaneously energy is released. Fission runs all commercially operating nuclear reactors. During **fusion**, isotopes of some light element (or elements) are fused together to make a heavier element (e.g., heavy isotopes of hydrogen bonding to form helium); in the process, energy is also released. The sun generates light and energy through fusion reactions, but on Earth, controlled fusion is still only in the developmental stage.

Fission

In fission, a **neutron** penetrates the nucleus of a fissionable atom (easily split by neutron penetration, such as uranium-235 [$^{235}_{92}U$]) and causes the nucleus to split into two or more smaller nuclei (these smaller nuclei, although nonfissionable, are highly radioactive). In the process, a tremendous amount of energy is released. The newly formed nuclei are positively charged and repel each other; as a result, they travel apart at high speeds and correspondingly high kinetic energies. In addition, if other fissionable atoms are nearby, they may be induced to fission by the extra neutrons released by the first fissioning atom; thus, a **chain reaction** takes place (**FIGURE 7-9**).

Modern nuclear power plants use a nuclear reactor, with a sustained but controlled fission chain reaction, to generate tremendous amounts of heat. The heat is then used to boil water, producing steam that powers a turbine that turns a generator and produces electricity. The typical nuclear reactor has a **core** containing the uranium (or other fissionable element, such as plutonium) fuel (**FIGURE 7-10**), the **moderator** (used to slow neutrons and help control the fission process), and **control rods** of some substance that readily absorb neutrons, such as cadmium or boron. When the control rods are fully inserted, they absorb enough of the neutrons to halt the nuclear chain reaction. By selectively inserting the control rods, the number, or flux, of neutrons can be regulated and thus the rate of

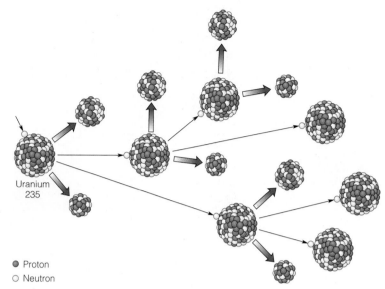

● Proton
○ Neutron

FIGURE 7-9 Diagram illustrating a nuclear chain reaction.

fission and energy release controlled. In some reactors, if the control rods are removed completely and excess heat is not removed quickly from the reactor core, the core may get so hot that it will begin to melt.

Water typically circulates around the core to cool it and is heated in the process. Because the water exposed to the core potentially becomes radioactive, it is recycled and recooled in the primary water loop. The primary water loop, in turn, heats a secondary water loop that produces steam to operate a turbine. The turbine is connected to a generator that

FIGURE 7-10 The interior of the containment structure of the Trojan nuclear power plant, Rainier, Oregon. In the center of the photograph is the reactor core positioned under 10.7 meters (35 ft) of water. The plant was shut down in 1992.

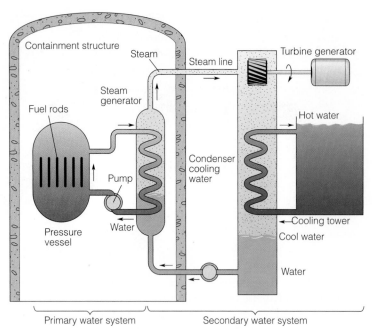

FIGURE 7-11 A schematic diagram of a typical pressurized light water reactor (LWR) power plant.

produces electricity (**FIGURE 7-11**). The reactor core and primary water loop are housed within a thick steel tank, or pressure vessel, that is designed to contain all radioactive traces. In most reactors, the **reactor vessel** is housed in a **containment structure** composed of thick steel-reinforced concrete. The containment structure is designed to protect the outside environment from major radioactive contamination if the nuclear reactor should fail.

The principal type of reactor used in the United States is the **light water reactor** (**LWR**).

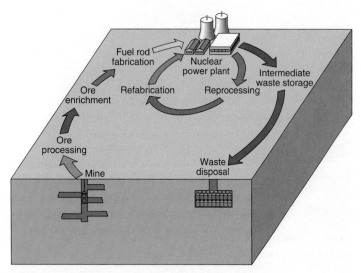

FIGURE 7-12 The fuel cycle for electricity generation using a fission reactor.

Currently, all commercial nuclear power plants in the United States and about 75% of the nuclear plants worldwide are LWRs. LWRs are named after their moderator, which is ordinary or light water (as opposed to heavy water, which contains an abundance of the hydrogen isotope deuterium). Chernobyl-type reactors use a graphite moderator and have other basic design differences from American-type LWRs (see the discussion later in this chapter).

The fuel used in typical LWRs is uranium that has been slightly **enriched** in fissionable U-235. The uranium is made into fuel pellets, which are placed into fuel rods. The fuel has to be enriched because ordinary water, while slowing the neutrons, actually captures too many of the neutrons to sustain a chain reaction with uranium composed of its natural isotopic abundances. Typically, an LWR's fuel contains about 3% U-235 and 97% U-238. The uranium must first be extracted from uranium ore (**FIGURE 7-12**). Most uranium ore contains a relatively small percentage of actual uranium—typically only 0.1% uranium metal by weight. After being mined, ore is concentrated during the **milling process**, and the result is the formation of the substance U-3O8, referred to as **yellowcake** (because of its color). Yellowcake is composed of 85% pure uranium by weight, but the uranium is still in its natural isotopic mixture of 99.3% nonfissionable U-238 and only 0.7% fissionable U-235.

U-238, although not normally fissionable, is a fertile isotope that, by absorbing a neutron, can give rise to plutonium-239 (Pu-239), which is a fissionable element. **Breeder reactors** take advantage of the ability of nonfissionable U-238 to be converted to fissionable Pu-239. With a breeder reactor, the 99.3% of uranium that did not fission can be put to use, thus greatly extending the amount of energy that can be generated from uranium ore deposits. In addition, **thorium**-232 (Th-232), an element that is several times more abundant in the Earth's crust than is uranium, can be used in a breeder reactor to produce fissionable U-233.

To obtain fissionable plutonium produced in a breeder reactor, used fuel from conventional reactors must be reprocessed at a **reprocessing facility**. In reprocessing, the fissionable materials are isolated and concentrated into fuel, but this same fissionable material could be used to make an atomic bomb. A terrorist group or hostile country would need to steal or produce only a small amount of more concentrated (weapons-grade) fissionable material to produce a weapon. In addition, given the high concentrations of fissionable material that could accumulate in a breeder reactor, and the fact that Pu-239 is fissionable by fast neutrons, some people fear that a breeder reactor might even have the potential to explode. In contrast, light water nonbreeder reactors theoretically never have high enough concentrations of fissionable materials to produce an actual atomic explosion, although an uncontrolled chain reaction could result in a **meltdown**, inducing nonatomic explosions (such as huge quantities of steam breaking containing pipes and vessels or water being converted to hydrogen and oxygen, which could then explode).

Even if we could ensure that no fissionable material would be lost or stolen from a reprocessing facility, reprocessing spent nuclear fuel presents other problems:

- It is dangerous and expensive.
- The spent fuel is so radioactive much of the reprocessing must be done by remote control.
- The fuel must be dissolved in strong toxic acids to extract the uranium and plutonium.
- The plants must be meticulously designed to avoid radiation leakage.

For these reasons, along with concern over nuclear weapons proliferation, reprocessing is not being actively pursued in the United States.

Fusion

The sun's energy is the result of nuclear fusion reactions within its pressure-cooker core. During fusion, the nuclei of light elements, in this case hydrogen, are fused together to form a heavier element, helium, releasing an enormous amount of energy in the process. As you might guess, extremely high temperatures are necessary to initiate fusion reactions. The first artificial fusion reaction was attained in 1954 when the United States successfully detonated a hydrogen bomb.

Controlled and sustained fusion reactions are not yet feasible as a commercial energy source. Researchers are working on two approaches to produce the extremely high temperatures needed for fusion and to contain the fusion reactions after it gets started: (1) magnetic confinement and (2) high-energy lasers and particle beams. Both methods are based on the fusing of hydrogen isotopes— either two deuterium atoms or a deuterium and a tritium atom, to form a single helium atom plus a neutron. Deuterium has a neutron as well as a proton in the nucleus, whereas tritium has two neutrons and a proton in the nucleus.

In magnetic confinement, powerful electromagnets surround the fusion chamber and repel the fusion materials as they approach the walls of the chamber. Various types of magnetic confinement reactors have been designed and tested experimentally; the most common and successful design is known as a **tokamak.** A tokamak is a large (the size of a house) machine with a doughnut-shaped vacuum reactor vessel that magnetic coils surround.

Fusion reactions can also be initiated by bombarding a fuel pellet containing deuterium or a deuterium-tritium mixture with laser or particle beams from all directions. For a split second, the fuel pellet will remain suspended within the reaction chamber and implode with tremendous force and pressure, attaining enormous temperatures. If the pressure and temperature are great enough, fusion will occur.

In many ways, fusion would be an ideal energy source, especially if deuterium or deuterium-tritium fusion reactions can be harnessed. Deuterium makes up only a small percentage of all hydrogen atoms (about 1 of every 6,700 hydrogen atoms), but hydrogen is the most abundant element in the universe. The deuterium contained in just 100 liters (about 26 gallons) of seawater, if totally fused, would release as much energy as approximately 33,500 liters (approximately 8,850 gallons) of gasoline. Some researchers have estimated that if the deuterium in the top 3 meters (10 ft) of water in Earth's oceans were used as a fuel in fusion reactors, it could satisfy the energy needs of all humans for

The Arctic National Wildlife Refuge

VISIT

http://environment.jbpub.com
/mckinney/5e/
for more information

The Arctic National Wildlife Refuge, often abbreviated ANWR, covers about 8 million hectares (20 million acres) in the northeastern corner of Alaska (**FIGURE 1**). The area is approximately the size of South Carolina and is about 5% of the area of the state of Alaska. ANWR is at the center of a major national environmental controversy: should oil drilling be allowed? This question has been debated in the U.S. Congress for many years, seemingly with each jump in the price of gas.

The amount of oil underground in ANWR that is economically feasible to extract is estimated to range between 3.2 and 7.8 billion barrels. People who favor drilling in the ANWR argue that this large supply of oil could be used to reduce U.S. energy dependence on oil imports. They also argue that the environmental impacts are minimal because the land in ANWR is tundra, with much less biodiversity than is found in rain forests and other highly productive ecosystems. Proponents note the total area affected by drilling is much smaller than the entire refuge. In addition, they point out extraction technology exists that will minimize the "footprint," and considerable money will be spent to reduce the impact of the actual drilling and the roads and large pipelines needed to distribute the oil.

(a)

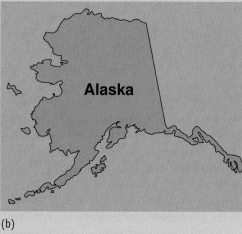

Alaska

(b)

FIGURE 1 (a) Summer in the Arctic National Wildlife Refuge (ANWR). (b) The ANWR is in the northeastern corner of Alaska.

about 50 million years. Fusion energy has another advantage as well; it does not produce fissionable materials that could be incorporated into the production of bombs.

On the other hand, a sustained fusion would produce some radiation, but much less than fission. The extra neutron released when hydrogen atoms fuse to form helium atoms would cause the internal parts of the reactor to become highly radioactive. The neutron bombardment would weaken the fusion reactor over time, as would the intense heat generated by the fusion reaction.

Some people predict that fusion power could solve many of the world's fundamental material problems. The sun-like high temperatures generated during the fusion process could be used to vaporize any and all waste

Opponents to the drilling argue that the amount of oil from ANWR is not that large compared with the U.S. demand. The United States consumes about 7.3 billion barrels per year, so the refuge contains enough oil to supply the nation's needs for somewhere between 6 months to 1 year. These opponents argue that if the drilling and construction money were instead spent promoting energy conservation and alternative energy, this would not only improve the environment but also would lead to a permanent reduction in oil imports.

Drilling opponents also contend that, although the land within ANWR is tundra, it is one of America's richest areas for unique wildlife. Thousands of musk ox, arctic foxes, arctic hares, wolves, and other wildlife make the refuge their permanent home. Half of the polar bears in Alaska den in ANWR. It is where the eastern, western, and Rocky Mountain flyways converge for 130 species of migrating birds each summer. It is a calving ground for an immense caribou herd. Changes in the caribou herd would affect the local indigenous populations that rely on the caribou for much of their food supply.

To what extent oil drilling will harm the wildlife is a topic of much scientific interest. A 2002 study by biologists in the Department of Interior concluded that the caribou herd, which uses the coastal plain for calving each summer, "may be particularly sensitive to development" because it has little quality habitat elsewhere. Oil development most likely will result in restricting the location of concentrated calving areas and will lead to fewer calves being able to survive and, in turn, possibly a decline in the herd. Similarly, snow geese, among the millions of migratory birds on the coastal plain, may be displaced because of increased activity, including air traffic. It cannot be assumed the geese will find adequate feeding areas elsewhere, the study says.

There is also concern over the impact of drilling on native Inupiat that live in the region. The only permanent human population in ANWR is the town of Kaktovik, with a population of just 300. The Inupiat in the town own shares, granted to them from native-rights settlements, in the Arctic Slope Regional Corporation. The corporation would benefit from oil drilling in the area. The people of Kaktovik have made it clear they would support oil exploration and development but only if they are given the authority and the resources to ensure that it is done properly and safely. The Inupiat's neighbors to the south, the Neets'aii Gwich'in, live in Arctic Village, which sits on the southern boundary of the reserve. The Gwich'in people strongly oppose any oil development in the coastal plain. Their cultural tradition is linked to the caribou, which they believe will be endangered by development.

Questions

1. List both the advantages and disadvantages of drilling in the Arctic refuge. Do you think the advantages outweigh the disadvantages? Why or why not?
2. Name three other areas of the world where oil drilling is common. Do you think that drilling in these areas is less harmful to the environment? Explain.

materials; wastes would be broken down to their constituent atoms. Piles of waste—garbage, sewage, refuse—could be vaporized and sorted into piles of carbon atoms, oxygen atoms, iron atoms, gold atoms, copper atoms, and so on to be reused—the ultimate recycling. The raw atoms could be used to produce whatever is needed. We are unlikely to exhaust the supply of hydrogen; it is estimated that more than 90% of the atoms in the universe are hydrogen atoms.

Uranium Resources

Uranium resources, like fossil fuels, are a nonrenewable source of energy. Mining and processing uranium can have a major impact on the environment. A typical light water fission reactor with a 1,000-megawatt capacity may require more than 140,000 metric tons of uranium ore to produce the fuel it consumes in a year. The mining operation to obtain this raw ore may extend over 7 hectares (18 acres) and displace as much as 2.5 million metric tons of rock and earth.

Assessing potential uranium resources is very difficult, but clearly, some high-grade deposits of uranium ore (as in the western United States) are already being depleted. Like other mineral resources, concentrated uranium deposits are not spread evenly over the globe (**FIGURE 7-13**). Based on current consumption, a world dependent on fission power, and not using breeder reactors, might deplete these high-grade, currently economically minable uranium resources in only a couple of centuries.

Other sources of uranium exist. For example, seawater contains uranium, but it is so diffuse that large amounts of energy would be required to extract it. On a more positive note, uranium that would be uneconomical to mine for its own sake is recovered as a by-product during the mining and refining of other minerals. By-product uranium has been produced from phosphate plants and copper operations. With the end of the Cold War, we have been able to reprocess weapons-grade uranium from nuclear warheads into fuel for power plants. Some lower-grade deposits of uranium also have the potential to be used, but it would take significantly more deposits than would be needed if higher grade ore were used.

7.5 Advantages and Disadvantages of Nuclear Power

The Safety Record of Nuclear Power

Advocates of nuclear power rightfully claim that, in practice, nuclear power generation has thus far proven to be the safest form of large-scale commercial power generation. Nuclear power plants, so far, are much safer (in terms of human lives lost), despite several widely publicized accidents, and are less damaging to the environment than are fossil fuel-burning power plants, and they are also safer than hydroelectric power plants. Other sources of electricity generation also have many accidents, although they do not seem to attract as much attention from the media and the public as accidents at nuclear power plants. In addition, the pollution generated by coal-fired power plants causes fatal health problems for workers and nearby residents on a routine basis. The air pollution, acid rain, and greenhouse gases they release are causing untold property damage (estimated at billions of dollars a year in the United States alone), disrupting ecosystems, and contributing to changing the global climate (see this text's discussion of global air pollution). Hydroelectric power plants, although apparently clean and natural, cause untold environmental havoc by flooding upstream areas and reducing water flow downstream. The disruption they cause to natural water flow can encourage the proliferation of disease-bearing organisms, and a large dam failure could conceivably kill hundreds of thousands of people and cause billions of dollars worth of property damage.

Drawbacks of Nuclear Power

Two major categories of negative environmental impacts are associated with the use of conventional nuclear fission reactors. The first are the types of impacts that occur with any large power plant that uses a bulky fuel. Enormous amounts of energy, land, and materials must be used to build the plant. Mining the enormous amounts of uranium ore needed to feed the plant involves substantial energy and potential environmental degradation. A typical light water fission reactor with a 1,000-megawatt capacity may require more than 140,000 metric tons of uranium

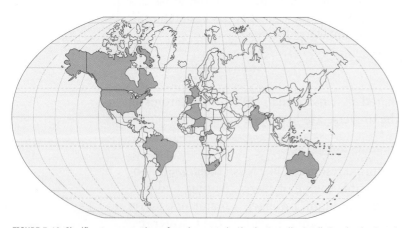

FIGURE 7-13 Significant concentrations of uranium occur in Algeria, Australia, Brazil, Canada, the Central African Republic, Gabon, France, India, Niger, South Africa, Spain, and the United States.

ore to produce the fuel it consumes in a year. The mining operation to obtain this raw ore may extend over 7 hectares (18 acres) and displace as much as 2.5 million metric tons of rock and earth. During routine operations, large quantities of water are used for cooling, and disruptive amounts of waste heat are dumped into the environment (such thermal pollution can be damaging to the natural flora and fauna). Eventually, after 30 or 40 years or less, an aging nuclear power plant must be decommissioned, which also requires enormous amounts of energy and materials.

One of the chief advantages of nuclear power is that it does not produce harmful, carbon-based greenhouse gases such as carbon dioxide; it also does not spew particulates, sulfur dioxide, and similar harmful substances into the environment. However, looking at the entire process, and not just at the actual operation of a nuclear power plant, reveals that greenhouse gases are produced. The building and later decommissioning of nuclear power plants involve the use of energy derived from fossil fuels, as do the mining and processing of uranium ore and the transportation and storage of the uranium fuel and spent fuel. It is sometimes suggested that nuclear power helps to decrease the U.S. dependence on foreign oil, but in fact, many processes associated with nuclear power are driven by oil. In addition, nuclear power is used almost exclusively to generate electricity, and at most, a mere 6% of the oil used in the United States goes toward generating electricity. The real competitor with nuclear power for electricity generation is coal. The pollution from coal burning is known to be causing severe environmental degradation. Nuclear advocates rightfully point out that the large-scale substitution of nuclear power plants for coal-burning plants would avoid many of the problems inherent in coal-burning technology. However, nuclear power generation contributes its own set of wastes and attendant problems.

The second category of disadvantages of nuclear power is inherent and specific to this technology: the dangers of **radioactivity**. The very basis of nuclear power production involves radioactivity. A typical modern nuclear power plant contains within its walls radiation equivalent to that of a thousand Hiroshima bombs. The presence of this much radioactivity leads to fears that radioactive wastes may inadvertently leak into the environment, accidents may occur (perhaps resulting in a reactor meltdown or explosion), or fissionable isotopes may fall into the wrong hands. Compounding these concerns are a number of widely publicized mishaps and the fact that there is still no long-term, satisfactory method of disposing of the radioactive wastes that nuclear power plants generate. Many people are concerned that a single "worst-case" nuclear accident could nullify all the potential benefits of nuclear power. A 1982 study by Sandia National Laboratory in New Mexico predicted that a major accident in the United States might cause 50,000 to 100,000 immediate deaths and as many as another 40,000 subsequent deaths from radiation-induced cancer; monetary damages would amount to at least $100 billion. Even the Chernobyl accident was not a "worst-case" scenario—indeed, some argue it was relatively minor compared with what could happen.

Chernobyl and Three Mile Island
The Chernobyl Disaster

In April 1986, the Number 4 reactor at the Chernobyl nuclear plant exploded, creating the single worst nuclear power plant disaster in history (**FIGURE 7-14**). The Chernobyl plant is located on the Pripet River about 130 kilometers (80 miles) from Kiev in Ukraine. The reactor explosion released an estimated 185 to 250 million curies of radioactivity into the environment. Within the first few days and weeks, according to official reports, 31 people died as a direct result of the explosion and the release of radiation (28 died of acute radiation poisoning, including a dozen firefighters), but the subsequent reports of mortalities are much higher. The International Atomic Energy Agency estimates that there will be 4,000 fatal radiation-induced cancers among the 600,000 people most highly exposed to the radiation. The less-exposed individuals who ultimately will die prematurely because of the incident could number in the tens of thousands. Immediately after the accident hundreds of people were diagnosed as having radiation

FIGURE 7-14 The Chernobyl Number 4 reactor as it appeared after the explosion.

reindeer were declared unfit for human consumption because of radioactive contamination. In Italy, thousands of tons of vegetables were unusable. In Corsica, as a result of eating local contaminated cheeses and other milk products, children were found to have large accumulations of radioactive iodine-131 in their thyroid glands (the gland in the neck that regulates growth). In all, more than 100 million Europeans were subjected to voluntary or mandatory food restrictions during the years after the Chernobyl accident.

Three years after the Chernobyl accident, the official reports stated that the severely contaminated area around the nuclear power plant was much larger than first estimated. Over an area of approximately 10,000 square kilometers (3,900 square miles), contamination was at a level of 15 curies or higher per square kilometer—a very unsafe level. According to some estimates, one quarter million people still lived in badly contaminated areas. Four years after the accident, 627,000 Soviets were under permanent observation for symptoms and effects of radiation poisoning. As of 2003, registries of persons exposed to the Chernobyl radiation in Ukraine, Belarus, and Russia listed 3 million people.

Four years after the incident, the financial costs of the Chernobyl accident totaled about $13 billion, and no one knows the ultimate costs. The direct costs are expected to be greater than the Soviet government's total investment in nuclear power before the accident.

Operator error caused the accident, but various design defects (such as inadequate control systems and the lack of a containment vessel) made matters worse. The accident occurred during a test of the backup electrical system. The operators were supposed to slow the reactor so that it would run at a low level during the test, but they found they had slowed it too much. To speed up the reaction again, they pulled out too many control rods and reduced the amount of water cooling the reactor. The reaction sped up but quickly went out of control. An intense build-up of steam caused an explosion that blew the concrete roof, weighing about 1,000 metric tons, off the reactor and shot radioactive elements into the atmosphere. Burning materials rained down on the plant's roofs and started a number of

sickness, and more than 100,000 people living within 30 kilometers (18 miles) of Chernobyl had to be evacuated. It has been estimated that as many as 40,000 to 70,000 additional cancer deaths (many of them occurring outside the former Soviet Union) may result from the effects of Chernobyl over the next several decades. Dr. Vladimir Chernousenko, a nuclear physicist who was the scientific supervisor of the emergency damage control team sent into Chernobyl after the accident, has suggested that the high levels of radiation released from Chernobyl have damaged perhaps 35 million people.

When the Chernobyl reactor exploded, a thick cloud of radioactive gases rose about 1.5 kilometers (1 mile) into the atmosphere. These gases very quickly covered much of the Northern Hemisphere, affecting areas more than 1,600 kilometers (1,000 miles) away. In Scandinavia, Germany, and Great Britain, plants and animals were contaminated by radioactivity for years and declared unsafe for human consumption. In Lapland, the

fires outside the containment building. Inside, studies indicate that the actual reactor core at Chernobyl underwent a meltdown, burning through a steel and gravel barrier 1.8 meters (6 feet) thick; it remained exposed and burning for 10 days, releasing radiation the entire time.

Only the gallant but ultimately suicidal work of firefighters and other emergency workers kept the fire from spreading to an adjacent nuclear reactor. Thousands of tons of boron, lead, and other radiation-absorbing materials were dropped onto the reactor, but unfortunately, much or all of this material may have missed the actual core. The helicopter pilots could not see through the toxic smoke billowing up as they tried to hover over the exposed core. Liquid nitrogen was pumped under the reactor vessel to cool it, and finally, the damaged reactor was entombed in reinforced concrete to try to contain the remaining radiation (**FIGURE 7-15**). No one is sure how long this tomb will hold; the radiation may begin to destroy it and leak through. Already

there are reports that this concrete tomb is riddled with holes, is structurally unstable, and could collapse at any time. Approximately 2,600 square kilometers (1,000 square miles) of land around Chernobyl will remain contaminated with high levels of radioactivity into the indefinite future.

More than a dozen reactors of the same design type are still operating in Russia, Ukraine, and Lithuania. The other three reactors at Chernobyl were finally shut down in 2000, 14 years after the Number Four reactor melt down. Despite the early predictions of nuclear advocates that the probability of a major catastrophe was negligible, we now know that catastrophes do occur. Chernobyl was not even the first major nuclear accident.

Three Mile Island

In March 1979, a nuclear meltdown nearly occurred in the United States. At the Three Mile Island nuclear power plant (**FIGURE 7-16**) near Harrisburg, Pennsylvania, a minor problem developed in the plumbing of Unit 2. As a result, cooling water drained away from the reactor, and the core began to partially melt—the worst commercial nuclear power plant accident in the United States. Operator errors, a stuck valve, faulty sensors, and design errors are all partially to blame for what happened at Three Mile Island. Some radioactive gas was released because of the accident (only one thousandth as much as at Chernobyl), but fortunately the containment structure around the reactor held in most of the radioactivity. If the

FIGURE 7-15 A view of the entombed Number 4 reactor at Chernobyl.

FIGURE 7-16 The Three Mile Island nuclear power plant near Harrisburg, Pennsylvania.

7.5 Advantages and Disadvantages of Nuclear Power

structure had failed, about 18 billion curies of radioactivity could have been released (the Chernobyl accident released a mere 185 to 250 million curies). Nuclear advocates hailed the Three Mile Island accident as "proof" that a more serious accident would not occur in the United States—for the containment structure did hold. However, many experts concluded that luck more than anything else kept the Three Mile Island accident from being worse. Some scientists calculated that the core, which reached temperatures in the range of 2,700°C to 2,800°C (4,900°F to 5,000°F) or higher, was just short of becoming hot enough to totally melt down. If emergency measures had not been initiated when they were, given another 20 to 30 minutes, the core might have completely melted through the steel reactor vessel and containment unit, releasing all 18 billion curies of radioactivity.

No one knows for certain what the overall negative health effects of Three Mile Island are, in large part because no one is certain how much radiation was actually released. A few days after the accident, the then governor of Pennsylvania, Richard Thornburgh, evacuated all young children and pregnant women from within an 8-kilometer (5-mile) radius of Three Mile Island as a safety precaution. There is evidence that radioactive-induced damage to immune systems contributed to the premature deaths of some older people in the affected area. Dairy farmers reported that many animals died shortly after the accident. Local residents have experienced leukemia and other cancers, and one study suggested that an increase in infant mortality and severe thyroid disorders in babies born after the accident was due to the effects of radiation exposure.

Safer Nuclear Reactors

The typical nuclear reactor is very large and complex. In the United States, the average nuclear reactor has about 40,000 valves, compared with about 4,000 valves for a coal-burning power plant that generates approximately the same amount of electricity. Ten times as many valves means 10 times as many chances for a valve to go bad or be turned the wrong way, perhaps leading to an accident. The complexity of nuclear reactors makes them difficult to build; documented errors include improper weldings, air bubbles in cement, unstable foundations, and even a reactor vessel that was installed backward (fortunately, someone noticed the mistake before the plant was completed).

This complexity makes nuclear reactors very expensive to build. Commercial utilities tend to build large reactors to generate large amounts of energy that can be sold to cover construction costs and still see a profit. Thus, commercial nuclear reactors in the United States typically have a capacity of 1,000 megawatts or greater. Of course, the larger the reactor, the larger the potential catastrophe if anything malfunctions.

Addressing these sorts of concerns, a new series of nuclear fission reactors is being designed. These reactors probably will be smaller and include such features as passive safety systems and modular designs. Passive safety systems depend on natural forces, such as gravity, to cool and control the reactor, especially in the event an accident occurs. Rather than relying on pumps and complex mazes of pipes to bring water to an overheating core, a passive design might have cooling water located in storage tanks around and directly above the reactor vessel. Under normal conditions, the water would not be released, but in an emergency, valves would open, and the water would be dumped directly on the core to cool it. The core itself might be buried in a cavity underground to protect it from terrorist attacks. Researchers have built and successfully tested small experimental reactors incorporating such passive safety systems.

Building standardized reactors made of modular components (series of identical subunits) could reduce the risks of construction errors. Modules prefabricated in a factory would streamline assembly procedures. Standardization and modular construction would also reduce costs and make it more practical to build smaller nuclear power plants. Such smaller reactors might also run at lower temperatures than the reactors currently in use; lower temperature reactors would probably mean overall safer reactors. A reactor design developed in Great Britain has a planned output of only 320 megawatts.

Disposal of Nuclear Wastes

Although nuclear reactors have been online since the 1950s and massive amounts of radioactive wastes have been generated by nuclear reactors and nuclear weapons-manufacturing facilities, no satisfactory plan has been implemented to dispose of these wastes. Nuclear wastes include spent fuel, radioactive products generated in the core of a reactor during operation, contaminated materials and clothing, and radioactive mining wastes and tailings. (Wastes from the use of radioisotopes in medicine, smoke alarms, and the like are also radioactive, but pose a relatively minor problem.) The amount of irradiated uranium fuel waste has approximately tripled since 1989 and is continuing to accumulate at a fast pace (**FIGURE 7-17**). Currently, it amounts to more than one quarter million metric tons.

In the United States, this irradiated uranium is less than 1% by volume of all radioactive wastes but accounts for 95% of the radioactivity. The typical American commercial reactor discharges about 30 metric tons of irradiated fuel each year. This will continue to emit significant quantities of radiation for thousands of years. Currently, more than 30 countries are operating nuclear power plants, yet nowhere in the world has a permanent disposal system for high-level nuclear wastes been fully developed, much less put into place. Instead, the spent fuel rods and other high-level wastes are generally stored "temporarily" onsite in pools of water (**FIGURE 7-18**). So much high-level toxic nuclear waste is now stored in this manner that many reactors are running out of temporary storage space. Low-level nuclear wastes, which occur in much larger volume, generally have been placed in steel drums and the like and buried in shallow landfills. At numerous such sites, the radioactive wastes have leached into the ground, contaminating soils and water. From the 1940s through the 1960s, barrels of radioactive wastes often were dumped directly into the oceans. Now a number of these barrels are leaking and contaminating the seas.

Nuclear waste experts and many nonexperts have suggested a number of ideas for dealing with nuclear wastes. Schemes to bury

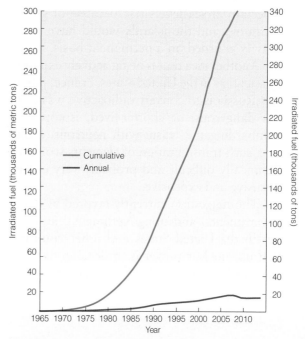

FIGURE 7-17 World generation of irradiated fuel from commercial nuclear power plants, 1965–2005. (*Source:* Based on data and projections from Worldwatch Institute. Vitals Signs, 1995. New York: W. W. Norton, 1995, p. 89. Data extrapolated to 2005.)

nuclear wastes at the bottom of the oceans or inside glaciers in Antarctica have been proposed; however, both the oceans and the ice are delicate ecosystems, and nuclear waste contamination could cause irreparable damage. The heat radioactive wastes give off might even begin to melt unstable Antarctic ice sheets.

Another suggestion is to store radioactive wastes on otherwise deserted islands, at least temporarily, until more permanent disposal facilities are established. Again, such a scheme poses grave risks to the biosphere,

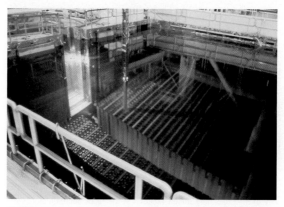

FIGURE 7-18 A fuel rod storage pool at a nuclear power plant.

especially if sea levels rise because of global warming, and the islands would have to be heavily guarded on a permanent basis.

Another idea that is being actively explored by scientists in the United States, France, Japan, and Russia is to convert radioactive wastes to less dangerous, or shorter-lived, isotopes by bombarding the waste with neutrons. However, such transmutation of elements would be technically difficult and probably very energy intensive and expensive.

The suggestion currently favored by many governmental and nongovernmental authorities in the United States and other countries that use nuclear power is to develop disposal facilities buried deep underground in geologically stable regions, perhaps in old mines. The U.S. National Research Council has concluded that geological burial is the "best, safest long-term option" for disposing of high-level radioactive waste. The problem is to find an underground formation that is isolated from the general environment (e.g., groundwater cannot be flowing through the site) and is not in an area of geological activity, such as earthquakes or volcanic eruptions. One must be able to guarantee that the site will remain stable for tens of thousands to millions of years—a virtual impossibility given the active nature of the Earth's crust. Even if a geologically appropriate site is located, social and political considerations can be a problem. When a nuclear waste disposal facility is proposed, local residents tend to give a NIMBY ("not in my back yard") response.

Old salt mines have been proposed as storage facilities, but some mines have water percolating into them that forms a very **corrosive** brine. This solution will damage steel drums or other containers holding the radioactive wastes and allow the contents to leak. In February 2002, Secretary of Energy Spencer Abraham formally recommended to President Bush that Yucca Mountain be developed as a long-term national repository for high-level radioactive waste (**FIGURE 7-19A**). Yucca Mountain is located about 136 kilometers (85 miles) northwest of Las Vegas, Nevada, on land owned by the Shoshone Native Americans (Figure 7-19b). From a political perspective, Yucca Mountain has advantages as a nuclear repository site: the population of Nevada is fairly small, which limits potential opposition, and the site is part of the former Nevada test site where nuclear weapons were tested, thus setting a precedent for nuclear-related activities in the area. Yet it is unclear if Yucca Mountain is appropriate from a geological perspective. Various U.S. Geological Survey geologists and engineers who have evaluated the site have raised questions about it. Geological faults and volcanoes surround Yucca Mountain. Since 1857, eight major earthquakes have occurred within a 400-kilometer (250-mile) radius of Yucca Mountain. Although none of the volcanoes

(a)

(b)

FIGURE 7-19 (a) Interior view of the Exploratory Studies Facilities at the Yucca Mountain proposed national nuclear waste repository. (b) A view of Yucca Mountain showing coring activities.

in the immediate vicinity have erupted in historical times, estimates are that one volcano erupted a mere 5,000 years ago. Yucca Mountain's geological stability for even the next 10,000 or 20,000 years seems questionable.

Predicting the climate of the Yucca Mountain area for the next 10,000 years or so is another significant problem. Currently, Yucca Mountain is located in a desert, and the water table is very deep below the surface. However, given the potential for climatic change (see this text's discussion of global air pollution), there are no guarantees that Yucca Mountain will not experience significantly increased levels of precipitation in the future. Rainwater could not only seep into and leach the radioactive waste, resulting in radioactive runoff, but also could cause the water table to rise and flood the nuclear waste. Possible contamination to groundwater fueled citizen organizing and related legal actions. As of this writing (2012), repository plans have ceased.

Decommissioning Nuclear Reactors

A nuclear reactor cannot last forever. The high temperatures and radiation bombardment, especially the neutron bombardment of the reactor vessel, cause it to weaken and become brittle over time. Eventually, radiation will contaminate and weaken the entire plant. Pipes may begin to break. Instruments may fail, and structural components collapse. The reactor vessel could even rupture. For these reasons, any nuclear power plant, even if it has a flawless record of operation, must eventually be taken out of service—it must be decommissioned.

Many researchers consider the appropriate life span of a good nuclear reactor to be in the range of 30 to 40 years. Until recently, nuclear power plants in the United States were licensed for only 40 years, after which they would have to be decommissioned. Because more than 60 nuclear power plants in the country began operating in the 1960s, we are now at a time when many of these plants should begin planning for decommissioning. However, in June 1991, the Nuclear Regulatory Commission approved a plan that allows plants to apply for a renewal of their operating license for as long as 20 more years.

This ruling may push the costs and problems of massive decommissionings into the future, but many people worry that it may also increase the risks of major accidents as the aging plants continue to operate.

A fully operational large-scale commercial nuclear power plant has never been fully decommissioned in the United States, and scientists are not in complete agreement on what decommissioning will involve or how much it will cost. The spent fuel will have to be removed from the reactor site and eventually stored in some permanent storage facility. As far as the reactor complex itself, several options have been suggested. The plant could simply be closed up, fenced off, and put under guard. Such mothballing is perhaps the cheapest solution in the short run and also allows the plant to be reopened in the future. The plant could be further dismantled in the future, after the levels of radiation have died down. However, in some ways mothballing is also the most dangerous option because the radioactivity is still on site; if not guarded carefully, it could be disturbed purposefully or inadvertently, perhaps resulting in radiation leaking into the surrounding environment. A second option is to entomb the plant—seal it off by encasing it in concrete and steel so that nothing, including radioactivity, can get in or out. Finally, a decommissioned power plant could be totally dismantled. All components and structures would be taken apart or cut up. All materials contaminated with radioactivity, including soil and water, would be removed to a safe facility for the permanent storage of radioactive waste.

Estimated costs to decommission and dismantle a nuclear power plant completely are $3 billion or more. For comparison, building a nuclear plant costs approximately $4 to $9 billion and takes about 15 years. If the plant runs at full capacity for 40 years, it will generate about $12 billion worth of electricity. However, in reality typical U.S. power plants operate at only about 60% capacity.

Global Nuclear Power Today and in the Future

The world's electrical generating capacity of nuclear power plants increased from 1 gigawatt

Barrel. A measure of volume of petroleum oil, containing about 42 gallons or about 159 liters. Depending on the quality of the particular oil, a typical barrel of petroleum may produce between approximately 5 and 6 million BTUs of energy when completely burned (see quad and petajoule).

British thermal unit (BTU). The basic unit of energy in the English system—the energy that, when converted to heat, will raise the temperature of 1 pound of water 1°Fahrenheit. One BTU is equal to 0.252 kilocalories.

Calorie (cal). A basic unit of heat energy in the metric system. A calorie is the quantity of energy that will warm 1 gram of water 1°Celsius. One calorie equals 4.184 joules.

Cubic feet. A measure of volume that is used to measure natural gas. A cubic foot of dry natural gas can produce approximately 1,031 BTUs of energy.

Exajoule (EJ). 10^{18} joules. An exajoule is equal to 1,000 petajoules or 1 million terajoules.

Gigawatt (GW). Energy conversion of 1 million (1,000,000) kilowatts.

Horsepower (hp). A somewhat obsolete (except for cars and mechanical engines in some countries) term for power that is approximately equivalent to 750 watts or three quarters of a kilowatt.

Joule (J). The basic unit of energy, heat, or work as measured using the metric system—equal to a force of 1 newton times 1 meter. From a macroscopic human perspective, this is a very small amount of energy. The burning tip of a wooden match gives off an estimated 1,000 joules of energy.

Kilocalorie (kcal or Cal). 1,000 calories, or 4,184 joules.

Kilowatt (kW). Energy conversion at the rate of 1,000 watts (1,000 joules per second).

Kilowatt-hour (kWh). A measure of energy based on the concept of the kilowatt—the energy converted or consumed during an hour if the energy is being converted or consumed continuously at the rate of 1 kilowatt. A kilowatt-hour equals 3.6 million joules, 860.4 kilocalories, and 3,413 BTUs.

Kilowatt-year (kW-yr). The energy converted or consumed during a year if the energy is being converted or consumed continuously at the rate of 1 kilowatt. A kilowatt-year is approximately equivalent to the amount of energy given off when 1,050 kilograms (approximately 1 metric ton) of coal are burned. If people consume 1 kilowatt-year of energy per year (1 kW-yr/yr), their average rate of energy consumption is 1 kilowatt.

Megawatt (MW). Energy conversion of 1,000 kilowatts (1 million joules per second), or 1 million watts. Many conventional nuclear power plants have generating capacities of 1,000 megawatts or more; that is, they can produce 1,000 million (1 billion) joules of energy per second (enough energy to power 10 million 100-watt light bulbs).

Megawatt-year (MW-yr). The energy converted or consumed during a year at the rate of 1 megawatt. A megawatt-year equals 3.156×10^{13} joules, 7.542×10^9 kilocalories, and 2.993×10^{10} BTUs.

Metric ton. 1,000 kilograms (2,204.6 pounds or 1.10231 English "short" tons). This unit is commonly used to measure amounts of coal (see quad and petajoule). A standard metric ton of coal may produce 27 to 28 million BTUs when completely burned, but the energy that an actual ton of coal produces will vary considerably depending on its grade and quality. Oil is also often measured in terms of metric tons. Depending on the type of oil, the exact number of barrels per metric ton will vary, but it is generally about 7.33 barrels of oil per metric ton.

Metric ton of oil equivalent (toe). Equal to 41.868 gigajoules. Thus, 1,000 metric tons of oil equivalent (1,000 toe) is equal to 41,868 gigajoules or 0.041868 petajoules, and 1 million metric tons of oil (1,000,000 toe) is equal to 41.868 petajoules.

Petajoule (PJ). One quadrillion joules. There are approximately 1,054 petajoules to a quad. One petajoule is approximately equivalent to the energy contained in 163,400 "U.N. standard" barrels of oil (see barrel) or 34,140 "U.N. standard" metric tons of coal (see metric ton).

Quad (Q). One quadrillion (10^{15} or 1,000,000,000,000,000) BTUs of energy. One quad is equivalent to the amount of energy contained in approximately 171 to 172 million barrels of oil, 36 million metric tons of coal, or 1 trillion cubic feet of natural gas.

Terajoule (TJ). 10^{12} joules; 1,000 terajoules is equal to 1 petajoule.

Terawatt (TW). Energy conversion of 1 billion (1,000,000,000) kilowatts.

Terawatt-year (TW-yr). The energy converted or consumed during a year at the rate of 1 terawatt. A terawatt-year is approximately equivalent to the energy given off by burning 1 billion metric tons of coal. The world as a whole consumed approximately 10 terawatt-years worth of energy in the single year 1980; thus, for 1980, we can express the total world energy consumption rate as approximately 10 TW (10 TW-yr/yr equals 10 terawatts). One terawatt-year is equal to approximately 31.6×10^{18} joules, 31,600 petajoules, or 31.6 million terajoules.

Therm. A measure of natural gas, equal to 100,000 BTUs. A therm of gas is equivalent to approximately 97 cubic feet of natural gas.

Watt. A measure of power, energy consumption, or conversion. It is commonly used in connection with electricity. One watt is equal to 1 joule of work or energy per second. A 100-watt light bulb consumes 100 joules of energy per second, or 360,000 joules per hour.

in 1960 to 366 gigawatts in 2004 (**FIGURE 7-20**) (see Fundamentals 7-2 for an explanation of gigawatts and related units).

The heyday of nuclear reactor construction was in the 1970s and 1980s. In 1970, global nuclear electrical generating capacity was only 16 gigawatts, but 20 years later (1990), it was 328 gigawatts. In the last decade, increases in nuclear generating capacity have been small. The role that nuclear power will play in the global energy mix of the future is uncertain. In many areas, there is strong opposition to nuclear power; thus, in North America and Western Europe, no new reactors are being built. For instance, Germany plans to phase out the use of nuclear power and slowly shut down its 18 reactors; by 2025, there may be no commercial reactors operating in Germany. Still, in the early 21st century about two dozen nuclear reactors are under construction. Most of this construction is located in Asia: China, India, Japan, South Korea, and Taiwan. However, with the March 2011 earthquake, tsunami, and nuclear power plant disaster in Japan, the advisability of nuclear reactors in at least some regions may have to be rethought.

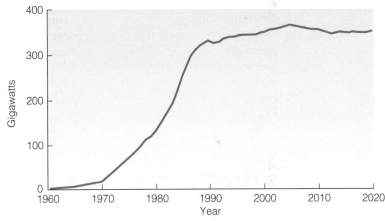

FIGURE 7-20 World electrical generating capacity of nuclear power plants, 1960–2020. (*Source:* Based on data from Worldwatch Institute. Vital Signs, 2005. New York: W. W. Norton, 2005, p. 33.)

■ study guide
SUMMARY

- Global energy consumption has increased dramatically during the last half century.
- There are major discrepancies in energy use around the world—among nations, the United States is both the largest producer and the largest consumer of energy, although the United States consumes considerably more energy than it produces.
- Globally, the big five energy sources are petroleum (oil), coal, natural gas, hydroelectric power, and nuclear power.

- Power plants convert various chemical or nuclear fuels into electrical energy.
- Hard energy technologies rely on large, centralized plants and complex distribution systems.
- Soft energy technologies are on-site, localized systems that are usually more environmentally friendly.
- Major advantages of fossil fuels are their historic abundance and relative ease of distribution and use; disadvantages are their growing scarcity and contributions to environmental

problems, such as local and regional pollution and global climate change, as they emit large quantities of carbon dioxide.

- Nuclear power, based on fission reactions, was developed commercially after World War II (after 1945) as a means of generating electricity.

- A chief advantage of nuclear power is the lack of greenhouse gas emissions; drawbacks include massive amounts of radioactive wastes that must be disposed of and the fear of a major nuclear accident that could far exceed the Chernobyl disaster.

KEY TERMS

blackouts
breeder reactors
brownouts
BTU
calorie
chain reaction
coal
containment structure
control rods
core
corrosive
daughter products
decommissioning
efficiency
electricity
energy
enriched
entropy
fission

fluidized bed combustion
fossil fuels
fusion
generator
hard technologies
hydroelectric power
joule
light water reactor (LWR)
megawatt (MW)
meltdown
milling process
moderator
natural gas
neutron
nuclear power
oil
oil shale
peak load
petajoule

petroleum
plutonium atom
power
power plants
quad
radioactivity
reactor vessel
reprocessing facility
soft technologies
strip mining
terawatt (TW)
thermodynamics
thorium
turbine
tokamak
uranium
watt
work
yellowcake

STUDY QUESTIONS

1. Why is energy consumption important to society, and what concerns does it raise?
2. Name the basic laws of thermodynamics. Why are they important when considering energy transformations?
3. Summarize the current global energy situation. Which countries use the most energy? The least?
4. What is a power plant? How do most power plants operate?
5. Distinguish between "hard" and "soft" energy technologies, citing examples of each.
6. Why is modern industrial society so dependent on fossil fuels? Can we maintain this dependence in the future?
7. What does it mean to say that fossil fuels are a nonrenewable resource?
8. Discuss the pros and cons of each of the major types of fossil fuels: oil, coal, and natural gas.
9. Describe the environmental degradation attributed to the use of fossil fuels.

10. Discuss the potential consequences of a nation being overly dependent on foreign oil supplies.
11. Describe the current role of nuclear power in the global commercial energy budget.
12. Is nuclear power generation a "hard" or "soft" technology? Why?
13. Explain the principles behind modern nuclear fission power plants.
14. What is fusion? In what context have human-induced fusion reactions taken place on Earth?
15. What are some of the advantages of nuclear power? Some of the disadvantages?
16. Briefly describe the Chernobyl and Three Mile Island accidents. How did these incidents affect the public's perception of nuclear power?
17. What methods have been suggested to dispose of the high-level and low-level radioactive wastes that result from nuclear power generation?

18. What steps are being taken to produce safer nuclear reactors?

19. Why is nuclear power so controversial? Is this controversy justified or simply the result of some people's ignorance of the technology involved?

20. Some people have suggested that we get nuclear waste off the planet completely, perhaps by loading it onto rockets and shooting it into the sun. What do you think of this idea? Consider such factors as the costs of such an undertaking, the amount of energy and materials that would be required, the consequences of an accident.

WHAT'S THE EVIDENCE?

1. The authors assert that the burning of fossil fuels has created environmental havoc on a global scale. What are they referring to? Do you agree with the authors' assertion? Are they simply being environmental alarmists who too quickly blame modern Western technology for any ills, real or imagined? Cite evidence to defend your position.

2. The authors suggest that there is still no long-term, totally satisfactory method of disposing of radioactive wastes from nuclear power plants. However, the authors do discuss a number of options that have been suggested to deal with radioactive waste. What evidence do the authors present to bolster their claim that radioactive waste from nuclear power plants is a major problem? Taking the waste generated into consideration, what is your position on nuclear power? Cite evidence to defend your position.

CALCULATIONS

1. Plutonium-239 has a half-life of approximately 24,000 years. If we start with 7.0 ounces avoirdupois (198.4 g) of plutonium-239, how much plutonium will remain after 120,000 years? (Express your answer in both ounces and grams.)

2. Radon-222 has a half-life of approximately 3.8 days. If we start with 9.0 ounces avoirdupois (255.1 g) of radon-222, how much radon will remain after 19 days? (Express your answer in both ounces and grams.)

ILLUSTRATION AND TABLE REVIEW

1. Referring to Figure CS7 1-1, world energy consumption increased by what percentage between 1970 and 2002? Based on the projections given in Figure 7-1, it is estimated that world energy consumption could increase by what percentage between 2002 and 2025?

2. Referring to Figure CS7 1-1, how much more energy did the United States consume than it produced in 1965? Express U.S. energy consumption as a percentage of production in 1965. How much more energy did the United States consume than it produced in 2003? Express U.S. energy consumption as a percentage of production in 2003.

3. Referring to Table 7-1, list the top three countries in terms of their percentage of electricity supplied by nuclear generation. List the top three countries in terms of total amount of nuclear energy supplied.

http://environment.jbpub.com/mckinney/5e/

Connect to this text's website at http://environment.jbpub.com/mckinney/5e/. This site features eLearning, an online review area that provides quizzes, chapter outlines, and other tools to help you study for your class. You can also follow useful links for more in-depth information.

a is
ilding
add
ower to their
by 2016. South Africa's
Department of Energy
reports that the country's
total energy supply will
more than double to
approximately 85 gigawatts
during the next 20 years
with the added expansion
of solar energy. In addition,
hundreds of jobs will be
created, including at least
100 permanent positions
for ongoing maintenance
operations.

8

Renewable and Alternative Energy Sources

Chapter Objectives

After reading this chapter, you
should be able to explain or describe
the following:

- The concept of renewable
  and alternative energy
  sources
- The principles of
  hydroelectricity
- The pros and cons of
  hydropower
- Advantages and
  disadvantages of biomass
  energy
- Ways that humans can use
  solar energy
- The advantages and
  disadvantages of solar energy
- Wind power
- Geothermal energy
- Ocean energy
- Types of energy storage
  systems
- Principles of energy
  conservation and efficiency

Hydroelectric power is most closely associated with the **alternative energy sources**. Along with hydropower, the alternative forms of energy, such as wind-generated electricity or solar thermal power, geothermal power, and combustion of biomass, often are thought of as renewable, sustainable (or at least potentially sustainable), and environmentally benign. These alternative energy sources all provide significant amounts of energy in local settings, but they currently account for only a small percentage of the global energy budget. In the long run, it may be inevitable that these alternatives will have to make up a greater percentage of the energy produced after nonrenewable fossil fuel supplies are exhausted, but before they are used on a larger scale, they must become cost competitive with traditional methods of generating electricity. The most recent Department of Energy statistics (2009) indicate that a little more than 8% of total U.S. energy consumption and 10.5% of electricity generation come from renewable resources. President Barack Obama has made the use of renewable resources a priority of his administration, so that number is expected to increase during the next 2 years.

Solar power, geothermal power, wind power, and biomass combustion (such as wood, dung, urban waste, methane produced from garbage, or alcohol fermented from plants) and even more minor sources of energy (such as ocean tidal power) present alternatives to the **"big five"** energy sources: coal, oil, natural gas, nuclear power (as seen in a review of the fundamentals of energy, fossil fuels, and nuclear energy), and hydropower (discussed in this chapter). We see here that there are trade-offs; some potential environmental problems are associated with all of these sources of energy, and geothermal energy in particular is not strictly a renewable or sustainable energy source. Despite the fact that hydroelectricity is generated from the movement of water, which is a renewable resource, overly large and mismanaged hydroelectric dams can lead to environmental abuse and long-term damage.

Alternative energy sources have received a major boost with the rise of **independent power producers (IPPs)**. IPPs construct electricity-generating plants and then sell the electricity to the large utilities. In many cases, the IPP plants use alternative energy technologies and are relatively small, innovative, and environmentally friendly. Such independent power generation is supported in the United States by the Public Utility Regulatory Policies Act of 1978, which allows smaller, independent, relatively unregulated power companies to build generating plants that use renewable fuels (such as wind or solar power). However, IPPs are not just an American phenomenon; they are appearing around the world and herald a new emphasis on alternative energy sources.

Along with the development of alternative energy sources, it is crucial that increased attention be paid to energy conservation and efficiency—doing more with less energy. Energy conservation and efficiency have a ripple effect. As a simplified example, less household electricity used means less coal needed to burn at the power plant, which means less air pollution, which results in fewer health problems and more productivity, which translates into greater economic growth, and on and on. Already, great strides are being made in this area, and it is likely to attract greater interest in the future.

8.1 Hydropower

Hydroelectricity (**hydropower**) is the fourth largest source of commercial energy production and consumption globally—trailing far behind oil, coal, and natural gas. Nuclear power comes in a close fifth by most current estimates but has been catching up quickly in recent years. Unlike fossil fuels and nuclear power, hydropower is a renewable resource. The basic principle behind this age-old source of energy is the damming of rivers to create artificial waterfalls (**FIGURE 8-1**). Natural waterfalls can be used, but the water flow must be reliably constant. In fact, the first major hydropower plant was completed at Niagara Falls in 1895. The falling water is used to turn turbines that drive electrical generators. The great advantage of hydropower is that after the dam is built and the turbines are in operation, it is a relatively cheap and very clean source of electrical energy. An operating hydropower plant produces no waste; it does not release carbon dioxide or other pollutants into the atmosphere, as do fossil fuel-powered plants, and no toxic, hazardous, or radioactive waste is generated (as occurs with nuclear power plants). An operating hydropower plant can be safe and efficient. For this reason, some people strongly support the construction of more hydropower plants to displace the use of fossil fuels and nuclear power. Nevertheless, the United States seems unlikely to build any more large-scale hydroelectric power plants in the near future because dams for the most cost-effective

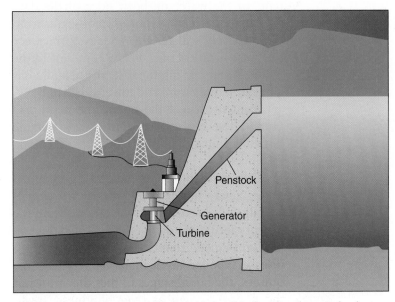

FIGURE 8-1 A cross-section of a typical hydroelectric dam. Water flows down the penstock and turns the turbine blades. The generators, rotated by the turbines, produce the electricity.

regions have already been developed. However, scientists are working on more efficient turbines, and hydroelectric power is growing internationally. Hydroelectric power currently produces 6% of all the energy in the United States; however, of all energy produced through renewable sources, hydroelectric power is responsible for more than 70%. In the United States, hydroelectric facilities can supply 28 million households with electricity. This saves nearly 500 million barrels of oil a year.

Environmental Disadvantages to Hydroelectric Energy

Despite its advantages, hydropower also has drawbacks. The construction of a major hydroelectric power plant requires large quantities of fossil fuels to power machinery and move materials. Large-scale hydroelectric operations purposefully flood huge tracts of land and seriously disturb natural ecosystems. Forests and other natural habitats are destroyed, along with local villages (often inhabited by indigenous peoples). Farmland is lost. People are displaced. Erosion rates are increased. Water may become polluted and choked with particulate matter. Water flow is affected downstream, and severe unanticipated flooding can occur.

After a hydroelectric plant is built across a river, the mineral and dissolved oxygen contents of the water on both sides of the dam are often affected. Evaporation of water from the relatively still reservoir tends to increase the concentration of salts and minerals in the water. Water leaving the reservoir usually has a higher temperature and lower dissolved oxygen content than it did before the plant was built. These changes can radically affect the flora and fauna that live in the river and along its banks.

Decomposing plant material submerged under newly flooded reservoirs releases carbon dioxide locked up in the biomass and the potent greenhouse gas methane. The dams can also create conditions that favor the spread of disease. The larvae of malaria-carrying mosquitoes, the larval stages of flies that transmit the disease known as river blindness, freshwater snails that are the hosts for the parasites (schistosomes) that cause bilharzia, and other disease vectors can be favored by water changes that artificial dams bring about.

Large dams can also affect coastlines and coastal fisheries by obstructing the passage of nutrients and particulate matter to the sea. The construction of the Egyptian Aswan Dam resulted in the trapping of nutrients that previously flowed into the Mediterranean, resulting in damage to the Mediterranean sardine fishery. Similarly, by trapping sand and silt, dams can cause coastlines to flood. Normally, the new sand and silt would replace material that is eroded from the coastal beaches; when the dam cuts off this supply of new material, the beaches continue to erode, and the ocean creeps landward. After the Volta Dam was built in West Africa, some 10,000 people on the coast of Togo reportedly lost their homes to the Atlantic Ocean in this manner.

In many ways, giant hydroelectric power plants share the problems of other large power plants fueled by coal, nuclear energy, or some other energy source. The setup expenses of a major hydropower plant are enormous. The plant must be carefully maintained and operated. The consequences of a major accident or failure, such as the dam cracking and giving way, could be catastrophic. At 55 meters (180 feet) high, the St. Francis Dam, a reservoir dam 80 kilometers (50 miles) north of Los Angeles, collapsed in 1928. It is estimated that at least 470 people died in the resulting flood. In contrast to popular belief, most large hydropower projects have only a limited life span. The critical water reservoirs behind the dam often fill up with the sand and silt that are not allowed to pass downstream. Such silting can occur very rapidly, especially if the upstream watersheds are deforested and/or subjected to poor agricultural practices that lead to accelerated erosion. After the reservoir fills with silt, the usefulness of the hydropower plant is greatly diminished. The deep reservoir is necessary for storing as much water (energy) as possible during periods of high rainfall and runoff and for regulating production so that a steady stream of electricity can be produced when needed, especially during dry seasons. Without the reservoir, the plant becomes an intermittent source of electricity at best.

In the United States, most dams built in the early and mid-20th century were very large. The U.S. Bureau of Reclamation constructed numerous large plants during this era. The Grand Coulee Dam in Washington is the largest (**FIGURE 8-2**). Built between 1933 and 1941, it has an electrical energy output capacity of approximately 6,800 megawatts. Hoover Dam, built in 1935, has an output of just more than 2,000 megawatts; however, many operating hydroelectric plants in the United States are much smaller. Currently, the United States has approximately 2,400 hydropower facilities and 79,000 dams, with an estimated 78,000 megawatts of hydropower generating capacity. The remaining dams are located in areas that do not lend themselves to easy hydroelectric power production. Globally, 17% of total electricity generation and nearly 90% of renewable electricity generation comes from hydropower. Central and South America generate nearly 70% of all their electricity through hydropower.

A Federal Energy Regulatory Commission study identified another 5,000 sites that have the potential to sustain hydropower facilities. If all of these sites were developed, they would contribute another 74,000 megawatts of capacity, doubling the United States' use of hydropower. Such statistics can be misleading, however. Some of the best sites for development of hydropower capacity are found in scenic areas, parks, and wildernesses. Federal, state, and local legislation and regulations (especially the federal 1968 National Wild and Scenic Rivers Act) will hamper or stop the building of hydropower plants on many prime sites. Studies of the remaining sites indicate that most would be uneconomical to develop given the current relatively inexpensive cost of energy in this country, but the primary hindrance to additional development of hydropower seems to be the environmental and social disruption caused by large dams. By one estimate, 30 to 60 million people worldwide during the last century have been forcibly displaced from their homes to make way for large hydroelectric and reservoir projects, and many of these persons belong to indigenous groups or are ethnic minorities with little political

(a)

(b)

FIGURE 8-2 (a) The Grand Coulee dam is the largest hydroelectric power plant built on the Columbia River. It produces enough electricity to supply the entire Seattle metropolitan area. (b) A turbine is lowered into place in the new, third powerhouse at the dam.

clout. In the United States, current research is addressing small-scale hydroelectric facilities that would supply much smaller communities with more limited environmental impact because building a dam is not needed.

The world's largest hydroelectric dam, located in the Three Gorges region of the Yangtze River in China, stretches over 2.3 kilometers (1.4 miles) across the river and stands 185 meters (607 feet) high. To make room for the more than 660-kilometer (410-mile) long reservoir, an estimated 1.2 to 1.9 million people were relocated, with enormous social disruption and great expense. Reported costs indicated a $24 billion price tag; many analysts believe the actual cost was much more than that. More than 100 workers lost their lives during the construction of the dam. A reported 632 km² (244 mi²) of land was flooded to create the reservoir, which resulted in flooding thousands of villages and towns that held great cultural significance for the displaced. The history and artifacts of the Ba people who lived in the region 4,000 years ago are among the archeological treasures now below the 175-meter (574-foot) deep reservoir. Although the government promised compensation for those forced from familial lands, reports of people not receiving promised compensation have been documented. Factories, waste dumps, and other industrial buildings were also submerged. Humans were not the only species to suffer because of the dam construction.

Annual fish harvests are down 50–70%. The Yangtze River basin is home to more than 30% of all China's freshwater fish species. Three of China's ancient fish species, the river sturgeon, the Chinese sturgeon, and Chinese paddlefish, are experiencing record decline and are threatened or endangered. The saddest case is of the freshwater dolphin, which at this writing (2011) is believed to be extinct. Scientists had been trying to secure a captive breeding program, but it failed. This is the first human-driven extinction of a large marine vertebrate in 50 years; the freshwater dolphin was the last of its evolutionary line. Numbers had been declining (to about 150 before the dam's construction) because of a number of management issues, but the loss of habitat after the dam was built is what scientists believe ended the species' battle for survival.

The Chinese government insists constructing the largest hydroelectric dam in the world was necessary to prevent future loss of human lives caused by flooding. In the 20th century, an estimated 300,000 people in the region of the three rivers lost their lives because of flooding. In addition, the shipping industry is growing, building the economy of the region, with approximately 15 million more tons of cargo per year shipped through the region as a result of the dam. The dam has a generating capacity of more than 18,000 megawatt (MW). By 2012, the Chinese government hopes to bring six more turbines online, for an unprecedented capacity of 22,500 MW. Because of silt build-up, several more dams are in the works upstream.

Although the dam held up to major flood pressure in the record floods from May to August of 2008, there were many landslides during the storms, depositing thousands of tons of sediment. In 2010, the Chinese government admitted persistent issues with landslides in some places along the banks, resulting in more human displacement. Many scientists warn that this structure might not hold up to a major earthquake; the dam was built in a geologically active area. In 2010, small-scale earthquakes were documented leaving visible damage to places along the banks. People who relocated to these areas when the dam was built were again displaced. Also of ongoing concern is an inadequate waste management system and algal blooms caused by natural water flow disruption. Despite all these challenges, Chinese officials deem the project a success because at capacity, the project saves an estimated 50 million tons of coal a year and decreases CO_2 emissions by 100 million tons. China media reports celebrated the fact that in 2008 the Three Gorges Dam did what it was meant to do: held up to the record flood waters of 2008 and saved countless lives. Internationally, controversy continues surrounding the dam as China continues to lay the groundwork for future construction there and abroad.

Small-scale hydropower plants may be more compatible with the natural environment and entail fewer negative social impacts. Some small-scale plants do not even use dams. Located on fast streams and rivers, such damless facilities direct water flow directly through the turbines located in the middle of the channel or off to one side. Damless facilities can avoid many of the undesirable environmental impacts associated with dams, but they require a fast rate of water flow and are suitable for only a very small number of sites (**FIGURE 8-3**).

Most types of small-scale hydropower facilities use dams, but many take advantage of smaller dams with lower heads (the distance the water drops from the top of the reservoir to the bottom of the dam). It has been suggested repeatedly that many already existing small dams could be fitted with electrical generators. In the United States alone, there are approximately 75,000 to 80,000 dams and reservoirs on rivers and streams, mostly small dams that are not currently used to generate electricity. These dams were built for navigation, flood control, irrigation, and other uses. Some proponents of hydroelectric power advocate developing this potential resource by fitting these small dams with the equipment to generate electricity. However, such small-scale hydroelectric facilities can provide power to people only in the immediate vicinity of the plant or dam. Such small operations may prove to be relatively unreliable, intermittent energy sources. In addition, given the continued presence of cheap fossil fuels in the United States, the costs of converting these dams to energy-producing facilities may be prohibitive. Instead, the recent trend has been to decommission and remove dams to restore natural waterways and ecosystems. In the last 30 years, hundreds of small dams have been demolished in the United States.

Another way of increasing the hydropower capacity of the country is to upgrade and refurbish existing hydropower plants to increase their capacities by as much as 10%. Upgrades to existing power plants, an ongoing process since the late 1970s, have added the equivalent of another Hoover Dam and Power Plant to the energy production pipe-

FIGURE 8-3 A 1,200-kilowatt hydroelectric plant in Kankakee, Illinois. The hydroelectric plant provides all of the electricity for the city's wastewater treatment plant. Excess electricity is sold to a local power company at a profit to the city.

line at less than 20% of the cost of generating the same amount of energy from a new coal-powered facility. An added advantage of refurbishing and upgrading is that additional environmental deterioration can often be minimized. For instance, replacing old turbines with new, more efficient versions may place no additional pressure on the environment. However, reconstructing and enlarging a dam might.

Globally, the prospects for additional hydropower development look more promising. In particular, the developing world continues to increase its hydropower capacity, using smaller facilities as well as huge centralized hydropower plants. Localized, small-scale energy production may be more feasible and compatible with local economies and energy use patterns in some developing countries than it is in many heavily industrialized nations that are wedded to centralized grid systems of power production and transmission. For instance, China has more than 86,000 small hydroelectric power plants in operation. There are hundreds of thousands of dams already in existence around the world that are not currently used to produce hydropower. Even though smaller plants tend to be somewhat more expensive per kilowatt of capacity than large plants, converting these dams to small-scale hydroelectric power plants may be relatively efficient and inexpensive and may have minimal detrimental impacts on the environment.

Since the middle of the last century, worldwide hydroelectric generating capacity has increased by a factor of 15 or more. Currently, the United States and Canada each contain about 9% and 12% of the world's hydroelectric capacity, respectively. Russia maintains about 5%, Brazil about 11%, and China currently 18%. Experts estimate that the current global capacity represents about 15% to 25% of the world's theoretical hydroelectric potential, but certainly 100% of the global potential will never be reached because of environmental and social concerns about hydropower. Although energy generated by hydropower will continue to increase globally during the next few decades, it is doubtful if in the near future it will ever contribute as much power as oil, coal, or natural gas currently do.

8.2 Biomass

The burning of biomass (**FIGURE 8-4**), especially fuelwood, has served as a major source of energy for most of recorded history. Even today, billions of people, perhaps more than half of the world's population, burn wood or other plant or animal products (such as animal waste) as their principal source of energy. Biomass provides as much as 10% to 15% of all energy consumed worldwide; in some developing countries, it provides as much as 90% of the energy consumed.

In the United States, biomass supplies approximately 4% of energy demands. The paper and pulp industry, which burns large quantities of wood and paper milling wastes to supply energy for its needs, accounts for much of this. Other substantial consumers of biomass in the United States include households that burn wood or wood products (such as fuel pellets made of sawdust) as a primary source of heat (about 5% fall into this category, and another 20% occasionally burn wood in a stove or fireplace), commercial industries and establishments that burn wood as a source of energy (in some cases, simply for space-heating purposes), and **waste-to-energy** facilities that burn municipal solid waste (as a review of municipal solid waste and hazardous waste shows). Biomass-derived fuels, such as ethanol, currently account for a very small percentage of the U.S. energy budget, although approximately 8% of the gasoline sold in the United States contains ethanol.

Raw Sources of Biomass Energy

Sources of **biomass energy** can be classified into three major categories: wastes (that is, material that is considered "waste" from other perspectives), standing forests, and energy crops.

Wastes

Wastes include such things as wood scraps, unusable parts of trees, pulp residue, paper scraps that the wood and paper industry generate, and municipal solid waste. During lumbering operations in timber forests, many sticks, branches, leaves, stumps, and roots are left over. Potentially, these could be collected and burned for energy (however, they are important sources of nutrients for regrowth in forest ecosystems). Likewise, agricultural wastes from food crops, such as leaves, stems, stalks, and even surplus or

FIGURE 8-4 The Tracy Biomass Plant in the San Francisco Bay area provides renewable electricity from wood residues discarded from agricultural and industrial operations. The plant is rated at 21 megawatts.

damaged food items, can serve as a source of biomass energy. In the United States, surplus corn is the principal raw material for the ethanol industry. A final large source of biomass energy is the organic material found in municipal solid waste (as shown by a review of municipal solid waste and hazardous waste).

Standing Forests

Natural standing forests afford a vast source of biomass energy, but unless harvested in a low-impact sustainable manner, they are a nonrenewable resource. Furthermore, diverting wood from forests for the purpose of energy use may compete directly with interests in other forest products, such as lumber for building and the furniture industry. Of course, clear-cutting forests for any purpose presents major environmental concerns (as shown by a review of land resources and management). All too often, timbering results in soil erosion and ecosystem destruction.

Energy Crops

Many advocates of increased use of biomass energy envision crops grown specifically for fuel purposes. Fast-growing varieties of trees, as well as grasses and other crops, could be raised commercially on large monoculture energy farms. The biomass would be harvested and the fuel burned directly or converted into other types of fuels (such as biogas, methanol, or ethanol). **Coppicing trees** (ones that will regenerate from stumps left in the ground) and perennial grasses may be particularly well suited to energy farming. The crop could be clear-cut, but no further planting would be necessary because a new crop would grow from the stumps and roots already in the ground. Such dedicated biomass plantations require vast amounts of land—land that is fertile and could potentially be used for other purposes such as food crops. Indeed, growing crops for energy generally requires more land than any other energy-generation technique (TABLE 8-1).

One country in particular, Brazil, has had great success in growing energy crops. In the 1970s, spurred by the uncertainties and volatility of the oil market, the policy makers of Brazil decided to put a major effort into a sugarcane-based ethanol program as an alternative to gasoline. Thirty years later the result is that many Brazilians drive "flexible fuel" cars that can run off of ethanol, gasoline, or a mixture of both. Furthermore, Brazil is exporting ethanol, and ethanol can be substituted for many fossil fuel uses.

TABLE 8-1	Land Requirements for Various Electric Generation Technologies	
Technology Land Requirement	Square Mile-Years per Exajourle*	Square Kilometer-Years per Exajoule*
Dedicated biomass plantation	48,250–96,500	125,000–250,000
Large hydro	3,204–96,500	8,300–250,000
Small hydro	66–6,562	170–17,000
Wind†	116–6,562	300–17,000
Photovoltaic central station	656–1,274	1,700–3,300
Solar thermal trough	270–1,158	700–3,000
Bituminous coal	259–1,274	670–3,300
Lignite coal	2,586	6,700
Natural gas-fired turbine	77–259	200–670

A square mile-year is a square mile used for 1 year, and a square kilometer-year is a square kilometer used for 1 year. One exajoule is equal to 1,000 petajoules, which is equal to approximately 0.949 quadrillion BTU (quads); in 2002, the total world commercial energy consumption was approximately 434 exajoules or 412 quads (see review of fundamentals of energy, fossil fuels, and nuclear energy). In addition, many alternative energy sources, such as wind and solar power, actually require less land than some conventional energy sources, such as coal and hydroelectric power. For instance, large hydroelectric power plants can require the flooding of vast tracts of land, and strip mining for coal can devour huge land areas (often leaving the land devastated and virtually unusable once the coal has been mined).

*End-use energy figure averaged over assumed 30-year life cycles for power plants, mines, and so on.

†The lower range for wind includes only land occupied by turbines and service roads, whereas the higher number includes total area for a project.

Source: Data are adapted from L. R. Brown, et al. State of the World, 1995. New York: W. W. Norton, 1995, p. 73. Reprinted with permission of Worldwatch Institute, Washington, D.C. Copyright © 1995.

Environmental Advantages and Disadvantages of Biomass Energy

Biomass energy has become a rather controversial subject among environmentalists, in part because of the diverse sources and technologies that fall into this category. The environmental appeal of biomass energy is that, in theory, fuelwood or other biomass fuel is a renewable energy source; however, in practice, it often is not. To be truly renewable and cause no deleterious environmental impact, every time a tree from a standing forest was cut down and burned, a new tree would have to be planted and assured of surviving. During growth, the new tree theoretically would absorb the carbon dioxide that the burning of the old tree gave off.

The reality is that forests are often harvested for fuelwood in an unsustainable manner, leading to their destruction and ultimately to desertification. For example, one of the world's poorest countries, the island nation of Haiti, has just 1% of its forest remaining because Haitians have little else to use as fuel. In fact, harvesting wood for fuel is one of the major causes of the world's deforestation problems. Burning forest or agricultural residues, dung, or other animal wastes can deprive the soil of nutrients that should be recycled back to the earth, reducing the soil's fertility and making it much more difficult for the land to support either natural vegetation or human-planted crops. Loss of forest cover can mean major soil erosion, landslides, and other disasters, especially when the rains come. The burning of fuelwood and charcoal derived from fuelwood not only contributes to the global greenhouse problem, but also can be a major component of local and regional air pollution. Nairobi has often found itself blanketed in a smelly haze because of the burning of wood and charcoal.

Fuelwood gathering, biomass burning, and their attendant problems are causing an environmental nightmare in many developing countries. However, approximately half of the world's population uses fuelwood or charcoal on a daily basis. At the close of the 20th century, the U.N. Food and Agriculture Organization estimated that approximately 80% of wood harvested in developing countries was for fuelwood, compared with more than 80% of wood harvested in developed countries, which went toward industrial uses. About a quarter of total energy in Asia is estimated to be supplied by fuelwood (in Nepal and Bangladesh, it may be more than 90%), and other parts of the developing world, such as Africa and Central America, are highly dependent on fuelwood as well (**FIGURE 8-5**). In some African countries, such as Ethiopia, the Congo, and Sudan, fuelwood may supply 75% to 90% of energy needs. There are currently major fuelwood shortages in parts of Africa, the Middle East, Asia, and Latin America. As many as 3 billion people may be experiencing a shortage of fuelwood.

Traditional biomass use, especially in developing countries, is typically very inefficient: from 3 to 10 times more energy may be used than is actually needed to accomplish the task at hand. For example, most of the heat is lost when food is cooked over an open fire. A simple but more efficient stove can cut the wood used by one third to one half. In some cases, wood use could be cut even further by introducing solar box cookers. These cookers use no wood or animal waste at all, but they are not so useful on a cloudy day. **Biogas digesters** convert manure and other organic wastes to methane for use in cooking, heating, lighting, and other purposes.

In industrial countries, some people are strong advocates of the waste-to-energy approach, essentially burning garbage in specially designed incinerators to generate electricity. The major problem is that this approach discourages reuse and recycling, and the facilities can generate large quantities of air pollutants and hazardous waste.

Energy farms where crops such as fast-growing trees or grasses are raised specifically for use as fuel also have their advocates, but critics argue that these farms will suffer from the same problems as traditional food farms, including soil erosion and massive pesticide, herbicide, and fertilizer use. Although improving management techniques could reduce soil damage and losses, raising energy crops would create another problem by displacing food crops in a world where there is already a shortage of

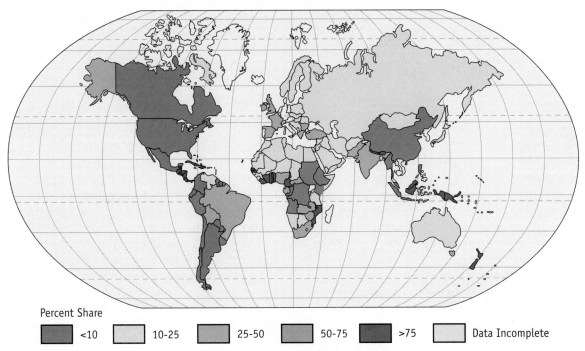

Percent Share

| <10 | 10-25 | 25-50 | 50-75 | >75 | Data Incomplete |

FIGURE 8-5 Percentage of overall energy consumption derived from fuelwood (including charcoal derived from fuelwood). (*Source:* Data from International Energy Agency.)

food in many areas. Raising energy crops on marginal lands that are not already cultivated for food is not a solution either, for those lands may be prone to environmental degradation and perhaps should not be planted.

It is unlikely that energy crops, even under the best of circumstances, could ever supply the majority of the energy requirements of a major industrial nation. For instance, even if the entire annual U.S. corn crop were converted to ethanol, it would supply only one fifth of the country's motor-vehicle fuel needs. To make matters worse, in some cases, the biomass conversion processes are very inefficient. In some instances, converting corn to ethanol actually requires more energy than the energy value of the final ethanol produced, but this inefficiency occurs largely because corn is not really a good energy crop (it is used because there is a surplus left over from subsidized food production, as shown by a review of food and soil resources). It would be more efficient to use ethanol made from fast-growing grasses; some varieties may contain five times as much energy as was used to grow and process the plants.

A major disadvantage of any biomass burning is the air pollution it can generate.

Biomass burning or combustion of biomass-derived fuels can emit carbon monoxide, nitrogen oxides, and particulate matter (such as ash and soot) into the air. The burning of biomass found in municipal solid waste can be extremely dangerous from this perspective, potentially releasing known carcinogens and heavy metals into the environment. Some have suggested that alcohol-based biofuels may release significant quantities of formaldehyde and other aldehydes (known carcinogens) when burned, but some studies indicate that such emissions from biofuels are less than those produced from comparable fossil fuels. Biomass burning can also contribute to the greenhouse effect when done unsustainably (i.e., when replacements are not planted for all the materials burned).

The long-term role of biomass in the energy mix of modern developed countries is uncertain. Biomass energy, in the form of biofuels such as ethanol, may be most important as a transitional measure between our virtually total dependence on fossil fuels and the full-scale introduction of a truly clean and environmentally benign energy system, such as a **solar–hydrogen economy** (see **CASE STUDY 8-1**).

The Natural Gas Transition to a Solar-Hydrogen Economy

It is very clear that in the future the world will have to stop burning fossil fuels at the high rate of the past few decades. Nevertheless, the world will continue to require large amounts of energy. The size of the human population continues to grow, and the developing nations rightfully want to improve the living conditions of their peoples. Sooner or later, a post-fossil fuel energy economy will be necessary.

What energy path to follow in this post-fossil fuel world has been the subject of much discussion, but two major routes have been advocated:

1. A strong reliance on nuclear power supplemented by a mix of other energy sources, such as hydroelectric, solar, wind, and geothermal.
2. A strong reliance on solar and other "clean" and renewable energy forms, combined with hydrogen gas as a way to store and transport energy conveniently—a so-called solar–hydrogen economy.

The idea behind a solar–hydrogen economy is that it would be based on renewable energy from solar power plants (using photovoltaics or thermal solar technologies) and wind- and water-driven power plants, supplemented by such technologies as geothermal power, tidal power, and combustion of biomass to a limited (and sustainable) extent. The primary form of power generation would be electricity, but it would be dependent on the sun shining brightly or the winds blowing hard. Peak electricity production and peak demand would not necessarily correspond to the weather conditions in many areas, which would create a problem because electricity is notoriously difficult to store. Surplus electricity (not used directly by private consumers or industry) would be used to subject water to electrolysis to form hydrogen gas. Essentially, an electrical current passed through water will break the water (H_2O) into hydrogen (H_2) and oxygen (O_2). The hydrogen can then be stored and used when needed to generate electricity or heat, in a manner very similar to natural gas. When hydrogen gas is burned, it releases energy as it recombines with oxygen in the atmosphere to produce water.

Burning hydrogen gas is very clean (small amounts of nitrogen oxides [NOX] may form, but it is hoped that with proper technology, this hazard can be minimized). Furthermore, water to form hydrogen gas is abundant, and the water can be recycled as the hydrogen combines with oxygen to form water once again. Advocates of a solar–hydrogen economy point out that with the proper technology, much of which already exists, hydrogen gas can be used in place of virtually all fossil fuels. Hydrogen could be used to heat buildings, run appliances, power motor vehicles, and fuel airplanes. Alternatively, hydrogen-burning fuel cells, in which the hydrogen is once again chemically combined with oxygen to produce an electric current, could run electrical appliances or motors (**FIGURE 1**). Automobiles that use electric motors and hydrogen-burning fuel cells could become standard in the next 30 years.

However, hydrogen is bulky relative to the amount of energy to be derived from a cubic unit, and it is dangerously explosive. Still, these drawbacks seem relatively minor. After all, fossil fuels, especially natural gas, can be very dangerous, explosive substances, too. As for the bulk of hydrogen gas, this consideration may be negligible if, as some suggest, many hydrogen-powered vehicles may be as much as twice as energy efficient as their fossil fuel-powered equivalents.

Even strong proponents of a solar–hydrogen economy realize that the world cannot switch instantaneously to this type of energy mixture. They envision a transitional stage in which the world switches first to a relatively abundant and relatively clean fuel—natural gas (although underground hydraulic fracturing, commonly known as "fracking," has come under heavy criticism as causing tremendous environmental damage, including the contamination of underground water supplies by benzene, a carcinogen, which can literally cause water to ignite when exposed to a flame). Natural gas use and technology are already familiar, and natural gas burns much more cleanly than do oil or coal (substituting natural gas in many situations decreases carbon dioxide emissions by 30% to 65% and decreases other air pollutants by 90% to 99%). Natural gas is also relatively abundant. As natural gas technologies and infrastructures are developed and put into place, they will form the basis of future hydrogen use. Much

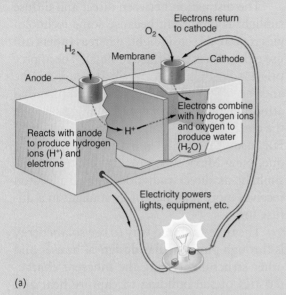

VISIT

http://environment.jbpub.com
/mckinney/5e/
for more information

(a)

(b)

FIGURE 1 (a) A hydrogen fuel cell produces electricity from hydrogen and oxygen. (b) An example of a fuel cell, produced by Ballard Power Systems Inc. of Burnaby, British Columbia, Canada. (*Source:* Photo courtesy of Ballard Power Systems Inc.)

of the hardware (piping systems, furnaces, engines, and so on) that is used for natural gas can easily be converted to hydrogen use when the time comes.

In summary, the steps to a sustainable solar–hydrogen economy would include the following:

1. Substitution of natural gas for other fossil fuels and the development of a natural gas-based technology and infrastructure;

2. The concerted development of solar-powered electrical generating facilities (supplemented by other renewable energy sources) to replace other types of electric power plants;

3. Continued research and development in hydrogen production, storage, transportation, and use; and

4. Finally, the complete substitution of hydrogen for natural gas.

8.3 Solar Energy

Much of the energy we use currently is indirectly **solar energy**. Fossil fuels, wood, and other biomass combustibles are the result of organisms that trapped the sun's energy into a form that we can conveniently employ. Hydropower and wind power also derive from the sun, in that the sun differentially heats the atmosphere, causing the winds, and evaporates water and recycles it as rain, which lets our rivers flow. Potentially, the sun provides the Earth with more energy every day than humans would ever be able to use. Although humans currently harness energy at a rate of about 13 terawatts (TW), the light and heat of the sun deliver energy to the Earth's surface at a rate of about 80,000 TW. In just 20 days, the Earth receives energy from the sun equal to all of the energy stored on Earth as fossil fuels.

People have long been aware that they can harness solar energy directly. For thousands of years, people have known they could build

FIGURE 8-6 This greenhouse in Cornwall, England, where the annual maximum temperature averages 12.7° Celsius (55°F) can sustain a humid tropics biome.

their homes to face the sun and take maximum advantage of incoming solar energy to heat the living quarters. Similarly, gardeners have long known how to trap the sun's energy in greenhouses, even simulating relatively tropical conditions in temperate climates (**FIGURE 8-6**). Much of the modern technology for solar energy was developed during the space race in the United States. Modern solar technology encompasses old and new themes: (1) simply trapping solar heat and light and redirecting it to purposes useful for people and (2) directly or indirectly converting solar energy into electricity. The Million Solar Roofs Initiative is a unique public–private partnership, aimed at overcoming barriers to market entry for selected solar technologies. The goal of the initiative is practical and market driven: to facilitate the sale and installation of one million "solar roofs" by 2016. By the end of 2010, California was generating more than 1 MW of energy from solar power.

Direct Use of Solar Energy and Passive Solar Designs

Sunlight striking the surface of the Earth at any one spot may be either direct or diffuse, although there is a continuous gradation between the extremes. Direct sunlight has traveled through the atmosphere with very little scattering or diffraction; generally, direct sunlight is bright and leaves sharp shadows. Direct sunlight can be concentrated with mirrors and lenses to form very energetic, focused beams. Diffuse sunlight has been scattered and diffracted by the atmosphere, clouds, and haze before reaching the surface of Earth. To the naked eye, diffuse sunlight does not appear as bright as direct sunlight, and it leaves soft shadows.

The distinction between direct and diffuse sunlight is important because some technologies require direct sunlight, whereas others can use either direct or diffuse sunlight. The appropriate solar technology will depend on the area because the amount of cloud cover and atmospheric turbulence experienced through the course of a year varies from place to place. For instance, in the United States, Albuquerque, New Mexico, receives about 70% of its sunlight as direct radiation, whereas Boston receives more than half of its sunlight in a diffuse form.

The time-honored way to use solar energy is through **passive solar design** of houses and other structures—using the inherent characteristics of the building to capture heat and light from the sun. Such architecture has been practiced for thousands of years but to a certain extent has been lost or at least greatly de-emphasized with the coming of the industrial age. With cheap, convenient fossil fuels readily available, architects and builders no longer worried about orienting houses toward the south (in the Northern Hemisphere) to capture the heat and light during the winter months. Likewise, windows were no longer required as a primary source of light after electric lights were widely introduced; in many buildings today, electric lights are used for illumination during daylight hours because some rooms do not even have a single window.

With passive solar design, the entire building becomes a solar collector. Passive solar strategies include the installation of large, south-facing windows and few or no north-facing windows. During winter, the windows allow incoming sunlight and trap heat. New types of glass with better thermal and light properties are available, including superinsulated glasses and electrochromic windows that can change their optical properties in response to small electrical currents and either block or admit sunlight as necessary to maintain a comfortable interior temperature. Windows and skylights and light pipes also provide natural lighting that can reduce the need for energy-burning electrical lighting. Even simple methods such as mirrors and brightly painted surfaces can be used to illuminate the deep recesses of a building during the day.

Shade trees situated around a small building can keep it remarkably cool during even the hottest summers. Inside the building, adjusting windows and skylights to admit or block solar light and heat and natural ventilation will contribute greatly to efficient cooling, and setting out pans of water to evaporate can cool the interior further.

On the other hand, a complete retrofit to an existing building so that it becomes an efficient passive solar system is usually very difficult. The key to passive solar techniques is to design the building with such considerations in mind from the beginning. Adding passive solar systems to the initial design of a new house often raises the cost of construction by only 3% to 5% (usually less than 10%). Accordingly, some advocates of passive solar design believe that local building codes and ordinances should be modified to encourage passive solar design in new construction.

A major drawback of passive solar design is that it is not easily applied to larger, more compact buildings, such as large office buildings, skyscrapers, and major apartment complexes. However, even in these cases foresight can lead to the successful implementation of passive solar techniques. Buildings can be oriented relative to the sun and windows placed appropriately. Interior atriums can reduce the need for artificial lighting and cooling. Allowing windows to open for natural ventilation would help; in many large buildings today, the windows cannot be opened, and a mechanical system must pump cool air throughout the building year-round. The bottom line is that even large structures can benefit from passive solar techniques.

Active Solar Techniques

Using **active solar techniques** can also reduce a building's energy consumption. The major use of active solar collectors is to produce hot water and secondarily to provide space heating for building interiors. (Nearly 25% of all fossil fuel energy consumed in the United States goes toward space heating and cooling of buildings and heating hot water.) One type of basic active solar system uses a **flat-plate collector**. In the simplest case, this device consists of a black metal plate that absorbs heat from the sun when it is struck by sunlight (either direct or diffuse). The heat is transferred to a liquid (such as water or alcohol) carried in pipes in contact with the metal plate. The heat of the hot liquid can then be used as desired (**FIGURE 8-7**).

Active solar collectors are usually placed on the tops of buildings, facing the predominant direction of the sun in that part of the world. Putting solar collectors on roofs keeps them from taking up space that could be used for other purposes. Some systems collect solar heat during daylight hours and store a portion to heat the building at night or on a rainy day when the sun is not at full intensity. Large solar collectors can be built to serve a community; for instance, special ponds filled with a combination of salt and water can trap and store solar energy.

The Economics of Active Solar Collectors

In the United States, active solar collectors are not used as much as they could be. With low energy prices and the large upfront costs

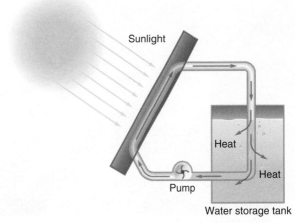

FIGURE 8-7 In an active solar heating system, flat plate collectors absorb heat. The heat is transferred to the fluid in the pipes, and the fluid is pumped to a storage tank, where the heat can be transferred again.

of installing even a simple solar water heater, most Americans are unwilling (or unable) to make the necessary capital investment. Flat-plate collectors are relatively simple to manufacture and install, but their prices will remain high as long as there is little market for them. Some people predict that demand would increase if the systems were less expensive, but until demand actually does increase, manufacturers cannot justify investing in the equipment that will allow mass production and bring prices down. It seems to be a vicious circle that will be broken only by sharp increases in conventional fuel prices or government intervention (such as tax incentives to install solar collectors).

Heat collected by active solar collectors potentially can be used to cool buildings as well as to heat them. To actively cool a building using solar energy, a system similar to that used in a gas-powered refrigerator might be used (where heat is used to bring a coolant, such as ammonia, to the boiling point; subsequently, the ammonia is condensed and then allowed to evaporate, producing cold temperatures—all of this is done in a closed loop with no moving parts). Currently, such systems are still in the developmental stage,

but they could become widespread in the future.

Electricity from the Sun

Active **solar thermal technology** (i.e., using the sun to heat substances) is not limited to flat-plate collectors. In areas with a high incidence of direct sunlight, mirrors, reflectors, or lenses can be used to concentrate the sun's rays to superheat a liquid; then the hot liquid can be used to generate electricity. For example, between 1984 and 1990, LUZ International Limited built nine solar thermal electric power plants in the Mojave Desert and sold electricity to the Southern California Edison Company. (Unfortunately, LUZ went bankrupt in 1991, putting a halt to its plans to build additional solar power plants. The plants it built are now owned by private investors and continue to operate. One reason LUZ went bankrupt was a tax structure that requires solar power plants to pay much heavier taxes than conventional power plants.) In the solar thermal trough system that LUZ used, trough-shaped mirrors focus the sunlight on steel pipes encased in glass tubes (**FIGURE 8-8**). The sun's rays heat synthetic oil inside the pipes to more than 400°C (750°F). The hot oil is used to turn water into steam, which drives a turbine connected to an electric generator. Through advanced computer technology, light-sensitive instrumentation, and microprocessors, the mirrors track the sun throughout the day to maintain the high temperatures necessary to produce the steam.

Other forms of solar thermal technologies include parabolic dish collectors that focus sunlight onto a single point; the heat can be used to generate electricity. In another method, mirrors focus sunlight on a solar receiver mounted on top of a tower, which may be as tall as 200 meters (660 feet) (**FIGURE 8-9**). The sunlight heats fluids, such as molten salts, in the receiver. The fluids are used to generate steam and power a turbine connected to an electric generator.

Photovoltaics

Photovoltaics use semiconductor technology to generate electricity directly from sunlight

FIGURE 8-8 Kramer Junction Power operates the world's largest solar power facility in the Mohave Desert near Kramer Junction, California, where it covers more than 1,000 acres (400 hectares) and has a capacity of 354 megawatts.

FIGURE 8-9 Solar Two, a solar thermal demonstration project in the Mojave Desert near Barstow, California, operated between 1996 and 1999. Solar Two showed that an advanced molten salt system (a mixture of sodium and potassium nitrate), instead of a water-steam system, could more efficiently collect solar energy and dispatch electricity as needed. The mirrors focused sunlight on a solar receiver mounted on top of a tower. After the salts were heated to approximately 565°C (1,050°F), they were used to produce steam that drove a turbine connected to an electric generator, or they were stored and used to produce steam even after sunset.

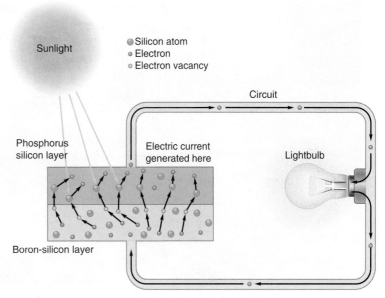

FIGURE 8-10 A photovoltaic panel generates electricity when sunlight strikes the silicon layers. The silicon loses electrons, which flow into the electric wires.

(**FIGURE 8-10**). Currently, various silicon substances are most commonly used for this purpose. A variation on simple photovoltaic systems, which produce electricity and send it to consumers directly, is photovoltaic systems that use some of the electricity they generate to split water into its component gases, hydrogen and oxygen. The hydrogen can then be used as fuel.

Most of the photovoltaic cells you see are in small consumer products, such as portable calculators, but the real potential lies in the application of photovoltaics to power homes, factories, and other commercial ventures. Photovoltaic systems are convenient because they can operate alone in remote areas with either direct or diffuse sunlight. They are already used for telecommunication and signaling devices, in space applications (such as satellites and space missions), and to provide power to recreational equipment, homes, and small villages that are far from other sources of electricity (**FIGURE 8-11**). In the United States and other developed countries, some utility companies already use photovoltaic systems to supply extra electricity to grids during peak hours. The use of photovoltaics is expected to increase in the coming decades, especially in developing countries.

The Economics of Photovoltaics

The main constraint on the use of photovoltaics at present is their cost. Currently, generating electricity using certain photovoltaic systems is several times as expensive as using fossil fuels (of course, such comparisons are misleading because fossil fuels may not be available in a certain area and cause environmental damage that is not accounted for in their cost, and the cost of photovoltaic systems is primarily a capital expense). However, since the mid-1970s, the manufacturing cost of photovoltaic cells has been decreasing steadily, from more than $70 per watt in 1975 to between $2.00 and $3.50 per watt in 2006 (price depends on the exact type of cell). Their efficiency at converting sunlight into electricity has been steadily increasing, and some analysts believe that in the not-too-distant

(a)

(b)

FIGURE 8-11 (a) Rooftop photovoltaic modules provide electricity for a village health center in Satyanarayanpur, West Bengal, India. (b) The Mars Rover uses solar panels to power its jaunts across the Martian landscape.

future prices may drop to $0.50 per watt or less. As the price has dropped, production and sales of photovoltaic systems have increased sharply from only 0.1 megawatts worldwide in 1971 to about 742 megawatts in 2003. Leaders in the production of photovoltaic units are Japan (where the government subsidizes rooftop solar systems), the United States, and Germany. As these trends continue, photovoltaic systems will become increasingly competitive, and expectations are that they will play an ever-larger role in our global energy budget. Although most existing photovoltaic systems are currently installed in developed countries, they are also increasingly being used in off-grid applications in developing countries in Asia, Latin America, and Africa. For example, 30,000 solar home photovoltaic systems are being installed in rural South Africa, 15,000 in the Philippines, and tens of thousands in rural Sri Lanka.

Environmental Advantages and Disadvantages of Solar Energy

In many ways, sunlight is the ultimate and perfect renewable, and virtually limitless, source of energy for humankind. Of course, this energy from the sun may be dilute, dif-fuse, and intermittent at any one point on the surface of the globe, but if even a small fraction of this energy was harnessed, it potentially could solve many of our current energy problems. The main obstacle to using solar energy appears to be its dispersed nature. Because of the diffuse nature of sunlight, solar systems will have to cover large areas even in areas of constant sunshine (see Figure 8-8). A major solar-generating facility might cover a circular area 8 kilometers (5 miles) in diameter with photovoltaic cells or mirrors and reflectors to capture sunlight. The wide-open spaces of the deserts and semiarid lands are the most logical place to build these systems, but that choice has its own obstacles. The desert ecosystem is very fragile, and solar power plants and the human activity they entail could be very disruptive to the local biota. Many types of solar systems also need large amounts of water, a substance already in short supply in desert environments. Furthermore, although sunlight (the "fuel" of a solar energy plant) is free and clean, any large power plant is a capital-intensive investment and initially resource-intensive. It has been estimated that the production and deployment of solar cell arrays (and wind turbines) can use about 3% of the fossil fuel that would be burned in a coal-burning power plant to generate approximately the same amount of electricity. This is clearly a drastic improvement, but it means that solar energy is not absolutely pollution free. Various solar technologies also generate toxic waste. Producing solar cells on a large scale requires large quantities of highly poisonous chemicals, such as hydrofluoric acid, boron trifluoride, arsenic, cadmium, tellurium, and selenium compounds. Photovoltaic cells and other solar energy systems have a life expectancy of 30 years or so, which means large portions of systems must be replaced or refitted every few decades. Doing so entails manufacturing new systems, generating poisonous wastes and pollution in the process, and disposing of the used components.

Another major challenge to using solar technology for electricity generation is that the power source is intermittent. Sunshine is not available at night or on rainy or overcast days; the flux of sunshine also varies seasonally (two or three times more solar energy is received during the summer than in the winter in some areas). Yet we use electricity 24 hours a day throughout the year. One solution to this problem is to collect energy when the sun is out and put it into storage. However, with current technology, the direct storage of electricity is notoriously difficult. Batteries are large, bulky, expensive, and dangerous (they contain many poisonous chemicals). Any battery system that is regularly subjected to charging and discharging degrades and must be replaced and disposed of periodically.

An alternative is to pump water (using sun-generated electricity) uphill into a reservoir during the daylight hours and subsequently release the water downhill as needed, using it to drive a generator that produces electricity. Essentially, this is a form of "artificial" hydroelectric power, and all of the problems associated with hydroelectric dams (flooding large areas, disrupting natural water flow, high cost of construction and maintenance, the chance of the dam breaking) are applicable to such a system. Another possibility is to compress gas into tanks and then release it again to drive generators when the sun is not available. However, as described in Case Study 8-1, perhaps the most popular concept at the moment is to use electricity to produce hydrogen gas, which is clean burning and easily stored and transported.

Another way to deal with the intermittent availability of solar energy is to supplement solar systems with backup systems using proven conventional means for generating electric power, but backup systems need to be able to start and stop on a regular, short-term basis; thus, they would almost inevitably have to burn fossil fuel. Except on a very small scale, backup systems do not make sense environmentally or economically. Essentially, the whole country would have to be covered with both a solar and a nonsolar electrical grid. A complete complement of fully operating nonsolar (i.e., fossil fuel or nuclear) power plants would be needed to supply electricity at night, at considerable cost. A large power plant cannot simply be turned off during the day and turned on at night. Indeed, many people believe that at least in the near future, solar energy, and not conventional power, will be the backup. Solar energy can serve as a supplement to conventional power sources, especially during the summer when electricity to run air conditioners is in high demand.

If it comes to pass, the large-scale implementation of solar systems will entail a whole new set of environmental concerns. If past experience with other new technologies is any indication, many of the problems associated with solar technologies will not be anticipated—they will manifest themselves only after years of large-scale operation.

Although large, centralized solar power plants are feasible in some areas, such as desert regions, solar energy also lends itself to small-scale, decentralized, dispersed applications, such as passive and active lighting and heating of buildings, localized small-scale electrical generation, the use of individual solar-powered ovens for cooking, solar hot water heaters, and so on. The widespread use of solar power in such situations could entail enormous energy savings.

8.4 Wind Power

People have been using **wind power** for thousands of years to operate windmills and sail ships. In the first quarter of the 20th century, wind-powered irrigation pumps were very common in the United States, but their use gradually declined as inexpensive oil and gasoline became more readily available (**FIGURE 8-12**).

In the last 30 years, fluctuations in oil prices and the realization that our fossil fuel reserves are limited have helped rekindle interest in wind power. Today, wind generates approximately 1.5% to 2.0% of the world's electricity, but its use is growing quickly (**FIGURE 8-13**). Germany, the United States, Denmark, and Spain are leaders in the large-scale commercial application of wind power. Other countries, including China, are implementing the large-scale use of wind power.

Wind farms—vast tracts of land covered with wind-powered turbines—currently generate billions of kilowatt-hours of electricity per year at competitive prices. Indeed, like

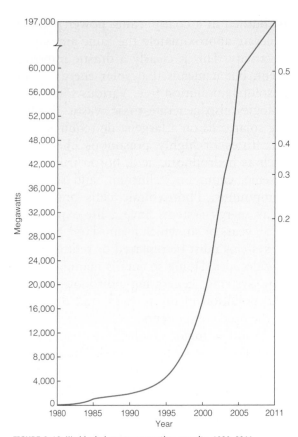

FIGURE 8-13 World wind energy generating capacity, 1980–2011. (*Source:* Adapted from Worldwatch Institute. Vital Signs, 2002. New York: W. W. Norton, 2002, p. 43, with updates from various sources.)

FIGURE 8-12 The use of wind power to pump irrigation and livestock water was once very common on farms and ranches.

the costs of solar thermal and photovoltaic power, the costs of wind-derived electricity have decreased dramatically in the last 10 years. Conservatively, some researchers estimate that wind power could supply about 12% of the world's electricity needs. Other analysts have optimistically (but perhaps unrealistically) suggested that as much as 100% of electricity needs could be met by wind power in some areas, such as Europe.

In 2008, the EPA put forth a plan to meet 20% of the demand for electricity from wind by 2030. Regionally geographic challenges differ. Wind technology cannot be developed in some areas because sufficient winds do not blow consistently. Wind machines typically are erected along coasts (some can even stand in water offshore) or in mountain passes, where a constant and steady wind supply is assured. In general, the most suitable areas for wind turbines in the United States are along the west coast, along the northeast coast, along ridges in the Appalachian and

Rocky Mountains, on the rims of mesas in Texas, and in a large area of the Great Plains. As of the end of 2010, the United States had 40,180 megawatts of wind capacity in place and operating—California accounted for 3,177 megawatts, Texas for 10,085 megawatts, and Iowa for 3,675 megawatts. Projects totaling thousands of more megawatts are planned throughout the country.

Modern wind turbines are relatively simple machines that consist of blades and a rotor connected to an electrical generator, along with a control system, all mounted on top of a tall tower. In the larger machines, the towers can be as tall as 120 meters (400 feet) (wind speed increases with height above the ground), and a single blade can measure up to 90 meters (300 feet). In intermediate- and large-sized wind turbines, a transmission and gears connect the rotor shaft to the generator; in some smaller wind turbines, the rotor is connected to the generator directly. Most wind machines have rotors with two or three blades mounted on the end of a horizontal shaft and look like huge fans or airplane propellers, but a few wind machines have blades that look like the end of an eggbeater mounted on a vertical shaft. Conventional propeller-like wind machines are controlled from the ground so that they face into the wind.

Currently, in the United States, the emphasis has been on intermediate-size machines with capacities of 50 to 500 kilowatts. The machines stand on towers about 30 to 50 meters (100–165 feet) tall, and a single wind farm may contain thousands of machines. Such wind farms can feed electricity into the local utility grid.

Small wind machines with capacities to 50 kilowatts are also becoming more popular in the United States and other countries. Small wind turbines are well suited to supplying power in rural locations that are not on a major grid, especially in developing countries. In some cases, small wind turbines can be combined with a photovoltaic system to increase the reliability of the power supply. Wind power offers many advantages. It is very safe and generally environmentally benign. The lands used for wind farms are often not in high demand for other purposes.

On farms, cattle can graze and crops can be planted around wind turbines. In the Great Plains, agricultural farms and wind farms can coexist peacefully on the same land (**FIGURE 8-14**).

Nevertheless, wind power has faced some heated local opposition. Wind farms have been subjected to the NIMBY (not in my back yard) syndrome. Some people complain that wind machines are noisy and say they are ugly and destroy the aesthetic qualities of the landscape. However, perhaps the most serious concrete environmental problem with wind machines has been bird deaths. Birds may collide with the blades or be electrocuted if they land on certain parts of the machine. In Altamont Pass in California, where a number of wind machines are located, wind turbines have killed several dozen rare golden eagles and scores of other large birds of prey. Researchers are working to devise systems that will solve this problem.

FIGURE 8-14 Wind turbines can operate effectively on agricultural land.

Bird deaths aside, the major practical drawback of wind power currently is the intermittent nature of the energy source. Although prevailing wind patterns for any given location can be determined and predicted overall (as mariners did for centuries), on a seasonal and even daily basis, wind speeds and directions can be subject to rapid and wide fluctuations. Many large utilities are understandably wary of relying too heavily on the unknown "whims" of the wind. On a seasonal and daily basis, the strongest and steadiest winds may not blow during the time of peak demand for electricity. For these reasons, wind power has little chance of becoming a major or dominant source of electrical power until large-scale electrical storage facilities become readily available.

Despite its drawbacks, wind power capacity on a global scale continues to expand. In 1980, the global wind energy-generating capacity was a mere 10 megawatts; by 2005, it had grown to nearly 60,000 megawatts, by mid-2010, growth had reached 175 gigawatts, and the growth shows no signs of abating. Indeed, although wind power supplies only about 2.0% of the world's electricity, it is one of the fastest growing sources of commercial energy.

8.5 Geothermal Energy

The interior of the Earth is very hot. This heat, which is primarily generated by radioactive decay within the Earth, reaches the surface by such means as molten rock, erupting volcanoes, and hot geysers and springs. The principle behind geothermal power is to tap and harness this natural heat from the Earth's interior. **Geothermal energy** can be used either directly for such purposes as heating buildings or indirectly to produce electricity (generally, either by using naturally vented steam directly or by heating water to produce steam to drive a turbine generator). Geothermal energy has already found widespread use in some areas such as California, which produced approximately 4.5% of its electricity geothermally in 2007 (**FIGURE 8-15**). In the United States as a whole, geothermal energy constitutes about 4% of the primary energy consumed (coming in after the fossil fuels, hydroelectric power, nuclear power, and biomass energy). Approximately one third of world geothermal use is found in the United States. The six nations with the highest geothermal use in 2010 were the United States, the Philippines, Indonesia, Mexico, Italy, and New Zealand. Globally, the use of geothermal energy has increased dramatically during the last 50 years and accounts for more than 40 million metric tons of oil equivalent each year. Many researchers predict that geothermal energy use will continue to expand in the future.

Geothermal power is not practical everywhere; it is most easily used along plate margins and at other points where hot magma comes close to the surface (**FIGURE 8-16**). Thus, some of the most likely areas for developing geothermal power lie along the Pacific Rim from New Zealand to New Guinea, the Philippines, Japan, the western coast of the United States, Mexico, Central America, and the Pacific coast of South America. Iceland, which lies on the Mid-Atlantic Ridge, already heats some 80% of its houses and many of its other buildings with geothermally heated hot water. Hawaii is also well suited to the utilization of geothermal power.

There are four basic types of geothermal deposits: **hydrothermal fluid reservoirs**, geopressured brines, magma, and hot dry rock. Hydrothermal reservoirs (the only deposits currently used commercially) are basically areas of the **crust** where hot rock occurs at

FIGURE 8-15 The Geysers is a steam-dominated geothermal field in Sonoma.

CHAPTER 8 Renewable and Alternative Energy Sources

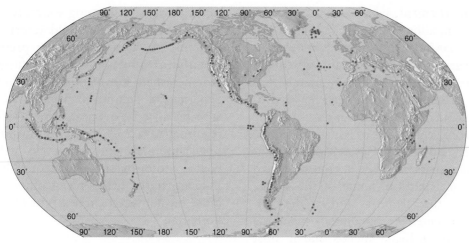

FIGURE 8-16 Global locations of geothermal sites.

relatively shallow depths and natural ground-water is heated, sometimes to extremely high temperatures. Such reservoirs may naturally manifest themselves on the surface as hot springs or geysers. The most useful, but also by far the most rare, are the steam-dominated hydrothermal reservoirs. At The Geysers in northern California, the naturally produced and vented steam fuels more than two dozen power plants. Hot water-dominated hydro-thermal deposits can also be used to run electricity-producing power plants; with lower temperature deposits, the hot water is often used to heat homes and other buildings in the vicinity directly.

Geopressured brines are naturally occurring deposits of hot, salty water (brine) under pressure at depths of 3,000 to 6,000 meters (10,000–20,000 feet). Such brines may contain significant quantities of dissolved gases, including methane. Potentially, these brines could be tapped for their heat content, dissolved gases, and pressure (which could be converted to useful forms of energy). However, like fossil fuel, after a geopressured brine is exploited, it cannot be expected to renew itself on a human time frame.

Magmas, or hot molten rock, are also a potential source of geothermal energy. Where magmas approach the surface, such as around active or not-long-dormant volcanoes (e.g., on the Hawaiian Islands), pipes could be inserted into the magma and water circulated through the pipes. The heated water or steam could then be used to drive turbines.

Although progress has been made in finding ways to prevent the pipes from melting—the magma is typically at temperatures of 600°C to 1,300°C (1,100–2,400°F)—this power-generating technique is still a long way from commercial application.

The long-term future of geothermal power may lie in the developing field of **hot dry rock technology**. Although the temperature of the Earth increases with depth virtually everywhere, this geothermal gradient is much more pronounced in some areas, reaching 70°C per kilometer (200°F per mile) of depth in some places. In hot dry rock technology, a hole is drilled some 3 to 10 kilometers (2–6 miles) into the subsurface rock until a sufficiently hot and thick layer of hard, dry rock is found. After such a layer is located, two wells placed relatively close together are drilled in the rock. Next, the rock between the two wells is fractured. Then cold water is pumped down one well. The hot rock heats the water, and it is withdrawn from the second well for use in generating electricity or for direct heating applications. The United States, Great Britain, and Japan have successfully tested hot dry rock techniques, but no commercial plants based on this principle have yet been constructed.

The Disadvantages of Geothermal Energy

Although geothermal energy production is often considered relatively clean compared with traditional fossil fuels, it is not without environmental hazards. By its nature, it tends to expel excessive quantities of heat into the environment, which may cause thermal pollution, killing plants and animals and disrupting natural ecosystems. Some of the hot underground water contains dissolved salts, other minerals, and heavy metals that are toxic pollutants on the surface. Hydrogen sulfide and other dangerous gases may be released from vents. Tons of hydrogen sulfide are released daily at The Geysers, despite that efficient scrubbers are used at the plants to minimize air pollution. Some of the sludge produced by the scrubbers and by the condensation of vented steam contains such high concentrations of heavy metals that it must be treated as a hazardous waste. Such waste could potentially be reinjected deep underground, but care must be taken to avoid contaminating freshwater aquifers.

Geothermal operations also require large quantities of water. Geothermal power plants currently in operation use water as a cooling and condensing agent, and hot dry rock techniques will also use considerable quantities of water.

In a sense, geothermal energy is an undepletable energy source, at least on a human time scale, because the Earth is not going to cool anytime soon. However, on a small scale this resource can be depleted—and in some cases, quite rapidly. Naturally circulating hot water flow can be inadvertently diverted over time, thereby disrupting easy access to the energy. Hydrothermal fluid reservoirs often contain "fossil water" under pressure that is withdrawn by a power plant faster than it is replenished. For instance, The Geysers has experienced significant drops in steam pressure because of overproduction since 1987, resulting in a 25% reduction in power output. Ultimately, any hot rocks will cool if heat is withdrawn from them quickly enough. Like any natural resource, geothermal energy must be managed carefully to maximize its potential.

8.6 Ocean Energy

The oceans are a vast, if sometimes diffuse and seemingly uncontrollable, source of energy. Researchers have been exploring various ways of exploiting this **ocean energy** for decades. The waves, tides, and currents all have potential as energy sources. In 1995, a small electrical power station that runs on wave energy was installed off the coast of Scotland.

The regularity of the tides can be used to drive mechanical systems, and this mechanical power can be transformed into electricity (**FIGURE 8-17**). A **tidal power** station operates by allowing water at high tide to flow into a reservoir or bay through sluice gates; then the

(a)

(b)

FIGURE 8-17 The La Rance River Tidal Power Plant in St. Malo and Dinard, France. (a) Water from the Atlantic Ocean (upper part of photo) is trapped at high tide, and then the water is released back into the sea at low tide. (b) The water passes through 24 turbines, like the one shown. These turbines are connected to generators that produce enough electricity for a city of 300,000.

CHAPTER 8 Renewable and Alternative Energy Sources

gates are shut, forming a dam. At low tide, the water is released through a turbine and used to generate electricity. Ultimately, this energy comes from the gravitational pull of the moon and sun.

A 240-megawatt tidal plant has been operating in France since 1966, and a 20-megawatt facility is currently operating in the Bay of Fundy, Nova Scotia. In most areas, the difference in height between high and low tide is not enough to efficiently drive a turbine, but because of local topographic conditions, the Bay of Fundy is an exception: there the difference between high and low tide can be as much as 16 meters (53 ft). A disadvantage of tidal power is that it is very intermittent. The times of the tides change daily, and the power is generated only after high tides. A final potential drawback is that the artificial damming and releasing of water may be disruptive to local marine life.

8.7 Energy Storage

A major obstacle to the widespread use of many alternative fuels is the problem of **energy storage** and transportation. One of the reasons that the fossil fuels in general and oil in particular came to dominate the world energy mix is because of their ease of storage and transportation. Petroleum products pack a high quantity of energy in a small amount of space, and the energy is easily obtainable by combustion. Clean-burning hydrogen gas (see Case Study 8-1) offers some of the same advantages, but being a bulky and light gas, it is much more difficult to store. Electricity is one of the most versatile energy sources, but it is difficult to store in large quantities. Most electricity is shipped through wires as soon as it is produced, and the consumer uses it almost immediately.

Of the big five energy sources powering the modern world (oil, coal, natural gas, hydroelectric power, and nuclear power), the only one that actively stores newly transformed energy and then releases it as needed is hydropower. The artificial lake created by the dam at the power plant effectively stores energy. When people need energy, water is released through the dam to power turbines connected to electrical generators.

Electrical Energy Storage

The most common way to store electrical energy is with **batteries**. Alessandro Volta developed the first primitive battery around 1800. The device was made with alternating disks of zinc and copper. A cloth soaked in brine separated each pair of disks. A wire attached to either end produced a low electrical current. All batteries work on basically the same principle—a chemical reaction creates energy. When a battery is charged, electrical energy (essentially, a stream of electrons) is converted into chemical energy in the form of atomic arrangements. During discharge, the chemical process is reversed. Today, common large batteries, such as conventional car batteries, are typically of the lead–acid variety, using lead plates and dilute sulfuric acid. Electronics equipment typically uses dry cell batteries.

The problem with current battery technology is that the batteries tend to be heavy and bulky relative to the amount of energy they store. This is not a major problem for small applications, but for larger scale, longer term applications, it is a serious limitation. In addition, batteries do not last forever; for instance, typical lead–acid batteries can be run through only 500 to 2,000 charge/discharge cycles.

Some futurists anticipate an increasing demand for medium-size batteries such as those that can be used to power electric cars. Electric cars in existence today, using conventional battery technology, typically have driving ranges of only about 112 to 160 kilometers (70–100 miles) before recharging is necessary, and recharging can take 3 hours. Such a car might be appropriate for short commutes but would be unsuited to long-distance travel. In addition, although an electric car in isolation may give off zero emissions, it must be recharged from some energy source. If it is recharged by electricity that photovoltaic cells or a hydroelectric power plant produce, then it may indeed truly be a zero-emission vehicle. However, if it is recharged by electricity generated by a coal-burning electrical power plant, it is ultimately still responsible for the emission of greenhouse gases and pollutants, although perhaps considerably less than a standard gasoline-powered car.

Currently, a new generation of lighter batteries that can store more energy per unit weight for longer periods of time is under development. These newer batteries use nonconventional combinations of elements as electrodes and electrolytes, such as zinc-bromine, zinc-chlorine, or hydrogen-nickel oxide. The next generation of batteries also promises to be longer lasting than current conventional batteries, but so far, they are also much more expensive.

Somewhat similar to batteries are **fuel cells** (see Case Study 8-1). In a typical fuel cell, hydrogen gas (H_2) is passed by a catalyst that separates the hydrogen atoms into hydrogen ions and electrons. The electrons, carrying energy, can be used to power an electric motor or virtually any other electric device. The hydrogen ions and electrons ultimately combine with oxygen (O_2) to form water. Electric cars powered by fuel cells would carry refillable tanks full of hydrogen and would emit only water. In addition, some fuel cells can use various hydrocarbon compounds, such as gasoline, natural gas, or methanol, as a source of hydrogen. In such cases, there may be undesirable emissions, but they are much less than is the case with conventional fuel combustion systems.

Fuel cell technology is being aggressively pursued by a number of automobile makers, including Ford, DaimlerChrysler, BMW, Volkswagen, Volvo, Toyota, Hyundai, and many more. For instance, Ford has introduced its P2000 Prodigy, a fuel cell-powered sedan that uses hydrogen and is designed to meet the same performance standards as the once popular Ford Taurus. Many people in the industry believe that the future of motor vehicles, both private cars and public transportation vehicles lies in hydrogen-powered fuel cell vehicles. Such vehicles include buses (the Chicago Transit Authority ran a pilot test of three fuel cell buses between 1997 and 2000, and similar vehicles are being tested around the world), as well as trucks and other commercial vehicles. Until the infrastructures are put into place to produce, transport, and deliver pure hydrogen gas per se, some futurists believe that service stations may use on-the-spot machinery to generate hydro-gen gas from natural gas (which is already being used widely, although not primarily for motor vehicles). Such natural-gas-to-hydrogen equipment will pave the way for a pure system of hydrogen storage and distribution. It is possible that by the year 2020, a large percentage of cars on the road will be running off of electric drives powered by hydrogen-consuming fuel cells.

8.8 Energy Conservation

Most authorities agree that the simplest and cheapest way of stretching our energy resources and mitigating energy-related problems is through first improving energy conservation (decreasing the consumption of energy) and second increasing energy efficiency (increasing the usable output per unit of energy). Energy conservation and efficiency go hand in hand; increasing efficiency is often the best way to conserve energy. A simple anecdote can help clarify the relationship of conservation to efficiency. Suppose that a room is illuminated at night using four standard 100-watt incandescent bulbs. One way to conserve energy would be to turn off three of the bulbs. However, that strategy would sacrifice illumination. Another way to conserve energy, through increased efficiency, would be to replace the incandescent bulbs with four "natural light" compact fluorescent lamp (CFL) bulbs. Each bulb draws only 25 watts of electricity, yet produces the light equivalent of a 100-watt incandescent bulb. Thus, the energy consumption has been cut from 400 to 100 watts without sacrificing quality, but perhaps we should also rethink our standards when considering energy conservation. Perhaps a room is over-lighted with four bulbs. We would conserve even more energy if we use only two 25-watt CFL bulbs and turn them off when no one is in the room (**FIGURE 8-18**).

A relatively small investment in energy efficiency can pay off handsomely in the long run. A car that gets better gasoline mileage may cost slightly more, but the gasoline that is saved will more than make up the difference in cost and the environmental degradation prevented. Likewise, even though CFL bulbs may cost considerably more than incandescent bulbs, over their lifetime, they may save $35 to $55 per bulb in electricity costs. Because of their advantages, in the last decade, the global sales of CFLs has skyrocketed, increasing from 45 million in annual sales in 1988 to more than one half billion annually by the end of the century. Fueling the rise are stronger exports of fluorescent bulbs from China, which currently account for about 54% of China's total exports of lighting products. In 2004, outbound shipments of fluorescent bulbs were worth $778 million. Just as investing in CFLs can save money in the long run, many other energy efficiency investments—from more efficient appliances and cars to utilities—can yield financial benefits. One study of Brazil concluded that investing $10 billion in making the country electrically more efficient could save approximately $44 billion in projected electrical needs.

Reductions in energy consumption need not result in a decrease in living standards—in many cases, living standards are actually increased. However, increasing energy efficiency ultimately may entail a qualitatively different lifestyle, as more labor-intensive activities replace energy-intensive activities. Some analysts predict a **decentralization** of energy supplies and sources, especially as alternative energy sources continue to make inroads in the global energy budget. Residences and business may become partially or wholly energy self-sufficient as they consume less energy and use local energy-generating options, such as photovoltaic cells mounted on the roof.

Reducing Consumption of Fossil Fuels

To many people, "energy conservation and efficiency" means more than reducing the energy used per se; it also means reducing the consumption of nonrenewable fossil fuel resources. The aim is to reduce the environmental degradation that the big three energy sources (oil, coal, and natural gas) engender, especially the release of atmospheric carbon, which is responsible for a major component of global climate change (as shown in a discussion of global air pollution). Studies of the **carbon efficiency** of the world economy

FIGURE 8-18 Replacing each of these incandescent lights with a CFL bulb would reduce the use of electricity and save money.

show that from an economic point of view, the world has managed to raise its output per unit of carbon released. In 1950, one kilogram of carbon dioxide was released for every $0.64 worth of world economic output, as measured by gross domestic product (or a pound of carbon dioxide for every $0.29); currently, about $1.28 of output is generated for every kilogram of carbon dioxide released ($0.58 per lb). This increase in carbon efficiency during the last half-century is the result of many factors working together: improved energy efficiency, spurred on by the oil crisis of the 1970s; increased emphasis on energy sources that produce virtually no carbon (the only carbon emitted is during construction and maintenance), such as solar, wind, hydroelectric, geothermal, and nuclear power; shifting from the use of oil and coal to natural gas (which emits less carbon per unit of energy); and increasing emphasis on communications and various high-tech industries that use relatively little energy per economic contribution versus traditional heavy, energy-intensive industries.

Interestingly, the carbon efficiency of developing countries used to be higher than that of industrialized countries, but in the last few decades, this situation has generally been reversed. Formerly, the developing countries relied primarily on manual labor, but as they have begun to industrialize, many have become very energy intensive. In many cases, these countries invest in less energy-efficient technologies simply because the upfront costs are lower. In contrast, many of the developed countries have been expanding their economic output while keeping carbon emissions stable. One study estimated that Japan's economic output is about $4.61 per kilogram ($2.09 per pound) of carbon dioxide; in contrast, the United States has an economic output of $1.46 per kilogram ($0.66 per pound) of carbon dioxide, and China has an economic output of only $0.25 per kilogram ($0.11 per pound). Theoretically, there is no limit to how high carbon efficiencies could climb. The exclusive use of virtually carbon-free energy sources would cause the carbon efficiency index to approach infinity.

Energy Efficiency at a National Level

At a national level, some countries are much more energy efficient than others. Energy efficiency can be measured in various ways: total energy consumption per year for a particular country, energy consumption per person per year in a particular country (per capita energy consumption), or primary energy consumption per unit of gross domestic product (i.e., how energy efficient a country is in producing goods and services) (TABLE 8-2). For example, the People's Republic of China uses about half the amount of energy used by the United States, but China has more than 4.5 times as many people. The per capita annual consumption of energy in China is just more than one tenth of that in the United States. Nevertheless, when we compare China's economic output to that of the United States, we find that China uses more than 4.5 times as much energy to produce the same amount of gross domestic product (calculated in terms of the 1995 U.S. dollars exchange rate). In other words, in this respect, China is much less energy efficient than the United States.

The ratio of a country's energy consumption to its economic output is the country's **energy intensity** (FIGURE 8-19). In general, developing countries have higher energy intensities than do developed countries—that is, they are less energy efficient even though they may use considerably less energy overall and per capita than do the developed

TABLE 8-2	Primary Energy Consumption, Energy Consumption Per Million Persons, and Energy Intensity for Selected Countries, Late 1990s			
Country	Energy Consumption	Energy Consumption per Million Persons	GDP/EC	Energy Intensity
United States	85.89	0.31	90,409 (90,657)	11.1 (11.0)
Russia	23.52	0.23	27,390 (14,378)	36.5 (69.6)
China	43.65	0.034	87,922 (19,129)	11.4 (52.3)
Japan	20.45	0.16	148,434 (263,284)	6.7 (3.8)
Germany	13.79	0.17	126,536 (180,347)	7.9 (5.5)
Canada	9.45	0.30	72,055 (64,764)	13.9 (15.4)
France	9.83	0.17	131,319 (162,237)	7.6 (6.2)
United Kingdom	9.05	0.15	135,199 (129,539)	7.4 (7.7)
Ukraine	5.96	0.16	18,645 (12,544)	53.6 (79.7)
India	18.31	0.02	87,911 (20,634)	11.4 (48.5)
Italy	6.49	0.11	179,870 (171,285)	5.6 (5.8)
Brazil	6.83	0.04	155,193 (109,338)	6.44 (9.1)

Primary energy consumption is in quads per year. Energy consumption per million persons is in quads per year per million persons. GDP/EC is gross domestic product in millions of dollars per quad of primary energy consumed. Energy intensity is billions of BTUs consumed per $1 million of gross domestic product = thousands of BTUs per $1 gross domestic product. For GDP/EC and energy intensity, two numbers are given: the first number was calculated using an estimate of purchasing power parity, which can be rather controversial and is subject to interpretation, whereas the second number (in parentheses) was calculated using the 1995 U.S.$ exchange rate; it is the latter number many people refer to when they discuss these issues. The numbers in this table are for comparative purposes among countries; the figures do not necessarily agree precisely with energy data given elsewhere in this text because of rounding and standardization among various countries in terms of minor energy sources (such as various types of domestic, traditional, or home use of energy) that are included or excluded from this table.
Source: Figures are calculated from data published in World Resources, 2000–2001. Washington, D.C.: World Resources Institute, 2000.

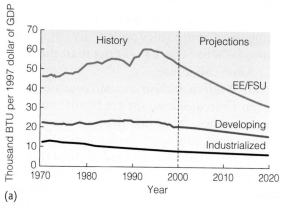 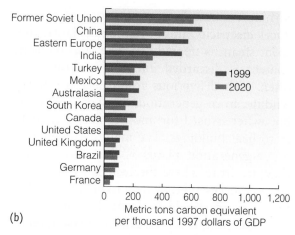

(a)　　　　　　　　　　　　　　　　　　　　　　(b)

FIGURE 8-19 (a) World energy intensity by region, 1970 to 1999 with projections to 2020, as estimated by the Energy Information Administration of the United States Department of Energy. In this graph, "Industrialized Countries" refers basically to North America, Western Europe, Japan, Australia, and New Zealand; "EE/FSU" is Eastern Europe and the former Soviet Union. (b) Because much energy used in the world today is from fossil fuels and other carbon-emitting sources, energy intensity is related to the concept of "Carbon Intensity" (the amount of carbon released, such as carbon dioxide into the atmosphere, per unit of economic output). This graph is the Energy Information Administration's estimate of carbon intensity for selected countries in the years 1999 and 2020 (projection). (*Source:* Adapted from the United States Department of Energy, Energy Information Administration, International Energy Outlook, 2002.)

countries. By becoming increasingly energy efficient over the last several decades, various industrialized nations (such as Japan, France, Italy, Germany, and other western European nations) have been able to decrease their energy intensities while maintaining strong economic growth and high standards of living. A country can increase its economic output while simultaneously decreasing its relative energy consumption—a concept virtually unimaginable before the 1970s.

Improving Energy Efficiency

Energy efficiency (reduced consumption) can be improved at every level. For instance, electricity distributed through power grids tends to be very inefficient. It is estimated that the average American electric power company loses about 2.5 times as much energy as it delivers. Energy is lost in the conversion process and as the electricity is transmitted through power lines. Rather than turning the energy stored in coal into electric power and using it to run heaters that warm a building, it would be more efficient from an energy perspective—it would take much less coal—to heat the same building by burning the coal directly in a stove or furnace within the building.

Much energy could be saved, which is the same as gaining energy, by simply

1. Not unnecessarily converting energy from one form to another;
2. Increasing the efficiency at which energy is converted from one form to another;
3. Not unnecessarily transporting energy; and
4. Increasing the efficiency by which energy is transported.

Efficient planning and decentralizing energy reserves and networks can decrease unnecessary energy conversions and transport. Sometimes several widely scattered, smaller, self-sufficient power plants may be more efficient than one large, centralized plant. Readjusting and fine-tuning old equipment can increase energy conversion and make transmission more efficient. However, more important in this area is the development of new technologies. For instance, superconductor technology may dramatically increase the efficiency of all aspects of electrical energy use.

Co-generation, an old, but newly resurrected technology, can sometimes more than double the usable energy produced by a power plant. The basic principle behind co-generation is that a power plant simultaneously produces several types of energy, such as electricity and heat that can be used locally. Remember the electrical power plant heats water to produce steam to power a turbine connected to a generator that produces electricity. In conventional plants, the spent steam is treated as waste, and its heat content is discarded into the environment, often with deleterious results for the local wildlife. In co-generation, the spent steam or hot water is used for another purpose, such as to heat buildings. The main disadvantage to co-generation plants is that the farther they are from where the heat is to be used, the less successful they tend to be; ideally, the power plant should be located right within the community it serves. For this reason, relatively small-scale co-generation plants appear to have more potential for community acceptance and energy efficiency than do large-scale versions.

Heavy and light industry can save much energy and increase energy efficiency in many ways. Using recycled materials can entail tremendous energy savings in this sector. In general, recycling saves energy. For example, recycling steel uses only 14% of the energy that it would take to produce a ton of steel from the raw ores, and recycling aluminum saves about 95% of the energy that it would take to produce the same amount of aluminum from the raw ore bauxite.

In the United States, more than one third of all energy consumption is used in residential and commercial buildings, primarily for lighting, heating, or cooling the interior space and running small appliances. Switching to more efficient heating, cooling, and lighting systems would reduce energy consumption. Many existing buildings can be insulated much better than they are, through various green-building techniques, and new construction materials and techniques enable new buildings to be superinsulated. In a well-designed, well-insulated, and sealed modern building, the interior will retain the heat given off by lights, human bodies, and various appliances during the cold season; ideally, little additional heat should be needed. During the warm months, the insulation will keep heat out and cool air in the structure. Of course, a superinsulated building should also be well ventilated to prevent the buildup of harmful indoor pollutants. Such buildings should be designed and situated to take maximum advantage of passive solar heating during the cold months and to minimize such effects during the warm months. Homes using these approaches potentially could consume 50% to 90% less energy than the average American house.

In Western industrialized countries such as the United States, the transportation sector consumes about one third or more of all energy used, and worldwide, almost 25% of all energy used goes toward transportation. Automobiles, trucks, and other road vehicles account for about 80% of energy consumption in transportation. Fossil fuel, especially oil, products are used to fuel the great majority of all transportation needs.

Currently, more than 700 million vehicles are on the roads worldwide, and more than 75% of these are cars. Of these cars, slightly more than one third are in the United States (although we have less than 5% of the world's population). Slightly more than one third of cars are in Europe, and the remainder are spread over the rest of the globe. The vehicle growth rate is very high, and based on current rates, there may be 1.1 billion vehicles on the roads by about 2020. Given the number of automobiles and other vehicles, it is imperative that something be done to control the amount of fossil fuel energy they consume and the amount of pollution they produce. The growing popularity of vehicles with **hybrid engines** is a good start (see **CASE STUDY 8-2**).

Encouraging Energy Savings—Voluntary Versus Mandatory Measures

How do we encourage a reduction in energy consumption? Will individuals, companies, and nations do it voluntarily? Or must conservation and efficiency measures be actively encouraged with subsidies and tax incentives and disincentives (e.g., high taxes on gasoline) or even mandated by law? These are difficult questions, and they elicit diverse reactions from various special interest groups.

Perhaps the most effective way to encourage voluntary energy conservation is through education. Sometimes a very practical approach that appeals to the individual's basic instincts can be effective. Improving the

Improving the Fuel Economy of Gasoline-Powered Vehicles

VISIT

http://environment.jbpub.com
/mckinney/5e/
for more information

Despite continued advances in the development of cars and trucks powered by electricity, natural gas, or biomass fuels, automobiles and light trucks that burn some form of conventional gasoline or diesel are likely to be the mainstay of the private transportation sector for many years to come. Thus, it is imperative that the fuel economy of the vehicle fleet be increased. The more fuel efficient we can make our vehicles, the greater the reduction in environmental deterioration caused by petroleum burning and the longer our fossil fuel supplies will last. We know we can improve fuel efficiency. The U.S. government first imposed fuel efficiency standards during the oil crisis of the mid-1970s, when the average fuel economy of a passenger car was 18 miles per gallon (13 L/100 km); by the first half of the 1990s, it had risen to a standard of 27.5 miles per gallon (8.5 L/100 km) for new cars sold in the United States. In 2006, the National Highway and Transportation Safety Administration established new standards of 24 miles per gallon for light trucks (which includes most SUVs). According to the administration, the new fuel economy standards for light trucks will save 250 million gallons of fuel a year after the largest sport utility vehicles are required to meet the standards in 2011. The standard for passenger cars has not changed in the last 10 years; nevertheless, prototype automobiles have been built using existing technologies that attain average fuel economies of better than 78 miles per gallon (3.0 L/100 km).

What are some of the technical options that can be used to increase fuel efficiency? To begin, as much as 70% to 80% of the energy in gasoline can be lost because of inefficient engines. Newly developed engines, which burn "leaner" because of delicate electronic oxygen sensors, can increase fuel efficiencies by 20%. The development of efficient two-stroke engines, as compared with the standard four-stroke engines used in cars today, may help increase engine efficiencies. In addition, work is proceeding on ceramic materials that could be used in the manufacture of engines, particularly advanced diesel engines. The new ceramics can withstand much higher temperatures, thus reducing the need for cooling and decreasing the amount of energy lost.

In addition to improvements in engine design, fuel efficiencies can be improved in many other ways. When power is transmitted from the engine to the wheels, some usable energy is lost. Some automobile manufacturers offer more efficient transmission systems. Most of the hybrid automobiles available now use continuously variable transmissions that allow the engine to operate at its maximum efficiency, even as the driver of the car increases or decreases speed.

Another area of promise is the increased streamlining of automobiles and the reduction of excess weight. Well-designed streamlined profiles also reduce aerodynamic drag and increase fuel efficiency. Likewise, excess weight adds to fuel consumption. Increasingly, automobile manufacturers are substituting composite plastics and lightweight metals, such as aluminum and magnesium, for the traditional steel in cars.

Currently, the most practical development along the lines of increasing fuel efficiency, decreasing emissions, and storing energy that would otherwise be lost is the marketing of new hybrid gasoline–electric cars. Toyota's Prius is the world's first mass-produced hybrid car; the Prius was made available to consumers in Japan in December 1997, and in 2000, it was introduced to the American market (**FIGURE 1**). Honda was not far behind Toyota in the race to produce the first commercial mass-produced hybrid cars and actually introduced the Honda Insight hybrid vehicle

FIGURE 1 The world's first mass-produced gasoline/electric hybrid car, the Toyota Prius. In addition to getting an average of 48 miles to the gallon, the car qualifies as a super ultra low emissions vehicle in California.

Improving the Fuel Economy of Gasoline-Powered Vehicles

to America in 1999. A few American automobile manufacturers started selling hybrid sports utility vehicles in 2005. A typical hybrid car such as the Prius has a gasoline engine and gas tank, an electric motor(s) and batteries, a generator (in some cases, the electric motor can also serve as an electric generator when not being used as a motor), and a power-split device or transmission that will send engine power to where it is needed at a certain time—to the wheels driving the car, to a generator powering the electric motor, to the generator to recharge the batteries, or to some combination of the various options. Hybrid cars also typically are outfitted with a regenerative braking and coasting system. When the driver brakes or coasts in the car, an electric motor acts as a generator capturing and storing energy that would otherwise be lost and wasted.

From the point of view of the consumer, a hybrid gasoline–electric car is fueled with only gasoline. The battery for the electric motor is charged either by the gasoline engine driving a generator or by otherwise "wasted" energy, mentioned previously, generating electricity; the batteries never need to be charged from an external source. Either the gasoline engine driving the wheels, the electric motor, or both, depending on the most efficient method in a particular situation (automatically monitored and regulated by an onboard computer), propel the vehicle itself. In some cases, such as during braking and stop-and-go traffic, the gasoline engine automatically shuts down, resulting in zero exhaust emissions and no use of fuel. The electric motor powers the car, and during braking, the generator recaptures energy. When the gasoline engine is running, temperature and other parameters are constantly regulated to keep it and the catalytic converter running at the lowest level of emissions. As a result, hybrid cars such as the Prius can qualify as "super ultra low emissions vehicles" under California's strict guidelines. Hybrid cars also attain remarkable fuel ratings. The fuel efficiency of the American version of the Prius is listed at 52 miles per gallon in the city and 45 miles per gallon on the highway; the Japanese version of the Prius, developed primarily for heavily congested, slow driving, is rated at an extraordinary 66 miles per gallon. These efficiencies are not only remarkable but the reverse of what is expected of most standard gasoline or diesel vehicles, where higher mileages are attained on the highway and lower mileages in stop-and-go city traffic, but a hybrid car thrives in heavy, slow, congested traffic. Thus, such hybrid cars may be ideal for city situations, both in terms of fuel economy and ultra-low emissions.

However, many environmentalists advocate not just increasing the efficiency of automobiles, but also eventually eliminating them. Although the automobile fleet continues to increase, it is not growing as fast as it might. Even as tens of millions of new cars are produced each year (44 million automobiles were produced in 2004), tens of millions of others are taken out of service. As of 2005, there were more than 700 million passenger cars, trucks, and buses in service worldwide, and this may grow to 1.1 billion in 2020. To offset this increase, conscientious citizens are restricting their auto use, and many cities, especially in Europe and Japan, are placing restrictions on the use and ownership of cars.

- More and more emphasis is being placed on efficient public transportation and environmentally friendly modes of transportation, such as walking and bicycling. For instance, Amsterdam is phasing out the use of motor vehicles except for public transportation and delivery purposes. London charges a "daily use" fee for vehicles entering the central downtown area.
- Car sharing can help decrease the number and use of automobiles. As of 2004, 15 U.S. car-sharing programs claimed 61,651 members sharing 939 vehicles, and 11 Canadian car-sharing programs claimed 10,759 members sharing 528 vehicles. These programs are growing in popularity. Members do not own private vehicles but pay by the hour and kilometer (plus in most cases a refundable deposit and perhaps an annual membership fee) to use a car as necessary from the car-sharing fleet. The fees that the car-sharing subscribers pay are used to cover the costs

of maintenance, gasoline, insurance, and so forth. Car sharing works best in situations in which participants do not need a car on a regular basis; for normal commuting, grocery shopping, and so forth, members use other modes of transportation.

Bicycles are very important as a major mode of transportation in many countries, such as China and India. Currently, bicycles are being produced at a rate of about 90 to 100 million per year globally. This means that environmentally sound bicycles are being added to the transportation fleet much more quickly than are automobiles. Indeed, in many parts of the world, bicycles still greatly outnumber cars. For instance, in China there are about 250 bikes for every automobile (**FIGURE 2**).

http://environment.jbpub.com /mckinney/5e/ for more information

FIGURE 2 An example of a "bicycle culture," in Shanghai, China.

insulation in one's home or buying a more fuel-efficient car will not only repay the investment and then begin to line the consumer's pocketbook, but will be good for the global environment as well. In addition, advertising and popular culture can change our perceptions and instincts radically. In the past, wasting energy was a status symbol or considered "sexy." For instance, we can point to the 1960s, when "muscle cars" that guzzled gasoline and generated tremendous amounts of horsepower for no practical reason were popular. Today, there is a trend among some people to desire bigger and bigger sport utility vehicles (SUVs), which have poor gas mileages and appear to serve as status symbols the same way that other gas-guzzlers did in earlier decades. With proper education (starting very early), perhaps being energy efficient could become a mark of status. Celebrities are now getting publicity for driving around Hollywood in their hybrid vehicles.

Across the United States, electric utilities are adopting strategies that are both good for the environment, in that they save energy, and good for business, in that they reduce demand that cannot be readily met by the utility. This new approach is commonly referred to as "demand-side management" (also known as "negawatts"). Demand-side management aims to improve the efficiency with which electricity is used, rather than stressing the production of ever-increasing amounts of electricity. Currently, U.S. utilities are investing about $2 billion a year in this approach; it is an investment that is well worthwhile. Demand-side management saves the company both short-term operating expenses and the considerable long-term expenses of building new power plants to increase production.

Much to the surprise of consumers, some utilities have found it cost effective to promote energy savings actively, sometimes even offering to pay for home improvements. Utilities have helped customers improve the insulation of their homes, distributed high-efficiency CFL bulbs, and promoted the installation of high-efficiency air conditioners and heaters. In the long run, the utility's savings can be split between stockholders and the consumers, who benefit from lower rates than would be the case otherwise.

Energy conservation can be promoted at every governmental level—local, state, and national. A first step is for government to set a positive example by using the most up-to-date energy-efficient technologies. The single largest consumer of energy in the United States is the federal government. The government can help fund the research and development of new products and then be the first consumer. In this way, the government can not only save energy but also help create a market for new, energy-efficient technologies. For instance, if all government vehicles had to meet very stringent fuel-efficiency requirements, automobile manufacturers would have an automatic incentive to build cars to these specifications. After the cars were under production, they could be marketed to the general public relatively easily.

Governments could simply mandate energy efficiency measures within their jurisdiction, but such heavy-handedness is often unpopular. Through direct funding, the federal government in particular can encourage the research and development necessary to produce energy-efficient alternatives. Tax breaks and subsidies can be given to industries and individuals that are more energy responsible, and traditional energy consumption can be taxed heavily so that the price consumers pay reflects the real cost of the product. For example, many people contend that gasoline should be much more heavily taxed in the United States. In many European countries and Japan, gasoline costs considerably more than the average price in the United States because of the heavy taxes imposed. These taxes are not just a way to discourage gasoline consumption and raise funds for the government, but also reflect and help to cover the real costs of burning a gallon of gasoline. The cost of gasoline in the United States is not fully reflected in the price. The real cost involves much more than the cost of pumping the oil out of the ground, refining it, and transporting it to the gas station; the real cost also includes such items as road building and maintenance, traffic regulation, health care needed because of traffic injuries, and the environmental degradation caused by burning that gallon of gasoline and releasing pollutants into the atmosphere. In the United States today, these additional costs, not covered by the price paid at the service station, are covered by government subsidies using money derived from taxes. In addition, the gallon of gasoline the average American burns to drive less than 30 miles (or 1 liter to drive less than 13 kilometers) took nature untold numbers of organisms and millions of years to produce.

People have suggested that the direct consumer, the person driving the automobile, should pay a larger share of the true cost of a gallon of gasoline, and this could be accomplished through higher gasoline taxes. In response, other people argue that stiff gasoline taxes are regressive—they unfairly tax those who can least afford it because gasoline is a "necessity" (one has to drive to work) and not a luxury. Theoretically, a system could be devised that would exempt or otherwise subsidize low-income persons or impose different rates so that recreational driving would be more costly than driving to work. However, such a system would result in a complicated and costly bureaucracy and lend itself to cheating and abuse. Many people argue that what we need instead is more mass transportation systems, such as buses and trains.

An alternative is to tax cars according to their fuel efficiencies. If gas-guzzler taxes were set high enough to pay for all the environmental damage a gas-guzzler causes over its operating life, the average fuel efficiency of new automobiles would almost certainly increase quickly. Another suggestion is to tax the energy content of all fuels (not just gasoline), but then fuels that we might want to promote, such as clean-burning hydrogen gas, also would be taxed. In addition, various "carbon taxes" have been suggested that could be imposed on all fuel sources. The basic idea is that a particular fuel, no matter how it is consumed, would be taxed in proportion to the amount of carbon dioxide and other greenhouse gases and pollutants it emits into the atmosphere per unit of energy it produces. Thus, if in a given context coal produces about twice the amount of greenhouse gases as natural gas does, then coal would be taxed twice as heavily per unit of energy generated. A very clean fuel, such as pure hydrogen, might not be taxed at all. The idea is to discourage the burning of some fuels while simultaneously encouraging the use of cleaner fuels that are less damaging to the environment. Some people argue that this policy could end up discouraging burning coal, our most abundant fuel, and encouraging the use of fuels in shorter supply, but others feel that the use of coal should be discouraged. There are no easy solutions to these problems.

■ study guide
SUMMARY

- Alternative energy sources, such as solar power, geothermal power, wind power, and biomass power, are alternatives to the "big five" energy sources: oil, natural gas, coal, hydropower, and nuclear power.
- Along with hydropower, most alternative energy sources are renewable (an exception in some cases is geothermal power), sustainable, and relatively environmentally friendly.
- Hydropower uses dammed water directed through generators to produce electricity.
- Although dams are "clean" and provide renewable power, their construction can cause havoc to natural ecosystems and have negative social impacts, such as when large numbers of people are exposed to water-borne disease vectors or must be displaced to make room for a reservoir.
- Biomass burning (wastes, wood, crops) produces greenhouse gases but can be renewable and sustainable; however, such use of biomass may work against recycling efforts and cause environmental problems.
- Solar energy takes many different forms—electrical solar power plants, active and passive solar techniques, and photovoltaics; solar energy systems have only a limited life span and may involve hazardous materials that are difficult to dispose of.
- Wind power has been used for thousands of years; some people consider modern wind farms "ugly," and wind farms have been implicated in bird deaths.
- Geothermal energy draws heat from hot rocks and brines far beneath the surface of Earth but may also bring toxic pollutants to the surface. Locally, high-quality geothermal reservoirs can be depleted.
- A major problem with many alternative energy sources is that the energy is intermittent (the sun is directly useful only during daylight hours), and energy storage becomes a problem.
- Using electrical energy to produce hydrogen gas (from water) may be the way to conveniently store energy in the future—the hydrogen can then be burned or used in a fuel cell to produce electricity when needed.
- Energy conservation and increasing energy efficiencies are important ways of stretching energy resources and mitigating energy-related environmental problems.

KEY TERMS

active solar techniques
alternative energy
batteries
"big five" energy sources
biogas digesters
biomass energy
carbon efficiency
co-generation
coppicing trees
crust
decentralization
energy conservation
energy efficiency
energy farms
energy intensity
energy storage
flat-plate collector

fuel cells
geothermal energy
hot dry rock technology
hybrid engines
hydropower
hydrothermal fluid reservoirs
independent power producers (IPPs)
ocean energy
passive solar design
photovoltaics
solar energy
solar–hydrogen economy
solar thermal technology
tidal power
waste-to-energy
wind farms
wind power

STUDY QUESTIONS

1. Define alternative energy. How do alternative energy sources differ from traditional energy sources?
2. List some of the important advantages and disadvantages common to many alternative energy sources.
3. Briefly describe the principles of hydroelectric power generation.
4. Discuss the pros and cons of hydroelectric power.
5. Describe the various types of biomass fuels and technologies.
6. What are the pros and cons of "waste-to-energy" techniques?
7. Describe some of the means currently being used to harness solar energy. Distinguish between active and passive solar techniques. How can electricity be generated from solar energy?
8. Describe the basic concept behind wind power.
9. What is geothermal energy? Is geothermal energy a renewable energy source?
10. What attempts are being made to harness ocean energy?
11. List the pros and cons of various renewable and alternative energy sources. Do you personally favor one over the other? Why? Are all such sources equally feasible in all areas? Explain.
12. List some of the major ways of storing electrical energy. Which are already in use today?
13. What is the distinction, if any, between energy conservation and energy efficiency?
14. Describe some ways that energy efficiencies can be improved.
15. What is a "solar-hydrogen economy"? Do you think it is feasible or just a pipe dream? Justify your answer.

WHAT'S THE EVIDENCE?

1. The authors suggest that it may be inevitable that hydroelectric power and alternative energy sources will make up a greater percentage of the energy mix in the future. Why do they suggest this? What evidence do they base their suggestion on? Do you find their evidence persuasive? What energy mix nationally and globally do you envision in 50 years? Support your answer with appropriate evidence.

2. The authors seem to imply that decreasing the energy intensity of a country is a positive move. Discuss the concept of energy intensity (Table 8-2). How does energy intensity relate to a nation's overall energy consumption? How does it relate to a nation's overall economic production? How does energy intensity relate to the per capita energy consumption of a nation's citizens? How does it relate to the per capita economic production? Some thinkers believe that all nations should strive to decrease their energy intensities, even if this means using more energy overall. What are some of the economic assumptions that underlie such a belief? Do you agree or disagree? Justify your answer.

CALCULATIONS

1. Referring to Table 8-1, calculate the amount of land required to generate 36.25 exajoules of electricity (the approximate amount of electricity generated in the United States during 1999) using solar thermal trough technology. Express your answer in both square kilometer-years and square mile-years [one square kilometer equals 0.386 square miles]. Your answer will consist of a lower and upper range.

2. Referring to Table 8-1, calculate the amount of land required to generate 36.25 exajoules of electricity using large hydroelectric power plants. Express your answer in both square kilometer-years and square mile-years. Your answer will consist of a lower and upper range.

ILLUSTRATION AND TABLE REVIEW

1. Referring to Figure 8-13, world wind-generating capacity increased by approximately how much between 1990 and 2005? Express this in terms of an approximate percentage increase during this period.

2. Referring to Table 8-1, which electric generation technologies have the largest land requirements? The smallest?

3. Referring to Table 8-2, in terms of energy intensity, where does the United States fall relative to other nations listed in the table? Name a couple of countries that have lower energy intensities than the United States. Name a couple of countries that have higher energy intensities than the United States.

http://environment.jbpub.com/mckinney/5e/

http://environment.jbpub.com/mckinney/5e/

Connect to this text's website at http://environment.jbpub.com/mckinney/5e/. This site features eLearning, an online review area that provides quizzes, chapter outlines, and other tools to help you study for your class. You can also follow useful links for more in-depth information.

According to the latest UNICEF data, more than 4,000 children die each day from contaminated water and poor sanitation. Worldwide, 2.5 billion people live without proper sanitation, 1.1 billion of whom also lack secure access to a consistent drinking water supply. In developing nations, children like this young child in the photo often travel barefoot to retrieve water for the family, especially during times of drought when help is needed and school attendance is low. The average distance of a water source to a residence is 4 miles (6.4 km). Around the world, students lead 6k events to raise awareness and funds to mitigate this global challenge. See: http://www.unicefusa.org/work/water and http://www.h2oforlifeschools.org

9 Water Resources

Chapter Objectives

After reading this chapter, you should be able to explain or describe the following:

- Properties of water and the hydrologic cycle on Earth
- Water usage in the United States
- Global water usage and supply
- Types of water resources
- Major groundwater problems
- Ways to increase effective water resources
- Benefits and drawbacks of large dams
- Ways to deal with water scarcity

ater is one of the most remarkable materials on Earth. Despite its chemical simplicity—H2O—water's properties make it absolutely essential for all life. Water covers over 70% of the planet's surface. If water were evenly distributed, it would cover the entire Earth to a depth of 3.2 kilometers (2 miles). However, more than 97% of this is saltwater and is not usable by land life. About 30% of the world's renewable freshwater supplies are already being used; this is about eight times the yearly flow of the Mississippi River. Even so, theoretically by some estimates enough freshwater is available to support more than 20 billion people, although the quality of freshwater is not always sufficient for all uses, such as ingestion by humans. Unfortunately, because of variable climatic and geologic conditions, freshwater is not uniformly distributed, so many areas of the world experience severe water shortages.

Such natural shortages are greatly aggravated by mismanagement of local water supplies, especially (1) lack of water conservation and (2) water pollution. Lack of water conservation is most evident in agriculture, which accounts for more than 70% of water consumed worldwide and 80% consumed in the United States (90% in several western states). For example, farmers in the western United States waste vast amounts of water because they do not pay its true cost. The government heavily subsidizes the water cost, so farmers do not conserve during irrigation, when much water evaporates. Water-short urban areas such as Los Angeles could use this wasted water.

In addition, the farmers in the drier western states often grow water-intensive crops, such as rice, tomatoes, and many fruits, that could easily be grown elsewhere where water is more plentiful.

Water pollution is an especially large problem in developing countries. Indeed, a main cause of human misery in the world is not thirst, which kills very few people, but death and disease from polluted drinking water. Although estimates vary, approximately 1.1 billion people around the globe do not have access to safe drinking water, and nearly 3 billion people (approximately 40% of the world's population) lack access to adequate sanitation services; 3,900 children a day die of waterborne diseases (see **CASE STUDY 9-1**).

On the other side of the equation, a surplus of water can result in flooding. Rivers can overflow their banks during heavy rains or rapid melting of snow. Often human activities can cause or exacerbate flooding. Man-made dams can break, releasing significant amounts of water in a short period of time. Farming, overgrazing, deforestation, and community development, such as widespread pavement, reduce the ability of water to penetrate the ground's surface. This leads to increased runoff, which promotes flooding and soil erosion. In addition, activities such as mining and construction can clog stream channels with sediment, impeding water flow and promoting flooding. For additional details, see review the dynamic earth and natural hazards.

9.1 Water and the Hydrologic Cycle

Water: A Most Unusual Substance

Water is so common that we often assume that it is a typical liquid. In fact, nearly all of its chemical and physical properties are unusual when it is compared with other liquids, such as alcohol, oil, ammonia, benzene, vegetable oils, or mercury. These unique properties account for why water is the major component of most cells and living tissue. Some of these properties and the reasons they are so important to the Earth and its life include the following:

- *Density.* Water is the only common liquid that expands when it freezes. This causes ice to float. If water were like most liquids, ice forming on lakes would be denser than liquid water and sink to the bottom. Lakes would freeze solid each winter, killing fishes, plants, and other familiar life forms. In addition, much of the activity that shapes our landscapes derives from this property: much weathering of rocks occurs when water freezes in the cracks and expands, breaking the rock apart.

- *Boiling point.* If water were similar to most other liquids, it would boil at Earth's normal surface temperatures and thus exist only as a gas. This would render it useless to life, which needs water in the liquid state.

- *Specific heat.* Specific heat is the amount of energy required to raise the temperature of a given mass of a substance by 18° Celsius (C). The specific heat of water is higher than any commonly known liquid except ammonia. It takes 1 calorie of heat to raise the temperature of 1 gram of liquid water by 18°C. The heat capacity of water plays a crucial role in the Earth's climate. Because water takes a long time to absorb heat and a long time to release it, or cool down, large bodies of water such as the oceans serve as moderating influences on climate. It also explains why coastal climates have less extreme temperature variations than do those areas located far from the oceans or large lakes. For example, without the northward-moving Gulf Stream, Great Britain would experience much colder winters.

- *Solvent.* Water is a better solvent than any other common liquid, which means that it dissolves more substances. This property makes water the most effective liquid for transporting dissolved nutrients and other materials. Indeed, water is crucial for transporting nutrients such as phosphorus throughout the biosphere. This dissolving ability also explains why water is the most important agent in eroding the landscape and why tropical areas are much more eroded than deserts. Unfortunately,

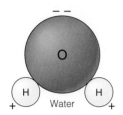

FIGURE 9-1 The water molecule, consisting of two hydrogen (H) atoms and one oxygen (O) atom.

this ability is also the reason water is so easily polluted and often stays polluted for such a long time.

The water molecule's structure easily explains most of water's unusual properties. Most important is that, although the H_2O molecule as a whole is electrically neutral, it contains electric charges (**FIGURE 9-1**). These charges are distributed in a "polar" manner, meaning that the water molecule has a positive charge at one end and a negative charge at the other end. These charges create an attraction between water molecules called hydrogen bonds. This phenomenon explains why ice melts and water boils at such high temperatures: energy is required to break the hydrogen bonds. Without such bonding, water would boil at −200°C (−32°F). The electric charges are also important in increasing the chemical reactivity of a substance, which explains why water is such a good solvent.

The Hydrologic Cycle

Although the "blue planet," Earth is rich in water, 97.4% of this water is saltwater, which contains about 35 parts per thousand (equal to 3.5%) dissolved substances (**TABLE 9-1**). Sodium chloride (NaCl), or "table salt," is the most abundant of these substances, with sodium and chlorine atoms making up about 86%.

TABLE 9-1	Major Dissolved Substances in Seawater	
Ion	Parts Per Thousand	Percentage
Chloride	19	55
Sodium	11	31
Sulfate	3	8
Magnesium	1	3
Calcium	0.5	1
Potassium	0.5	1
Total	35	99

Magnesium, calcium, and other atoms in lesser amounts make up the rest. Indeed, saltwater in the sea contains dozens of elements, including gold, in trace amounts. However, right now no one will be getting rich on seawater. The majority of the precious metals occur in such small quantities and so much energy would be needed to obtain them that they are not economically recoverable. In addition, saltwater generally is not usable for drinking, agricultural, or industrial purposes. Drinking water requires dissolved substances of no more than one part per thousand before it is considered potable. Industry requires even less because salts destroy machinery.

Therefore, we must turn to the 2.6% of Earth's water that is freshwater to satisfy most of our needs. About three-fourths of this water occurs as ice and is inaccessible (**FIGURE 9-2**).

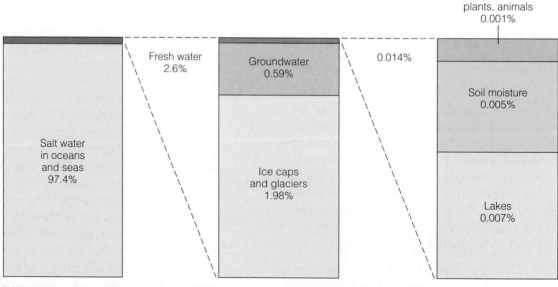

FIGURE 9-2 Most of the world's water is saltwater. Of the 2.6% that is not saltwater, most is ice. (*Source:* Data are from D. Speidel and A. Agnew. The world water budget. In D. Speidel D., et al., eds. Perspectives on Water Uses and Abuses. New York: Oxford University Press, 1988, p. 28.)

Ninety percent of this ice occurs in Antarctica. Only about 0.6% of Earth's freshwater is in the readily available liquid state. The majority of this occurs as groundwater. Only a fraction (less than 0.014%) of Earth's water occurs as the freshwater lakes and rivers that we usually associate with water.

All of Earth's waters, whether freshwater or saltwater, are connected through the **hydrologic cycle** (see **FIGURE 3-3**). The hydrologic cycle is the "great pump" that circulates water through the atmosphere, land, and oceans. The circulation is powered by energy from the sun. This global water cycle involves two main processes: (1) evapotranspiration and (2) precipitation. **Evapotranspiration** is the transfer of water into the atmosphere (as the gas, water vapor) by evaporation and transpiration. Evaporation occurs from the heating of liquid water by the sun, and transpiration is the release of water vapor by plants. **Precipitation** occurs when water falls to the ground as rain, snow, sleet, or hail. Essentially, evapotranspiration removes water from the liquid state, and precipitation puts it back.

Let us follow this cycle by starting with the evaporation of ocean water, which amounts to about 425,000 km^3 (102,000 mi^3) of water per year. Most of this precipitates back into the oceans, but about 40,000 km^3 (9,600 mi^3) falls on land. The cycle is completed when about the same amount of water returns to the oceans as runoff via rivers or groundwater flow. Thus, the cycle is in equilibrium in that the same amount of water taken from the sea by evaporation and precipitated onto the land is returned to the sea by runoff from the land. The most important part of the hydrologic cycle to people is this 40,000 km^3 (9,600 mi^3) of runoff because it provides most of our water supply, as discussed later.

How fast does water move through this system? Each year, evapotranspiration removes an amount of water equivalent to a 1-meter (39-inch) thick layer around the globe. It takes about 40,000 years to recycle all the water in the oceans. In contrast, the much smaller water reservoir of the atmosphere recycles in 9 to 10 days (depending on seasonal variation). Stream and river water is fully renewed about every 2 weeks.

FIGURE 9-3 Agriculture consumes more water than any other activity, largely because so much water is lost through evaporation.

9.2 Water Demand

The United States uses about 1.54 trillion liters (408 billion gallons) of freshwater per day, or about 4,330 liters (1,143 gallons) per person; when estimates are corrected for the difference between commercial and personal use, each person in the United States actually uses closer to 379 liters (100 gallons) of water a day, which is still an exorbitant amount when compared with a Kenyan, who uses 45 liters (12 gallons). This is more than any other nation and more than twice the average usage of water in Europe. Nevertheless, people require only about 3.75 liters (approximately 1 gallon) of water per day for biological needs. Why do we use so much? In the United States, 41% of the water is used by agriculture, 38% to cool power generators, and 11% for industrial manufacturing. **TABLE 9-2** shows some examples of how much water is required to produce various agricultural and industrial products.

The public uses only about 10% of the water, or 530 liters (140 gallons), per day per person. Some of this is for fire hydrants and other municipal uses. **TABLE 9-3** shows the daily indoor domestic water use of an average American family of four; each person uses about 230 liters (61 gallons). All of these figures largely reflect the amount of water withdrawn from the water supply. **Withdrawn water** is water that is taken from its source

TABLE 9-2	Water Used to Make Various Agricultural and Industrial Products				
Agricultural Products	**Gallons**	**Liters**	**Industrial Products**	**Gallons**	**Liters**
Egg, 1	40	151	Refine 1 gallon of crude oil	10	38
Milk, 1 glass	100	380	Sunday paper	280	1,060
Flour, 1 pound	75	285	Aluminum, 1 pound	1,000	3,800
Rice, 1 pound	560	2,120	Automobile, 1	100,000	380,000
Beef, 1 pound	800	3,030			
Source: Based on U.S. Geological Survey data, 1992.					

(such as a river, lake, or aquifer), but it may be returned to its source after use. For example, a power plant may withdraw water from a river to cool generators but return the water to the river when done.

Consumed water is water that is withdrawn but not returned to the original source: it is "lost" to the local part of the hydrologic cycle, usually by evaporation. Approximately a quarter of all water withdrawn in the United States is consumed, 80% of which agriculture consumes (Figure 9-3). California agriculture alone accounts for about one-third of the total U.S. water consumption. Of all the water consumed in agriculture, 85% is used for irrigation. Naturally, crops cannot be grown without water, but most irrigation systems are extremely inefficient. On average, only about 37% of the source water is actually absorbed by the plants being irrigated. The rest is mostly evaporated from the drainage ditches as the water travels to the crops. When irrigation

pipes spray water upward, most of that water can also be lost to evaporation. Irrigation is particularly problematic in dry climates. Not only does the dry atmosphere increase evaporation, but also the hard soils in such climates tend to absorb water less quickly, so even more water is lost to evaporation and runoff. In addition to agriculture, irrigation is used extensively in landscaping, including golf courses, commercial landscapes, and domestic lawns and gardens.

Since 1950, total world water withdrawal has tripled (**FIGURE 9-4**) as the global population has increased. Because water conservation has led to a relative slowing of demand growth in industrialized countries in recent years, increasing withdrawal in developing nations causes the rapid growth of global demand. Developing countries are using more water for the same basic reasons that their use of all resources is increasing: (1) increasing population growth (more people) and (2) increasing per capita demand (more water per person). Per capita demand increases with industrialization because water is needed to

TABLE 9-3	Indoor Domestic Daily Water Use for an American Family of Four People	
	Gallons	**Liters**
Toilet flushing	100	380
Showers and baths	80	303
Laundry	35	132
Dishwashing	15	57
Bathroom sink	8	30
Utility sink	5	19
Total	243	921
Source: U.S. Environmental Protection Agency, 1993. Latest date for which accurate data are available, but it is believed that per capita domestic indoor water usage has not significantly changed in the last 2 decades.		

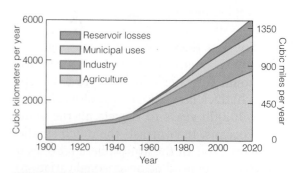

FIGURE 9-4 Estimated annual world water use, 1900 to 2020. Agriculture is the largest user by far. (*Source:* Based on data from L. R. Brown, ed. State of the World 1983. Washington, DC: Worldwatch Institute, 1993.)

produce water-intensive consumer items such as electronics.

As in the United States, agriculture is the main user of water worldwide, accounting for about 40% to 75% (differs geographically). Industry uses about 19% to 21%, and domestic use is about 6% to 9% (estimates vary). Use varies from country to country, however. In Nepal, for instance, 96% of all water withdrawal goes toward agriculture; whereas in Germany, 68% goes toward industry, and in Uganda, 43% goes toward domestic use.

Industry is the greatest "withdrawer" of water, but agriculture is the greatest consumer. Unfortunately, this water returned by industry is not always clean and must be treated. As a result, even water that is not technically "consumed" is often rendered unusable for other purposes such as drinking.

9.3 Water Supply

Water shortages occur when supply does not meet demands. For now at least, such shortages are mainly regional problems that occur because the hydrologic cycle distributes water very unevenly.

Regional Water Shortages: Inequalities in the Hydrologic Cycle

FIGURE 9-5 illustrates the inequalities in the hydrologic cycle. **Surplus areas** receive more precipitation than is needed by well-established vegetation, including crops, and the local biota (including humans). In contrast, **deficit areas** receive less precipitation than well-established vegetation and the local biota ideally need (including human occupants).

Figure 9-5 exhibits two main patterns:

1. Severe deficit areas are deserts found mainly in Australia, the western United States, and especially Africa and the Middle East. These correspond strongly to rainfall patterns: most deserts tend to occur at about 30-degrees north and south latitude.
2. Major surplus areas are especially prominent in South America, Asia, and eastern North America. About two-thirds of stream flow in the United States is east of the Mississippi River. Most of the one-third found west of the Mississippi is confined to the Northwest.

Perhaps the most surprising aspect of these patterns is that so much of Africa is water-deficit

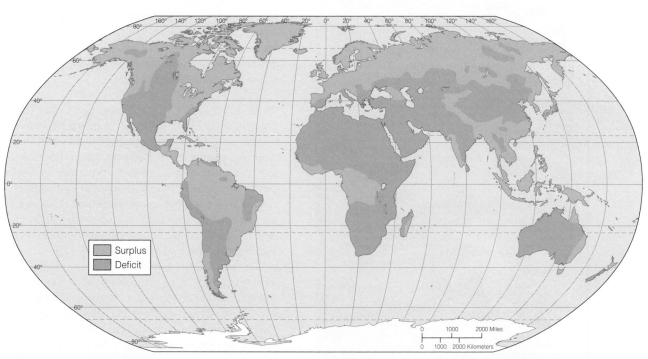

FIGURE 9-5 Regions of water surplus (green) and deficiency (brown) in the world. Deficit areas receive less precipitation than is needed by well-established vegetation and organisms; surplus areas receive more. (*Source:* Adapted from M. Falkenmark. Water and mankind. Ambio 1977;6:5.)

The Global Safe Drinking Water Crisis

VISIT

http://environment.jbpub.com
/mckinney/5e/
for more information

Many of the cities in developing countries do not have an adequate supply of safe drinking water. Limited water resources, inadequate facilities for treating and distributing water, and the absence of proper sewerage all contribute to the shortage (**FIGURE 1**). Despite efforts over the last half century by industrialized countries and international agencies, the situation in these cities is actually worsening.

The poor in developing countries have long been perceived as being unable to pay for household water service. In reality, because the urban poor of these countries generally are not provided with public water service, they have to buy water, often of questionable quality, from private, unregulated vendors. Buying small amounts of water from vendors is more expensive by far than obtaining water from municipal sources.

A study of the water-vending system in Onitsha, Nigeria, found that these private vendors were responsible for more than 95% of water sales to the city's residents. The poor were annually paying water vendors twice the operational and maintenance costs and 70% of the annual capital costs of the new municipal water system. The new municipal water system in Onitsha was planned without any participation from the city's residents. As a result, they were unsure of its reliability and quality and were reluctant to connect to the system. Instead, they continued to pay high prices for a much inferior service.

Visitors to these cities may see an impressive skyline with modern hotels, offices, and apartment buildings. What they do not see is the absence of physical infrastructure, such as sewers and water lines, to serve these buildings. In cities where the public water supply is inadequate and developers sink wells, which lower the water table, land subsidence often results, and in coastal areas, saltwater intrusion ultimately fouls the wells. The absence of sewage systems is even more serious. The wastewater is discharged with little if any treatment into drainage channels and urban streams so that the surficial groundwater and the soil become badly contaminated. In Bangkok, for example, household wastewater is discharged into the klongs, or canals. Pipes containing water destined for homes run through the klongs, and when the water pressure in the pipes is low, wastewater in the klongs seeps into the pipes.

The lack of basic water and sanitation services leads to a number of problems. For example, although water-borne diseases are manageable in the industrialized world, they exact a heavy toll in developing countries. In addition, many hours each day are spent fetching water—time that might be put to more socially and economically productive uses. The water problem is worsening. Eighteen of 22 "giant cities" (urban areas with more than 10 million inhabitants) are in developing countries, compared with only 1 of 4 such cities in 1960.

A lack of technology or funds cannot alone explain the failure of some developing countries to provide an adequate water supply and sanitation services. Even in cases in which funds are available and the required technology involves only well-established practices, projects have not been sustained. As this suggests, solving the water supply problems of these countries will not be a simple matter and will require more fundamental changes than merely providing funding.

FIGURE 1 A floating market in Thailand illustrates the demands placed on water in urbanized areas of developing nations.

ridden. Although we often think of Africa as a continent rich in lush, tropical forests, this is true of only the central portion. North Africa and the Middle East, for instance, withdraw a huge portion of available water.

A Global Water Shortage?

Regional and local water shortages have always existed because of the inequalities of the hydrologic cycle. Even small bands of hunters and gatherers long ago experienced droughts and other serious shortages. For example, the mysterious Anasazi "cliff-dwellers" of the American southwest are thought to have abandoned their dwellings because of a long-term drought hundreds of years ago (**FIGURE 9-6**). With the rapid growth in population and industrialization (and consequent per capita water use), increasing demand has, of course, increased the extent of regional and local shortages.

Does this mean that countries with surplus water from the hydrologic cycle will not suffer water shortages? Not necessarily! A global view indicates that the entire hydrologic cycle is nearing the limits of use. If this occurs, even surplus areas will experience shortages. The hydrologic cycle annually creates about 40,000 km³ (9,600 mi³) of freshwater runoff from the land to the sea. It is estimated that about 9,000 km³ of this runoff is accessible in the form of a stable, predictable supply, such as rivers, that can be captured by dams, canals, and other methods (**FIGURE 9-7**).

Currently, people use about half of the 9,000 km³ (see Figure 9-4), but crude projections based on Figure 9-7 suggest that we are nearing the limits of the accessible water. Countries can be categorized according to the amount of water they have available for their populations. Nations with less than 1,000 m³ of renewable water supplies per year per

FIGURE 9-6 Native Americans, often called "cliff dwellers," lived in these cliffs in the southwestern United States. Research suggests an extended drought forced the people to abandon the area.

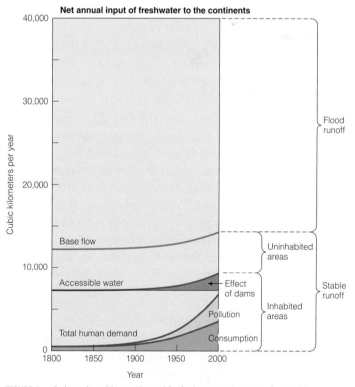

FIGURE 9-7 Estimated total human demand for freshwater and amount of accessible freshwater, 1800 to 2000. Of the total input of freshwater to the continents, only about 9,000 cubic kilometers was accessible in inhabited areas as of the year 2000. This barely exceeds the total amount of water used by humans either through direct consumption or by rendering it unusable through pollution: 4.1655 km³ equals approximately 1 mile³. (*Source:* Adapted from Meadows, Meadows, and Randers. Beyond the Limits. White River Junction, VT: Chelsea Green Publishing Company, 1992.)

person are said to be "water scarce," and nations with 1,000 to 1,700 m³ per person per year are "water stressed." Various studies have demonstrated that as countries fall into the water stress and water scarce categories, their imports of food in particular must increase dramatically and the standard of living may decline. Today, an estimated 2.2 to 2.5 billion people live in water-scarce or water-stressed countries. The group Population Action International projects that depending on varying population growth rates, this number may rise to between approximately 1.2 and 2.9 billion by 2020. Particularly affected are peoples in the Middle East; north, south, and east Africa; and southern and western Asia. Today, an estimated 4 million deaths each year (mostly of infants and young children) are the result of a lack of safe freshwater, and in many regions, social and political tensions are increasing because of the scarcity of freshwater. Although water conflict has been documented since biblical times, the current water conflicts are occurring more frequently and often are violent. Since 2000, there have been more than 50 violent conflicts related to water access, distribution, and control.

Types of Water Resources

Water resources are usually categorized as either (1) surface waters or (2) groundwater. Surplus areas are rich in water resources. Deficit areas also usually have these two types of water resources, although in shorter supply of course. Even the arid areas of the western United States have a few major rivers, such as the Colorado River (see **CASE STUDY 9-2**), and groundwater supplies. However, because precipitation is lower in those areas, these resources are replenished much more slowly when used.

Surface Waters

People have traditionally used surface waters where possible because they are more accessible than groundwater. Surface waters include both (1) flowing waters, such as streams and rivers, and (2) basinal waters, such as ponds and lakes.

About 20% of the precipitation that falls on land flows over the surface, first as sheet wash, then as riverlets, and then as streams.

The streams (often called "tributaries") eventually merge together into a large river. As the running water flows from the riverlets through the rivers, it carves out progressively larger channels, creating a tree-like pattern (**FIGURE 9-8**).

Stream and river valleys, from the Nile to the Mississippi, have always been among the most heavily populated areas. The reasons are many:

1. Soil on the floodplain is rich with nutrients deposited when the river overflows.
2. The water provides a ready means of transportation and shipping.
3. Moving water carries wastes away more readily than do lakes and other standing supplies of water.
4. The river provides a source of clean drinking water.

However, these various uses often conflict, and the conflicts have intensified as growing numbers of people have placed ever more pressures on stream and river resources. For example, rivers used for extensive waste disposal are not useful for drinking water or even recreational activities. In 1969, such large amounts of in-

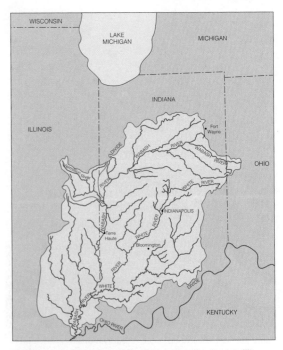

FIGURE 9-8 Drainage basin of the Wabash River showing tree-like pattern of tributaries that drain into larger channels.

dustrial waste were dumped into the Cuyahoga River, which flows through Cleveland, Ohio, that the oily waste on the river caught fire. Such incidents led to the passage of the U.S. Clean Water Act and other legislation in the early 1970s, which has succeeded in keeping flowing waters available for many uses (for additional information see a review of water pollution). Forty years later, in 2009, the Cuyahoga community celebrated the hard work to save the river and set ongoing goals. Restoring a river once it is dead is a long and arduous undertaking; despite ongoing conservation and the return of invertebrate, fish, and bird populations, the river has still passed only 4 of the 14 points needed to deem the river healthy.

Basinal freshwaters, such as ponds and lakes, are more readily polluted than flowing waters. Ponds and lakes form where water accumulates in basins (local depressions) in the land. Wastes are not carried away, but tend to accumulate in the basin. The water supply in even very large basins such as the Aral Sea in central Asia can be depleted if enough demands are made on it.

Groundwater

The majority of accessible freshwater occurs as groundwater (see Figure 9-2). Indeed, groundwater supplies the drinking water for about half the U.S. population and likely will provide much more in the future. **Groundwater** is a general term referring to water beneath the Earth's surface. It occurs because about 10% of the water falling as precipitation ultimately enters the ground, infiltrating down through the soil and rock (**FIGURE 9-9**). As the water is pulled downward by gravity, it fills the rock pores, called voids, until a layer of impermeable rock stops it. As the infiltrating water "backs up," it forms the **zone of saturation**, which consists of voids that are fully saturated with water. Above this is the **zone of aeration**, which consists of voids that may be moist but are not saturated. The **water table** is the boundary between the zones of saturation and aeration. The rock that contains the zone of saturation is known as an **aquifer** (*aqua* = water; *fero* = containing). Practically speaking, an aquifer must be a relatively permeable rock, such as sandstone,

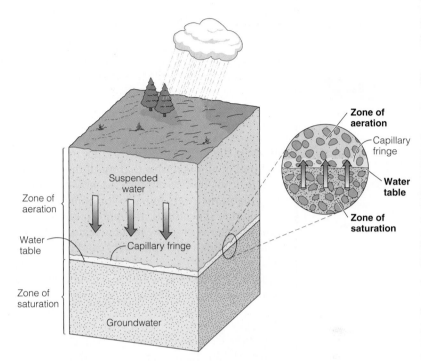

FIGURE 9-9 Rainfall infiltrates the soil, moving downward to form the zone of saturation. The top of this zone is the water table.

gravel, fractured limestone, or fractured granite, and economically accessible.

Impermeable rocks that obstruct water flow are called **aquicludes.** Shales and other clay-related rocks are common aquicludes. Aquicludes can separate the zone of saturation into one or more aquifers, often called **confined aquifers** (**FIGURE 9-10**). These confined aquifers are supplied with water from one or more **recharge areas**, from which rainfall infiltrates into them. If aquicludes sufficiently confine the aquifer so that pressure is built up from the recharge, an **artesian well** can result. The pressure causes water to rise in the well, possibly to the surface where it can flow freely, without the pumping required by most aquifers. Waste-well injection into deep confined aquifers was once a common method of waste disposal because it was erroneously thought that wastes would not leak into the higher aquifers that often provide drinking water. However, aquicludes often have fissures or other openings that allow water to move between aquifers; "confined" aquifers are rarely completely confined. Consequently, waste-well injection has been banned in many areas.

The Colorado River

Even in very dry regions, there are usually a few rivers that are of vital importance to the livelihood of all organisms that dwell there. Water is needed not only to support aquatic organisms that inhabit the rivers but also for humans to use in industrial, agricultural, and municipal activities. Because water is so scarce in these regions, the demand placed on individual rivers can be devastatingly high.

One of the largest and longest of the rivers in the arid southwestern United States is the Colorado River. The Colorado River begins its journey in the Rocky Mountains of northern Colorado, just west of the Continental Divide. It travels approximately 2,330 kilometers (1,450 miles) through the American southwest into northern Mexico, emptying into the Gulf of California. Along its path, the river serves as habitat for a variety of aquatic organisms, many of which are endemic to the Colorado River. It is also a source of water for humans living in the region, primarily for irrigation and municipal purposes.

Human use has drastically affected nearly every aspect of the Colorado River system. To harness enough water for the growing populations in

the southwest, numerous dams have been built along the length of the river. The northernmost, the Glen Canyon Dam, was completed in 1966 (**FIGURE 1**). After completion, it began blocking the flow of the river, flooding Glen Canyon, and forming a reservoir called Lake Powell. The dam has been the object of much environmental controversy since the beginning. Environmental groups have been calling for the slow increase in the amount of water released, thus slowly draining Lake Powell and restoring a more natural ecosystem.

Years before the completion of Glen Canyon Dam, the Hoover Dam was built on the Colorado, on the border between Arizona and Nevada. It was completed in 1936 for the purposes of controlling flooding, as well as creating a reservoir for irrigation and domestic water use. It also harnesses hydroelectric power as the river flows through turbines in the dam. Today, the Hoover Dam provides water to more than 25 million people in southwestern states. The electricity generated at the dam is transmitted as far away as Los Angeles. The reservoir, Lake Mead, is a popular recreational destination in the southwest, drawing several million visitors per year for swimming, boating, and fishing.

There are five large dams on the Colorado River in Arizona alone (**FIGURE 2**). The adverse effects of such dams are numerous. The building of dams drastically alters the sediment flow in a river. Before the formation of dams and reservoirs, the Colorado River carried large amounts of sediment along its length and deposited it in the Gulf of California, supplying the Colorado River Delta. Today that sediment is trapped behind the dams, not only rapidly filling the reservoirs, but also starving the downstream areas of sediment. It is estimated that Lake Powell will be filled within 100 to 300 years, and various heavy metals (including arsenic, selenium, mercury, boron, and lead) and other toxins are trapped and concentrated in the reservoir, rather than flowing harmless in dilute quantities to the sea. These high concentrations of toxins in the sediment and water may pose health risks to wildlife and humans.

FIGURE 1 The Colorado River seen here behind the U.S. Bureau of Reclamation's Glen Canyon Dam is one of the few large sources of water in the southwestern United States.

Dams not only starve downstream areas of sediment, they also deprive them of water. Except in years of exceptionally high precipitation, the Colorado River no longer makes it to its original destination in the Gulf of California. Rather, it dries up just south of the border between the United States and Mexico. What is not stored or diverted by the dams and canals evaporates in the Sonoran Desert heat. The Colorado River Delta, once flourishing with freshwater, brackish, and marine animals and plants, has been reduced to about 5% of its original nearly 20 million acres of wetland.

Native species of plants, fishes, and other aquatic life are severely endangered by the changes wrought in the river's ecosystem. Damming affects not only the amount of water available downstream but also its quality. Diversion of clean upstream waters and evaporation of water from reservoirs causes increased concentrations in pollution and salts downstream. By the time the Colorado River reaches Mexico, its salinity is 20 times higher than at its source. People cannot easily use water at this salinity without dilution or desalination, and it is also toxic to many species of plants and animals. Also problematic is the year-round release of cold water from the bottoms of reservoirs. It lowers the average water temperature downstream, negatively affecting the spawning patterns of native fishes.

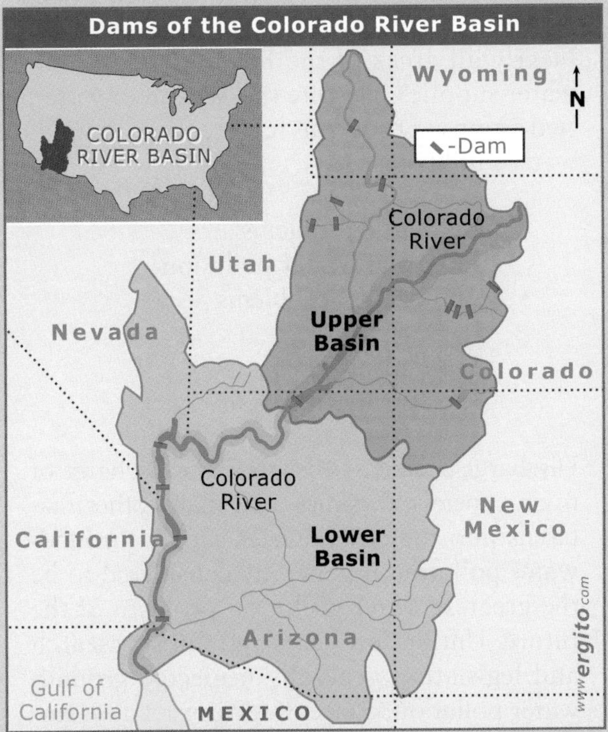

Several species of fish native (and possibly endemic) to the Colorado River are endangered, such as the humpback chub, the Colorado pike minnow, the razorback sucker, and the bonytail sucker. The various stresses on the ecosystem have provided an opportunity for the invasion of nonnative species. Dozens of species of fishes have been introduced, either accidentally or on purpose (e.g., for recreational fishing); currently, nonnative species outnumber native species in the Colorado River. Many of these introduced species are better adapted to the disturbed conditions of the river and thus can out-compete native species for resources such as habitat and food. A similar situation has arisen on the delta, where invasive plant species such as salt cedar and arrowweed have out-competed native species such as cottonwoods and willows.

Efforts are under way to restore environmental conditions and advocate for the sustainable use of the Colorado River. Increased water flow to the delta, as well as to other parts of the river, will be critical to the restoration and protection of endangered species. Several groups, such as the Sierra Club and the Glen Canyon Institute, are promoting the decommissioning of the Glen Canyon Dam. Other key issues are the reduction of water consumption and an increase in programs that promote water recycling to relieve the pressure on this overextended ecosystem. It is hoped that with attention to these problems, as well as those of pollution, salination, sedimentation, and others, the vital Colorado River will continue to support life in the southwestern United States.

VISIT

http://environment.jbpub.com/mckinney/5e/
for more information

Dams of the Colorado River Basin

COLORADO RIVER BASIN

N

⬦ -Dam

Wyoming

Utah

Colorado River

Upper Basin

Colorado

Nevada

Colorado River

California

Lower Basin

New Mexico

Arizona

Gulf of California

MEXICO

www.ergito.com

FIGURE 2 Dams on the Colorado River.

281

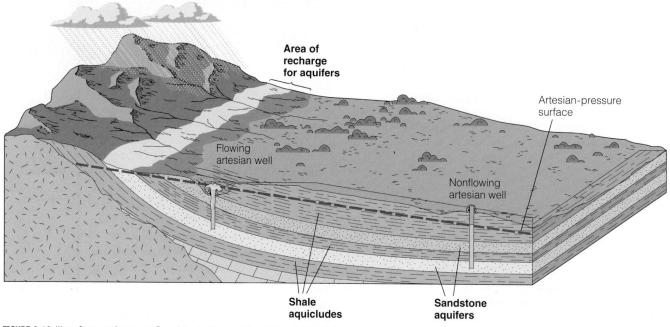

FIGURE 9-10 Water from a recharge area flows into sandstone aquifers that aquicludes enclose.

In areas of high rainfall, the water table tends to be near the surface. This is especially true in low-lying places such as Florida, which is near sea level. In such areas, lakes and rivers are especially abundant because they occur where the land surface dips below the water table (**FIGURE 9-11**). Where the ocean is nearby, groundwater closest to the ocean likely will be composed of saltwater. The saltwater groundwater will migrate inland if humans withdraw too much water.

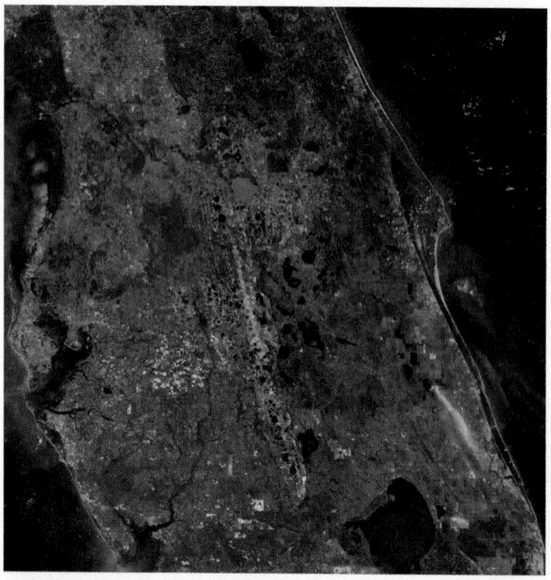

Groundwater Problems

Nearly all areas of the Earth have groundwater supplies. Despite this wealth of water, two groups of problems have reduced groundwater's utility:

- Discharge problems
  1. Groundwater pollution
- Withdrawal problems
  1. Depletion
  2. Land subsidence
  3. Saltwater intrusion

Discharge problems arise from the discharge of toxins, metals, organics, and many other materials into the groundwater supply. Groundwater pollution is generally considered to be the greatest water pollution problem of the future. Until a few years ago, most research and legislation generally neglected groundwater pollution, concentrating instead on surface waters. However, groundwater's slow movement (an average of about 15.2 meters or 50 feet per year) causes pollutants to stay in the water for long periods of time. Particularly problematic sources of groundwater con-

FIGURE 9-11 Florida's many lakes are caused by a combination of high rainfall and low-lying areas, including some created by sinkholes. This satellite image of central Florida shows hundreds of lakes and thousands of sinkholes. Limestone underlies the area.

tamination include leaky underground storage tanks (containing gasoline, oil, hazardous wastes, or chemicals), improperly installed or maintained septic tanks, landfill seepage, and profligate use of road salt and agricultural chemicals such as pesticides and fertilizers. These sources of pollution have created many long-term groundwater problems that will take much time and money to clean up (which is why the precautionary principle should always be applied when possible).

Withdrawal problems arise from the removal of groundwater from aquifers. Depletion occurs when groundwater is withdrawn faster than the aquifer can be recharged. Aquifers can be depleted even in areas of very high rainfall because human-made pumps can remove groundwater at a much greater rate than the aquifers recharge. Recharge water must infiltrate through the sediment and rock, which usually takes years. Geologists estimate that the average aquifer takes 200 years to fully recharge. In such cases, groundwater becomes a nonrenewable resource that is "mined," in terms of a human lifetime. For example, much of central and south Florida is already experiencing severe groundwater shortages despite its very high rainfall. Paving over **recharge areas** often greatly aggravates depletion, preventing water from infiltrating. An example is the destruction of wetlands. Wetlands are major recharge areas because they retain water for long periods, purifying the water and allowing it to infiltrate the aquifer. For this reason, wetlands could be called the "kidneys" of the hydrologic cycle. When they are paved over or filled in, the aquifer's recharge ability is reduced. The United States has already lost about half its wetlands to urban and agricultural development.

Recharge time is longest in areas with lower rainfall. Deserts and other water-deficit areas tend to have aquifers at great depth, and deep wells must be drilled to reach them. This groundwater is often called "fossil" water because it was deposited thousands of years ago when the climate was wetter in the area. As a result of the water's age, its quality is often poor, and it may be highly saline. Even more important, when water is withdrawn in such arid areas, it often takes thousands of years to

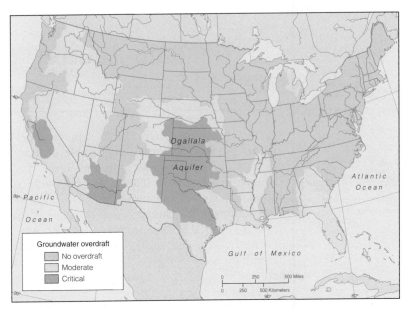

FIGURE 9-12 Areas of groundwater overdraft where groundwater is being pumped out faster than it recharges. Note the Ogallala (or High Plains) Aquifer overdraft in the central United States. (*Source:* U.S. Geological Survey.)

recharge. When high demand occurs in areas of low rainfall, such as southern California with its dense population or the midwestern United States with its intensive agriculture, groundwater supply problems are especially acute (**FIGURE 9-12**). The great Ogallala Aquifer of the Great Plains provides the water for the "breadbasket" of the world. Nevertheless, this aquifer is perhaps half depleted, with recharge rates that are much slower in today's drier climate: most of the Ogallala Aquifer formed many thousands of years ago during the wetter Ice Age climate (see **CASE STUDY 9-3**).

The rate of groundwater depletion can be estimated by the rate that the water table falls in an area. The global extent of groundwater depletion is apparent in this small sampling of global rates:

Water Table Fall Rate	
North China plain	2.3 meters per year
Guam, Mexico	1.5–3.5 meters per year
Beijing, China	0.91 meters per year
Manila, Philippines	9.1 meters per year
Northeastern Iran	2–3 meters per year
Various states in India	1.3 meters per year

The second withdrawal problem, land subsidence, is more localized than is depletion.

The Ogallala Aquifer

The Ogallala (or High Plains) Aquifer underlies most of Nebraska and sizable portions of Colorado, Kansas, and the Texas and Oklahoma panhandles (**FIGURE 1**). This area is one of the largest and most important agricultural regions in the United States. It accounts for about 25% of U.S. feed-grain exports and 40% of wheat, flour, and cotton exports. More than 5.7 million hectares (14 million acres) of land are irrigated with water pumped from the Ogallala. Yields on irrigated land may be triple the yields on similar land cultivated by dry farming (no irrigation).

The Ogallala's water was, for the most part, stored during the retreat of the Pleistocene continental ice sheets (the last Ice Age, which ended about 10,000 years ago). Current recharge is insufficient over most of the region (see Figure 1). Each year, farmers draw from the Ogallala more water than the entire flow of the Colorado River. In 1930, the average thickness of the saturated zone of the Ogallala aquifer was nearly 20 meters (66 feet); currently, it is less than 3 meters (9 feet), with the water table locally in various areas dropping by amounts ranging from 15 centimeters (6 inches) to 1 meter (3.3 feet) per year. Overall, it is believed that the Ogallala will be effectively depleted within the next few decades. In areas of especially rapid drawdown, it could be locally drained in less than a decade. Furthermore, even if all groundwater pumping stopped today, it would take approximately 1,000 years for the aquifer to recharge to the level it was in the mid-20th century.

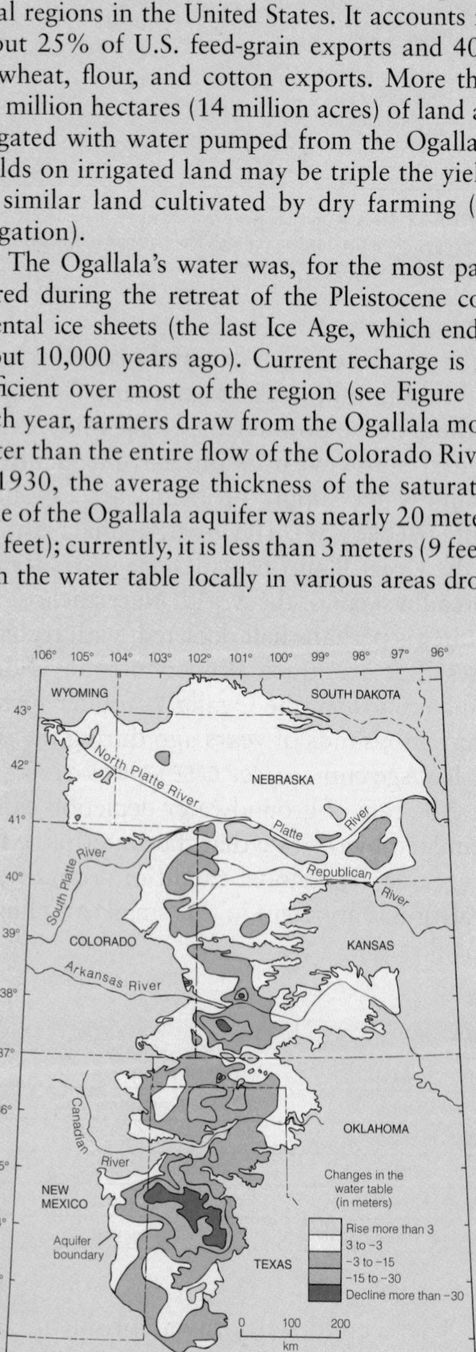

Reversion to dry farming, where possible at all, will greatly diminish yields. Less vigorous vegetation may lead to a partial return to pre-irrigation, Dust Bowl-type conditions. Alternative local sources of municipal water are not at all apparent in many places. Planners in Texas and Oklahoma (regions where water withdrawal from the Ogallala is currently greater than other areas of the reservoir) have advanced ambitious water-transport schemes as solutions. Texas is considering alternatives that would involve transferring water now draining into the Mississippi River from northeastern Texas across the state to the panhandle. Oklahoma's Comprehensive Water Plan would draw on the Red River and Arkansas River basins. Such schemes would cost billions of dollars, perhaps tens of billions, and could take a decade or longer to complete. However, before the projects are completed, acute water shortages can be expected in the areas of most urgent need. Even if and when the transport networks are finished, the cost of the water may be 10 times what farmers in the region can comfortably afford to pay if their products are to remain fully competitive in the marketplace. There seem to be no easy solutions. Perhaps because of this, progress toward any solution has been slow. Meanwhile, the draining of the Ogallala Aquifer continues.

It is noteworthy that in the short term, farmers have little incentive to exercise restraint, in large part because of federal policies. Price supports encourage the growing of crops such as cotton that require irrigation in the southern part of the region. The government shares the cost of soil-conservation programs and provides for crop-disaster payments, thus giving compensation for the natural consequences of water depletion. In fact, federal tax policy provides for groundwater depletion allowances (tax breaks) for High Plains farmers using pumped groundwater, with larger breaks for heavier groundwater use.

FIGURE 1 The Ogallala Aquifer.

Changes in the water table (in meters)

- Rise more than 3
- 3 to -3
- -3 to -15
- -15 to -30
- Decline more than -30

Nevertheless, subsidence is very costly to those who live in areas where it occurs. Land subsidence means that the land "sinks," causing buildings, roads, and other surficial structures to sink with it. Subsidence occurs where depletion of groundwater causes the water table to fall. At first, this lowering is localized around the well, where water table drawdown forms a **cone of depression** (**FIGURE 9-13**). As more wells are drilled to draw water from the aquifer, the entire water table will lower. In many cases, a lower water table does not lead to significant land subsidence. This is very fortunate because water tables have been lowered in most parts of the United States and many areas of the world from rapid withdrawal, as we have just seen. However, in some cases, the grains of sediment composing the aquifer have very large voids or are otherwise supported by the water pressure of the groundwater. Removal of that water causes the voids to close up as the sediment or rock grains crowd closer together. This loss of rock volume underground causes the overlying land to subside.

Beneath the fruitful farmlands of California's San Joaquin Valley is a large reserve of groundwater held in a loosely packed, porous layer of fine-grained sediment, including clays. Excessive withdrawal of groundwater for agricultural and municipal purposes since the 1920s has caused significant subsidence of the area because the clay is compacted. The subsidence is less severe in areas where considerable surface runoff recharges the aquifer more quickly; however, areas with slower recharge in the western part of the valley have experienced as much as 30 feet of subsidence in the past 80 years.

Subsidence tends to be especially bad in urbanized areas (high withdrawal) with aquifers of sediment that are highly saturated with water. A notable example is Mexico City. It was built on an ancient lake bed, and massive groundwater withdrawal has caused the ground to sink by many feet. Some buildings must now be entered by the second floor. Houston, Texas, New Orleans, Louisiana, Venice, Italy, and other coastal cities are experiencing significant subsidence. Efforts have been made to reverse subsidence by pumping

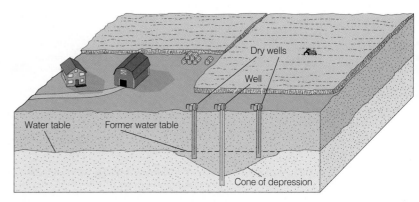

FIGURE 9-13 A cone of depression is created where overpumping causes the water table to drop below its former level.

water back into some aquifers. However, aside from the high cost, the ability of the aquifer to hold water is often permanently reduced because the void space, once diminished, cannot be recreated.

Sinkholes are a special type of land subsidence caused by water withdrawal. These are depressions in the ground that occur when the thin layer of rock overlying a previously existing underground cavern collapses into the cavern (**FIGURE 9-14**). Sinkholes are especially common in areas underlain by limestone, such as much of the southeastern United States and the Yucatan of Mexico. This limestone is often dissolved by groundwater to form underground caverns. Water pressure from the groundwater helps support the overlying land surface, so a drop in the water table from heavy withdrawal removes that support, causing the ground to collapse.

Another withdrawal problem is saltwater intrusion. Normally, the groundwater underlying coastal regions has an upper layer of

FIGURE 9-14 Collapse of the underlying limestone caused this sinkhole in Winter Park, Florida.

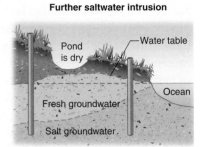

| No saltwater intrusion | Saltwater intrusion | Further saltwater intrusion |

FIGURE 9-15 Saltwater intrusion into groundwater supplies is caused when rapid removal of freshwater allows saltwater to migrate farther inland, where it can reach freshwater wells and contaminate valuable groundwater.

freshwater, underlain by saltwater. This layering occurs because rain, falling as freshwater, is less dense than saltwater; it therefore tends to form a lens, floating above the saltwater. This is also why oceanic islands are inhabitable: islands the size of several acres or more have their own underground freshwater supply from the rains. Unfortunately, rapid pumping of either coastal or island fresh groundwater can cause the underlying saltwater to migrate upward, "intruding" into the freshwater (**FIGURE 9-15**). This degrades the quality of the freshwater for drinking, agriculture, and other uses. Saltwater intrusion is a significant problem in many coastal cities. Sections of the Gulf Coast and some parts of California are especially affected. Globally, such cities as Manila, Philippines, and Lima, Peru, have major problems with it.

These different types of groundwater problems are not exclusive: Some areas are plagued by all of them. For example, some urban areas of the Gulf Coast, such as Tampa, Florida, have problems with groundwater pollution, depletion, land subsidence (and sinkholes), and saltwater intrusion.

9.4 Increasing Our Water Resources

Although much of the United States and many other parts of the world are facing critical shortages of water, these shortages are not insurmountable. Indeed, the main reason that most shortages exist is simply that, until recently, water resources were taken for granted. On the positive side, this means that water resources can often be extended with relatively little sacrifice, simply by cutting past wasteful uses and overconsumption.

This section discusses six ways that water resources can be extended. The first three are familiar as methods of reducing the input of any resource: increased efficiency, recycling, and substitution. In the case of water, recycling refers to recycling wastewater. Desalination is one of the many ways of substituting plentiful seawater for naturally occurring freshwater. Two methods of extending water supplies are dams and canals, which do not reduce total demand. They shift water supplies from one region to another where they are needed more, extending water supplies locally but reducing them elsewhere. The last method has to do with protecting, restoring, and enhancing agricultural and forestland for the purposes of protecting and enhancing water quality and quantity for beneficiaries (such as citizens, water utilities, and businesses).

Increased Efficiency

A major theme in environmental science is that input reduction (reducing consumption) is the best all-around solution to most environmental problems. The reduction of resource inputs by conservation not only prolongs the time that the resource will last, but also reduces the pollution and other disturbances generated when the resource is extracted and used. Water resources provide many opportunities for conservation because they have been used so wastefully and inefficiently in the past. Water-rich areas have had no incentive to conserve.

Agriculture in particular and industry account for the large majority of water withdrawn and consumed in nearly all modern societies, so more efficient use in these areas is especially important. This will involve both technical and social changes. A classic example of a technical change is **microirrigation** (sometimes called "drip" irrigation), whereby water is transported to crops via pipes instead of open ditches that promote evaporation. The water is then dripped onto the plants from tiny holes in the pipes installed on or below the soil. This increases the water that reaches the plant from 37% to as high as 95%. Microirrigation saves nearly 300% of the water previously used. This method has been very effective in water-poor countries such as Israel. Only 3% of U.S. irrigation is by microirrigation, largely because of its expense; in contrast, Israel has about half of irrigated lands under microirrigation, and in Cyprus, better than 70% of its irrigated lands use microirrigation. The initial cost is about $1,500 to $3,000 per 1 hectare (about 2.5 acres), so it is used mainly on highly valued fruit and vegetable crops. However, the use of microirrigation is rapidly growing in the United States, especially in the West, where incorporation of the true environmental costs of food production is raising prices (and some states are subsidizing the cost of irrigation). Similarly, industrial engineers have found it relatively easy to design production technologies that use much less water to produce everything from newspapers to aluminum. Recycling the same water during production is just one example of a design change. Although such technical changes may temporarily increase costs, they often ultimately lower costs compared with what would be paid for a steadily decreasing water supply.

Conserving water often requires social changes in addition to technical change. Changes in diet can reduce agricultural water loss. Much more water is required to raise domestic animals for food than to produce the same amount of food as crops; in general, we use fewer resources when eating "lower on the food chain."

The average person can usually do much to conserve water. These changes require little personal sacrifice and may even save money. Simply shaving with a basin instead of running water, or with an electric razor, reduces water use by 2,000% **(TABLE 9-4)**. Installing low-flush toilets and taking shorter showers are especially important because flushing and showers are, by far, the two greatest personal uses of water (Table 9-3). Home lawns and gardens often use much water; landscaping changes such as switching to native plants that have low water needs can save water. **Xeriscaping** (pronounced "zeriscaping") is landscaping designed to save water. It includes planting drought-tolerant plants, such as buffalo grass, sage, juniper, rosemary, verbena, forsythia, pomegranate, cypress, and oleander, and many other changes, such as weed control, watering times, and height of mower settings.

Desalination

Because 97.4% of Earth's water is saltwater, removal of dissolved salts, called desalination, would obviously create a huge supply

TABLE 9-4	**Ways to Conserve Water**		
Activity	**Normal Water Consumption**	**Water-Saving Methods**	**Consumption**
Bathing in a full tub	36 gallons	Regular shower or Wet down, soap-up, rinse off	25 gallons 4 gallons
Washing hands with the water running	2 gallons	Fill the basin	1 gallon
Brushing teeth with the water running	10 gallons	Wet brush & quick rinse	½ gallon
Each toilet flush	5–7 gallons	Minimize flushing	
Leaking faucet	25 gallons a day	Fix as soon as possible	

of freshwater. Unfortunately, desalination is expensive, even with the latest, most efficient technology. Thus, it is economical only where other sources of water are even more expensive or are not available. More than half of the world's desalination capacity is located in the Middle East and North Africa, where more than 1.5 billion cubic meters of freshwater is produced each year.

The primary reason for the high cost of desalination is the cost of energy. On average, seawater has about 35 parts per thousand of dissolved solids, which must be reduced to less than one part per thousand for drinking and industrial uses. About 90 kilograms (approximately 200 pounds) of salt is produced for every 3,800 liters (approximately 1,000 gallons) processed. Because a desalination plant often processes many millions of gallons per day, millions of pounds of salt must be removed daily. Disposing of all this salt is another major cost of desalination. Landfill costs are rising fast, and the concentrated salt cannot be dumped into the ocean with-

out damage to ecosystems. Some economic uses, such as de-icing roads, have been found for this salt, but without costly treatment, it is generally not clean enough for many purposes. In some cases, producing water that is still too salty for drinking and industry but that is nevertheless useful for such purposes as watering salt-tolerant plants may reduce the costs of desalination.

The membrane and the distillation desalination methods have proven to be the most effective in minimizing energy costs. The **membrane method** (also called the filter and reverse-osmosis method) removes dissolved ions by passing the liquid through a membrane at high pressure (**FIGURE 9-16**). Many homes have water purifiers based on reverse osmosis installed in kitchen faucets. This method is generally cheaper and faster than distillation. A large municipal plant can process several billion gallons per day. The membrane method is generally unable to desalinate normal saltwater on a large scale because it rapidly clogs the tiny pores.

FIGURE 9-16 At the desalination facility on Culebra Island, Puerto Rico, seawater is forced through a series of membranes in each blue tube. By the time the water is forced through the last tube and reaches the tank in the back, it is fresh enough to drink.

The method has proved highly effective for areas that need to desalinate brackish (moderately salty) water. Many cities near the ocean pump brackish water from underground when using this method. An example is Santa Catalina Island off the coast of southern California.

The second desalination method, **distillation,** relies on heat: when saltwater evaporates, the dissolved solids are left behind. Such a device can be designed in many ways, but the most common is the **multistage flash distillation** (MSF) method (**FIGURE 9-17**). This pipes cold seawater through a series of coils in chambers that become progressively hotter; the heated seawater is then routed back underneath the coils (Figure 9-17). Freshwater condenses on the outside of the coils from the boiling seawater at the bottom. This method is more than six times more efficient than simply heating a single chamber but is still costly in terms of energy and thus money. Locating MSF plants alongside electric power plants and using the heat they produce as a by-product often partially offsets the high costs of MSF. Solar energy is used for desalination on a small scale, but sunlight is too dilute to make this method practical at large scales: even under ideal conditions, only about 8 liters per m^3 (approximately 0.2 gallons per $foot^2$) can be collected per day. Many square miles of land would be required to rival the output of an average MSF plant.

Wastewater Reclamation

Most of us do not like the idea of drinking wastewater, especially reclaimed sewage. Nevertheless, wastewater can be safe to drink if it is properly treated. Greater detail about wastewater treatment is provided in a review of water pollution. Here we simply note that most municipalities treat their wastewater and then release it into nearby rivers, lakes, or oceans. People downstream or elsewhere often drink and use this water. Many natural processes such as dilution help reduce pollutant levels, even if the water has traveled only a few miles, but constant testing is needed to guarantee safety.

Recently, many cities have begun to consider reusing their own wastewater directly. This so-called **closed loop reclamation** involves treating the wastewater to the level needed before direct reuse. For drinking water, this means using advanced—and very expensive—treatment to render the water safe. Even if this is done, the water must be stored in tanks or ponds for a few days before it is safe to drink. A number of cities, especially in arid parts of South Africa, have successfully produced drinking water in this way. Denver, Colorado, has a well-known plant that produces water to irrigate the city's parks. (It also harnesses methane to heat the facility.) However, it is often much more economical to use closed loop reclamation to produce water for nondrinking purposes and save natural sources

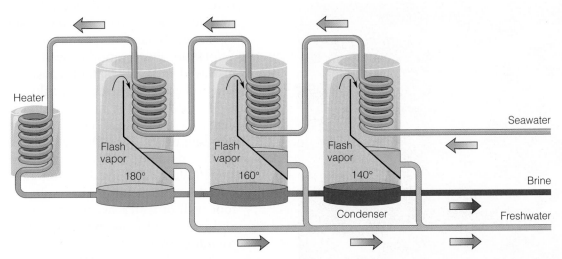

FIGURE 9-17 In multistage distillation, saltwater is pumped through a series of coils that produce condensation of freshwater. (*Source:* U.S. Department of the Interior.)

for drinking. **Gray water** is untreated or partially treated wastewater that is used for watering golf courses and lawns instead of using cleaner water of drinkable quality.

Despite our emotional reactions to wastewater, it generally is much cheaper to treat wastewater than to desalinate seawater, especially if we need to produce only gray water or some other type of partially treated water. It typically costs eight times as much to produce drinking water from desalination as to reclaim moderately polluted water. Some experts suggest that the city of the future will have closed loop reclamation engineered into the natural environment through the "3 Rs": return, repurify, and reuse. In this system, treated wastewater is returned to irrigate crops; the water is renovated (repurified) by percolation as groundwater and is then reusable as drinking water.

Dams and Reservoirs

All of the ways of extending water resources discussed so far have a major advantage: they use water more efficiently and produce more for society in general. In contrast, canals and dams and reservoirs simply redistribute water from water-surplus areas with low populations to water-deficient areas with high populations. This results in no net gain and often a net loss for society in general because evaporation, soil infiltration, and other processes lose water during storage and transport. Nevertheless, redistribution is often the most economical way to meet society's water needs and can have minimal environmental impact if properly designed.

Dams: The Benefits and Costs

Dams are structures that obstruct river or stream flow to make lakes (**FIGURE 9-18**). By keeping water from flowing downstream, dams allow people upstream to have preferential use. There are more than 45,000 major dams in the world (and hundreds of thousands of smaller dams). Dams are built for one or more of three reasons: (1) provide better control of water flow to minimize flood damage, (2) create a reservoir that can serve as a water source for nearby populations, and (3) provide hydroelectric power. Even well-designed dams have some environmental impacts. These include the following:

1. *Sediment accumulation.* Dams trap sediment. Unless the sediment is removed, the reservoir will fill up in a few decades to a few hundred years.
2. *Scouring downstream.* The water that escapes the dam has little sediment. Such "sediment-starved" stream flow scours out the river bottom on the other side of the dam.
3. *Water loss.* Water evaporates from the lake. The scale of this loss is vast: In Texas, about three times as much water is lost from evaporation of reservoirs as is used by industry and cities. Attempts to reduce the loss by spreading a "skin" of chemicals have not proven effective.
4. *Salination.* Evaporation of water leaves dissolved minerals behind, causing the lake to become progressively saltier.
5. *Dam breaks.* Heavy rainfalls can cause dams to burst, leading to disastrous flooding down river. On December 14, 2005, a dam on the Black River in Missouri failed, releasing 1 billion gallons of water in 12 minutes. The resultant 20-foot wall of water swept away vehicles and inundated numerous homes in surrounding areas.
6. *Biological disturbances.* Dams can disturb both land and aquatic life. An obvious ef-

FIGURE 9-18 A flood control dam on the Shenango River in Pennsylvania.

fect is when dams block migration patterns of fishes, such as spawning salmon in the northwestern United States. Fish ladders built to allow migration around the dam are often ineffective. A second cause of biological disturbance is loss of both land and aquatic habitats. Land habitats can be flooded upstream; downstream aquatic habitats can suffer from decreased water flow. One of the most infamous examples is the Aswan Dam on the Nile River, which has had many biological impacts. For example, the rich silt that used to fertilize the floodplains is now piling up behind the dam, forcing the farmers to use millions of tons of chemical fertilizers to replace natural nutrients. The dam has also adversely affected the fertility of the coastal Mediterranean waters. Once nutrient-rich water flowing into the Nile Delta during the flood season fed great quantities of phytoplankton, which in turn fed sardines and other pelagic fishes. The Aswan Dam now prevents this from occurring, and the sardine population is a fraction of its former size. An overgrowth of plant life in the canals leading from Aswan's reservoir, Lake Nasser, has been implicated in massive outbreaks of schistosomiasis, a disease caused by parasitic flatworms that freshwater snails carry. The abundance of plants in the canals provides increased habitat for the snails, thus increasing human exposure to the disease. In recent years, there has been increasing pressure to dismantle unnecessary dams, and in the United States, hundreds of dams (mostly small dams) have been removed during the last few decades. Decommissioning of these dams has occurred largely because their Federal Energy Regulatory Commission licenses expired and were not renewed.

7. *Social disturbances.* Large dams in particular can impose heavy social damage, displacing people, forming reservoirs where stagnant water can breed disease vectors, such as insects that afflict the human population, and hindering the natural ability of a river to break down and process human-dumped sewage. An example of the massive displacement of people by a single dam is the Three Gorges dam in China. By the time it was complete, more than 1.2 to 1.9 million people were relocated and the reservoir reportedly submerged 140 towns and 326 villages, totaling an area larger than 632 km^2 (244 mi^2).

One often-proposed solution to the many environmental and social problems caused by dams is to make more use of natural methods of flood control, such as wetlands and local geological conditions. This suggestion runs counter to the traditional policy that the U.S. Army Corps of Engineers often uses, which is to alter natural conditions with dams, channelization, and other means according to economic and engineering criteria, without much consideration to environmental impacts. This alternative has received renewed attention since the great flood of 1993 in the Mississippi River system. Many environmental groups noted that dam breaks and many kinds of damage will occur again unless more consideration is given to conforming to natural conditions instead of trying to alter them. Wetland mitigation banking, established in 1983, with guidance from the U.S. Fish and Wildlife Service (FWS), has evolved as one approach to preserve, enhance, and restore wetlands in the United States.

Canals

Canals, or aqueducts, are artificial channels built to transfer water over long distances. As urban areas grow, their water needs often exceed local supplies, and canals become necessary for growth. For example, New York city has imported water for more than 100 years, moving to sources progressively farther away, such as upstate New York. Many long canals are even more important for large cities in dry areas. In California, two-thirds of the water runoff is north of San Francisco, whereas two-thirds of the water use is south of San Francisco. Since the early 1900s, the growth of cities in southern California, especially Los Angeles, has led to the construction of many canals to import water from the north and from the Colorado River. Population growth in southern California has continued to increase in recent years, causing

Forests and Water

VISIT

http://environment.jbpub.com
/mckinney/5e/
for more information

Forests play an important role in naturally regulating water flows, water purity, and erosion. A clean and reliable water supply is one of the most important benefits of well-managed forests and is a resource that generates immense economic value for communities all over the world.

Water Flow Regulation

Forests and forested wetlands affect the timing and magnitude of water runoff and water flows. Some forest ecosystems act as sponges, intercepting rainfall and absorbing water through root systems. Water is stored in porous forest soils and debris and then is slowly released into surface waters and groundwater. Through these processes, forests recharge groundwater supplies, maintain baseflow stream levels, and lower peak flows during heavy rainfall or flood events.[i] According to one study, less than 5% of rain falling on a forest is converted to runoff, while 95% of rain falling on impervious surfaces, such as concrete, is converted to runoff (Cappeilla et al., 2005).

The water flow regulation services that forests provide can yield economic benefits to communities. By reducing water runoff during rainstorms, forests reduce the volume of water that a municipal stormwater containment facility or retention pond must store. Communities, therefore, do not need to invest as much in constructing stormwater control infrastructure. Based on this avoided cost of stormwater storage, one assessment estimated that forests near Atlanta saved the city $420 per forested acre per year (American Forests, 2001).

Water Purification

Two-thirds of the nation's clean water supply comes from precipitation that is filtered through forests and ends up in streams (Smail and Lewis, 2009). Forests help prevent impurities—mostly from nonpoint source pollution[ii]—from entering streams, lakes, and groundwater in numerous ways. Root systems of trees and other plants keep soils porous and allow water to filter through various layers of soil before entering groundwater. Through this process, toxins, nutrients, sediment, and other substances can be filtered from the water. Leaves and other debris on the forest floor play a role, too. Through the process of denitrification, for example, bacteria in wet forest soils convert nitrates—a nutrient that can lead to harmful algal blooms if too much of it enters bodies of water—into nitrogen gas, releasing it into the air instead of into local streams (Sprague et al., 2006).

The water purification benefits of forests are economically valuable. The water purification service can reduce drinking water treatment costs. Studies conducted by the American Water Works Association and the Trust for Public Land of 27 different water supply systems found that drinking water treatment costs decrease as the amount of forest cover in the relevant watershed increases (Figure 1).[iii] They found that 50–55% of the variation in operating water treatment costs can be explained by the percentage of forest cover in the water source area (Ernst, 2004).

Erosion Regulation

Forests help keep soil intact and prevent it from eroding into nearby bodies of water in several ways. By intercepting rain, a forest canopy reduces

---

[i]Forests, however, typically result in lower surface flows to nearby waterways because of infiltration and the transpiration of water into the atmosphere through leaves. Therefore, reducing forest cover and density generally increases surface water yield from watersheds, although these changes can be short-lived and depend on climate, soil characteristics, and the percentage and type of vegetation removal. For instance, streamflows increased 28 percent following a clear-cutting experiment in a southern Appalachian watershed. *Source:* McGuire, Kevin. "Water and Forest Cover Literature Review." Virginia Water Resources Research Center & Dept. of Forest Resources & Environmental Conservation, Virginia Tech. Citation in literature review taken from Swank et al. 2001.

[ii]According to the U.S. Environmental Protection Agency, nonpoint source pollution from agriculture, urban development, and suburban development accounts for more than 60 percent of impairment in U.S. waterways, including many drinking water sources. *Source:* Barten and Ernst 2004.

[iii]A "watershed" is the area of land above (in terms of elevation) a given point on a stream, lake, river, or estuary that contributes water to that waterbody. A watershed is also referred to as a "drainage basin."

the impact of heavy rainfall on the forest floor, reducing soil disturbance. Leaves and natural debris on the forest floor can slow the rate of water runoff and trap soil washing away from nearby fields. Tree roots can hold soil in place and stabilize stream banks. This also allows other plants, such as native grasses, to help further purify run off. In addition, coastal forests and forested wetlands protect coastlines by absorbing some of the energy and impact of storm surges, thus reducing erosion and other onshore impacts.

The erosion control provides numerous benefits to people. For instance, it can help reduce the deposition of sediment behind hydroelectric dams—the Tennessee Valley Authority alone has 30 dams (Tennessee Valley Authority 2003)—and thereby reduces the need for expensive dredging through investing in sustainable forest management upstream.

Sources

American Forests. 2001. *Urban Ecosystem Analysis Atlanta Metro Area: Calculating the Value of Nature.* Washington, DC: American Forests.

Cappeilla, K., T. Schueler, and T. Wright. 2005. Appendix A: Effect of Land Cover on Runoff and Nutrient Loads in a Watershed. *Urban Watershed Forestry Manual, Part 1: Methods for Increasing Forest Cover in a Watershed.* NA-TP-04-05. 94. Ellicott City, MD: USDA Forest Service.

Ernst, C. 2004. *Protecting the Source: Land Conservation and the Future of America's Drinking Water.* Water Protection Series. San Francisco, CA: The Trust for Public Land and American Water Works Association.

Hanson, C., L. Yonavjak, C. Clark, S. Minnemeyer, A. Leach, and L. Boisrobert. 2010. *Southern Forests for the Future.* Washington, DC: World Resources Institute.

Hanson, C., Talberth, J., and Yonavjak, L. 2011. *Forests for Water: Exploring Payments for Ecosystem Services in the U.S. South.* Southern Forests for the Future Incentives Series: Issue 2. Washington, DC: World Resources Institute.

Smail, R. A., and D. J. Lewis. 2009. Forest Land Conversion, Ecosystem Services, and Economic Issues for Policy: A Review. PNW-GTR-797. Portland, OR: U.S. Department of Agriculture, Forest Service, Pacific Northwest Research Station.

Sprague, E., D. Burke, S. Claggett, and A. Todd, eds. 2006. *The State of Chesapeake Forests.* Arlington, VA: The Conservation Fund.

Swank, W. T., J. M. Vose, K. J. Elliott. 2001. Long-term hydrologic and water quality responses following commercial clearcutting of mixed hardwoods on a southern Appalachian catchment." *Forest Ecology and Management* 143 (1–3): 163–178.

Tennessee Valley Authority. 2003. Dams and hydro plants. Online at: http://www.tva.com/power/pdf/hydro.pdf.

VISIT

http://environment.jbpub.com
/mckinney/5e/
for more information

ever-greater water demands. This has led to increasing legal and social confrontations as people in other areas, especially farmers, object to the diversion of their water to urban areas. Such disputes illustrate the disadvantage of redistribution. There is no net gain for society, whereas reduced consumption, desalination, and wastewater recycling result in greater amounts of available water. California cities such as Santa Barbara and San Diego are turning to all of these solutions and moving away from water importation.

In Florida, a vast artificial system created in the 1960s and earlier containing 2,250 kilometers (1,400 miles) of canals drained away much of the water feeding the Everglades. The diverted water goes to urban centers and especially to agriculture industry (**FIGURE 9-19**)

use. Whereas the canals led to shortages and disputes over water rights in California, those in Florida illustrate another potential hazard of canals: habitat destruction and pollution. The Everglades ecosystem is one of the largest and most unique wetland habitats in North America. It was drying up and becoming polluted as canals diverted water. At the same time, straightening or **channelization** of the Kissimmee River was causing more pollution to flow into the Everglades. This pollution not only caused species extinction but also affected water quality because the Everglades serve as a major recharge area for water supplies in South Florida. To help correct this situation, in 1992, the United States Congress authorized the Kissimmee River Restoration Project to restore, at a cost of tens of millions

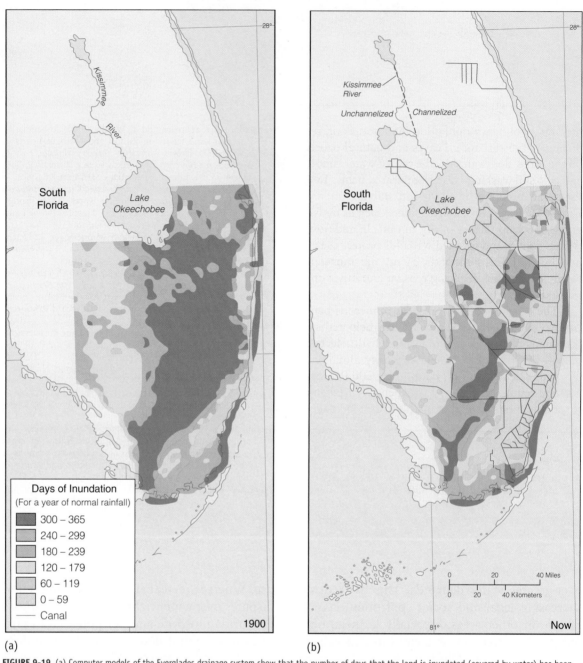

Days of Inundation
(For a year of normal rainfall)

■	300 – 365
■	240 – 299
■	180 – 239
□	120 – 179
□	60 – 119
□	0 – 59
—	Canal

1900

Now

(a)   (b)

FIGURE 9-19 (a) Computer models of the Everglades drainage system show that the number of days that the land is inundated (covered by water) has been greatly reduced since 1900. (b) Canals now drain rainwater away before the land can be saturated for long periods, especially the "permanent" wetlands, shown in deep blue. (*Source:* Modified from South Florida Water Management District data.)

of dollars, 100 square kilometers (40 square miles) of riverine and floodplain ecosystems, including the restoration of 69 kilometers (43 miles) of meandering river channel that had been "straightened." The first phases of this project have now been completed, and ultimately, it is planned that some 7,800 square kilometers (3,000 square miles) of floodplains, wetlands, and river channels will be restored at a cost of more than $5 billion.

Despite enormous economic and environmental costs, there probably will be increasing social pressures for water importation. Such pressures will grow not only because of increasing population growth in California and other parts of the West, but also because of agricultural needs. Recall that agriculture is the greatest consumer of water, and the West and Midwest are vast dry agricultural areas. Added pressure to import will come later this century because of the depletion of the Ogallala Aquifer underlying the Midwest.

Conserving and Managing Land for Water Resources

Forested watersheds and lands in agricultural use provide a number of benefits to the nation's citizens, communities, and businesses, including water flow regulation, flood control, water purification, erosion control, and freshwater supply (see **CASE STUDY 9-4**). The loss and degradation of these lands can reduce their ability to provide these watershed-related ecosystem services. Land conservation, restoration activities, and management techniques can help these lands continue to provide necessary watershed services. One approach to achieving these outcomes is for landowners to receive payments for the role their lands play in improving watershed benefits, which makes conservation, restoration, and management more economically viable. This often requires a payment scheme whereby downstream water beneficiaries (such as companies, water utilities, or citizens) pay for the upstream watershed benefits they receive from conserved, restored, or sustainably managed lands. These payments can be voluntary, minimize the cost of regulation, or be made to generate public benefits. These payment schemes are designed to invest in "green infrastructure" instead of "gray infrastructure." They are investments in natural open space instead of human-engineered solutions to address watershed service problems. In many instances, these investments in green infrastructure can be more cost effective than investments in gray infrastructure.

One example of a pilot phase payment for watershed services project is the Florida Ranchlands Environmental Services Project (FRESP), which is under way in the Lake Okeechobee watershed in Florida. In this project, the World Wildlife Fund and other environmental partners are working to offer ranchers in the Lake Okeechobee watershed an opportunity to provide environmental services such as improved water quality and water retention in ways that save taxpayers money, preserve rural communities, enhance wildlife habitats, and provide additional income revenue.

9.5 Social Solutions to Water Scarcity

We have discussed many ways that society can extend water supplies, from conservation to redistribution by canals. However, like all technical solutions to environmental problems, these solutions have important social aspects. For instance, conservation is strongly influenced by economic considerations, such as the price of water and, in the case of upstream land protection for water resources, the price of implementing the best management practices on private lands, purchasing easements, and in some cases, actual land acquisition. Similarly, laws governing water use, especially in the western United States, have enormous influence on where dams and canals can be built.

Economics of Sustainable Water Use

When a resource is cheap, it will be wasted. Only when the resource is costly do consumers have an incentive to conserve or reduce resource input by more efficient use, recycling, or substitution (**FIGURE 9-20**). In the case of water resources, wastewater reclamation is an example of recycling. Substitution could include switching to desalinated saltwater instead of

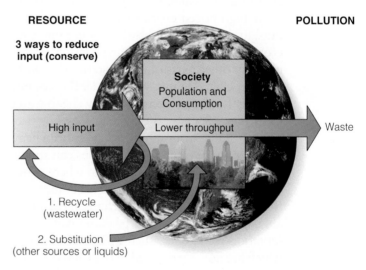

FIGURE 9-20 Three ways to conserve water (reduce water input to society), with examples: more efficient use, recycling, and substitution. If water resources become more costly to the water users, they will have an incentive to use these three methods of conservation. Conservation also tends to reduce water pollution because less wastewater is produced and water flow into natural waters is reduced.

depleting groundwater or using liquids other than water.

The low cost of water in the United States has offered little incentive to conserve. Most water supply systems charge only a fraction of the total cost of obtaining, storing, treating, and delivering water. The average family pays less than 1% of its income for water. As a result, many states and local areas are increasing water prices to provide more incentive to reduce water use: for each 1% increase in cost, water use in homes initially decreases by about 0.5%. As costs rise, cities find it economical to monitor water supplies more closely, reducing waste.

Subsidizing Water for Agriculture

As the largest consumer of water, agriculture is an especially important target for conservation. California provides a good example. Farmers use 85% of California's water, much of it for the irrigation of water-intensive crops. Because of California's agricultural history, farmers not only have "first rights" to water but also pay far less (often less than 10% of the price urban users pay) because the government subsidizes their water. In the early 1990s, urban areas began to rebel against this practice and urged their state legislators to reduce subsidies to farmers. As expected, this has led to farmers using less water. One response has been to grow high-priced crops such as certain fruits, nuts, and vegetables instead of water-intensive crops such as rice. A similar situation occurs in many other parts of the western and southwestern United States, where federal taxes historically funded projects to supply cheap water to farmers. Society as a whole does not necessarily suffer from refusal to subsidize cheap water because water-intensive crops can still be grown in other parts of the nation where water is more common. They can also be imported from other countries, or perhaps most desirably, the higher costs will promote water conservation techniques, such as microirrigation, which are now rarely used in the United States.

Taxing Water Use in Industry and Households

Although water for industry and household use is less heavily subsidized, it is still very cheap. As a result, state and local governments have found that increasing the price of water via taxes or other means can greatly reduce the amount of water used. Often these taxes are levied as **effluent charge**s, which is the cost of disposing of industrial wastewater. This illustrates again how input management not only conserves resources but also reduces pollution. For example, water withdrawals by U.S. manufacturing industries in the year 2005 (latest data available) were estimated to have been 60% less than in 1977 and 11% less than in 1995 and 8% less in 2005 because of increased taxes on wastewater disposal. Similarly, increased prices on domestic water encourage people to conserve by using low-flush toilets, taking shorter showers, and employing other means of conservation. In some countries, higher prices also encourage domestic water recycling and reclamation. For example, Japan and Singapore have long used reclaimed wastewater in flush toilets.

Increasing the cost of a resource to its true or "environmental" cost will discourage both the rapid depletion and pollution that occur when a resource is very cheap. The improved efficiency of resource use often reduces not only overall environmental costs but also economic costs to society. Industrial recycling of water is a good example. Cities can also gain water efficiency and reuse infrastructure from this. Beijing, China, recently invested in more modern plumbing, leakage reduction, and water recycling. This will save billions of gallons of water per year, at a much lower dollar cost than the water development project originally planned to import water from farther away. There are also a number of examples where cities are investing in upstream land conservation in conjunction with, or in lieu of, water treatment plant upgrades or other human-made infrastructure improvements.

Legal Control of Water Use

As one of the most important components of living things, water probably has more laws governing its use than any other resource. In the United States, a hierarchy of laws extends from the federal to the local level; as with all resources, federal law takes precedence when there is conflict between levels. However, most direct regulation is found at the state level for reasons that date to homesteading and the use of the land when it was originally settled. This also explains why the laws for the control of water are well developed for surface water, which has been intensively used through the years. In contrast, the rapidly growing use and pollution of underground waters have led to a confusion of legal activity in recent years. These difficulties are magnified because groundwater is much more difficult to monitor than surface water, and it moves so slowly that depletion and pollution often take years to detect and remedy (more information is provided in the review of water pollution).

Surface Waters

Laws regulating surface waters can be divided into two basic types: riparian law and prior appropriation law. **Riparian law** dictates that all landowners have the right to withdraw water that is adjacent to the land, such as from a river or lake. (Riparian means "bank," as in riverbank, in Latin.) The laws vary by state, but generally, a landowner cannot take so much water that it adversely affects the reasonable uses of other owners. In times of drought, all landowners must cut back on their water use. A key provision is that water returned must be returned in a relatively unpolluted condition so that owners of adjacent land, such as downstream, do not have their water rights violated. Riparian law is a common-law idea, meaning that it evolved through practical use, as opposed to being dictated from theory or administrative decree.

Riparian law is practiced mainly east of the Mississippi River. It works well in those states, where water is relatively plentiful and areas were settled gradually. In contrast, water laws in the western states are based to some degree on the principle of appropriation, where use takes precedent over adjacent land ownership. **Appropriation law** basically establishes that the first to put the water to beneficial use owns "first rights" to the water and can take as much as needed to a limit (the principle of "first in time, first in line"). Beneficial use originally was defined as water used for irrigation, watering livestock, and domestic uses. The reason for this type of law in the western United States is that water has often been scarce so it has to be closely protected. The next right holders are allowed to get the maximum they need and on down the order of ar-

rival. If someone cheats or there is a drought, the next in line may end up getting no water at all. In addition, the Western states have used the beneficial use concept to write appropriation laws allowing states to take the water from landowners or deny the right to withdraw water if a more beneficial use is found. For instance, water from the Colorado River is tightly regulated by state and federal agencies (**FIGURE 9-21**). The greater state control of water under appropriation law makes it better suited for implementing environmentally optimal management. For example, if a study shows that an industry is being exceptionally wasteful, the state can threaten to appropriate the water if the industry does not take conservation measures. In contrast, under riparian law, if the industry owns the land along the

FIGURE 9-21 This satellite image shows the northern portion of Lake Powell and the Colorado River winding its way through the dry Colorado Plateau Region.

river, the state may have to file suit and undergo expensive, prolonged litigation to make the industry change its ways.

Groundwater

Until recently, groundwater was virtually unregulated. Owners of land above the water were considered to have rights to the water, much as in riparian law. Even today, people in many rural areas generally are free to drill holes and withdraw water. In areas with rapid groundwater recharge or low population, this practice created few problems. The rapidly falling water tables of many areas have required much stricter laws on drilling and withdrawal. Most of these are local, municipal ordinances that prohibit withdrawal inside city or county limits without a permit. Similarly, many cities regulate water use, such as prohibiting lawn sprinkling during the afternoon when evaporation is high. Groundwater depletion is easier to regulate than groundwater pollution (a review of water pollution provides greater information). There is increasing pressure on the U.S. Congress to enact a single coherent set of laws regulating groundwater depletion and pollution to replace the piecemeal regulation now in place because of the many local and federal statutes.

Which Are Better for Enabling Sustainable Use: Economic or Legal Solutions?

The type of water resource problem determines whether economic or legal solutions are better for producing sustainable use. Most economists agree that market conditions provide the optimal allocation of resources when many parties want access to them. For example, in California, where many farmers and city dwellers want access to water, one effective solution is to make all parties pay a price for the water that incorporates the environmental costs. This could include taxing water, removing subsidies, or otherwise increasing the costs to all parties. Such an economic solution not only tends to distribute costs more equitably, but also is often the only practical way of controlling water use: regulating a resource that is being used by many parties is often very costly because of the equipment and personnel needed to monitor use.

Thus, legal solutions are often most applicable where only a few parties use the resource, making monitoring possible. In addition, legal regulation often is preferable where a resource is in very short supply so that waste is minimized through stricter control than a market approach allows.

■ study guide
SUMMARY

- Water is a remarkable substance that is essential to life as we know it—unusual properties include its density, boiling point, specific heat, and solvency, which are all affected by the bipolar structure of the water molecule.
- Despite the abundance of water on Earth (mostly in the form of the oceans), there is only a limited supply of freshwater.
- The hydrologic cycle acts as a pump, cycling water through evapotranspiration and precipitation.
- From a human perspective, water is very unevenly distributed across the continents and nations; many areas are currently water stressed.
- Withdrawn water is water removed from its source (such as a lake, river, or aquifer); consumed water is withdrawn but not returned to its original source.
- Worldwide, the majority of withdrawn water goes toward agriculture; industry, uses the next highest amount, followed by domestic use.
- Types of water sources include surface waters and groundwater; in many areas, groundwater is being polluted and depleted.
- Water resources can be stored and redirected using dams and canals, or effectively stretched and increased through efficiency measures (e.g., microirrigation), desalination, and wastewater reclamation.
- Taxes, subsidies, and various legal controls have been used to address issues of water scarcity.

KEY TERMS

appropriation law
aquiclude
aquifer
artesian well
channelization
closed loop reclamation
cone of depression
confined aquifer
consumed water
dams

deficit area
distillation
effluent charge
evapotranspiration
gray water
groundwater
hydrologic cycle
membrane method
microirrigation
multistage flash distillation (MSF)

precipitation
recharge area
riparian law
sinkhole
surplus area
water table
withdrawn water
xeriscaping
zone of aeration
zone of saturation

STUDY QUESTIONS

1. Why is water an unusual substance? List and explain its properties. How does the structure of the water molecule explain water's unusual properties?

2. What percentage of the Earth's water is readily available in the liquid state as groundwater? As surface water?

3. The global water cycle involves what two main processes? Define these two processes. How much water returns to the oceans as runoff?

4. What are the major deficit areas of the world?

5. Why do hundreds of millions of people experience water shortages if we are not using all readily available freshwater?

6. Why have streams and river valleys, and areas near them, usually been among the most populated regions?

7. What are three withdrawal problems associated with groundwater? How does wetland destruction promote withdrawal and pollution problems in groundwater?

8. Why is agriculture a greater consumer of water than industry? How can microirrigation reduce this consumption?

9. What are the two main desalination methods? Explain how they function.

10. Why are dams and reservoirs less efficient and more harmful than other ways of extending water resources?

11. What happens when the cost of a water resource is increased? Name some ways to do this.

12. Discuss the economics of water use, focusing on ways to pay the true cost of water. Why is the true cost often underpaid? Give examples.

13. Name and define the two types of laws regulating water resources in the United States. Where are they used?

14. Discuss various ways of reducing water inputs to society, including increased efficiency of use, economic aspects, and legal aspects.

WHAT'S THE EVIDENCE?

1. The authors assert that human mismanagement of water supplies can greatly aggravate natural water shortages. What evidence supports this contention? Do you agree that people often mismanage water supplies? If so, how can we better manage water resources? Give specific examples. If you disagree with the idea that humans have historically mismanaged many supplies of water, cite evidence to support your position.

2. The authors appear to be critical of the large amounts of water used by the United States, including more water per person in the United States than among most other nations. Is such criticism well founded? Is the United States, in general, excessive in its water use? Being arguably the only superpower, is such water usage justified? Is it necessary? State your position, and support your views with relevant evidence and persuasive arguments.

CALCULATIONS

1. Suppose a river carries 10 cubic miles of water into the ocean over 10 years. Convert this amount of water to cubic kilometers, given that 1 cubic mile = 4.168 cubic kilometers.

2. Given the river in question 1, how much water would be carried to the ocean on average each day during the decade under consideration? Give your answer in both cubic feet and cubic meters.

ILLUSTRATION AND TABLE REVIEW

1. Referring to Figure 9-4, calculate the estimated world water use in 1960. In 2000. (Express your answers in units of cubic kilometers and cubic miles.) Estimate the percentage of total world water use that industry used in the year 2000.

2. Referring to Figure 9-7, calculate by approximately what percentage the estimated total human demand for freshwater increased between 1900 and 2000. During this same time, by approximately what percentage did the amount of accessible freshwater increase?

http://environment.jbpub.com/mckinney/5e/

http://environment.jbpub.com /mckinney/5e/

Connect to this text's website at http://environment.jbpub.com/mckinney/5e/. This site features eLearning, an online review area that provides quizzes, chapter outlines, and other tools to help you study for your class. You can also follow useful links for more in-depth information.

In the 1960s and 1970s, some environmentalists predicted that by 2000, our finite supply of many minerals and metals would be nearly exhausted and the process for any remaining reserves would be astronomical. Yet today, due in part to more sophisticated detection and more extensive extraction methods, some minerals seem to be "more common" than they were five decades ago. Mineral and metal resources really are finite, however. The Bingham Canyon Copper mine shown here measures 1.2 km (0.75 miles) deep, 4 km (2.5 miles) wide, and covers approximately 7.7 km² (1,900 acres). It is the single largest human-made excavation in the world and is visible from space with the naked eye. Its reserves are expected to last until 2019, and with permit extensions, reserves may last into the 2030s.

Mineral Resources

10

M ineral resources are a classic form of resource needing careful management. Mineral supplies are finite and nonrenewable. The extraction and process-ing of minerals can be terribly damaging to the environment. Yet modern technological society is heavily dependent on minerals (see the diagram for Section 3). As the world population continues to grow and more nations become increasingly "developed," the need to extract and use mineral resources will increase.

Yet mineral and metal prices sometimes drop even as the world's population is increasing. Many factors affect prices of resources. New deposits may be discovered, and new technologies may make it possible to recover minerals and metals economi-cally from previously known but lower grade deposits. Recycling has decreased the need for new sources of metals and minerals, as has the continued substitution of rela-tively abundant materials for traditional but scarcer materials (such as the use of glass optical fibers in place of traditional copper wire).

Even so, the trend of increasing mineral and metal consumption over time cannot be ignored. Remember the equation $I = P \times C$ (Impact = Population × Consumption) from the overview of environmental science. Many researchers contend that our immedi-ate concerns should not be the eventual depletion of our mineral and metal resources, but rather the environmental damage that mineral exploitation causes. Such thinkers

Chapter Objectives

After reading this chapter, you should be able to explain or describe the following:

- The basic types of mineral resources
- Relative demands and values of mineral resources
- Environmental degradation linked to mineral extraction and use
- Overall historical trends in mineral use
- Ways to deal with mineral scarcity

301

emphasize the development of ever more efficient and environmentally friendly ways of mining, processing, using, and disposing of mineral and metal resources—a strategy that ultimately should benefit everyone. For example, the owners of the Bingham Canyon copper mine, shown in the opening photograph, point out they have the cleanest smelter in the world.

Mineral resource refers not only to **minerals** in a strict sense (such as quartz, diamond, mica, or graphite), but also to material substances composed of minerals or extracted from minerals. Thus, mineral resources include ores, metals, gravel, sand, marble, granite, phosphate rock, and so on, including elements and compounds that can be extracted from earth materials. These are usually inorganic compounds, although life forms may concentrate or accumulate some deposits; limestone composed of the calcium carbonate of the shells of sea creatures is a classic example. As used in some contexts, the term *mineral resource* includes fossil fuels (as seen in a review of the fundamentals of energy, fossil fuels, and nuclear energy).

10.1 Types of Mineral Resources

Minerals can be divided into different categories. **Metallic minerals** include iron, aluminum, copper, zinc, lead, and gold. Metallic minerals are often referred to as **ferrous** (iron and related metals, such as chromium, manganese, nickel, and molybdenum, which are commonly alloyed with iron) or **nonferrous** (gold, copper, silver, and so forth). **Nonmetallic minerals,** a second major category, can be divided into two groups: **structural materials,** such as building stone, sand, gravel, and components of cement and concrete, and **industrial materials,** such as fertilizer components (phosphorus and potassium), salts, sulfur, other nonmetals used in manufacturing, asbestos, and abrasive minerals (such as emery, or corundum, and industrial diamonds). The **gemstones** and semiprecious minerals that serve no pragmatic function constitute a third category; they are purely of ornamental and aesthetic value (of course many materials can serve multiple purposes; for instance diamonds are used both in-

dustrially for grinding, cutting, and drilling, and in a different context are valued ornamental/aesthetic purposes). Finally, uranium and the fossil fuels (coal, oil, and natural gas—as seen in a review of the fundamentals of energy, fossil fuels, and nuclear energy) are sometimes regarded as a fourth category of **energy minerals**.

10.2 Relative Demands and Values of Mineral Resources—An Overview

Of all nonrenewable geological resources, including the fossil fuels (**FIGURE 10-1** and **TABLE 10-1**), the greatest demand in terms of total bulk quantity is for basic sand, gravel, crushed stone (road aggregates), and building stone. These are the raw bulk materials used in constructing roads, bridges, buildings, and so forth. However, despite the demand, they make up only a small percentage of the total value of nonrenewable mineral materials (approximately 4–10% of the total value, including the fossil fuels) because of their worldwide abundance. Virtually all of the cost involved with these materials is for their extraction and shipment—the material itself may be almost "free." No one is concerned that we will run out of sand and gravel in the near future; indeed, it is doubtful that we could ever exhaust the supply of these materials, although certain fine building stones, such as high-grade marble from a particular quarry, may be exhausted, and clean sand to fill holes (natural or human made) may become scarce locally.

In terms of bulk, fossil fuels are the class of nonrenewable, mined resources that are next highest in demand. Because of their increasing scarcity and modern society's dependence on them, the fossil fuels are currently the most valuable sector of our mined materials (fossil fuels account for more than 50% of the total value of nonrenewable mineral materials). Fossil fuels are discussed in the review of the fundamentals of energy, fossil fuels, and nuclear energy.

The remaining mineral resources that are significant on a global scale are metals (iron, aluminum, copper, zinc, and so on) plus certain other elements and compounds (such as

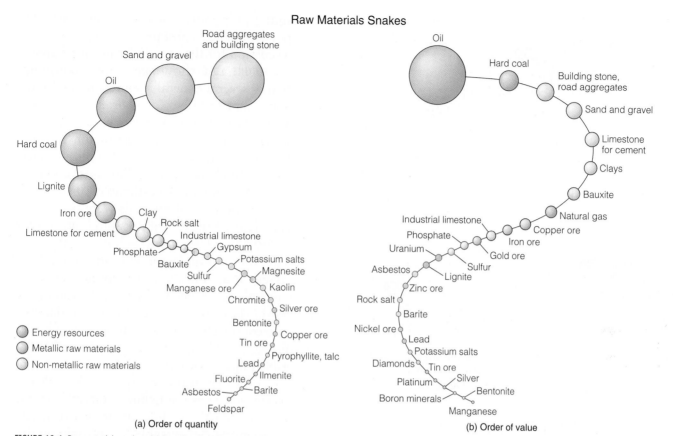

FIGURE 10-1 Raw materials snakes: (a) in order of world production by quantity and (b) in order of economic value. Although the relative ordering of the various materials is accurate, the relative sizes of the circles are merely diagrammatic. (*Source:* A. A. Archer, et al., ed. Man's Dependence on the Earth. Paris: UNESCO, 1987, p. 68.)

TABLE 10-1	Representative Annual World Production of Selected Minerals		
Mineral	**Production (Thousands of Metric Tons)*†**	**Mineral**	**Production (Thousands of Metric Tons)*†**
Metals		*Nonmetals*	
Steel	1,050,000	Stone	11,000,000
Aluminum	24,461	Sand and gravel	9,000,000
Copper	14,676	Clays	500,000
Manganese	8,600	Salt	191,000
Zinc	8,922	Phosphate rock	141,589
Chromium	3,784	Lime	135,300
Lead	3,038	Gypsum	99,000
Nickel	1,107	Soda ash	32,000
Tin	216	Potash	28,125
Molybdenum	543		
Titanium	102		
Silver	15		
Mercury	6		
Platinum	0.162		
Gold	0.2		

*All data exclude recycling.
†One metric ton equals approximately 1.102 English tons.
Source: J. E. Young. Mining the Earth. In: L. R. Brown et al. State of the World, 1992. New York: W. W. Norton, 1992, p. 102. Updates are from L. Mastny, et al. Vital Signs, 2005. New York: W. W. Norton, 2005, p. 53. (Steel) and Mining, Minerals and Sustainable Development Project (MSND). London: Earthscan Publications, 2002, p. 36 (Aluminum, copper, zinc, lead, nickel, molybdenum, platinum, and phosphate rock).

10.2 Relative Demands and Values of Mineral Resources—An Overview

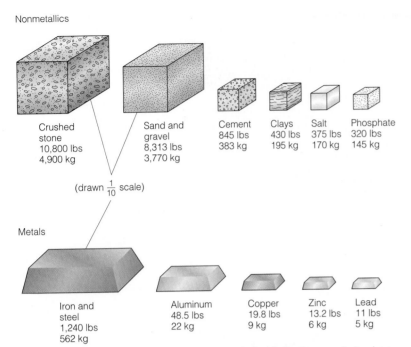

Nonmetallics

Crushed stone
10,800 lbs
4,900 kg

Sand and gravel
8,313 lbs
3,770 kg

(drawn 1/10 scale)

Cement
845 lbs
383 kg

Clays
430 lbs
195 kg

Salt
375 lbs
170 kg

Phosphate
320 lbs
145 kg

Metals

Iron and steel
1,240 lbs
562 kg

Aluminum
48.5 lbs
22 kg

Copper
19.8 lbs
9 kg

Zinc
13.2 lbs
6 kg

Lead
11 lbs
5 kg

FIGURE 10-2 Approximate annual per capita consumption of selected mineral resources by Americans.

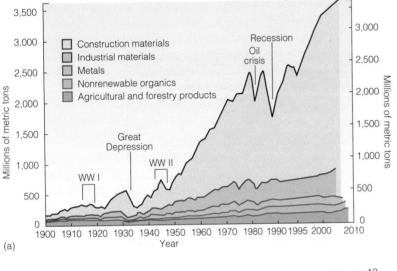

Construction materials
Industrial materials
Metals
Nonrenewable organics
Agricultural and forestry products

Millions of metric tons

WW I
Great Depression
WW II
Oil crisis
Recession

(a)

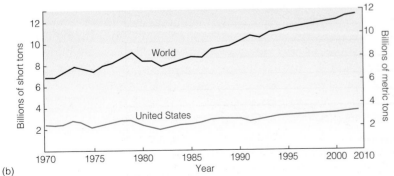

Billions of short tons

World

United States

Billions of metric tons

(b)

sulfur, phosphorus, salt, lime, soda ash, and potash). These substances form the basis of discussion for the remainder of this chapter.

FIGURE 10-2 shows per capita consumption of several mineral resources in the United States that have increased dramatically since World War II. For example, world steel production was 190 million metric tons in 1950 but had increased to 1,050 million metric tons in 2004 and 1,414 million metric tons by 2010. Worldwide, an estimated 10 to 11 billion metric tons of minerals were being used each year by the end of the 20th century. Our industrialized economy is built on the consumption of ever-larger amounts of materials (**FIGURE 10-3**), and the consumption of various minerals is unevenly distributed around the world (**TABLE 10-2**). Some of this material goes into constructing **infrastructure**, such as roads, bridges, and hospitals, or producing other durable objects. However, a large part of it consists of materials that are quickly discarded as wastes—such as the **tailings** from mining (the residue after the high-grade ore is extracted), inexpensive "disposable" electronic equipment, batteries, and cars (which are typically discarded after several years). Although some of this material can be recycled (as discussed in this chapter), mineral resources are inherently nonrenewable. After a particular copper or gold deposit is mined to exhaustion, it is gone forever; it does not "grow" back.

10.3 Mineral Deposits, Ores, and Reserves

Mineral resources are finite and exhaustible, just like fossil fuel resources. In addition, these resources are not distributed evenly over the face of the globe. Various types of **mineral deposits**, including fossil fuel deposits, form only under special geological conditions.

FIGURE 10-3 (a) Nonfood and nonfuel raw materials consumed in the United States, 1900–2010. (b) Nonfood and nonfuel raw materials use for the world and United States, 1970–2010. The data in these graphs do not correspond exactly to the data in Table 10-1 because of discrepancies in reporting and estimating around the world; the information in the table and these graphs is presented to give a general picture of the magnitude of materials use and trends through the 20th century and into the 21st century. (*Source:* United States Geological Survey.)

TABLE 10-2	Consumption of Selected Metals and Minerals on Different Continents						
Material	North America	South America	Europe	Former Soviet Union	Asia	Africa	Other
Aluminum	7,291	823	6,632	612	8,819	294	421
Lead	1,294	212	1,854	179	1,866	118	47
Zinc	1,714	352	2,572	280	3,563	162	240
Copper	3,649	534	4,551	270	5,868	116	176
Nickel	165	24	416	25	449	31	2
Steel (million tonnes)	170	33	206	25	377	18	9
Gold (tonnes)	306	83	906	42	2,423	179	7
Coal (million tonnes of oil equivalent)	613	37	241	197	767	123	158
Phosphate rock	44,580	6,298	11,008	8,965	43,210	23,087	2,718

Thousands of metric tonnes except where noted.
Source: Mining, Minerals and Sustainable Development Project (MSND). London: Earthscan Publications, 2002, p. 48.

On a global scale, a few elements (namely oxygen, silicon, aluminum, iron, calcium, sodium, magnesium, and potassium) are abundant, making up more than 99% of the Earth's crust by weight. However, even for these abundant elements, the practically obtainable supply is not inexhaustible. To obtain most mineral products economically, we must locate mineral deposits that natural geological processes formed, but not all mineral deposits are usable. An **ore deposit** is a mineral deposit that can be economically mined at a certain time and place with a certain technology. Depending on supply, demand, and other factors, a mineral deposit that is not considered an ore deposit one year may be considered an ore deposit the next year (perhaps because the price of the mineral has risen enough to offset extraction costs).

Reserves are identified ore deposits that have yet to be exploited. Reserve estimates for a particular mineral will vary over time, increasing as new mineral deposits are located, improved technology and more efficient methods of extraction decrease the costs of exploiting deposits, or economic factors drive prices up, causing lower grade deposits to become ore deposits. Reserve estimates will decrease as known deposits of a particular mineral are exhausted or if market prices fall, making it economically unfeasible to mine lower grade deposits.

Iron and aluminum illustrate the concept of reserves. Given the abundance of these minerals in the Earth's crust, the supplies seem virtually inexhaustible. According to estimates, given current consumption rates, the iron and aluminum in the Earth's crust could last more than 1 million years and 100 million years, respectively, without recycling; however, when only proven, economically viable world reserves (in the form of ore deposits) of iron and aluminum are considered, estimates are that at present-day production and consumption rates (again, without recycling) there will be a severe shortage of these metals within a few hundred to a couple thousand years. Which view is more realistic will be determined by the amount of energy and the kinds of sources that will be available in the future. With unlimited amounts of energy, we could extract virtually unlimited amounts of aluminum from very low-grade, diffuse deposits, but extracting from low-grade ores, even with unlimited energy, will surely cause environmental disruption and degradation on a scale even more massive than that of current mining.

Many substances are much more rare than aluminum or iron. For instance, copper makes up on average only 55 to 63 parts per million (ppm) of the Earth's crust and tin only 2 ppm. To have a minable ore of most metals, the metal must occur in the rock at a concentration of tens to thousands times higher than in ordinary rock. Even with unlimited amounts of energy, there is an upper limit to the quantity of minerals that can be produced. In the real world, where energy is limited and minerals must be produced from ore deposits, taking world production, consumption, and reserves into account, it has been estimated that our known **virgin supplies of ores** for copper, tin, lead, zinc, and various other metals will be exhausted within a century.

10.4 Environmental Degradation Due to Mineral Exploitation

Today, vast amounts of land are devastated because of the direct effects of mining activities. Around the globe many millions of acres (millions of hectares) of land have been laid waste. Trees have been destroyed. Earth and rock have been churned up, and billions of tons of air and water pollutants and solid waste in the form of tailings have been generated (**FIGURE 10-4**).

The largest human excavations in the world are open-pit mines. Metal smelters typically produce enormous quantities of air pollution, including sulfur dioxide (causing acid deposition) and many toxic heavy metals—lead, cadmium, and arsenic among them. An estimated 8% of worldwide sulfur emissions into the atmosphere come from smelters. **Dead zones**, in which no vegetation or animal life can survive, cover thousands of hectares (thousands of acres) around many large smelters. The mining industry routinely uses hundreds of tons of mercury, cyanide compounds, and other very toxic substances to remove metals from ores. Acid drainage flows from abandoned mines and tailing piles. Removed overburden (rock and soil removed during mining), tailings, and other wastes pile up in massive amounts. On a global scale, mining operations have become a force as powerful as natural erosional processes. It has been estimated that more than 23 billion metric tons (25 billion tons) of nonfuel minerals are removed from the Earth annually (less than half of this is actually "used" or "consumed"), and including the overburden, the total amount of material that humans artificially move in mining each year may be twice the amount of sediment that all of the world's rivers carry annually.

The mineral industry is one of the largest consumers of energy worldwide. Approximately 1% of all the world's energy goes toward aluminum production each year, and more than 5% of the world's energy goes toward steel making annually. Much of the energy used in the mining industry is acquired from burning fossil fuels, charcoal, or fuelwood (causing all of the problems associated with massive energy use).

We can get a glimpse of the magnitude of the total destruction by focusing on the United States. As of the early 1990s, abandoned and operating coal and metal mines covered an estimated 9 million hectares (22 million acres) of land; for comparison, 16 million hectares (40 million acres) of land in the United States are covered by pavement (roads, parking lots, and so forth). Each year, nonfuel mineral mining in the United States produces an estimated 1.0 to 1.3 billion metric tons (1.1 to 1.4 billion tons) of waste. This is about five to six times the amount of municipal solid waste that the United States produces annually. Worldwatch Institute reports from 2010 indicate that between 1950 and 2005, worldwide metals production grew sixfold; oil consumption grew eightfold, and natural gas consumption 14-fold. Annually, a total of 60 billion tons of resources are now extracted. This represents a 50% increase from just 30 years ago. In 2010, a person living in

FIGURE 10-4 Tailings from an open-pit copper mine, Morenci, Arizona.

Europe used 43 kilograms of resources daily, whereas a citizen of the United States used 88 kilograms.

Much of the volume of the waste from mining is overburden—the rock and dirt above the ore deposit that must be removed to obtain the ore. Overburden can be relatively safe and may be used in reclaiming land after mining, so it need not pose a major environmental problem. However, if overburden is allowed to erode and thus be dispersed, it may have a major negative impact on the environment: after flourishing land loses its topsoil and associated flora, streams become clogged with silt, and wildlife is destroyed.

In generally, tailings (the leftovers from processing the ore) are more hazardous. These may leach and produce acids, and the leachate may contain heavy metals or other toxic substances. Leaching and drainage from mines and piles of mining wastes have damaged an estimated 16,000 kilometers (10,000 miles) of streams in the western United States (**FIGURE 10-5**). **Smelting** and refining metals from minerals also release pollutants into the atmosphere, such as heavy metals (for example, arsenic and lead) and sulfur oxides. A 2009 Worldwatch Institute report underscored this fact, indicating that in the production of the 10 most common metal commodities, 3 billion tons of waste material is also produced. This amount is four times the weight of the metals that are extracted. If current rates of mining continue, the Environmental Protection Agency estimates that by 2013, the United States will lose 3,860 kilometers (2,400 miles) of streams.

Mining can contribute directly to deforestation. Charcoal, a necessary fuel in many of the mining and refining operations in less developed countries, is created by partially burning wood. For example, in Brazil, 123,500 acres (193 square miles, or 50,000 hectares) of forest will be destroyed each year to produce charcoal to fuel the Grande Carajas iron ore mining and smelting project. If a proposed bauxite mine and smelter are also constructed, the devastation will be even worse. In the northwestern portion of the Brazilian Amazon, diamonds, uranium, and gold have been discovered in territory where the Yanomami

FIGURE 10-5 Acidic water drains from an old mine near Cooke City, Montana.

Indians live. Miners are now invading the area and threatening both the forest and the native populations. **CASE STUDY 10-1** describes how mining has affected another native population. Recent estimates indicate that humans have degraded almost 80% of all the Earth's forests.

Many mineral products have no ready substitutes. Petrochemical products, such as plastic compounds, can be substituted for some mineral uses, but these products have their own environmental drawbacks. The use of metals and other minerals is not in itself generally harmful to the environment; the great majority of harm is done during the initial mining and processing of the minerals. In the

CASE STUDY 10-1

Trading a Mountain for a Hole in the Ground

Massive mining has the ability to obliterate natural landmarks. Goldsmith and colleagues tell the story of the obliteration of Mt. Fubilan in New Guinea (**FIGURE 1**):

For the Ok people of the highlands of central New Guinea, Mt. Fubilan was a sacred mountain, sitting on top of the land of the dead. In the late 1960s they were persuaded to lease their mountain to a mining company. To the utter astonishment of the Ok, the company began systematically to scoop away the peak of Fubilan. Within the next two decades, the 2,000 meter (6,560 ft) peak will have ceased to exist. In order to exploit Mt. Fubilan's reserves of copper and gold, the mining company intends to remove the sacred mountain altogether. When the mine is finally exhausted, all that will be left is a hole in the ground, 1,200 meters (3,900 ft) deep.

Questions

1. Is it "right" for the mining company to do what it is doing? On the one hand, everyone seems to benefit—the mining company makes a profit, the Ok profit monetarily from their mountain, and the world can use the minerals derived from Mt. Fubilan. On the other hand, it can be argued that we have a moral and aesthetic obligation to preserve natural wonders such as Mt. Fubilan. The mining company can be viewed as simply exploiting the Ok and their land. What do you think?

2. When the Ok leased the mountain to the mining company, do you believe they realized that it would be destroyed? What responsibility, or moral obligation, does a mining company have to explain all of the "fine print" to people who may not be familiar with legal contracts?

3. Considering that the mountain was sacred to the Ok, what does its destruction say about the mining company's regard for the Ok religion and culture? Should primitive beliefs be allowed to stand in the way of modern technological progress?

FIGURE 1 Mt. Fubilan in the process of being mined for gold and copper, 1984.

Source: E. Goldsmith, N. Hildyard, P. McCully, and P. Bunyard. *Imperiled Planet: Restoring Our Endangered Ecosystems.* Cambridge, MA: MIT Press, 1990, p. 201.

past, the easiest, most convenient way to obtain more mineral products was usually considered to be the mining of more virgin ore. With reuse and recycling of already extracted mineral products, the mining of virgin ore could become a rare event. New materials that have less environmental impact, such as using vegetable starches to create plastic-like disposables for the fast food industry, are an example of how humans can reduce the use of virgin ore and apply creative problem-solving to pressing challenges.

10.5 Trends in Mineral Use

In global terms, mineral use has tended to increase over time. Before the 19th century, the use of minerals was relatively insignificant compared with the abundance of minerals in geological deposits. However, since the Industrial Revolution and its associated technological developments and burgeoning population, the use of mineral resources has increased at very high rates. Between roughly 1750 and 1900, the world's population doubled, but mineral use increased by a factor of 10. Between 1900 and the present, mineral use increased by 13-fold or more. The use of certain metals has increased even more dramatically. Current annual world production of crude pig iron is approximately 550 million metric tons (600 million tons), tens of thousands of times the production of 3 centuries ago, and steel production totals some 800 million metric tons a year (900 million tons). Copper and zinc production have increased by factors of 560 and 7,300, respectively, in the last 200 years. As noted, mineral consumption in the industrialized countries is much higher than in the developing nations. Americans use about 25 times as much nickel per person as do citizens of India and four times as much steel and 23 times as much aluminum as the average Mexican. According to one estimate, the United States consumed more minerals between 1940 and 1976 than did all of humanity to the year 1940 (**FIGURE 10-6**). Humans worldwide have consumed more resources in the past 50 years than in all of recorded history to now. Looking at total material consumption, a variety of studies indicate that from the 1970s to the present, people in the United States consumed more than 50% more now than they did 30 years ago—this rate of consumption is at least two times more than the rate of population growth.

Yet since the 1970s, the use of certain raw materials in the United States and Western Europe has leveled (see Figure 10-3) or at least not grown as rapidly as in the past. One reason for this trend is that Western industrial economies are placing more emphasis on high-technology goods and consumer services that use fewer raw materials; in addition, the infrastructure of most industrialized nations

FIGURE 10-6 No other industry uses more materials by weight than the construction industry. For example, the average high-rise building uses thousands of tons of steel and other metals in addition to concrete and glass. The Wills Tower (formerly the Sears Tower) in Chicago, now the tallest building in North America, used 76,000 tons of steel for its frame.

is firmly in place—major construction and public works projects are less common than they once were.

Nevertheless, in absolute terms, the demand for minerals remains very high, and the developing countries will need even more minerals as they industrialize. In addition, within the next few decades, some industrialized countries may need to replace major segments of their aging infrastructures, thus once again sharply increasing the demand for mineral resources. Also, as new technologies are developed, such as high-speed railway systems, new infrastructure components may have to be built.

The bottom line is that on a global scale, people are depleting their mineral resources very quickly. However, it is unclear whether mineral scarcity will ever reach the point that we actually run out of most minerals. The effects of environmental degradation and attendant pollution that mineral extraction and refining cause may force us to curtail the exploitation of mineral deposits before they are exhausted.

10.6 The Economics of Mineral Consumption: What Accounts for the Prices of Minerals?

Minerals, including the precious metals, are artificially cheap. Since the middle of this century, the overall trend in the prices (using stable [noninflated] dollars) of various metals and other minerals generally has been downward. Prices have tended to decline even though mineral deposits are a nonrenewable resource, mineral consumption generally has been increasing worldwide, and the environmental costs of mineral extraction and production are rapidly mounting. Given these factors, common sense might suggest that mineral prices should be increasing rapidly. Why are they falling?

One reason is that known reserves for many metals and other substances have grown at least as fast as production during the last few decades. Given this record, most experts do not foresee imminent shortages even when known reserves seem to be sufficient for only a few decades. Consequently, in the near future, the absolute limits to and relative scarcity of mineral resources will have little impact on the price of most minerals.

One exception might be copper. In 2006, mine closures and declining ore grades cut copper stockpiles to the equivalent of about 3 days of global consumption. As a result, copper prices reached record highs over speculation that the copper production would not increase fast enough to meet demand driven by Chinese economic growth. Currently, although most reports indicate copper reserves will peak by the year 2025, mining rates and pricing increases are led by Chile, for whom copper is the cornerstone of its economic boom.

A second reason prices have declined is that the current price of most raw mineral products reflects only the immediate costs of their extraction. In many cases, such as when the minerals are on public land obtained from the government, the source of the minerals, the ore body, is virtually free. In addition, the price the consumer pays does not include externalities. Externalities are environmental, social, and other costs that are not included in the prices of products that cause the costs (as a review of environmental economics shows). In this sense, the public heavily subsidizes mineral production, for the public ultimately must bear the burden of the environmental devastation caused by mineral production. However, even more immediately and directly, the mining and mineral industry is heavily subsidized. In the United States and many other countries, mining companies receive major tax exemptions and deductions from their gross income; the lost taxes are equivalent to subsidizing the industry by billions of dollars. Mining companies are often de facto exempt from many environmental regulations, including laws governing pollution emissions and land reclamation. Low-interest loans, loan guarantees, and direct investments from governments and financial institutions such as the World Bank contribute to maintaining mineral prices at artificially low levels.

Governments traditionally have supported mineral production for several reasons. Mineral production historically is associated with national security concerns—metals are needed to build weapons. Minerals are also important for the economies of many countries. The developed nations tend to think a healthy economy is synonymous with economic growth, and traditionally, economic growth has been closely linked to manufacturing and heavy industry, which require large supplies of mineral products. Consequently, governments have tried to ensure that the manufacturing sector received a constant flow of inexpensive raw materials. Historically, opening up virgin mineral deposits was a way to attract people to frontiers that "needed" to be developed, or so politicians thought. For many present-day developing countries, mineral exports are the primary means of earning currency to pay their international debts. Such nations have an incentive to open, subsidize, and maintain large mining operations, no matter what the environmental costs.

10.7 Dealing With Mineral Scarcity

There are two basic strategies for coping with mineral scarcities: increasing the supply (by locating new ore deposits or recycling old materials) or decreasing the demand (by finding alternatives or substitutes for or simply eliminating consumption through sustainable technological developments and changes in lifestyle).

Expanding the Resource Base

In the past, humans relieved the demand for more raw minerals by simply going out and discovering new mineral deposits, but the days of finding significant new ore deposits are almost surely over. The major geological provinces on land where deposits are found have for the most part been heavily explored. The number of new discoveries of large and/or high-grade deposits has been declining for years. The high-grade deposits that remain to be discovered will be smaller, less accessible, and more expensive to exploit. Significant new mineral deposits may be found under the sea, but the technical difficulties and expense of mining in deep water have yet to be resolved. In general, drilling deeper into the crust is not a solution either, for most rich mineral deposits are found in the upper regions of the crust, and the cost of mining rises exponentially as one penetrates extreme depths.

Using progressively lower grade ores as high-grade deposits are depleted may sound feasible in theory, but in practice, it does not work for all minerals. For a few of the geologically common minerals (namely iron, aluminum, manganese, magnesium, chromium, and titanium), higher grade ores are rare, but the lower the grade of the ores, the more abundant they are. Of course, as lower grades of ore are worked, the extraction costs increase dramatically, as do the energy requirements, concomitant pollution, and other environmental costs associated with mining, processing, and disposing of huge volumes of rock. In addition, the amount of waste produced for every ton of metal extracted rises exponentially, as progressively lower grades of ore are mined. For instance, in the United States, the amount of waste generated by copper mining has increased dramatically as the average grade of copper ore has steadily dropped from about 2.5% in 1905 to approximately 0.5% to 0.6% today. The increase in waste simply reflects the fact that more raw rock must be mined, crushed, sorted, and processed to extract the metal. As more rock is mined and processed, more energy is consumed, and more of the landscape is devastated. Most of the energy consumed in mining is derived from fossil fuels; thus, mining ever lower grades of mineral ores contributes to the quickening pace of fossil fuel depletion and the resulting pollution generated when the fuel is burned.

Most metals are geologically scarce and are not found in a continuous gradation from high-grade ores to lower-grade deposits. Metals such as copper, tin, and zinc have been geologically concentrated into the high-grade, minable deposits that people are currently exploiting. After these deposits are depleted, the next lower grade of "ore" is essentially ordinary rock that may contain only approximately 0.1% to 0.001% of the amount of metals found in the high-grade ores. Enormous amounts of rock "ore" would have to be processed to obtain relatively small amounts of metal. Likewise, some people have argued that an abundant supply of metals is dissolved in seawater, but huge amounts of seawater would have to be processed to extract these metals because their concentrations are so low. For example, to extract a year's supply of zinc for the world, one would have to process approximately 75 billion metric tons (or 7.5×10^{13} kg) of ordinary rock or 503,400,000 billion liters (5.034×10^{17} liters, or 133,000,000 billion gallons) of seawater—assuming that the extraction processes were 100% efficient.

"Mining" ordinary rock or seawater for scarce minerals and elements not only would be energy and cost prohibitive, but also would cause untold environmental devastation on a much larger scale than the considerable damage already caused by the mining of high-grade deposits.

Recycling

As we have seen, in an absolute sense, the stock of any mineral on and within the Earth's crust is finite and can only decrease. Yet, in a fundamental way, we never appreciably decrease the stock of any element on Earth (fissionable elements, such as uranium used in nuclear processes would be the exception to this rule). As we use materials, the atoms of which they are composed physically still exist; they are simply in altered arrangements. Metals and many other substances can be recycled. In effect, recycling can be viewed as increasing the effective or usable amount of a substance that we have at our disposal. Moreover, **recycling** often consumes less energy than extracting raw, virgin material. Producing aluminum from recycled goods uses 90% to 97% less energy than is required to produce the same aluminum from raw ore.

However, recycling has practical limits. During the recycling process, the **recovery ratio**—how much of the original material can actually be recovered and recycled—is always less than 100%. For instance, if we begin with 1 ton of aluminum and manufacture 1 ton of aluminum cans and other products, use these products for a number of years, and then attempt to gather them all up and melt them down into pure metal, we will end up with something less than 1 ton of aluminum. Normal wear and corrosion will result in some loss of metal—perhaps microscopic—that is unrecoverable. In the very process of recycling and producing new products, there will be more loss. Heating the metal, cutting it, or crushing it will result in less than 100% recovery (**FIGURE 10-7**). A recycling recovery ratio of 90% is difficult to achieve for most metals. The only metals for which a higher ratio of recovery is normally possible are precious metals, such as gold, that are relatively nonreactive (so they do not corrode) and are used primarily in jewelry or other products not subject to heavy physical wear. In fact, the current recycling efficiency for many common metals can be 30% or even less.

As a hypothetical example, assume that we have a recycling ratio of about 80%. If we begin with 1 ton of aluminum, use it in products, and subsequently recycle it entirely (i.e., as best we can—not all of the metal will even necessarily be returned for recycling) once a year, then the next year, we will have only 0.8 ton of the metal. After 2 years, we will have only 0.64 ton of the metal. After 3 years, we will have only 0.512 ton, and so on. If we keep doing this and then sum all of the metal used from year to year, we will find that, in effect, the 1 ton of aluminum was stretched into 5 tons (or what would have been 5 tons had virgin metal been used each year). Turning 1 ton into 5 tons through recycling may help, but given increasing demands for metals, it can only slightly delay the inevitable exhaustion of the supply. In a practical sense, recycling alone, without the input of new raw materials, cannot satisfy current demands indefinitely.

In addition, even if recycling extends the effective amount of a substance over a number of years, it does not increase the amount of the substance at any one time. Returning to

FIGURE 10-7 Recycling metal, such as aluminum shown here, always results in some amount of material loss.

our 1 ton of aluminum, no matter how meticulously we recycle, we can only manufacture 1 ton's worth of goods from 1 ton of aluminum at any one time.

Recycling's Dark Side

In some cases, recycling used products that were not designed with recycling in mind can be difficult and labor intensive. For example, the metals used for many manufacturing and industrial purposes must be quite pure—they cannot be contaminated by small amounts of other metals. One cannot simply take a junk car and "melt it down." The various metals must be carefully separated, and even then, large amounts of energy may be required to purify the resulting metal. It is for this reason that various electronic devices, such as obsolete computers, can be incredibly difficult to recycle.

It is virtually impossible to recycle some substances in certain uses. Many uses of materials are dissipative—under normal use, they are degraded and dispersed. Examples of such materials include dyes, paints, inks, cleansers, solvents, cosmetics, fertilizers, and pesticides. For instance, metals used in dyes and paints cannot be recovered. As the relative need for certain metals (e.g., copper) decreases and demand remains relatively steady, a greater proportion of consumption may well be for dissipative uses. Even when consumption is relatively low, if the uses are dissipative, the metals cannot be recycled—after they are consumed, they are gone for good. Eventually, the supplies of such metals, if subjected to constant consumption without recycling, must run dry.

Substitutability

As the Earth's population continues to grow, recycling may not be enough; we could conceivably reach a point where there is simply not enough copper, for instance, to produce all of the copper piping and wiring demanded. The absolute amount of copper is limited, no matter how carefully we recycle. We can cope with these situations by finding substitutes (**substitutability**). By manufacturing household pipes from plastics, making wires out of metals other than copper, and using fiber optics in place of wires, we cannot increase the supply of copper, but we can lower the demand, at least relative to what it might have been. However, that even with substitutes, real demand may still increase as the number of consumers increases. Thus, the demand for copper is increasing by approximately 2.7% per year despite many substitutes.

Certain metals are valued for their physical properties and characteristics, which can sometimes be found in other substances. To use an often-cited example, the metal tin is relatively scarce from a global perspective. At current consumption rates, the known reserves of tin will be exhausted in 20 to 30 years. Tin is desirable because it is durable, malleable, and light, but each of these properties can also be found in other substances. Thus, tin cans have been largely replaced by aluminum and steel cans and by glass jars and bottles. Replacements can also be found for many other increasingly scarce metals, such as lead, zinc, and mercury.

However, substitution deals only temporarily with the scarcity of one or a few metals (or other substances). It is not a solution to the generally increasing demand, and therefore scarcity, of all minerals and other exhaustible resources because ultimately the reserves of the substitutes will themselves become depleted. Some people have suggested that plastics could substitute for metals, for example (and indeed great strides have been made in this direction), but most plastics are derived from petroleum products—which are also nonrenewable, rapidly diminishing resources. However, after plant-based plastics become common, the situation may be different.

In some cases, certain elements or compounds have unique properties, and there is little realistic hope of finding a ready substitute. Platinum and other metals are used as catalysts in industrial processes. Mercury is the only metal that is liquid at room temperature, and small but critical amounts of certain elements (such as copper, lead, zinc, mercury, nickel, tin, manganese, chromium, cobalt, and titanium) are absolutely necessary in modern metallurgy. The prospects of developing a totally new system of metallurgical techniques seem remote.

A few substances that have no ready substitute and cannot be recycled very efficiently also happen to be in very short supply and high demand; they are also critical to life in a direct way. Perhaps the best example of such a substance is phosphorus. Phosphorus is essential to all living organisms: it is used in DNA, cell membranes, and the bony tissue of vertebrates, among other things. Without phosphorus, life as we know it could not exist. Not surprisingly, phosphorus is a key component of fertilizer. Phosphorus has no known substitute. It is produced commercially from nonrenewable deposits of phosphate rock, and these deposits are relatively rare and being depleted very quickly. Estimates vary, but assuming current conditions and consumption rates, the practically minable phosphorus supply could be depleted in less than 100 years or, most optimistically, in about 1,300 years.

Phosphorus is difficult to recycle because it is used in a diffuse form (e.g., spread on the ground as fertilizer), which makes it very difficult to recover. One way of recycling phosphorus would be to crush bones and extract the phosphorus from them, but the **recovery rate** would probably be fairly low. Another possibility is that wastewater treatment plants are required to remove phosphorus from wastewater, and possibly this phosphorus could be recycled.

Reducing Consumption:
Conservation and Durability

Just as conserving energy can reduce the demand for energy, so too can **conservation** reduce our demand for mineral resources. We can simply use less.

A very simple way to reduce the demand for mineral resources is to produce **durable goods** (objects that are designed to be used over time, such as cars, appliances, and furniture) that are designed to last as long as possible. This can be done by building better quality products to begin with and designing products so that they can be repaired, rebuilt, modified, or refurbished as they grow old—rather than simply discarding and replacing them. The old Checker cabs in New York were a good example. The cabs' fenders were bolted on; if the driver of such a cab got into an accident, the fender was unbolted. A usable fender was bolted on, and the damaged fender was repaired and used when the next cab came into the shop needing a fender. In some cases, reusing goods, rather than discarding them, can accomplish considerable energy and mineral savings. Food and drink packaging is a classic example. The refillable glass bottle can typically be reused (washed and refilled) dozens of times, producing less packaging during the manufacturing process. Compared with recycling, which entails breaking down the product and using the raw materials to make a new product, **reuse** conserves valuable mineral resources and saves energy and money.

Another strategy is to reduce the size of products where possible, a concept sometimes referred to as **dematerialization**. A smaller, lighter car uses a correspondingly smaller amount of mineral resources and may serve the function equally well. Indeed, a smaller car may be more fuel efficient. Miniaturization, as in electronics, can save substantial amounts of raw materials and energy—and miniaturized electronic components can be faster, more dependable, and more powerful. Take computers, for example: computations that in the 1950s required tons of computer hardware housed in its own building can now be accomplished with a system that fits into a briefcase or even a palm-sized machine.

Specific cases of dematerialization must be analyzed carefully; in some cases, first impressions can be deceiving. If a smaller, lighter product lasts only 75% as long as the product that is 25% heavier, the heavier, longer-lasting product will in this case use less material over time. Environmentally, sometimes it may be worthwhile to use a more durable product if it lasts significantly longer, rather than a dematerialized product.

Another consideration is that although dematerialized products may require substantially less material to produce, because of the complexity of their designs, they may be virtually impossible to disassemble and recycle once they have come to the end of their useful lives. Such products are often intended to be permanently discarded once their useful life is over. This is the case with many modern electronic devices such as cell phones. In practice, it is not always cost-effective or energy effective to separate and refine the small quantities of various valuable metals that occur in electronic equipment (although recently advances have been made in recycling metals from electronic circuit boards and other computer hardware). The metals used in manufacturing electronic devices may be lost forever.

However, in other cases, a dematerialized product may last even longer than the heavier product.

■ study guide

SUMMARY

- Modern technological society is dependent on mineral resources, but of course, humans have used minerals since prehistory.
- Mineral resources can be classified as metallic minerals and nonmetallic minerals; the latter category includes structural materials (such as building stone), industrial minerals (such as fertilizer components and nonmetallic minerals used in manufacturing), gemstones, and energy minerals (fossil fuels, not technically minerals, and uranium ore, which is actually a metal).
- An ore is a mineral deposit that can be economically mined.
- Mineral reserves are ore deposits that have not yet been exploited.
- Mining and smelting operations can cause severe environmental degradation; in the United States, nonfuel mineral mining produces more than 1 billion tons of waste each year.

- Since the Industrial Revolution more than 200 years ago, mineral use has increased by a factor of 100 or more, and for some specific minerals, the increases have been by factors measured in the thousands.
- Prices of many minerals are low because they do not include externalities—environmental, social, and other costs that are associated with procurement of the mineral; mining may be subsidized by the government, or the ore may be virtually free to the miner.
- Mineral scarcity can be dealt with by expanding the resource base (discovering new mineral deposits), recycling, substitution of one mineral or substance for another, and conservation of mineral resources (including the manufacturing of more durable goods that last longer).

KEY TERMS

conservation
dead zones
dematerialization
durable goods
energy minerals
ferrous
gemstones
industrial materials
infrastructure

metallic minerals
minerals
mineral deposits
mineral resource
nonferrous
nonmetallic minerals
ore deposit
recovery rate
recovery ratio

recycling
reserves
reuse
smelting
structural materials
substitutability
tailings
virgin supplies of ore

STUDY QUESTIONS

1. What is a mineral resource?
2. List some of the major types of mineral resources. Which are most important in terms of quantity used? Which are most important in terms of economic value?
3. Why are mineral resources said to be non-renewable?
4. Distinguish between mineral deposits, ores, and reserves.
5. Summarize some of the ways that mineral exploitation and use can lead to environmental degradation.
6. What are some of the factors that determine the price of a particular metal or other mineral? Have mineral and metal prices generally increased or decreased over the last couple of decades? What accounts for these trends?
7. Describe the broad historical trends in mineral and metal use.
8. Why has the mining industry received special treatment from governments, including the U.S. government, in the past?
9. Discuss the advantages and disadvantages of the following strategies to deal with mineral scarcity: recycling, substitutability, conservation, and durability.
10. What is dematerialization?
11. Do you believe we will ever run out of mineral resources (will demand for minerals ever outstrip supplies)? Justify your answer.

WHAT'S THE EVIDENCE?

1. The authors assert that the prices of minerals are, in general, artificially cheap (with a few exceptions, such as diamonds and other precious stones whose values may be artificially inflated). What evidence do the authors present to support their assertion that the prices of many mineral resources fall short of their real costs and values? What factors might be included in the prices of minerals to bring the amount the consumer pays more in line with their actual values and the expenses of procurement?
2. The authors maintain that recycling alone cannot satisfy current demands for metals and other minerals indefinitely. What lines of evidence do the authors put forth to support this contention? If recycling alone is not enough, what else might be done to increase the effective supply of mineral resources?

CALCULATIONS

1. If a certain ore body of tin contains 0.7% tin by weight, how much ore will have to be mined and processed (assuming complete recovery of all the tin) to produce one metric ton (1,000 kilograms, or 2204.623 pounds, or 1.1023 English tons) of tin? (Express your answer in both metric tons and English tons.)
2. If a certain ore body of copper contains 0.4% copper by weight, how much ore would have to be mined and processed (assuming complete recovery of all the copper) to produce 1 metric ton of copper? (Express your answer in both metric tons and English tons.)

ILLUSTRATION AND TABLE REVIEW

1. Referring to Figure 10-3a, calculate by approximately what percentage the consumption of metals increased in the 55 years from 1945 (the end of World War II) to 2000. During this same period, by what percentage did the consumption of industrial materials increase? By what percentage did the consumption of construction materials increase from 1945 to 2000?
2. Referring to Figure 10-3b, calculate by what percentage world materials use increased between 1970 and 2000. How does this compare with the increase in use by the United States during the same time period?

http://environment.jbpub.com/mckinney/5e/

http://environment.jbpub.com/mckinney/5e/

Connect to this text's website at http://environment.jbpub.com/mckinney/5e/. This site features eLearning, an online review area that provides quizzes, chapter outlines, and other tools to help you study for your class. You can also follow useful links for more in-depth information.

Conserving Biological Resources

11

Biological diversity, or **biodiversity**, has many definitions, all of which involve the variety and variability among living organisms and the ecological complexes in which they occur. One reason for the lack of a precise definition is that life is hierarchical. For example, genes occur in cells. Cells occur in organisms, and organisms occur in ecosystems (a review of the biosphere and its populations, communities, ecosystems, and biogeochemical cycles provides more information). At what level do we measure the diversity of life in an area? The number of genes? Organisms? Ecosystems? All can be used.

People depend on the diversity of biological resources for food, energy, shelter, clothing, good health, medicine, and in many other ways. In this chapter, we focus on the current rapid loss of natural biological resources, often called the "extinction crisis." We examine why biological diversity is being lost, why it is worth saving, and how to go about saving it. Few people truly want to see any species become extinct, and the effort to slow the loss of perhaps thousands of species each year is gaining national and international support. One indication is the very rapid growth of **conservation biology**, a new subdiscipline of biology that draws from genetics, ecology, and many other fields to find practical ways of saving species.

Chapter Objectives

After reading this chapter, you should be able to explain or describe the following:

- Why biodiversity measurement is important
- Different types of extinction
- The causes of extinction
- Traits that make some species prone to extinction
- The benefits and problems of the Endangered Species Act
- The values of biodiversity
- Ways to promote biodiversity preservation

317

11.1 Measuring What's at Risk

Even if we simplify the definition of biodiversity to describe just the number of species, or **species richness**, no one knows how many species live on Earth. As seen in a review of the distribution of life on Earth, taxonomists have classified about 1.8 to 2 million species, but this sample of 2 million species is highly biased toward the northern hemisphere and may not reflect the true species richness of the world's biosphere. The tropics, the areas with most species diversity, are still not well known, and it is safe to say that the oceans contain an unknown number of species that we have never seen. Of the 2 million species classified, 56% are insects, 14% plants, and just 3% vertebrates such as birds, mammals, and fishes.

Estimating the Unknown

It is important to understand species diversity and the abundances of individual species so that we have some idea of what we are losing and what we are trying to conserve. As discussed in a review of the distribution of life on Earth, biologists have devised a number of ways to develop biodiversity estimates from limited information, such as rain forest insect samples, ecological ratios, and species-area curves. These methods are based on small samples. All methods of estimating biodiversity involve a certain amount of extrapolation; so naturally, there is disagreement about how many species exist. Estimates range from as low as 4 million land species to as high as 100 million land species, and from 1 to 10 million species in the ocean. Whatever the exact number of species today, the fossil record indicates that we live in a special time of Earth's history when global biodiversity is at or near its all-time peak.

11.2 Biodiversity Loss

Biodiversity loss is the result of **extinction**, which is the death of a group. In most cases, people mean species extinction when they refer to extinction, but other kinds occur. Local extinction, or **extirpation**, means that a species has died out in a local area. For example, grizzly bears once inhabited at least part of every state in the western United States, as far east as Kansas and as far south as Texas. They were hunted into local extinction in most areas but persist in Canada and in a few areas totaling 2% of their original U.S. range. **Ecological extinction** means that a species has become so rare that it has essentially no role or impact on its ecosystem. Many of these species are called the "living dead" because their rarity dooms them to eventual total extinction in the wild. The Florida panther, which has a population size of less than 80 individuals, including only 20 breeding females, cannot possibly survive without much help because of inbreeding and other problems of small populations discussed here (**FIGURE 11-1**). Tigers, rhinos, and many other animals suffer similar problems and likely will survive only if direct action is taken soon. Even then, such species may survive only in zoos and parks outside their original habitat.

Biological impoverishment by local or ecological extinction is much more common than species extinction, but because local and ecological extinctions are rarely reported, most of the media and the public do not notice this kind of impoverishment. This may be one reason the public does not seem to perceive the extinction crisis as urgent.

Extinction is the ultimate fate of all species. The fossil record indicates that since life began more than 3.5 billion years ago, more than 95% of all species that existed are now extinct. Some of these past extinctions occurred during **mass extinctions** (see Figure 4-7 in the

FIGURE 11-1 The Florida panther has a population of fewer than 80.

discussion on the distribution of life on Earth). These were catastrophic events that possibly killed more than 60% of all species on Earth; by some estimates, the Permian extinction of approximately 250 million years ago may have involved the loss of 90% of marine species living at that time and 70% of terrestrial species. The causes behind major mass extinctions continue to be debated. Some apparently were caused by climatic disruption of habitat, possibly associated with global cooling, glaciations, or sea level changes, and were at least in some cases triggered by unusual events, such as asteroid activity and/or significant tectonic activity. The last major mass extinction, which occurred approximately 65 million years ago, included the disappearance of the dinosaurs. Evidence strongly indicates that a large meteorite impact on the Yucatan Peninsula caused or accelerated it.

Despite widespread interest in past catastrophic mass extinctions, careful analysis of the fossil record shows that more than 90% of all extinct species died out during normal background times, not during mass extinctions. **Background extinctions** are those that occur because of small changes in habitat, climate, predation, evolution of competing species, localized catastrophes such as volcanic eruptions or floods, or other conditions that would require a species to adapt. If a species does not adapt to the new conditions, it can become extinct. Common estimates are that between 2 and 10 species per year died out from natural causes.

Current Biodiversity Loss: Another Mass Extinction?

The European Age of Expansion in the 15th and 16th centuries initiated a wave of extinction that has continued to accelerate. Some people have suggested that this current wave of extinction is comparable to some of the major mass extinctions of the geologic past, but while acknowledging the seriousness of present extinctions, this comparison should be taken only so far. Major mass extinctions that the fossil record documents consisted of not only the loss of numerous species, but also the loss of larger taxonomic groups (types of organisms). Thus, the entire group of trilobites (a diverse group of fossil arthropods) was extinct by the end of the Permian, about 250 million years ago, and all dinosaurs were extinct by the end of the Cretaceous, about 65 million years ago. We are not witnessing the comparable loss of major taxonomic groups today. In addition, although current extinctions are severe and likely add up to thousands of species, the trends do not equal the magnitude of past major mass extinctions as indicated in the fossil record. Current data from the International Union for Conservation of Nature (IUCN) indicates that approximately 19,000 of the 52,000 species documented are threatened with extinction.

TABLE 11-1 shows the number of recorded species extinctions among vertebrates and plants since the year 1600. In the last 4 centuries, 1.9% of the known mammal species and 1.4% of bird species on Earth have gone

TABLE 11-1	Recorded Extinctions of Vertebrates and Plants, 1600 to 2000				
	Species Extinct	Species Threatened	Total Species	Percentage Extinct	Percentage Threatened
Mammals	87	1,130	4,600	1.9	24.6
Birds	131	1,183	9,500	1.4	12.5
Reptiles	22	296	6,300	0.3	4.7
Amphibians	5	146	4,200	0.1	3.5
Fishes	91	750	19,100	0.5	3.9
Plants	90	5,611	250,000	0.04	2.2

Many additional species have presumably gone extinct without scientists having recorded them. Many more species are now threatened with extinction in the next few years. Except for mammals and birds, most threatened percentages are probably greatly underestimated.
Source: Data from International Union for Conservation of Nature and Natural Resources, The World Conservation Union Red List 2000.

extinct. Although these numbers may not seem especially alarming, they almost certainly underestimate the true nature of the extinction rate, for a number of reasons:

1. The extinction rate of birds and mammals has been increasing rapidly. Between 1600 and 1700, the extinction rate was only about one species per 10 years. Between 1850 and 1950, the rate was one species every year. Current evidence indicates that the extinction rate has increased yet again since 1950.

2. Some species of birds and mammals became extinct before they were described, so the true extinction rate probably is higher than that recorded in Table 11-1.

3. Species are not officially recorded as extinct until they have not been seen for 50 years. This lag also means that the true extinction rate of birds and mammals probably is higher than is shown in Table 11-1.

4. Many species of birds and mammals have too few individuals remaining alive to successfully breed. These species are not yet recorded as extinct but will be in the near future.

As birds and mammals are among the best-described groups, the extinction rates for other groups, including invertebrates (not shown in Table 11-1), are likely to be underestimated even more significantly. To solve such problems, we can use other methods to estimate the rate of species extinction. TABLE 11-2 shows a variety of extinction rate estimates based on the species area curve and other methods. These indicate that between 1% and 11% of all species on Earth are lost each decade. The average estimate is around 5% per decade, which translates into an extinction rate of at least 50 species per day. This is much higher than the recorded rates of Table 11-1 because many extinctions have gone unrecorded. This is many hundreds of times higher than the normal average extinction rate of about 2 to 10 species extinctions per year, as estimated from the fossil record. Such currently high rates lead many, perhaps most, biologists to predict that as many as half the world's species could be extinct sometime in the 21st century.

Perhaps a better way to assess impending extinction is to examine the number of species that we know to be at immediate risk of extinction. Table 11-1 shows the number of vertebrate and plant species that have been designated as threatened with extinction in the next few decades or even years. These are species that have documented declines in population size over the last few decades. They often have population sizes in the range

TABLE 11-2	Estimated Rates of Extinction	
Global Estimate	Loss per Decade (%)	Method of Estimation
One million species between 1975 and 2000	4	Extrapolation of past exponentially increasing trend
15% to 20% of species between 1980 and 2000	8 to 11	Estimated species area curve; forest loss based on U.S. government projections
12% of plant species in Western Hemisphere tropics	—	Species area curve
15% of bird species in Amazon basin		
2,000 plant species per year in tropics and subtropics	8	Loss of half the species in area likely to be deforested by 2015
25% of species between 1985 and 2015	9	As above
At least 7% of plant species	7	Half of species lost over next decade in 10 "hot spots" covering 3.5% of forest area
0.2% to 0.3% per year	2–3	Half of rain forest species assumed lost in tropical rain forests to be local endemics and becoming extinct with forest loss
5% to 15% forest species by 2020	2–5	Species area curve
2% to 8% loss between 1990 and 2015	1–5	Species area curve; range includes current rate of forest loss and 50% increase

Source: Adapted from W. V. Reid. How many species will there be? T. C. Whitmore and J. A. Sayer, eds. Tropical Deforestation and Species Extinction. London, Chapman & Hall, 1992.

of or below several hundred individuals. An alarming 22% of mammals and 14% of birds are now threatened or extinct. As with extinction percentages, mammals and birds are the best studied, so the threat percentages for the other groups are probably gross underestimates. In reality, reptiles, fishes, plants, and most, if not all, other groups probably have similar percentages of species that are threatened. Although extinctions are increasing at an alarming rate, scientists at the IUCN also point out that because of human conservation efforts, 16 species of birds have been spared extinction in the last 15 years.

11.3 What Causes Biodiversity Loss?

Environmental change causes extinction. In many cases, human activities induce or exacerbate environmental change. Species can adapt to environmental change if it is not too rapid (a review of the biosphere and its populations, communities, ecosystems, and biogeochemical cycles provides more information), but the changes in human population growth, activities, and technology have been too massive and too fast to allow many species to adapt.

For convenience, we can categorize all environmental changes that cause extinction into two basic categories: the **physical environment** and the **biological environment** (**TABLE 11-3**). Biological changes occur in three ways: introducing species, overhunting, and removing species.

It is usually difficult to identify which of the four causes in Table 11-3 was involved in a particular extinction. In most cases, species extinction has more than one cause.

TABLE 11-3	Four Ways That People Cause Population Decline and Species Extinction	
Change Physical Environment		**Examples**
1. Habitat destruction, change biological environment		Drain swamp, toxic pollution
2. Introduce new species		New Predator
3. Overhunting		Big-game hunting
4. Secondary extinctions		Loss of food species

For instance, all four causes have led to the decline of the Florida panther. Loss of habitat from development, overhunting, dozens of introduced species, and loss of prey species have combined to reduce the panther population drastically in the last 200 years.

Keeping in mind that extinction-causing environmental changes often occur together, we now discuss each of these four causes of extinction. We try to focus on examples where one cause has clearly played a major role.

Habitat Disruption

Habitat disruption refers to disturbance of the physical environment in which a species lives. Such disturbances can range from minor to drastic. Minor disturbances, such as mild chemical changes from air pollution, tend to affect only the most susceptible species.

In contrast, extreme physical changes in habitat can eliminate many species from the area. Deforestation of tropical rain forest by logging, burning, or other agricultural practices is one of the most publicized examples of massive habitat disruption. **FIGURE 11-2** shows the projected loss of this critical habitat if current trends continue. Recall that the tropics contain many more species than other areas (a review of the biosphere and its populations, communities, ecosystems, and biogeochemical cycles provides more information). By some estimates, rain forests contain more than 50% of the world's species, even though they cover only 7% of the Earth. As of the year 2009, more than half of the world's rain forests had been deforested, with as much as 90% destruction in some places such as West Africa. Using the species area curve, some experts estimate that this 50% rain forest loss could mean perhaps a 10% rain forest species loss thus far. However, the curve also shows that continued habitat loss will soon result in a rapid decrease in species numbers.

There has been an enormous loss of habitat in the United States. **FIGURE 11-3** shows that less than 25% of the natural vegetation covering most of the central and northeastern United States remains. Only in the more desert-like regions of the western United States is more than 75% of the area covered by natural vegetation.

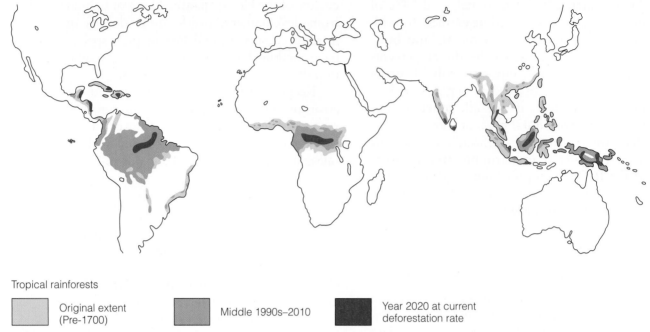

Tropical rainforests

| ▢ | Original extent (Pre-1700) | ▨ | Middle 1990s–2010 | ▉ | Year 2020 at current deforestation rate |

FIGURE 11-2 Past decline of tropical rain forests and projected decline by year 2010 if current deforestation rates continue.

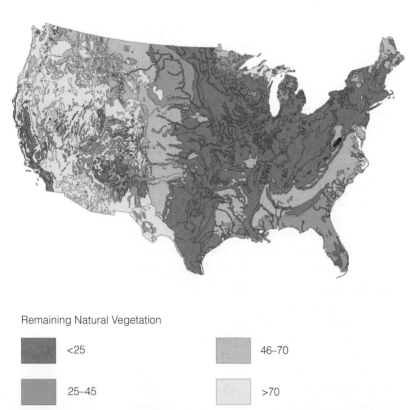

Remaining Natural Vegetation

| ▉ | <25 | ▨ | 46–70 |
| ▨ | 25–45 | ▢ | >70 |

FIGURE 11-3 Remaining natural vegetation in the United States. (*Source:* B. A. Stein, L. S. Kutner, and J. S. Adams. Our Precious Heritage: The Status of Biodiversity in the U.S. New York: Oxford University Press, 2000, p. 229.)

Removal of natural vegetation for farming or development not only destroys the plant species but also reduces habitat for the animals that require those native plants to survive. Devegetation is usually a piecemeal process, with the habitat being broken into progressively smaller and more isolated fragments. Such habitat fragmentation has occurred in nearly all places where modern agriculture and urban society have moved in. **Habitat fragmentation** is especially disruptive for two reasons. For one thing, remaining areas of habitat become separated and form islands of refuge in a hostile environment. If the islands are too far apart, members of a species may not be able to reach other members to reproduce. The species could die out if all of the isolated islands of habitat are so small that no island can maintain a self-sustaining population.

Second, habitat fragmentation leads to **edge effects** (**FIGURE 11-4**). Our pets, wind and temperature changes, and many other disturbances from the surrounding area can penetrate along the edges of the preserved area, resulting in loss of habitat. Edge effects may be very dramatic. For instance, roads that occupy only 2% of a certain area may actually dis-

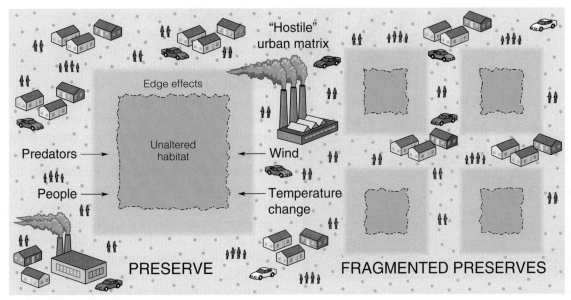

FIGURE 11-4 Even though the fragmented preserves on the right may seem similar in total area to the single preserve on the left, the amount of unaltered habitat is greatly reduced from edge effects.

rupt approximately 50% of the habitat. Habitat fragmentation is one of the main causes of decline in global and North American songbird populations.

Habitat disruption is not limited to land. Water pollution, dam building, and other human activities have a huge impact on fishes, invertebrates, and other species in lakes and rivers because such species are confined in limited areas. Wetlands loss is a particular concern because traditionally many wetlands, such as swamps, marshes, and bogs, have been viewed as "wasteland" that simply breeds mosquitoes and other vermin. Therefore, not only did people not bother to protect wetlands, but also sometimes actively tried to fill them in and destroy them.

The oceans are experiencing widespread disruptions as people pour raw sewage, toxic chemicals, sediments, and other pollutants into them. It is difficult to monitor extinctions and ecological damage in the oceans, but many studies of coral reefs have found signs of extensive damage. Reefs are good biodiversity indicators because they support a great variety of species. Many reefs are in critical condition; an estimated 10% to 15% of the world's reefs are already dead. At current rates, another 60% could be lost in the next 20 to 40 years. The major causes of death are global warming and sedimentation from logging, farming, mining, construction, and other coastal activities. The suspended sediment blocks sunlight, decreases available oxygen, smothers the reefs, and has many other harmful effects, such as increasing susceptibility to disease.

Introduced Species

In many parts of the world, populations of **exotic species** have been introduced into habitats that they have not occupied under natural conditions, sometimes with beneficial effects. Many of the food crops and virtually all of the livestock species, including horses, in the United States were imported. However, more often exotic species are imported with devastating effects. More than 1,500 species of nonnative insects, including fire ants, gypsy moths, Asian Longhorn beetles, and Africanized bees, have become established in the United States since 1860. Many other kinds of organisms have also been introduced. Well-known U.S. examples include kudzu, walking catfish, and English starlings. After an exotic species becomes established, it is virtually impossible to remove (**FIGURE 11-5**).

Florida, with its tropical climate, is especially vulnerable to the more exotic species. Animals that have made themselves at home in the state after pet owners released them include giant Burmese python snakes, Nile

FIGURE 11-5 The Old World climbing fern, shown here overtaking cypress trees in Florida, was native to East Asia and Australia before it was introduced to the United States intentionally as a landscape ornamental plant.

monitor lizards, and African rats the size of dogs. Indeed, South Florida is often said to have more exotics than any other part of the United States. More than 1,100 of the 4,000 plant species currently documented in Florida are nonnative. Most of the damage is from 30 species of plants, with a total of 130 considered a serious threat. More than 500 fish and other animal species found in Florida are nonnative, with more than 100 established as invasive. Estimates indicate that thousands of species of insects and agricultural-related pests have been introduced and have quickly died out. Florida's Fish and Wildlife Conservation Commission indicates that approximately 40 new potentially invasive agricultural-related species arrive in Florida every month. These species are new to the area, in part because the Florida peninsula had been isolated from other tropical regions, such as those in Latin America, and Florida is a major destination for goods entering the country from other nations.

The economic toll of exotic species is astonishing. The U.S. Office of Technology Assessment estimates that nonnative species caused more than $100 billion in damage during the 1990s, and the current yearly spending is approximately $120 billion. About 17% of the more than 1,500 insect species introduced into North America are pests requiring the use of pesticides for control. Many of these introductions were intentional. The English starling, perhaps the most common bird in the United States today, was introduced in a production of a Shakespearean play in the early 1900s. Game animals have been introduced for hunting or fishing (striped bass in Tennessee) and predators to reduce pests (mongoose in Hawaii). The kudzu plant now overgrows and smothers many acres of land in the southeastern United States at a yearly eradication cost exceeding $6 million dollars; this does not include the loss of agricultural income due to lost acreage. Nevertheless, in the 1930s, the federal government paid farmers to plant it to control soil erosion.

Although purposeful introductions have been sharply curtailed, accidental introductions continue to increase, as international trade and travel expand. The very destructive zebra mussel is an example of a species that hitchhiked from one nation to another (see **CASE STUDY 11-1**). Studies show that the large majority of new species fail to become established, but because so many introductions occur, some of them eventually succeed. People promote this both by providing transportation for nonnative species and by disturbing native ecosystems. By upsetting natural interactions, we make it easier for new species to become established.

Exotic species are especially important as a cause of extinction on islands. Because of their isolation, native island species are often poorly adapted to cope with new species. As a result, new species not only have an easier time becoming established on islands but also have a more devastating impact on native island species than on continental species. Predators and competitors have driven many native prey and competitor species to extinction. Others survive only in carefully protected preserves. The Nature Conservancy reports that invasive species have directly contributed to the decline of more than 40% of all species on the U.S. threatened and endangered list.

An example is found in the ground birds that evolved on many islands, such as New Zealand and Hawaii, in the absence of large predators. The introduction of domestic cats

Invasion of the Zebra Mussels

VISIT

http://environment.jbpub.com
/mckinney/5e/
for more information

An oceangoing ship teems with alien life. The organisms come aboard when the ship takes on ballast water from the harbor. A ship will take ballast for several reasons. One is stability: a lightly loaded vessel may ride too high and is thus more liable to capsize. Another reason is propulsion: the propeller of an under-loaded ship may rise half out of the water. To avoid these problems, a ship may distribute ballast water through a network of tanks inside the hull.

In the mid-1980s, a ship leaving a freshwater European port for North America began to take on ballast. As it did, several members of the harbor's population came aboard along with the water. Unfortunately for North America, one of them was a hardy bivalve named *Dreissena polymorpha*, the zebra mussel, which over a 200-year period had spread from the region of the Caspian Sea into much of Europe, plugging water pipes and clamping onto the hulls of boats (**FIGURE 1**). The unknown ship that carried the zebra mussel across the Atlantic was headed for the St. Lawrence Seaway. Past Quebec City it sailed, past Montreal, past Toronto, and on through Lake Erie. When the ship finally reached its destination and flushed its ballast somewhere above Detroit, a founding population of zebra mussels tumbled into Lake St. Clair (where it was discovered in 1988)—and a new continent.

Biologists are hastening to quantify the effects of this invasion, as the introduction of any new species is called, but already the costs have been

staggering. Propelled by relentless fertility and a talent for spreading themselves abroad, zebra mussels are clogging the intake pipes of power stations and water treatment plants; colonizing navigation buoys in such numbers that they drag them under; fouling fishing nets, marine engines, and hulls of boats; and displacing spawning grounds that are the mainstay of a commercial and sports fishery valued in the billions. In North America, the zebra mussel seems to have found a very hospitable environment. They quickly swept down the length of Lake Erie and penetrated Lake Ontario. By 1990, they could be found in all of the Great Lakes, and within 2 years, zebra mussels were found in the Hudson and Illinois rivers and then made their way into the Mississippi River and its drainage area. By 1992, zebra mussels were in the Arkansas, Cumberland, Ohio, and Tennessee rivers. The zebra mussels spread so quickly in the United States that by 1994 they had been reported in or along the borders of Alabama, Arkansas, Illinois, Indiana, Iowa, Kentucky, Louisiana, Michigan, Minnesota, Mississippi, Missouri, New York, Ohio, Oklahoma, Pennsylvania, Tennessee, Vermont, West Virginia, and Wisconsin. As of 2002, they were found in Connecticut and Virginia as well—and they continue to spread.

The zebra mussel is a tiny but troublesome mollusk. It is typically no more than half an inch (1.27 cm) long (although it can sometimes grow to 2 inches [50 mm]) and comes in a handsome shell marked by alternating bands of light and dark. In addition to the tongue-like "feet" that mussels use to push themselves along the bottom, zebra mussels, alone among freshwater mussels, possess thread that allows the mussel to attach itself to hard surfaces such as rocks, steel, or other mussels. Unlike native mussel larvae, which disperse themselves by hitching rides on fishes, zebra mussel larvae are veligers, meaning that each possesses cilia, tiny hair-like fibers that enable them to suspend themselves in water. These features allow zebra mussel larvae to spread with remarkable swiftness in a current.

There seems to be no doubt that the zebra mussel will continue to spread throughout much of North America. They can survive almost anywhere in a range that covers about two-thirds of the United States and the southern part of

FIGURE 1 Zebra mussels clog pipes, causing billions of dollars in damage.

Invasion of the Zebra Mussels

VISIT

http://environment.jbpub.com
/mckinney/5e/
for more information

Canada. In terms of temperature, only Canada's cold northern lakes and the warmer waters of the American South will not sustain them. Their reproductive rates are amazingly high. A female zebra mussel can produce 40,000 eggs a year and the male a similar amount of sperm. Even if only a small percentage of these eggs are fertilized and advance to maturity, the rate of proliferation can be impressive. For example, on Hen Island Reef in Lake Erie, the density was 3,500 zebra mussels to the square meter (2,900 to the square yard) on the first reading; 5 months later, the count was 23,000 per square meter (19,230 per square yard). Densities as high as 700,000 per square meter (585,000 per square yard) have been reported at a Michigan power plant.

Zebra mussels appear to be driving out the native North American mussels, which could have disastrous effects on local ecosystems. The zebra mussel also threatens native fishes with its insatiable appetite for phytoplankton, the microscopic green plants at the very bottom of the aquatic food chain. The mussels' filtering of phytoplankton (as well as contaminants) from the water seems at first to be a positive thing. It removes toxins from the water and makes the water look clean and clear, but because many fishes feed on the zooplankton that eat phytoplankton, tinkering with phytoplankton is tinkering with one of nature's building blocks. If the mussels deplete the phytoplankton, the population of zooplankton could crash—with disastrous consequences for certain herbivorous fishes, for the predators of those fishes, for piscivorous (fish-eating) waterfowl such as ducks, and for commercial and recreational fisheries. In addition, zebra mussels also take in contaminants from the water, which build up in their tissues. When predators consume the mussels, the contaminants accumulate in the predator, a process called biomagnification. Accumulation of toxins can negatively affect the fitness of organisms in higher trophic levels. In this way, the zebra mussel, in what amounts to a nanosecond of evolutionary time, could alter the ecosystem of the entire Great Lakes and, ultimately, most freshwater ecosystems in the United States.

How can the spread of the zebra mussel be controlled and its range contracted? One possibility is **ozonation**, an environmentally benign oxidant that chews away at the soft parts of the organism. Ozonation could be helpful to power and water treatment plants that are trying to prevent zebra mussels from clogging their intakes, but it is very expensive. Even then, ozonation can alleviate only specific problems here and there; it cannot do anything to halt the overall proliferation of the zebra mussel. Various chemical, thermal, electrical, and other methods have been suggested as ways to control zebra mussel populations, but any such method may also have adverse effects on the native organisms and ecosystems. Manual and mechanical removal and destruction of the bivalves is tedious and expensive.

Biological methods have been suggested, such as using predators, parasites, or disease agents that will attack the zebra mussels. Of course, such methods could end up releasing another organism that could cause havoc in the local ecosystems. A bottom-feeding fish called the drum feeds on the mussel, but unfortunately, this benefits only the lowly drum itself. Because it is not exactly a tasty fish, it would not help establish a profitable fishery. At one site in Lake Erie, a diving duck has begun to prey on the zebra mussels. Unfortunately for the Europeans, it is only the overwintering waterfowl that have an effect on mussel control.

and dogs alone has driven dozens of island ground bird species to extinction or near extinction (**FIGURE 11-6**). The introduction of the brown tree snake to Guam has effectively eliminated 11 of 18 native bird species on the island. In some areas of Guam, more than 1,900 snakes per square kilometer (more than 5,000 per square mile) have been counted.

Although large continents experience relatively fewer species extinctions from introduced species, they certainly experience major impacts. The kudzu plant has driven out many local plant populations where it grows. Freshwater lakes and rivers are especially susceptible to destructive invasions, perhaps because they are confined environments, leaving no refuge for native organisms. Exotic aquatic plants such as hyacinth and hydrilla now cover vast areas of many of Florida's lakes, choking out other plants and blocking sunlight. No less than 50 of the 133 fish species sampled in a recent California study were found to be introduced. Many extinctions have occurred from such introductions. Native fishes, clams, and other freshwater groups are among those most severely affected by exotics, such as the zebra mussel, that are out-competing the native species. The American Fisheries Society estimates that approximately 79% of the nearly 300 species of freshwater clams in North America are either extinct or in decline.

The most dramatic devastation that an introduced fish caused is in East Africa's Lake Victoria, where a single species, the Nile perch, has exterminated more than 35 species of native fishes in just a few years. If the situation continues as expected, the perch will cause the extinction of several hundred more native fish species, setting a record for the greatest number of extinctions from a single introduction.

Overhunting

Overhunting is the unsustainable hunting, capturing, or collecting of organisms that causes species decline. It is very difficult for people to cause the extinction of pest species, such as roaches or mice, in this way because the pests are so abundant and reproduce so rapidly; however, overhunting has often caused the decline of organisms that were initially rare and/or that reproduce slowly. A review

FIGURE 11-6 The New Zealand kakapo once numbered in the hundreds of thousands, but this flightless bird was no match for dogs, cats, and other introduced species. The 86 or so remaining birds survive on two outer islands, which are carefully protected sanctuaries.

of the biosphere and its populations, communities, ecosystems, and biogeochemical cycles shows that most species are naturally rare, so many can be eliminated this way. This is especially true of most large animals and plants, which tend to have much lower abundances and slower reproduction rates than small organisms.

Large animals are also the species most often killed for sport or economic motives. Economic motives include killing for food, to protect domesticated animals from predators, or to sell parts of the animal. Elephant ivory, leopard skins, rhino horns, and tropical bird feathers are a few examples.

This preference for killing large animals, combined with their low abundance and reproductive rate, means that overhunting has caused drastic reductions in large species throughout the world. Such a process may have begun with prehistoric people contributing, along with natural climate change, to the extinctions of the woolly mammoth, saber-toothed cats, and other large animals. As killing technology has improved, so has the effectiveness of overhunting. A classic example is the North American bison, which declined from many millions to fewer than 1,100 in just

11.3 What Causes Biodiversity Loss?   327

FIGURE 11-7 Slaughter of Buffalo (bison) on the Kansas Pacific Railroad, from Plains of the Great West (1877). (*Source:* Courtesy of the John Hay Library, Brown University.)

a few decades (**FIGURE 11-7**). Today, virtually all large species are in decline, including herbivores such as elephants, rhinos, and pandas, and carnivores such as large cats and canids (wolves and coyotes). Ocean-dwellers are not spared; most whale species are in decline, as are commercially extracted fishes such as shark, bluefin tuna, swordfish, and cod.

To subsistence hunters, nearly all animals are potential food or a source of income. As hunting grounds shrink because of development, even more pressure is put on the remaining animals. The economic reality for indigenous people whose tribes have survived off the land for generations offers them little other choice, as far as they are concerned, than to continue to hunt, even if it means driving some animals to extinction locally.

Rare plants used for herbal medicine have also long been at risk for overharvesting to the point of extinction. For instance, in ancient Greece, the *silphion* plant was so valued as a contraceptive, in addition to other uses, that coins minted in the city of Cyrene were embossed with an image of the plant. However, it seems that the Greeks used it too much. Botanists have been unable to locate any living specimens. The gutta-percha tree, once endemic to the mountainous regions of China and sought after for its medicinal bark, was harvested into extinction in the wild. The tree still exists, but only as a cultivated species. In the United States and Europe, many wild plant harvesters are sensitive to overharvesting and have organizations that adopt codes of **ethics** to sustain threatened species.

Secondary Extinctions

Secondary extinctions occur when the extinction of one group causes the extinction of another. Often this involves the loss of a food species. For example, the familiar panda of China subsists largely on a diet of bamboo. Bamboo is being destroyed, so the panda may become extinct from that cause alone. Other examples are subtler, reflecting the complex and unpredictable interactions among organisms that make the effects of human disturbances so difficult to predict. For instance, the well-known extinction of the dodo bird has caused the Calvaria tree to become unable to reproduce. When the dodo ate the seeds of the tree, it digested the outer seed covering and excreted the seeds without their coverings, allowing them to germinate and grow.

Minimum Viable Populations

Extinction can occur even if the population is not reduced directly to zero. Even if many individuals survive the disturbances, the population may never recover if it becomes too small. The species will fall into an **extinction vortex**.

There are two basic causes for this extinction vortex. One is that small populations may have breeding problems. Too few females may be left, or if the population is too dispersed, individuals may not be able to locate each other to mate. Even if there are enough locatable mates, genetic inbreeding is a major problem in small populations. It leads to decreased diversity in the gene pool, which increases chances of congenital (hereditary) disorders and many other physical and health defects. For example, many males in the remaining population of less than 80 Florida panthers have testicular malformations.

Aside from breeding difficulties, the second cause of the extinction vortex is that small populations are much more easily extirpated by random environmental fluctuations, such as an abnormally harsh winter, that would not significantly affect larger groups. The chance of being wiped out increases exponentially with decreasing population size.

The smallest population size needed to stay above the extinction vortex is often called the **minimum viable population** (**MVP**). If a population falls below this size, it is said to be no longer viable, and long-term breeding problems and environmental fluctuations eventually will finish off the population. This is what we meant earlier by the living dead—when habitat fragmentation leaves too few individuals in each isolated fragment.

In the past, calculations of MVP have been oversimplified. For instance, some ecologists initially suggested that an MVP of perhaps 500 individuals would permit almost any species to survive for many years. Most ecologists now acknowledge that there is no such magic number for MVP that applies to all species. Organisms vary widely in their ability to rebound from low numbers. For example, organisms that breed and grow rapidly often will recover from lower population sizes than more slowly growing organisms and thus often have lower MVPs. In general, most population viability analyses, which seek to estimate how long a certain population will persist, indicate that at least a few thousand individuals must remain alive to ensure long-term survival of most species. In this case, long-term survival means more than a few decades.

Community and Ecosystem Degradation

A growing criticism is that conservationists have focused too much on individual species. Although much money has been devoted to saving a few highly publicized species, entire ecosystems are rapidly being degraded or even destroyed. As a result, there is increasing interest in saving endangered ecosystems as well as endangered species.

A review of the biosphere and its populations, communities, ecosystems, and biogeochemical cycles shows that a community consists of all populations of species that inhabit an area. An ecosystem is a community plus its physical environment—all living and nonliving components of the area. A community or ecosystem becomes extinct when the populations comprising it die out. For example, a forest fire or a development may completely destroy large areas of a certain forest community so that it becomes extinct.

Community Degradation

Communities and ecosystems are more likely to be degraded by human activity than destroyed. By "degraded" we mean that some species within the community or ecosystem experience major decreases in abundance. A classic type of disturbance is pollution of a community's water, soil, or air. A basic effect of disturbance is **ecosystem simplification**, meaning that the number of species in the ecosystem declines. In addition, some of the remaining species become superabundant. These are the species that thrive in the polluted environment, such as certain sewage-eating bacteria or "trash" fishes such as garfishes that have adapted to low-oxygen waters.

In addition to simplifying structure, a disturbance can disrupt the functioning of communities (and ecosystems) by altering the flow of matter and energy through them. The flow of excess nutrients into the ecosystem can cause eutrophication and other disruptions often involving rapid population growth and decay. Conversely, slash and burn agriculture and other disturbances often lead to "leaky" ecosystems, with excess nutrient flow out of the system (a review of the biosphere and its populations, communities, ecosystems, and biogeochemical cycles provides more information).

Indicators of Ecosystem Health

Some species in a community or ecosystem are more susceptible to disturbances than others. These are often called **indicator species** because they indicate the health of an ecosystem. A decline in the abundance of indicator species is evidence that the entire ecosystem may soon decline. For example, some trees, such as the Fraser firs in the Smoky Mountain National Park, are more sensitive to air pollution than others and will be affected by it first. Similarly, trout and freshwater "jellyfish" are among the organisms most sensitive to water pollution.

Are Diverse Communities More or Less Easily Disturbed?

Until the early 1970s, many ecologists believed that the more species a community had, the

FIGURE 11-8 A sea otter lunches on an urchin. Courtesey of NOAA.

species for food, reproduction, or some other basic need. If the keystone species are removed, many parts of the community can be affected drastically. For example, in the giant kelp forests along the Pacific coast of North America, sea otters are the keystone species, keeping the sea urchin population in check (**FIGURE 11-8**). If predation or overhunting decreases the otter population, the number of urchins increases, with harmful consequences. They feed voraciously on kelp beds, not only damaging the kelp, but also depriving many kelp-dwelling fish species of habitat. The removal of a keystone predator seems to be especially damaging in some highly diverse tropical communities. Because they have been undisturbed for so long, they have built up highly intricate interactions that are easily disturbed.

11.4 Stopping Extinctions

Why worry about extinctions? Some people argue that people have little need of wildlife. They view elephants, exotic tropical insects or plants, and the like as curiosities that have no immediate value, so they are not very concerned about the prospect of the loss (**FIGURE 11-9**). Others take the opposite view and want to preserve all of nature for its own sake. They see people as the intruders and insist that all extinctions must be stopped as soon as possible. Between these two extremes are many practical realities.

Reasons to Stop Extinctions: The Many Values of Biodiversity

People who worry least about extinction tend to place only direct values on biodiversity. A review of people and natural resources shows that direct values are based on the immediate economic gain made when a resource is harvested destructively. Examples include whaling, logging, and illegal trade in endangered species.

Illegal trade is an especially good example because the entire value of an organism is reduced to the price it brings, either dead or alive, in the marketplace. Recent data from the World Wildlife Fund and Wildlife Conservation Society estimates that illegal trade in wild

more difficult it was to disturb. Although there is still much ongoing research and debate, mathematical models and field data show that the relationship between diversity (number of species) and stability (ease of disturbance) in a community is much more complicated than ecologists formerly believed.

Most ecologists now agree that stability is lowest in communities with very low diversity. For example, in monoculture, only one plant, such as corn, is grown for many acres. Such simple communities are easily disturbed; an invading pest can wipe out much of the community if the corn plants are susceptible. In communities that have more plant species and thus support more kinds of animals, it is very unlikely that all the plant species will be susceptible to a single pest. Thus, increasing diversity leads to redundancy: if one food species is lost, other species will still exist to support at least part of the community.

However, after a certain point, increasing diversity may make a community easier to disturb. Increasing diversity means that more species are interacting, creating more complex food webs (a review of the biosphere and its populations, communities, ecosystems, and biogeochemical cycles provides more information). Many mathematical models show that such complex webs can become very sensitive because minor changes can cascade through the community. For example, many species can become dependent on certain **keystone**

animals globally produces $2 to $6 billion per year. For example, rhino horns are used as aphrodisiacs, dagger handles, and other adornments. Many other species, such as the imperial Amazon macaw and mountain gorilla, are popular pets or attractions.

However, such direct short-term valuation of an organism omits many other potential values, called indirect values (a review of people and natural resources provides more information). When these are ignored, the true value of a species is underestimated.

Indirect Values

One could argue that an ethical reason to save species is that people have no right to harm or destroy other species. According to this view, animals have intrinsic rights of their own, outside extrinsic human needs. This includes the right to live and to live relatively free from pain.

There are esthetic (aesthetic) and emotional reasons to save species. Biological diversity can make life more enjoyable and enriching. Given the mental and physical rejuvenation many of us experience after a long hike or other outdoor recreation, this argument alone would seem to have much validity. Such recreation might be less stimulating if we had only the same few species to experience all the time.

Indirect economic values include the many nondestructive ways that people can use species. Ecotourism and sustainable harvesting of exotic foods, medicines, and many other materials are two examples of practical social incentives for providing goods and services while preserving species. When measured over many years, such sustainable economic benefits of biodiversity generally are greater than the short-term economic gain from selling an organism for its skin, horns, or other attributes.

Environmental services refer to the value of biodiversity in providing us with many of life's essentials. Ecosystems are environmental support systems that supply us with things that we now take for granted: oxygen to breathe (from plants), drinkable water (purified by microbial activity), and many other natural chemical cycles that occur via ecosystem functions (**FIGURE 11-10**). If we remove too many species from an ecosystem, ecosystem function will be impaired. If we lose too many

FIGURE 11-9 What is the value of this endangered plant? This is the Eureka Valley evening primrose, known from fewer than five occurrences in Inyo County, California.

species in too many ecosystems, environmental services of the entire biosphere could be impaired.

The last indirect value is the evolutionary value of species. This is the value of today's species to future generations. Rather than being constant entities, species often change to adapt to new environmental conditions;

FIGURE 11-10 Wetlands, the "kidneys of the water cycle," illustrate many ecosystem functions by filtering groundwater, producing oxygen, and decomposing organic wastes. Seen here is part of a restored wetland in the prairie pothole region of northern Iowa.

however, the raw material of evolution is variation. As biodiversity disappears, the biosphere will become less able to adapt to change because fewer species will be alive. An impoverished biosphere today means an impoverished biosphere for future generations.

A practical example of the loss of evolutionary value is food production by people. Today only about 30 plant species provide 95% of the world's nutrition. Just four of these, wheat, corn, rice, and potatoes, provide most of the world's food, and all four have been subjected to centuries of inbreeding. Do we really want our diet to be this bland? Botanists estimate that at least 75,000 edible plant species exist, many superior in flavor and nutrition to those commonly eaten in the United States.

The ability of species to adapt to change also has other important implications for our future food supply. Low diversity makes organisms more susceptible to extinction. Inbreeding of most crop species makes them notoriously susceptible to diseases and insects. By interbreeding wild species with closely related cultivated forms, resistance to disease and general hardiness can often be improved, and genetic engineering now allows us to combine useful traits from species that cannot interbreed naturally (a review of food and soil resources provides additional information).

It is thus economically advantageous to maximize diversity by growing a variety of crops and by having as many different crop species at our disposal as possible. Nevertheless, if current rates of extinction continue, some 25,000 plant species may die out in the next few years, before we have a chance to study them. Some biologists have proposed that gene banks be set up as a reservoir to save seeds, spores, sperm, and other genetic materials for species that cannot be saved in the wild. Currently, seed banks contain only a small fraction of most plant varieties, and many wild species have yet to be collected and described.

We have focused on plants, but similar arguments can be made for domesticated animal species. Breeding cattle with buffalo ("beefalo") to improve the stamina of cattle breeds and increase their resistance to predation is one of many examples for which wild genes have been useful. Similarly, many wild animal species are potentially tasty and healthy food sources, even without being interbred with existing food species.

Which Species to Save?

So many species are at risk that they cannot all be saved from extinction. In 2002, more than 50% of all crayfishes and freshwater clams (mussels) in the United States were at risk (**FIGURE 11-11**). At least 30% of flowering plant, amphibian, and fish species and more than 15% of mammal species were at risk. Some species may survive with adequate attention, but the cost of saving all of them far exceeds the limited monetary and human resources available for saving species. Thus, painful decisions will have to be made concerning which species we want to save the most.

Conservation biologists often use the concept of **species triage** to refer to the difficult process of selecting which species to try to save from extinction. Triage is a medical term used at accident scenes and in emergency rooms: some patients are beyond care (fatally wounded), and some are walking wounded; thus, most attention should be focused on those in the middle who are likely to survive if they get immediate care. So it is with species. Weedy species (species that are opportunistic and reproduce rapidly), such as squirrels and deer, are in no danger of extinction, whereas

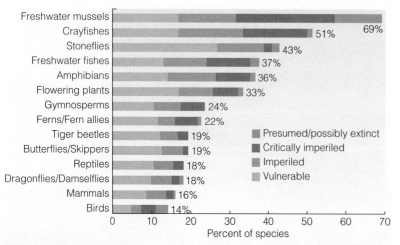

FIGURE 11-11 Freshwater organisms are among the most imperiled species in North America. These data include only species at risk, not those already extinct. (*Source:* Adapted from B. A. Stein, L. A. Kutner, and J. S. Adams. Our Precious Heritage: The Status of Biodiversity in the U.S. New York: Oxford University Press, 2000, p. 102.)

other species are among the "living dead," with so few individuals left that they will likely go extinct. Thus, many experts argue that conservation efforts should be directed at those species that are at risk and can still be saved. Unfortunately, the number of species at risk far outweighs the social resources currently allotted to preserving them.

Which Species Are at Risk?

Species are not equally likely to become extinct. For many reasons, some species are more able than others to survive environmental change. A well-known example of resilience is the cockroach family, which has existed for more than 300 million years and probably will be alive for millions more. TABLE 11-4 identifies nine characteristics that make some species more susceptible to extinction than others. Island species are sensitive to the introduction of new species because they have been isolated for a long time and are very specifically adapted. Indeed, most of the bird and mammal extinctions until now have been of island-dwelling species. Species with limited habitats become extinct easily because human activity can quickly eliminate the small space available. For instance, one species of tropical insect will often be adapted to only one part on one kind of local plant. When the plant is eradicated, so is that insect species, along with others that are adapted to other parts of the plant.

Species with large territories die off quickly because they need lots of area to support them. This has been shown repeatedly when apparently substantial game reserves are set up and wide-ranging species (such as large predators) still die out. When a single individual needs many square miles to forage, it takes a sizable reserve to support enough individuals to maintain the species. Many species are naturally rare, even without human disturbance; such species are prone to extinction because they have few individuals to begin with. Low reproductive rates make it difficult for a species to rebound from habitat disturbance, hunting, or other causes of population declines. Economic and sporting values cause species to be sought by hunters. Predators are high on the food pyramid, so they are relatively less abundant than organisms that eat lower in the food chain. Sensitivity to pollution is another trait leading to extinction. Finally, the interaction between human invention and animal behavior can contribute to population decline and ultimately extinction. Birds can fly into windshields. Boat injuries such as hull and propeller strikes are the leading cause of mortality for the endangered Florida manatee, causing the deaths of 40 to 50 individuals each year (FIGURE 11-12). Recent data indicate that although manatees have excellent perception of significant biological sounds in their environments, their sensitivity has not yet adapted to low frequency sounds, such as those that boat motors

Characteristics	Reason Characteristics Tend to Cause Extinction	Examples
1. Island species	Unable to compete with introduced species	More than half of the native plant species in Hawaii
2. Species with limited habitats or breeding areas	Some species found in only a few ecosystems	Woodland caribou, Everglade crocodile, red-cockaded woodpecker
3. Species that require large territories to survive	Widespread habitat destruction	California condor, blue whale, Bengal tiger, Florida panther
4. Species with low reproductive rates	Many species evolved low reproductive rates because predation was low	Blue whale, California condor, polar bear, rhinoceros, Florida manatee
5. Rare species	Few individuals to replenish population	Tropical insects, rhinoceros
6. Species that are economically valuable or hunted for sport	Hunting pressures by people	Snow leopard, blue whale, elephant, rhinoceros, tiger
7. Predators	Often killed to reduce predation of domestic stock	Grizzly bear, timber wolf, Bengal tiger
8. Species that are susceptible to pollution	Some species are more susceptible than others to industrial pollution	Bald eagle (susceptible to certain pesticides), pelicans
9. Species with inadaptive behaviors	Behaviors promote death in human environments	Manatees swimming close to motorboats

TABLE 11-4 Characteristics of Extinction Susceptibility

FIGURE 11-12 The Florida manatee is a threatened species that suffers from ocean pollution, collisions with motor boats, and many types of human activities. Manatees swim very close to the surface, and it is likely this manatee's scars are from a motorboat's propeller.

produce. Manatee deaths also occur from entanglement in and ingestion of fishing gear.

These nine traits can co-occur: many species have more than one of the traits, increasing their likelihood of extinction. For example, large animals tend to be more rare, require larger territories, and have lower reproductive rates than small animals. If the large animal is also a predator and is hunted for sport, then it has at least five of the traits that promote extinction. Similarly, the Florida manatee not only has destructive interactions with human-made equipment, but also has low reproductive rates.

Are All Species Equally Important?

With so many species at risk, triage decisions cannot be made on the basis of risk alone. Thus, conservation biologists often ask whether one species is more important than another. Ethically, perhaps one could argue that all species are equally valuable; an insect may have as much right to live as a panther.

However, in other ways, particularly in ecological and evolutionary importance, all species are not equal. Ecological importance reflects the role a species plays in its ecological community. Keystone species play large roles because they affect so many other species. For example, large predators often control the pop-

ulation dynamics of many herbivores. When the predators, such as wolves, are removed, the herbivore population may increase rapidly, overgrazing plants and causing massive ecological disruption. Similarly, certain plants can be crucial food sources for many animal species. The extinction of keystone species often has cascading effects on many species, even causing secondary extinctions. Many people argue that saving keystone species should be a priority.

Evolutionary importance varies among species because all species do not have the same potential for contributing to future biodiversity. **Unique species** are not closely related to any other living species (**FIGURE 11-13**). Unique species represent unusual gene pools, with many genes and traits not found in other species. The loss of a unique species often represents a much greater loss of genetic and evolutionary potential than the loss of a species with many living close relatives. For example, pandas are a unique species of bear. Unlike most bears, which eat many kinds of food, the panda is specialized for eating bamboo. Recent genetic studies indicate that the panda diverged from the rest of the bear family relatively early in its evolutionary history. The panda is a result of evolutionary events that are unlikely to recur, and the Earth may never again see anything similar to it.

Rather than concentrating on species of ecological and evolutionary importance, to the regret of some biologists, conservation efforts

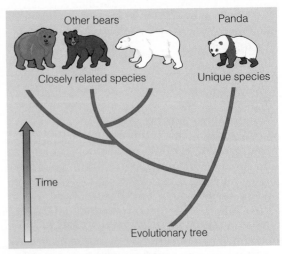

FIGURE 11-13 Unique species have no living closely related species.

have often focused on high-profile **charismatic species** that attract public support. The Florida panther and the bald eagle are examples of charismatic species. Focusing on such species is not necessarily bad because they are sometimes ecologically and evolutionarily important. In addition, because they are usually large animals and often predators, substantial areas must be set aside for them. These sizable preserves often contain the habitat of many other species in danger of extinction. For this reason, large charismatic species such as the Florida panther are also called **umbrella species**. Many other species are also protected under the umbrella set aside to preserve the panther.

How to Save Species

Because extinction is caused by change in a species' environment, such as habitat disruption or introduced species, the cheapest and most efficient way to save any species is to preserve its natural environment. This requires establishing protected areas such as wilderness preserves; however, protected areas are relatively small in size and number, and many threatened species also occur in areas that are not protected. In the United States, the Nature Conservancy estimates that more than half of all federally threatened species occur on private lands.

Thus, conservation also requires approaches that protect species living in areas that people often intensely use. This entails legal and economic incentives that reduce the human impact on threatened species. In the rest of this chapter, we focus on the laws that attempt to achieve this goal.

Laws Protecting Endangered Species

Early laws to protect endangered species focused on the trade and sale of their skins and other products. In recent decades, the statute that has been most widely used to protect species and their habitats is the **Endangered Species Act** of 1973. This Act directs government agencies, especially the U.S. Fish and Wildlife Service, to maintain a list of species and populations (subspecies, varieties, or distinctive populations) that are endangered or threatened. Endangered species are in immediate danger of extinction. Threatened species

are likely to be endangered soon. The Service is also directed to produce a recovery plan for each listed species in the United States and protect its designated "critical habitat" needed for survival.

The Endangered Species Act is very controversial, and Congress often debates whether the act should be reauthorized in its present form. Many people want to weaken or even eliminate the act, whereas many others want to strengthen it. Opponents claim that the act has cost billions of dollars and violates the Fifth Amendment's ban on taking property without compensation by preventing property owners from developing their land when it contains habitat of an endangered species. Landowners, logging companies, and other organizations have often brought suit against the government, claiming that the act violates the right to own private property.

Opponents also point out that the act has been relatively ineffective. The number of listed species has grown much more rapidly than the number of delisted species, and certainly the list of endangered and threatened species and populations, which focuses primarily on the United States, is but a small sample of the actual number of species worldwide that is endangered and threatened. As of late May 2006, there were 1,880 species and populations worldwide (1,311 of these occur in the United States) listed by the U.S. Fish and Wildlife Service as threatened and endangered (more than one-third of these are plants; see **TABLE 11-5**). In contrast, only 41 species and populations (33 of these in the United States) had been delisted (**FIGURE 11-14**). **Delisted species** include both species that are no longer endangered or threatened, such as the brown pelican and the American alligator, and species that have gone extinct, such as the blue pike and the Santa Barbara song sparrow. So even some of the delistings indicate a failure, not a success. In fact, only 15 (8 in the United States) of the 41 species were delisted because they recovered. This means that for every 100 species added during the last 30 years, fewer than 1 species has recovered. Overall, only about 10% of listed species are classified as improving. The rest are only stabilized or are declining, but still many may have been saved from final

TABLE 11-5	Number of Species and Populations in Different Taxonomic Groups That Are Endangered and Threatened Worldwide, May 2006		
Taxonomic Group	**Endangered**	**Threatened**	**Total**
Mammals	324	33	357
Birds	251	21	272
Reptiles	79	39	118
Amphibians	21	11	32
Fishes	87	62	149
Clams	64	8	72
Snails	25	12	37
Insects	51	10	61
Arachnids	12	0	12
Crustaceans	19	3	22
Flowering plants	572	143	715
Conifers and cycads	2	3	5
Ferns and allies	24	2	26
Lichens	2	0	2
Total	1,533	347	1,880

Source: Data are from U.S. Fish and Wildlife Service, 2006.

extinction thus far, although more will need to be done to ensure their long-term survival. In addition, hundreds, perhaps thousands, of candidate species that are already rare are waiting to be listed, but there is inadequate funding to pay for the listing process.

Supporters of the Endangered Species Act argue that it has not worked well because it is too weak and that far too little money has been spent on enforcement, less than $100 million annually in the federal budget. State and federal agencies combined spend about $200 million per year to protect endangered species; this is less than 4% of the money spent annually on U.S. lawn care and less than $1 per person. In addition, most of this money is spent on just a few species. In any given year, just a few popular and charismatic listed species receive more than half the money. For example, many millions are spent each year on the bald eagle and Florida manatee. Meanwhile, many less popular species receive a few hundred dollars or less. Another sign of imbalance is that as of late May 2006, only 61 insect species worldwide (57 in the United States) were listed as endangered or threatened compared with 357 mammals worldwide (81 in the United

States), despite that there are thousands more species of insects than mammals. Supporters of the act also note that the listing process takes too long because by the time species are finally listed, many are too close to extinction to be saved. In most cases, it takes many years for a species to become listed and protected.

A potential solution may be to protect entire ecosystems before species are on the verge of extinction, instead of focusing on "one species at a time" and waiting until a species is in imminent danger. Such **habitat conservation plans** have been relatively successful thus far (see **CASE STUDY 11-2**).

Other laws also protect species. In the United States, each individual state has its own "endangered species" act that protects species that are threatened with extinction within the state boundaries. In most cases, these species are not federally threatened but are disappearing from the state. State wildlife agencies are usually responsible for promoting the recovering of these species.

International law states that it is illegal to trade, transport, and sell products made from endangered species. In 1975, 81 countries signed the **Convention on International Trade in Endangered Species** (CITES; as of May 2006, 169 countries had signed), which outlaws trade in endangered species products. Although CITES has helped reduce illegal trade in some areas, illegal wildlife trade continues to proliferate (**FIGURE 11-15**). Unfortunately, thousands of animals die each year, especially tropical birds, in the process of being smuggled into the United States and other nations that are the main markets for illegal wildlife.

Species Recovery: Breeding and Reintroduction

In addition to protecting species in the wild, most wildlife laws also have the goal of **species recovery**, which refers to increasing the population size of a threatened species. One important method for doing this is to breed species in captivity. This allows for their protection from people, better medical care for individual organisms, and control over which individuals breed (to reduce genetic defects in offspring from inbreeding).

However, breeding in captivity generally is used only as a last resort to save a species.

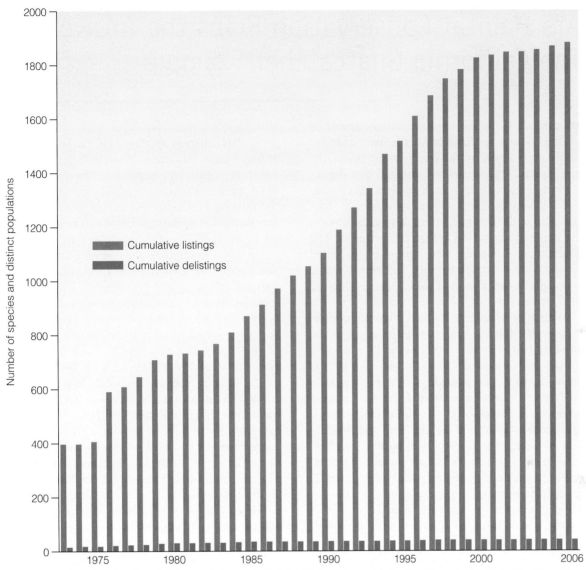

FIGURE 11-14 The number of species and distinct populations worldwide being listed as endangered or threatened is growing much faster than the number being delisted. (*Source:* Data are from U.S. Fish and Wildlife Service, 2006.)

Captive breeding is almost always much more expensive and less effective than preserving habitat and saving species in the wild. Animals tend to be unhappy outside their natural state, especially in older zoos where conditions are poor (**FIGURE 11-16**). Even under the best of conditions, many animals, such as pandas, do not readily breed in captivity. In addition, there are far too few zoos in the world to sustain sufficient populations of all the world's threatened species, even if only large charismatic mammal species are considered. Yet because habitat continues to disappear at alarming rates, captive breeding is a necessity

to save many species. As conditions become very desperate, as with the now-extinct dusky seaside sparrow, genetic material (sperm and eggs) is being frozen for future use as a gene bank.

The ultimate goal of most captive breeding is for the animals and plants to proliferate and attain sufficient population size for some to be reintroduced into the wild. A **reintroduction** is the release of plants and animals back into habitat that they formerly occupied. For example, the American bison was extirpated throughout most of the West but is now being reintroduced in many areas. The controversial

Are Habitat Conservation Plans the Answer? The California Gnatcatcher Example

In 1982, Congress passed an amendment to the Endangered Species Act that was intended to meet some of the main criticisms of the original act. Opponents had charged that the act (1) interfered with economic growth, (2) emphasized saving species rather than whole ecosystems, and (3) waited until a species was on the verge of extinction before protecting it. The amendment created a new approach designed to make the act more flexible, reduce economic costs, and protect many species before they are near extinction.

The new approach is called a habitat conservation plan (HCP). Under an HCP, some of the habitat of an endangered species can be destroyed (called an "incidental take") as long as a plan is drawn up to reduce future losses. An HCP usually evolves as a compromise from discussions among landowners, developers, environmental groups, local governments, and the U.S. Fish and Wildlife Service, which eventually must approve the HCP. Ideally, an HCP will protect all current or potentially endangered species in an area while simultaneously permitting human use of nearby lands as deemed necessary by social consent.

By 1994, seven HCPs had been approved, and another 60 were under discussion from California to Key Largo, Florida. Each HCP is unique,

and some are more successful than others. California has the most approved HCPs. The first was at San Bruno Mountain, south of San Francisco, which emerged as a compromise between housing developers and environmentalists wanting to save the mission blue butterfly. In fact, the conflict at San Bruno led to the legislation creating the HCP concept.

A good example of the complexities an HCP can encounter involved the coastal sage scrub habitat of southern California. This habitat includes some of the nation's most expensive real estate in prime locations around Los Angeles, San Diego, and nearby areas (FIGURE 1), but it is also home to a rare songbird called the California gnatcatcher (FIGURE 2), with 70% to 90% of the habitat already destroyed. In 1993, after 3 years of discussion among developers, environmentalists, and government officials, then Interior Secretary Bruce Babbitt announced that the gnatcatcher would be listed as "threatened," instead of "endangered" as the U.S. Fish and Wildlife Service had proposed. The less-urgent "threatened" status allowed officials to work out the details of an HCP that permitted development on some of the remaining scrub habitat, but the HCP also called for establishing as many as 12 reserves that will benefit as many as 40 other coastal sage scrub species that are also in jeopardy from this disappearing habitat. In 2000, as part of the Natural Communities Conservation Program, the U.S. Fish and Wildlife Service designated 13 critical habitat units for the California gnatcatcher, covering an area of more than 513,000 acres. The species is still listed as threatened.

As of 2006, 446 HCPs have been approved. The most recent as of this writing was a plan to conserve the Florida scrub jay. Community development has fragmented and decreased the habitat of the scrub jay so that its numbers have declined by at least 50% in the past 100 years. Most recent Nature Conservancy data combined with other monitoring sites indicate that, between 2,000 and 3,000 individuals survive. Because the birds live on private and public land (particularly in several national forests), efforts are focusing on preserving as much of the pri-

FIGURE 1 Encroachment of gnatcatcher habitat.

FIGURE 2 The California gnatcatcher has generated national controversy.

vately owned scrub habitat as possible, through recommendations for, and agreements with, developers and landowners. Preservation of the scrub lands is the key to the continued survival of this threatened bird.

Questions

1. Many environmentalists strongly dislike the HCP concept because they believe it is "giving away" species and habitat to development. Can you think of a better way to resolve habitat versus development conflicts? Explain.

2. Because "extinction is forever," why would anyone ever approve of the HCP concept? How could development ever be justified over the irreversible loss of species? Explain.

3. Would you approve of an HCP that permitted one species to go extinct, but allowed five other species to survive? If it allowed 10 others to survive? Twenty others? Explain your reasoning.

VISIT

http://environment.jbpub.com/mckinney/5e/
for more information

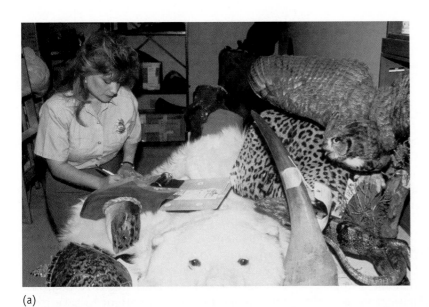

(a)

(b)

FIGURE 11-15 Trade in illegal wildlife continues to grow globally. (a) Seized wildlife parts. (b) Seized illegal shipment of Eclectus parrots and cockatoos.

FIGURE 11-16 Many animals in zoos live in poor conditions that bear no resemblance to their natural habitat. The result is often a lifetime of boredom.

reintroduction of wolves into some parks and other areas in the United States is another example. Some species can be introduced into habitats they did not formerly occupy. For instance, many African savanna species, such as lions, antelope, and rhinos, have been successfully introduced onto game preserves in Florida and Texas.

Captive breeding can also be used to raise plants and animals for sale. Llamas bred in captivity are one of many exotic species popular in U.S. breeding farms. Other examples include tropical fishes and birds. Some people oppose breeding exotic species for pets and food for a variety of reasons, including **animal rights**, and because it creates markets for captive animals, but supporters point out that such breeding helps reduce the market for the illegal wildlife trade by satisfying the demand for exotic animals and plants without taking them from their native habitats.

Sustainable Uses of Biodiversity

Conservation biologists agree that governmental decrees alone will not solve the extinction crisis. Establishing preserves and prohibiting trade in endangered species will not work as long as people, especially local inhabitants, are poor and must rely on destructive uses of biodiversity to stay alive.

Instead, sustainable uses of biodiversity must be developed and encouraged. Sustainable uses of natural ecosystems, such as tourism and sustainable harvesting of the rain forest, provide economic incentives to save species while also respecting the right of all people to support their families and have a decent quality of life (a review of land resources and management provides additional information).

■ study guide
SUMMARY

- Biodiversity is often measured as the number of species.
- It is estimated 5 to 100 million species may now inhabit our planet, but only 1.8 to 2 million have been described.
- Five major mass extinctions occurred in the past.
- Extinction rates are best known for mammals and birds.
- Since 1600, approximately 2% of mammal species have gone extinct; 25% are currently threatened.
- Four causes of extinction are habitat disruption, introduced species, overhunting, and secondary extinctions.

- Biological communities can be degraded, as in ecosystem simplification.
- The health of an ecosystem can be inferred from indicator species.
- Biodiversity has many potential values: ethical, esthetic, emotional, economic, environmental services, and evolutionary values.
- The best way to save species is to preserve natural habitats.
- Large, rare, specialized, and predatory species are among those most at risk of extinction.
- Unique species, with few living close relatives, are among the most important to save from extinction.

- Species are listed much faster under the Endangered Species Act than they are delisted.
- A way to improve the Endangered Species Act may be to use more flexible habitat conservation plans.

- Species recovery in captivity is an expensive last resort, but it provides protection and ways to increase reproduction.
- The ultimate goal of species recovery is to reintroduce plants and animals back into their natural habitat.

KEY TERMS

animal rights
background extinctions
biodiversity
biological environment
biological impoverishment
charismatic species
conservation biology
Convention on International Trade in
    Endangered Species
delisted species
ecological extinction
ecosystem simplification
edge effects
Endangered Species Act
ethics
exotic species
extinction

extinction vortex
extirpation
habitat conservation plans
habitat fragmentation
indicator species
keystone species
mass extinctions
minimum viable population (MVP)
ozonation
physical environment
reintroduction
species recovery
species richness
species triage
umbrella species
unique species

STUDY QUESTIONS

1. How many major mass extinctions have occurred before now? What was the average extinction rate before people? How much higher is the rate now?
2. Name and describe two key reasons habitat fragmentation is one of the most destructive ways of disrupting habitat.
3. Name the four main causes of extinctions. Which are biological? Do most extinctions involve just one cause?
4. What is an extinction vortex? What are two basic causes of an extinction vortex? What is the MVP?
5. Where are exotic species an especially important cause of extinction? Why? Give examples.
6. Are diverse communities more easily disturbed? Explain.

7. Are all species equally important? Give two major examples. Where do unique species fit in?
8. Why is biodiversity important? Discuss some of its many values and indicate the ones you favor the most.
9. Are marine species being threatened? By what? Discuss possible solutions.
10. Discuss the pros and cons of the U.S. Endangered Species Act. Is it a failure? A success? How should it be improved?
11. What are the traits that make a species vulnerable to extinction?
12. Why are extinction rates almost certainly underestimated?
13. What is the ultimate goal of captive breeding? Why is it best used only as a last resort?

WHAT'S THE EVIDENCE?

1. The authors suggest that the Earth may currently be undergoing a major period of extinction. What evidence do they present to support this contention? If you agree that the Earth is currently experiencing a major period of extinction, do you think people are causing it, either partly or entirely? Cite evidence to support your opinion.
2. The authors note that the Endangered Species Act is very controversial. What evidence is given to support this statement?

CALCULATIONS

1. If you estimate that 50 species per day are going extinct, how many species will be extinct in a year? In 100 years? What percentage of all species will be extinct in 100 years, if there are 10 million species on Earth?

2. In 2006, there were 1,311 species and populations listed as threatened or endangered in the United States and only 8 species that had been delisted because of recovery. What is the ratio of listed to delisted species? If this ratio continues, how many delisted species will occur if 2,000 species are listed in the future?

ILLUSTRATION AND TABLE REVIEW

1. In Table 11-1, which group of organisms has the greatest number of threatened species? Which has the lowest? Which group has the highest proportion? The lowest proportion?

2. In Figure 11-14, which is greater, the number of listed or delisted species? Is the gap between listed and delisted species increasing with time?

http://environment.jbpub.com/mckinney/5e/

http://environment.jbpub.com/mckinney/5e/

Connect to this text's website at http://environment.jbpub.com/mckinney/5e/. This site features eLearning, an online review area that provides quizzes, chapter outlines, and other tools to help you study for your class. You can also follow useful links for more in-depth information.

These satellite images document a 39% increase in developed land in the Washington, DC, metropolitan area (which surrounds the Chesapeake Bay) between 1986 (top) and 2000 (bottom). These Landsat images are created by coding data with an algorithm that illuminates changes in low-density residential land use, exemplifying sprawl. Declining water quality in the Chesapeake Bay estuary system is directly related to increased urbanization. Scientific modeling using images like this can help citizens, scientists, and policy makers create a more sustainable development plan. Considering that the Chesapeake Bay is home to more than 3,900 distinct species of organisms and approximately 17 million *Homo sapiens*, preserving, protecting, and restoring the vitality of the Chesapeake Bay should be a top priority. However, as with many areas, ecosystems and ecosystem services are often undervalued, and land use is mainly determined by immediate economic, social, and political factors. The fate of the largest estuary in the United States, and one of the most productive bodies of water in the world, rests in the actions of citizens who must work together to create a more sustainable development plan if the biodiversity of life along the Chesapeake is to remain viable throughout future generations. For more information about Chesapeake Bay restoration projects, see: http://www.chesapeakebay.net.

Land Resources and Management

12

Land resources in the United States, and most other nations, typically are managed under two general categories: public and private lands. Public lands include national parks, national forests and rangelands, and other lands that are not available for citizens or businesses to purchase. State, county, and city governments have similar public land holdings, including state parks and state forests. In the United States, most public lands are managed for many uses but especially for resource extraction and recreation. Recently, there has also been growing public pressure for more wilderness preservation on public lands.

Private lands, owned by individuals and businesses, are both a major challenge and a major opportunity for solving current environmental problems. On one hand, private lands are where most intensive land modifications occur, such as farming and building shopping malls. Developers buy land and convert it into commercial or residential real estate that might produce needed services and homes for a growing population but at the same time might produce large amounts of pollution and might provide very little habitat for native species of plants and animals. On the other hand, local land conservation groups can purchase or manage private lands and thereby make them available for wilderness recreation and critical habitat for species that would otherwise be excluded from highly developed areas. Aside from conservation groups, individual landowners anywhere can make enormous contributions by being

Chapter Objectives

After reading this chapter, you should be able to explain or describe the following:

- Why land-use planning is important
- How public lands are managed
- Ways to improve nature preservation on public lands
- How private lands are managed
- Why urban and suburban sprawl are harmful to the environment
- What causes sprawl
- Ways to curb sprawl
- Legal and economic incentives to curb sprawl

343

ecologically concerned stewards of their land. For example, farmers can leave riverbank vegetation intact to reduce run-off and flooding, and landscapers and homeowners can leave some native vegetation intact, and so on. Such "bottom-up" or grassroots stewardship is the key to long-term solutions because it allows landowners to have a good quality of life while promoting a healthy sustainable environment.

12.1 Managing Public Lands

The largest landowner in the United States is the federal government (**FIGURE 12-1**). The National Park System includes more than 50 major parks and many smaller recreation areas that total more than 84 million acres. The National Forest Service oversees 155 national forests and 20 grasslands totaling more than 193 million acres. The National Wildlife Refuge System consists of 550 refuges and 38 wetland areas, for a total of more than 150 million acres. The Bureau of Land Management oversees vast areas of the western United States, about 886 distinct areas totaling more than 27 million acres,

much of which is used for grazing and mining. Federal land ownership totals 650 million acres, which is about 30% of the total land area of the United States. The western states are especially rich in federal land. The federal government owns more than 40% of California, Nevada, Utah, Idaho, Wyoming, Arizona, New Mexico, and Alaska. This federal land includes federal parks, national monuments, Native American reservations, and military bases. It does not include the large amount of public land owned by state and local governments.

The management of public lands has been a matter of controversy since the beginning of their establishment more than a century ago. On one side have been **preservationists**, who argue that public natural resources should be preserved as much as possible as pristine wilderness. On the other side have been **conservationists**, who argue that public resources should not be allowed to go unused; the lands should be open to mining, logging, grazing, and other human uses (as long as used judiciously). This debate peaked in the early 1900s, when those seeking access to

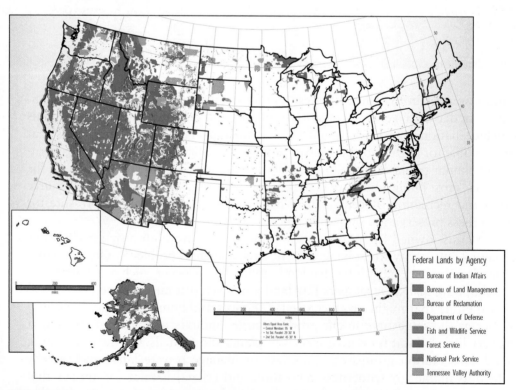

Federal Lands by Agency
- Bureau of Indian Affairs
- Bureau of Land Management
- Bureau of Reclamation
- Department of Defense
- Fish and Wildlife Service
- Forest Service
- National Park Service
- Tennessee Valley Authority

FIGURE 12-1 Some major systems of public lands that contain some natural habitat. (*Source:* Data from U.S. Geological Survey.)

the resources essentially won it. Since then, the conservationist philosophy and the multiple-use principle (a review of people and natural resources provides more information) have dominated much public land use policy. Many historians note that this policy probably was inevitable given the rich resources, sparse population, and rapidly growing industrial economy of the United States.

An End to Subsidized Abuse of Public Land?

In recent years, support for more preservationist policies toward public land use has been growing. The multiple-use approach historically has been employed to justify very high levels of resource extraction, which often have resulted in widespread and highly visible environmental damage to public lands. In the western United States, active and abandoned mines now scar many areas of public land where mineral deposits once existed. Each year, many of these mines produce tons of toxic acid and heavy metal drainage that contaminates the soil and water. Similarly, overgrazing of public lands by privately owned cattle damages native vegetation, destroys wildlife habitats, and encourages widespread growth of nonnative weeds. Logging on public lands is a major reason the remnants of ancient old-growth forests are rapidly dwindling.

Much of this destructive resource extraction has been accelerated by subsidies from federal tax money. Mining, grazing, logging, and other activities have been conducted at very cheap prices, far below market cost, because taxpayers pay much of the cost. The 1872 Mining Law, which allows miners to lease public land for about $5 or less per 0.4 hectare (1 acre), is a good example. Grazing is also much cheaper on public land than on nearby private land. The World Resources Institute estimates that these subsidies cost more than $1 billion per year in public funds spent and taxes not collected from land users. Although public resistance to this subsidized damage is growing, the opponents of subsidies have had difficulty achieving political change. Intense lobbying and the opposition of senators and representatives from the western states have generally defeated the efforts. Some of the leading lobbyists have been advocates of the "**Wise Use Movement**," which consists of a variety of individuals and organizations that seek to maximize private utilization of public lands for mining, logging, recreation, and other purposes.

The National Park System

Historians generally credit the artist George Catlin with originating the concept of the national park. In 1832, he was in the Dakotas on a painting expedition and was troubled by the potential impact of America's westward expansion. The wildlife, wilderness, and Indian civilization might be preserved, he wrote, "By some great protecting policy of government . . . in a magnificent park . . . a nation's park, containing man and beast, in all the wild and freshness of their nature's beauty!" The first major national park in the United States was Yellowstone National Park, established in 1872. In a foreshadowing of the coming debate, the preservationists, with their idealistic motives to save more of these spectacular areas, often found themselves lobbying alongside western rail barons who wanted to promote tourism to boost their passenger business.

As the parks were created, usually through acts of Congress, it became apparent that an organization was needed to manage them. Accordingly, the U.S. National Park Service was founded in 1916. In 2005, the National Park System covered more than 34 million hectares (83 million acres) and included 368 sites ranging from major parks such as Yellowstone to urban recreation areas, battlefields, trails, rivers, prehistoric ruins, and homes of presidents. Areas in the park system often are designated as national rivers, national monuments, national lakeshores, and such to specify their significance. California has 25 individual parks and another 7 it shares with other states, the most of any state. The idea of national parks has been very successful. The spread of such parks worldwide has led some observers to comment that national parks are one of America's great exports. Through the collaborative efforts of thousands of people from all sectors of the population, the International Union for Conservation of Nature (IUCN) reports that there are more than 100,000 protected

FIGURE 12-2 Bumper-to-bumper traffic is a common sight in the more popular national parks. This scene shows a traffic jam caused by visitors stopping to view wildlife in Yellowstone National Park, Wyoming.

areas across the globe, measuring 49 million square kilometers (19 million square miles); marine habitats are less represented, accounting for only about 1% of all protected areas. Globally, many of these protected areas are in politically unstable locations, making enforcement of protection problematic.

In recent years, the National Park System has experienced increasing problems. One is popularity, or overcrowding (**FIGURE 12-2**). The National Park Service reports that the number of annual visitors to the national parks has risen dramatically, from 22 million in 1946 to 133 million in 1966 and more than 428 million in 2004. Attendance is increasing faster than the U.S. population and reflects the public's growing awareness of the value of nature. People who want to camp or rent cabins at the more popular parks, such as Yellowstone, Yosemite, and Great Smoky Mountains, generally must make advance reservations. For example, in 2004, more than 20 million people visited the Great Smoky Mountains National Park. The parks often turn away hundreds of potential visitors for lack of space. Traffic jams, air pollution from cars, and crime have become problems.

The Funding Problem

Funding to maintain the national parks has not kept pace with the increasing attendance. The 2001 National Park Service budget of $1.8 billion forced the service to delay maintenance,

reduce staff and programs, and allow environmental problems to continue. The service had a $5 billion backlog of high-priority construction projects, $2 billion of unfunded land acquisitions, and $800 million worth of unfunded repairs. Lack of maintenance leads to problems ranging from rusting cannons at Gettysburg in Pennsylvania to complaints of raw sewage at Mammoth Cave in Kentucky and Sequoia National Park in California. Roads in many parks have gone unrepaired. The Grand Canyon alone needs an estimated $370 million worth of road, sewer, and other repairs. Park rangers suffer, too. More than 50% of their housing units have problems such as rotting floors, faulty wiring, or insect infestation.

One reason for the underfunding has been efforts to reduce federal government spending. However, a number of observers, including some prominent park officials, argue that part of the problem is that Congress has purchased new national parks at the expense of the parks already in existence. Between 1975 and 1999, Congress established 88 new parks. Because of the growing interest in preserving the environment and historical places, establishing parks is very popular with many voters, so money for purchasing land has been relatively easy to obtain. In contrast, funding for sewage treatment, road repairs, and other maintenance needs has less popular appeal. As use of the parks increases during economically troubled times, and voters express their desire to preserve the parks, politicians have responded with an increase in park spending during the last two fiscal cycles. In 2009, the National Park Service budget was $2.92 billion, and in 2010, it was $3.16 billion. Because the National Park Service budget is approved by the Congress, voters and community organizers for the parks can influence funding through their elected officials. Despite this recent increase in budget spending, the parks still struggle to keep up with maintenance. In Arizona, 13 state parks were closed in 2010 because of budget strain. The parks collectively have about a $2 million carryover from previous years. If it were not for the efforts of many local citizens through volunteer groups, more parks would be negatively affected.

Two possible solutions to the funding problem have attracted considerable support: (1) making the parks more self-sufficient by relying on visitor fees and other private sources and (2) removing some of the less popular or less unique parks from the park system. Advocates of increased self-sufficiency point out that park visitors spend more than $10 billion annually, but much of this money does not go toward park maintenance. Instead, it goes to concession stand operators in the parks and especially to general government funds used for a variety of purposes. Increasing entrance or activity fees is another possibility. The National Park Service spends approximately $900 million per year on visitor services but collects only about $190 million in entrance and activity fees, $60 million in concession fees, and $1.2 million in special project fees (such as when a movie studio pays to film in a park).

Removing some less-visited parks from the National Park System is another option. Many critics point out that numerous sites attract few visitors but are costly to maintain. For instance, the Yukon-Charley Rivers National Preserve in Alaska has an annual budget of more than $500,000 but receives fewer than 1,000 visitors per year. Critics argue that such wilderness can be preserved at less cost either by allowing private conservation groups to purchase it or by managing it outside the National Park System. Some groups are also interested in conducting mining or logging activities in parks in Alaska and the western United States that have few visitors. At the same time, despite the number of visitors, these parks contain natural resources, wildlife protection, and beauty beyond what visitor counts reveal. Congress is considering selling dozens of smaller urban parks, battlefields, and monuments to local governments or private groups. San Francisco's Golden Gate National Recreation Area and Cleveland's Cuyahoga Valley National Park are examples. Additions and removals of parks from the park system is not new. Congress has deauthorized 24 parks since 1916. Currently (2011) there are 394 parks in the National Park System with 3 added so far by President Obama, 7 authorized and 1 deauthorized by former President George W. Bush, and 19 new parks added by former President Clinton. The Oklahoma City National Memorial is an example of a positive example of deauthorization; the memorial is preserved, promoted, and maintained through a highly active nonprofit that has built substantial partnerships, creating an impressive educational and cultural experience that could not have been done with the budget allocated by the National Park Service. This is in contrast to the closing of Fossil Cycad National Monument in South Dakota, which was deauthorized by Congress in 1957 because of extreme loss of the fossil resource as a result of collecting activity. In other areas such as parks near sacred tribal lands, groups of Native Americans petition to close parks because of many visitors' lack of respect, the desecration of culturally important sites with graffiti, and diminished staffing because of lack of resources.

Recreation or Preservation? The Dilemma

Some critics do not agree that the solution to the overcrowding problem is increased funding for more facilities and staff. Instead, they argue that the number of visitors should be reduced. They contend that too many visitors will despoil the natural beauty and contribute to the extinction of endangered species in the parks.

At the heart of the debate is whether the main purpose of national parks is to serve the needs of people or to preserve natural areas. The **multiple-use philosophy** that has dominated since the early 1900s has tended to emphasize human uses; however, many visitors recognize the conflict as soon as they encounter the huge crowds, traffic jams, commercialism, air pollution, and other unpleasant aspects of urban life at the most popular parks. There is also controversy regarding what kinds of recreation should be allowed. For example, should snowmobiles be allowed in Yellowstone National Park (see **CASE STUDY 12-1**)?

Public support is growing for reducing the size of crowds in national parks and perhaps limiting public access to preserve the natural areas. One solution has been to designate some lands in the park system as wilderness areas. Under the **Wilderness Act of 1964**, a designated wilderness area is federal land that

Should Snowmobiles Be Allowed in Yellowstone National Park?

Yellowstone National Park is America's oldest park, established in 1872. It is also a unique park. Yellowstone's physical beauty contains a wide variety of pristine mountain views and even volcanic activity that produces elegant geysers such as Old Faithful. Elk, grizzly bears, and wolves are among the wildlife that are common in this popularly visited park.

However, in recent years, a major controversy has arisen that illustrates the conflicts inherent in the "multiple-use" approach to public lands. Should snowmobiles be allowed in the park? Each year there are more than 80,000 snowmobile visits to Yellowstone (**FIGURE 1**). Advocates who wish to see the park focus on wilderness preservation argue that snowmobiles should be banned. They note that the loud noises these machines produce often disturb the wildlife. For example, it is not uncommon for bands of snowmobiles to startle herds of buffalo. In addition, preservationists argue that the wilderness experience is ruined by the noise, fumes, and headaches produced by breathing snowmobile exhaust. Some visitors complain that they cannot hear the splashing of Old Faithful because of the buzzing noises. In addition to many wilderness groups, many and perhaps most of the park rangers seem to agree that some type of snowmobile restrictions are needed. In fact, surveys show that most of the general public is against snowmobiles in the park.

On the other side are advocates of snowmobile use in the park. Yellowstone contains approximately 8,987 square kilometers (3,470 square miles) of land with more than 1,600 kilometers

FIGURE 1 Snowmobiles in Yellowstone National Park have provoked considerable debate.

(1,000 miles) of established back woods trails and 965 kilometers (600 miles) of groomed snowmobile trails. Advocates assert that in a park this large, there is room for responsible snowmobile use. Aside from the users themselves are many businesses in nearby towns that rent snowmobiles and sell related items. For example, the mayor of West Yellowstone is an outspoken critic of a snowmobile ban, arguing that it would "devastate" the town's economy. The town of West Yellowstone has many snowmobile rental shops

is to be managed to retain its primeval character, with no commercial enterprise, no permanent roads, and no motorized vehicles. The amount of land set aside as designated wilderness has increased greatly since 1970, from 4.2 million hectares (10.4 million acres) that year to 43 million hectares (107.45 million acres) in 2005 (**FIGURE 12-3**).

These figures include only land in the National Forest and Wildlife Refuge Systems and the National Park System. In the National Forest System, the largest system by far in the lower 48 states, approximately 15% of the land is set aside as wilderness.

Interest in preservation has also led to two other trends: (1) increasing the size of

and thousands of hotel beds that snowmobilers use in the winter.

After 5 years of study and several rounds of public hearings, the Clinton administration proposed a ban on snowmobiles in the park. The snowmobile visits every winter were to be phased out over a few years; however, on the day he was sworn into office, President George W. Bush imposed a moratorium on the ban, and thus, snowmobiles continue to be widely used. The park is open to snowmobiles on approved trails from December 15 through March 15 annually.

In the meantime, several compromises have been proposed. One is the use of quieter and less-polluting four-stroke engine snowmobiles. However, because the machines are more expensive and less powerful than the traditional two-stroke models, renting them has not been easy. In addition, snow coaches, which can carry several people, have been suggested as an alternative, but they have limited appeal to those who want the freedom and exhilaration that come from a solo ride on a snowmobile (**FIGURE 2**).

VISIT
http://environment.jbpub.com
/mckinney/5e/
for more information

FIGURE 2 Snow coaches do not provide the same open-air experience of snowmobiles.

parks by acquiring adjacent land through purchases or donations and (2) restoring parks to a more natural condition. Donations often come from conservation groups. For example, in 1995, a local conservation group, The Foothills Land Conservancy, purchased a few hundred acres of important black bear habitat and donated it to the Great Smoky Mountain National Park (**FIGURE 12-4**).

Park restoration efforts include allowing some forested areas to return to the old-growth condition and reintroducing native species that were exterminated, such as the reintroduction of wolves into Yellowstone National Park. This project also illustrates the conflict

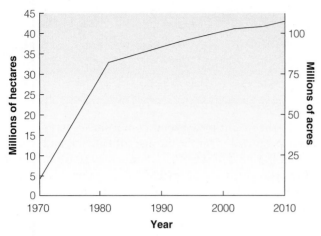

FIGURE 12-3 The amount of land set aside as designated wilderness since 1970.

that can arise from the multiple use of public lands. Many environmentalists welcome the return of the wolf, whereas some farmers argue that the wolves pose a threat to people and livestock. At some parks, there are plans to place a greater limit on visitation. In Yosemite, for example, some buildings have been torn down, and visitor access to some areas may be reduced.

FIGURE 12-4 The recent donation of critical black bear habitat to the Great Smoky Mountain National Park is a good example of how private citizens can enhance the preservation of public lands.

The trend toward increasing preservation is not without controversy. President Clinton was strongly criticized for designating public lands as national monuments. Many environmentalists applauded the efforts to protect what they regarded as unique but threatened natural treasures, but many other people, particularly in the western United States, resented what they saw as a federal tactic to interfere with the use of their lands.

National Forests: Land of Many Uses

Since the mid-1600s humans have cleared approximately 120 million hectares (300 million acres) of North American forests. The most comprehensive land census, completed in 2002, indicates that the United States contains some 263 million hectares (651 million acres) of forests. About two-thirds of this land is classified as commercial timberland, meaning that it is available for timber production. Companies directly involved in producing forest or paper products own only about 15% of this commercial timberland. Instead, the timberland is on federal, state, and other public land. Most of the federal forests are in the National Forest System and are managed by the U.S. Forest Service (Figure 12-1). Indeed, nearly half of the large commercially valuable timber trees in the United States are found on Forest Service land. This includes the last virgin stands of old-growth forests in the continental United States, which are found in the national forests of the Pacific Northwest.

Utilization or Preservation?
The Dilemma Continued

The dilemmas that arise from multiple use of public lands are especially common in national forest management because the National Forest System originally was designed to allow considerable use of natural resources. Although national parks were designed mainly for recreation and to protect wildlife, water, and all natural resources, national forests were designed not only to make available these resources but also to provide U.S. citizens with access to the land for livestock grazing and the extraction of wood and mineral resources. This difference is evident in the National Park Service and the Fish and Wildlife Agencies being in the Depart-

TABLE 12-1	Common Forestry Practices and Their Effects on Biodiversity	
Forestry Practice	**Purpose**	**Effect on Biodiversity**
Planting of exotics (nonnative species) or genetically "improved" tree species	Improved yield of commercial tree species	Replacement of native species
Pesticide spraying	Protection of forest tree species of commercial value	insect species; secondary effects on nontarget organisms; major disruption of ecosystem
Clear-cutting/reforestation	Maximum utilization of existing tree biomass/ maximum speed of new forest growth	Artificial cycle of disturbance; loss of species richness; loss of structural and functional diversity
Clear-cutting/even-aged management	"Efficient" regulation of the forest; maximum profit from growing trees	Shortened successional cycle; loss of forest structural diversity
Slash-burning	Site preparation for new forest; esthetics ("neatening up" the forest)	Major loss of structure, biomass, and nutrients from forest ecosystem
Tree thinning	Increased growth of commercial tree species	Reduced structural diversity in forest
Brush removal/herbicide spraying	Removal of species believed to be delaying reestablishment of commercial tree species	Truncated succession with loss of important successional processes; loss of species richness
	The Bottom Line	
Maximum production of commercial sawtimber and pulpwood	Maximum forest output; maximum profit	Reduced structural and species diversity of forest; deterioration of forest ecosystem

Source: Adapted from R. Noss and A. Cooperrider. *Saving Nature's Legacy.* Washington, DC: Island Press, 1994.

ment of the Interior, whereas the Forest Service is in the Department of Agriculture. As a result, national forest management historically has been less inclined toward preservation. Because the National Forest System is so large, this philosophy helps explain the relative lack of wilderness preservation in the United States.

With the growing public interest in environmental preservation, the traditional methods of forest management, such as the use of pesticides and nonnative plants (**TABLE 12-1**), often are criticized as unnecessarily destructive to wildlife and their habitat. Of special note is the controversial practice of clear-cutting, also called "even-aged" management (**FIGURE 12-5**). In this tree-harvesting system, an entire stand of trees is removed, and the site is prepared for planting a new "crop" of trees. Clear-cuts are unattractive because the land is stripped bare of trees and can contribute greatly to increased sedimentation and erosion, which not only deplete the soil but also often pollute local streams. Very large clear-cuts are no longer practiced in the United States; however, even patchworks of clear-cuts can be harmful and unattractive, especially because they often require large road networks to be built into the forest. In addition, to communities that rely on the attraction of their natural beauty to lure tourists, patchwork clear-cuts on hillside vistas can also negatively affect local economies.

An alternative to clear-cutting is **selective cutting**, also called "uneven-aged" management. In this method, only certain trees in the stand are cut down, so the land is not stripped bare. However, selective cutting generally is more expensive than clear-cutting and is not without its own environmental problems. Roads must often still be built, and selective cutting usually alters normal forest development because shade-intolerant trees such as most pines and the Douglas fir cannot grow. Nevertheless, there is a movement, called "New Forestry," to incorporate biodiversity

FIGURE 12-5 Clear-cutting strips the land bare of trees, promoting erosion and water pollution and devastating habitat. This photograph was taken in Willamette National Forest, Oregon.

preservation into forest management and use selective cutting instead of clear-cutting. New Forestry promotes replanting and careful restoration of harvested forest. In fact, the forest products industry is the largest planter of new trees in the country; one-third more wood is grown each year in the United States than is harvested or lost to fire, insects, or disease.

President Bush launched the **Healthy Forests Initiative** (HFI) in 2002 to promote the removal of trees in the national forests and other public lands with the stated goal of reducing the risk of wildfires. This initiative has been controversial. Logging and paper companies generally have promoted it, but many environmental groups, such as the Sierra Club, have criticized the HFI as an excuse to allow logging access to millions of acres of public forests. They also note that the revised rules in the HFI allow supervisors of each of the country's 155 national forests to approve logging, drilling, and mining and to ignore the forest plan's guidelines for protecting wildlife. The HFI also eliminates the need to monitor scientifically the effect of these activities on plants and wildlife and restricts public participation in the planning process. This topic continues to play out in the courts on a case-by-case basis. In 2002, federal courts halted the burn and subsequent timber sale of the largest trees that had survived a previously prescribed burn in 1999. In this particular case, the courts ruled that the Forest Service had ignored the environmental impact of the burn and in this case stopped the burn and timber sale.

In 2005, the Bush Administration overturned the **Roadless Area Conservation Rule** installed by President Clinton. This rule required that no roads should be built through wilderness areas currently lacking roads in the national forests. By overturning this rule, the Bush administration allowed individual states to decide whether new roads can be built into roadless areas. Some, such as Idaho, have decided to build roads.

Subsidized Abuse? The Tongass National Forest Example

We noted that taxpayer dollars are often used to subsidize rapid, and often harmful, depletion of natural resources. In the case of national forests, each year the Forest Service loses money on timber sales in most of its forests. If the cost of roads that the government builds for the logging companies is included, the Forest Service shows a net loss of millions of dollars, according to data from the Wilderness Society and other environmental groups. Such groups thus argue that the emphasis on tree production is outmoded at a time when forests often contain endangered species and shrinking, rare ecosystems. These shrinking forest ecosystems are not limited to the northern spotted owl and the old-growth forests of the Pacific Northwest but include many other areas containing dozens of endangered aquatic species (such as freshwater clams) and birds (such as the red-cockaded woodpecker of the southeastern United States).

The public lands of Alaska are especially important because they contain the largest natural areas remaining in the United States and also the largest amount of untapped resources. As a result, the stakes are high for both developers and environmentalists, and so is the controversy. For example, the Tongass National Forest in southeastern Alaska contains the last intact stretch of America's only temperate rain forest, which once extended down the Pacific Coast to California (**FIGURE 12-6**). Currently, timber companies are pushing hard for the right to log the heart of the Tongass Forest and have won consid-

FIGURE 12-6 The Tongass National Forest of southern Alaska contains the last large area of America's temperate rain forest.

erable support in Congress. Federal judges in 2010 supported the preservation of the Tongass, but the controversy continues, and the timber companies continue to fight for the right to harvest an area the size of West Virginia. With trees more than 500 years old, the Tongass is not only a national treasure, but also sacred to Native Americans in the area. Alaskans living near the forest point out that although past logging restrictions have cost jobs, they have also improved tourism and commercial fishing, which were harmed by the clear-cutting operations. A similar controversy is brewing over the Arctic National Wildlife Refuge, the largest wildlife preserve in the United States, where oil companies want to drill. Critics point out that the damage to wildlife, especially caribou, from the roads and development could be enormous. In addition, according to geological estimates, there may be little oil in that area. Whatever happens in the Tongass and Arctic National Wildlife Refuge, it seems likely that the debate over preservation versus extraction of Alaska's wealth will persist for many years to come, demonstrating the controversies inherent in the multiple-use concept.

12.2 National Environmental Policy Act and Environmental Impact Statements

In 1969, a fundamental federal law was passed called the **National Environmental Policy Act** (**NEPA**), which required that all major federal actions, including all activities on federal public lands, be reviewed for their environmental impacts before those activities could go forward. A key part of NEPA was the **Environmental Impact Statement** (**EIS**), which has become a basic tool to protect the environment on federal lands. Most states also require such a review on state public lands. The EIS requires that anyone proposing action on public lands, such as logging or construction of roads or buildings, first perform a study that documents the need for the project, reasonable alternatives to it, an overview of the natural environment affected, and the environmental consequences of the project.

Like many federal laws, NEPA was well received and is widely regarded as a good idea. However, in practice there have been many problems. One issue is the huge amount of paperwork that is often generated to produce the impact statement. As a result, in 1979, the process of **scoping** was introduced; this means that local government agencies and the general public are invited to participate in producing the EIS for a project. For example, members of the community near a proposed Forest Service logging project are invited to town meetings with service officials to discuss the pros and cons of its impact on the community. As a result, the EISs of many projects produce considerable controversy because nearly all projects have both harmful and beneficial impacts. For example, local environmentalists often complain that EISs prepared by private companies for work on federal lands are biased because the private companies hire the technical firms that produce the EIS and the data are often selectively used.

12.3 Managing Private Lands for Biodiversity Preservation

The future resolutions of many environmental problems rely heavily on how private lands will be managed. Environmentalists hope that many of them will be managed for the preservation of native species biodiversity. Because extinction is caused by change in a species' environment, such as habitat disruption or introduced species, the best way to save any species is to preserve its natural environment. Until recently, the establishment of preserves has occurred mainly on public lands, such as those discussed. However, in the United States and most of the world, such public preserves are currently far too inadequate to prevent the extinction of many species.

One inadequacy is simply that the amount of land (and ocean) protected is insufficient. One common estimate is that approximately 10% of any given area, such as the United States, would need to be preserved as unmodified habitat to save the large majority of threatened species. In 2003, slightly less than 6% of the world's total land area (excluding

Antarctica) is protected, far below the 10% goal established by the International Union for Conservation of Nature (IUCN, formerly the World Conservation Union). Currently, the *United Nations Environment Programme* (UNEP) and the IUCN report that 11.6% of the terrestrial surface area of Earth is protected. Much of the pledged land is in politically unstable areas where enforcement of protection laws is challenging and unreliable. Much more work is needed to protect marine environments. The IUCN's goal is to achieve a minimum of 10% protected land in each of the world's major biomes. This has already been achieved for 9 of 14 biomes.

Another inadequacy is that even where larger amounts of land are preserved, the habitat is of poor quality: it is often highly fragmented and/or located on mountains, in deserts, or in other habitats that were set aside simply because the land had little commercial value for people. Finally, many of the preserves are ineffectively managed. Especially in poor nations, there is little money available to train local residents and prevent poaching and habitat loss.

Fully aware of these inadequacies of preserved public lands, some environmental groups have safeguarded additional private lands. The past few decades have seen the rapid growth of private organizations in many nations, usually called nature conservancies or land trusts, which purchase, lease, or find some other way to legally protect private lands from development.

The Nature Conservancy is one group that has made a huge impact by doing this. Founded in 1951, the Nature Conservancy has a membership of approximately 1 million. By using a large fund of donations (almost $436 million in 2004 alone), the Nature Conservancy has created the largest system of private natural areas and wildlife sanctuaries in the world. As of 2010, the Nature Conservancy had secured the protection of 119 million acres (48,157,591 hectares) of land and 5,000 miles (8,046 km) of rivers and operated more than 100 marine conservation projects worldwide.

By focusing on buying species-rich threatened habitats, buffer zones, and corridors, the Nature Conservancy has greatly enhanced the survival chances of many species. Another example is the **Land Trust Alliance**, which is an association of more than 1,700 land trusts in the United States.

Managing land for preservation is actually quite complex and involves two key steps: (1) selecting and designing preserves on the basis of ecological principles and (2) using legal and economic principles to establish and maintain the preserves. The first step is largely scientific and the second largely social.

Selecting Preserves

Creating a preserve involves more than just setting aside a section of land or water. Preservation of habitats containing many nonnative species or abundant native species often does little to maximize regional or global biodiversity. Instead, preserves should be selected that save native species that are rare and in danger of extinction.

Historically, natural habitat preservation in North America and Australia usually has arisen from the desire to preserve unique or scenic landscapes. For example, recall that many public parks were established for this reason; however, many of these early parks were created on lands that were available simply because they were thought to have low commercial value (mountains and canyons have poor soils for farming and are isolated from population and transportation centers). In Europe, many of the first game preserves were created to preserve large game animals, whose numbers were decreasing. One of the earliest was established in Poland in 1564 to attempt to preserve wild cattle. Developing tropical nations also have established many preserves to protect disappearing species.

As many native species become rare at increasing rates, it is clear that simply preserving either a few unique habitats or the habitats of a few charismatic species will not suffice. If the goal is to save many threatened species, a more comprehensive approach is needed. A common approach has been to use popular charismatic species as **umbrella species** to preserve large areas, but this approach has its limitations. Many rare species are narrowly distributed in areas that are not currently occupied by umbrella species. In addition, the past emphasis on lands with low commercial value has disproportionately favored preserving some habitats and neglecting others. For example, national parks in the United States well represent mountainous and canyon habitats, whereas river valley and wetland habitats are poorly represented. River valleys and wetlands have been favored for human settlement because they have a high commercial value, with fertile soils and river access for industry and travel.

A way for future biodiversity planning to correct these past omissions is to identify species hot spots that should be preserved. **Hot spots** are areas of exceptionally high species richness, especially concentrations of localized rare species that occur nowhere else. Hot spots generally are areas of great geographic diversity that have promoted lots of evolutionary change. **FIGURE 12-7** shows a number of global hot spots. Ecologist Norman Myers has estimated that plant species hot spots cover just 0.5% of the Earth's land surface but contain the only habitat of about 20% of the Earth's plant species. Similar analyses of birds indicate that about one-fourth of the world's bird species are confined to just 2% of the Earth's land surface. Hot spots for birds do not necessarily occur in the same areas as

plant hot spots. Conservation International, which compiles data regarding biodiversity hot spots, reported in 2007 on 34 worldwide biodiversity hot spots, each of which hold at least 1,500 endemic plant species. When first identified, the 34 hot spots covered 15.7% of the Earth's land surface. The most recent report indicates that 85% of the hot spots' habitat has been destroyed. What remains covers only 2.3% of the Earth's land surface. In the United States, a number of hot spots for various groups have been identified (**TABLE 12-2**). By many estimates, most of the hot spots in Figure 12-7 and Table 12-2 will experience additional habitat loss within the next generation.

Designing Preserves

After a location is selected, the design of the preserve is critical. Species preservation is maximized by three key characteristics: size, shape, and connectivity. Larger size increases the number of species contained in the preserve. Rounder shape minimizes edge effects because the perimeter (edge) is smaller relative to the area inside than with other shapes. Connectivity is the opposite of fragmentation. Increasing the connections (corridors) between potential fragments allows members of the same species to immigrate and interbreed.

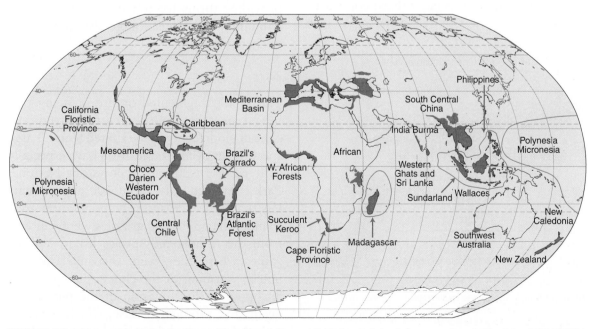

FIGURE 12-7 Global hot spots for biodiversity. (*Source:* Adapted from N. Myers, et al. Biodiversity hotspots for conservation priorities. Nature 2000;403:833.)

TABLE 12-2	Hot Spots of Species Richness in the Continental United States
Species	**Areas of Greatest Species Richness**
Vascular plants	California, followed by Texas, Arizona, Oregon, and Florida (in that order)
Trees	Southeastern Coastal Plain and Piedmont, northern Florida
Mollusks	Tennessee River system (Tennessee, Alabama) and Coosa River system (Alabama)
Butterflies	Western Great Plains and Central Rocky Mountains (Colorado)
Fishes	Cumberland Plateau in the Tennessee and Cumberland River drainages
Amphibians	Southern Appalachians and Piedmont
Reptiles	Gulf Coastal Plain (eastern Texas)
Birds (breeding)	Sierra Nevada, southeastern Arizona–southwestern New Mexico
Mammals	Sierra Nevada and, secondarily, southern Cascades and desert Southwest
Source: Adapted from R. Noss and A. Cooperrider. Saving Nature's Legacy. Washington, DC, Covelo, CA: Island Press, 1994.	

Buffer zones are another important preserve characteristic (**FIGURE 12-8**). A **buffer zone** is moderately used land that provides a transition into the unmodified natural habitat in the core preserve where no human disturbance is allowed. For example, campgrounds and limited cattle grazing may be permitted in the outermost buffer zone, with hiking in the innermost buffer zone. Buffer zones are a major departure from traditional preserves that were viewed as islands of natural habitat in a hostile matrix of agricultural or urban landscape.

Buffer zones are very important for both psychological and practical reasons. Sharp boundaries, as were used around traditional preserves, tend to promote the idea that nature should be fenced in and people fenced out; such boundaries suggest that people are separate from nature rather than being part of nature. On a practical level, conservation biologists have found that inhabitants of areas surrounding preserves must derive some benefits from the preserve. Many preserves have been established in developing nations only to have the endangered species poached to extinction because local inhabitants needed food or money. Buffer zones integrate some of the area surrounding the preserve into the local economy without harming the core preserve. By permitting moderate recreational, forestry, farming, and other activities, buffer zones provide jobs and income with no ill effects on species in the core preserve. Indeed, endangered species in the core preserve occasionally wander into the buffer zones, providing thrills for tourists, income for inhabitants, and larger forage area for the endangered species.

Preserve Networks

The small, fragmented nature of many preserved areas has contributed to a steady decline in biodiversity in the United States. Large species that need sizable areas have especially suffered. For example, Yellowstone National Park contained only about 100 grizzly bears less than a decade ago. The population grew to more than 500, and grizzlies were removed from the endangered species list in Yellowstone, where they were thriving. Unfortunately, warming of the climate has provided an opportunistic invasion of the pine forest beetle and a fungal disease, which have aggressively attacked the white bark pine (*Pinus*

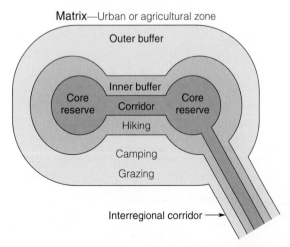

FIGURE 12-8 Buffer zones can be used to support economic activities around a core of natural habitat. (*Source:* Modified from R. Noss and A. Cooperrider. Saving Nature's Legacy. Washington, DC: Island Press, 1994, p. 148.)

albicaulis), which produces pine nuts, a major food source for the grizzlies. Because of food shortages and other considerations, such as genetic relatedness among individuals, the grizzlies were put back on the endangered list in 2008. This example illustrates the multifaceted relationships between species in the fight for survival on an ever-changing planet (**FIGURE 12-9**). Even if the bear population recovers, there is a limit to the number of grizzlies a park such as Yellowstone can support.

Many conservation biologists suggest that the solution is to (1) expand the size of existing preserves, (2) create buffer zones, and (3) create habitat corridors that connect the preserves. Such preserve networks would allow habitat preservation in the United States to have the desirable characteristics of preserve design discussed here. Many of the proposals for extensive networks have been criticized as unrealistic; one plan, for example, includes much of Florida (**FIGURE 12-10**). An even more extensive plan, called the **Wildlands Project**, outlines a similar goal with networks throughout the entire United States. However impractical such plans may seem, most wildlife experts agree that corridors, buffers, and other network concepts eventually will be necessary for wide-ranging species to survive and for many regional ecosystems to stay healthy. Indeed, in other parts of

FIGURE 12-9 Yellowstone National Park, like many parks, is far too small to sustain viable populations of large mammals such as grizzly bears for many years.

the world, preserves and habitat corridors are being established. As one example, The Critical Ecosystem Partnership Fund (joint initiative of Conservation International, the Global Environment Facility, the Government of Japan, the John D. and Catherine T. MacArthur Foundation, and the World Bank) is working to establish and protect habitat corridors in Nicaragua, Costa Rica, and

(a)

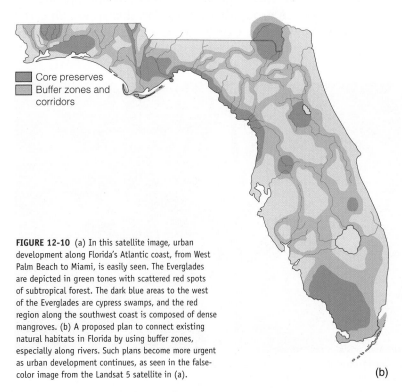

Core preserves
Buffer zones and corridors

FIGURE 12-10 (a) In this satellite image, urban development along Florida's Atlantic coast, from West Palm Beach to Miami, is easily seen. The Everglades are depicted in green tones with scattered red spots of subtropical forest. The dark blue areas to the west of the Everglades are cypress swamps, and the red region along the southwest coast is composed of dense mangroves. (b) A proposed plan to connect existing natural habitats in Florida by using buffer zones, especially along rivers. Such plans become more urgent as urban development continues, as seen in the false-color image from the Landsat 5 satellite in (a).

(b)

Panama. This area of southern Mesoamerica, the thin strip of land connecting North and South America, is a biodiversity hot spot for both endemic species and higher taxa (major types of organisms). Habitat corridors can also be established on a small, local scale. Thus, the city park in Nashville, Tennessee, has larger wooded areas that are purposefully connected by wooded corridors, allowing animals to have cover as they move from one section of the park to another.

Sustainable Management of Preserves

Conservation biologists agree that governmental decrees alone will not solve the extinction crisis. Establishing preserves and prohibiting trade in endangered species will not work as long as people, especially local inhabitants, are poor and must rely on destructive uses of biodiversity to stay alive.

Instead, sustainable uses of biodiversity must be developed and encouraged. Sustainable uses provide incentives to save species while also respecting the right of all people to support their families and have a decent quality of life. This is sometimes called the **bottom-up approach** to biodiversity conservation because it permits local citizens to play a role in planning and establishing preserves. A group that exemplifies this approach is the Nature Conservancy, which for half a century has been working with local communities in the United States and around the world to preserve natural ecosystems and biodiversity. To give just one of numerous examples, in Indonesia, the Nature Conservancy is working with indigenous peoples to sustainably manage endangered forest ecosystems of Borneo. Partnerships have been fostered between timber companies and local communities to sustainably harvest wood from the forests and provide appropriate financial benefits to the communities that own the land. In contrast, with the top-down approach, national and regional governments simply establish preserves regardless of the needs of local citizens.

Ecotourism

One very important sustainable use of preserves is **ecotourism**, which the Ecotourism Society defines as "responsible travel to natural areas that conserves the environment and sustains the well-being of the local people." Ecotourism uses the travel facilities that the native population created. It avoids international hotel chains in favor of properties owned and managed by locals. Adherents dine not on foods flown in from abroad, but on regional dishes, supporting the community's economy. Ecotourism respects the local environment and the native culture of an area. To be done correctly, ecotourism must not harm the local ecosystems in any significant way.

Examples of this fast-growing industry abound throughout the world. This growth is driven by widespread environmental concern combined with the fact that ecotours can be an enjoyable escape from the routine of industrialized society. Travel agencies arrange ecotours that range from a few days in an air-conditioned boat cruising the Amazon to weeks of hiking in the Himalaya Mountains in Nepal. Many of these tours do accomplish at least some of the goals of ecotourism.

A good example is the Orangutan Rehabilitation Centre in Bukit Lawang, Indonesia. This project rehabilitates orangutans held in captivity illegally or displaced by deforestation and returns them to natural habitats (**FIGURE 12-11**). Because of the great international interest in this disappearing primate species, the Indonesian government is promoting

FIGURE 12-11 Ecotourism can benefit animals such as these orangutans if it leads to money to buy quality natural habitat.

the Orangutan Rehabilitation Centre as a destination where tourists can view orangutans in their habitat. Consequently, it has become one of the most popular ecotours in Southeast Asia and is bringing in much revenue to help support the Centre.

Although ecotourism can help preserve natural habitat and promote a sustainable economy, it has limitations. One is that many natural areas are not suitable for ecotourism. Many ecosystems are simply not appealing enough to the general public to provide a large pool of visitors. Aesthetic sites, such as coral reefs, rain forests, and pristine mountain hikes, are popular, as are areas with charismatic species, such as elephants or grizzly bears, but many ecosystems lack the spectacular scenery or unique creatures necessary to draw sufficient visitors. In all likelihood, large portions of deserts, swamps, tundra, and many other areas cannot be preserved via ecotourism. Indeed, some experts suggest that only a few isolated, relatively unique ecosystems can be completely supported by ecotours.

Even where ecotourism is feasible, practical problems arise. Money that ecotours generates does not always contribute to the local economy, especially in poor developing nations. For example, some fees from the Orangutan Rehabilitation Centre have gone to government officials in Jakarta. In addition, in many cases, crowds of people and all of the activity have harmed native species and ecosystems. Motorized tundra buggies that take groups to see Hudson Bay polar bears have been criticized as seriously disturbing the bears. Similarly, the Manuel Antonio National Park in the heart of the Costa Rican rain forest now gets 1,000 visitors per day, and 300 once-wild monkeys have become garbage feeders. Even native cultural diversity can be affected. Some boat tours along the Amazon now make stops so that ecotourists can trade with native Indians who previously had little contact with civilization.

Despite these problems, ecotourism has enormous potential because the travel industry is the second largest industry in the world (agriculture is the largest), employing hundreds of millions of people and generating more than a quarter trillion dollars annually.

If properly done, ecotourism can reduce local poverty, conserve nature, and educate the public about both biological and cultural diversity.

Sustainable Harvesting

Another sustainable use of preserves is **sustainable harvesting**. Sometimes called extractive forestry, sustainable harvesting includes the harvesting of nuts, fruits, and many other products that can be extracted from an ecosystem without harming it. Specific examples include collecting Brazil nuts and tapping rubber trees in the rain forest to sell to industrialized nations. Sustainably harvested products can also be used for health care at home. Traditional healers using native plant remedies provide most of the primary health care in many developing countries. The World Health Organization estimates that nearly 3 billion people use wild-harvested medicines, with an annual value as great as $15 billion. With so many undescribed plants, many potentially useful medicines will be lost if the plants they are derived from become extinct before their potential value is discovered. This has led to a new field, called **bioprospecting**. Biologists and chemists are rapidly compiling a huge database of the commercial potential of various species. One of the most effective types of bioprospecting uses the knowledge of native healers, or shamans, about the medicinal properties of local plants, but bioprospecting is not limited to rain forest species or even land species. Many marine organisms have yielded biochemical compounds now widely used as medicines, foods, and other products.

Although bioprospecting has been very successful in finding potential uses for biodiversity, many local inhabitants still do not benefit from it. Typically, it is foreign-owned drug and agricultural companies and other businesses that profit from the genetic diversity of developing nations. This inability of the developing nations to profit from their own biodiversity was a major issue at the 1992 Earth Summit in Rio de Janeiro. After heated debate, 158 nations signed the Convention on Biological Diversity, which acknowledged that developing nations deserve a greater share of the profits generated from genetic resources by agriculture and genetic technology. In essence, this means that genetic resources are similar to intellectual property rights that can be patented, with profits returning to local populations where the genetic information originated. Such **genetic patent rights** would allow native people who conserve their biodiversity to receive some of the enormous profits often derived from pharmaceuticals, agriculture, and other uses of their genetic resources. An example of a group actively working toward such goals is the Australian Plants for People program (part of the Desert Knowledge Cooperative Research Centre). Plants for People is working with indigenous peoples of the Titjiaka region, Central Australia, to collect plant specimens and record traditional knowledge about the plants. The information is entered into an electronic database, but access to the data is restricted to preserve and protect the intellectual property rights of the indigenous peoples.

Studies often show that the long-term value of a rain forest is greater when it is sustainably harvested for medicines, rubber, foods, and other products than when it is logged, burned so that grass can be planted for grazing cattle, or otherwise destructively exploited for short-term profits. The long-term value from sustainable harvesting is even greater if the preserve can also be used for ecotourism. For example, the total revenue from Amboseli National Park in Africa has been estimated at about 40 times the revenue that would be produced from farming the same land. Each lion was estimated as worth $27,000 in tourist revenue, more than its value either dead or alive in the illegal wildlife market (**FIGURE 12-12**).

Sustainable harvesting of natural healthy temperate forests also is important. Private companies that grow trees for commercial logging own much of the forest area in the United States. Many rare species occur in these private forests, and many more could exist there if current forestry practices were discontinued (Table 12-1). An important result of wood production is the loss of old-growth, or virgin forests. More than 95% of such forests have been cut down in the lower 48 states. An old-growth forest takes many decades to grow because it is a late-successional, climax stage in forest growth (a review of the biosphere, its populations, communities, ecosystems, and biogeochemical cycles provides additional information). Old-growth timber generally is denser, higher-quality wood. Old-growth forests are also rich in species not found in early-growth forests. Because timber takes so long to grow, timber companies tend to raise only early-successional trees, such as certain pine species, that grow quickly. These can be cut down in a few years and replanted to start the cycle anew. This practice may maximize wood production, but it reduces overall biodiversity in U.S. forests.

12.4 Marine Preserves

Although the extinction of land species has attracted more attention than the extinction of species living in water, aquatic species, in both freshwater and marine environments, are also being decimated by widespread human disturbances. The protection of aquatic species has lagged behind that of land species for many reasons. One is that people cannot see the extent to which biodiversity loss is occurring there. In addition, in some ways, aquatic species are more difficult to protect. Large bodies of water, especially oceans, are classic commons and are not clearly owned

FIGURE 12-12 A lion is generally much more valuable as a source of income to local inhabitants if it is alive.

(a)

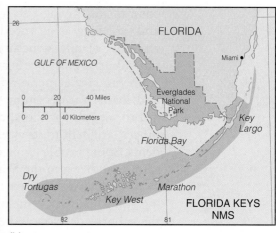

(b)

FIGURE 12-13 Florida coral reefs are threatened by shipping, development, water pollution from land, and many other activities. (a) A Spanish hogfish and coral in the Florida National Marine Sanctuary. (b) The Florida Keys National Marine Sanctuary preserves the only U.S. reef.

by any one nation. Consequently, regulating protected areas can be difficult. For example, pollution discharge in one area may travel hundreds of miles in an ocean current.

Nevertheless, there is growing interest in protecting these crucial marine habitats, especially nearshore bays and reefs that are being overvisited by tourists and contaminated by water pollution. Governments are establishing **marine protected areas**. In theory, fishing, construction, tourism, pollution, and other human disturbances are closely regulated and restricted in these areas, but the restrictions often are difficult to enforce, especially in poor developing nations with relatively few resources.

In the United States, marine protected areas are included in the **National Marine Sanctuary** program. This program was created in 1972, a full century after Yellowstone was dedicated as the first national park, illustrating how marine protection has lagged. Since 1972, 14 sanctuaries have been established under the jurisdiction of the National Oceanic and Atmospheric Association that encompass more than 388,000 square kilometers (150,000 square miles). One of the largest is the 354-kilometer long (220-mile) Florida Keys sanctuary, where fishing and coral col-

lecting are restricted because of widespread ecosystem disruption by people (**FIGURE 12-13**).

Many marine biologists argue that many more such sanctuaries desperately need to be established, particularly throughout the Caribbean and western Pacific, and especially around Southeast Asia, where poor nations have allowed massive destruction of coral reefs and other crucial marine habitats. Buffer zones, ecotourism, sustainable harvesting, and many other economic incentives, if properly applied, can be used to promote marine preserves just as effectively as with forest and other land preserves.

12.5 Managing Lands to Reduce Urban Impacts

The last 200 years have seen a strong global trend toward **urbanization**. In about 1800, only 3% of humanity could be classified as urban dwellers. By 1950, this number had risen to approximately 30%; as of 2005, about half of all people live in cities. If the current trends continue, it is projected that in 2030, the urban population will amount to 60% of the world's total population.

As a city's population grows, it spreads out along the edges into the surrounding countryside, giving rise to **suburban sprawl** and an increasing need for tremendous amounts of infrastructure. Slowly, residential suburbs push farther and farther away from the city center, and shopping centers and malls, theaters, and medical facilities, not to mention schools, churches, and post offices, are built to service the suburban residents. Soon, people living in the suburbs may rarely venture into the heart of the city; they can find everything they need without leaving suburbia. City centers tend to decline as the suburbs develop and expand. An important general rule is that lateral expansion increases at a much faster rate than population growth of a city. In fact, the population of many cities in the United States has declined, whereas the area occupied by sprawling development has increased! As the population moves away from the center, the core of the city may deteriorate until eventually low land prices attract businesses and developers once again. A large city may go through cyclical stages of growth, stagnation, and rejuvenation.

As the suburbs of large cities expand outward, adjacent cities and their suburbs may meet and begin to merge, forming a network of adjacent urban and suburban communities (sometimes referred to as a *conurbation*), until they become a single enormous urban area, the **megalopolis**. The term *megalopolis* was first coined to describe the almost continuous urban and suburban sprawl in the northeastern United States from Boston, through New York City and Philadelphia, down to Washington, D.C. Examples of megalopolises can now be found throughout the world; for instance, in the Netherlands, Amsterdam, Leiden, The Hague, Rotterdam, and Dordrecht all run together, and in Japan, the Tokyo–Yokohama complex forms a megalopolis.

In the United States, suburban sprawl is a very widespread problem. Currently, approximately half of the 310 million people in the United States live in the suburbs, whereas 30% lives in cities. A mere 20% of the population, the lowest percentage in U.S. history, lives in rural areas. More ominously, the amount of land being converted to "built-up" areas is increasing exponentially so that nearly 1.2 million hectares (about 3 million acres) per year are being developed in the United States. The rate of development is now much faster than the rate at which land is preserved as public or private land (**FIGURE 12-14**).

Causes of Sprawl

Surveys consistently show that most people tend to have a negative opinion of sprawl. Then why does it happen? Suburban sprawl has many causes. One of the most important is the rise of automobile use and the proliferation of roads that make commuting easy. Economics influences both of these factors in the United States, where gasoline is much cheaper than in nearly all other nations. In addition, road construction is encouraged by federal and state gasoline taxes that are strictly dedicated to the state and federal departments of transportation and provide enormous funds for the purpose of building roads.

Of course, ease of transportation is only part of the cause. Suburbanites must have some incentive to move to the outer fringes of cities. When asked, many suburban dwellers say that the main attraction to living in suburbia is the desire to own a parcel of land that can be gardened. (Gardening is the most popular outdoor activity in the United States.) Another reason often cited is the desire to escape the greater perceived crime rates of downtown areas.

Cheap land at the suburban fringe is yet another incentive because it allows developers

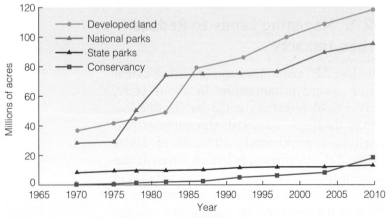

FIGURE 12-14 In the United States, acreage being developed is increasing much faster than acreage being preserved as public land (parks) or private land (conservancy). (*Source:* Data from U.S. Census Bureau, 2003.)

to build residential subdivisions at relatively low cost. This land is often cheap because it is agricultural land, and the farm industry in the United States is in decline in many areas (**FIGURE 12-15**). This is especially true of the small family-owned farms surrounding many urban areas. The expansion of large-scale corporate farming has made it very difficult for small farms to compete. As a result, it is usually much more profitable for individuals to sell farmland near cities for development into residential and commercial subdivisions than to farm the land. Thus, many homeowners move to the suburbs because the housing is most affordable there.

Consequences of Sprawl

Sprawl has many costs. These include economic, environmental, and human health costs. Starting with economic costs, sprawl is the most expensive way for people to live because it requires a widespread infrastructure of roads, sewers, telephone and television cables, police and fire units, and many other services. It is much more costly per person to provide these services to individual residences than to a few communities (such as large apartment buildings) that house many people. However, to be fair, apartment buildings are a major resource sink in terms of construction materials.

Sprawl is also the most harmful way to live, in terms of harmful impact per person on the environment. By widely dispersing human habitation on the landscape, the effects of each person in producing air and water pollution and consuming resources (such as gasoline and wildlife habitat) are greatly enhanced. For example, a single gallon of gasoline produces more than 20 pounds of carbon dioxide plus considerable amounts of smog and other harmful pollutants. Nevertheless, 80% of the automobiles on the road have only one person in them, most of them commuting to work, going to a store, or performing some other local chore that can be accomplished only in a car because walking is not feasible. Similarly, water pollution is greatly increased by the extensive pavement and concrete that much sprawl necessitates: at least 50% of the surface area of the aver-

FIGURE 12-15 There are no easy answers to the question of which is more desirable: affordable homes for a growing population or farmland to feed the growing population?

age suburb or commercial center is covered with buildings and pavement. In fact, 5% of the surface area of the lower 48 states is covered with pavement and concrete. Such "sealed surfaces" do not permit the infiltration of rainfall into the ground; thus, surface water runoff carries increasing amounts of litter, oil, and many chemicals into local streams. This also slows the recharge of groundwater supplies and increases flooding. Extensive areas of roads and concrete are the most devastating of all human activities to both plants and wildlife, producing what is essentially a "biological desert" in which nothing grows. If ugly landscapes can be considered visual pollution, then we should note the ugliness created by traffic, strip malls, and monotonous housing developments that sprawl causes.

Sprawl increases the consumption of resources per person in many ways. There is one car per person in the United States. The extensive driving of these cars consumes many billions of gallons of gasoline. The average American adult drives more than 200 miles per week. Nationally, Americans drive more than 1.5 trillion miles per year, which is equal to nearly 15,000 trips to the sun, which is 93 million miles away.

Sprawl also takes a very heavy toll on human health. Many thousands of people in the United States die prematurely of illnesses caused by smog that originates mainly from automobiles (a review of air pollution provides more information). Water pollution arising from sprawl contains sewage and many other harmful substances that take a human toll in waters used for drinking and recreation (a review of water pollution provides more information). Perhaps a less obvious health cost is the lack of exercise produced by spending so many hours in a car. The average American spends nearly an hour per day in a car, which adds up to nearly 3 years for a person who has a lifespan of 70 years or so. Instead of walking to the store or the neighbor's house, suburbanites drive everywhere and as a result get out of shape. In 2002, the American Medical Association issued a report stating that about 60% of Americans are overweight, and many Americans have heart disease and other health problems; the main cause is apparently a lack of exercise that their suburban lifestyle produced. Finally, there are mental health costs. Many studies by social scientists have concluded that the typical suburban lifestyle produces consid-

erable social isolation that leads to a "loss of community." Unlike true villages of past cultures, modern neighbors have little need for cooperative interactions, and they often do not even know the names of people living 100 feet away. These studies also show that the resulting social isolation generally has harmful effects on both mental and physical health.

Curbing Sprawl

The best solution to suburban sprawl is **compact development**. Instead of dispersing human habitation across the landscape, compact development minimizes the amount of space occupied by setting boundaries on city growth (**FIGURE 12-16**). You might think that compact development would lead to a lower quality of life for urban dwellers; however, many studies by urban planners show that there is an enormous amount of inefficiently used land within most cities. By designing dwellings, public transportation, and other human needs to more efficiently use this land, compact development can accommodate much population growth at a high living standard.

There are at least four ways that cities can increase the efficiency of land use to meet the needs of a growing population. Perhaps the most obvious is the use of **multifamily housing**. Instead of single-family homes on individual land parcels, many more people can be housed per acre in condominiums, apartment buildings, and other "high rises." A second way is the use of **clustered housing**, where developers cluster houses fairly close together but leave relatively large amounts of open or undeveloped land that can be used for a social commons, park, and wildlife habitat. Clustered housing contrasts with the standard approach of placing houses on uniformly separated land parcels, with very little land left for recreation, wildlife, or other nonresidential uses. A third way is called **multiple-use development**, where the same piece of land is simultaneously used for different purposes. For example, some cities now use riverside land for recreation, flood buffers, water filtration by plants, and wildlife habitat. Some recreational city parks have been designed with

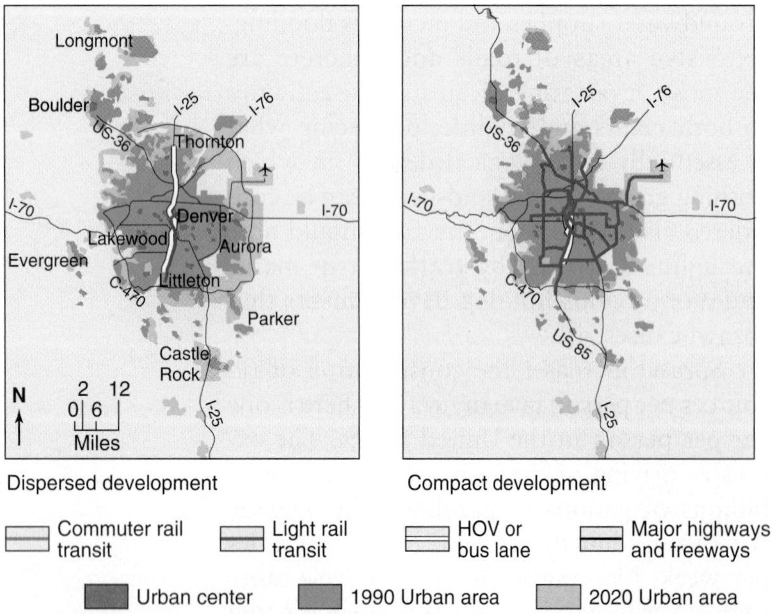

FIGURE 12-16 Compact development (using Denver, Colorado, as an example) is a more efficient way to accommodate population growth because it encourages development that builds "up" instead of "out," as seen in dispersed development. HOV = high occupancy vehicle.

native plants and natural features to provide much better wildlife habitat than the typical urban park designed only for people.

A fourth way of more efficient land use is **sequential-use development**, where land that is no longer needed in some capacity is used in another way. For example, old landfills have been converted into parks, residential neighborhoods, and shopping areas. In Knoxville, Tennessee, a local nature center is transforming an old mining quarry into a natural landscape to be used as wildlife habitat and also for recreation (see **CASE STUDY 12-2**). A very important kind of sequential-use development is **brownfield development**. A "brownfield" refers to old, run-down parts of a town, such as an old factory or a downtown area with empty stores. Brownfield development thus refers to improving such abandoned parts of a town by renovating them into new industries, stores, or multifamily housing (**FIGURE 12-17**). For instance, abandoned stores in many downtown areas of the United States are being modernized into comfortable condominiums that are bought by professionals who want to avoid commuting from the suburbs.

Even in regions where a city must expand into adjacent natural lands, there are ways to minimize the negative impacts of land use on the environment. One very important approach is to use **natural siting criteria** when determining where buildings, roads, and other structures will be constructed. This means that the location (siting) of a structure is selected to be where the geological and biological factors are most favorable. For example,

if you have a choice between building a new mall on the side of a mountain that contains many rare species versus building the mall in a flat area with only widespread common species, you make the latter choice because it is much less environmentally harmful. This is sometimes called the "working with nature" approach to designing and engineering human settlements. Unfortunately, this approach is very rare, especially in the United States. Many studies that urban planners have done show that human cultural needs, such as real estate values and commuting time, are the overwhelming criteria used to determine where a structure will be located. In the vast majority of the cases analyzed, geological and biological criteria played no role at all in determining where structures were built. As a result, erosion, landslides, and flooding have increased, unique natural vistas are lost, and many species are removed from the area.

Another way to minimize the impact of sprawl is sometimes called **wildscaping** (a contraction of "wilderness" and "landscaping"), referring to the retention of native vegetation, soil, and other natural features of the land when building on it. Most construction projects begin with the complete removal, using bulldozers, of the soil and native vegetation. A main reason for this is economic: the presence of vegetation, especially trees, impedes the extensive movement of people and machines during construction. When construction is finished, the area is then replanted, but usually with ornamental plants that are not native to the area and thereby are lower quality

(a)

(b)

FIGURE 12-17 (a) This brownfield development project turned a former blighted area into (b) an apartment and retail complex in Seattle.

12.5 Managing Lands to Reduce Urban Impacts   365

The Restoration of Mead Quarry

VISIT

http://environment.jbpub.com
/mckinney/5e/
for more information

The restoration of Mead Quarry by Ijam's Nature Center of Knoxville, Tennessee, is a good example of sequential land use. Mead Quarry was first established as a marble quarry in the early 1880s. It was then sold to Williams Limestone Company in 1945 and operated as a limestone quarry until the company filed for bankruptcy and abandoned the property in the mid-1970s. The site has played a significant role in Knoxville's history since it was established, and throughout the years, marble cut and polished at the quarry has been used in the construction of several notable buildings in downtown Knoxville, as well as other buildings and monuments in Maine, New York, Ohio, and Indiana. Since the mid-1970s, the site has been used as a dumping ground for everything from stolen cars to household garbage to toxic chemicals and worse.

In 2001, Knox County acquired the 50-acre Mead's Quarry property that is located adjacent to the Ijam's Nature Center. The property, which has been subsequently leased to the Center on a long-term basis, has provided an unprecedented opportunity for the preservation of spectacular natural beauty in South Knoxville. Alongside the Center's existing 100-acre park, the site's natural high bluffs, a 25-acre lake, an abundance of wildlife, and noteworthy geological features combine to provide unique recreational and educational opportunities for the local community.

Ijam's Nature Center has a long-standing history of environmental stewardship within the local community through considerable wetland protection and conservation initiatives. Through strong partnerships with other local organizations, the Center is able to make a difference in the quality of life of community residents. The restoration of the Mead's Quarry property will continue this trend by preserving stunning green space within 3 miles of downtown Knoxville.

Restoration Plans

To restore the property to a state of natural beauty for the enjoyment of the local community, the site will undergo significant cleanup activities to remove all toxic waste and other dangerous materials. The Center will coordinate a series of volunteer-led workdays to remove lightweight debris by hand. Larger piles of industrial waste, including potentially dangerous materials such as asbestos or glass that cannot be cleared by the volunteers, will be removed by industrial front-loading equipment.

wildlife habitat. Thus, wildscaping usually is more expensive because equipment use is restricted and special care is taken during construction of the houses or buildings; however, a growing number of cities, homeowners, and even golf courses are using wildscaping techniques to preserve natural habitat and reduce environmental impacts where construction is deemed necessary (see **CASE STUDY 12-3**).

A final way of minimizing sprawl impact is **mitigation**, which is the replacement of a destroyed ecosystem with a new one that people create. For example, if a developer is permitted to pave over 5 acres of wetland habitat, then state and federal law usually requires the developer to replace those 5 acres by creating or restoring a healthy wetland somewhere nearby. In fact, most laws require a ratio of 2 or 3 to 1 so that 2 or 3 acres must be created to replace each one that is destroyed. Although this sounds like a good compromise between social and ecological needs, many environmentalists and ecologists are critical of mitigation because it is very difficult to "create" an ecosystem that will function as well and persist as long as a natural ecosystem.

Legal and Economic Incentives to Curb Sprawl

The traditional method of regulating urban development has been zoning regulations. These regulations specify the type of development, such as commercial or residential, that

In addition to the removal of discarded waste, the major components of the restoration plan include the following:

Vegetation Removal and Native Plantings

The Center's education staff will undertake conservation activities to restore the environmentally damaged area and strengthen the wildlife habitat already in existence. The focus will be to increase biodiversity through exotic plant removal and native planting and to improve water quality through protection of riparian zones. The planting of forage and grasses will stabilize cleanup sites and excavation. Native trees and shrubs will be planted from the Center's stock to restore the site. Habitat improvements will also include ground leveling and refilling of disturbed areas. Landscape plantings in the lime-rich soils will greatly beautify and soften the site. Selective reforestation will be undertaken in appropriate areas and nest boxes will be installed for bird populations (**FIGURE 1**).

Trail Construction

The first proposed hiking trail will be half a mile in length. Commencing at the main entranceway, the trail will follow an existing roadbed along the southern perimeter of the property and ascend up a gradual incline at the west end of the lake. The trail will incorporate a natural overlook affording spectacular views of the property. Expansion of the trail to complete a circular walkway around the whole site will be undertaken at a later date.

VISIT
http://environment.jbpub.com
/mckinney/5e/
for more information

FIGURE 1 The property has already shown itself to be a source of neighborhood and community pride. Since the acquisition of the property, there has been a groundswell of support from within the community to assist with a volunteer-led cleanup operation.

is allowed in an area. County, and sometimes city, governments typically make zoning regulations based on the needs of the community. In the past, most zoning has not been aimed at restricting suburban growth. Rather, it has focused on keeping areas of commercial, residential, and agricultural uses from becoming intermixed. However, it would be possible to zone certain areas for "nondevelopment," where the land would be managed as wildlife habitat and light recreation. A common example of this is **greenbelt zones** that prohibit any removal of native vegetation around a city (**FIGURE 12-18**). These zones are especially important in cities surrounded by hills or mountains, such as Boulder, Colorado, because the vegetated mountains are such an important part of the scenic view visible from throughout the city. Currently, the use of zoning to regulate sprawl varies greatly among communities, so there is growing interest in producing state laws that provide a coherent set of standards. Some states, such as Tennessee, have already passed such laws, requiring cities to designate boundaries that limit their spread. However, as with all laws, it remains to be seen how effectively these state urban growth boundary laws are enforced.

As more cities struggle to control sprawl, they have devised a variety of economic incentives to slow its growth. One method is a gasoline tax, but the tax has to be high enough

Should We Rethink Lawns and Golf Courses?

VISIT

http://environment.jbpub.com
/mckinney/5e/
for more information

Lawns and golf courses are excellent examples of how small-scale activities that we take for granted can accumulate to cause widespread environmental harm. Both lawns and golf courses have become traditions in the United States and other developed countries, ironically because of our desire to enjoy nature (**FIGURE 1**). Yet, as so often happens, traditional activities can become environmentally damaging if individuals carry them out on a huge scale.

If all the lawns in the United States were assembled together, they would cover an area of 10 million hectares (25 million acres), which is equivalent to the combined area of Vermont, New Hampshire, Massachusetts, Connecticut, Rhode Island, and Delaware. This makes lawn grass the largest single crop in the United States in terms of area. We care so much for our lawns that we spend $25 billion on the lawn care industry each year.

Many psychologists believe that lawns, and our care for them, are a way to feel close to nature.

The smell and images of greenery are pleasant to many of us. Unfortunately, this particular way of staying close to nature is actually very "unenvironmental." Suburban lawns account for much of household water use, which is a serious problem in water-poor areas such as the southwestern United States and especially southern California. In addition, lawns average more than three times more pesticides per acre than an equivalent area of cropland. U.S. lawns absorb more synthetic fertilizers than the entire nation of India applies to all its food crops. Finally, mowing the lawn for 1 hour with a gas-powered mower can produce as much air pollution as driving a car for 544 kilometers (340 miles).

What can we do? Growing a variety of grass types in good topsoil and keeping the grass high discourages weeds and reduces the need for pesticides and herbicides. Many air quality management districts have sponsored buy-back events in which people exchanged their gas-powered mowers for a rebate coupon, typically worth $100.00,

FIGURE 1 Unless carefully managed, golf courses can use more pesticides and fertilizers and cause more water pollution per acre than most farms.

good toward the purchase of an electric mower from a specific manufacturer. The **Environmental Protection Agency** is phasing in stricter emission standards for new models. F. Herbert Bormann edited the book Redesigning the American Lawn: A Search for Environmental Harmony (1983). This book's answer is that if we wish to enjoy nature, we should stop trying so hard to cultivate the grasses and other plants that are not native to our backyards. The huge investment in water, pesticides, fertilizers, and mowing in the typical suburban lawn is needed only because we attempt to grow nonnative plants. Instead, it is suggested that we step back and let some of the native plants, such as crabgrass, chickweed, and many others, become established. We call such plants "weeds," but they are really just native plants that recur because they are better suited to the environment. Bormann and his Yale colleagues suggested that a totally environmental "Freedom Lawn" would consist entirely of native plants, require less care, and still provide the same feeling, if not more, of being close to nature.

On a per-acre basis, golf courses are even more harmful than lawns. Golf courses can use nearly seven times as much pesticide per acre as cropland, twice as much as the average for lawns. A single golf course can cover 81 hectares (200 acres) and usually requires even more fertilizer per acre than a lawn. The United States already has nearly 18,000 golf courses occupying an area larger than the state of Delaware, and the rate of new golf course construction is increasing. From 1995 to 2000, an average of 340 courses were built or older courses expanded each year. By January 2011, there were 15,890 golf courses in the United States, an increase of approximately 400 new golf courses from the year 2000. This is more than twice the rate of the mid-1980s.

Florida, California, and Michigan have the most courses, with more than 1,000 each.

All of these pesticides have had visible effects. Bird kills related to golf course pesticides have been reported: in some cases, hundreds of birds died. Human health effects could be very serious. A 1993 study by the University of Iowa Medical School examined data on the cause of death of 618 golf course superintendents between 1970 and 1992. It showed that the superintendents had exceptionally high frequencies of cancers of the lung, brain, intestine, prostate, and some lymphomas. These are often related to pesticides, and the study recommended more restricted use of pesticides and better protective clothing for golf course workers.

What can be done? As with lawns, there are more "environmentally friendly" ways to design golf courses. For new and existing courses, environmental planning, wildlife and habitat management, chemical use reduction and safety, water conservation, and water quality management by course superintendents can help courses serve as ecologically valuable green spaces.

Questions

1. Do you think lawns and golf courses should be redesigned? Perhaps you are like many Sierra Club members: one in every six members is a regular golf player. Some are not concerned, and others have had to rethink their hobby.

2. What about lawns? If you owned a home, would you have a traditional yard and mow the lawn on weekends?

3. What are some alternatives to green lawns? How would you redesign golf courses to be less environmentally harmful?

VISIT

http://environment.jbpub.com /mckinney/5e/ for more information

FIGURE 12-18 Even areas within suburban developments can serve as a greenbelt, prohibiting development and greatly increasing the visible beauty of the suburb. Shown here is a greenbelt buffer along a small stream in a suburb of Des Moines, Iowa.

to have an impact (even at gasoline prices of more than $3.50 a gallon in 2011, some people continued to drive gas-guzzling sport utility vehicles). Increasing gasoline taxes make long commutes less attractive, so there is more demand for less dispersed housing and commercial space. These taxes also cause drivers to take fewer unnecessary trips, buy cars that are more fuel efficient (and less polluting), and encourage commuters to carpool and use mass transit. Perhaps more important is that the funds collected by increased gas taxes can be used by the city to improve mass transit systems, for brownfield development, and to buy land (or the "development rights" to that land) that can be used as parks and other undeveloped spaces. Even without an increase in gasoline taxes, a current cause of sprawl is that much of the money now collected as gas taxes is automatically transferred to state and federal departments of transportation. As a result, these departments have an enormous amount of funding available for building roads and very little incentive to promote alternative transportation. In most states, education departments and most other state agencies must compete for dwindling funds every year, whereas the transportation agencies have almost sole access to the large and increasing funds created by gas taxes. Thus, a key solution is to see that more gas tax money is distributed for purposes other than road building.

Another option is to increase taxes on developed land to collect funds for brownfield development and buying undeveloped land. At the same time, the government could reduce taxes on land that is farmed or otherwise not used for development. A final option is the developer-pays approach, in which cities require the developer of a residential subdivision to pay for the costly infrastructure of roads, sewer lines, utility lines, and other services that are often paid for by taxpayers. Such taxpayer subsidies only encourage the growth of sprawl by making it relatively cheap for the developer to build housing and acquire very large profits in the process. If the developer is required to pay for infrastructure, there will be less incentive to build large subdivisions and more incentive to spend development money on compact development, brownfield development, and other ways of more efficiently using land that is already available. Under the **developer-pays** approach, housing in new suburban subdivisions will be more expensive because it will not be subsidized by taxpayers; this is another example of when consumers may be obliged to pay the "true environmental costs" of items if the free market system is to promote sustainability (a review of people and natural resources and environmental economics provides more information).

A final important option for controlling sprawl is by the actions of private preservation groups, especially the land trusts and conservancies noted. These groups do a great public service by preserving land before it is developed; however, often is too costly to own preserved land outright because of maintenance costs. Therefore, a more common preservation tool is the use of **conservation easements**, which are legal agreements with a landowner by which the development rights to the land are surrendered. In other cases, the preservation group leases these rights.

Such land trust agreements are often complex, involving expiration dates when development rights may be repurchased and problems when the owner sells the land; however, land trusts will be a fast-growing and very important method of preserving land for many decades to come.

SUMMARY

- Management of public lands must balance the needs of people versus the need to maintain healthy ecosystems for current and future generations.
- Public land management in the United States varies with the goal of the government agencies involved.
- There is a continuing debate over how many tax dollars should be spent on subsidizing resource extraction of public lands.
- A major tool for protecting public lands is the Environmental Impact Statement, which requires that planned activities be assessed for their impacts.
- Management of private lands presents both a challenge and an opportunity for conservation.
- Unlike the past, when preserved land was selected for its scenic beauty or low commercial value, future parks should include biodiversity preservation as a selection criterion.
- Preserves should be designed to have buffer zones, connecting corridors, maximum size, and other variables that promote biodiversity preservation.
- Ecotourism and sustainable harvesting are good examples of how preserves can be maintained for sustainability while providing income for local human inhabitants.

- Suburban sprawl is a rapidly growing problem in many parts of the world, especially the United States.
- Causes of suburban sprawl include cheap land prices at the urban fringe, a strong reliance on the automobile, and taxes and economic incentives that promote suburban development.
- Harmful impacts of sprawl include air and water pollution, habitat loss for wildlife, and human health costs from lack of exercise and social isolation.
- Compact development reduces sprawl and can be promoted by clustered housing, multiple-use development, sequential-use development, and brownfield development.
- Ways to minimize sprawl impacts include using natural siting, mitigation, and wildscaping.
- Incentives to control sprawl include greenbelt and other zoning, lower taxes on farmland, buying development rights, the "developer-pays" approach, and conservation easements.
- Lawns and golf courses are often not environmentally friendly, they disrupt local native flora and fauna and may require the excessive use of fertilizers and pesticides.

KEY TERMS

bioprospecting
bottom-up approach
brownfield development
buffer zone
clustered housing
compact development
conservation easements
conservationists
developer-pays
ecotourism
Environmental Impact
    Statement (EIS)
genetic patent rights
greenbelt zones

Healthy Forests Initiative
hot spots
Land Trust Alliance
marine protected areas
megalopolis
mitigation
multifamily housing
multiple-use philosophy
multiple-use development
National Environmental
    Policy Act (NEPA)
National Marine Sanctuary
natural siting criteria
preservationists

Roadless Area Conservation Rule
scoping
selective cutting
sequential-use development
suburban sprawl
sustainable harvesting
umbrella species
urbanization
Wilderness Act of 1964
Wildlands Project
wildscaping
Wise Use Movement

STUDY QUESTIONS

1. What are some controversies over the use of public land?
2. What is the bottom-up approach? How are buffer zones, ecotourism, and sustainable harvesting related to this?

3. What is an EIS? What does it do and what are some problems with it?
4. How are genetic patent rights important in promoting biodiversity conservation? Give specific examples. What is bioprospecting?

5. What is a hot spot? How can hot spots be used in selecting the location of preserves?
6. What are two widely suggested solutions to the national park funding problem?
7. Compare and contrast selective cutting and clear-cutting.
8. Discuss the problems of the U.S. Forest Service and those of the National Park Service. What differences and similarities do you see in the problems of the two agencies?
9. Name four ways that land can be more efficiently used to reduce sprawl.

10. What is ecotourism? Discuss its advantages and disadvantages in saving species.
11. What is wildscaping? What is mitigation? How are they useful in coping with sprawl?
12. What is a conservation easement? Why is it sometimes better than buying land to preserve?
13. Give some examples of sustainable harvesting.
14. Name some ways that traditional forestry can harm the forest ecosystem.
15. What is the Wildlands Project? Is it realistic? Why or why not?

WHAT'S THE EVIDENCE?

1. The authors make several assertions that suburban sprawl is bad for human health and the environment. What is the evidence to back this up?

2. The authors suggest that abuse of federal public land is sometimes subsidized by taxpayers. What is the evidence for this? Is this evidence objective?

CALCULATIONS

1. By what percentage did the amount of designated wilderness land increase between 1970 and 2005? (Refer to Figure 12-3.)

2. If an average American drives 200 miles per week, how many miles will that person drive in a year? In 10 years? In a driving lifetime of 50 years?

http://environment.jbpub.com/mckinney/5e/

http://environment.jbpub.com /mckinney/5e/

Connect to this text's website at http://environment.jbpub.com/mckinney/5e/. This site features eLearning, an online review area that provides quizzes, chapter outlines, and other tools to help you study for your class. You can also follow useful links for more in-depth information.

Food and Soil Resources

13

We are part of the global biosphere, and as such, we are ultimately dependent on other living organisms for virtually all aspects of our lives. Many of the ways we depend on other organisms are obvious. Trees shade us. Green plants and other photosynthetic organisms produce the oxygen (O_2) that we breathe and absorb carbon dioxide. Worms and microorganisms prepare the soil for us and recycle necessary nutrients, and insects pollinate our crops. We do not fully understand all of the complex interrelationships among bacterial and other microorganismal, plant, fungal, and animal species that make up the biosphere, and we have no way of knowing what roles certain seemingly insignificant species perform now or will play in the future.

Modern industry and medical technology are dependent on animals and plants for a plethora of important drugs and other substances. All of our fossil fuels were formed by organisms over millions of years, as were many commercially important rock and ore deposits (such as limestone and phosphate deposits). Biological resources provide a large percentage of the raw materials that we use (such as wood, natural rubber, and leather).

Perhaps most importantly, we are totally dependent on other organisms as a source of nutrition. We raise plants and animals to eat; we harvest wild plants and animals as food (for example, fishes). In this chapter, we focus on people's most fundamental use of biological resources. Because the majority of the world's food needs are met

Chapter Objectives

After reading this chapter, you should be able to explain or describe the following:

- What constitutes the major categories of human food sources on a global scale
- The concept of hunger
- Agricultural food production, both at present and projected into the future
- The differences between "traditional" and "modern" agriculture (including the "Green Revolution")
- Genetically modified food
- Aquaculture
- The nature and importance of soil

373

today by the cultivation of crops, we include a discussion of the soil resources that form the basis of **agriculture**.

13.1 Food as a Biological Resource

Virtually all of the food that people depend on is derived from other organisms. Although we eat many different types of plants and animals, actually only a very small number of species provide the majority of our food. Only 20 different species of plants supply 80% of the world's food supply, and just three kinds of plants constitute 65% of the food supply—rice, wheat, and maize (corn).

Hunger

The numbers are daunting. Recent United Nations studies estimate that 1.2 billion people do not eat enough every day for a normal, healthy life; malnutrition and hunger-related disease cause 60% of the more than 10 million deaths that occur yearly among children in developing nations. Even in the United States the Food Research and Action Center estimates that some 17 to 50 million people and close to 25% of all children do not have adequate and nutritional food sources.

Terms such as **hunger** and **malnutrition** can be hard to define, although their meaning can be painfully obvious when one sees a starving human. To maintain good health, a person must have adequate nutrition—proper amounts of protein and various vitamins and minerals—and an adequate supply of kilocalories as an energy source (in popular usage, kilocalories are often referred to simply as "calories," but a kilocalorie or Calorie is actually equal to 1,000 calories). The amount of kilocalories and nutrients that individuals require also depends on their age, gender, and other characteristics. Statistics of global hunger are often based on the number of kilocalories available or ingested per person per day. According to the United Nations, the recommended daily intake per person is 2,350 to 2,500 kilocalories. The U.S. National Research Council recommends a kilocaloric intake of 2,700 and 2,000 for the average adult male and the average adult female, respectively.

From a simple kilocaloric point of view, many people in the world do not consume enough to live and work actively. It is estimated that as many as 1.1 billion people are undernourished and underweight. In some countries, especially in Saharan Africa and Asia, the average daily kilocalorie intake is less than required for people to carry out productive work. In 2007, the United Nations reported that Zambia had a daily per capita supply of just 1,873 kcal. The United States had a daily per capita supply of 3,748 kcal. For comparison, the least developed nations had an average of 2,157 daily per capita kcal intake; all of North, Central, and South America together had a daily per capita supply of 3,216 kcal; and all of Asia had a corresponding supply of 2,668 kcal. Studies have demonstrated that in many countries, malnutrition is correlated with high death rates, particularly among children. Children with malnutrition are characterized by stunted growth, reduced mental functions and learning capacities, and lowered general activity levels. Damage caused by severe malnutrition may be irreversible, but one study found that increasing the per capita consumption of food from slightly under 2,000 kilocalories to 2,700–3,200 kilocalories could cut the death rate in half (**FIGURE 13-1**).

In addition to an adequate intake of kilocalories, a healthy person requires proper amounts of vitamins and minerals. Vitamin and mineral deficiencies can cause many symptoms, including general poor health; blindness (vitamin A deficiency); mental retardation; learning disabilities; decreased work capacity; decreased resistance to illnesses, diseases, and infections; and premature death. More than 1 billion people, primarily in developing countries, currently have vitamin and mineral deficiencies, whereas another 1 billion are at risk.

Feeding the World Today

Why are so many people in need of food, especially when some countries have grain surpluses? In part, the large inequities in food distribution are related to political and social problems; all too often, food is withheld as a weapon (**FIGURE 13-2**), especially in internal struggles in developing nations. However, political considerations are only part of the problem. Around the world, arable land is being lost to construction (for instance, build-

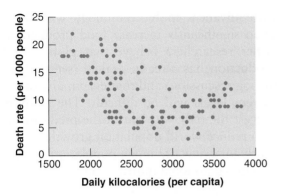

(a)

FIGURE 13-1 (a) The relationship between the daily energy supply per capita in 1989 and the crude death rate per 1,000 people in 1990 for 119 countries. Both too little and too much food, as measured using kilocalories, results in an increased death rate. (b) Severely undernourished children are more likely to die of childhood illnesses. (*Source:* (a) Adapted from H. R. Pulliam and N. M. Haddad. Human population growth and the carrying capacity concept. Bulletin of the Ecological Society of America 1994;75:150.)

(b)

ing more subdivisions and shopping malls), and cultivated land is being degraded through **erosion**. Furthermore, global climate change (a review of global air pollution, destruction of the ozone layer, and global climate change provides additional information) is displacing weather patterns, causing localized droughts and floods. Warming of temperatures also fosters insect infestations, and rising sea levels can flood low-lying coastal areas and cause saltwater contamination of groundwater supplies in coastal regions.

Still, today there is a remarkable amount of food being produced annually, but it is barely enough (**TABLE 13-1**). In fact, perfect management of the world's current food production might just barely feed the global population, but this would be only a temporary measure, for soon the global population will outrun food supplies.

Per capita **grain production** actually peaked in 1985 at 343 kilograms per person, and since then, the population has grown faster than grain production; lower global grain production and more people to feed brought this number to 299 kilograms in 2001. Based on a peak total grain harvest of approximately 1.9 to 2.4 billion metric tons (seen in 1997, 2001, 2004, and 2010), researchers have calculated the percentage of the world's population that could be fed using various diet models. In the United

FIGURE 13-2 A Palestinian security guard directs trucks bringing medicine and food from Jordan to the Gaza Strip in May 2006.

TABLE 13-1	Global Food Production in 2004
Food Source	**Metric Tons per Year**
Grain	2 billion
Meat	258 million
Wild fish	93 million
Aquaculture	40 million*
*Data for 2002.	

States, the typical citizen consumes about 25% to 30% of his or her kilocalories from animal products, such as meat and cheese—a notoriously inefficient way of deriving energy from foodstuffs. If the entire world followed the American dietary example, less than half of the current global population could be fed adequately. Typical Latin Americans consume about 10% of their kilocalories from animal sources. Using the Latin American diet as a model, approximately 4 billion people could be fed based on the harvest of 2010. Only with everyone maintaining a strictly vegetarian diet and assuming perfect food distribution systems (which is unrealistic given current political and transportation problems) could the current population of 7 billion be adequately fed. If we could considerably decrease the waste factor (for example, as much as 40% of all food typically spoils or is eaten by insects, rats, or other pests, and much food is thrown away as "leftovers," made into food for nonessential house pets, and so forth), perhaps as many as 8 billion people worldwide could be fed a subsistence diet.

Such food conservation measures would be of little consequence to the extra mouths that would need to be fed if the world's population reached 8 to 10 billion. There seem to be few options for dealing with this situation. One possibility is to increase global food production dramatically and to ensure that the food is equitably distributed; however, whether major gains in global food production will be possible is a point of heated debate. By 2050, the world population is estimated to top 9 billion people. To meet the basic dietary requirements of every person on the planet at that time, current food production would have to increase, by 70%. Some analysts argue that innovative agricultural techniques, combined with advances in biotechnology will allow us to significantly increase food production. Other researchers believe that global food production has already peaked (see the following discussion) and that even maintaining current levels will be difficult. The analysts suggest the only way out of the predicament is to reduce world population growth significantly (a review of human population growth provides additional information).

Food for the Future

If the world population continues to grow, we will need more food in the future. World food production could be increased through two basic, but certainly not mutually exclusive, strategies: increase the amount of land under cultivation or increase the yield per unit of land under cultivation. Today, slightly more than 1.5 billion hectares of land worldwide are under some form of cultivation—an average of approximately one-quarter hectare per person. According to various theoretical estimates, there are between 2 and 4 billion hectares of **cultivable land** on Earth (much of the variation in the estimates is attributable to the use of different criteria for defining "cultivable land"—how fertile the soil must be, how much water must be available, and so forth). However, much of the theoretically cultivable land realistically will never be cultivated, and doing so would either irrevocably damage natural ecosystems on which we depend or irrevocably disrupt human society. That is, cultivable land includes vast expanses of forests and grasslands (a review of the distribution of life on earth provides additional information), some of which are designated as National Parks or National Forests. In addition, much cultivable land has been paved over as highways, roads, parking lots, and shopping malls. Cities and suburbs are built on cultivable land.

Differing scenarios involving various combinations of land cultivation and yields on a global scale have been proposed (see **FIGURE 13-3**). If the population increases at the exponential rate seen in the 1990s (a review of human population growth and the term "unmodified extrapolation" provides additional information) and current average yields continue (that is, they do not increase), before the year 2050, all of the theoretical 4 billion hectares of cultivable land will have to be put into production, and by the end of the 21st century, the population will quickly outstrip its food supply. This scenario unrealistically assumes that the current high crop yields of prime land can be maintained, even on marginally arable land. However, it also assumes a higher population growth rate than is currently expected for the first half of this century. It also assumes that no cultivable land is lost during the next 100 years, which is also unrealistic because, as explained later, the absolute amount of cultivable land on Earth is declining because of various factors.

If the population increases according to various World Bank projections, which are probably more realistic at this point, we can sustain the Earth's population through the 21st century within the range of estimates of land available for cultivation. However, unless crop yields can be significantly increased or food losses and wastage significantly decreased, an additional 1 billion or more hectares may have to be placed under cultivation by the end of the century to feed all the globe's people. Such a dramatic increase in cropland will severely strain natural ecosystems and may be quite difficult pragmatically and politically.

Some optimistic studies have concluded that if we really wanted to, we could grow enough food to support a global population of 50 billion people. However, such conclusions are based on totally unrealistic assumptions. For one thing, they assume that all potentially arable land would be cultivated, including land occupied by forests and land that is of marginal fertility or is so arid that massive irrigation would be necessary. The human population would have to live in areas, such as the polar regions, where agriculture is totally impossible, whereas the potentially arable land beneath our current cities and towns would be put under the plow. In addition, these studies assume the yields on all this land would either match or (with technoagricultural advances) surpass those that have been attained under ideal conditions on the most fertile land in the past. Such super-high yields are only a pipe dream. Finally, these projections ignore the detrimental and nonsustain-

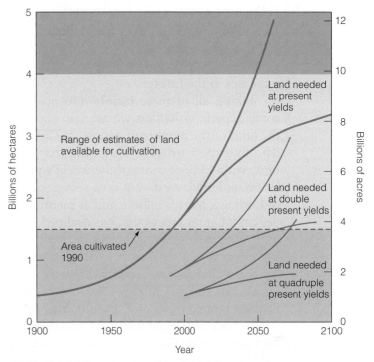

FIGURE 13-3 Possible land futures. The pairs of lines show the amount of land needed to maintain current per capita food production if the world population continues to grow exponentially at current rates (upper line of each pair) or if the world population grows according to early 1990s World Bank forecasts (lower line of each pair). The heavy solid pair of lines shows land requirements at current crop yields; the two lighter pairs of lines show land requirements at double and quadruple current crop yields. The yellow shaded area is the range of estimates of land that could possibly be cultivated for food; much of this land currently is covered with forests. (*Source:* D. H. Meadows, D. L. Meadows, and J. Randers. Beyond the Limits. Post Mills, VT: Chelsea Green, 1992, p. 51. Adapted from Beyond the Limits, © Meadows, Meadows, and Randers.)

able aspects of modern agriculture (discussed in a later section) and the consequences (climatic and otherwise) of destroying the world's remaining forests. The question of where the massive quantities of energy, **fertilizers, pesticides, herbicides,** and freshwater necessary for modern, intensive, high-yield agriculture will come from is not addressed, nor are the attendant problems of pollution caused by chemical use, **topsoil** (a mixture of mineral matter and humus, alive with microscopic and macroscopic organisms) loss and exhaustion, soil salinization, and waterlogging addressed. To suggest that we could feed close to 50 billion people is irresponsible—indeed, it is a lie.

A more conservative estimate is that we could potentially feed about 8 billion people, but even this estimate assumes better yields that occur today in many agricultural areas and also assumes essentially a subsistence diet. Although some increase in yields may be possible (for instance, the best farmers in Iowa can produce four times the world's average corn yield per acre), in recent years, global yields have shown signs of leveling, and we cannot necessarily count on additional large increases in the future.

Taking all of these factors into account, some experts insist that we are just about at the limit of the number of people we can realistically expect to feed. They note that even now, with 15% or more of the world's population underfed, we do not do a very good job of feeding a mere 7 billion, and as mentioned, per capita grain production has declined. Still, it must be pointed out that this is a very controversial topic. New crop varieties continue to be developed, and new methods of farming and more efficient equipment have yet to be used around the world. Even if crop yields have leveled in the last few years, this does not necessarily mean that they will remain at these levels for all time. They may begin to increase once again as successful new cultivation techniques spread around the world.

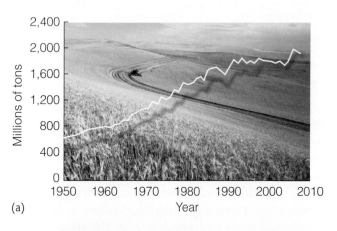

(a)

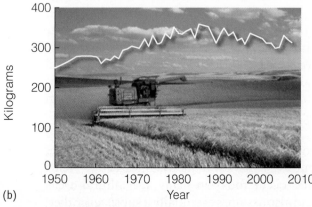

(b)

FIGURE 13-4 (a) World grain production, 1950–2005. (b) World grain production per person, 1950–2004. The background images show wheat and barley harvesting in Palouse, Washington. (*Sources:* USDA/FAS and U.S. Bureau of the Census.)

Agricultural Food Production and Supplies

Grain (rice, wheat, maize, sorghum, barley, oats, rye, millet, and so forth) forms the backbone of the world's food supply, so the global annual grain production is of utmost importance. Total world grain production nearly tripled between 1950 and the end of the century (**FIGURE 13-4**), in large part because of the "Green Revolution," which is discussed later.

Not all grain grown is used for human consumption. Since 1960, about one-third to more than two-fifths of the world's grain supply each year has been used to feed livestock and poultry. To raise livestock and poultry on grains, farmers must routinely include a protein supplement in the animals' diet. The most important such supplement is the protein-rich soybean, which is also a valuable source of oil and protein for people.

It is far more efficient to consume grains and other plants directly than to feed them to animals and then consume the animals. As the world population continues to increase and grain production remains stable or declines, it may become necessary to allocate less grain to animals if we want to ensure that all people are fed adequately.

Not all grain produced in a given year is necessarily consumed in that year. An important concept is that of **carryover grain stocks** from one year to the next. The size of the world's grain carryover stock is often used as an indicator of global food security. Often, if the stocks drop too low, as they did in the early 1970s, grain prices may fluctuate widely. In such cases, it becomes even more difficult for the poor to obtain adequate food until a good harvest is restored.

Some observers point out that historically there has always been grain to buy—as expressed in stockpiles of grain around the world. They contend that the primary reason hunger exists in the world today is because poor nations cannot afford to buy food. Some researchers maintain that the best way to ensure that all people are adequately fed is to expand food production in temperate regions (such as the United States) where advanced agricultural technologies and transportation systems are in place. In addition, there must be a global free trade policy for food. A worldwide free market system for food would discourage inefficient, often government-subsidized, food production in marginal areas. For instance, in India, overpumping of groundwater supplies using free government-provided electricity has resulted in lowered water tables, salinization, and waterlogged soils. In the name of "self-sufficiency," Indonesia has cleared 607,000 hectares (1.5 million acres) of tropical rain forest to grow soybeans for use as chicken feed. The problem is that the cost of Indonesian soybeans is higher than the price for soybeans on the world market. Likewise, India produces milk at a cost above world market prices.

Even those who espouse a "free market solution" to world hunger must acknowledge that the problems of debt (particularly on the part of developing countries), trade imbalances, and restrictions on free market policies are complex and not easily solved. Politically, it would be very difficult to have a genuine global free market system for food products. Even if free trade/free market policies could accomplish the equitable distribution of food around the world, there must be enough food to go around. Currently, the food supply appears to be adequate, but as we have discussed, if the global population continues to grow, it is far from certain that there will be enough food to feed everyone in the future.

Land, Fertilizers, and Water Devoted to Agricultural Production

An enormous amount of land is devoted to agricultural use. Globally, 3.3 billion hectares (8.15 billion acres) of land is used for grazing animals and slightly more than 1.5 billion hectares (3.7 billion acres) is devoted to cropland. Not all agriculture is devoted to food production—for instance, cotton and other crops are grown to manufacture textiles, and animals are raised for leather products. An estimated 275 million hectares (680 million acres) are artificially irrigated to grow crops.

Every year, new land is put under cultivation as forests are cut and dry areas are irrigated, but other land is removed from cultivation because of such factors as soil exhaustion, degradation, and the building of residences and shopping malls. Currently, there is a rough balance between land newly put

FIGURE 13-5 Genuine crop circles. Each of these circular features is an irrigated crop field in the desert southwest of Riyadh, Saudi Arabia.

under cultivation each year and land removed from cultivation, but some observers fear that within a few decades, the cropland area may begin to diminish substantially.

More importantly, because of global population growth and the dependence on grain crops, the amount per capita of land used to grow grain has steadily declined. For now, this decrease has been offset by an increase in yield. This was accomplished through intensive, often mechanized farming techniques using specially bred varieties of crops and massive doses of artificial fertilizers, pesticides, herbicides, and in some cases artificial irrigation (the "Green Revolution" is discussed later). In many regions, **irrigation** is essential to grow crops that would not otherwise survive (**FIGURE 13-5**).

13.2 The Effects of Agriculture

Most agriculture alters and manipulates natural ecosystems, transforming them into artificial ecosystems that are inherently unstable and can survive only with constant human attention. Maximum food production is the only goal.

In nature, during the process of **ecological succession**, successive groups of plants and animals will colonize a clear patch of land.

The first settlers generally will be smaller, fast-growing, pioneer plants. Then larger, slower growing, and longer lasting plants will progressively replace the original colonists. The final stage of succession is the **climax community**, which in many terrestrial areas consists of mature forest composed of large trees interspersed with younger trees and other plants and animals.

In clearing land for agricultural use, farmers essentially begin the cycle of succession anew (**FIGURE 13-6**). However, the farmer does not allow succession to follow its natural course and reach a climax stage. Instead, the land is artificially maintained at the pioneer

stage. The pioneer plants that are allowed to grow on the land are carefully picked, maintained, and managed. Corn (maize), wheat, or rice may be planted as the pioneer plant; other weeds are eliminated (many major food crops are essentially cultivated weeds). When the crop has matured, it is harvested; the next season the land is cleared, and the system begins again. Most agricultural systems emphasize the **pioneer stage of succession** because this is when an ecosystem is most productive (although not most efficient in energy use). In this stage, virtually all of the energy and nutrients that the plants use go into growth. In contrast, in the climax stage, much of the energy and most of the nutrients go into maintaining the system; the only new growth that occurs replaces plants that die.

However, the long-term arrestment of ecosystems at the pioneer stage leads to major problems. Pioneer ecosystems are inherently unstable, and this instability is exacerbated by the human habit of planting only one variety of plant per field at a time (**monoculture**). In a mature, climax ecosystem, the complex relationships and interactions among many species of plants and animals promote long-term stability. In the climax community, there are natural checks and balances on predator–prey relationships (including insect attacks on vulnerable plants), disease, population explosions of particular species, and so forth. Climax communities are also less susceptible to the ravages of climatic fluctuations such as droughts or floods. In the artificial environment of a crop field, human interventions must mitigate to a greater extent the pests, disease, and the vagaries of climate. Although there are more natural alternatives, pests might be controlled by applying poisonous chemicals to a field. Watering or irrigation may compensate for a lack of rain.

Pioneer stages also extract a heavy toll of nutrients from the soil without replenishing them. Replenishment occurs naturally during later stages of ecological succession, but when people harvest and remove their pioneer crops, the nutrients are lost from the land, and artificial fertilization must restore them. In contrast, complementary, even symbiotic, relationships between organisms character-

(a)

(b)

FIGURE 13-6 (a) Pioneers cut down climax stage forests to build their homes, grow crops, and generally "tame" the wilderness. (b) Strip cropping and woodlots in Leelanau County, Michigan.

ize the climax community. The nutrients that one organism extracted are eventually passed on to and restored by another organism. The cycle of growth, death, decay, and regrowth—all on the same parcel of land—ensures continued recycling of raw materials.

The characteristics of pioneer communities that make modern monoculture farming so productive and successful on a short-term basis cause continued environmental degradation in the long term. Rapid nutrient uptake (absorption) without recycling destroys the soil's fertility. A lack of a balanced vegetation—or no vegetation at all between harvesting and the next planting season—to hold the soil in place and absorb moisture can lead to massive erosion of valuable topsoil, flash floods, dust storms, and droughts. Monocultures are notoriously susceptible to attack by disease and pests uncontrolled by natural predators or other mitigating agents.

The Effects of Irrigation

In areas where irrigation is necessary, a whole new set of problems is encountered, particularly **salinization** and **waterlogging**. All soils contain various mineral salts. Under natural conditions, in areas that are characterized by relatively high rainfall and good drainage, these salts are washed out of the soil and travel, through water flow, to the sea. This is why the sea is salty—it is where salts from the land surface accumulate. In contrast, arid regions tend to have higher natural concentrations of salts in the soil and in any groundwater or standing bodies of water simply because there is not a constant flow of water to remove the salt. Irrigating arid land dissolves the salts in the soil, and as the water evaporates, the salts are drawn toward the surface. Many artificially irrigated lands are poorly drained, and as a result, the salts simply remain in the upper levels of the soil, rather than being flushed out and carried to the sea. Furthermore, the poorly drained land and soils themselves can become waterlogged, and the water table can rise over time, as the groundwater and soils become progressively saltier. Accumulated mineral salts are toxic to most plant life, and as land becomes increasingly salinized, it may reach a point where it can no longer support most crops or other plants (**FIGURE 13-7**).

Modern Agriculture's "Solutions": Fertilizers, Pesticides, and Herbicides

"Modern" agricultural techniques of the late 19th through 21st centuries have shunned many traditional farming methods as inefficient and unsuited to mass production. This has resulted in the "quick fix" of bumper

FIGURE 13-7 Salt-affected agricultural land near Katanning, Western Australia.

crops but has been at the expense of the land, nonrenewable mineral and energy resources, and long-term sustainability.

Much modern agriculture is synonymous with the circumvention of biological agents in the restoration of depleted soil fertility; crop diversity, crop rotation, and even manure use are abandoned. Instead, minerals and chemicals are mined, processed, and applied directly to croplands in the form of fertilizers; simultaneously, irrigation efforts are intensified. Sometimes this has been referred to as "force-feeding" the land.

The main nutrients applied to the soil are phosphorus, potassium, and nitrogen. Phosphorus and potassium are mined from mineral deposits; phosphorus in particular is potentially in short supply as high-grade, naturally formed phosphate deposits are exploited faster than new deposits are discovered (like the fossil fuels, there is only a finite supply of geologically formed high-grade phosphate deposits, many of which are the result of accumulations of ancient animal bones millions of years ago). Nitrogen was initially supplied from manure or from bird droppings known as "guano." Some isolated islands contain huge mountains of bird droppings, and these were mined for their nitrogen content. Now artificial nitrogen-bearing fertilizers can be manufactured synthetically using the abundant nitrogen of the atmosphere.

An increasing emphasis on monoculture (planting huge fields with a single variety of a single crop) accompanied the heavy use of fertilizers. Monoculture allows the farmer to tailor the fertilizers to the specific needs of the particular crop and increases efficiency in mechanical harvesting and processing of the crop. But monoculture brought with it increasing problems from pests and diseases that found a happy point of attack in the huge, ecologically unstable fields. This meant that such pests and diseases needed to be controlled. The preferred way to control them was through the use of more chemicals—synthetic pesticides and herbicides that could be designed to kill everything except the crop being cultivated. In the United States, pesticide use in agriculture rose from about 154 million kilograms (340 million pounds) per year in 1965 to about 404 million kilograms (890 million pounds) in the early 1980s. U.S. pesticide consumption then dropped slightly; it is currently 340 to 363 million kilograms (750 to 800 million pounds) per year. Globally, there has been a steady increase in pesticide use, with the average amount applied to agricultural land more than tripling—from about 0.5 kilogram per hectare in 1961 to between 1.5 and 2.0 kilograms per hectare at the end of the century. But such techniques led to obvious problems. Despite the massive addition of fertilizers, the soils slowly became exhausted. The major nutrients extracted by the plants were being temporarily restored, but many trace elements necessary for the ultimate sustainability of agriculture were not; examples of such trace elements include zinc, iron, boron, copper, molybdenum, and manganese. In addition, good healthy soil is more than just a handful of dry minerals and fertilizers. It is full of organic debris, humus, and living organisms, including worms, beneficial insects, fungi, and bacteria. These organisms help mix and aerate the nutrients. The texture, structure, and quality of the soil are necessary for the roots of plants to take hold and the soil to retain water—which helps minimize both droughts and waterlogging of the earth below. Dumping massive amounts of toxic substances (in the form of pesticides and herbicides) literally kills the soil (**FIGURE 13-8**). Dead soil

FIGURE 13-8 Spraying pesticide on leaf lettuce in Yuma, Arizona.

loses its structure and no longer functions properly. An unstable, dead soil may quickly erode away, perhaps further spreading the noxious chemicals that killed it.

The Green Revolution

Shortly after the end of World War II, modern, chemically based agriculture began to be used on a large scale in the industrialized countries. In the 1960s, the Food and Agriculture Organization of the United Nations began a massive program to increase world food production, especially in developing countries. This effort was based on the use of modern agricultural techniques applied to high-yielding, Western-designed crops. The resulting food production gains of the 1950s through 1980s (**FIGURE 13-9**) are often termed the **Green Revolution**.

The immediate, short-term gains of the Green Revolution were truly impressive. For instance, between 1950 and 1986, annual world grain production rose from approximately 631 to 1,664 million metric tons. This growth rate in grain production was greater than the rate of human population growth across the planet, with the net result that per capita grain production rose from 250 to 338 kilograms per person during the same period (see Figure 13-4). Given the increase in the Earth's population from just slightly more than 3 billion in 1960 to the current 6.6 billion, the Green Revolution may have staved off immediate starvation for billions, but this has come at a price. The massive application of "modern" agricultural techniques has resulted in numerous problems, and it is unclear whether current food production levels will be maintainable for much longer. The peak per capita annual production of grain in 1984 (343 kilograms) has thus far (through 2010) not been duplicated. In many areas, **soil fertility** is declining rapidly as nutrients are extracted from the soil but not returned in kind. Massive irrigation, even in countries where forms of traditional irrigation have been successfully carried out for millennia, is causing waterlogging and salinization at unprecedented rates. Ironically, in arid and semiarid regions, this

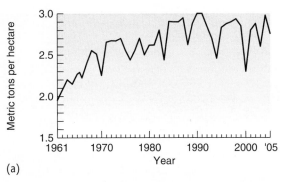

(a)

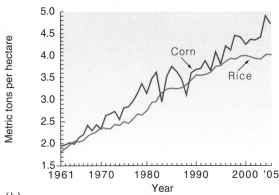

(b)

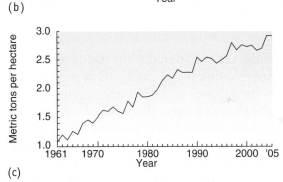

(c)

FIGURE 13-9 (a) World mixed grain yields, 1961–2005. (b) World maize (corn) and paddy rice yields, 1961–2005. (c) World wheat yields, 1961–2005. All yields are in terms of metric tons per hectare. (*Source:* United Nations Food and Agriculture Organization.)

often leads to **desertification** (**FIGURE 13-10**), the spread of desert-like conditions that human exploitation and misuse of the land have caused. In China, perhaps more than 1 million hectares of agricultural land have had to be abandoned since 1980 because of problems with salinization and waterlogging. Similarly, 2.9 million hectares were removed from use in the Soviet Union between 1971 and 1985, and in India, the newly irrigated land that is put into production each year is counterbalanced by damaged land that must be removed from production. In Egypt, irrigation has been a necessity since ancient

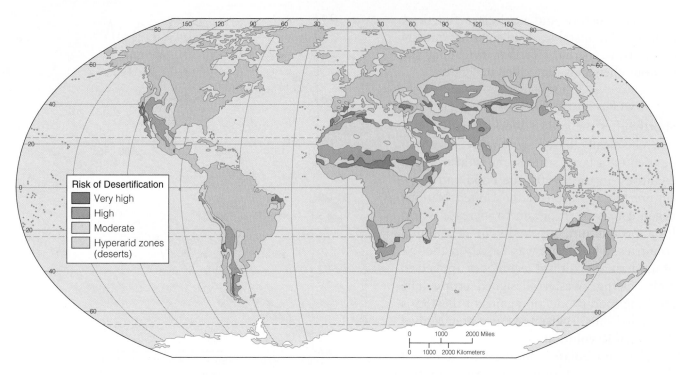

FIGURE 13-10 Deserts and areas at risk of desertification. (*Source:* Based on United Nations data.)

times, yet traditionally the fields were not used continuously; the fields were allowed to lie fallow, and thus the accumulating salts could be naturally washed out and the land rejuvenated. Since about 1960, intensive modern irrigation has caused salinization of about one-third of Egypt's cultivated land, and an estimated 90% is experiencing the effects of waterlogging.

Just as serious as nutrient depletion, salinization, and waterlogging are the effects of toxins (in the form of pesticides and herbicides; a review of the principles of pollution control, toxicology, and risk provides additional information) and soil erosion. Pollution from herbicides and pesticides is a global problem: It destroys not only the living organisms in the soil, but also other wildlife and vegetation and is directly harmful to the human population. Denuded soils quickly erode, and topsoil loss is a serious global problem (discussed later). A study in Tanzania found that land with natural vegetation cover experienced virtually no topsoil loss and absorbed almost all the rainfall compared with land that was

either artificially cultivated or left bare. It has been estimated that currently about 26 billion metric tons of topsoil are lost worldwide from erosion every year. It can take as long as 1,000 years to form a layer of soil 1 centimeter (0.4 in) thick, yet it can be lost in just a few years from poor agricultural management. Some researchers estimate that, at current rates of erosion, the once fertile land of the U.S. corn belt could be nearly depleted of topsoil before the middle of the 21st century.

The Green Revolution has also been held responsible for the contamination and depletion of groundwater supplies in many parts of the world. Agrochemicals—artificial fertilizers, pesticides, herbicides, nitrates (derived from fertilizers), and other chemicals (many of which are highly toxic and carcinogenic)—applied in abundance to fields have penetrated and polluted groundwater supplies. An estimated 50 million people in the United States are potentially exposed to pesticide-contaminated groundwater used for drinking. Contaminated groundwater often is virtually impossible to clean. Underground aquifers are cool, dark, and well protected; they have poor water circulation, contain little in the way of life forms, and thus form an ideal place for contaminants to be stored and remain stable (and toxic) for centuries or millennia. Deep in the aquifer there are no natural mechanisms to break down or neutralize contaminating toxins, so they remain indefinitely. Many authorities think that in most cases, a badly contaminated groundwater supply must be dismissed as a future source of freshwater.

The Green Revolution has also greatly stressed the supplies of freshwater. Many of the miracle crops of the Green Revolution were hybrid varieties that, although they may have been more productive, required much more water than traditional varieties of wheat and rice. In addition, in some areas crops for export were introduced that required even greater amounts of water; a case in point is sugarcane, which can require 10 times as much water as wheat. In addition, arid land was put under artificial irrigation using modern wells that tapped deep underground aquifers. In the past few decades, many areas have routinely withdrawn water from aquifers much faster than the aquifers are recharged by rain. The net result of these practices is that water tables are declining around the world.

Beyond the Green Revolution—Higher Yields Through Sustainable Agriculture

The acme of the Green Revolution, with its heavy dependence on synthetic chemical compounds—fertilizers, pesticides, and herbicides—and its use of water-consuming, genetically identical monocultures grown with the help of heavy equipment powered by fossil fuels, has now passed. The gains of the Green Revolution were impressive, but they were achieved unsustainably. The downside of the Green Revolution has been environmental damage to an extent previously unknown in recorded history. Fortunately, an increasing number of farmers around the world are using "new" (in fact, based on traditional practices), ecologically sound, and sustainable methods of growing food (see **CASE STUDY 13-1**).

Traditional and Sustainable Methods of Coping With Agriculture's Effects

A traditional and time-honored way of circumventing the problems inherent in agriculture is to occupy the land for a year or two, often using slash and burn or **swidden techniques** (cutting and burning the natural vegetation to clear the land and release the nutrients into the soil), and then move on. In this way, the natural ecological cycle of succession can occur once again; the land is allowed to regenerate and replenish itself. This method is feasible as long as the human population in any one area is relatively small and they are willing to pick up and move on a regular basis.

A variation on this theme is to maintain a permanent place of residence but use alternating fields in different years. In late ancient and medieval Europe, many farmers used a "two-field system" in which only half of their land was planted with crops in any particular year; the other half lay fallow. Native wild plants would colonize the fallow land, and farm animals were allowed to graze there; their manure helped to restore fertility to the soil. Fallow fields could also serve as a home for wildlife, such as birds, that could help keep insects and other pests in check. Each year, the crops would be planted on the previous year's fallow land. A related method is to rotate crops from field to field and season to season. With proper rotation, the next crop can restore nutrients that a previous crop used. For instance, periodically planting a field with legumes (members of the pea and bean family) will restore nitrogen to the soil. These plants' roots attract soil bacteria that have the ability to remove nitrogen from the air and produce nitrogen compounds on which other forms of life are dependent. Crop rotation also tends to decrease the threat of pests and disease. If the same crop is planted in the same field year after year, a colony of a harmful pest or disease agent (be it a rodent, insect, fungus, or other life-form) can take up permanent residence in or near the field. Such a colony has less chance of establishing itself if crops are rotated from year to year.

Subsistence Growers and Sustainable Agricultural Practices

For much of agricultural history, farming has been carried out on a subsistence level. That is, a subsistence farmer produced only enough food to feed himself or herself, his or her immediate family, and perhaps some of the farmer's close neighbors. A number of subsistence farmers might support a local village that included artisans and other workers who themselves did not farm. However, in such situations, farm produce does not travel far; all markets are nearby, local tastes are accommodated, and there is little issue of produce losses during transport or storage. Furthermore, in local subsistence farming, wastes and refuse tend to be returned to the local ecosystem, and the cycling of nutrients forms a relatively closed loop with little seepage or net loss from the immediate area (a review of the biosphere's populations, communities, ecosystems, and biogeochemical cycles provides additional information about matter cycles).

Under traditional subsistence farming, occasional surpluses might be sold outside of the local community, but such was not a primary goal of farming. In modern times and especially during the last century and a half, there has been a strong trend away from subsistence farming and toward commercialization of agricultural activities and the expansion of markets. Modern technology, including fertilizer and pesticide use, fast-growing and high-production strains of crops, mechanization, and modern transport systems, means that on a local, national, and global scale more food and other agricultural products (such as fibers for cloth production or energy crops) can be produced with reduced human labor demands—fewer and fewer people are directly, or even indirectly, involved with farms and agriculture. Thus, an American consumer today may think nothing of sitting down to a meal that includes kiwis from New Zealand, apricots from Turkey, and mangoes from Peru. Decorating the table may be a bouquet of flowers from Colombia.

A farmer today may grow crops or raise fowl and livestock exclusively for a market that is on the other side of the world with little consideration for local needs. The farming activities become purely a commercial enterprise embedded in the national or global market economy, and

this often leads to mass production, economies of scale (cheaper to specialize and grow one product in bulk), and the movement toward large agribusiness at the expense of the small-scale subsistence grower. Certainly modern agribusiness is in large part responsible for the development of modern civilization as we know it. For instance, huge cities would hardly be possible otherwise. However, if not carefully managed, modern agricultural developments may disrupt fragile local societies and economies, as well as adversely affect local ecosystems. Sustainable and self-perpetuating subsistence farming may give way to unsustainable practices that maximize short-term gains (whether the "short term" is measured in years or decades).

Problems associated with large-scale agriculture include damage to local ecosystems with loss or extinction of indigenous organisms and biodiversity, topsoil depletion, surface water and groundwater depletion and contamination, loss of indigenous human cultures and ethnographic diversity, and disruption and breakdown of local time-tested social and economic systems, which may result in increasing poverty, crime, and other societal evils.

Many analysts argue that, if we as a species are to survive at anything better than a marginal level well beyond the 21st century, we must re-institute sustainable agricultural practices. A key factor is that sustainable agriculture, unlike at least some concepts of large-scale mechanized agribusiness, is not a one-size-fits-all solution. Rather, it is extremely important, if sustainable agriculture is to be successful, that crops and techniques be carefully fitted to local environmental and cultural conditions. Indigenous resources, knowledge, conditions, and customs must be honored.

Some basic principles and themes of sustainable agricultural practices can be enumerated as follows:

1. The needs of the present must be met without compromising the ability of future generations to meet their needs and fulfill their potential.
2. Crops suitable to a particular setting, which are often local indigenous forms, are empha-

sized and fostered. Tolerance of local climatic and soil conditions, resistance to local pests, and similar factors are taken into account. Local biodiversity is maintained.

3. Careful stewardship of all resources (material, energy, organismal, land, ecosystems, and people and human institutions) is necessary. Nonrenewable resources must not be exploited unsustainably (thus reuse and recycling must be implemented), and ecosystems must not be degraded. Inputs to the system (such as water and nutrients) must be managed efficiently; low or no input of any synthetic fertilizers and pesticides is fostered. Practices such as rotation of crops, planting crops that will restore nutrients to the soil, and so forth are used. Local peoples and their cultures, lifestyles, and societal values must be treated with respect and dignity. Laborers are not a resource to be exploited or taken advantage of.

4. All participants in the system are important and deserve a decent standard of living and a fair share of the profits earned. The roles and value of laborers, farmers, managers, consumers, governing bodies, and policy makers must all be acknowledged. All participants, from a local to a global level, share in both the responsibilities and the benefits for making sustainability work. Social and economic equity is a necessary component of long-term sustainability.

5. Sustainability in agriculture and farming encompasses an interdisciplinary systems approach (a review of environmental science provides additional information). All components of the system are interdependent and affect all other components, from the smallest to the largest. Thus, for example, fertilizers or pesticides must not be applied indiscriminately; rather, they must be used judiciously with a clear understanding of their implications for long-term soil fertility, ecosystem health, and so forth. A chemical application, while increasing agricultural production, may cause damage to the health of the local citizenry. To give another example, it must be acknowledged that high-level global

policy decisions, such as those concerning import tariffs, can have major ramifications for local growers.

Even as we move from subsistence farming to national and global economies integrated with sustainable agricultural practices, age-old subsistence techniques can continue to be used to advantage and, in some cases, rediscovered. An example of the latter is the use of *albarradas* (Spanish for "earthworks") of the Santa Elena Peninsula of western Ecuador. This region, like other parts of Central and South America, is highly dependent on the fluctuations of El Niño (a review of El Niño and integrated Earth systems provides additional information) to bring rains; periods of drought are interspersed with deluges brought by El Niño. In pre-Columbian times, for thousands of years, the indigenous peoples built horseshoe-shaped *albarradas* to capture and store the rainwater brought by El Niño so that it could be used during dry times. Similar ancient stone and earthwork structures are known in Peru, where they were also used to collect water during the wet season (**FIGURE 1**).

FIGURE 1 Ancient earth and stonework structure in the Peruvian Andes used to collect water during the wet season.

Subsistence Growers and Sustainable Agricultural Practices

In some *albarradas*, crops were grown in the moist soil at one end, even as the structures served to recharge the local aquifers. In modern times, structures to trap rainwater are widely built, but many of these modern structures have been damaged by the ferocity of the strongest El Niño storms, whereas the ancient structures survive and continue to do their job. Modern researchers are studying the ancient structures to learn how to apply their successful design and construction techniques to modern local farming. Furthermore, ancient *albarradas* have yielded evidence of the plants that once formed the ancient Ecuadorian ecosystem but have been displaced or destroyed in modern times. In some cases, the *albarradas* have served as a living refuge for ancient genetic varieties that are not found anywhere else. By studying the ancient *albarradas*, progress is being made toward understanding and ultimately restoring the local biodiversity and fragile indigenous dry tropical forest ecosystems of western Ecuador. Here indigenous ancient knowledge, subsistence growing and wider markets (that is, surpluses can be marketed beyond the local area), and sustainable agricultural practices can come together to ensure a bright and self-perpetuating future for the region.

Another traditional way to avoid the problems inherent in some agricultural practices is to promote diversity. This can take many forms and is not unrelated to the concept of crop rotation. In many traditional aboriginal agricultures, numerous varieties of many different crops are planted each season; for instance, the aboriginals of Amazonia used at least 70 varieties of manioc (a group of tropical plants with edible roots, also known as cassava or tapioca). In some cases, many different types of plants are cultivated within a small area, even planted together in the same space—mimicking some of the characteristics of a climax community. In Central America, the farmers traditionally have interplanted maize (corn), beans, and squash. The three crops benefit one another, and the system leads to greater long-term productivity and sustainability than planting a single crop at a time. The more varied diet such interplanting promotes is also nutritionally preferable for people. In addition, using a variety of crops is a form of insurance—one does not put all of one's eggs in a single basket. Different crops and varieties have different tolerances for adverse pest, soil, disease, and climatic conditions (**FIGURE 13-11**). Even if unexpected rains or droughts occur or an abnormal fungus or insect plague strikes, it is less likely to destroy the entire harvest if a variety of plants have been cultivated.

Integrated Pest Management and Biological Controls and Organic Farming

The basic philosophy behind **integrated pest management (IPM)** and **biological control** is that the farmer does not try to eliminate pests, as was often the idea behind using massive

FIGURE 13-11 Planting two or more crops in alternating strips is referred to as strip cropping. Here, alternating strips of alfalfa with corn protect this field in Iowa from soil erosion.

amounts of poisons as part of the Green Revolution, but simply attempts to control pests so that they do not cause serious damage. IPM advocates "natural" controls, such as the use of the pests' biological predators. IPM systems also use cultural practices such as **crop rotation**, allowing fields to lie fallow periodically and interplanting to help control various pests and weeds.

To an increasing extent, farmers are returning to the use of natural fertilizers such as crop wastes that are plowed back into the soil or left to rot on top of the soil, natural compost, animal manures, and even human wastes. In some areas, farmers have taken up true **organic farming**, which avoids the use of any synthetic chemicals—fertilizers, pesticides, or herbicides.

IPM, biological control of pests, and organic farming are proving to be productive and economically feasible. In some cases, the yields have been slightly lower (although sometimes they are higher), but because the farmers did not have to purchase extra synthetic chemicals, their costs were lower and their profits the same as or higher than they would have been with the use of more conventional methods. In fact, one study of nine crops in 15 U.S. states found that the farmers using IPM systems had a collective profit of $579 million more than their projected earnings using other methods. Thus, these new techniques are economically viable; they will not drive farmers out of business or cause a dramatic drop in food production. Most importantly, however, they do not deteriorate the land and general environment to the extent that the techniques of the Green Revolution did. In fact, at their best, organic farming and IPM, combined with very limited use of synthetic chemicals, appear to be sustainable—a claim the Green Revolution could never approach.

Biotechnology and Genetically Modified Crops

The Green Revolution was based on many new "miracle" strains of crops that grew faster and produced higher yields. Many hope that we can continue to increase food production through **biotechnology** and **bioengineering**—the artificial use and manipulation of organisms toward human ends, including genetic manipulations that can in effect produce new types of organisms (see **CASE STUDY 13-2**). **Genetically modified (GM)**, transformed, or **transgenic crops** (transgenic varieties) are already a reality. Many people may not realize it, but sizable percentages, from one-third to one-half or more of such crops as soybeans, corn, and cotton, are composed of transgenic varieties in America. In 1994, the Flavr Savr tomato that Calgene, Inc. (since taken over by Monsanto Corporation), of Davis, California, developed became the first genetically engineered whole food product to hit the markets. Essentially, the Flavr Savr was designed so that an altered gene blocked production of a certain enzyme that controls ripening and softening. Normally, tomatoes are harvested before they are ripe, shipped, and then artificially ripened (such as by using ethylene gas) after they reach their destination. However, flavor is lost with such procedures, and an estimated 30% of the tomato crop is still destroyed by rotting or is damaged during shipping. The idea was that the Flavr Savr could be left on the vine longer so that it would ripen naturally and develop a better flavor, resist spoiling during the shipping process, and have a longer shelf life once it arrived at a supermarket or home; however, the Flavr Savr was not a success. Within a few years, it was off the market because of reported problems with taste, damage during shipping, poor growing in soils and climates outside of California where it was developed, and the general public wariness concerning GM foods. Still, many other fruits and vegetables are plagued by the same ripening and spoilage problems as tomatoes and might benefit from similar genetic engineering.

Genetic engineering can also change the taste or other properties of plants by modifying their sugar and starch content. Peas, corn, tomatoes, and other crops can be made sweeter, or the starch content of potatoes can be raised, making them more suitable for potato chips.

In the long run, genetic engineering's most important contribution may be to increase the resistance of crops to insect and disease vectors. The common bacterium *Bacillus thuringiensis* (abbreviated as Bt) naturally produces a substance that is toxic to certain types of pest caterpillars. For several decades, Bt and

Genetically Modified Foods and Crops

In less than a decade, GM crops have grown by orders of magnitude on the world stage. In 1996 (when the first edition of this book was published), a mere 1.7 million hectares were planted worldwide; however, by the turn of the century, some 50 million hectares were being planted with GM crops, and in 2005, it was estimated that more than 80 million hectares were dedicated to GM crops. The use of GM crops is concentrated largely in the United States (the leader in such operations), Argentina, Canada, and China. It is estimated that in the United States, between 60% and 70% of all food contains some GM component.

However, GM crops did not explode just in terms of dramatic increases in plantings; they also exploded in terms of worldwide controversy over the advisability of relying on such technologies. In many European countries, much of the populace is very wary of genetically modified foodstuffs. For example, it was not until early 2004 that Britain finally gave the go-ahead for the first GM crop for commercial growing. In Europe, generally there is much concern with the labeling of GM foods and regulation of their importation. In the United States, there is no requirement to label foods containing genetically modified components.

GM organisms are plants (or in some cases animals or microorganisms) that contain genes extracted from other types of organisms (viruses, bacteria, plants, animals, and so forth) inserted artificially into the subject organism. Thus, "transgenic organisms" is a more accurately descriptive term for organisms that contain genetic material transferred from another species. People have been selectively breeding, and thus artificially modifying, the genetic makeup of domesticated organisms for thousands of years, but transgenic organisms are different. Traditional breeding involves, by necessity, the crossing of organisms that are closely enough related that they can interbreed. The engineering of transgenic organisms involves the mixing of genes from organisms that are widely separated evolutionarily, such as the inserting of bacterial genes in a plant or animal, or even animal genes in a plant or vice versa. For this reason, sometimes transgenic food crops are referred to as "Frankenfoods" (after the fictional Dr. Frankenstein's monster, which was manufac-

tured from a combination of body parts originally belonging to different individuals).

Why engineer transgenic organisms? Direct desired benefits include increasing crop yields, developing more advantageous characteristics of crops (be it better taste, higher nutritional value, or longer shelf life), resistance to pests, and tolerance of herbicides and pesticides (so that the crop can be treated with a herbicide that will kill everything other than the desired crop). Through transgenic engineering, plants can also be developed that will yield precursors of plastics, vaccines, and other products not necessarily associated with the plants in nature, and it is not only plants that are genetically engineered. The "super salmon" has been genetically engineered with a growth hormone gene that causes the fish to grow extremely quickly (four to six times as fast as the original, unaltered fish) and reach very large sizes.

One early example of a genetically engineered plant is the GM soybean that the Monsanto Company developed. This variety of soybean was developed to be immune to the Monsanto herbicide known as "Roundup." Farmers could plant the genetically modified soybeans and then control weeds simply by spraying Roundup on the crop. All plants other than the soybeans with the immunity to Roundup would be eliminated. This, it was argued, would reduce the need to use other, more toxic and dangerous herbicides. Various herbicide-resistant varieties of soybeans, corn, cotton, and canola (rapeseed) are among the most common types of transgenic crops currently planted (in 2010, the latest year for which there are accurate statistics), accounting for 71% of the area worldwide planted with transgenic crops. Another major development was the insertion into corn, cotton, and other crops of bacterial genes (from the common bacterium *B. thuringiensis*) that produce a toxin poisonous to various insect species. Such genetically modified Bt crops (named after the initials of the bacterium) have a "built-in" resistance to insect damage, and as of 1999, Bt corn and Bt cotton accounted for 22% of the transgenic crop area planted globally. Plants can also be modified to both produce Bt toxins and be herbicide-resistant simultaneously, known as "trait-stacked" varieties that contain more than one genetic modification; in 1999, 7% of the transgenic crop area consisted of corn and cotton varieties that produced Bt and were herbicide-resistant.

GM crops such as those described have incredible potential, their proponents argue, to increase production and decrease costs (less money needs to be spent on pest control), so transgenic crops have been rapidly adopted in a few countries, led by the United States. However, critics argue there is a darker side to GM foods. One example often pointed to is the use of so-called terminator technologies. In the late 1990s, Monsanto and other companies were worried, in part, that their investments and what they regarded as their valid intellectual property rights would be lost if genetically modified plants that they had worked hard to develop could simply be purchased once and then regrown year after year by saving some seeds from the previous year's harvest, as is done by many traditional farmers. To eliminate such a possibility, they worked to engineer sterility into the seeds—that way, each year the farmers would be forced to buy more seeds from the parent company. However, such technology came under intense public scrutiny, and under pressure from the critics, Monsanto decided in 1999 not to commercialize terminator technology. Instead, Monsanto decided to threaten legal action against "seed savers" or "unauthorized" farmers found with the technology on their property.

This incident did not make for good global public relations when it came to GM products. It also brings up larger philosophical and ethical issues, such as should private individuals and companies be allowed to patent life forms or parts of life forms? On the one side, some people believe it is wrong to "play god" and patent organisms or simply unethical to develop a beneficial strain of a crop that could help feed the poor in developing nations, yet make it available only to those willing to pay the asking price, thus potentially eliminating those who could most benefit from such a development. On the other hand, it can be argued, why shouldn't a company reap the benefits from the risk and investment involved in attempting to develop new GM products? Indeed, in 1980, the U.S. Supreme Court cleared the way for the patenting of newly developed types of organisms,

which helped make investing in genetic engineering more attractive to private corporations.

Many questions have been raised, especially by environmental and consumer advocate groups in Europe, about the potential health and environmental risks of GM foods and crops. In contrast, according to advocates of the benefits of GM foods and crops, no such "hypothetical" risks to people have yet to be definitively demonstrated for GM organisms, and what minimal risks might be involved in the usage of genetically modified organisms are far outweighed by the benefits. Let us briefly examine both sides of the issue.

The Arguments For and Against Transgenic Foods

It has been suggested that transgenic foods may produce toxic or allergic reactions in people (although no studies to date have definitively demonstrated this to be the case), and at the least, critics argue that consumers should be allowed to choose whether they want to purchase and ingest GM foods. For instance, why should a strict vegetarian be unwittingly subjected to ingesting a GM product that might contain an animal gene? Of course, this would not only mean labeling all GM foods, but keeping separate and distinct GM crops and non-GM crops and processing so that the end products could be labeled correctly. (In the fields, non-GM crops planted too close to GM crops may become inadvertently pollinated with GM pollen, thus turning a non-GM crop into a GM crop, to the detriment of the farmer attempting to grow a GM-free product. This has already become a concern in some areas where GM crops are grown.) GM proponents argue that there are no demonstrated human health risks involved with commercially available GM foods, so such separation and labeling are unnecessary and only an added expense that will scare consumers. In opposition, GM opponents note that there have been very few studies of GM foods and their safety, and the few studies that have been carried out were aimed primarily at determining whether the GM food in question is "substantially equivalent" to its natural counterpart, and if it is "substantially equivalent" in composition to the natural form, then it is considered safe. However, "substantial equivalence" is a poorly defined concept, and potentially a very slight dif-

VISIT

http://environment.jbpub.com/mckinney/5e/ for more information

Genetically Modified Foods and Crops

ference in composition could have major health or environmental ramifications.

Indeed, depending on what level of evidence one accepts, there is at least some evidence that some GM foods may pose health risks. In a widely publicized and controversial study, Dr. Arpad Pusztai of the Rowett Research Institute in Aberdeen, Scotland, reported stunted growth, damaged internal organs, and damaged immune systems in rats after feeding them experimental varieties (not commercially grown) of genetically modified potatoes for several years. Part of the problem, at least according to critics of GM food, is found in the very techniques that are used to engineer the organisms. Viruses are typically used to insert foreign genes into an organism, and this process can result in other, unintended insertions of genetic material with unanticipated consequences either in the short term or the long term.

As a result of the disparate GM climates on various continents, different parts of the world have treated GM food very differently. In the United States, there is no labeling requirement. It is suggested that Europeans in general have a different attitude toward food than do Americans. Europeans traditionally have cared more about the food they eat and have been more "purist" concerning what they ingest. In addition, the mad cow disease (bovine spongiform encephalopathy) scares have eroded European public confidence in their health care officials, and GM food was also viewed as an imposition from America. Regardless of whether any harm could be demonstrated to be caused by GM foods, as a precautionary measure and to allay public concerns, beginning in the late 1990s, the European Union implemented the labeling of GM foods and restricted the import of GM crops. This had a severe negative effect on corn imports from the United States to Europe, given that much of American corn is composed of GM varieties. By 1999, the concerns of Europe had spread to many other large importers of American crops, including Japan and South Korea. Various food companies in Europe and Japan implemented policies of removing any GM ingredients from their products. These developments have cost U.S. agriculture hundreds of millions of dollars as exports dropped

and transgenic crops were devalued because of a lack of markets.

There are also many environmental concerns about GM organisms. For instance, laboratory studies have suggested that pollen that Bt corn produces can harm Monarch butterfly larvae, although these studies have been disputed and any risks from Bt crops under real-life conditions may be relatively low. Other studies have suggested that the toxins Bt corn produces can accumulate in soils and may have negative ecological effects. There is also the concern that using crops genetically engineered to produce their own "insecticide toxins" could, analogous to the overuse of antibiotics to treat diseases, induce the evolution of insects and other crop pests that are resistant to the toxins. Such super pests might then not only attack the crop, but also begin attacking other plants. If such toxin-resistant pests have not evolved yet, that does not mean there is no potential for their evolution; after all, GM crops have seen widespread commercial use for less than a decade. There is also concern that genetically engineered herbicide resistance could be spread to wild plants, such as wild relatives of the crops (many crops are essentially cultivated weeds), or the plants may even escape into the wild and ultimately produce "super weeds" that cannot be controlled by standard herbicides. Much damage has already been caused by the introduction of exotic species from one region of the world to another; the escape of GM organisms and their genes would effectively constitute additional cases of the introduction of exotic species, in this case from the laboratory to nature.

Another concern, not distinct from the above considerations, is the basic unpredictability of the effects of genetic engineering. A recent case in point involves potatoes. When potatoes were genetically modified to repel aphids, it was found that they actually attracted other pests, including the potato leafhopper that feeds on the plant's leaves. It has also been found that the stems of herbicide-resistant Monsanto soybean plants are more prone to cracking open in hot climates than are non-GM soybean plants.

The use of GM crops might not be "all or nothing." Grafting is common in certain types of horticulture, and genetically modified root stocks that are pest and disease resistant can have non-GM stalks and fruit- or nut-bearing portions grafted on to them. This has been done, for instance, with walnuts. Such hybrid plants may combine the best of both—the advantages of GM with a final non-GM product that will cause no concern to consumers.

GM organisms not only serve as foodstuffs, but also have many other purposes. For instance, microorganisms can be engineered to digest and get rid of toxic substances. Another potentially important application is to genetically engineer plants, such as corn (maize) to produce the substances that can be extracted and processed into "green" (that is, environmentally friendly), fully biodegradable, plastics as an alternative to using scarce petroleum resources to make plastics. Such plastics (organic plastics or bioplastics) are produced from a renewable resource (plants) and can be disposed of in the same manner as organic compost and other wastes (a review of municipal solid waste and hazardous waste provides additional information). The production and use of green plastics is in its infancy; however, a limited number of products using green plastics are available, and we can expect to see their use accelerate.

Despite their environmental appeal, drawbacks to bioplastics have been noted. In some cases, bioplastics are mixed with regular (petroleum-based) plastics, and at present, pure bioplastics generally are not suitable for durable goods. Rather, bioplastics are used for such purposes as packaging materials, hygiene products (such as diapers or sanitary napkins, toothbrushes, disposable plates, cups, and utensils), medical uses (suture threads and protective films), and agricultural and horticultural materials (seed trays and potting containers for plants, time-release capsules for fertilizers and pesticides). One company that has entered the bioplastic field in a major way is Toyota Motor Corporation as part of their biotechnology division (which is involved in many products other than cars, including flowers and garden products). Toyota is investing heavily in bioplastics and already uses them for the car mats and spare tire cover of one of its models sold in Japan.

To produce plant-based plastics with current methods requires large amounts of energy for extraction and processing—energy that typically comes from fossil fuels (although it could be derived from plant material as well, or from a renewable source). Based on one analysis, more fossil fuel will typically be used to produce plant-based plastic than to produce an equivalent amount of petroleum-based plastic, in which case it makes more sense environmentally to go with the petroleum-based plastic. Furthermore, given the problems of global climate change, biodegradability is not necessarily an advantage. If the plastic is allowed to break down, it will produce greenhouse gases such as carbon dioxide and methane. In addition, devoting crops to the production of plastics may detract from the world's food supply, and there are already many people who do not have enough to eat. An exception might be if plastic precursors can be simultaneously grown in an otherwise inedible part of an edible plant, such as in the leaves of corn (maize), in which case the same field can produce food and plastics simultaneously.

Passions run deep on both sides of the GM organism debate. Should farmers in Britain, France, Sweden, and India be forced to destroy their crops of rapeseed or cotton, as has been the case in some instances, because they were sowed using "illegal" or "unapproved" GM seeds? Irrational fear of GM crops can lead to nonsensical actions, contend GM proponents. Yet opponents of GM crops argue that it is better to be "safe than sorry" when we do not know the potential consequences of the widespread use and consumption of genetically modified organisms. For example, the president of famine-stricken Zambia refused to accept GM grain from food aid organizations to feed his people. Part of his concern was that European countries would not accept grain imports from Zambia (if contaminated with GM varieties) should the country ever recover its crop production. Some proponents of GM foods and crops suggest that the opponents of GM are hurting primarily the poor farmers in developing countries who may ultimately be the ones to most benefit from the "Gene Revolution" (following a half century after the "Green Revolution"). It has been suggested that wealthy European consumers can easily afford to indulge in non-GM products, but the poor and starving need all of the help they can get. These proponents say GM plants and

VISIT

http://environment.jbpub.com
/mckinney/5e/
for more information

Genetically Modified Foods and Crops

VISIT

http://environment.jbpub.com
/mckinney/5e/
for more information

technologies should be designed specifically to increase yields and decrease costs in tropical and developing countries and distributed at fair prices. In fact, over the last few years, this is exactly what has been happening in China. China currently has the second largest GM program (the United States is number one), but in China, the emphasis has been to engineer insect and disease resistance, rather than focus on herbicide resistance. As of 2002, it was reported that the Chinese had introduced more than 120 different genes into approximately 50 different species. Already either in use or in the late trial stage are Chinese GM varieties of rice, cotton, wheat, tomatoes, sweet peppers, potatoes, rapeseed, peanuts, cabbage, melons, maize, chilies, papaya, and tobacco. Literally millions of Chinese farmers benefit from GM crops, especially Bt cotton. Approximately 2 million Chinese plant Bt cotton on 7,000 square kilometers of fields, and since the introduction in 1997 of the Chinese version of Bt cotton, the use of toxic pesticides on these fields has dropped by 80%. Costs have decreased by 28%, and the farmers have accordingly increased their earnings. Naysayers suggest

it is only a matter of time before resistant insect strains evolve and all of the GM gains will be lost, but so far, the Chinese are benefiting from GM technology.

Questions

1. List some of the reasons GM organisms have been and are being developed. How are they enhancing food production? Or are they?

2. If you live in the United States, you probably eat some foods that are composed, at least in part, of GM organisms. Does this bother you? Are you aware of when you are eating foods with a GM component?

3. It has been suggested that GM crops and organisms might "escape" into nature, or crossbreed, with close natural relatives and cause damage to ecosystems. What is the evidence, thus far, for such a potential threat? What precautions do you think should be taken with GM crops to avoid such problems?

its derivatives have been used as a natural pesticide on crops with good results; it is relatively nontoxic to birds, mammals, and various nonpest insects. Through biotechnology, the Bt bacterial genes can be implanted into the crops themselves so that they produce the toxin. Such a transgenic organism is in effect mostly plant but also part bacterium. Spiders and other creatures also produce toxins that kill insect pests. Plants that resist various viral, bacterial, and fungal diseases can be designed by implanting genes from various viruses, bacteria, plants, and animals into crop plants.

Clearly, transgenic crops require reduced loads of standard pesticides and have an advantage in resisting diseases. However, researchers point out that genetically engineered crops

must be used carefully. Many insects and diseases can evolve very rapidly. If too much reliance is placed on one or a few types of transgenic crops, natural pest populations may rapidly evolve immunities to the toxins given off by the crops. Already there are reports of insects that can tolerate fairly high levels of Bt toxins. This situation is analogous to insect populations evolving the ability to withstand the assaults of standard insecticides. To prevent immunities from developing in the pest insects or disease vectors, IPM techniques can be used in conjunction with transgenic crops. For example, different types of transgenic and standard crops might be combined in the same field. The unaltered, nonresistant stands of plants would act as a feeding and breeding ground for insects that are not immune to the toxins engineered into the transgenic crops. Thus, the more damaging individuals—those carrying natural immunities—would never be allowed to dominate the population. By not overusing the resistant strains of crops, their effectiveness will be maintained.

Efforts are under way to alter plant crops genetically so that they will have increased tolerances to stresses such as drought, cold, heat, or high soil salinities. However, less progress has been made in this area than in developing insect- and disease-resistant strains. Stress-tolerant crops could be a real boon in developing countries that have only marginally arable lands, have soil salinization, or lack adequate irrigation systems. Some observers worry that in the long run, stress-tolerant crops could cause more harm than good by encouraging the continued cultivation of marginal, fragile, or already damaged lands until they are destroyed.

From a human dietary perspective, an important potential of genetic engineering is to improve the nutritional content of familiar foods. People whose staple is rice often experience vitamin A deficiency; thus, researchers engineered a rice variety containing substantial quantities of beta carotene, a precursor of vitamin A. (However, this "Golden Rice" has come under some criticism. It is suggested that the average person would have to consume 12 times the normal intake of rice to get the necessary amount of beta carotene, and beta carotene is best converted into vitamin A in a healthy person, not in the undernourished individuals targeted for Golden Rice. It might be more useful to help people with vitamin A deficiencies to grow various green vegetables that are rich not only in beta carotene but also in other nutrients that are lacking in rice, Golden or otherwise.) Likewise, levels of various proteins might be increased in crops that are eaten directly by people and in those used as feed for farm animals.

Genetic engineering is also being used to meet the specialized needs of consumers in industrial countries: coffee with a lower caffeine content and rapeseed (oilseed, canola) varieties that produce specialized oils for use as lubricants, in cosmetics, for soaps, and in cooking. A mustard family plant has been designed to produce the biodegradable plastic known as polyhydroxybutyrate, which is similar to polypropylene (derived from petroleum). One concern over this kind of research is that currently many specialty oils, waxes, and rubbers are derived from tropical forests and are among the major exports of the developing countries. Successful development of "oil crops" could restrict the market for these goods. However, by growing "plastic crops," the developing countries could produce their own plastics without relying on oil or petrochemical facilities. In addition, biodegradable plastics could alleviate many of the disposal problems associated with traditional nonbiodegradable plastics (a review of municipal solid waste and hazardous waste provides additional information).

Fisheries

It is sometimes suggested that we could feed extra billions of people by harvesting the natural, renewable biological resources that grow wild on land and in the seas. People have been hunting wildlife and collecting naturally growing edible plants for millennia, but the reserves of such sources are virtually depleted. Certainly, traditional hunting and gathering on land is not a viable option for feeding anything but an infinitesimally small proportion of the current global population.

However, the ocean is often viewed in a different light. Using modern techniques, tens of millions of tons of fish and other seafood are harvested from the sea each year. The oceans are so vast, couldn't we make a dent in our food shortages (especially protein shortages, for fish is high in protein) by drawing more from this resource?

Many people have a mistaken impression of the magnitude and abundance of life in the seas. They may be familiar with the productive shallow-water coastal and reefal areas, which are unrepresentative of the life in the oceans. Most of the open ocean is a "biological desert" that is very sparsely populated by life forms. Only certain areas where nutrients upwell and collect near the surface are highly productive. Because these fishing grounds are limited, we are already closely approaching, and have perhaps now surpassed, the maximum sustainable yield of fishing from the oceans: estimated at approximately 80 to 100 million metric tons of fish a year. In the last decade, the global fish catch from the oceans has been in the range of 80 to 95 million metric tons per year. The oceans are literally mined; fishes and other organisms are removed much more quickly than they can replenish themselves. Fishing on the open oceans has often been done on a massive scale using drift nets. These huge nylon nets—some are as long as 50 kilometers (30 miles) and 30 meters (100 feet) deep—indiscriminately catch everything in their path, including squid, fishes, dolphins, seals, sea turtles, and water birds. The carcasses of these unwanted animals, or by-catch, are simply thrown overboard as waste. As a result, there has been

FIGURE 13-12 The collapse of a commercial fish population almost always creates financial hardship for fishing communities; local fishermen lose their livelihood and their expensive boats sit idle.

mounting international pressure to limit the use of drift nets. Since 1992, the United Nations has imposed an international moratorium on the use of drift nets more than 2.5 kilometers (1.55 miles) long, but even if drift nets are banned, some fishing fleets will continue to use them illegally on the high seas. In addition, many broken pieces of nets or damaged and abandoned nets ("ghost nets") are floating unattended through the oceans. These ghost nets continue to catch and kill sea organisms indiscriminately.

Already, people have exploited certain species of ocean organisms, perhaps to the point of **commercial extinction**—that is, it is no longer economically viable to harvest them (**FIGURE 13-12**). Populations that have dropped to such low levels may never fully recover; indeed, they may become extinct. Classic examples of such overexploitation include the Peruvian anchovy fishery, the Alaskan king crab fishery, and the exploitation of whales. The Peruvian anchovy industry more than tripled its catch from 1960 to 1970, peaking at about 13 million metric tons in 1970. By 1973, it had collapsed to less than 2 million metric tons, possibly because of both overexploitation and adverse climatic conditions. It has never recovered to the levels of peak production. The Alaskan king crab story presents a similar scenario: Peak production in 1980 was 84,000 metric tons, but this dropped to 7,000 metric tons in 1985 and has not fully recovered. Whales have been hunted to commercial extinction over several centuries, beginning on a large scale in the 1700s and early 1800s and continuing into this century.

In addition to mining our oceans, we are polluting them at a tremendous rate. Some seafood species are being killed off altogether, and others contain such high levels of toxic chemicals that they are unfit to eat. Many coastal cities continue to dump their raw or inadequately treated sewage and waste directly into the oceans. This pollution is destroying the wildlife. In 1988, 10,000 seals died in the North Sea, apparently from a viral infection that they could not fight off because the pollutants in the water had weakened their immune systems. Beluga whales inhabiting the St. Lawrence River have been reported to contain such high levels of heavy metals, **polychlorinated biphenyls** (PCBs), and other pollutants in their flesh that their corpses are classified as toxic waste.

The surfaces of the oceans are manifesting the symptoms of the "pollution disease." Around the world, surface algal blooms, sometimes called "red tide," are appearing with increasing frequency. Currently, dozens of red tides occur each year in Hong Kong's harbor, where they were unknown before the mid-1970s. In a red tide, certain species of phytoplankton in the upper layers of the oceans proliferate out of control. Apparently, sewage, fertilizer runoff, and other pollutants that are released into the water are fit nutrients for the algae, which grow on the surface of the water. As the algae grow, they deplete the oxygen in the water that is necessary for the survival of other organisms. Shellfishes, crabs, shrimp, a variety of fishes, and numerous other organisms can be suffocated under the red tide (**FIGURE 13-13**). An algal bloom off the coast of Norway reportedly killed more than 609,000 kilograms (1.34 million pounds) of salmon and trout. Poisoned shellfish, if consumed, can cause food poisoning.

Perhaps even more pernicious than overfishing and pollution are the effects that global warming and the weakening of the ozone layer may have on marine life (a review of global air pollution, destruction of the ozone layer, and global climate change provides additional information). Abnormally warm ocean temperatures appear to be killing coral reefs; it is estimated that 11% of the world's reefs have been lost as of 2001, and 40% or more could be lost by 2015. In the Philippines, where the destruction is the worst, more than 70% of reefs have been destroyed, and only 5% can be said to be in good condition. If the current rate of destruction continues, more than 70% of the reefs will be obliterated in the next few decades. Reefs are particularly important because an estimated 500 million people live within a 100 kilometers (62 miles) of a reef and depend on the reefs and their biota for food and employment, either directly or indirectly. A 2008 comprehensive report published by **National Oceanic**

FIGURE 13-13 Red tide bloom near La Jolla, California.

and Atmospheric Administration (NOAA) indicated that one-third of all corals on the planet are threatened; more than one-half of the coral reef habitats are in poor or fair condition because of climate change and human-related causes. An estimated 25% of the fish catch in developing countries comes from coral reef areas, helping to feed 1 billion people.

In 1988, when ozone levels reportedly declined 15% because of the ozone hole over Antarctica, phytoplankton levels also decreased by 15% to 20%. Phytoplankton are small, photosynthetic organisms; they form the basis of the oceanic food chain and also help the oceans to absorb carbon dioxide (the prime greenhouse gas). If increasing global heating and destruction of the ozone layer adversely affect the phytoplankton, this will have a detrimental effect on the entire ocean ecosystem. Some researchers have even suggested that life in the oceans may collapse. As phytoplankton die, the oceans will take up less carbon dioxide, which will lead to increased global warming, which in turn will accelerate the destruction of the ocean ecosystem. Ironically, both too much of certain phytoplankton (those producing red tide) and too few phytoplankton (destroying the base of the marine food chain) are detrimental to oceanic ecosystems.

What about the possibility of increased "fish farming" through **aquaculture** (used to

Aquaculture

As catches from natural fisheries have stabilized or even decreased, production from fish farms (including freshwater and marine fishes, mollusks, crustaceans, and other aquatic edible animals) has skyrocketed over the last 2 decades (**FIGURE 1**). Today, aquaculture continues to be the fastest growing form of food production in the world; 30% of the world's food fish is produced by aquaculture, and this is bound to increase in years to come. China is the leader in fish farming, producing an estimated 70% of the world's output. In terms of volume, but not value of final product, the next largest producer is India, followed by Japan, Indonesia, and Bangladesh. Overall, nearly 90% of all aquaculture is done in Asia.

Fish farms produce not only fishes, such as flounder, salmon, trout, carp, tilapia, and catfishes, but also oysters, clams, shrimp, prawns, and many other aquatic organisms (**FIGURE 2**). More than 200 species are farmed using aquaculture, although the majority of production is only a dozen or so species. More than half of world production consists of relatively low-value freshwater fishes, such as carp and tilapia, which are primarily raised for local consumption. High-

value species, such as salmon, shrimp, and certain mollusks, are grown primarily for export. About two-thirds of fish farming activities take place along inland rivers and in lakes, ponds, and artificial tanks, whereas the remainder are located along the coasts, in bays, and sometimes even in the open ocean.

An important reason for the steady expansion of aquaculture is that fishes and other aquatic organisms typically are very efficient at turning feed from plants into animal meat. Fishes, crustaceans, and mollusks are cold-blooded, so they do not burn excess calories to keep warm. The water they live in helps support their body weight, so they expend less energy than do comparable terrestrial animals. As a result, only 2 pounds (or less) of feed is typically required to produce a pound of fish, far less than is needed to produce an equivalent amount of beef or pork.

However, there are numerous drawbacks to fish farming. Aquatic farming naturally requires tremendous quantities of clean water, a substance in increasingly short supply. Of course, the water is not consumed in the same way that water is when crop plants are irrigated, but fouling of the water environment can be a real problem. Excess organic wastes may pollute the water to the extent that all aquatic organisms suffer; for instance, excess wastes may induce algae blooms, resulting in oxygen depletion and suffocation of fishes, mollusks, and crustaceans. Like any animal farmers, fish

FIGURE 1 Global aquaculture production, 1950–2002, millions of metric tons per year. (*Source:* Adapted from L. R. Brown, ed. Vital Signs 2005. New York: W. W. Norton, 2005, p. 27.)

FIGURE 2 An aquaculture facility in Louisiana raises catfish. The color differences between ponds are due to the number and type of algae in each pond.

farmers generally must purchase grain products to feed their stock. This, on a global scale, means there is less grain available for other purposes (such as feeding terrestrial livestock or even people). Furthermore, some aquatic species, such as shrimp or salmon, are carnivorous or omnivorous. In aquaculture, such species typically are fed pellets with a high-protein content. But where does the protein come from? In some cases, it comes from plant-based proteins, but in many cases, the feed pellets used in aquaculture are made, at least in part, from fishmeal derived from relatively low-value fish caught in the wild, such as anchovies and herrings. In such cases, aquaculture is not supplementing and adding to the wild fish catch, but actually consuming part of the wild catch. In contrast to carnivorous species that must be artificially fed a rich diet, marine mollusks (such as oysters and clams) raised in pens along coasts and in bays can feed on nutrients that naturally occur in the water and thus require little in the way of artificial inputs.

Specialized and expensive equipment may be necessary in aquaculture operations, especially if the stock is being raised in artificial tanks, ponds, or holding areas. Hormones, antibodies, vaccines, and other medical supplies may be required. The dense populations of fish that are typical of modern fish farms are vulnerable to infectious diseases. Disease outbreaks among farmed aquatic species, especially in monoculture situations (where a single species is being raised in a small, confined area), are a constant threat. For instance, in 1999, farmed shrimp in Ecuador experienced an outbreak of white spot virus that resulted in a loss of nearly $500 million of product. Inbreeding, resulting in genetically weakened strains, can also be a problem, particularly if cultivated organisms escape and interbreed with a wild population.

Another controversy that has developed in recent years is the use, or potential use, of transgenic (bioengineered or genetically modified) species in aquaculture. For instance, transgenic Atlantic salmon (sometimes colloquially referred to as "Super Salmon") that have been genetically engineered with growth hormones reportedly can grow four to six times as quickly as the wild variety, reach larger sizes, and are more efficient

at turning feed into fish flesh. As of this writing, the transgenic salmon have yet to be approved by governing authorities in all of the countries where their use has been proposed, including Canada, Chile, New Zealand, and the United States. Concerns raised relative to the transgenic salmon are comparable to the concerns expressed more generally relative to genetically modified organisms: Are they totally safe for human consumption? Is there a possibility that the modified fish could have a negative impact on natural fish populations and ecosystems? If, or when, transgenic fish escape into the wild, will they outcompete their nonmodified relatives? Will they destroy natural prey populations? (Many critics suggest that it is inevitable that some will escape—farmed salmon have been known to escape into the wild, sometimes in large numbers. In December 2000, as many as 100,000 farmed salmon held in pens off the coast of Maine escaped when a storm damaged their cages.) Will transgenic fishes interbreed with nonmodified populations and essentially genetically destroy the wild populations? These are all issues that will increasingly come to the fore in future years. Currently, there are major efforts under way to derive the benefits of transgenic salmon and other fishes while protecting wild varieties. Examples include raising and keeping transgenic species in secure, self-contained, land-based facilities so that there is no possibility of escape into the wild, or developing strains where the final adults raised for food purposes are sterile so that even if they inadvertently escape into the wild they will not reproduce or mate with wild varieties.

Another major problem with fish farming is the space it requires. The best settings for fish farms are along coasts, rivers, and lakes, but these same areas are considered prime waterfront property, and land values are often very high. Furthermore, in some areas, coastal mangrove forests and other wetland areas have been cleared to build fish farms. These coastal wetlands are the breeding grounds for wild fishes, so clearing such areas often causes natural (wild) fish populations to decline.

Still, fish farming continues to expand and with good reason. It is one of the most efficient means of turning plant products into animal meat. With proper management, fish farming can

VISIT

http://environment.jbpub.com/mckinney/5e/
for more information

399

VISIT

http://environment.jbpub.com
/mckinney/5e/
for more information

have a very low impact on the environment. As the human population increases, we can expect that more and more of the animal protein in our diet will come from the cultivated aquatic realm.

Questions

1. How can aquaculture serve to protect wild populations of aquatic organisms?
2. What are some potential ways that fish farming can harm wild populations?
3. Do you believe that "Super Salmon" and other genetically modified fish varieties should be farmed? If so, under what conditions? If not, why are you opposed to the raising of such varieties?
4. List the advantages and disadvantages of aquaculture. Do you think that the advantages outweigh the disadvantages? Explain.
5. Do you believe that aquaculture operations should be promoted in the United States?

refer to aquatic organism farming in general, or freshwater "seafood" farming in particular) or **mariculture** (saltwater seafood farming)? Organisms such as salmon, shrimp, and edible seaweed are being raised under controlled conditions in many countries. Aquaculture systems can be very productive and efficient at producing animal protein—generally more efficient than terrestrial farms (see **CASE STUDY 13-3**). The drawbacks of aquaculture are that it is very labor intensive, it can involve very intricate management of delicate ecosystems, and it is not suited to all locations. One must have adequate water of the right purity, salinity, and so forth, and temperatures need to be maintained within close tolerances. The startup and maintenance costs of aquaculture can be relatively high. Some of the best locations for aquaculture are coastal areas that are being destroyed by pollution and development. For these sorts of reasons, many experts have little hope that aquaculture will ever significantly relieve the world's hunger.

13.3 The Soil of the Earth

Soil is one of our most precious commodities: It is vital for the health and well-being not only of human civilization, but also of most terrestrial ecosystems. Without soil, we could not grow food, our single most important activity.

We are quickly squandering our natural inheritance of soils. Various estimates suggest that we are losing soil to erosion at a rate of 25 to 75 billion metric tons a year globally. When land is cleared for agriculture, very little plant material is left to reinforce the soil when it rains. The soil washes away. The annual loss of topsoil from agricultural lands averages about 17 metric tons per hectare in the United States and Europe and as much as 30 to 40 metric tons per hectare in parts of Asia, Africa, and South America. In contrast, erosion rates in undisturbed natural forests are on the order of 0.004 to 0.05 metric ton per hectare per year; in nature, soil loss generally is more than offset by soil formation. It has been estimated that soil erosion costs the United States some $44 billion a year in direct damage to agricultural lands and indirect damage to infrastructures, waterways, and health (**FIGURE 13-14**). Globally, the direct and indirect costs of soil erosion may be close to $400 billion a year.

As soils are depleted on prime agricultural lands, crop yields decrease. For every 2.5 centimeters (1 inch) of topsoil lost, average corn and wheat yields drop by about 6%.

Generally, at least 15 centimeters (6 inches) of topsoil are needed to grow crops. After the layer of topsoil has become too thin, the land is no longer useful for agricultural production.

We can sustain some soil loss, for soil is continually produced on the surface of the Earth. The generation and maintenance of soils are functions that healthy natural ecosystems perform for "free." However, under the best of conditions, soil formation is a very slow process. Some scientists have estimated that soil is forming in the United States at an average rate of only about 2.5 centimeters (about 1 inch) per century, which is equivalent to about 3.7 metric tons per hectare per year; other researchers suggest that the average rate may be closer to 1 metric ton per hectare per year. Around the world, some studies indicate that average rates of topsoil formation may be as little as 2.5 centimeters (about 1 inch) in 500 or 1,000 years. The inescapable conclusion is that we are losing our soils more quickly than they are forming.

Currently, modern civilization is living off—and eroding into—the capital of the past (the soils accumulated over many thousands of years), rather than using the soils in a sustainable manner. We need to learn to live off income—that is, to deplete the soils no faster than they are forming under natural conditions.

What Is Soil?

Soil is a combination of weathered, disintegrated, decomposed rocks and minerals (technically known as regolith) plus the decayed remains of plants and animals (organic matter or humus), small living animals, plants, fungi, bacteria, other microscopic organisms, water, and air. Typical soil is about 50% mineral and organic matter by volume and about 50% water and air. There are literally thousands of different types of soils around the world, but they all serve the same vital functions in the ecosystems in which they are found. Soils hold nutrients and water in place such that surface fauna and flora can grow and thrive. Without healthy, porous soils, most rainwater quickly runs off the surface of the land instead of soaking in. Soils supply

FIGURE 13-14 Lack of growth (the light areas) for this field of new wheat in Palouse, Washington, indicates topsoil loss.

the vital nutrients, such as usable nitrogen, phosphorus, sulfur, carbon, hydrogen, oxygen, and various trace elements and important compounds, to the plants that grow in the soil. As organisms die and are decomposed in the top layer of the soil, the nutrients are recycled back to the above-ground organisms.

Healthy soil is alive, a complex ecosystem unto itself. Without its living components, soil lacks its characteristic properties: texture, fertility, and the ability to dispose of wastes and recycle nutrients. An amazing number of organisms can live in a handful of soil. Larger animals that live in soils include earthworms, mites (relatives of spiders and ticks), millipedes, and insects. In a square yard (less than 0.85 m^2) of pasture in Denmark, researchers found 40,000 small earthworms and related organisms, almost 10 million roundworms, and more than 40,000 mites and insects. This is not even taking truly microscopic organisms into account. One ounce (approximately 28 g) of good forest soil can contain

- More than 28 million bacteria
- Approximately 3 million yeast cells
- 1.4 million individual fungi organisms

An ounce of good agricultural soil can contain

- Billions of bacteria
- 11 million fungi
- 1.4 million algae organisms
- 850,000 protozoa

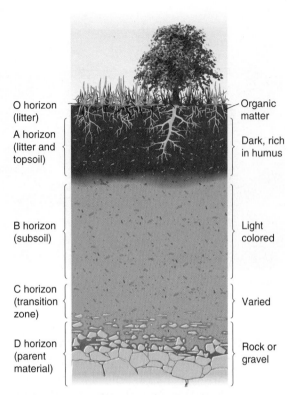

O horizon (litter) — Organic matter

A horizon (litter and topsoil) — Dark, rich in humus

B horizon (subsoil) — Light colored

C horizon (transition zone) — Varied

D horizon (parent material) — Rock or gravel

FIGURE 13-15 A typical soil is composed of five horizons.

A typical well-developed soil is not a homogeneous mass. It consists of layers or **soil horizons** that are approximately parallel to the surface of the Earth (**FIGURE 13-15**). The horizons have different biological, physical, and chemical attributes, such as the amount of living and dead organisms and organic matter they contain, water and air content, texture, structure, color, and mineral content. In any particular part of the world, the soil horizons develop characteristics based on the underlying bedrock and the influence of the climate, flora, and fauna over time. In some places, many horizons develop, whereas elsewhere only one or two horizons are distinguishable. From top to bottom, a typical soil profile (a vertical section through the soil at a particular locality) may exhibit the following basic soil horizons: the uppermost organic matter and humus (heavily decomposed organic matter), the topsoil, the subsoil (composed mainly of minerals), a layer of partially disintegrated rock, and the underlying bedrock.

Global Assessment of Soil Degradation

For many years, the amount and degree of **soil degradation**, leading to land degradation, has been a topic of intense controversy. For several decades, certain environmentalists have enumerated cases of **deforestation**, overgrazing, desertification, and clear destruction of once-fertile lands. At the same time, other experts pointed out that crop yields and livestock production have increased significantly since World War II and concluded that land degradation is not a major global problem. This dispute was essentially unresolvable without a global database on soil degradation.

A major study, the Global Assessment of Soil Degradation, was sponsored by the United Nations Environment Programme and coordinated by the International Soil Reference and Information Centre in the Netherlands. Global Assessment of Soil Degradation, unlike many earlier studies, looked only at soil/land degradation that has occurred because of human intervention since World War II, specifically, from 1945 to 1990. Hundreds of soil scientists measured the degree, area, and causes of land degradation since 1945, and the results were compiled by continental regions and globally. The findings were alarming. Globally, approximately 2 billion hectares (4.8 billion acres), or 17%, of the vegetated land surface of Earth has been degraded by people to some extent in less than half a century.

On a worldwide basis, livestock overgrazing, deforestation, and agricultural activities account for more than 90% of the soil degradation since 1945. Overgrazing is responsible for 35% of land degradation, deforestation for 30%, and agricultural activities for 28%. Of course, these percentages vary greatly by continent, but the damage from soil and land degradation is still occurring. As **FIGURE 13-16** shows, every inhabited continent has areas that are at serious risk.

The U.S. Midwest and Great Plains, the breadbasket area, have experienced various degrees of soil degradation. The Soil Conservation Service has determined that approximately one-fourth of all U.S. cropland is eroding faster than is sustainable. Central America is experiencing extreme soil degradation and is an area of serious concern for the future. Most of this damage is a result of deforestation, overgrazing, and mining, but improper agri-

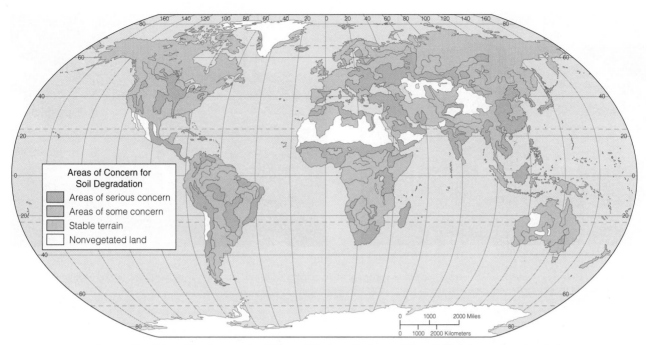

FIGURE 13-16 Areas of concern for soil degradation. (*Source:* Adapted from World Resources Institute. World Resources 1992–1993. New York: Oxford University Press, 1992, p. 117.)

cultural practices are also an important factor. In South America, areas of intensive deforestation are particularly in danger of continued soil degradation, as is the mountainous region on the west coast.

Europe, particularly the middle and eastern portions, is experiencing extreme soil degradation that is predicted to continue into the future. Much of this is attributable to pollutants, including industrial and urban wastes and pesticides that have damaged the soil. Asia is also experiencing severe and continuing soil degradation, especially in India, China, and Southeast Asia. Deforestation, agriculture, and overgrazing primarily cause Asian soil degradation. Africa has continued soil degradation and desertification, especially along the north coast, in the sub-Saharan Sahel region and in South Africa. Overgrazing, wind erosion, and poor agricultural practices are to blame for much of Africa's problems. Compared with the rest of the world, Australia is an area of only moderate soil degradation. Most of Australia's problems with soil deterioration result from overgrazing.

Stopping Soil Degradation

As the world population continues to increase, stabilization and restoration of soil resources will become increasingly important. Since World War II, modern agricultural technologies have masked the soil deterioration by increasing yields even as the soil has become degraded. Crop yields would have been even higher with healthier soils. If unsustainable agricultural practices continue, the soil will become so degraded that despite our fertilizers, pesticides, and high-yield crop varieties, we will still not be able to produce a good harvest. After the soil is dead or eroded, the land becomes barren.

Techniques, such as no-till sowing of crops, drip irrigation, crop rotation, and leaving land fallow, can mitigate or prevent soil degradation, but many farmers often do not practice them for simple economic reasons (**FIGURE 13-17**). The farmers find that it does not make short-term economic sense to invest very heavily (if at all) in soil conservation, preservation, or restoration efforts. Of course, in emphasizing the maximization of short-term profits, the farmer is destroying the capital upon which the business depends, but in a fertile area, the soil may take half a century to become severely degraded—longer than the individual farmer may stay in business. In the past, the next generation would move on to new land, but the world is

(a)   (b)

FIGURE 13-17 (a) No-till planting in the residue of the previous crop reduces erosion and returns nutrients to the soil. This field is in northwest Iowa. (b) Drip irrigation delivers water directly to plants (like the grapes shown here) and helps prevent erosion by reducing water runoff.

now running out of new land to place under the plow.

Aside from economic considerations, even if an isolated farmer wants to practice soil conservation techniques, often this is impossible because the action required to stop soil degradation may go beyond the scale of a single farm. As we have discussed, many factors are causing soils to degrade around the world. Soil conservation measures, such as watershed management and river and catchment basin maintenance, may be required on a regional or national level. Only governmental authorities on a local, national, or even international level can implement such projects. An isolated farmer may be virtually helpless if the government does not support and implement sound conservation policies.

Ultimately, the problem of global soil degradation can be solved, but it must happen through a variety of actions addressing a multitude of causes on every level from the individual to the international community. The long-term needs of society, which essentially means sustainability, must take precedence over all other concerns, be they personal short-term economic gains, debt payment on the part of a poor government, or political jockeying in the international arena. Without healthy soil, civilization as we know it cannot survive.

■ study guide

SUMMARY

- Virtually all food for humankind comes from other organisms, with just three kinds of plants (rice, wheat, maize [corn]) constituting 65% of the global food supply.
- Depending on estimates, globally at least 500 million people are chronically hungry, and more than 1 billion are undernourished and underweight.
- Globally, the number of people who can be fed adequately depends on the components of the diet (grain versus animal products, for instance);

how many people can be fed in the future is a topic of controversy.
- Modern agriculture of the last 50 years, including the "Green Revolution," was heavily dependent on monocultures, mechanization, irrigation (where needed), and the use of fertilizers, pesticides, and herbicides.
- Newer, ecologically sound, and environmentally friendly farming methods are now increasingly being used, such as Integrated Pest Management and crop rotation.

- Genetically modified foods and crops potentially hold great promise for the future but are also surrounded by controversy.
- Aquatic organisms (both wild and farmed) are an important component of the human diet, and aquaculture (aquatic organism farming) is increasing at a fast pace.
- Soil is vital for terrestrial food production and also for the well-being of most terrestrial ecosystems.
- Globally, primarily because of the intervention of people, soil is being lost at an alarming rate—much faster than it is being produced by natural processes.

KEY TERMS

agriculture
aquaculture
bioengineering
biological control
biotechnology
carryover grain stocks
climax community
commercial extinction
crop rotation
cultivable land
deforestation
desertification
ecological succession
erosion
fertilizers
genetically modified (GM)
grain production
Green Revolution
herbicides

hunger
Integrated Pest Management (IPM)
irrigation
malnutrition
mariculture
monoculture
organic farming
pesticides
pioneer stage of succession
salinization
soil
soil degradation
soil fertility
soil horizons
swidden techniques
topsoil
transgenic crops
waterlogging

STUDY QUESTIONS

1. Describe the current global food situation. Is everyone fed adequately?
2. What is hunger? Approximately how many hungry people are there in the world?
3. What three plant species supply most of the world's food?
4. What was the "Green Revolution"?
5. Describe the effects of modern intensive agriculture.
6. Given that the world population continues to grow, more food will be needed in the future. What are two basic strategies that can be pursued to increase world food production?
7. Do you believe that the world could adequately feed a population of 10 billion? Justify your answer.
8. What are the high and low estimates of the number of people that could be fed in the

future? What types of assumptions are these estimates based on?
9. What is the major nonagricultural food source? Is it being used sustainably?
10. How can biotechnology, bioengineering, and the increasing use of genetically modified organisms help us deal with increasing food scarcity?
11. What are some of the criticisms of GM organisms?
12. How does Integrated Pest Management attempt to control crop pests?
13. What is soil?
14. Discuss the types and extent of soil degradation that are occurring globally.
15. Why are soil and potential soil degradation such important issues?
16. What is being done to help stop global soil degradation?

WHAT'S THE EVIDENCE?

1. The authors state that perfect management of the world's current food supply could adequately feed the world's population. Do you agree? What is the evidence for this statement? Is "perfect management" realistic or possible?

2. The authors contend that global soil loss and depleted soil fertility, in part the result of modern agricultural techniques, are vital issues that must be addressed in this century. Are you convinced they are as important as the authors suggest? Cite evidence to support your answer.

CALCULATIONS

1. Currently, about 1.5 billion hectares (3.7 billion acres) of land worldwide is under cultivation. Assuming that currently just enough food is produced to feed everyone on Earth, if the world population increases from 7 billion to 8 billion and yields per unit area of land remain constant, approximately how much more land will need to be cultivated to feed everyone?

2. Using the same assumptions as in Question 1, how much more land will need to be cultivated if the world population increases from 6.6 billion to 10 billion?

ILLUSTRATION AND TABLE REVIEW

1. Study carefully the projections shown in Figure 13-3 (possible land futures). If current crop yields per area of land remain the same and the world population grows according to World Bank estimates, approximately how much land will need to be cultivated in 2050 to feed the world's population? What percentage increase is this over the amount of land currently under cultivation?

2. If current crop yields per area of land double and the world population grows according to World Bank estimates, approximately how much land will need to be cultivated in 2050 to feed the world's population? How does this compare with the amount of land currently under cultivation?

3. Referring to Figure 13-9a (world grain yields), beginning in 1961, approximately how many years did it take to increase by 25% average world grain yields?

4. Referring to Figure 13-16 (areas of concern for soil degradation), a continuous band of desert and areas at moderate to very high risk of desertification stretches from the coast of Western Africa east into the Asian heartland. About how many thousands of miles long is this continuous band?

http://environment.jbpub.com/mckinney/5e/

Connect to this text's website at http://environment.jbpub.com/mckinney/5e/. This site features eLearning, an online review area that provides quizzes, chapter outlines, and other tools to help you study for your class. You can also follow useful links for more in-depth information.

California vineyard in Spring.

Dealing with Environmental Degradation

"Pollution and waste are symptoms, not causes, of the environmental crisis."

Paul Hawken, businessman and environmentalist

Population | Consumption

1. The Environment
and Humans
(Chapters 1–2)

Lithosphere | Biosphere

Hydrosphere | Atmosphere

2. The Environment
of Life on Planet
Earth (Chapters 3–5)

Lithosphere | Biosphere

Hydrosphere | Atmosphere

3. Resource Use and
Management
(Chapters 6–13)

Lithosphere | Biosphere

Hydrosphere | Atmosphere

4. Dealing with
Environmental
Degradation
(Chapters 14–18)

Lithosphere | Biosphere

Hydrosphere | Atmosphere

5. Environmental Issues: Social Aspects and
Solutions
(Chapters 19–20)

Nearly all human activities produce waste that will find its way into the air, water, and soil. To mitigate this challenge, domestic and international citizen groups are working in their communities to make a positive impact on the infrastructure and behaviors that will increase societal sustainability. Even if you do not have the time to volunteer or the financial means to contribute to a cause, the small thoughtful actions made daily in our ordinary lives can have impressive global consequences. For example, composting kitchen waste properly as a society could reduce annual landfill waste by at least 17 to 20%. Furthermore, recent studies suggest that through a "zero-waste" commitment throughout our whole country, collectively we could reduce greenhouse gas emissions that would amount to closing 21% of all U.S. coal-fired power plants. Additionally, establishing a backyard habitat for song birds or creating a butterfly garden on a balcony all help to conserve species in small ways, especially during migratory times. When it comes to the best way to conserve Earth's resources for future generations, being thoughtful of everyday impacts can truly make a big difference. For more information, see: http://zerowasteinstitute.org and http://www.nwf.org/Get-Outside/Outdoor-Activities/Garden-for-Wildlife.aspx.

14 Principles of Pollution Control, Toxicology, and Risk

Chapter Objectives

After reading this chapter, you should be able to explain or describe the following:

- What is meant by "pollution"
- How pollution is produced
- How pollution is controlled
- Toxic chemical testing
- How society copes with risk
- The costs and benefits of pesticides

To this point, we have focused on environmental problems caused by resource depletion or excess inputs to society. We now turn to the second major category of environmental problems, those caused by too many outputs by society (see the diagram at the beginning of the discussion of dealing with environmental degradation). It is also important to know which chemicals in our daily lives impose the greatest risks to our health. Despite many problems (both of an ethical sort and questions involving the applicability of animal data to humans), data gathered from laboratory animals provide the most widely used way of making social decisions about whether chemical products, including pesticides, should be used.

14.1 What Is Pollution?

Pollution generally refers to society's excess outputs into the environment. In this case, "excess" means something produced in amounts high enough to be harmful to us, other life, or valued objects, such as cars and buildings. Almost anything can be harmful if it is concentrated enough in a particular context, so all matter and energy can

410

cause pollution if locally produced in sufficient amounts (carbon dioxide, for example). Because pollution is such a widespread environmental problem, many aspects of it deserve special consideration:

- *Pollution as matter cycling and energy flow.* All of the environment, including land, sea, air, and life, ultimately consists of matter cycles and energy flows. Pollution represents local concentrations in the matter cycle or energy flow. For instance, the rapid burning of fossil fuels releases into the atmosphere tons of carbon that was stored underground as coal and petroleum. Similarly, concentrations of energy can be a form of pollution. Heat pollution is a serious form of air and water pollution. Astronomers speak of "light pollution" from nearby cities that disrupts their view of the stars.

- *Pollution as an accelerated natural process.* We associate pollution with belching factories, but many natural processes have been causing "pollution" for billions of years. For example, volcanoes release gases that are harmful to life, affect global climate, and cause acid rain. Many organisms produce highly toxic chemicals.

However, people cause pollution at a much greater rate than does nature for two reasons, which are related to both the quantity and the quality of our waste:

1. The quantity of waste produced is staggering. For example, the more than 6.5 billion metric tons of carbon we release annually from combustion, especially the burning of fossil fuels, far exceeds what would be released from forest fires and other natural combustion sources.

2. The quality of our waste accelerates pollution because it includes so many new substances. Although often they are not more harmful than natural materials, these substances are not easily decomposed by natural processes and thus are more persistent. This quality aspect of waste began in 1856, when the first human-made industrial chemical was created: an artificial dye. In 1895, the first bladder cancer associated with artificial dyes was reported. Between 1957 and 2003, the American Chemical Society recorded more than 15 million new chemicals. Now the society registers new chemicals at the rate of more than 70 per hour. Only approximately 500 of the chemicals invented each year ever reach a wide market. Still, more than 70,000 chemicals are in everyday use worldwide, yet less than 1% of the chemicals on the market have ever been completely evaluated as potential health or ecological hazards. Most have been subjected to very little, if any, testing. In fact, the Toxic Substances Control Act of 1976 requires that only chemicals registered after 1976 be checked.

- *Pollution as a stepwise process.* A primary reason people produce so much waste is that large concentrations of matter and energy are produced as by-products ("waste") during each step of many human activities (**FIGURE 14-1**). Because "everything must go somewhere" as either energy flowing or matter cycling, waste often ends up polluting water, air, or land. Increasingly, engineers are using **life cycle analysis (LCA)** to pinpoint steps that can be eliminated or made more efficient. Both measures reduce waste generation. LCA analyzes the entire "life cycle" of a product (such as an appliance), from mining of raw materials to its eventual disposal by the consumer, and measures how much energy and matter are used at each step.

Sometimes called "**cradle to grave**" analysis, LCA is most widely practiced in Germany, where laws require that cars and many appliances be recycled. For example, all of the various parts in a car or appliance have their own manufacturing steps. To recycle the entire car or appliance, the life cycle of each part must be studied. In Germany, the parts are labeled (with computer bar codes) when they are assembled to make it easier to dismantle and recycle them when the car or appliance is discarded.

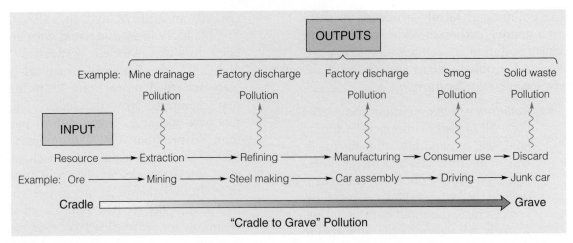

FIGURE 14-1 "Cradle to grave" pollution refers to the many ways that a single product, such as a car, can pollute during its lifetime. Each step, from mining through final disposal, often results in the release of dozens of toxic pollutants into the air, water, and land (life cycle analyses elucidate such processes). Nearly all products, including food and other agricultural products, create such stepwise pollution.

- *Pollution = population × consumption.* Recall from the overview of environmental science that environmental impact (I) = population (P) × consumption (C). In this case, the amount of pollution that society produces depends on two basic factors: the number of people (P) and the amount of waste produced per person, mainly determined by consumption.

Thus, total pollution increases as both population and waste per capita (per person) grow. In the past, waste per capita has been correlated strongly with the rapid growth of resource-intensive industry and technology. "Green"

technologies emphasize source reduction as a main goal. The need for reduction is especially great in the United States, which has among the world's highest per capita air, water, and land pollution. Each U.S. citizen accounts for more than 907 kilograms (2,000 pounds) of waste per day when transportation, agricultural, industrial, and household wastes are included. As a result, with 5% of the world's population, the United States produces perhaps 50% of the world's solid waste and 20% to 35% of most other pollutants.

History of Pollution

Natural pollution is as old as the Earth. For instance, the earliest volcanoes emitted vast amounts of gases, which helped form the early atmosphere of the earth. Gases from the eruptions of less ancient volcanoes also affected global climate, causing at least some of the mass extinctions that devastated the biosphere in the past. In addition, people are not the only animals that pollute. Huge herds of grazing animals have been known to pollute rivers with urine and feces; droppings from large flocks of birds can contaminate the local environment near nesting areas.

Early people also polluted the environment. The remains of the first "landfills" are among the most common finds in caves and other sites of fossil humans. Often these contain the bones of hundreds of fossil animals (**FIGURE 14-2**). The first written records indicate that Egyptians, Greeks, and other early civili-

FIGURE 14-2 A shell mound in Georgia where prehistoric Native Americans discarded shellfish remains and other wastes.

CHAPTER 14 Principles of Pollution Control, Toxicology, and Risk

zations sometimes had to cope with polluted drinking water. As cities grew and population density increased, so did pollution. Roman laws prohibited dumping in certain areas. One of the earliest references to air pollution dates back to 1307, when air pollution in London was so bad that King Edward I banned coal burning. Modern wastewater treatment and the flush toilet did not become common until the 1920s in Europe and the United States. Until that time, pollution of lakes, rivers, and groundwaters from human waste posed enormous threats to drinking water quality. Not until 1912 was the first federal law, The Safe Drinking Water Act, enacted to set standards for drinking water in the United States.

Although pollution has always occurred, the scale and rate of pollution in the last few decades are vastly greater than in the past. In today's world, large human populations and fossil fuel-driven technologies are daily producing huge volumes of many different kinds of pollutants on regional and even global scales. These pollute all parts of the environment, from land to water to air to living things.

14.2 Controlling Pollution

We often associate environmental deterioration with smokestacks and pipes pouring out toxic substances (**FIGURE 14-3**). However, controlling pollution involves far more than just passing laws that require factories to stop toxic emissions. Realistic efforts to control pollution must consider the many complexities involved.

Myths of Pollution Control

We begin by examining some of the misconceptions about pollution and how it can be controlled. These myths are widely held and often lead to costly political and economic decisions.

The Myth of Purity in Nature

We often hear phrases, such as "pristine waters," that imply a certain purity in natural places untouched by humans. In reality, virtually nothing is "pure" in nature (**FIGURE 14-4**). A single drop of water contains approximately 500 trillion molecules. Even before people existed, no drop of rain or river water would have contained 500 trillion molecules of only

FIGURE 14-3 Smokestacks billowing pollution are an icon of environmental degradation.

H_2O. Millions or, more likely, billions would have been molecules of various dissolved gases, organic solids, mineral solids, and many other natural "pollutants." Similarly, the "pure" country air we breathe contains a huge number of gases and particles. Even highly polluted water or air usually contains only tiny concentrations, far less than 1% of toxins or other pollutants. This is why pollution is measured in parts per million or billion. The raw sewage entering treatment facilities usually is at least 99.9% water.

FIGURE 14-4 Even a clear mountain stream has some amount of "natural" pollutants.

Myth of Zero Pollution

Although eliminating all pollution sounds ideal, zero pollution is an unrealistic goal for four reasons:

1. Modern society cannot exist without producing pollutants. Everything must go somewhere, so human activities inevitably will produce waste matter and energy.
2. As we see later, the costs of removing all pollutants from any given activity increase exponentially after a given point, making total purity economically impossible in nearly all cases.
3. The benefits of pollutant removal decrease exponentially after a point, making total purity unnecessary.
4. Not even nature is totally pure. Even if zero pollution were economically feasible, should we necessarily seek an environmental standard higher than nature?

Myth of Zero Risk

Most of us have many misconceptions about risk, including that of a risk-free existence. In reality, every activity, from eating peanut butter to taking a walk, involves some risk. Therefore, in all activities, including control of pollution, we can at best seek to minimize, not eliminate, the many risks we face. However, our efforts to minimize risk are greatly hindered by our inaccurate perceptions of risk.

Deciding How Much Control: Being Realistic

The pollution myths have had a strong influence on policy, leading to some unrealistic and costly decisions. Pollution control in the 1970s and early 1980s emphasized "rights-based" and "technology-based" approaches. Rights-based control assumes that all individuals, even the most pollutant sensitive, have a right to be exposed to the least pollution that society can provide. For example, the Clean Air Act of 1970 set pollutant concentrations at levels low enough to protect even the most sensitive citizens. Technology-based control sets pollutant levels according to technological ability; technology-based laws often specify that companies reduce pollution by using the "best available technology."

In the past, rights- and technology-based approaches have been very successful in improving many aspects of air and water quality. However, many people have argued that such laws tend to be too stringent and costly because they strive for zero pollution and zero risk, which are unattainable ideals. Determining pollution control based only on the most sensitive individuals or the best technology is a very narrow approach that can lead to overprotection for most of society at an unnecessarily high cost, according to critics.

Rise of the Benefit–Cost Approach

Concerns about a declining U.S. economy have led to increasing interest in lowering the costs of pollution control in recent years. As **FIGURE 14-5** shows, the portion of U.S. businesses' total expenditures that goes for pollution control has been climbing since the early 1980s; the 2001 percentage of 19% represented more than 2% of the gross national product. By 2010, the total amount spent on pollution control in the United States amounted to 1.6% of the U.S. gross national product. Some critics argue that much, or all, of these costs were a needless drain on the U.S. economy.

Others argue that the money spent on pollution control has yielded many benefits to human health and the environment. Nonetheless, the trend toward increasing costs, combined with general economic problems, has encouraged a search for less costly approaches. A sustainability movement, beginning in the early 1980s, has gradually replaced the environmentalism of the 1960s and 1970s. One of the traits of the **sustain-**

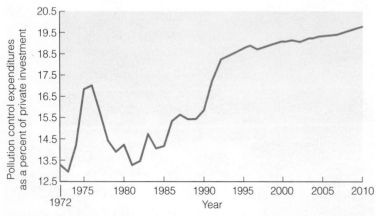

FIGURE 14-5 The cost of pollution control increased from approximately 13% of business investment in 1972 to approximately 19% at the beginning of the 21st century. (*Source:* U.S. Commerce Department.)

ability movement is its focus on addressing fundamental economic changes, as opposed to the legal mechanisms that environmentalism often emphasizes (as seen in an overview of environmental science). In the case of pollution control, this has meant replacing the rights- and technology-based approaches with a benefit–cost approach.

Benefit–cost analyses address the benefit–cost ratio. With pollution control, this means finding the greatest relative health and environmental benefits for the least amount of economic cost. The cost of pollution control increases slowly at first so that the initial reduction in pollution is relatively cheap (**FIGURE 14-6**). After a certain point, costs increase rapidly as increasingly refined equipment and more energy must be applied to extract the more problematic pollutants, such as finer particles or less soluble gases.

Fortunately, society usually does not have to pay the very high costs of further removal because the detrimental effects of pollution tend to follow a curve that is the inverse of the cost curve. As pollution is initially removed, rapid benefits accrue, but the "law of diminishing returns" sets in as pollution removal continues, and continued health benefits are much less. The point where the cost and benefit curves intersect can be considered the point where benefit:cost (ratio = 1) is optimized, denoting maximum relative benefit for the cost incurred. Ratios greater than 1 indicate that additional controls could result in a net overall gain to society. Ratios less than 1 indicate "overcontrol," for which controls are costing society more than the benefits it receives.

Although benefit–cost studies are extremely useful in social decisions about pollution control, they are not without problems. Benefit–cost results may be very misleading, or even worthless, if some benefits or costs are omitted or inaccurately estimated. This is very likely when dealing with environmental problems because it is very difficult to estimate the value of many environmental resources. Society historically has tended to underestimate the true environmental costs (externalities) of most activities. For example, we can measure the health costs of particulate matter relatively easily using medical data, but we should also include the aesthetic damage done to wilderness areas if plants are killed. What is the value of your favorite fishing lake? How much would you pay to keep it from being polluted? Such environmental benefits of pollution control are not only difficult to place a dollar value on, but people value them differently. To some people, a wilderness area has less value than a shopping mall (**FIGURE 14-7**). Others place no value on the mall.

This "valuation problem" is one of the most fundamental obstacles to solving pollution and other environmental problems. A partial solution that economists often use is called the **contingent valuation method**; it attempts to measure "objectively" the dollar value of changes in environmental quality, often by using questionnaires and other surveys that ask people what they would pay for various environmental improvements. For example, one survey found that U.S. citizens would be willing to pay $1.30 to $2.50 per household for air pollution controls on a power plant that was reducing air quality in the scenic Grand Canyon. The contingent valuation method has been used for more than 30 years and has considerably improved our understanding about environmental values, such as how much people are willing to pay to remove pollutants. Even so, many hurdles remain because personal and social values are inherently subjective, change often, and are difficult to measure.

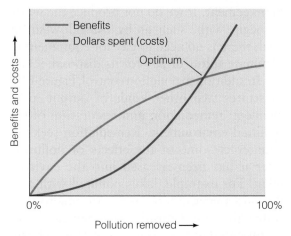

FIGURE 14-6 As pollution is removed, the relative benefits of additional pollution control decrease, and the economic costs of control increase.

(a)   (b)

FIGURE 14-7 Which do you value more—the (a) shopping mall or (b) the wilderness? Or do you value them the same? Determining how much people value something is extremely difficult, yet it is essential if natural resources are to be conserved and managed.

Input versus Output Controls

The flow of matter and energy through society can be reduced in two basic ways: input reduction or output management. Input reduction is usually the most desirable way to reduce pollution because it also slows resource depletion and usually saves money, such as by increasing efficiency and reusing or recycling materials (**FIGURE 14-8**).

In contrast, output management deals with the flow of matter and energy only after society is finished with it—when it becomes waste. This rarely promotes resource conservation. It also is generally very expensive compared with input reduction. Wastewater treatment facilities and scrubbers that trap particles and gases in smokestacks are examples of pollution control through output management. Whereas input reduction saves money by increased efficiency and reuse/recycling, output management increases costs even further because the pollution must be cleaned up after it is produced, and many businesses see it as taking away from profitability. In the case of pollution control, these added costs include (1) cost of the pollution control equipment, or **abatement devices**, which are relatively expensive, and (2) cost of waste disposal. For example, an average coal-burning electricity generating plant produces many tons of toxic ash, sulfur, and other waste that is very expensive to remove and dispose of in landfills or incinerators. This demonstrates why input reduction has much higher benefit–cost ratios than output management. If we do not produce the waste to begin with, such as by using low-sulfur coal, there is no need to pay for equipment to eliminate sulfur or pay for its disposal.

In addition to pollution control, Figure 14-8 illustrates two other kinds of output management: remediation and restoration of the polluted environment. **Remediation** seeks to counteract some of the effects of pollution after it has been released into the environment. For example, lakes polluted with acid rain are sometimes treated with lime to neutralize the acid. Consider how expensive this is compared with stopping acid rain by burning low-sulfur coal or even trapping the sul-

SOURCE

Input reduction (conservation)

1. Efficiency improvements

High input

2. Reuse and recycle

3. Substitution

Society
Population and Consumption

Lower throughput

SINK

Output management

4. Remove (pollution control)

5. Remediate

6. Restore

FIGURE 14-8 Input reduction, or conservation, can be accomplished in three ways. Green technologies are used to achieve efficiency improvements, reuse/recycling, and substitution.

fur before it is released. Remediation of toxic waste dumps has proven so expensive that the amount of Superfund money that Congress originally estimated is much too small. A rapidly growing type of remediation that promises to lower costs in some cases is bioremediation, which uses microbes to digest pollutants in groundwater, oil in oil spills, and many other toxic substances.

Restoration is more ambitious than remediation. Rather than just counteracting some of the effects of pollution in the environment, restoration seeks to reinstate the environment to its former condition. Restoration of the Kissimmee River and Everglades in Florida and restoration of the tall-grass prairie in the Midwest are examples. Restoration is, not surprisingly, the most expensive of the three output management methods. When the original Everglades restoration plan was adopted in 1998, the total projected cost was $7.8 billion, with another $182 million a year needed to maintain each stage of the restorations. In 2000, the plan was updated with a projected expenditure of $10.9 billion. As of 2010, when the latest comprehensive data was released a total of $16.8 billion dollars had actually been spent since the 1990's. Despite ambitious plans and impressive monetary investment, complete restoration is almost never accomplished in any project. After the environment is harmed, it is nearly impossible to return it fully to its former state.

Despite the high costs, pollution reduction attempts traditionally have emphasized output management. Most of the state and federal laws passed during the 1970s and early 1980s, such as the Clean Air and Water Acts and the Superfund Act, mandated pollution controls and pollution cleanup, but as noted previously, the early 1980s saw increasing interest in reducing costs, such as by the use of benefit–cost analyses. Such analyses repeatedly indicate that pollution can be reduced much more cheaply by greater application of input reduction methods. Reducing the need for output control with increased efficiency, reuse/recycling, or other input reduction methods can result in more pollution removed for less money. Growing numbers of businesses are recognizing this.

Implementing Pollution Controls

Implementing pollution controls is often the most difficult step because it usually entails an initial monetary cost and/or change in behavior. Although society as a whole may decide that the benefits of controlling a certain pollutant outweigh the costs, the average citizen is often reluctant to pay the added costs and sacrifices necessary to bring this about. Even the cheapest forms of pollution control, such as increased energy efficiency, require an initial cost, perhaps buying a new, smaller car that yields benefits only after payment.

Social scientists tell us that there are three basic ways to promote change in human activities such as those that pollute:

1. *Persuasion*. Ask people to change polluting behaviors.
2. *Regulation*. Pass laws requiring less pollution.
3. *Incentive*. Reward behavior that reduces pollution.

Persuasion can be an effective way of changing people's behavior, especially if it is accompanied by education. People are much more likely to change if they have information about the harmful consequences of their actions. An example would be an advertisement urging people not to pour household chemicals into the sewer system. Persuasion is also relatively inexpensive compared with regulation and incentive methods, but because it has no means of enforcement, persuasion alone is rarely sufficient. By itself, it is generally the least effective of the three methods, but it can be used to supplement the others.

Legal Aspects of Pollution Control

TABLE 14-1 lists the major federal laws regulating pollution in the United States. Most of these laws are discussed in greater detail this text, but key generalizations are noted here. First, these laws can be subdivided into two basic categories: those using the input approach and those using the output approach. The input approach is used in laws that control the production of toxic chemicals, such as **Federal Insecticide, Fungicide, and Rodenticide Act (FIFRA)** and **Toxic Substances Control Act (TSCA)**. These laws try to restrict the creation

TABLE 14-1	Some Federal Laws Regulating Chemicals in the United States		
Year First Enacted	**Regulatory Law**	**Agency**	**Focus of Regulation**
1938	Food, Drug, and Cosmetics Act	FDA	Food, drugs, and cosmetics, medical devices, veterinary drugs
1947	Federal Insecticide, Fungicide, and Rodenticide Act	EPA	Pesticides
1960	Federal Hazardous Substances Act	CPSC	Household products
1970	Occupational Safety and Health Act	OSHA	Workplace chemicals
1970	Clean Air Act	EPA	Air pollutants
1972	Clean Water Act	EPA	Water pollutants
1974	Safe Drinking Water Act	EPA	Drinking water contaminants
1976	Toxic Substances Control Act	EPA	Industrial chemicals not covered elsewhere
1980	Superfund Amendments and Reauthorization	EPA	Contaminants at waste sites
1990	Clean Air Act Amendments	EPA	Reduction of pollutants via incentives
1990	Pollution Prevention Act	EPA	Contaminant source reduction
1996	Food Quality Protection Act	EPA	Stricter pesticide regulation
1996	Safe Drinking Water Act Amendments	EPA	Increased source to tap protection
2011	Cosmetics Safety Act (in house for approval at this writing)	FDA	Limits use of carcinogenic and hormone disrupting chemicals in cosmetics

CPSC = Consumer Product Safety Commission; EPA = Environmental Protection Agency; FDA = Food and Drug Administration; OSHA = Occupational Safety and Health Administration.

of materials that can become pollutants. Other laws use the output approach, trying to reduce pollution after it has been produced. Examples include the Clean Air and Clean Water Acts, which limit pollution discharges from factories and other places, and the Superfund Amendments, which mandate pollution cleanup. The laws are administered by different agencies (although the Environmental Protection Agency [EPA] administers most of them).

Regulation by laws is most useful where polluters are few in number and the pollution can be easily monitored. For example, hazardous waste disposal and large factories are relatively easy to oversee. Regulation by laws is also important when the pollution is very dangerous and must be tightly controlled; again, this includes toxic and hazardous wastes.

Economic Aspects of Pollution Control

The United States has seen a rapid increase in regulatory laws at the local, state, and federal levels. Many businesses and citizens complain that there are too many laws and that they are too complicated, with environmental lawyers and administrators being the main benefactors. In addition, laws are not very effective at controlling activities that involve many people acting independently. It is difficult and costly to regulate pollution when there are many

polluters, such as people littering or disposing of used oil. In such circumstances, incentive methods are often better.

Rather than using force and other legal threats, incentive methods provide economic rewards for nonpolluting activities. Tax incentives and subsidies for renewable energy and other green technology will stimulate the market and often provide enough incentives for businesses to make changes. These control pollution at much lower costs because society does not have to pay for many law enforcement officials to police the many individual polluters. Instead, individuals carry out nonpolluting activities on their own because they are monetarily rewarded for doing so. Two common examples are (1) deposits paid when potential waste is purchased and (2) "pay as you throw" schemes that make the polluter pay for waste discarded.

Requiring deposits for such items as aluminum cans greatly increases the rate of return for can recycling. Currently, Maine, Vermont, Connecticut, Delaware, Hawaii, New York, Massachusetts, Iowa, Oregon, Michigan, Oklahoma, and California have deposits for cans and bottles, such as glass and plastics. Some laws, such as those in Vermont, prohibit selling certain containers in the state made from materials that do not recycle or biodegrade. Generally, the larger the deposit, the higher the rate of return. The use of deposits has been so successful with cans that it is being increasingly applied to other waste items. For example, discarded car engine oil is a hazard to water supplies in many areas. Initial studies show that requiring a deposit when new oil is purchased and reimbursing the buyer when the waste oil is returned can drastically reduce oil waste. "Pay as you throw" schemes are also becoming common, especially with solid waste disposal. Both industry and households are being charged for the amount of waste that is picked up by waste disposal companies.

In summary, regulatory means of pollution control are most effective when polluters are relatively few in number and can be readily monitored. Incentive means are better for pollution from many sources and generally are less costly than regulation. Persuasion is generally a weak but cheap form of social implementation that can supplement the other two.

International and National Aspects

Pollution in both ex-communist countries and developing nations illustrates the economic principles just discussed. Ex-communist countries generally are the most polluted in the world, largely because the old communist governments subsidized fossil fuels, thus encouraging pollution. This kept prices very low, so people had no incentive to conserve. Studies show that increasing energy prices to western European levels will decrease air pollution by more than 50%. This again demonstrates the effectiveness of input reduction.

Developing countries illustrate input management of pollution in another way, that of investment in "clean" (green) technologies. Such technologies emphasize hydropower and other renewable resources for electricity generation. Clean technologies not only decrease pollution as those countries industrialize, but according to studies by the World Bank, the overall cost of these technologies also is lower than the cost of current technologies because of improved efficiency and less fuel use.

National Pollution

As a leading industrialized country, the United States is the world's leading polluter in many respects. For instance, the United States pro-

duces more than 25% of the world's greenhouse gas emissions. Nevertheless, some types of water and air pollution in the United States have been reduced considerably since the early 1970s because of more stringent efforts at all levels of society; however, the amount of solid waste in the United States continues to increase exponentially, driven by a steady increase in consumption.

The **Toxics Release Inventory** (**TRI**) provides an annual overview of toxic pollution in the United States. The Emergency Planning and Community Right-to-Know Act of 1986 requires industries to report to the EPA the amounts of their toxic releases. The EPA uses these reports to produce the TRI. Several hundred toxic substances must be reported. Nevertheless, environmental groups have criticized the TRI for not including many kinds of toxic substances. Another criticism has been that some polluters, such as coal-burning power plants and sewage plants, are not required to file reports.

Despite these shortcomings, the TRI data have proved useful in tracking the flow of toxic substances in the United States. For example, the 2003 TRI data show that the majority of toxic substances were released by the electrical utilities and mining, chemical, and metal industries (**FIGURE 14-9**). Direct releases into all parts of the environment have declined significantly due at least in part to community efforts and the publicity generated by the release of the TRI data (**TABLE 14-2**).

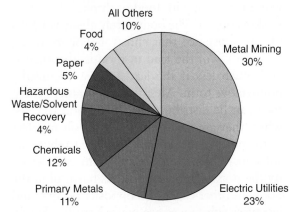

2003 TRI Total Disposal or Other Releases
4.44 billion pounds

FIGURE 14-9 Percentages of disposal and releases reported in the 2003 TRI by major industries and sources. (*Source:* Environmental Protection Agency, 2005.)

TABLE 14-2	TRI Total Reported Disposal or Other Release by Industry 2001–2010
Textiles	−58%
Coal mining	−50%
Primary metals	−37%
Electrical utilities	−34%
Metal mining	−30%
Petroleum	−18%
Source: Environmental Protection Agency.	

14.3 Toxicology: The Science of Poisons

Toxicology studies the effects of chemicals that are harmful or fatal when organisms consume them in relatively small amounts. A conventional pollutant may be deadly in parts per million, whereas a toxic pollutant can be deadly in parts per trillion or even less. Toxic pollutants include not only pesticides, which are purposely made to be toxic to certain target organisms, but also thousands of other chemicals that have toxicity as a side effect. Tens of thousands, probably more than 100,000, chemicals are in everyday use worldwide. An estimated 2% of the thousands of new chemicals available each year are sufficiently harmful to be considered toxic.

Effects of Toxic Substances

Toxicologists trace the path of a toxic substance through the body in five steps identified by the acronym **ADMSE**: absorption, distribution, metabolism, storage, and excretion. Absorption of toxic substances can occur in three ways: ingestion with food or drink, inhalation, or skin contact. The body rarely absorbs 100% of any substance that enters it. For example, about 50% of the lead in food is absorbed—and much less if the lead is in a chemically nonreactive form. After toxic substances enter, they can be distributed via the blood to cells, where they are metabolized. Metabolism refers to the body's biochemical response to the toxic substance. Sometimes metabolism can detoxify the poison, rendering it less harmful in the body. In other cases, the opposite occurs, and harmless substances are made harmful. Sometimes the chemical is neutral.

A **carcinogen** is a cancer-producing substance. Cancer is the uncontrolled multiplication of cells, often forming tumors and spreading throughout the body. Estimates vary, but overall cancer causes about 23% of U.S. deaths annually. There are numerous kinds of cancers, some of which can be inherited. However, in many cases, environmental factors, such as diet and chemicals in the workplace, can contribute to or directly cause cancer. Tobacco and diet are the two most important environmental factors, accounting for perhaps two-thirds of avoidable (not inherited) cancer deaths (**TABLE 14-3**). These, like most causes of cancer deaths, are related largely to our personal lifestyle choices. Pollution and food additives, which are cancer causes often associated with environmental problems, are responsible for a very small percentage of cancer deaths. Such estimates are very approximate because it obviously is difficult to do controlled experiments with humans. Nevertheless, similar results from many rigorous statistical studies lead most cancer experts to consider these figures valid as rough estimates of noninherited cancer causes.

Two types of **chronic toxicity** can cause birth defects. A **mutagen** is a substance that causes genetic mutations in sperm or egg cells. A **teratogen** is a substance that affects fetal development, such as when the pregnant mother drinks alcohol or has other harmful chemicals

TABLE 14-3	Estimated Causes of Cancer Deaths
Factor	**Percentage of Total Cancer Deaths**
Tobacco	30
Alcohol	3
Diet	35
Reproductive and sexual behavior	7
Occupation	4
Food additives	1
Pollution	2
Industrial products	1
Sunlight, ultraviolet light, other radiation	3
Medicines, medical procedures	1
Infections or inherited factors	13
Total	100
Source: Data are adapted from R. Doll, R. Peto. Avoidable risks of cancer. U.S. Journal of the National Cancer Institute 1981; 66:1191.	

in her body. Teratogens are common because development is a very complex sequence of cellular and biochemical interactions and thus is easily disturbed. A general rule is that sensitivity to chemicals decreases with age, at least until old age is reached. Very young embryos are most sensitive to chemicals because cell multiplication, organ formation, and many other delicate growth processes are occurring. This is why pregnant women are discouraged from smoking, drinking alcohol, and taking drugs. In addition, males who consume illegal drugs and alcohol can also raise the risk of genetic defect in a potential fetus as harmful mutations can take place as sperm cells mature and can in turn effect fetal development. In fact, because of the far greater number of cellular divisions in the male germ cell line, the general mutation rate occurring in otherwise healthy sex cells is much higher in males and increases with age. A male in his 20's has a sperm mutation rate 8 times higher than a female's sex cell mutation rate. By age 40, the male sex cell mutation rate is nearly 100 times that of a female's.

Toxic Risk Assessment

One of the most important changes in environmental policy in the 1980s was the incorporation of risk measurements into environmental decision making. This was based on the realization that everything involves costs (including risk) and that the benefits of substances must be weighed against the risks to people and other organisms. During the 1990s, interest in risk grew even more.

For this policy to work, we must have some way to measure the risks. One way is through the use of **epidemiological statistics**, which examine correlations between environmental factors and the health of people and other organisms. For instance, the strong relationship between smoking cigarettes and developing lung cancer is statistically highly significant. This approach has two disadvantages. First, such statistical correlations examine harm "after the fact" so that much damage has already occurred by the time we determine that a chemical was harmful. Second, the best evidence in science is usually based on controlled experiments, in which certain variables are held constant. Thus, information on toxicity often

FIGURE 14-10 Animals are used to test the effects of substances on humans.

requires the direct testing of a chemical's effect on organisms under controlled conditions. Of course, people are not usually tested directly, so animals ranging from bacteria to primates are used (**FIGURE 14-10**). Animal testing raises the issue of animal rights (see **CASE STUDY 14-1**).

Toxic risk assessment in the United States is a four-step process:

1. *Toxic identification.* Determines if a chemical is toxic, usually by testing microbes.
2. *Dose–response assessment.* Determines the strength of toxicity, often by testing rodents.
3. *Exposure assessment.* Determines how often people are exposed to the chemical.
4. *Risk characterization.* Considers the scientific data from steps 1–3 along with societal values to reach a final decision.

The first two steps use animals; the dose–response assessment may test higher animals, including dogs or monkeys.

14.3 Toxicology: The Science of Poisons   421

Should We Use Animals to Test Products? Which Animals?

VISIT

http://environment.jbpub.com
/mckinney/5e/
for more information

Many people question whether animals should be used to test products. Animal activist groups, such as People for the Ethical Treatment of Animals and the Animal Liberation Front, oppose animal experimentation. Although some argue that no animals should be used for experimentation, others oppose the use of only certain kinds of animals (such as monkeys and apes). Others agree that animal experimentation is needed but call for more humane treatment: The amount of pain inflicted, which is often ignored, should be considered.

Although most animal rights advocates take a nonviolent approach, others have acted violently. The National Association for Biomedical Research documented 107 attacks on institutions conducting animal research between 1981 and 1994. The monetary damage from these attacks was estimated at more than $7.7 million, which does not include the loss of scientific data that could have saved human lives or even animal lives, in the case of data used for veterinary medicine. The Federal Bureau of Investigation reports that, since 2005, terrorist attacks in the name of animal rights are on the rise, with more than 40 separate events since 2008.

Perhaps the most desirable solution to the animal-testing controversy is to find ways of testing products that provide the same toxicological information but do not use "higher" animals (such as mammals) or at least do not inflict pain on or kill them. Fortunately, several new methods are becoming available, although none is fully satisfactory for all purposes. These methods include computer models, in vitro tests of cell cultures, and the use of microbes and invertebrates. Computer models based on previously gathered data can be very powerful predictors if properly used. For instance, knowing the biochemical pathways of the way one substance is metabolized can allow us to predict how a similar substance will affect an organism; however, the quality of the data and the accuracy of the assumptions made always limit computer models. Just one inaccurate assumption out of hundreds can nullify the model and produce invalid results. Consequently, many toxicologists are skeptical that models will ever replace high-quality data gathered under controlled laboratory conditions. In vitro tests of cells or tissues can avoid animal pain or death when the cells can be grown in the lab, but such tests are very incomplete compared with whole-animal data because they provide little information on how organ systems in the animal will interact if exposed to the chemical.

Fewer objections are raised to the use of microorganisms and invertebrates, such as shellfishes or insects, but chemical testing on them is often criticized for two reasons. One is scientific: The less closely related a species is to humans, the less similar to people are its biological responses to chemicals. Thus, vertebrates, especially monkeys and apes, provide the closest analogs to humans. In contrast, the genes, biochemistry, and physiology of microbes and invertebrates are often so different from our own that the data produced are useless for human toxicology. The other problem is ethical: Can we say that a mammal deserves more humane treatment than a shellfish? If so, is this just because we are mammals? Or is it because mammals are "smarter" or feel more pain? Where do we draw the line? Is it valid to subject some mammals, such as mice, to pain, but not others, such as monkeys?

These questions cut to the very heart of the animal-testing issue, and there are, as we often find in environmental issues, no simple answers. Perhaps a "technosolution" eventually can be found, such as the use of very complex computer models or cloning of cell tissues that will provide precise toxicological information, but until that time, animal testing seems certain to remain controversial as the debate over the needs of people versus the rights of animals continues.

Questions

1. Do you think that all animal testing should be banned? Why?
2. Should apes be used for testing? Why or why not?
3. How can computer models reduce the use of animal testing? Give specific examples.
4. What are two criticisms of chemical testing of invertebrates?

Toxic Identification

In the initial step of toxic identification, researchers seek evidence if a certain substance is toxic. Given the number of new chemicals developed each year, it is impossible to thoroughly test each chemical for toxicity. Instead, government agencies and manufacturers use hierarchical sequences of tests, called **bioassays**, to test each chemical. A relatively short, inexpensive initial test is administered to search for any possible indications of harmful effects. If harmful effects are indicated, then progressively more thorough tests are carried out until statistical analysis shows that the chemical is harmful (or not) with a high degree of certainty.

An example is the test sequence for cancer, which begins with a short-term test to see whether the chemical may cause mutations. The most widely used mutagenicity test, called the Ames test, subjects certain strains of bacteria to the chemical. These bacteria are specially bred to be incapable of normal reproduction by cell division. If the bacteria begin to multiply and form visible colonies after the test chemical has been applied, it is likely to be mutagenic because only bacteria that have mutated back to a form capable of normal cell division can reproduce. Because some mutagens can cause cancer, a series of more thorough tests for carcinogenesis is begun. These involve longer term tests (several months), using mice and rats, to see whether tumors develop in specific organs. If these tests also show signs of carcinogenic properties, the chemical is subjected to an even lengthier and more costly set of tests, the chronic carcinogenesis bioassay. This involves hundreds or thousands of animals over a period of several years.

These procedures for identifying toxic substances have two basic problems: (1) they do not test people directly, and (2) they cannot detect very small risks. With animal tests, we can never be totally certain that a chemical is safe for humans. In addition, these chemicals react with one another in the environment in complex ways. The complex biochemistry of even closely related species often differs in subtle ways, called **intrinsic factors**, so that a chemical that has no effect on one species may have acute or long-term effects on another. The inability to detect very small risk arises because even the long-term testing of thousands of animals cannot generate statistically certain evidence on chemicals that have slight effects.

Dose–Response Assessment

If a chemical shows evidence of toxicity in the bioassays, the next step is to measure the strength of toxicity. **Supertoxins** in very tiny amounts can cause death; less than a drop can kill an adult human. Conversely, the chemical may be slightly toxic, with only very large doses being harmful. Toxic strength is essential to know because the benefits may outweigh the harm if the substance has nontoxic uses and is toxic only in amounts much greater than normally encountered, as is the case with aspirin. Any substance can cause harm in sufficient amounts, even vitamins and other nutrients that we require in moderation.

The dose–response assessment is based on the application of various doses of the test toxic substance to a number of organisms, usually animals, to see what the response is: no harm, impairment, or death. The statistics generated are then used to make probabilistic statements on the chemical's effect on humans. The usual pattern in any population of organisms is a bell-shaped curve. Some individuals are much more sensitive to the toxic substance than are others. Some are more tolerant, and most are somewhere in-between (**FIGURE 14-11A**). Lethal dose curves, which examine the dose needed to cause death in the test organism, are among the most common types of statistics (**FIGURE 14-11B**). The midpoint represents the dose that kills 50% of the population, called **LD-50**, meaning lethal dose 50%. LD-50 has become the standard reference for summarizing the toxicity of substances.

LD-50 testing has the same two problems as toxic identification: These tests use non-human data and cannot detect very small effects that accumulate over a long time. The problem of small effects is most pronounced with cancer-causing substances. For noncarcinogenic substances, it is usually assumed that there is a **toxic threshold** below which no harm is done. For carcinogens, it is often conservatively assumed that exposure to any amount will create some increased chance of

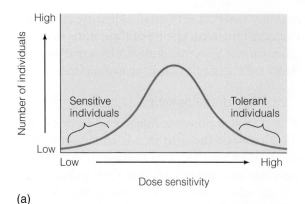

(a)

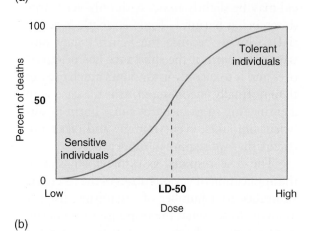

(b)

FIGURE 14-11 (a) Sensitive individuals react to small doses, but a few tolerant individuals can sustain large doses. The majority of the population, or the average person, reacts to an average dose size. (b) The LD-50 is the dose at which 50% of the population dies.

cancer in a population, so there is no threshold. For example, a study of mice may indicate that only 100 cancers may be produced in a population of 1 million mice over the next few years. Although the calculated risk is small (0.01%), it may be considered serious enough to make a chemical product unmarketable.

Another obstacle to extending animal results to people is differences in body sizes, often called **scaling factors**. For example, **dosage** generally is measured as the amount of a substance administered to the organism divided by the body weight of the organism. When LD-50 data on mice are applied to humans, the data must be adjusted for body weight differences between mice and humans.

Exposure Assessment

Although much public attention is given to identifying and measuring the toxicity of chemicals, the amount of exposure is equally important. Even very toxic substances can do no harm if we are not exposed to them in sufficient amounts. The first step in exposure assessment is to determine possible pathways by which toxic chemicals may reach humans. The three most important pathways are inhalation (air transport), eating (food transport), and drinking (water transport). Pathways can involve many steps, and many substances have a number of different, often complex pathways.

In addition to mobility, the **persistence** of a substance is important. Persistence refers to whether the substance remains intact long enough to be transported long distances. In the case of exposure by food, a crucial property is the tendency toward bioconcentration. **Bioconcentration** (sometimes called bioaccumulation or biomagnification) is the tendency of a substance to accumulate in living tissue; it involves two steps. First, an organism takes in a substance, but does not excrete or metabolize it. Then a predator eats that organism and in the process ingests the substance that its prey already accumulated (**FIGURE 14-12**). This second step is common in fishes, shellfishes, carnivores such as eagles and cats, and other organisms that prey on smaller organisms so that the chemical is concentrated by passage through the food pyramid. Mollusks, such as scallops, are exceptionally prone to bioconcentration because they feed by filtering millions of tiny plankton from the water.

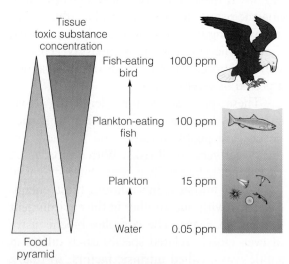

FIGURE 14-12 Organisms occurring higher on the food pyramid tend to have increasingly greater concentrations of toxic substances in their tissue (ppm = parts per million).

Heavy metals, such as mercury, attack the nervous system and are among the many substances that bioconcentrate in living tissue.

Recently, a relatively unstudied impact of toxic pollutants on wildlife and people has been arousing concern. Often called **environmental hormones**, these pollutants send false signals to the complex hormonal systems that regulate reproduction, immunity, behavior, and growth. The result can be infertility, behavioral and growth abnormalities, and low resistance to disease. Some ecologists suspect that hundreds of chemical products may have caused population declines in many species of birds, fishes, and other wildlife, but the effects are usually so subtle and complex that they are difficult to prove (see **CASE STUDY 14-2**).

A final key aspect of exposure assessment is that we can be exposed to many chemicals simultaneously. Consequently, we need to understand how substances can affect us when they interact with one another. There are three basic kinds of interactions: *antagonistic*, *additive*, and *synergistic*. **Antagonism** occurs when substances work against each other and cancel each other's effects. For example, chemical antidotes that destroy the toxic molecules can neutralize deadly snake venoms. **Additivity** occurs when the toxic effects are added together. If chemical A harms the liver and chemical B harms the kidneys, then exposure to both will harm the liver and kidneys. **Synergism** is perhaps the most important because it occurs when the toxic effects are multiplied through interaction, making this the most dangerous type of interaction of all. For example, smoking cigarettes increases your chance of getting lung cancer, as does inhaling radon gas, but a smoker who is exposed to radon gas has a much higher chance of developing lung cancer than one would predict by adding the chances of getting cancer from smoking and radon gas. Tobacco ash particles in the lungs serve as attachment areas for radioactive radon particles, greatly enhancing the chance of getting lung cancer.

Risk Characterization

The fourth step of toxic risk assessment is risk characterization, which combines scientific data from the first three steps with societal values to reach a decision about what to do with a new chemical. Is it too dangerous to be marketed? Do the risks outweigh the benefits? Because human emotions, economics, and many other social variables are involved, this is often the most complex step of all.

Risk refers to any potential danger. The first three steps of toxic assessment provide data on the risk of newly invented chemicals. However, even when data are available, our perception of risk is often distorted and does not accurately reflect the actual risk. Consequently, our efforts to try to guard ourselves against perceived risks can lead to underprotection or overprotection. The average American is much more likely to die from eating peanut butter than from being struck by lightning (**TABLE 14-4**), yet most of us worry much more about lightning.

There are two basic reasons for such misperceptions of risk. One is that our emotions play a large role in how we interpret facts. The most "dreaded" (highly perceived) risks are hazards that are neither observable nor controllable, such as radioactive waste

TABLE 14-4	Annual Risk as Calculated by Lifetime Odds of Death by Various Causes in United States Using Average Life Span and 2010 Census Bureau Population Data
Activity/Exposure	**Annual Risk**
Heart disease	1 in 6
Cancer	1 in 7
Stroke	1 in 30*
Motorcycle accident	1 in 90
Intentional self-harm	1 in 120
Accidental poisoning/Exposure to noxious substances	1 in 140
Falls	1 in 180
Accidental drowning	1 in 1070
Air and space travel	1 in 5.900
Exposure to excessive heat	1 in 6,200
Cataclysmic storm	1 in 51,200
Bees and wasps	1 in 63,000
Bitten or struck by dog	1 in 120,000
Earthquake	1 in 153,600
*Numbers rounded.	

Environmental Hormones: A Growing Threat?

VISIT

http://environment.jbpub.com
/mckinney/5e/
for more information

We are exposed to thousands of synthetic chemicals every day. We eat them, drink them, breathe them, and use them at work, at home, and in the garden (**FIGURE 1**). They are present in soil, water, air, and food. Direct environmental exposure occurs through the foods we eat and the products we use, including detergents, drugs, lubricants, cosmetics, pesticides, and plastics. Chemicals and their breakdown products found in industrial discharge and sewage effluent may also contaminate drinking water. A study released in 2002 of streams and rivers in the United States found surprisingly high levels of caffeine in most samples. When caffeine is urinated, it is very stable and remains in natural waters for quite some time, serving as a chemical tracer for human waste products. Indirect exposure to chemicals occurs when chemicals are released into the air and water. Examples include airborne particles from industry, which can land on grass or hay, be eaten by livestock, and be passed along to humans.

Some environmentalists warn that this constant exposure to synthetic chemicals may be posing significant health risks for people and wildlife. Usually the first health risk that comes to many people's minds is cancer, but in fact, the most serious risks, some contend, could be the loss of fertility and problems of fetal development. The problems occur because many synthetic organic chemicals disrupt natural hormones, such as estrogen, and when organisms are exposed to even extremely small amounts at critical times in their development, these chemicals can produce severe damage, especially in unborn fetuses. Sometimes called endocrine-disrupting chemicals, they alter hormonal functions in several ways:

1. They can mimic or partly mimic the sex steroid hormones estrogens and androgens by binding to hormone receptors.
2. They can alter the production and breakdown of natural hormones.
3. They can affect the manufacture and function of hormone receptors.

Perhaps the best-documented example involves diethylstilbestrol (DES), an artificially synthesized mimic of estrogen. DES was given to many pregnant women in the 1940s through the 1960s to prevent miscarriages. Then it was discovered that DES can interfere with the development of the fetus, particularly causing female sterility and other fertility problems, and a rare form of vaginal cancer in young adulthood in the children of mothers who took DES.

Although some natural plant-produced chemicals (called phytoestrogens) can alter hormonal function, the main focuses of concern are the many endocrine-disrupting synthetic chemicals that are

FIGURE 1 The next time you decide to linger over a meal at a sidewalk café you might want to pick a quiet street where the levels of carbon monoxide, sulfur, benzene, and heavy metals from vehicle exhausts are not too high. Shown here is a café in the Little Italy section of New York City.

commercially manufactured. Examples include pesticides (insecticides such as dichlorodiphenyl trichloroethane [DDT] and chlordane and various herbicides and fungicides), products associated with plastics such as bisphenol A, pharmaceuticals (drug estrogens—birth control pills), ordinary household products (such as breakdown products of detergents), industrial chemicals (polychlorinated biphenyls [PCBs], dioxin, and benzopyrene), and heavy metals (lead, mercury, and cadmium).

In recent years, such endocrine-disrupting chemicals have been blamed for reduced levels of intelligence, increases in violence, hyperactivity, and birth defects, an epidemic of prostate and breast cancers, and even increasing neglect of children on the part of their parents. However, such assertions are extremely controversial, and scientists really are not sure how many of these environmental hormones we are exposed to and how that exposure affects us. Specifically, there is no consensus among researchers as to the strength of the evidence, the validity of many studies, how much (or how little) of an endocrine-disrupting chemical may produce significant results, or whether studies performed on laboratory animals or wildlife are also applicable to humans.

A classic example is a study of the alligator population in Lake Apopka, Florida. Significant quantities of a pesticide were spilled into the lake, and several years later it was found that 60% of the male alligators had undersized penises and only about 25% of the normal level of testosterone. The conclusion was that the pesticide was affecting their hormonal levels, essentially "feminizing" the males. Dozens of other studies, from Lake Michigan minks to Baltic Sea fishes, have purported to show similar results, including reduced penis sizes in males, "feminized" male behaviors, female sterility, and other problems. Although such studies on wildlife document a direct impact of environmental hormones on ecosystems, it is unclear how well these data can be extrapolated to humans.

Another large group of published studies focuses on the impacts of environmental hormones on sperm counts around the world. Since the early 1990s, these studies have been reporting that human sperm counts have decreased by as much as 50% in cities and other areas of high exposure

to synthetic chemicals during the last half century. Early studies were justifiably criticized for many sampling errors, such as combining data from different areas at different time periods and failing to take potential geographic variations in sperm counts into account; however, more recent studies continue to suggest an impact on fertility. For example, a study released in 2002 by Lynn Fraser of Kings College in London showed that even small concentrations of three kinds of environmental estrogens had a much stronger (approximately 100 times greater) effect on sperm behavior in mice than did natural estrogens found in female mice. Fraser hypothesized that this could reduce fertility by reducing the sperm's ability to fertilize the egg.

Again, critics of such studies note that differences between mouse and human reproduction may invalidate their application to humans. To help resolve the debate, the United Nations, the World Health Organization, and many other international organizations sponsored a global comprehensive review by many scientists of the publicly available scientific literature on environmental hormones. This review was released in 2002 as the report *Global Assessment of the State-of-the-Science of Endocrine Disruptors*. The report found there is sufficient evidence that adverse effects have occurred as a result of exposure to environmental hormones in some wildlife species. The report also concluded that although the evidence supporting the contention that human health has been adversely affected by exposure to environmental hormones is generally weak, additional research and information are needed; therefore, because of continuing concerns and scientific uncertainties, studies on the potential effects posed by these chemicals should remain a global priority requiring coordinated and strengthened international research strategies. There is, in particular, an urgent need for studies of vulnerable populations, especially infants and children, because exposure during critical developmental periods may have irreversible effects.

Until such studies are completed, public concern over the hormonal impacts of synthetic chemicals will continue to have many social ramifications. For example, in 2002, the EPA released a pesticide assessment report saying that more than 6,000 pesticides had been successfully assessed for their

Environmental Hormones:
A Growing Threat?

impacts on people and animals. The EPA was sued by the National Resources Defense Council, which stated that the EPA's assessment was incomplete because the testing focused on pesticide **acute toxicity** and failed to test whether the pesticides harm the body's hormone controlled endocrine system.

Questions

1. How great is your concern over synthetic chemicals in the environment? How about in the food you eat? What would make you more concerned than you currently are? Do you think such issues need to be addressed more forcefully than they currently are by government and/or industry?

2. One large group of synthetic organic chemicals are the "organochlorines," including chlorinated pesticides (such as DDT), PCBs, dioxins, chloroform, chlorofluorocarbons (CFCs), vinyl chloride, carbon tetrachloride, and so forth (see also Case Study 14-3).

Although organochlorines have often been considered the mainstay of the chemical industry and chlorine is even used to kill pathogens in drinking water, many chlorinated organic compounds are known to be carcinogenic or otherwise toxic, whereas many others are suspected of having harmful effects. In 1991, Greenpeace launched a campaign to phase out the industrial use of chlorine. Do you think such a chlorine ban is called for? Should organochlorines perhaps be evaluated one by one for harmful effects and then their use restricted if necessary? How practical can such an approach be, given that there are more than 10,000 different organochlorines being manufactured? Phasing out the use of chlorine in most industrialized processes could cost at least tens of billions of dollars. How well documented should the potential harmful effects of organochlorines be to justify such an expense?

(**FIGURE 14-13**). Observable and controllable hazards, such as motorcycles, are perceived as less risky than nuclear power, although statistically the risk associated with motorcycles is much higher. The second reason for risk misperception is the way we receive information about environmental hazards. Nearly all of the news media tend to focus on dramatic spectacles, such as oil spills, rather than long-term problems because the media have found that people tend to be more interested in dramatic events. Although oil spills do represent short-term local risks, the most dangerous risks to society as a whole are often more long term in nature, such as global warming and species extinctions.

Public misperception of risk results in wasteful and inefficient spending of money on environmental problems. Much more money is spent on problems of lower overall risk, such as sewage and hazardous wastes, than on

high-risk problems, such as global warming and indoor radon. Some have proposed that such spending inefficiencies could be avoided by using benefit–cost analyses that incorporate accurate risk information. However, in practice this has proven very difficult largely because so many of the benefits and costs, including risk, are very difficult to measure. For example, the statistical risks of death in Table 14-4 omit many other crucial considerations needed for a complete benefit–cost analysis. What about the risks of nonlethal injuries, which are omitted from simple "death count" tabulations? What about the risks to wildlife and ecosystems? What about the social groups that will experience the highest risk and obtain the least benefit from a proposed activity? This last consideration is crucial because risks, costs, and benefits are almost never equally distributed. As a result of such complexities, society has spent enormous resources in legal, political,

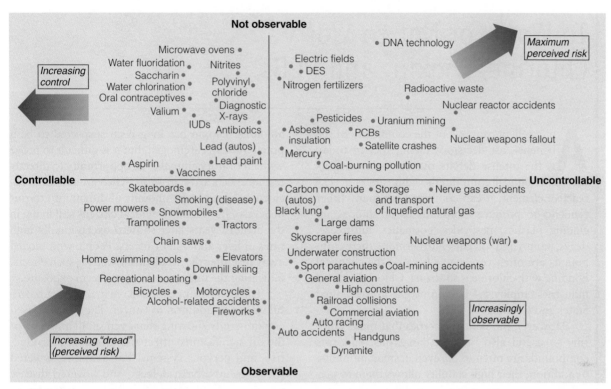

FIGURE 14-13 The most dreaded threats (maximum perceived risk) are often those that are least observable and least controllable. The least dreaded may be higher actual risks, but we dread them less because our control and observability are greater. (*Source:* Adapted from M. Granger Morgan. Risk analysis and management. Scientific American July 1993:41.)

and public debates over the costs and benefits of many substances (see **CASE STUDY 14-3**).

14.4 Pesticides: Pollutants Made to Kill

A **pesticide** is a chemical manufactured to kill organisms that people consider to be undesirable. Specific kinds of pesticides refer to the targeted pests: insecticides, herbicides, rodenticides, and fungicides. Herbicides are the most commonly used, followed by insecticides. More than 100,000 chemicals are recorded as potential pesticides.

More than 500 million kilograms (1 billion pounds) of pesticides are applied each year in the United States, at a cost of around $10 billion. It is estimated that this saves $64 billion in crops. Household pest eradication (mainly mice, termites, and other insects) is a $5-billion-a-year industry. Worldwide, more than $40 billion per year is spent on pesticides. However, chemical pesticides have many disadvantages, ranging from harming people and ecosystems to the evolved immu-

nity of pests. Approximately 45,000 accidental pesticide poisonings occur in the United States each year, resulting in about 50 deaths; however, many more people may die of cancer caused by pesticide exposure. A National Academy of Sciences study estimated that in the United States as many as 20,000 people per year get cancer because of pesticide residues in foods; even if only 10% of these die, the death rate would be 2,000 per year.

In addition, because pests can develop immunities to pesticides and new pests are introduced by human transport, the increasing use of new pesticides has not reduced the number of pests. Farmers still lose approximately one-third of their crops to pests, just as they did 50 years ago, before advanced pesticides were invented (**FIGURE 14-14**). Therefore, the current trend in the United States and some other parts of the world is toward nonchemical controls, such as using natural enemies. In undisturbed ecosystems, natural enemies control pest populations 5 to 10 times more effectively than do chemical pesticides, with little or no harm to other organisms.

To Worry or Not to Worry?
Chlorine, Dioxin, and PCBs

http://environment.jbpub.com
/mckinney/5e/
for more information

An excellent example of the complexity of the benefit–cost decisions facing society today is the intense debate over chlorine and its derived chemical compounds. Chlorine is a very **reactive** element. It can combine with many other elements to produce thousands of substances, including plastics, pesticides, cosmetics, antifreeze, drugs, paints, and solvents. Furthermore, these compounds are often very stable because the chemical bonding with chlorine is so strong. Chlorine-related industries employ 1.4 million people in the United States and Canada.

However, the same properties that make chlorine so useful also make it dangerous. Chlorine compounds are often toxic, even in small amounts. In addition, their high stability allows them to persist for a long time in the environment and often bioaccumulate in living tissue. CFCs are chlorine-based chemicals that pose a threat to the ozone layer because they persist for many decades in the atmosphere. An especially controversial group of chemicals is the organochlorines, which are formed from chlorine and carbon. Many are considered to be potentially toxic, and they are extremely common. Half of the 362 synthetic chemicals found in a recent survey of the water, sediment, and aquatic organisms of the Great Lakes were organochlorines.

Two of the best known organochlorines are dioxin and PCBs. Dioxin is a family of more than 75 related compounds, some of which are extremely toxic to laboratory animals in very small amounts. Dioxin has long been suspected to be a potent human carcinogen, but it is difficult to prove with laboratory animals. An opportunity to directly test its effects arose in 1976, when an industrial accident released large amounts of dioxin into the air in Seveso, Italy. Studies of local citizens still living in the area 15 years later showed exceptionally high rates of liver and other cancers. People who moved away just after the accident showed no such effects, indicating the importance of long-term exposure.

Dioxin also appears to have many other harmful effects in addition to cancer. The EPA released a major study showing that even small amounts of dioxin significantly affect the immune, reproductive, and nervous systems. For example, lowered sperm counts, birth defects, and lowered disease resistance may occur. This study provoked so much debate that a $4 million reassessment study was done, with contributions from more than 100 scientists from universities, governments, and private laboratories around the world. The reassessment confirmed many previous results and provided insight into how dioxin harms humans: It disrupts cell and tissue growth and function by binding to cellular proteins. The reassessment was unsuccessful in specifying just how much dioxin exposure is needed to harm humans, but most data indicate that people are in the middle range of sensitivity; some organisms are much more sensitive than people to dioxin's effects, and others less so. Continued research was called for and will no doubt occur for years to come.

The History of Pesticide Use

The history of pesticide use by people has occurred in three basic stages.

Inorganic Pesticides

The first stage of pesticide use began more than 4,500 years ago when the Sumerians used sulfur to kill insects. Other inorganic chemical pesticides include copper, arsenic, and lead. These substances were used in many parts of the world until the early 1900s but are now generally banned as pesticides because of their high persistence and especially their nonspecific toxicity. They can readily harm or kill nearly all forms of life, including humans.

Synthetic Organic Pesticides

The second stage, synthetic organic pesticides, revolutionized pest control when they were first widely used around the time of World War II. Initially, they were heralded as a "cure-all" that would result in the total elimination of

Although complete understanding of dioxin's impacts remains elusive, we have a fairly good understanding of how people are exposed to dioxin and thus how to reduce exposure. Although dioxin does occur in nature, such as from volcanoes and forest fires, most of it is a product of modern industry. Although just 13.6 kilograms (30 pounds) of dioxin is released each year in the United States, it is so pervasive that everyone has been exposed. Most exposure comes from eating meats and dairy products from animals that have eaten plants contaminated by dioxin emissions. What kind of industries emits dioxins? Waste combustion, especially medical and city waste, accounts for about 95% of all dioxin emissions. This is one reason that many communities oppose waste incinerators. Other sources of dioxin are the chemical manufacture of chlorine compounds, such as pesticides, and pulp and paper mills. Bleaching paper with chlorine is an especially important source. In response to growing pressure, many paper companies are switching away from chlorine to other substances, such as chlorine dioxide, which produces less dioxin. Oxygen compounds such as ozone that produce no dioxins can also be used, but they are more costly.

Another well-known organochlorine group is PCBs. These were once common in electrical insulation, paints, and many other industrial products. Their properties and effects on people are similar in many ways to those of dioxin. In fact, combustion of PCBs will produce dioxin. Manufacture of

PCBs in the United States ceased in 1978. They have been slowly disappearing since but are still common because of their persistence. Industrial waste sites and other old dump sites often have high concentrations of PCBs; they are one of the main challenges of hazardous waste cleanups.

In response to the problem of chlorine compounds, Greenpeace and other groups have campaigned to ban all use of chlorine by industry. The evidence of potential harm to people and other life can be used to support such a ban, but society also has to consider the overall benefits and costs. The costs of banning chlorine in industry are very high; according to some estimates, a switch to other compounds would cost consumers more than $91 billion. For instance, chlorine-based materials comprise more than half the pharmaceuticals on the market. Similarly, cleanup of dioxin and PCBs is exacting a huge cost right now. As much as $100 billion may ultimately be spent in the United States to remove PCBs from the environment. Is it worth it? Many people argue that although no one can deny that such compounds do some harm, the huge monetary costs to society will greatly outweigh the health and ecological benefits of removing these compounds. Once again, we see the problem of measuring true environmental costs. It is difficult to make monetary decisions when we cannot measure such things as just how many human lives are lost and how many ecosystems are seriously harmed by these chemicals.

pest species, but this has decidedly not been the case. Organic pesticides have caused a growing number of environmental and health problems and have steadily declined in effectiveness as pest species have become immune to them.

There are three basic kinds of synthetic organic pesticides (TABLE 14-5). **Chlorinated hydrocarbons** include chlordane, DDT, and other familiar pesticides. These are nerve toxins that cause paralysis, convulsions, and death. They are broad-spectrum toxins, attacking

any organisms with central nervous systems. Although these are the cheapest groups of pesticides, they also cause the most long-term harm to biological communities because they are easily bioconcentrated in the food pyramid. Chlorinated hydrocarbons are "organochlorines," which often have high persistence and tend to be concentrated in tissue. DDT, in particular, has one of the highest fat solubilities ever measured, and the body tissue of organisms often becomes concentrated to more

FIGURE 14-14 A gallery of agricultural pests. (a) The squash bug eats watermelon, cantaloupe, and squash. (b) Gypsy moth caterpillars feed on the leaves of more than 500 types of trees and shrubs. (c) The tarnished plant bug damages fruits, vegetables, cotton, and many seed crops. (d) The alfalfa plant bug feeds on alfalfa in the eastern United States and Canada.

than 1,000 times the surrounding levels. For this reason, DDT and most of the other chlorinated hydrocarbons have been banned in the United States. As a result, the concentration of such persistent pesticides in human tissue has steadily decreased in recent years.

The other two groups of organic pesticides, **organophosphates** and especially **carbamates**, are less persistent, lasting only a matter of days, weeks, or months in the environment, compared with decades for chlorinated hydrocarbons. These two groups are also more expensive and much more toxic to humans. An adult human would have to ingest more than 1 pound (0.45 kg) of DDT in one sitting to cause death but much smaller amounts of organophosphates or carbamates, which are rapidly absorbed by human skin, lungs, and the digestive tract. Like chlorinated hydrocarbons, organophosphates and carbamates attack the nervous system of the pest, but they do so in a

different manner. They attack an enzyme that controls nerve impulses, setting off a stream of uncontrolled impulses that cause the organism to twitch violently before death from organ failure. Common examples of organophosphates are malathion and parathion; Sevin is the brand name of a common carbamate (Table 14-5).

Biochemicals: Bacterial Toxins and Synthetic Hormones

The third and most recent stage of pesticides is the commercial production of chemicals that are ordinarily found in nature and can be used to control pests. Such naturally occurring biochemicals have the advantage of being very nonpersistent. Unlike many artificial chemicals, biochemicals are readily degraded into nontoxic forms by microbial digestion and other natural processes.

Bacterial toxins are pesticides that are made by such bacteria as *Bacillus thuringi-*

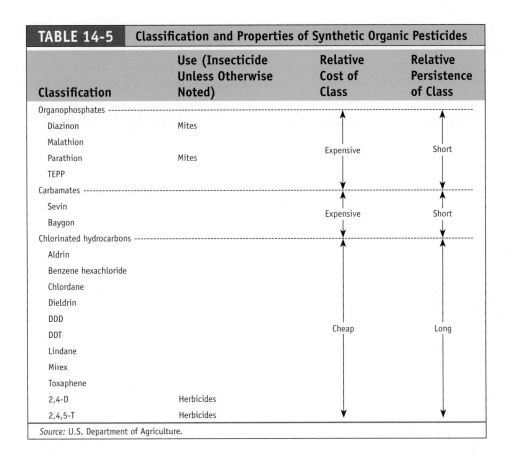

TABLE 14-5 | **Classification and Properties of Synthetic Organic Pesticides**

Classification	Use (Insecticide Unless Otherwise Noted)	Relative Cost of Class	Relative Persistence of Class
Organophosphates			
Diazinon	Mites		
Malathion		Expensive	Short
Parathion	Mites		
TEPP			
Carbamates			
Sevin		Expensive	Short
Baygon			
Chlorinated hydrocarbons			
Aldrin			
Benzene hexachloride			
Chlordane			
Dieldrin			
DDD			
DDT		Cheap	Long
Lindane			
Mirex			
Toxaphene			
2,4-D	Herbicides		
2,4,5-T	Herbicides		

Source: U.S. Department of Agriculture.

ensis. These are toxic to a number of insect pests, such as Japanese beetles. Synthetic hormones are biochemicals that mimic hormones of pests, especially insects. Two basic types are especially important. One type interferes with growth-regulating hormones. For example, insects must periodically shed, or molt, their hard outer skeleton to grow (**FIGURE 14-15**). Synthetic growth hormones inhibit the hormones that control molting. Being unable to shed their outer skeleton, the juvenile insects die. **Pheromones** are the second type of synthetic hormone; these are the biochemical scents used by males or females to attract the opposite sex. Artificial pheromones are used to attract individuals to traps, where they are killed. Pheromones are so powerful that only a drop can attract hundreds of individuals from many miles away.

Synthetic hormones have two major advantages over the nonpersistent synthetic organic pesticides. First, they are generally more specific and more likely to affect only the target species. For example, mammals and birds usually are unaffected by insect hormones. Second,

FIGURE 14-15 Biochemicals that prevent insects from molting or shedding their outer skeleton are effective at reducing reproduction in a certain insect species without harming other species. Seen here is a cicada emerging from its nymphal case.

14.4 Pesticides: Pollutants Made to Kill   433

only small doses are needed because they have such a powerful effect on the target species.

Problems with Chemical Pesticides

In 1962, Rachel Carson published *Silent Spring*, the first book that warned against the ecological and health effects of widespread pesticide use. She brought national and international attention to the toxic effects of DDT. Since then, many studies have confirmed much of what she said. Specifically, three main problems have emerged that severely limit the long-term utility of pesticides: *nontarget toxicity*, *secondary pest outbreaks*, and *increasing immunity*. These problems are especially severe with synthetic organic pesticides, which continue to be widely used because biochemical pesticides are not yet available for many pests. In addition, biochemical pesticides are often less effective. Trapping pests by attractants is not only much more labor intensive, but it generally reduces the pest population less than would a potent poison. A heavy poison dose will kill more than 90% of the pest individuals, whereas dozens of traps will kill only a few percent. In addition, the biochemical pesticides are not completely free of any of the following problems.

Nontarget Toxicity

The widely used organic pesticides can have devastating effects on nontarget species. Just one of countless illustrations occurred in 1958, when massive doses of the chlorinated hydro-carbon dieldrin were applied on Illinois farmland to eradicate the Japanese beetle. Cattle and sheep were poisoned. Ninety percent of local cats died. Twelve species of wild mammals and 19 species of birds were decimated, with robins, starlings, pheasants, and other birds virtually eliminated. Even synthetic hormones can affect nontarget species. For instance, synthetic growth hormones can affect nonpest insects because many insects have hormones that are biochemically similar. Nontarget toxicity is enhanced in the more persistent and fat-soluble organic pesticides, such as chlorinated hydrocarbons. This often results in increased toxicity in organisms high on the food pyramid from bioconcentration. Pesticides come in many different forms, such as dusts and sprays, which can be carried great distances by air and water. More than 11,600 kilometers (7,200 miles) of U.S. rivers and 1 million hectares (2.5 million acres) of lakes have restricted or banned fishing, largely because of pesticide contamination.

Nontarget toxicity is also directly hazardous to humans. Pesticides cause an estimated 10,000 deaths and 1 million injuries each year worldwide (**FIGURE 14-16**). These occur from a variety of toxic effects, ranging from immediate to long-term tissue accumulation. Immediate effects are most likely from exposure to the highly toxic organophosphates and carbamates. Temporary symptoms include nausea, blurred vision, and convulsions. Permanent symptoms from exposure to higher

(a)   (b)

FIGURE 14-16 Pesticides and herbicides pose considerable health risks to the farm workers who must use them. In less-developed countries, workers (a) often cannot read the warning labels, or they (b) lack access to the protective clothing that should be worn when handling the chemicals.

CHAPTER 14 Principles of Pollution Control, Toxicology, and Risk

concentrations include damage to major organs (heart, lungs, and especially liver and kidneys) and brain and other nervous system damage. Chlorinated hydrocarbons tend to cause chronic toxicity, leading to cancer, birth defects, and mutations. An example is Agent Orange, a chlorinated hydrocarbon herbicide used in Vietnam to defoliate forests. Now believed to cause cancer and birth defects, it has been banned in the United States since 1985.

How much pesticide do we ingest in our food? Unfortunately, a majority of the tested fruits and vegetables eaten in the United States contain pesticide residues. One special cause of concern is that many of these residues include pesticides banned in the United States, such as DDT and many other chlorinated hydrocarbons. Despite the ban on use in the United States, they are still manufactured here and shipped to developing countries, where they are widely used because of their low cost. U.S. citizens continue to be exposed to these pesticides because much of the produce grown in these developing countries is imported by the United States in what is sometimes called the **circle of poison**. Most of the residues on supermarket produce occur in such low concentrations that many experts are not concerned. Nevertheless, efforts to regulate pesticide use can fail, and increasing numbers of consumers are willing to pay more for foods grown organically without pesticides.

Secondary Pest Outbreaks

Biological interactions such as predation and competition constrain natural populations. When a pesticide strongly reduces key predators or competitors, the species they had kept in check often experiences a "population explosion." Cotton pests in Central America are a prime example. In 1950, synthetic organic pesticides were applied to reduce the boll weevil, resulting in much greater crop yields. However, by 1955 a number of secondary pests that were less affected by the pesticide, such as the cotton aphid and cotton bollworm, began to reduce crop yields again. These pests had been limited by competition from the boll weevil. The typical response to such secondary pest outbreaks is to apply a different kind of pesticide. By the 1960s,

farmers were combating eight pest species instead of the original two, with an average of 28 applications of various pesticides per year.

Evolved Immunity

Pests become immune to pesticides with striking speed. In the last few decades, insects, plant diseases, and weeds have shown an exponential increase in the number of pesticide-resistant species (**FIGURE 14-17**). For example, in 1970, there were no known weeds with herbicide resistance, whereas now there are more than 273 known herbicide-resistant weeds. (A "herbicide-resistant weed" is not necessarily resistant to all herbicides but is resistant to at least some.) It is estimated that across all kinds of agricultural pests and diseases, approximately 1,000 species are resistant to at least one pesticide. Some pests are now resistant to multiple pesticides. At least 10 species of rodents are immune to some brands of rat poison. Perhaps the most impressive pest in this regard is the common housefly, which has become immune to almost every pesticide invented, even a synthetic growth hormone designed for it.

The rate of adaptation is so high largely because pests are generally fast-growing organisms that reproduce often and are adapted to a wide range of conditions. In fact, these "weedy" traits are why they are pests. Their fast growth ensures that tolerant traits are

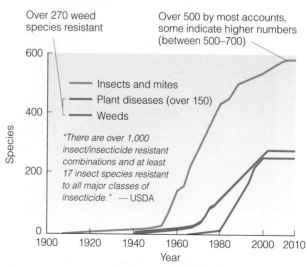

FIGURE 14-17 Pesticide-resistant species increased exponentially between 1900 and 2010. (*Source:* Food and Agriculture Organization of the United Nations.)

passed on to offspring quickly, so evolution occurs quickly. Their broad adaptations ensure that at least a few tolerant individuals likely will occur somewhere in the population. The result is what is often called the **pesticide treadmill**: To keep up with the rapid pace of evolving pesticide immunity, new pesticides constantly must be developed. As of the beginning of this century, pest resistance to pesticides was estimated to cost U.S. agriculture approximately $1.5 billion in increased pesticide costs and decreased yield.

Reducing Chemical Pesticide Use

The problems with chemical pesticides have caused a recent shift in agriculture toward **integrated pest management (IPM)**, which seeks to reduce the use of chemical pesticides by relying on other methods of pest control, such as natural predators and breeding pest-resistant crops (**FIGURE 14-18**). IPM is becoming increasingly common in the United States and many parts of the world. In fact, since 1980, U.S. pesticide use has leveled and begun to decrease (although some of this decrease can be attributed to more lethal pesticides so that less is needed).

Integrated pest management is yet another example of the efficiency of source reduction.

Reducing the amount of pollutant produced is much easier than trying to control pollutant dispersal or cleaning up the pollution after it is released into the environment. Less than 1% of sprayed pesticides actually reach pests. The rest is carried off by water and other transport methods to pollute groundwater, soil, and other parts of the environment. Source reduction—in this case, spraying less pesticide—is the best way to reduce such pollution.

In addition to reducing pesticide pollution, IPM often is cheaper than chemical pesticides. Furthermore, the use of nonchemical methods actually improves the efficiency of chemical pesticides when they must be used. Immunity to chemicals develops much more slowly when the pests are exposed to the chemicals less often. Most experts predict that these benefits will cause IPM to be increasingly used in the United States and worldwide. Indeed, the main obstacles now are a shortage of teachers trained in IPM and a lack of funds to finance the education of farmers in IPM.

Pesticide reduction through IPM often is associated with farms and agriculture, but cities and households are significant contributors to pesticide pollution. U.S. suburban lawns use more chemical pesticide per acre than U.S. cropland. Similarly, golf courses pour tons of

(a)

(b)

FIGURE 14-18 Integrated pest management can include the careful use of an invasive pest's natural predators or parasites to help reduce the pest's damage. (a) A female Catolaccus grandis wasp, a native of Mexico, homes in on a boll weevil larva. The wasp will deposit an egg in the larva-rearing cell, and soon the larva will become a meal for the newly hatched wasp. (b) An Aleiodes indiscretus wasp is parasitizing a gypsy moth caterpillar.

pesticides, fertilizer, and other pollutants into local waterways. Should homeowners and golf courses reduce pesticide pollution by substituting more natural plants native to the area for lawns?

14.5 Legal Aspects of Toxic Substance and Pesticide Control

Three major federal statutes are of special importance to toxic substance input reduction because they limit the actual production of toxic substances. In contrast, most other federal laws limit the release and environmental exposure of toxic substances after they are produced.

The Toxic Substances Control Act (TSCA) was designed to regulate toxic substances not covered under other laws. Its basic goals are to require industry to produce data on environmental effects of chemicals and to prevent harm to people and the environment, while not creating unnecessary barriers to technology. These objectives exemplify the ambiguity inherent in many regulations because the protection of people and the environment often conflicts with the promotion of technology. For example, what is an "unnecessary" barrier?

The centerpiece of the act is the "premanufacturing notice," which requires the chemical industry to notify the EPA 90 days before a new chemical is to be manufactured for sale. This notice must contain information on the chemical identity, molecular structure, and test data obtained from the toxic identification, dose–response assessment, and risk characterization procedures discussed earlier. The EPA uses this (and other) information to decide whether the chemical can be produced, should be banned, or requires additional testing. The TSCA has caused hundreds of chemicals to be withdrawn that otherwise would have been sold.

More than 60,000 chemicals were already in use in 1976 when the act was enacted. This was far too many to test, so nearly all of these chemicals were simply "grandfathered," with no requirement for testing. They continue to be sold without any testing, and these pre-1976 chemicals make up more than 99.9% of the approximately 2.7 trillion kilograms

(6 trillion pounds) of chemical products made in the United States each year.

The Federal Insecticide, Fungicide, and Rodenticide Act (FIFRA) requires anyone wishing to manufacture a pesticide to register that product with the EPA. More than 25,000 pesticides have been registered. As with other toxic substances under the TSCA, the EPA may ban the pesticide, permit manufacture (sometimes under limited use), or require further additional if it is deemed to present an "unreasonable risk" to people and the environment. This mandate is ambiguous: What is an "unreasonable" risk? The importance of ensuring the safety of pesticides before they are used is clear when (1) less than 1% of all fruits and vegetables are inspected by the Food and Drug Administration, (2) some pesticides are not tested for, and (3) tests take an average of 28 days, so that the produce is often sold before the tests are completed. However, because about 25% of U.S. produce is imported from countries that use EPA-banned pesticides, we are still exposed.

Even if manufacture and use in the United States are permitted, each pesticide must be reregistered every 5 years and may be banned if information has emerged showing it is exceptionally harmful. DDT and many other chlorinated hydrocarbons were taken off the market in the United States. More recently, diazinon, one of the most widely used lawn and garden insecticides, has been banned in the United States. Shipment to retailers stopped in August 2003. A major problem with FIFRA is that pesticide manufacturers are not required to list "inert" chemicals when they label the pesticide for consumers. Such "inert" chemicals often constitute more than 80% of the pesticide and sometimes are very harmful in their own right. The manufacturers insist that they must withhold this information to prevent competitors from discovering their formulas. The National Pesticide Information Center (800-858-PEST) has been established to obtain pesticide information.

The oldest statute regulating production of toxic substances is the **Food, Drug, and Cosmetics Act**, which authorizes the Food and Drug Administration to test products to see whether they are safe to market for human use.

The Food and Drug Administration is often criticized for being overly cautious and taking too long to test medical products that could save lives. On the other hand, the agency has a heavy responsibility because it cannot risk approving a highly toxic product that could take many lives.

Toxic Torts

The complexities of regulating toxic chemicals are enormous. Errors inevitably will be made in a large economy. When errors happen, the next line of protection—one that is increasingly being used—is the so-called toxic tort, a lawsuit by individuals against manufacturers for harm that has occurred. This area of law is very chaotic and varies from state to state because state law governs torts. Basically, the harmed individual(s) must prove that the company was negligent in not providing adequate warning or in marketing a product that was unnecessarily dangerous.

For example, one can sue a chemical manufacturer for negligence if it produces toxic chemicals that cause harm to unknowing consumers. As we have noted, in practice, it is very difficult to prove that a certain chemical causes cancer or other chronic illness. Nevertheless, many individuals have won millions of dollars in compensatory damages (for medical bills and other expenses to recover their original quality of life) and, less often, punitive damages, which are awarded to punish the company. An important environmental benefit of toxic torts is that chemical companies are now taking special care to police the toxic properties of the chemicals they market. Even if a toxic chemical gets by the EPA review process, the companies know that they are liable to lawsuits if negligence or exceptional toxicity can be shown.

Industrial Accidents

Accidental releases of industrial chemicals into the environment form a special class of toxic torts because of their huge potential for catastrophic ecological and human harm. This was made evident in the "**Bhopal** disaster" in India in December 1984, when approximately 15 metric tons of deadly pesticide gas were accidentally released (**FIGURE 14-19**). The gas covered more than 77 square kilometers (30 square miles) and exposed as many as 800,000 people. Local inhabitants estimate that at least 7,000 people died immediately, but much suffering continues. Medical reports indicate that at least 50,000 people experience a variety of serious ailments, and perhaps one person dies every 2 days from the effects. Union Carbide, which owned the pesticide plant, was held legally responsible for the accident and paid various types of compensations to some individuals. Such cases often are the subject of class-action lawsuits, where a single group, such as the citizens of Bhopal, sues the party responsible for the accident. However, in this case, many thousands of sick Bhopal citizens have complained that they received no compensation.

Despite the warning of Bhopal, the potential for similar catastrophes is growing as chemical industries expand in developing nations and industrial accidents are statistically increasing. In the 8 years after Bhopal, there were 106 major accidents, compared with just 74 in the 8 years before. In 2000, cyanide used in a Hungarian gold mine overflowed into a major river, the Tisza, and killed all aquatic life for dozens of miles downstream. Even the famous Danube River, which receives water from the Tisza, was affected by this spill.

The United States is not exempt from such events. More than 4.5 billion kilograms (10 billion pounds) of extremely hazardous chemicals are stored in the United States.

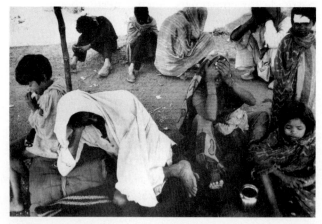

FIGURE 14-19 Victims of the Union Carbide Bhopal disaster, December 1984.

Between 1990 and 2000, several incidents of industrial accidental gas releases that exceeded Bhopal in quantity and toxicity occurred in the United States. For example, on December 13, 1994, there was an explosion in the ammonium nitrate plant operated by Terra Industries, Inc., in Port Neal, Iowa. Four persons were killed as a direct result of the explosion, and 18 were injured and required hospitalization. The explosion released approximately 5,700 tons of anhydrous ammonia into the air; approximately 25,000 gallons of nitric acid to the ground, lined chemical ditches, and sumps; and liquid ammonium nitrate solution into secondary containment basins. The chemicals released as a result of the explosion contaminated the groundwater under the facility.

The Precautionary Principle Revisited

We introduced the precautionary principle in the overview of environmental science, and its aphorism, "better safe than sorry," has been a guiding rule of thumb for centuries. In some ways, the saying "nothing ventured, nothing gained" states the opposite view. When it comes to introducing new chemicals into the environment, these two opposing views can lead to radically different actions and consequences. Suppose that a newly developed fertilizer has the potential to greatly increase crop yields, but it also might contaminate soils and groundwater. Here "might" is the key term. Do we go ahead and use the new fertilizer and worry about any negative consequences only if they occur? Do we attempt to solve the problem, such as by the cleanup of resultant pollution, after the fact? Do we avoid using the new fertilizer just in case there could be problems? How high must the likelihood of problems be before we decide not to use the fertilizer? What if in reality there are no negative consequences of using the fertilizer and it can produce great good by increasing crop yields for a hungry world? Do we reject the use of the fertilizer simply because there might be negative consequences?

In January 1998, there was a historic meeting of scientists, scholars, activists, and negotiators held at the Wingspread Conference Center, Racine, Wisconsin. The participants gathered explicitly to discuss the precautionary principle. By 1998, this principle had become increasingly invoked in various situations involving the use of toxic substances and other activities that can potentially threaten the environment. The precautionary principle was regularly discussed during the negotiation of international treaties. After much thought and discussion, the participants of the Wingspread Conference agreed to the following Consensus Statement on the Precautionary Principle:

The release and use of toxic substances, the exploitation of resources, and physical alterations of the environment have had substantial unintended consequences affecting human health and the environment. Some of these concerns are high rates of learning deficiencies, asthma, cancer, birth defects and species extinctions; along with global climate change, stratospheric ozone depletion, and worldwide contamination with toxic substances and nuclear materials.

We believe existing environmental regulations and other decisions, particularly those based on risk assessment, have failed to protect adequately human health and the environment—the larger system of which humans are but a part.

We believe there is compelling evidence that damage to humans and the worldwide environment is of such magnitude and seriousness that new principles for conducting human activities are necessary.

While we realize that human activities may involve hazards, people must proceed more carefully than has been the case in recent history. Corporations, government entities, organizations, communities, scientists and other individuals must adopt a precautionary approach to all human endeavors.

Therefore, it is necessary to implement the Precautionary Principle: When an activity raises threats of harm to human health or the environment, precautionary measures should be taken even if some cause and effect relationships are not fully established scientifically.

In this context the proponent of an activity, rather than the public, should bear the burden of proof.

The process of applying the Precautionary Principle must be open, informed and democratic and must include potentially affected parties. It must also involve an examination of the full range of alternatives, including no action.

A decade after Wingspread, the precautionary principle continues to be prominent and likely will be of increasing influence throughout the 21st century. Even if the strictest versions of the precautionary principle are not always implemented (for instance, to decide not to release a new product onto the market unless it is known to be *absolutely* safe, and how can one ever know *absolutely?*), the precautionary principle certainly provides an important baseline for discussion. Such discussion will benefit humanity and the environment as we face future challenges and may allow us to avoid mistakes similar to those made in the past.

■ study guide

SUMMARY

- Pollution refers to excess concentrations of substances.
- Life cycle analysis can eliminate waste and pollution at many steps in product use.
- Pollution = population × consumption.
- Three myths of pollution control are: purity of nature, zero pollution, and zero risk.
- The benefit–cost approach seeks to reduce pollution to acceptable levels while minimizing economic cost.
- Input reduction is more efficient than output control in reducing pollution.
- Persuasion, regulation, and economic incentives are three basic ways to implement pollution controls.
- Economic incentives are more effective than laws where many polluters are involved.
- The United States produces more pollution per person than any other nation.
- The Toxic Release Inventory provides an annual overview of toxic pollution.
- Toxicology studies how chemicals harm living organisms.
- Bioassays determine if a chemical is toxic.
- Dose–response assessment determines the strength of toxicity; a standard measure is LD-50.

- Inaccuracies in using dose–response data for people include the use of test animals and scaling factors.
- Exposure assessment examines the pathways by which toxic substances can reach humans, including bioconcentration.
- Risk characterization helps scientists decide if the benefits of a chemical outweigh the costs (risks).
- People widely misperceive risks.
- Benefit–cost analyses are rarely simple to carry out in practice, despite the widespread desire to apply them.
- Pesticides have evolved through three stages: inorganic, synthetic organic, and biochemical.
- Problems with pesticides include nontarget toxicity, secondary pest outbreaks, and evolved immunity.
- Integrated pest management minimizes pesticide use by relying on natural enemies and pest-resistant crops.
- The major U.S. laws regulating toxic chemicals include the Toxic Substances Control Act, the Federal Insecticide, Fungicide, and Rodenticide Act, and the Food, Drug and Cosmetics Act.

KEY TERMS

abatement devices
acute toxicity
additivity
ADMSE
antagonism
benefit–cost analyses
Bhopal
bioassays
bioconcentration
carbamates
carcinogen
chlorinated hydrocarbons
chronic toxicity
circle of poison
contingent valuation method
"cradle to grave"
dosage

environmental hormones
epidemiological statistics
Federal Insecticide, Fungicide, and Rodenticide Act (FIFRA)
Food, Drug, and Cosmetics Act
heavy metals
integrated pest management (IPM)
intrinsic factors
LD-50
life cycle analysis (LCA)
mutagen
organophosphates
persistence
pesticide
pesticide treadmill
pheromones

pollution
reactive
remediation
restoration
risk
scaling factors
supertoxins
sustainability movement
synergism
teratogen
toxicology
Toxic Substances Control Act (TSCA)
toxic threshold
toxic tort
Toxics Release Inventory (TRI)

STUDY QUESTIONS

1. What is life cycle analysis (LCA)? How does it reduce waste generation?
2. Why is nothing in nature truly pure?
3. What are some problems associated with benefit–cost studies? What do benefit–cost ratios over 1 indicate?
4. How does output management increase pollution control costs compared with input management?
5. What method uses questionnaires to try to objectively estimate the dollar value of changes in environmental quality? Explain how this works.
6. Name the three basic ways to promote change in human behavior, such as reducing pollution.
7. Name the three common myths of pollution control.
8. What is toxicology?
9. Name three kinds of synthetic organic pesticides. Which kind has the longest persistence and is easily bioconcentrated? Give two examples.
10. What is bioconcentration? What kinds of substances tend to bioconcentrate? Give examples.
11. What is an environmental hormone? How can these hormones harm organisms?
12. Define risk. How does perceived risk vary from actual risk? Why do people sometimes perceive risk inaccurately?
13. Name and describe the three basic kinds of chemical interactions among substances in the environment.
14. What causes secondary pest outbreaks?
15. What causes evolved immunity?
16. What does the Toxic Substances Control Act require industry to do? What is the centerpiece of the act? What happened to the chemicals already being produced before the act was enacted in 1976?
17. What is a toxic tort? Is this area of law a simple one? Explain.
18. Describe the Precautionary Principle and how it can be applied to the regulation and use of toxic substances.

WHAT'S THE EVIDENCE?

1. The authors state that benefit–cost analyses are difficult to implement properly. List the evidence presented to reach this conclusion. Do these reasons seem valid to you? Why or why not?
2. The authors indicate that pesticide use imposes many problems on society. What are these problems and what is the evidence for them?

CALCULATIONS

1. It is estimated that pesticides cause up to 1 million injuries to people each year. What percentage of the human population is this, if there are 6.6 billion people on Earth?

2. Less than 1% of sprayed chemical pesticides usually reaches pests. If someone sprayed 1 million gallons (3.785 million liters) on a field, how many gallons (liters) would reach the pests, assuming that 1% of the spray did? (Recall that 1 gallon = 3.785 L.)

ILLUSTRATION AND TABLE REVIEW

1. Table 14-2 shows percent changes in releases of toxic substances between 1998 and 2003. What kind of toxic release had the greatest decline? What kind had the smallest decline?

2. Figure 14-5 shows that after 1975 the percentage cost of pollution control in the United States reached its lowest point in what year? What has been the general trend since then?

http://environment.jbpub.com/mckinney/5e/

http://environment.jbpub.com /mckinney/5e/

Connect to this text's website at http://environment.jbpub.com/mckinney/5e/. This site features eLearning, an online review area that provides quizzes, chapter outlines, and other tools to help you study for your class. You can also follow useful links for more in-depth information.

In the early 1900s, Boise, Idaho, started using the Boise River for waste disposal. By 1962, untreated wastes, including rotting carcasses from slaughterhouses, grease, and raw sewage, clogged the river. Such environmental degradations helped instigate landmark environmental laws during the 1970s. One of the most successful has been the Clean Water Act of 1972, which provided federal funding for wastewater treatment and spurred many other improvements in "end-of-pipe" waste removal. The payoff for the improved surface water quality has been increased economic activity and a cleaner environment. In Boise, for example, thousands of people now swim and canoe in the river thanks to the thousands of volunteers over the years who have all played a part in its successful restoration. See, http://www.idahorivers.org and http://www.americanrivers.org.

Water Pollution

15

People have long dumped their wastes into natural waters because the waste was quickly diluted and dispersed. If the water was a river, the waste was conveniently swept away. Such natural processes of purification are effective as long as human population density is relatively low compared with the amount of water available. The rapid increase in human populations has so overwhelmed the natural purification processes that pollution of both fresh and oceanic waters has become a global crisis. The scarcity of clean water is one of the greatest problems in the world today, as seen in a review of water resources.

Potable water is water that is safe to drink. **Polluted water** is water rendered unusable for its intended purpose. Even if water is unfit for human use, it is not considered polluted if that water is intended for uses such as watering city flower beds, filling fountains, or watering golf courses. **Contaminated water** is water unusable for any purpose: People cannot drink it. Farmers cannot water crops with it, and it is too corrosive or otherwise harmful to machinery for industrial use. Disease organisms, heavy metals, and many other toxic substances causing death or illness can contaminate water.

15.1 Damages and Suffering

Water pollution is highly costly in the effects it has on the health of people, other organisms, and ecosystems and in the economic damage it inflicts on industries, including agriculture. The effects on human health are greatest in developing nations that do not

Chapter Objectives

After reading this chapter, you should be able to explain or describe the following:

- The concept and risks of polluted water, including the common major categories of water pollution
- Natural water purification processes
- Sources of water pollutants
- Where most water pollutants ultimately end up
- Technical ways to decrease and treat water pollution
- Legal methods used to address water pollution

443

yet have widespread modernized purification or wastewater treatment. Although the data vary, in 2010 an estimated 1.1 billion people lacked access to improved water sources; this figure represents 17% of the global population. An improved water supply can be defined as a household connection, a public standpipe, a protected well or spring, or rainwater collection. In 2010, nearly 2.6 billion people, or 42% of the world's population, lacked access to improved sanitation; it is estimated that 50% of developing world population lacks adequate sanitation. Inadequate sanitation (**FIGURE 15-1**) and poor water supplies lead to deadly outbreaks of cholera, typhoid, hepatitis, and many other debilitating infections. Every day, 3,900 children die of waterborne diseases; they make up almost 90% of the 1.8 million people who die every year of diarrheal diseases, including cholera. Approximately 500 million people are at risk of trachoma, which causes blindness in approximately 6 million annually. About 200 million people are infected with schistosomiasis (also known as bilharziasis—infection with parasitic trematode worms), with 10% of them experiencing severe suffering. All told, the global consequences of poor water supplies and inadequate sanitation are staggering.

Modernized societies have a much lower incidence of waterborne diseases. In the United States, improvements in sanitation beginning around 1850 have led to a steady decline in waterborne diseases. By 1920, outbreaks of cholera and typhoid were rare, but problems can still occur. For example, in 1993, a breakdown of the Milwaukee water supply system caused about 403,000 people to become ill and contributed to the deaths of at least 9 people (and possibly as many as 50).

Aside from its effects on human health, water pollution in the United States causes widespread damage to aquatic ecosystems, such as rivers, lakes, wetlands, and coastal marine areas. An Environmental Protection Agency (EPA) survey found that more than 17,000 (or about 10%) of the nation's streams, rivers, and bays are significantly polluted. In addition, about 25% of the nation's usable groundwater is contaminated. Groundwater is the only source of drinking water for about half of the U.S. population. The EPA estimates that total economic damages from water pollution in the United States exceed $20 billion per year. Currently, the United Nations is leading a worldwide campaign, called the "Millennium Development Goal for Water and Sanitation," which is aimed at increasing access to safe water and proper sanitation by 50% by the year 2015. The World Water Council estimates the cost to be $6.7 billion a year spread out over world governments, a figure less than what Europeans and U.S. citizens spend in 1 year on pet food.

FIGURE 15-1 Inadequate sanitation is a major cause of disease for at least one-third of the world's people. Seen here is an open sewer and slum housing in Belize City, Belize.

15.2 Water Purification in Nature

The water cycle, driven largely by the energy of the sun, purifies water in many ways. Thus, rain, streams, lakes, and groundwaters tend to be relatively clean of dissolved matter, and such water usually is drinkable in the natural state. Water is purified in nature through a number of processes. For convenience, these can be grouped into two basic categories: physical and chemical processes.

- Physical Processes. The four physical processes involve forces such as gravity. These processes are important in removing the larger particles and debris and thus often account for the removal of the bulk of the material in water (**FIGURE 15-2**).

  1. **Dilution** is the reduction in concentration of a pollutant when it is discharged into water. Dilution increases as one moves away from the point of pollutant discharge (Figure 15-2). Not surprisingly, the faster the water moves, the more effective the dilution is because more water flows past the point of discharge in a given period of time. Indeed, fast-moving streams can remove as much as five times more waste than can slow-moving streams.

  2. **Sedimentation** is the settling out of suspended particles (Figure 15-2). Sedimentation varies with particle size and water velocity, occurring most readily with larger particles and in slow-moving waters. Very fine particles (such as clay) can stay suspended for long periods and travel great distances.

  3. **Filtration** is the percolation of water through sand and other settled sediment to remove suspended particles (Figure 15-2). Even very fine particles can be filtered out if the water percolates through fine sediment.

  4. **Aeration** is the release of gaseous impurities into the atmosphere (Figure 15-2). Water, like all liquids, contains many substances in the form of dissolved gases. Aeration is greatly accelerated when more water is exposed to the air, such as when the water trickles over rocks in shallow streams. It is also more effective in warm and fast-moving waters.

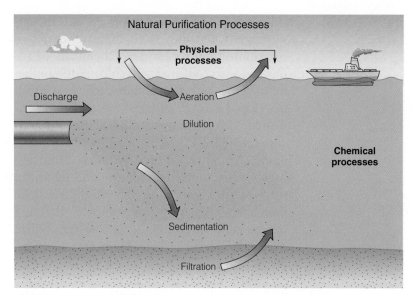

FIGURE 15-2 Five natural chemical and physical purification processes.

- *Chemical Processes.* Natural waters contain many dissolved minerals and gases that chemically interact in complex ways. Most of these reactions are biochemical, involving enzymes and many other chemicals produced by organisms, especially microbes. People have put these microbes to work as major degraders of waste in sewage treatment plants. A number of inorganic (nonbiological) chemical reactions are also important in purifying water. Many elements, such as iron and phosphorus, may react with one another to form molecules that are insoluble in water. After reaction, the molecules precipitate out and settle on the bottom.

15.3 Pollution: Overwhelming Natural Purification

As human populations and per capita use of water have increased, the quantity of wastes released into natural waters has increased exponentially, overwhelming natural processes for removing wastes from the waters in many areas. In addition to this massive increase in quantity, the quality of the waste has become increasingly toxic in recent decades. In the early part of the Industrial Revolution, large quantities of heavy metals, acids, sewage, and other pollutants were discharged into the environment, but they were, and are, easier to remediate than the organic chemicals of today's industry, such as polychlorinated biphenyls (PCBs) and chlorofluorocarbons (CFCs). The rapid growth of chemical and other industries leads to thousands of new chemicals created every year in the United States alone. Moreover, these chemicals often are highly resistant to degradation and persist in the water for some time. The result has been increasingly severe water pollution affecting increasingly larger bodies of water. Entire ocean basins, such as the Mediterranean Sea, are now becoming significantly polluted.

Water pollution can be classified by

- Its composition (what it is)
- Its source (where it originates)
- Its fate (where it goes)

The next sections examine these three categories in that order.

Composition and Properties of Water Pollutants

The U.S. Public Health Service classifies water pollutants into eight general categories by composition, as shown in **TABLE 15-1**. In reading about these pollutants, recall that such pollutants often interact synergistically to become exceptionally toxic and also may undergo bioaccumulation in living tissue. For example, influx of two weak toxic substances into a lake may lead to a massive fish kill when their joint effects are combined.

Oxygen-Demanding Wastes

Oxygen-demanding wastes include materials that plants and animals produce, such as body wastes in sewage and unused body parts from food preparation. That something as "natural" as body waste is a highly destructive pollutant is a good illustration of how anything is harmful in too great a quantity (as shown by a review of the principles of pollution control, toxicology, and risk). In this case, low levels of plant and animal matter provide food for the microbes in natural waters that live by decomposing such matter. However, too much plant and animal waste leads to a rapid increase in the rate of decomposition. Because decomposition requires oxygen, the **biological oxygen demand (BOD)**—defined as the amount of oxygen needed to support biological activity, including oxygen absorbed or used ("demanded") by organisms and chemical processes (such as natural oxidation)—in a particular stream, lake, or other body of water increases. The absorption, dissolution, and movement of oxygen through water take much longer than through air (the oxygen content of water can be increased by bubbling air through water or by allowing water to "fall" through air, as when water runs quickly down a rocky stream or over a waterfall). Thus, when there is a very large and sudden demand for oxygen in an aquatic system, the local oxygen supply may be depleted faster than it is replenished. When BOD levels from decomposition are higher than the local dissolved oxygen content in the water, there is not enough oxygen left for other organisms, such as fishes, causing them to die (**FIGURE 15-3**). Dead fishes and other organisms may then further add to the BOD as their bodies decompose, causing in turn the deaths of even more organisms deprived of oxygen. Unfortunately for fishermen, game fish such as trout tend to be less tolerant of low oxygen content than other fishes.

TABLE 15-1	Classes of Water Pollutants and Some Examples
1. Oxygen-demanding wastes	Plant and animal material
2. Infectious agents	Bacteria and viruses
3. Plant nutrients	Fertilizers, such as nitrates and phosphates
4. Organic chemicals	Pesticides, such as DDT, detergent molecules
5. Inorganic chemicals	Acids from coal mine drainage, inorganic chemicals such as iron from steel plants
6. Sediment from land erosion	Clay silt on stream bed, which may reduce or even destroy life forms living at the solid-liquid interface
7. Radioactive substances	Waste products from mining and processing of radioactive material, radioactive isotopes after use
8. Heat from industry	Cooling water used in steam generation of electricity
Source: U.S. Public Health Service.	

Infectious Agents

Four major kinds of disease-causing agents are involved in water pollution: bacteria, viruses, protozoa, and parasitic worms. These are usually found in association with oxygen-demanding wastes in sewage and other waters carrying plant, animal, and especially human wastes. Before there were modern sewage treatment plants, outbreaks of cholera, typhoid fever, and diphtheria were major causes of human death, and as noted, they still are in areas with poor sanitation. Modern sewage treatment plants can easily remove most of these pathogens simply by adding chlorine, which is highly toxic to many life forms. However, some pathogens, including the protozoan *Giardia* and a number of kinds of hepatitis viruses, are resistant to chlorine. These are often associated with human feces and can cause diarrhea, cramps, and other illnesses if wastewater treatment plants are ineffective. The illness of more than 400,000 people in Milwaukee in 1993 was caused by a combination of factors, including a contaminated water supply, poor water treatment, and an especially resistant parasite called *Cryptosporidium* (see **CASE STUDY 15-1**). Increasingly, ozone is being used at some water treatment plants in place of chlorine. *Cryptosporidium* is still a prevalent challenge. The Centers for Disease Control report that in the U.S. in 2006 there were more than 6,000 cases; in 2007, that number had increased to more than 11,000, and in 2008 there were 10,500 cases.

Plant Nutrients

Plant nutrients are another example of how too much of a "good thing" can be a major pollutant. Plants in aquatic ecosystems (such as algae) are adapted to a fairly narrow set of environmental conditions, including the chemical nutrients in the water. Ecologists have found that there generally is one nutrient, called the "limiting nutrient," that limits plant growth. If large amounts of this limiting nutrient suddenly become available, plant populations will increase rapidly. The result is sometimes called an "algal bloom," which appears as a green or red scum on the water (**FIGURE 15-4**). As the plants die, microbial decomposition increases the BOD, and animal life suffocates. This

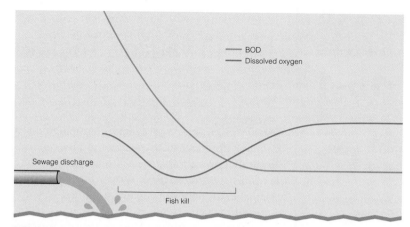

FIGURE 15-3 Biological oxygen demand (BOD) decreases away from the sewage discharge point. Fish kills, caused by a lack of oxygen, can occur many miles downstream.

process of abundant plant growth is called **eutrophication** (Greek *eu* = good, *trophikos* = food, so "good food" is a rough translation).

In freshwater lakes, rivers, and streams, the limiting nutrient is usually phosphorus. In many marine waters, it tends to be nitrogen. Fertilizers and human sewage are rich in both phosphorus and nitrogen, so pollution of fresh or saltwaters waters by either of these causes eutrophication. Eutrophication (and BOD increase in general) is usually greatest in standing bodies of water, where water circulation is slow.

Organic Chemicals

Some organic (carbon-containing) chemicals are natural, such as petroleum or coal, and

FIGURE 15-4 The discharge of nutrients into natural waters can lead to the rapid growth of algae and other plants.

CASE STUDY 15-1

The Milwaukee Incident

VISIT

http://environment.jbpub.com
/mckinney/5e/
for more information

In 1993, a series of events culminated in a major outbreak of a waterborne illness, commonly known as "traveler's diarrhea," among the residents of Milwaukee, Wisconsin. The culprit was the small parasitic organism named *Cryptosporidium*, which assembles along the inner membrane of the intestines. This organism infected half of the 800,000 individuals served by the city's water system. The symptoms varied in severity and generally lasted 2 weeks. Individuals with compromised immune systems, including those with AIDS or cancer, and older people experienced the worst symptoms. More than 50 deaths (or more than 100, according to some estimates) were attributed to the outbreak, including deaths directly and indirectly linked to the parasite.

The source of the outbreak, namely the municipal water system, was identified within several days, and Milwaukee's mayor, John O. Norquist, recommended that all users of city water boil their water before drinking or washing food. Although the contamination was remedied 8 days later, questions remained. How did the drinking water become contaminated with this infectious agent? Why did the municipal water treatment facilities fail to remove this organism?

After investigation, it was found that the *Cryptosporidium* most likely came from (1) infected cattle on farms upstream, whose runoff flowed into the Milwaukee River and then into Lake Michigan, the source of the city's water, or (2) Milwaukee's own sewage system discharge, which also flowed into Lake Michigan just a few miles from the city's source water intake.

In either case, the water treatment facilities should have been able to eliminate the problem. In fact, one facility was functioning adequately, but the second facility, which processed the greater volume of infected water, was not. In an effort to correct an earlier problem with the quality of its water, this facility was using polyaluminum chloride to lower the corrosiveness of the water, but this measure inadvertently decreased the effectiveness of the sand filters used to eliminate *Cryptosporidium* and other solids from the treated water.

Although incidents of drinking water contamination affecting large numbers of people are not common in the United States, the Milwaukee incident (by far the largest on record) was not the first case or the last. Between 1984 and 2001, a dozen significant outbreaks in the United States occurred. For instance, in 1987, 13,000 residents of Carrollton, Georgia, were affected by a *Cryptosporidium* infestation of the local municipal water system even though all state and federal standards were reportedly met at the time.

One issue that threatens water quality is out-of-date plumbing and the fact that many cities around the United States rely on pre-World War I era delivery systems and treatment technology. This aging infrastructure often breaks, leaching contaminants into the water supply. In addition, old-fashioned water treatment also often fails to filter out all particles in the water and does not kill or remove various parasites, bacteria, and contaminants, such as pesticides, industrial chemicals, and arsenic.

What should be learned from these events? They reveal the imperative need to address and deal with nonlocalized sources of contamination of our water supplies. They also illustrate the delicate balance of water treatment processes. The complex intricacies of these processes should not be altered randomly or haphazardly. Instead, thorough evaluations must be performed to determine any possible ramifications from system changes.

Questions

1. Does the Milwaukee incident scare you? Do you ever worry about the safety of your local water?

2. Do you know where your drinking water originates? Is your town or city dependent on one or more water treatment facilities?

3. It has been suggested that a prime way for a terrorist organization or enemy nation to attack the population of the United States is through the drinking supplies of major cities (contaminating the supplies and/or damaging water treatment facilities). Do you think this is a realistic assessment? How important and vulnerable are our water supplies?

these can be toxic to many species. Many other organic chemicals are human made. In fact, most of the many thousands of chemicals invented each year are organic because of carbon's unique bonding abilities. As a result, thousands of synthetic organic chemicals are now used in industry, agriculture, and other activities. Synthetic organic chemicals are among the most abundant and most toxic pollutants. They also tend to bioconcentrate. Among the most common organic pollutants are organo-chlorines, including some pesticides.

Inorganic Chemicals

The inorganic chemicals, which are not carbon based, form a broad group. Two of the most destructive water pollutants in this category are metals and acids. Metals are naturally occurring elements such as arsenic, zinc, lead, and mercury. In high concentrations, metals often have a very toxic effect on living organisms because metal atoms chemically bind to protein molecules such as enzymes, interfering with their functioning. The well-known example of mercury poisoning affecting more than 3,500 people near Minimata, Japan, in the 1950s was the result of the residents eating fish from the waters into which a local industry was discharging mercury; chronic illnesses, even deaths, and birth defects resulted from the contamination. A more common metal that can cause health damage is lead. It is often carried in tap water in homes with old plumbing. In ancient Rome, lead in piping systems and cooking vessels may have harmed or killed people. Like mercury, lead attacks the nervous system.

A second major type of inorganic pollution is the release of acids into natural waters. This can occur through "acid rain" (as shown by a review of local and regional air pollution) or from discharge of acidic waters, including drainage of coal and metal mines. Aquatic organisms, such as different kinds of fishes, vary greatly in their tolerance to acidity.

Sediment Pollution

Sediment from erosion is the sixth major type of pollution (see Table 15-1). We do not often think of sediment as pollution, but human activities can cause thousands of times more sediment to be introduced into natural waters than would otherwise occur. This sediment has many detrimental effects to human use, causing the water to be murky and aesthetically unappealing. In economic terms, it fills in channels and reservoirs and damages power-generating equipment. It is also very detrimental to aquatic organisms. For example, sediment can impair gills and other organs of fishes and shellfishes and reduce the amount of sunlight available to underwater plants, thereby slowing production at the base of the food pyramid. For instance, dredging operations that stir up large amounts of sediment and kill the tiny coral animals destroy many coral reefs in the Florida Keys, the Caribbean, and other parts of the world.

Radioactivity

Radioactive substances may be discharged into natural waters by nuclear power plants. Radioactivity has many detrimental effects on living organisms, ranging from nearly immediate death to cancer, mutations, and sterility. The radioactivity of the substances routinely released by power plants is generally considered to be so limited as to be harmless, but this practice remains controversial.

Thermal Pollution

Heat as a water pollutant often comes from hot industrial waters, such as those discharged by power-generating plants. These heated waters can seriously disturb aquatic ecosystems. Fishes have distinct temperature tolerances for spawning, egg development, and growth. When the water temperature changes, certain organisms may be eliminated and others may prosper. In many cases, a temperature change of only a few degrees can completely exclude a species from breeding in the area. Coral reef systems are especially susceptible to incremental temperature changes; 1998 was an exceptionally warm year and also a year that coral populations all over the world showed signs of definite damage. As with low-oxygen waters and acidity, game fishes tend to be more sensitive to temperature changes than are most other fishes. Plants, plankton, and shellfishes are also highly sensitive to temperature changes. For example, green algae grow best between

30° and 35°C (86° and 95°F), whereas blue-green algae grow best between 35° and 40°C (95° and 104°F). Because blue-green algae are a poorer food source than green algae, an increase in water temperature and the resulting shift from green to blue-green algae will have a big impact on animals higher on the food web. Heated water can also affect large organisms. The endangered manatees of Florida often crowd into the warm discharge waters of power plants during cold winter periods.

In addition to exceeding the temperature tolerances of organisms, heated water affects the oxygen supply. Warmer water is able to hold less oxygen than cooler water. The ability to dissolve oxygen ("solubility") decreases exponentially with increasing temperature. Even if some organisms are not directly affected by increases in temperature, they may be affected indirectly because they cannot tolerate the lower oxygen content of the water.

Sources of Water Pollutants

Virtually all human activities produce some kind of environmental disturbance that can contaminate surrounding waters with the eight pollutants just discussed. Eating (body wastes), gardening (pesticide and sediment runoff), and many other activities create by-products that can find their way into the water cycle. For convenience, we can assign the large majority of the sources for the eight pollutant groups to three broad categories of waste:

1. Industrial
2. Agricultural
3. Domestic waste

FIGURE 15-5 Pesticides are a major source of pollution of surface waters and groundwater.

Industrial and domestic wastes are often **point sources**, which discharge pollutants at easily identified single locations, such as discharge pipes. Agricultural sources of water pollution tend to be **nonpoint sources**, which discharge pollutants from many locations; an example is pesticide runoff from crops (**FIGURE 15-5**).

Industrial Wastes

Wastes from industry serve as major sources for all eight types of water pollutants. Pollutants released into wastewater systems inevitably find their way into natural water systems. Many major industries contribute significantly to water pollution, but some of the most important are as follows:

1. Manufacturing
2. Power generating
3. Mining, logging, and construction
4. Food-processing industries

Manufacturing industries contribute many of the most highly toxic pollutants, including a variety of organic chemicals and heavy metals. In many cases, both the product, such as paint or pesticides, and the by-products from the manufacturing process are highly toxic to many organisms, including people. A key problem with such toxic wastes is not just the different kinds produced, but also the sheer volume of each kind. Many billions of pounds of waste are produced each year, especially by the chemical and metal industries, which are by far the largest producers of toxic and hazardous wastes.

Power-generating industries are the major contributors of heat and radioactivity. Nearly all power plants, whatever the fuel, are major sources of thermal (heat) pollution. Radioactivity from nuclear power plants can pollute waters in a variety of ways, including discharge of mildly radioactive wastewater and groundwater pollution by buried radioactive waste.

The mining, logging, and construction industries are major contributors of sediment and acid drainage. Sediment pollution occurs because these industries can denude the land of vegetation. Construction in particular results in a drastic rise in the rate of land erosion and transportation of sediment into streams. Acid drainage is mainly a product of mining coal and metallic ore minerals. More than 19,300 kilometers (2,000 miles) of streams in the United States have been seriously affected by acid drainage from mining operations.

Food-processing industries include slaughterhouses, canning factories, and many other plants that produce large amounts of animal and plant parts that become oxygen-demanding wastes in nearby waters. These are also sources of waterborne diseases.

Agricultural Wastes

The cultivation of crops and animals generates agricultural wastes. Globally, agriculture is the leading source of sediment pollution, from plowing and other activities that remove plant cover and disturb the soil. Agriculture is also a major contributor of organic chemicals, especially pesticides.

The other three major agricultural pollutants have biological aspects. Oxygen-demanding wastes are largely body wastes produced by livestock, which produce five times as much waste per pound as people and are the major cause of this type of pollution (**FIGURE 15-6**). Infectious agents are nearly always found in body wastes, so livestock are also major producers of this type of pollutant. Agriculture is the chief source of plant nutrient pollution from runoff carrying fertilizers.

Domestic Wastes

Domestic wastes are those produced by households. Most domestic waste is from sewage or septic tank leakage that ends up in natural waters. In the past, some cities dumped untreated or barely treated sewage directly into rivers, lakes, or coastal waters. The bulk of domestic waste pollution consists of body wastes and other oxygen-demanding wastes. In addition, domestic sources may be a major contributor of infectious agents and plant nutrients. Infectious agents are a common hazard of all human waste. Plant nutrients occur in the form of nitrogen and phosphorus. These come not only from human waste, but also from fertilizers used extensively in household lawns and gardens.

Stormwater Runoff

Stormwater is rain and snowmelt. According to the EPA, **stormwater runoff** from pavement

FIGURE 15-6 Runoff from livestock yards such as this one is a source of water pollution.

(roads, highways, and parking lots) and rooftops is the most common cause of water pollution. The impervious surfaces used in urban and suburban development replace vegetation and soil that used to absorb rain and snowfall. Because there is less area for the water to be captured, stormwater runoff becomes an environmental issue for several key reasons.

This runoff can pick up pollutants such as trash, oil, fertilizers, pesticides, and animal waste. The untreated runoff then flows into streams, lakes, rivers, and wetlands. Most of these water bodies eventually connect, so runoff pollutants can severely damage the water sources, including aquifers, on which life relies. Flooding frequency and intensity, and severe erosion, can also increase because of unchecked stormwater runoff.

Although most stormwater is considered nonpoint source pollution, Congress declared in 1987 that the stormwater discharges from certain industries and municipalities were point-source pollution. These stormwater dischargers (including oil and gas exploration, sewage treatment plants, and municipal solid waste facilities) must obtain a National Pollutant Discharge Elimination System (NPDES) permit, or water quality discharge permit; these permits are administered by the EPA. The permitting process helps the EPA work toward locating and managing point-source pollutants. Another mitigation strategy is to use pervious pavements on roads and parking lots so that the stormwater is soaked up. In addition, buffers can be used at construction sites and by industry to prevent excess runoff. Homeowners can prevent unnecessary runoff by planting native plants in their yards, or storing captured rainwater from their roofs and using it for such tasks as watering grass and gardens during dry times.

Fate of Pollutants

Ultimately, most pollutants find their way into natural waters. This is inevitable given the dissolving power of water and its tendency to flow toward rivers and basins. The natural waters that ultimately absorb the pollutants can be divided between freshwater and marine water. The freshwaters, in turn, can be surface water (rivers, streams, or lakes) or groundwater.

Freshwater: Rivers and Streams

Rivers and streams drain water that falls on upland areas. Moving water dilutes and decomposes pollutants more rapidly than does standing water. Then why are more than 10% of U.S. rivers and streams significantly polluted? A primary reason is that all three major sources of pollution—industry, agriculture, and domestic (cities)—are concentrated along rivers. Industries and cities historically have been located along rivers because the rivers provide power, transportation, and a convenient place to discharge wastes. The very fact that fast-moving waters carry away wastes and purify themselves more readily has encouraged people to pollute them in ever-greater amounts. Agricultural activities likewise have tended to be concentrated near rivers. River floodplains are exceptionally fertile because of the many nutrients that are deposited in the soil when the river overflows, and the water can be channeled to irrigate the crops.

One of the unique aspects of river pollution, as compared with other types of water pollution, is that "everyone is downstream from someone else." Bodies of water that border countries are often areas of dispute. For example, the Rio Grande, the river that borders Texas and Mexico, is an issue because what the United States does with its water usage and pollutants inevitably affects "downstream" what Mexico can do with the water. If a city or factory discharges waste, it may find its way into someone's drinking water downstream (**FIGURE 15-7**). This is not so bad if the next water user is far enough downstream that the natural purification processes have time to act; however, many types of pollution are discharged into rivers, and the purification processes remove them at varying speeds. For example, some heavy metals are removed relatively quickly because suspended clay and organic particles have a slight electric charge and adsorb ("attach") the metal atoms. When the clay or organic particles settle out of the water, they take the metal atoms with them. Other pollutants that are very persistent in the water can actually accumulate downstream, causing great hazards for those living at the "end of the line." For instance, New Orleans draws much of its water from the Mississippi River, just before it empties into the Gulf of Mexico (**FIGURE 15-8**). In the past, the city has had many problems with water quality; some of the hundreds of industrial chemicals in the water were detected in amounts high enough to be harmful. Consequently, many residents and visitors drink only bottled water. The floods caused by Hurricane Katrina (August 2005) did not help matters. Contaminants spilled into the drinking water supply; water and wastewater treatment plants were damaged. Fortunately, it appears that the damage to the drinking supply was not as bad as some had feared initially. The massive volumes of water involved helped to dilute contaminants, and as of the summer of 2006, water quality in New Orleans was satisfactory.

The **Clean Water Act (CWA)** of 1972, along with other environmental laws, such

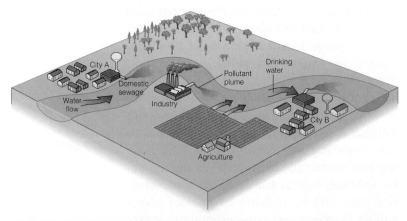

FIGURE 15-7 One city's wastewater is another city's drinking water because "everyone is downstream from someone else."

as those regulating pesticide use, have led to a substantial improvement in water quality in many U.S. rivers, especially the larger ones. Heavy metals and many synthetic organic compounds, including pesticides, have on average decreased substantially since the early 1970s.

Despite these improvements, many U.S. rivers remain in serious trouble, and some are declining rapidly for various reasons. One is nonpoint pollution. Some rivers, such as the Everglades river system, suffer from agricultural and other sources of polluting water runoff. The second is the problem of physical alteration of rivers. Alterations can take four forms, often called the "four horsemen" of river destruction:

- Dams
- Diversion of water (canals)
- Channel alteration
- Land development

These alter flow patterns, increase sediment pollution, and vary water temperature, among many other changes that reduce water quality and destroy aquatic ecosystems. Although many of the problems of point source pollution have been addressed, rivers (and streams) in many areas will continue to decline until nonpoint pollution and physical alteration of rivers are addressed.

Freshwater: Lakes

Lakes are subjected to many of the same pollutants as rivers, but lakes are more easily polluted than rivers for at least three reasons. First, lake waters circulate much more slowly. Deep waters in many lakes are circulated only during seasonal temperature changes. Second, lakes are often "dead ends," being basins into which water flows. Thus, pollutants accumulate in lakes, having nowhere else to go. Third, lakes often contain less water than rivers, especially if you consider the amount of water that flows through a river over time. Smaller lakes, of course, are polluted even faster than larger ones.

Lake Erie is a classic example of a polluted lake. Because of its very large size, Lake Erie (like the other Great Lakes) was thought to be quite resistant to pollution. Nevertheless, the

FIGURE 15-8 The Mississippi River contains many kinds of pollutants by the time it reaches the "end of the line" at New Orleans.

close proximity of many large industries and cities began to overwhelm the lake's natural purification systems. By the late 1960s, sewer systems serving more than 9 million people were emptying into the lake on the U.S. side alone. Tons of industrial waste from steel mills, paper mills, auto plants, and many others were added daily. By the early 1970s, the lake was declared "dying, if not dead" by many experts, based on the fact that game fishes and many other organisms were absent from large parts of the lake and increasingly rare in the remainder. Much of the problem was the result of widespread eutrophication, with rampant algal growth and decay causing oxygen to be depleted in the lake. A major contributing factor was the depth of the lake. Although Lake Erie is nearly 485 kilometers (300 miles) long, it is very shallow, with an average depth of less than 20 meters (65 feet). The lake is now recovering, thanks to concerted action taken since the early 1970s. Pollution from industry and sewers has been greatly reduced by water treatment before discharge (FIGURE 15-9). Game fishes have begun to reappear in areas where they had disappeared. Oxygen levels are up, and there are many other hopeful signs that water pollution problems are being corrected; however, increasingly severe biological damage from dozens of introduced species offset many of these gains.

FIGURE 15-9 Lake Erie may cover a large area, but it is relatively shallow and was readily polluted by surrounding cities and industries. Seen here is Sandusky Harbor, Ohio.

TABLE 15-2	Major Sources of Groundwater Pollution	
Unintentional		**Intentional**
Surficial percolation		
1. Landfills		1. Deep-well injection
2. Waste disposal ponds		
3. Spills		
4. Agriculture and land use		
Underground leakage		
5. Septic tanks		
6. Buried wastes		
7. Underground storage tanks		
Overpumping		
8. Saltwater intrusion		

Freshwater: Groundwater

Most freshwater is groundwater. As human populations increase and industrialize, the demand for this resource is rising. At the same time, there has been an accompanying rapid increase in pollution of groundwater. About half of the U.S. population relies on groundwater as their only source of drinking water, yet it has been estimated that approximately 25% of the usable U.S. groundwater is in danger of contamination or is already contaminated.

A primary reason groundwater is so easy to pollute is that it moves so slowly. The water must migrate through pores in the aquifer rock. Groundwater flow rates vary, but the average is only a few inches (several centimeters) per day. Compare this with a river in which water usually moves many hundreds or thousands of feet (up to thousands of meters) per day. This slow movement may cause water shortages because of the long times required to recharge aquifers. The slow movement also leads to pollution for the same reasons that lakes are more easily polluted than rivers: Natural purification processes such as aeration and dilution are slowed. After pollution occurs, groundwater tends to stay polluted for a long time, often for decades. Some aquifers will take many centuries to become purified if the only processes used for purification are natural ones.

TABLE 15-2 lists nine major sources of groundwater pollution. Only one of these is intentional: deep-well injection, which involves pumping waste into a deep aquifer. Ideally, this aquifer is separated from higher "drinking water" aquifers by an impermeable barrier, such as a shale layer. However, the barrier may be fractured, permitting leakage upward. The eight major sources of unintentional groundwater pollution in the table can be subdivided into three basic categories. One of these is saltwater pollution from overpumping the aquifer.

The other two categories are easily distinguished: surficial percolation and underground leakage. Some examples are shown in **FIGURE 15-10**. The four surficial sources involve downward percolation of rainwater that has percolated through landfills, waste disposal ponds (such as where industries put toxic chemicals), spills, and various agriculture and land-use products, including fertilizers, animal waste, and salt sprinkled on roads to melt ice. The three underground sources include leakage from septic tanks, buried wastes, and underground storage (such as gasoline and fuel oil tanks). Septic tank pollutants are primarily oxygen-demanding wastes, whereas buried wastes tend to be highly toxic chemical by-products of industry.

Petroleum product leakage from underground fuel tanks is one of the most common types of groundwater pollution. The EPA has estimated that more than 50% of such tanks leak before they are removed. Volatile organic compounds and petroleum products, both mainly from leaking petroleum storage tanks, rank third and fourth among groundwater pollutants in the United States (**FIGURE 15-11**). Nitrates (fertilizers) and pesticides rank first and second, respectively, illustrating that non-point sources are also a problem for groundwater; both enter the aquifer mainly as runoff from farms and other cultivated areas.

Much progress has been made in reducing many sources of groundwater pollution. Disposal of waste by deep-well injection has been greatly curtailed in most areas by strict regulations. Reduction of solid waste through recycling and other means has slowed the use of landfills. Even where landfills are being used, they are often designed with linings and other barriers that reduce percolation into the groundwater. Federal legislation passed in the early 1990s requires many gas station owners to replace their older underground tanks with new tanks designed to minimize leakage. Some of the most dangerous sources are the tens of thousands of hazardous and toxic waste disposal sites in the United States, such as disposal ponds, buried wastes, and some landfills. Tens of billions of gallons of water percolate through these sites each day. Like polluted rivers and lakes, nonpoint sources will remain a major challenge for years to come.

Marine Waters: Chemicals and Sediments

In 1990, a group of scientists appointed by the United Nations issued a major report that concluded that the oceans have contamination and litter that can be observed from the poles to the tropics, from the beaches to the deep sea. This pollution is not evenly distributed; most coastal areas are polluted, but the open oceans are relatively unpolluted. The National Oceanic and Atmospheric Agency (NOAA) explains that one reason for this distribution is that 80% of all ocean pollution comes from the land in the form of runoff. In 2008, The United Nations Joint Group of Experts on the Scientific Aspects of Marine Pollution (GESAMP) reported that 80% of the marine pollution is now plastic. Runoff, atmospheric pollution (primarily from factories and other land-based sources), and other

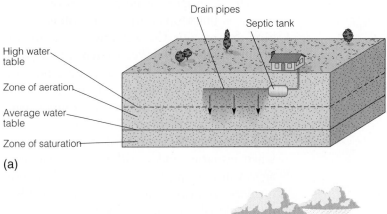

(a)

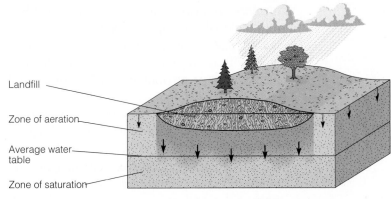

(b)

FIGURE 15-10 (a) Leaking or clogged septic tanks can soon pollute local groundwater. (b) Nearly all landfills eventually form leachate that diffuses into the groundwater.

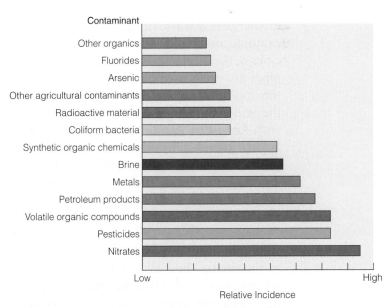

FIGURE 15-11 Major contaminants of U.S. groundwater: nitrates (from fertilizers) and pesticides (mainly from agriculture). Volatile organic compounds and petroleum products come from oil and chemical storage tanks underground. The rest come largely from hazardous waste sites and leaking sewage and septic tanks. (*Source:* Environmental Protection Agency.)

nonpoint source contaminants make up the rest of the waste.

The disproportionate pollution of coastal waters is unfortunate for both ecological and economic reasons. Coastal waters contain some of our most productive ecosystems. Because of rich nutrient runoff from land, nearshore estuaries are the "nurseries" for any number of young organisms. Furthermore, young organisms, such as larvae, are the most sensitive to pollution. The widespread destruction caused when estuaries and coastlines are developed aggravates this problem. The economic consequences are also significant. More than 99% of the global catch of marine fish comes from within 322 kilometers (200 miles) of shore; much of the decline in the worldwide fishing industry is attributable to loss of coastal habitat.

Agricultural runoff is the major source for sediment, plant nutrients, and pathogens. Industry is the main source for synthetic organic chemicals, such as toxic chemical waste. Domestic (city) waste is the main source of oxygen-demanding waste, especially through discharge of sewage. In some parts of the world, city sewage is often not treated before being discharged into the ocean. To save money, many coastal municipalities simply discharge raw waste directly into the sea. For example, much of the sewage discharge from cities in the Mediterranean is untreated. Long pipes are often used to transport the waste a few miles out to sea, so the local impacts of the waste were thought to be negligible. As with Lake Erie mentioned earlier, the rapid growth of coastal populations has begun to overwhelm the purifying abilities of many local marine waters, and currents often bring polluted waters back to shore.

Marine Waters: Oil and Litter

Oil spills probably receive more attention in the media than all the other ocean pollutants combined. However, most oil pollution does not come from spills. According to one estimate, globally nearly half comes from land-based sources, such as atmospheric pollution, city and industry waste, and runoff. Marine transportation provides another 45%, but less than one-third of this

is from oil spills. Most comes from routine ship operations. For example, tankers often discharge oily bilge and ballast water at sea. According to a 2002 National Academy of Sciences report, 110 million liters (29 million gallons) of petroleum are released into U.S. coastal waters annually, and nearly 85% of this oil originates from land-based runoff, polluted rivers, airplanes, ships, jet skis, and small boats. Less than 8% is from tanker or pipeline spills, and about 3% results from offshore oil exploration and drilling. Interestingly, the same study estimated that 178 million liters (47 million gallons) of oil enters U.S. ocean waters each year naturally, such as through cracks in the sea floor. Still, a major oil spill can be significant, and one such incident can greatly affect these annual averages. For instance, the grounding of a tanker off the Shetland Islands in 1993 spilled an estimated 98 million liters (26 million gallons); the grounding of the Exxon Valdez in Prince William Sound, Alaska, in 1989, spilled more than 38 million liters (10 million gallons); and the 2010 Deepwater Horizon/BP oil spill (from a well head blowout) released an estimated 779 million liters (205.8 million gallons) of crude oil into the Gulf of Mexico. Estimates of the amount of oil spilled during the Persian Gulf War in 1991 range from 95 to 490 million liters (25 to 130 million gallons).

Despite their relatively modest contribution to overall oil pollution, oil spills can cause massive devastation to the local environments where they occur. This is why they receive so much media coverage. Long-term events that continue for many years include tar balls washing up on local beaches and scarcity of marine life. In some cases, species that are highly sensitive to oil have never returned. Long-term economic effects include a decline in tourism and fishing. Fishing losses occur not only from death and a decline in breeding but also because the fishes and shellfishes that survive may contain high concentrations of oil in their tissue, often for many years.

Marine litter is a major form of water pollution (**FIGURE 15-12**). Such litter includes materials of many kinds, usually discarded from boats. In the past, it was common practice to dump trash at sea; municipalities used boats to routinely dump tons of garbage. They assumed that the sea is so vast that the garbage would effectively "disappear," but the rapid growth of human populations and material goods worldwide has begun to overwhelm the ocean's ability to absorb and decompose the solid waste. Millions of tons of shipboard litter alone are tossed into the sea each year. This great volume is made worse by the widespread use of plastics, which are very durable, often taking many decades to decompose. The impact of plastic is evident in the litter cleared from the beaches each year. As an example, a brief 3-hour cleanup of 253 kilometers (157 miles) of Texas coastline was typical of many U.S. coastlines: It collected 279 metric tons of litter. The litter collected in a typical beach cleanup is about two-thirds plastic; much of this is plastic cigarette filters, followed by other pieces of plastic. Metal, glass, and paper make up most of the remainder. Plastic is not just a problem for popular beaches. Plastic's durability, combined with its ability to float, allows plastic litter to travel very long distances. Recent surveys of remote locations, thousands of miles from large landmasses, have found increasing amounts of plastic debris.

In addition to the obvious ugliness of marine litter, it is harmful to marine life. It is estimated that as many as 2 million seabirds and 100,000 marine mammals die each year from eating or being entangled in plastics. The extent of this problem is evident from a study on Midway Island in the Pacific, which found that 90% of the albatross chicks had plastic in their digestive tracts. A random global sampling found that 25% of the world's seabirds have such undigested plastic particles. Sea turtles and marine mammals (such as porpoises and seals) often die from eating plastic, especially bags that they apparently mistake for jellyfish. In addition, birds, turtles, and mammals can be killed when discarded plastic bands encircle their bodies and strangle them as they grow, which is why cutting them before they enter the waste stream is important. Discarded fishing nets have also become a major cause of death, entangling and strangling the animals.

FIGURE 15-12 Tons of litter accumulate each year on an average beach.

15.4 Slowing Pollution: Reduction, Treatment, and Remediation

A review of the principles of pollution control, toxicology, and risk shows the various ways of slowing pollution. Most of these methods are currently used to slow water pollution, including

1. Source reduction of waste
2. Treating wastewaters before they are discharged into natural waters
3. Remediating (cleaning up) natural waters after they have been polluted

These are listed in order of generally increasing cost. It is usually much more costly to clean up a polluted lake than to produce less pollutant at the source. Currently, most of society's efforts are directed at the second method—treating polluted wastewaters—so most of our discussion focuses on that.

Source Reduction: Efficiency, Recycling, and Substitution

Source (input) reduction decreases pollution by producing less waste. Reducing the use of water-polluting materials is the least costly way of slowing pollution because it saves the cost of treating polluted discharge waters or cleaning up polluted natural waters. In addition, it is the only effective way of slowing pollution from the many nonpoint sources, such as agricultural runoff.

Efficiency

One way of reducing the introduction of polluting materials into natural waters is improved efficiency, or using less of a polluting product. An example is oil. Between 1974 and 1986, the total number of oil spills worldwide decreased from 1,450 to 118 per year. Although much of this decrease was from better transportation methods (such as improved oil containment), some of the decrease was from oil conservation. After the Organization of the Petroleum Exporting Countries (OPEC) embargo in the early 1970s, world exports decreased about 25% between 1977 and 1986 as higher oil prices led to increased energy efficiency. In other cases, concerns over the water-polluting effects of materials have led to improved efficiency. This is one reason for the reduced growth of pesticide use in the United States. Soil conservation efforts are another example. These not only promote sustainable agriculture, but also reduce sediment pollution into nearby waters; pollution from fertilizer is often reduced as well because natural soil fertility is retained.

Recycling

Recycling plastics, chemicals, and many other materials helps to reduce litter and nearly all types of water pollution because less waste is produced. Recycling of city and industrial wastewater, as in "closed loop reclamation," is a rapidly growing method of reducing water pollution while also extending water supplies.

Substitution

Another method of source reduction is the substitution of other materials for water-polluting materials in manufactured products. A classic example is the reduction of the amount of phosphate in detergents. As the limiting plant nutrient in many aquatic ecosystems, phosphate pollution is a primary cause of eutrophication. Until the early 1980s, a typical detergent contained about 9% phosphorus by weight, and more than 225 million kilograms (500 million pounds) of phosphorus was dumped into U.S. wastewaters yearly. The manufacturers' use of phosphate substitutes has significantly reduced phosphate pollution from domestic and industrial wastewaters. Other examples are biodegradable packaging (instead of plastic) and pesticides that decompose before they reach natural waters.

Treating Wastewater

Until recently, most efforts at pollution control have neglected the generally more effective and less costly methods of source (input) reduction just discussed. Instead, pollution control has emphasized output controls. For water pollution, this has meant an emphasis on wastewater treatment by industries, cities, and households.

Wastewater Treatment

Countries that lack it illustrate the overwhelming importance of proper wastewater treatment. Inadequate water sanitation accounts for much of the world's sickness and death. Even early civilizations realized that the human waste that accumulated under crowded conditions led to rapid disease transmission. The first known sewer systems were built in Mesopotamia more than 5,000 years ago. They were made of clay pipes that were used to carry wastes away from the cities. Some major Roman sewers are still used today. The flush toilet was invented in the late 1800s, removing the need to carry the waste to the sewer canals in buckets. Carts that picked up and carried the buckets were called "honey wagons," a term still used to refer to garbage trucks in many regions.

Although the early sewer systems greatly relieved the health problems of local urban populations, they increasingly damaged the health of the local aquatic ecosystems and contaminated drinking water downstream because they dumped the raw sewage directly into rivers or lakes. Therefore, scientists began to develop ways of treating the wastewater; by the 1920s, modern treatment had come into widespread use in North America and Europe.

Treatment today is often carried out by municipal treatment (sewage) plants, which serve as the endpoint for waters carried by city sewer systems. These sewer waters often carry both domestic and industrial wastes. The composition of these highly polluted sewer waters varies from city to city, depending largely on what industries are allowed to dump waste into the sewer system and how much they are required to treat it before they discharge it into the sewers. But even the most highly polluted wastewaters are usually more than 99.9% water. After the polluted sewer water reaches the municipal plant, its treatment is generally divided into three basic stages: primary, secondary, and advanced (or tertiary). Each stage is progressively more expensive, and only areas with special needs or problems use advanced treatment.

Primary and Secondary Treatment

Primary treatment uses physical processes, especially screening and settling, to remove materials (**FIGURE 15-13**). A typical primary treatment sequence has three basic steps:

1. A bar screen removes branches, garbage, and other large objects.
2. The grit chamber holds the wastewater for a few minutes while the sand and other coarse sediments settle out.
3. The primary settling tank holds the wastewater for about 2 to 3 hours, allowing finer sediments and organic solids to settle out (or float, after which they are skimmed). These settled sediments and solids, called sludge, contain large numbers of bacteria, fungi, protozoa, and algae. These are piped into a sludge digester that uses bacteria to decompose the sludge. After removal of water in the sludge drying bed, the digested sludge is usually taken to a landfill.

As landfills become full (review a discussion of municipal solid waste and hazardous waste), alternatives for sludge disposal are being explored, such as incineration and use as a soil conditioner. However, sludge usually has high concentrations of toxic synthetic

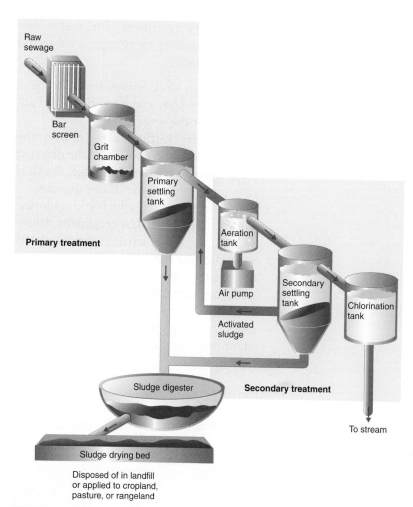

FIGURE 15-13 Conventional wastewater treatment plants use a two-step process of primary and secondary treatment. Note the disposal of sludge.

organic chemicals, heavy metals, and pathogens that require costly treatment before the nutrient-rich sludge can be used for crops. The toxic chemicals and metals also cause problems when the sludge is incinerated.

Primary treatment typically removes about 60% of the suspended solids and 35% of the oxygen-demanding waste. As recently as the early 1970s, this is all of the treatment received by the sewage of about one-fourth of the U.S. population. However, primary treatment alone is inadequate to prevent many long-term health and ecological problems, so secondary treatment has been required by U.S. law since 1977. When it follows primary treatment, **secondary treatment** uses biological (microbial) processes to remove as much as 90% of the suspended solids and oxygen-demanding waste.

In secondary treatment, the water discharged from primary treatment is subjected to three basic steps:

1. An aeration tank mixes together oxygen, wastewater, and bacteria. The bacteria digest the sewage.

2. A secondary settling tank allows the many fine organic particles in the digested wastewater to form more sludge, which is then piped back to the sludge digester.

3. A chlorination tank, which adds chlorine to the water, is the last treatment. Chlorine is a highly reactive chemical and is very effective for killing disease-causing organisms, although it produces potentially toxic chemicals that have become a matter of great controversy. In some cases, ozone processes are used in place of, or to supplement, chlorine treatments.

Advanced (Tertiary) Treatment

Despite its effectiveness in removing oxygen-demanding waste and suspended solids, secondary treatment is relatively ineffective in removing plant nutrients, toxic chemicals (especially synthetic organics and metals), and some pathogens. Secondary treatment may be adequate if treated water is discharged into natural waters where natural purification eventually eliminates these pollutants before the water is used for drinking. However, some situations require advanced treatment to remove some of these pollutants beforehand. As much as 50% of the nitrogen and 70% of the phosphorus often remain in the wastewater after secondary treatment. Therefore, treatment plants that discharge the treated waters into such sensitive ecosystems as Lake Tahoe use advanced treatment to remove the nutrients.

Dozens of advanced treatments are available. They use a wide variety of methods, from physical processes such as microfiltration, heating, electricity, and evaporation, to chemical processes, such as oxidation and precipitation. An example of a physical process is carbon adsorption, which is commonly used to remove synthetic organic molecules, such as PCBs or pesticides. The wastewater goes through filters of fine carbon particles. The carbon is so finely ground that a single handful has the surface area of an acre, allowing the organic molecules to adsorb to the carbon. An example of a common chemical process is chemical precipitation, which is often used to remove metals. Lowering the acidity of the wastewater, for instance, can cause metals to precipitate out of the water into a solid mass that can be easily removed. Advanced treatment is often used as a part of closed loop wastewater reclamation.

Septic Tanks

In areas without a sewer system (mostly rural areas), individual homes use septic tanks. A **septic tank**, which is usually made of watertight concrete and often has a capacity of about 3,785 liters (1,000 gallons), is buried underground to receive household sewage. Waste from the house (such as from the flush toilets) undergoes settling in the tank for several days, and sludge forms at the bottom. The more liquid part of the flow passes across the top of the tank into the absorption field where the polluted water diffuses outward into the soil (**FIGURE 15-14**). The waste is decomposed by bacteria in the tank and by soil bacteria in the absorption field.

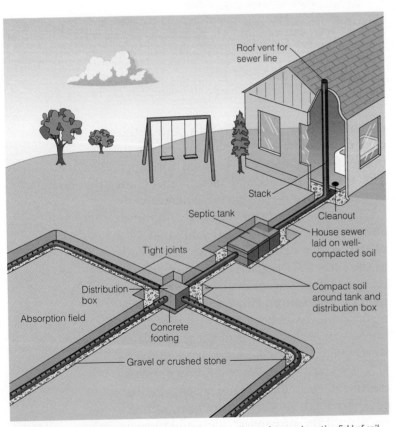

FIGURE 15-14 Septic tanks permit sewage to settle and then disperse into an absorption field of soil where bacteria digest it.

Soils with finer sediment (such as fine sand or silt) are most suitable for septic tanks because they slow the rate of diffusion, giving bacteria more time to work. They also provide more surface area for bacteria. In fact, septic tanks should not be built on land with coarse sediment (such as gravel) or underlain by fractured rocks (such as limestone) because the rapidly moving wastewater will pollute the groundwater. On the other hand, if the soil is too fine or the water table is high, the septic tank will back up because outward diffusion in the absorption field is impeded. After about 3 years, the sludge must be pumped out of even properly working septic tanks. The absorption field usually becomes saturated in a decade or so, causing persistent problems of backing up. This is why septic tanks are generally unacceptable for houses on lots of less than half an acre (one-fifth hectare) of land; the absorption field becomes saturated in just a few years.

Remediation: Cleaning Polluted Waters

Although cleaning polluted waters is much more expensive than reducing pollutant input or treating waters before discharge, it is sometimes the only option. Groundwater can be treated with microbes to decompose organic pollutants. This is accomplished either by (1) injecting "seed" microbes into the polluted groundwater or (2) injecting nutrients and oxygen into the groundwater to stimulate growth in microbes already there. This is often used to treat groundwater contaminated with gasoline. Groundwater can also be pumped to the surface, treated, and pumped back underground.

Polluted surface waters (lakes, rivers, oceanic) can be dredged to remove contaminated sediment. Lake acidity caused by acid rain can be treated with chemicals. Chemical bases such as lime are disseminated in the lake to neutralize acidity.

15.5 Legal and Social Solutions

Legal Solutions

Major federal legislation aimed at water pollution control began in 1899 with the **Refuse Act,** which prohibited dumping of waste into major waterways. Through the 1960s, Congress passed various laws that gradually expanded federal responsibilities for maintaining water quality and conducting research. Many of these laws were ineffective and were so widely ignored that water quality in the United States continued to deteriorate. The increased environmental awareness that led to many landmark laws in the early 1970s produced two federal laws, the Clean Water Act and the **Safe Drinking Water Act** (**SDWA**), which were far more effective than earlier laws for reasons discussed shortly. These acts are reauthorized every few years.

Many other laws have also played key roles in improving water quality. These include the Coastal Zone Management Act; Federal Insecticide, Fungicide, and Rodenticide Act; Toxic Substances Control Act; and the Ocean Dumping Ban Act.

Cleaner Rivers and Lakes

The Clean Water Act of 1972, along with its 1977, 1987, and 1996 amendments, has been very successful in its goal of improving the water quality of lakes and rivers. The act

1. Provided for enforcement by the EPA with stiff penalties
2. Created a system for identifying new point sources
3. Established water quality standards for discharged wastewaters
4. Set pretreatment standards for industrial wastes prior to discharge
5. Provided federal funding to build wastewater treatment plants

Although all five provisions have been important to the CWA's success, the funding for treatment plants and enforcement by the EPA have been especially crucial. Before the act was passed, many local communities could not afford wastewater treatment and often dumped highly toxic material into local waters despite local and federal laws against it. Through federal subsidies, the CWA provided as much as 75% of the cost of building treatment plants. As a result, many thousands of sewage treatment plants were constructed in the 1970s; more than 10,000 plants have been built since 1981. In 2004, more than 12,000 wastewater treatment plants were in operation processing more than 130 billion liters (34 billion gallons) of waste (**FIGURE 15-15**). In addition, the EPA was authorized to (1) establish specific water quality criteria (pollutant concentrations) that had to be met, (2) monitor surface waters to ensure compliance, and (3) punish violators. The EPA has established limits for more than 100 priority pollutants.

Cleaner Drinking Water

The CWA focused on improving water quality of rivers and lakes by reducing pollution in wastewater discharge. The SDWA of 1974 was explicitly aimed at improving drinking water. It set more rigorous standards that applied to more than 60,000 public water supply systems with 25 or more customers.

Public Water Supplies

Approximately 12% of the water used in the United States is distributed by the public water systems as the "tap water" found in all major U.S. cities and towns. This water is usually withdrawn from local groundwaters, rivers, or lakes and is often stored in reservoirs that allow sediment to settle out and other pollutants to decompose through natural processes. Before distribution to public taps, the water is sometimes treated, especially if it is from a river or lake. A variety of treatments are used, including coagulation to cause suspended particles (including most bacteria) to settle out, filtration by clay or sand filters to remove remaining suspended particles and chemicals (see **CASE STUDY 15-2** for an example in which a city attempts to avoid the need for costly filtration), and disinfection, usually by chlorine.

These methods are very effective in removing most pollutants, and tap water is generally safe. Nevertheless, the increasing age of treatment facilities, the growing number of chemical pollutants, the potential dangers of chlorine compounds (ozone treatments are being increasingly substituted for chlorine), and other problems have caused many citizens to opt for sources of drinking water other than tap water, as we discuss later.

Safe Drinking Water Act

Like the CWA, the SDWA authorized the EPA to monitor drinking water supplied by the public systems and other drinking water sources. Since 1974, standards have been added for more contaminants as they become progressively more common. For instance, a 1988 study listed more than 2,100 contaminants found in U.S. drinking water since the SDWA was passed in 1974. In 1986, a series of strengthening amendments was added to the SDWA. These amendments

1. Required the EPA to speed the creation of standards for 85 new contaminants
2. Increased civil and criminal penalties for violations of all standards
3. Required the use of lead-free solder in plumbing pipes

Before 1986, water pipe solder often contained about 50% lead, which resulted in widespread lead contamination of tap water in many homes.

The drinking water standards required by the SDWA fall into two categories. **Primary standards** specify contaminant levels based on health-related criteria. **Secondary standards** are based on nonhealth criteria. These improve the quality of water in aesthetic and other ways. For example, taste, color, odor, and corrosivity have secondary standards. Because secondary standards do not directly relate to

FIGURE 15-15 Until recently, construction of conventional wastewater treatment plants has been heavily subsidized with government funds. Seen here is the Hill Canyon Wastewater Treatment Plant in Thousand Oaks, California.

The New York City Watershed Protection Plan

VISIT

http://environment.jbpub.com
/mckinney/5e/
for more information

During the 19th and most of the 20th centuries, New York City was supplied with clean and tasty drinking water from upstate watershed areas covering more than 5,100 square kilometers, or nearly 2,000 square miles, in the Catskill Mountains and Hudson River Valley (**FIGURE 1**). Forests, soils, vegetation, and wetlands provided natural filtration and purification of the water that was collected and brought by aqueduct to the millions of residents of the city. No treatment of the water was necessary other than standard chlorination to ensure that there would be no contamination by living waterborne diseases.

In the last decades of the 20th century, the quality of New York City's water began to deteriorate as the watershed area became increasingly developed. More people living in the watershed area, with more septic and sewage systems and wastewater treatment plants, badly strained the natural filtration and purification processes of the land and its ecosystems. Increasing amounts of substances such as petroleum products, road salts, fertilizers, herbicides, and pesticides contained in the runoff from lawns, farms, roads, golf courses, parking lots, and other human-made constructions exacerbated the situation. By 1990, the EPA made it clear that either the watershed would have to be protected or New York City would have to build and operate a water filtration system. At the time, it was estimated that to build such a water treatment plant would cost anywhere from $3 to $8 billion, but to implement a watershed protection program would cost only around $1.5 billion.

On one level, it seemed obvious to some people at the time that theoretically the less expensive option, to protect the watershed areas, was the best. Not only would it save money, they argued, but it would also help protect natural ecosystems and open space, improving the quality of life in upstate New York.

However, generally speaking, protecting watershed areas is in many ways more difficult, at least politically, than simply building a water filtration/treatment plant through which the water can be processed. Protecting watersheds involves working with numerous individuals, industries, organizations, municipalities, local and regional communities, and state and federal governments. Developing and implementing a plan that will be accepted by all of the parties involved is a major undertaking, and at the outset, success is by no means certain. In fact, some parties involved may resent being asked to limit development or change their lifestyles or ways of doing business for the benefit of an enormous, wealthy city of which they are not a part. In the middle 1990s, the city of New York owned less than 7% of the critical areas of the watersheds, and the state of New York only owned about 20%; thus, some 73% of the critical watershed areas were out of the city and state's direct control.

Still, after many years of intense and sometimes contentious negotiations, in 1997, a watershed management and protection plan was agreed on that involved the city of New York, state and federal officials, and a variety of villages, towns, environmental groups, and organizations in the watershed areas. Based on this plan, the EPA waived immediate filtration requirements for New York City water. Aspects of the plan included New York City spend-

FIGURE 1 The Catskill Mountains watershed and the Hudson River Valley supply New York City with drinking water. This satellite image was taken during the autumn. Note the Catskills are southwest of Albany.

The New York City Watershed Protection Plan

ing $250,000,000 directly to acquire and preserve land in the watershed areas and development of new regulations to protect the watersheds and the quality of the water. New York City also agreed to provide funding for upgrades to wastewater treatment plants in the watershed areas, funding for farmers to implement voluntary measures to reduce runoff and contamination of waters, and payments to communities in the watershed areas to subsidize water protection procedures and environmentally friendly development that will help preserve the quality and integrity of the watershed areas.

Whether the program will actually accomplish its goals of avoiding the construction of a filtration system and protecting New York City's water quality is still an open question. Critics point out that it is dependent in part on the cooperation of numerous individuals, villages, and towns, as well as voluntary measures to be taken by farmers and others. In addition, there is no guarantee that the city and state will be able to acquire enough acreage in the critical watershed areas to ensure adequately acceptable water quality. Critics also suggest that the plan does not adequately limit development in watershed lands and may place too much emphasis on pollution control rather than more effective (at least in the long run) pollution prevention. Ultimately, some critics suggest, filtration will still be needed; however, the measures already being put into place to help protect New York City's watershed areas are having a positive benefit in terms of protecting open space, water bodies, and natural ecosystems—even if ultimately the measures are not enough to avoid filtering the water. As an added benefit, dialogue over environmental concerns has been created among many officials at all levels (local, state, and federal), communities, industries, and organizations. Even if the plan to protect New York City's watershed areas is not fully successful, it is hoped that the lessons learned from the experience can be used to develop watershed protection programs for other communities.

Questions

1. Do you agree with the initial decision that was made, to try to protect the watershed areas rather than build and operate a water filtration system?
2. What are some of the obstacles that must be overcome politically, socially, and technologically to protect adequately the watershed areas upon which New York City is dependent?
3. If, despite spending hundreds of millions of dollars attempting to protect the watersheds, it is found that a water filtration system still needs to be built, will the money have been wasted? Justify your opinion.

life-threatening characteristics, they generally are not legally enforced. They represent guidelines that water systems try to follow when possible. Another familiar secondary standard is **hardness**, which is the amount of calcium, magnesium, and other ions in the water. "Hard water," which has high amounts of these ions, causes two annoying problems for water users. One is that these ions react with soap to produce gummy deposits, such as a "bathtub ring." These deposits make all cleaning, from dishes to baths to laundry, more difficult. In addition, when hard water is heated, it forms rocklike "scales" (mineral deposits) that can clog pipes. Adding water softeners to hard water removes the ions and can reduce these problems.

Primary standards are actively enforced because human lives are directly affected. Contaminants that are covered by primary standards are classified into four basic categories: inorganic chemicals, organic chemicals, radioactive matter, and microbes (pathogens). These are also four of the eight main classes of water pollutants discussed previously. The remaining four—heat, sediment, oxygen-demanding wastes, and plant nutrients—are harmful to ecosystems but usually are not toxic to people

TABLE 15-3	Maximum Contaminant Levels (MCLs) for Certain Inorganic Chemicals	
Contaminant	Principal Health Effects	Maximum Containment Levels (mg/L^2)
Arsenic	Dermal and nervous system toxicity effects	0.01
Barium	Circulatory system effects	2.0
Cadmium	Kidney effects	0.005
Chromium	Allergic dermatitis	0.1
Fluoride	Skeletal damage	4.0
Lead	Central and peripheral nervous system damage; kidney effects; highly toxic to infants and pregnant women	0.015
Mercury	Central nervous system disorders; kidney effects	0.002
Nitrate and nitrite	Methemoglobinemia (blue-baby syndrome)	10.0/1.0
Selenium	Gastrointestinal effects	0.05
Thallium	Liver/kidney effects	0.002

Source: Environmental Protection Agency, 2002.

in small amounts and thus are omitted from drinking water standards.

TABLE 15-3 lists **maximum contaminant levels (MCLs)** for some inorganic chemicals. MCLs are the highest concentration allowed by the EPA; water with concentrations exceeding this level is considered harmful to at least some people if they consume it over a lifetime. The EPA has similar listings of MCLs for organic chemicals, radioactive matter, and microbes. Examples of common organic chemicals with MCLs are pesticides, petrochemicals, and chlorine by-products. The most common radioactive contaminant is dissolved radon gas, which enters groundwater from surrounding rocks. Radon is colorless, tasteless, and odorless, and it is potentially very dangerous where it occurs (for additional information, review a discussion of local and regional air pollution).

Microbial contaminants, the fourth category, include disease-causing microorganisms, mainly certain bacteria and viruses. Because running separate standard tests for each of the many species of disease-causing microbes would be too expensive, a simple test for one class of microbes, coliform bacteria, is carried out. **Coliform bacteria,** especially *Escherichia coli,* live in the human intestine in huge numbers, where they aid digestion and perform other functions essential for our health (**FIGURE 15-16**). Because they have a short life span, large numbers of dead coliform bacteria occur in human feces, about 50 million individuals per gram (1.4 billion per ounce) of feces.

Thus, the abundance of coliform bacteria in water is an excellent indication of how much human feces has recently entered the water. Because human feces are the main source of disease-causing microbes in many waters, the abundance of coliform bacteria also provides an estimate of the abundance of disease-causing microbes. Raw sewage contains about 1 virus for every 92,000 coliform bacteria. Similar ratios have been calculated for disease-causing bacteria, worms, and other pathogens. Fortunately, most of these disease-causing organisms have a much lower survival rate than coliform bacteria outside the human body.

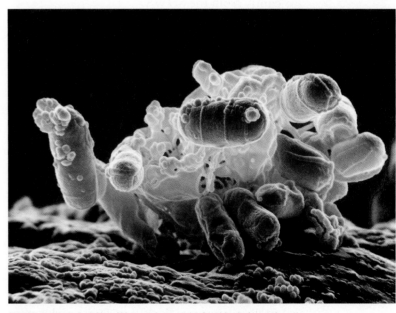

FIGURE 15-16 Escherichia coli is a common species found in the human intestine.

15.5 Legal and Social Solutions   465

TABLE 15-4	Treatment Methods and Alternative Water Sources: Advantages, Disadvantages, and Costs		
Treatment Method/ Alternative Source	**Contaminants Removed/Advantages**	**Disadvantages**	**Estimated Cost**
Carbon filter	Removes halogenated and some other organics, radon, and residual chlorine	Bacteria held in carbon may wash into water	Countertop under $50 and up. Under-sink $300 and up plus maintenance costs. Whole house under $1,200
Distillation	Removes heavy organics, inorganics, metals, and microbiological contaminants.	Not effective on volatile organics compounds, trihalomethanes, and radon. May corrode pipes.	$200–$2,000 plus maintenance costs
Reverse osmosis	Best method of removing inorganics; also reduces the level of many organics and metals	Wastes water; 75 to 90% of water is lost in straining process	$90–$700 and maintenance costs
Bottled water	Natural spring water from an isolated source could be water of good quality		$10–$15 per week depending on the size of the household.
Drilling a new well	Possible location of an uncontaminated nearby source		Averages $10–$15 per foot plus casing and pump costs nationwide but can range anywhere from $5 to $50 per foot, depending on the area

Although the EPA has oversight authority, each individual state is largely responsible for monitoring and enforcing the standards on a day-to-day basis. As a result, the effectiveness with which water quality is maintained varies considerably from state to state. Water quality in Maine, Massachusetts, and New Jersey often ranks highest in yearly lists that environmental groups produce. Maine devotes a large part of its drinking water budget to training water system operators and reportedly allows no community to deviate significantly from the SDWA.

Social Solutions

For many reasons, increasing numbers of U.S. citizens are concerned about their drinking water. Annual polls by the Gallup Organization regularly show that many Americans have a "great deal" of concern about their drinking water and are willing to take personal action.

Point-of-Use Treatment, Bottled Water, and Alternative Sources

The public's concern about their drinking water has led to a rapid increase in the use of point-of-use treatment, bottled water, and alternative sources such as drilling home wells.

Point-of-use treatment refers to a variety of home treatment devices that alter water quality as it enters the home. Three of the most

common are listed in **TABLE 15-4**: carbon filters, distillation, and reverse osmosis. The advantages and costs vary, depending on the area and the quality of the device. Countertop carbon filters are the most common. Not listed in the table are water softeners, which are home treatment devices that replace with sodium the calcium and magnesium atoms that make water "hard." Hard water is particularly common in the midwestern states. Limestone and other alkaline rocks are major contributors to hardness in surface and ground waters.

The use of bottled water (Table 15-4) in the United States has surged in recent decades. In 2004, the U.S. bottled water market sold 25,760 million liters (6,806 million gallons) of bottled water, an 8.6% increase from 2003 at a cost of $9,169 million, a 7.5% increase from 2003. A 2009 market report from the International Bottled Water Association indicates 32,000 million liters (8,454 million gallons) sold for a total of $10,595 million. The sales had peaked at $11,552 million for 33,228 million liters (8,778 million gallons) in 2007 and decreased approximately 5% during the 2007–2009 time frame, which analysts attribute to the economic times. Is bottled water worth it? Some people buy bottled water for improved taste, smell, or color compared with their tap water, but others buy it in the belief that it is more healthful than tap water. This is not always true; a recent study carried out by the Natural Resources Defense Council found that bottled water is not necessarily cleaner or safer than most tap water. Although most tested bottled water was found to be of good quality, at least some bottled water was found to be contaminated with such substances as bacteria, synthetic organic materials, or arsenic. A primary reason for such discrepancies in bottled water quality is that at a national level, the Food and Drug Administration regulates bottled water, whereas the Environmental Protection Agency regulates tap water quality. The Food and Drug Administration has considerably less rigorous standards, both in terms of testing and purity, for bottled water than the EPA has for tap water. In addition, state authorities, not the Food and Drug Administration, regulate the many bottlers selling in only one state, and one of five states fails to adequately regulate bottled water.

Another possibility is to find an alternative source to the local municipal water by drilling a private well. Costs of drilling vary considerably, depending on how deep clean groundwater is to be found, the type of rocks to be drilled through, and other variables. In some regions, groundwater is of such high quality that it requires no treatment at all for safe, tasteful drinking, but because of possible unseen contamination, professional laboratories should routinely test such water. The Centers for Disease Control reports that, in 2010, approximately 15 million people in the United States relied on private wells or springs for their household water supply.

Future of U.S. Water Pollution Control

The Clean Water Act of 1972 is reauthorized every 5 years or so. The main purpose of the reauthorizations is to rectify deficiencies in the laws that have become apparent over time. The CWA, along with the SDWA of 1974, has played an enormous role in improving water quality, and these laws and their modifications will largely determine how water pollution is controlled in coming years.

One of the most widely discussed deficiencies of water pollution control in the United States and nearly all industrialized nations is the emphasis on "end-of-pipe" methods. Hun-

dreds of billions of dollars have been spent on water pollution control since the CWA was established. This includes money spent on cleanup methods, the building and upgrading of urban sewage treatment plants, and the treatment of industrial and other wastes not directly associated with city sewer systems.

Although the end-of-pipe approach has been successful in reducing localized sources of water pollution (often called "point" sources because the pollution is discharged at a single point), many problems such as pollution of native ecosystems and groundwaters arise from nonlocalized or nonpoint sources, such as runoff from farm fields and urban areas that contains pesticides, oils, and other pollutants. The EPA has estimated that runoff from agricultural lands accounts for more than two-thirds of the pollution in "impaired" rivers. Thus, a major focus of future water pollution control has been on reducing runoff and other nonlocalized sources (**FIGURE 15-17**). This is much more difficult, but it can be accomplished through input reduction, such as using less pesticide, fertilizer, and many other substances that end up in our waters. Instead of concentrating on output reduction by end-of-pipe methods, it is often much more effective to stop pollution at the source. Input controls are cheaper

FIGURE 15-17 This waste management system helps control runoff from a 900-head hog farm.

than end-of-pipe methods, so this approach also will address the problem of the increasing costs of water pollution control.

Other key problems the United States must address include

1. The obsolescence and age of many urban wastewater treatment plants (many were built in the 1970s)
2. Increasing concern over toxic chemicals, such as chlorine-related compounds (resulting from the use of chlorine to disinfect water; ozone is now being used for the same purpose to avoid the problems associated with chlorine) and many newly invented chemicals
3. The lack of an integrated network of water testing facilities to provide comprehensive data on water quality
4. Pollution of waters that have been largely neglected until recently

Although cleanup efforts have focused on lakes and rivers, groundwater, marine waters, and wetlands have continued to experience increasing pollution and need immediate protection.

Clean Water Outside the United States

Poor water quality is a major cause of death in many developing nations. Providing modern water treatment is a primary goal of many development programs of the World Bank, the United Nations, and many other organizations. Currently, India is in the process of borrowing $1 billion from the World Bank for projects to clean up the 2,400-kilometer (1,500-mile) Ganges River. However, the sewage treatment plants, public toilets, and crematoriums this money purchased has scarcely dented the tons of sewage, industrial waste, pesticides, and other pollutants that flow into the Ganges daily.

Water quality is also very poor in many countries in Eastern Europe and in the territory of the former Soviet Union. The Baltic Sea is deteriorating rapidly. Dioxin concentrations are so high that once-popular Baltic cod liver is no longer edible; many fish have large tumors. The main reason is that surrounding cities have no or only primitive wastewater treatment. The sediment in

St. Petersburg's harbor has a thousand times the normal concentration of many toxic metals such as lead and cadmium. The Finnish Baltic Marine Environmental Commission estimates the cost of cleaning up the Baltic at $1 billion per year for 20 years. Each year Russia's 3,700-kilometer (2,300-mile) Volga River receives billions of tons of sewage and industrial waste, including toxic waste from more than 3,000 factories. Currently, 70% of the fish in the Volga contain mercury. Much of this pollution ends up in the Caspian Sea, which is also rapidly becoming polluted. A growing environmental movement, Save the Volga, has appeared but, as in other former communist countries, the lack of money seriously impedes cleanup efforts.

Global Water Security

Increasing shortages of clean, potable water around the world are now major concerns of government officials, policy makers, business leaders, humanitarians, national security advisors, and other experts and leaders. Major national and international conferences have been held on the issue, and in 2005, the Sandia National Laboratories (part of the National Nuclear Security Administration) and the Center for Strategic and International Studies jointly released the important position paper "Addressing Our Global Water Future." One of the fundamental conclusions was that the availability of clean water is critical to human well-being and prosperity, yet freshwater resources are limited in many areas. A lack of water resources can cause famines, declines in economic output, displacement of peoples, conflicts over limited and dwindling resources, and the weakening and collapse of once-stable governments. The United States (as well as other countries, and in particular the developed nations) has a strategic security interest in ensuring stable governments around the world, and this means that the United States should promote policies and technologies globally that not only provide adequate supplies of freshwater in the short-term, but also sustain the long-term management of such water supplies. Good water management is beneficial from a geopolitical and military perspective. In the case of freshwater supplies, humanitarian interests, environmental concerns, local socioeconomic and geographic factors, global security issues, and the promotion of long-term sustainability are all mutually compatible. Working together, diverse parties with many different backgrounds and agendas can agree on a common point: Freshwater is important for all peoples, and it is in our best interest to make certain that there are adequate supplies both locally and globally. The world as we know it depends on clean water.

■ study guide

SUMMARY

- Polluted water can be defined as water that is rendered unusable for its intended purpose.
- Currently, more than 1 billion people globally lack an improved water supply, and more than 2 billion lack improved sanitation facilities, leading to outbreaks of cholera, typhoid, hepatitis, and many other diseases.
- Water is purified in nature by physical processes (dilution, sedimentation, filtration, aeration) and chemical processes (biochemical reactions, many produced by microbes, and precipitation).
- Basic categories of water pollutants are oxygen-demanding wastes, infectious agents, plant nutrients, organic chemicals, inorganic chemicals, sediment pollution, radioactivity, and thermal pollution.
- Sources of water pollutants are: industrial, agricultural, and domestic wastes.
- Ultimately, water pollutants can end up in natural waters, such as rivers and streams, lakes, groundwater, and marine waters.
- Water pollution can be reduced and mitigated by source reduction, wastewater treatment, and remediation (cleaning polluted waters).
- In the United States, the Clean Water Act and Safe Drinking Water Act have been instrumental in addressing water pollution issues.

KEY TERMS

aeration
biological oxygen demand (BOD)
Clean Water Act (CWA)
coliform bacteria
contaminated water
dilution
eutrophication
filtration

hardness
maximum contaminant levels (MCLs)
nonpoint source
point-of-use treatment
point source
potable water
polluted water
primary standards

primary treatment
Refuse Act
Safe Drinking Water Act (SDWA)
secondary standards
secondary treatment
sedimentation
septic tank
stormwater runoff

STUDY QUESTIONS

1. What is polluted water?
2. Approximately how many people in the world are without an improved water supply? Without improved sanitation? What types of problems can this cause?
3. Describe the basic water purification processes as they occur in nature.
4. What are the eight basic categories of water pollutants that the U.S. Public Health Service has classified?
5. What is the effect of oxygen-demanding wastes as water pollutants?
6. Give an example of an effect caused by high concentrations of a metal in living organisms.

7. Give examples of how sediment from erosion can be a pollutant.

8. What are two harmful effects of increased temperatures on aquatic organisms?

9. What are the major sources of water pollution? Specify the water pollutants produced by each activity, such as agriculture.

10. What are two reasons that cities historically have been built along rivers?

11. Give three reasons that lakes are more easily polluted than rivers.

12. Why is groundwater so easily polluted? What are some of the major sources of groundwater pollution?

13. Discuss why ocean pollution is a growing concern. Describe some of the major types of ocean pollution and their sources, and suggest potential solutions.

14. What are the three basic ways of addressing water pollution concerns?

15. Describe the concept of source reduction and how it applies to water pollution issues.

16. Describe the process of wastewater treatment. What do primary treatment and secondary treatment refer to in this context?

17. The Safe Drinking Water Act of 1974 specifies maximum containment levels for what four classes of health-threatening water pollutants?

18. Under the Safe Drinking Water Act, what distinguishes a primary standard from a secondary standard? What is hard water, and what problems does it cause?

19. Distinguish between localized (point) sources of water pollution and nonlocalized sources of water pollution. Which has been a primary emphasis of water pollution control efforts in the past? Why is it important to address both classes of water pollution?

20. How are adequate supplies of freshwater related to global security issues?

WHAT'S THE EVIDENCE?

1. Many people buy bottled water thinking that it is better quality water. What does "better" mean in this context? More healthful? Better tasting? What is the evidence that bottled water is superior to local tap water? How are the answers to such questions locally specific?

2. The authors state that remediation—cleaning up polluted waters—can be more expensive than reducing pollutant input or treating waters before discharge. Are you convinced that this is the case? Can you cite general examples to support the authors' contention? (Suggestion: You might also consider the lost economic and other benefits of polluted waters, the increased "value" that clean waters can have, and the costs of treating disease caused by polluted waters.)

CALCULATIONS

1. If a beach cleanup gathered 307 tons (278.6 metric tons) of litter along 157 miles (253 kilometers) of Texas coastline, how much litter per mile was this? How much litter per kilometer? How many tons of plastic were collected if 64% of the total was plastic?

2. If a beach cleanup gathered 600 tons (544.5 metric tons) of litter along 200 miles (322 kilometers) of Florida coastline, how much litter per mile was this? How much litter per kilometer? How many tons of plastic were collected if 58% of the total was plastic?

ILLUSTRATION AND TABLE REVIEW

1. Referring to Table 15-3, which contaminant would you conclude is the most dangerous at very low levels? Which does the EPA allow at the highest concentrations?

2. Referring to Table 15-3, the maximum allowable concentration of barium is how many times that of mercury?

http://environment.jbpub.com/mckinney/5e/

Connect to this text's website at http://environment.jbpub.com/mckinney/5e/. This site features eLearning, an online review area that provides quizzes, chapter outlines, and other tools to help you study for your class. You can also follow useful links for more in-depth information.

Strict legislation has significantly reduced air pollution in many cities, including New York City (seen here on a clear evening). In fact, the American Lung Association's 2012 annual report, *State of the Air*, indicated nearly a 25% decrease in both particulate pollution and ozone in New York from 2001 to 2010. Although overall air quality is improving throughout the state, New York City still experiences an alarming number of code orange and code red days, especially in summer months. While stricter enforcement of existing laws played an important part in recent improvements, sustained citizen action focusing on improving air quality is needed to perpetuate positive gains. For air quality information in your state, see http://www.lung.org.

Air Pollution: Local and Regional

16

A ir pollution might be most visible around our congested cities, but the thin layer of air that envelops the Earth is polluted at many scales: It is contaminated indoors, locally, regionally, and globally (**FIGURE 16-1**). Just as everyone is downstream from someone else, everyone is downwind from someone else, making air pollution one of our most widespread environmental problems. Regional and global air pollution is especially difficult to control because polluters are often very distant from those damaged by their emissions.

Based on Environmental Protection Agency (EPA) statistics (2009), 107 million tons of pollution annually was emitted into the air from the United States. Although this is down considerably from years past, this number reflects data only from the pollutants most commonly regulated. In addition, 80 million people in the United States live in areas where air pollutants remain above acceptable levels. Air pollution remains on the forefront of global environmental challenges, and the resulting costs are distressing. Each year, outdoor air pollution causes tens of billions of dollars worth of direct damage in the United States, including crop and livestock damage, forest degradation, coastal ecosystem disruption, weathering of statues and buildings, and cleaning costs to clothes and other personal property. In 2005, ground-level ozone alone caused $500 million worth of reduced yields to agriculture and commercial forests. Many billions of dollars more are lost because of increased health care expenses. In 1995, the EPA launched their

Chapter Objectives

After reading this chapter, you should be able to explain or describe the following:

- Why it is important to be concerned about air pollution?

- The six basic criteria pollutants that the U.S. Environmental Protection Agency recognizes and their effects on health and the environment

- The major sources of air pollution

- Historical trends of various air pollutants

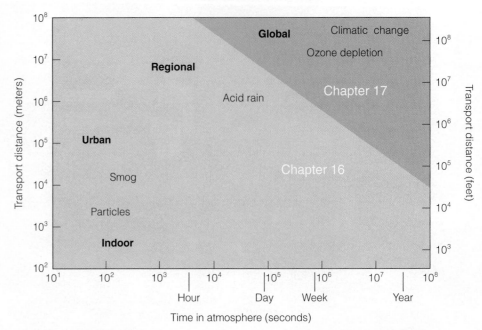

FIGURE 16-1 Global air pollution occurs when so much pollution, such as carbon dioxide, is released into the atmosphere that global changes occur. Urban and regional air pollution are more local; pollutants travel less distance and stay in the air for shorter time periods. Acid rain, smog, and other examples are discussed in the text. (*Source:* Modified from P. Mackenzie, J. Mackenzie. Our Changing Planet. Englewood Cliffs, NJ: Prentice-Hall, 1995, p. 237.)

model Acid Rain Program, which in 2010 was in its second phase. The results continue to be successful. In 2002, acid rain emissions were down more than 30% and have continued to decline by about 1 million tons of sulfur-based pollutants alone. Despite these efforts, more needs to be done, especially in the most challenged areas, such as cities along both coasts. Globally, approximately 1 million premature deaths are attributable to particulate air pollution yearly.

Indoor air pollution is potentially an even greater health hazard than outdoor air pollution. Radon and especially cigarette smoke cost many billions of dollars. In terms of overall impact, global air pollution is perhaps the greatest single environmental challenge that people will face in the next few hundred years. Climate changes caused by pollution could lead to massive alterations in agriculture, large-scale migration of people and other organisms, flooding of the world's coastal cities, and many other changes. Yet all is not bleak. The air we collectively breathe has improved considerably in the United States during the last 4 decades (see **CASE STUDY 16-1**). Since 1970, when enforcement of the **Clean Air Act** was put in place, the levels of six principal pollutants have dropped by 53% despite increasing population, increasing energy consumption, and increasing vehicle miles traveled (**FIGURE 16-2**). In 2005, the EPA reported that although many cities were still polluted by too much smog and ozone, a number of cities had seen major reductions in the number of days per year when the air quality index was greater than 100 (below 100 is considered acceptable or safe; see discussion in Legal and Social Solutions in this text).

Here we discuss local and regional air pollution and indoor air pollution. Global air pollution, including ozone depletion and climate change, is discussed in the review of global air pollution. We will see that local and regional air pollution is largely a problem of cities and industrialization: Large concentrations of people and machines produce large inputs into the atmosphere, especially by combustion of the fossil fuels that we now use to run civilization. Thus, alternative fuels, such as solar energy, would alleviate much air pollution. In contrast, global air pollution is not restricted to cities; the burning of tropical forests is an example.

Smog: Causes and Cures

VISIT

http://environment.jbpub.com
/mckinney/5e/
for more information

Smog as we know it today and air pollution more generally have been significant issues for more than a century. In London, bad air has been a concern since the 1200s (**FIGURE 1**). The witches in Shakespeare's Macbeth chant, "Fair is foul, and foul is fair: Hover through the fog and filthy air." London became known for its "pea-soup fog," an acrid combination of natural fog and coal smoke. Popular images of Victorian-era steam trains puffing out of stations and soot-smudged chimney sweeps dancing across endless rooftops romanticize an unhealthy city. In 1879, one fog lasted from November to March, 4 long months of sunshineless gloom. In 1892, 1,000 people died in London as a result of a "killer fog" (smog), and in 1952, some 4,000 people succumbed to another London smog. In the latter case, the smog and smoke were so dense that traffic was stopped and buses had to be led by a person on foot carrying a lantern.

In the United States, Los Angeles became well known for its high levels of air pollution and smog in particular. As early as the summer of 1943, there was a major smog incident in the Los Angeles area. Residents experienced irritation of the eyes and nose, nausea, vomiting, and breathing disorders. Visibility was limited to three blocks. At the time, the problems were blamed on a local butadiene (a chemical used in the manufacture of synthetic rubber) plant, but although it may have contributed to the problem, it was not the sole cause. Even after the plant shut down, the problems continued. In 1945, Los Angeles responded with its first pollution control program, and in the late 1940s, the state of California also began addressing air pollution issues through legislation. Still, the air quality in the Los Angeles area remained unhealthful. In the early 1990s, Southern California's air quality was the worst in the United States. Air pollution in the region reached unhealthful levels on half the days each year, and it violated four of the six federal standards for healthful air—those for ozone, fine particulates, carbon monoxide, and nitrogen dioxide. In 1991, the South Coast Air Basin exceeded one or more federal health standards on 184 days.

As bad as it was in the 1990s, the air quality had improved dramatically since the 1970s. From 1955 to 1992, the peak level of ozone—one of the best indicators of air pollution—declined from 680 parts per billion to 300 parts per billion. The California Air Resources Board documented

Smog: Causes and Cures

that population exposure to unhealthful ozone levels was halved between 1984 and 1994, when the lowest level on record was registered. All of these improvements were achieved at a time when the greater Los Angeles area population was growing to some 16.9 million people (depending on how the area is defined). The area now has approximately 10 million automobiles and other vehicles on the streets. Despite the growth in population, emission control standards remain aggressive with an overall state goal of ozone readings not to exceed 84 parts per billion in any eight hour period. However, in 2010 some areas in California had over 100 days in which ozone level exceeded this safer level. Overall, ozone levels in the state continue to improve as industry adjusts to meet more stringent guidelines.

So what causes smog? The hazy days of the Los Angeles area are caused in part by a natural weather phenomenon known as an inversion layer (see Figure 16-12). Such inversions often form off the coast of Los Angeles as the Pacific Ocean cools the atmosphere just above it. After ocean breezes blow the air mass inland, the layer of cool air may slip underneath a layer of warmer air. The inversion layer traps air pollutants in the cool air near the ground where people live and breathe. The mountains that surround the region compound the problem by preventing the pollutants from dispersing.

Natural materials, such as dust, pollen, fibers, and salt, are important components of haze, but industries and motor vehicles contribute to the problem by adding carbon particles, metallic dust, oil droplets, and water vapor. Automobiles, factories, and other sources also release such raw pollutants as hydrocarbons, water vapor, carbon monoxide, and heavy metals. When these chemicals are exposed to intense sunlight, they react to yield a vast number of secondary pollutants.

Over the years, regulations were gradually developed that concentrated on reducing the major sources of air pollution: particles from trash incineration, emissions from industry, and pollutants from motor vehicles. The EPA prepared standards for airplanes, trains, and ships that travel through the region. Local governments tried to reduce traffic by improving the transportation infrastructure and expanding mass transit. Automotive emissions in particular decreased because of more stringent tailpipe standards and programs encouraging carpooling and use of public transportation. Since the middle 1990s, the dramatic pace of improvements in Los Angeles air quality seen in previous decades has slowed. For instance, through 2005 (the last date for which data are available), peak ozone levels have been creeping upward. Still, there is good news because even as peak recorded ozone levels have increased slightly, the number of days during the year when ozone (and other) standards as established by the EPA have been exceeded has dropped. In 1992, the Los Angeles area failed to meet acceptable air quality standards on 175 days of the year, but this number has been reduced every year. In 2010, Los Angeles reported 45% clean air days but still remains under acceptable levels overall. Regionally, the area is making better strides, especially at the port of Los Angeles, which implemented a carbon reduction plan with the results of lowering carbon emissions by more than 40% in 2009. Los Angeles is certainly not alone in their struggles. The American Lung Association reports that 6 of 10 Americans live in an area where air quality endangers lives. In a 2009 American Lung Association report, Stephen J. Nolan, National Board Chair, remarked: "When 60 percent of Americans are left breathing air dirty enough to send people to the emergency room, to shape how kids' lungs develop, and to kill, air pollution remains a serious problem."

Questions

1. Is there significant local air pollution where you live, work, or attend college? Do you keep track of the local air quality? Are there days that you can "feel" that the air is unhealthy?

2. Some people are resigned to the "fact" that there is bound to be more air pollution in large metropolitan areas, yet they would rather put up with the issue in exchange for the opportunities and conveniences of the city. How do you feel about this matter? Must cities necessarily have unhealthful air? How does the example of Los Angeles influence your opinion on this issue?

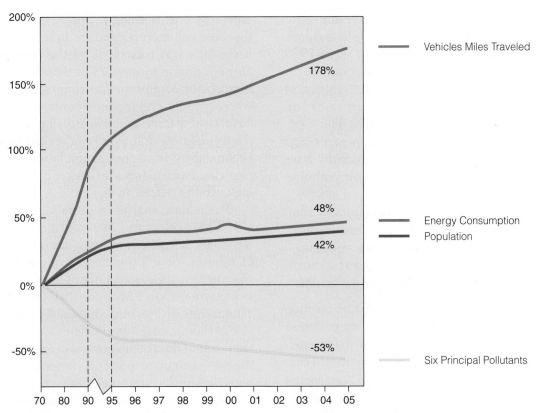

FIGURE 16-2 The levels of six principal pollutants have dropped in the United States despite increasing population, energy consumption, and increasing vehicle miles traveled. (*Source:* Environmental Protection Agency.)

16.1 Local and Regional Air Pollution

Until the late 1980s and early 1990s, local and regional pollution generally received more attention than global or indoor pollution. The United States and most Western industrialized nations have significantly reduced most types of local and regional air pollution since 1970, when strict amendments to the Clean Air Act were passed. Reducing the small group of so-called criteria pollutants (also known as "principal air pollutants") that are the source of most local and regional air pollution achieved this.

The six basic **criteria pollutants** as currently recognized by the EPA are particulate matter (PM), sulfur dioxide (SO_2), nitrogen dioxide (NO_2), ozone (O_3), lead (Pb), and carbon monoxide (CO). Carbon monoxide, lead, nitrogen dioxide, and sulfur dioxide are emitted directly into the air from various sources. Ozone and particulate matter are somewhat more complicated. Particulate matter (a num-

ber of different types of which can be distinguished) can be emitted directly, or it can form when various pollutants such as sulfur dioxide, nitrogen oxides (NO_X), ammonia, or other gases react in the atmosphere. Ozone forms as a result of the reaction of volatile organic compounds (VOCs) and nitrogen oxides in the presence of sunlight. A volatile organic chemical by definition is a gas that enters the air from a solid or a liquid; the fumes you smell in paint are a common example of a VOC. For this reason, it is often very important to monitor the levels of VOCs in a given area. These criteria pollutants have two main sources: (1) burning of fossil fuels by motor vehicles and stationary sources such as power plants and (2) industrial processes other than fossil fuel combustion. We will discuss each of these six criteria pollutants in turn, but first we should note some key patterns in the United States:

1. Most pollutants, with the exception of nitrogen oxides and ground-level ozone

in some areas (formed as a result of increased emissions of nitrogen oxides), have decreased significantly since 1970 due to increased pollution controls.

2. Transportation is the largest source of carbon monoxide, volatile organic compounds, and nitrogen oxides. This category includes motorboats, mowers, and small motors, which until recently have been largely unregulated by air pollution controls.

3. Stationary fuel combustion (mostly coal-burning power plants) is the largest source of sulfur oxides, which cause acid rain.

4. Industrial processes other than fuel burning are the largest source of particulates.

Take special note that fossil fuel burning is the main cause of much of the air pollution that we face. This is one reason many environmentalists advocate switching to alternative fuels, such as solar or wind. These energy sources are not only renewable, but also would greatly reduce many air pollution problems.

Particulates

Particulates are any particles of dispersed matter, solid, or liquid that are larger than individual molecules. This category is very complex because particulate matter varies widely in composition and size. Soot, which arises from incomplete fuel combustion in cars and especially coal-burning factories, is generally the most common particulate in urban areas. It comprises at least 50% of the particulate air pollution in most cities of the world. Soot is especially common in developing countries, where air pollution is poorly controlled. If you have visited large cities in developing countries, you know that white clothes can quickly become dirty from the accumulation of dark particles. Particulates also scatter light, reducing visibility (**FIGURE 16-3**).

Particulate matter is often classified by size. $PM_{2.5}$ refers to "fine" particulate matter that is 2.5 microns or smaller in diameter (1 micron, or micrometer, is one-millionth of a meter). PM_{10} particulates are "coarse," being larger than $PM_{2.5}$ particulates but still 10 microns or less in diameter. As a general rule of thumb, coarse particulates are primarily the result of direct emissions, whereas much of the fine particulate matter is composed of secondarily produced particles (as when various gaseous pollutants interact in the atmosphere).

Aside from their effects on clothes, machinery, buildings, and visibility, particulates are the single most damaging air pollutant to lungs. Larger particles are usually trapped in the hairs and lining of the nose and throat and then coughed or sneezed out. Small particulates can penetrate deep into the smaller canals and pockets of the respiratory system, where they cannot be easily coughed up or

(a)

(b)

FIGURE 16-3 (a) Smog in Beijing, China. Particulates and other pollutants greatly reduce visibility. (b) A satellite image of smog covering Beijing.

otherwise removed. Many grade school models of the lungs liken them to balloons that inflate and deflate, but a better example would be for you to imagine a sponge made of a fine material such as silk. We absorb air into our lungs; in the tiniest internal layers deep in our lungs, the lining is one cell thick. In this area, called alveoli, oxygen and carbon dioxide are exchanged. Imagine if you are trying to absorb water with a sponge and the nooks and crannies are clogged with dirt. This is what happens when particulates lodge in the delicate lining of our lungs. Although we have two lungs (the right has three lobes and the left has two lobes), there is still a finite surface area for oxygen absorption. This is why lung damage is cumulative over a lifetime, beginning in utero. We move around in the world literally marinating in the air as a fish swims in water. We breath in approximately 11,360 liters (3,000 gallons) of air a day (7.6 liters or 2 gallons a minute). Particulates are so small that several thousand pieces of particulate matter could fit on the period at the end of the sentence. The amount we take in during one breath is minimal, but because the effect is cumulative, it is like dropping half a drop of oil or a tiny pinch of dust into a 2-gallon fish tank every minute; the pollution has nowhere to go, and the filters can hold only so much, so eventually the filter clogs, oxygen cannot flow, and life cannot sustain itself. In terms of people, the World Health Organization reports that in 2008, 2 million people worldwide lost their lives prematurely because of unacceptable air quality (**FIGURE 16-4**). Accumulation of sufficient amounts of these fine particles will cause impaired breathing from blockage and irritation, resulting in long-term illnesses. Examples include "black lung" of coal miners, asbestos fibrosis, and urban emphysema. Many statistical studies show that increases in particulate concentration in the atmosphere are correlated with increased visits to hospitals for respiratory problems, cardiac disorders, bronchitis, and many other illnesses associated with

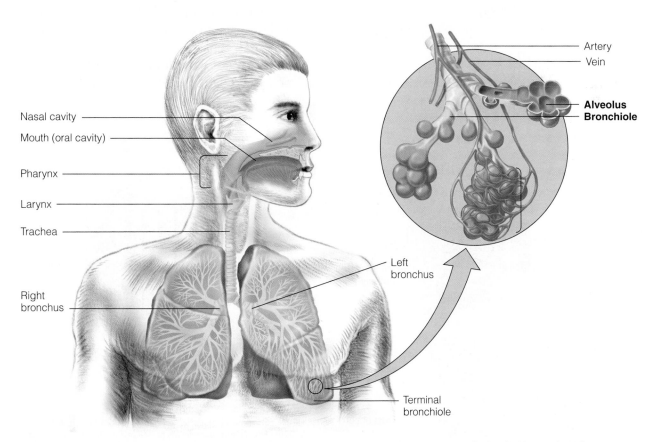

FIGURE 16-4 Anatomy of the lung. Fine particulates can penetrate deep into the bronchioles and alveoli, where oxygen and carbon dioxide are exchanged.

breathing. The Asthma and Allergy Foundation of America reports that every day in the United States 5,000 people visit an emergency room with an asthma attack, 1,000 people are admitted to the hospital with asthma-related illness, and such 11 people die. One in every 20 children has asthma.

An additional problem is that toxic chemicals may become attached ("adsorbed") to the surface of soot particles. These chemicals, such as the toxic metals cadmium or nickel, can be absorbed by the blood and cause heavy metal poisoning, usually affecting the nervous system. Organic chemicals, such as those used in pesticides or manufacturing, are also often adsorbed to soot; these can cause cancer. Because particulate inhalation is cumulative, as are the toxic effects of adsorbed chemicals, the effects are usually most pronounced in the elderly. Particulate matter can also have detrimental affects on plants, animals, and natural ecosystems. In addition, PM can cause damage to painted surfaces and building materials. Despite these effects, particulates as a whole are generally not considered as detrimental to the environment as most other forms of air pollution. For one thing, particulate pollution caused by people is actually small compared with pollution from natural sources. By various estimates, more than 50% to perhaps more than 90% of the particulates in the air globally come from salt spray, dust storms, and especially volcanoes. Second, unlike gaseous air pollutants, particulates settle out of the air in a few days because of gravity and rain. Thus, they tend to be local problems. Third, particulates are among the easiest and cheapest air pollutants to remove at the source.

Reducing Particulate Pollution

As with all pollutants, input reduction is always the cheapest and most effective method. In the case of particulates, this means burning coal, for instance, that has a low ash content; however, even low-ash coal produces tons of emissions. These are usually removed from the air flow after combustion in a stepwise fashion. Large particles are removed first by a so-called **cyclone collector** (**FIGURE 16-5A**). This spins the emissions in a vortex, causing the heavier particles to collide with the sides and slide down into a collecting bin at the bottom. This step typically removes more than 90% of the largest particles and is relatively cheap and maintenance free.

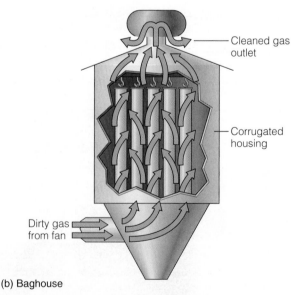

FIGURE 16-5 (a) A typical cyclone collector for particulates. (b) A typical baghouse filter for particulate pollution. (*Source:* U.S. Department of Health, Education, and Welfare.)

Zone of inlet interference — Outer vortex — Inner vortex — Gas outlet — Body — Gas inlet — Inner cylinder (tubular guard) — Outer vortex — Inner vortex — Core — Dust outlet

(a) Cyclone

Cleaned gas outlet — Corrugated housing — Dirty gas from fan

(b) Baghouse

Next, the smaller particles are removed by using either electrostatic precipitators or baghouses. **Electrostatic precipitators** take advantage of the small electric charge that most particles carry. A high voltage is created between metal walls, causing the particles to collect on the walls, where they are washed or shaken off into a bin. These are very efficient, removing as much as 99.9% of small particles. **Baghouses** are a series of fabric bags that act as filters (**FIGURE 16-5B**). In efficiency and cost, they are similar to precipitators. Despite certain disadvantages, such as fire and explosion potential, baghouse filters are used almost as frequently as precipitators.

The Downward Trends in Particulate Pollution

It can be difficult to get a precise handle on particulate emissions because there are several different types and sizes of particulate matter involved; some are emitted directly, and some particulate matter forms secondarily. In addition, in the late 1990s, the EPA changed its procedures for estimating and monitoring different types of particulate emissions. Still, it is clear that particulate emissions in the United States have decreased dramatically in the last few decades; it is estimated that emissions of PM_{10} declined by 31% and $PM_{2.5}$ declined by 10% during the 1998 to 2003 period. From 1990 to 2008, overall emission from the six most common pollutants decreased a total of 41%. Much of this reduction is the result of the Clean Air Act Amendments of 1970 and subsequent revisions, which, among other things, required the use of control devices to limit coal particle emissions. However, across the United States, there can be major regional discrepancies in the levels of particulate matter. The eastern United States, particularly the southeast and Mid-Atlantic states, tends to have higher levels of $PM_{2.5}$ than do the western states because of the sulfate concentrations resulting from coal-burning power plants.

Particulate air pollution has also decreased during the last few decades in nearly all of the other wealthier nations. In contrast, the former communist countries of Europe and the developing nations continue to have high levels of particulate pollution, largely because they lack the money to invest in controls. Historically, as per capita income increases, particulate emissions decrease because nations have more money to invest in pollution controls.

Sulfur Oxides

Sulfur oxides are produced when fossil fuels (in particular, coal and oil) that contain sulfur are burned or during metal smelting and other industrial operations. This oxidizes the sulfur to form SOX compounds, such as SO_3 or, most commonly, SO_2. Coal generally contains much more sulfur than do the other fossil fuels. Consequently, the majority of SO_2 air pollution in the United States is generated by coal-burning electric power plants.

In terms of overall damage to people and the environment, SOX may be the most serious local and regional air pollutant. Sulfur dioxide, or SO_2, is a gas that is toxic to living things. Plants are especially sensitive, exhibiting stunted growth, discoloration, and reduced crop yields. Plant damage has been recorded more than 80 kilometers (50 miles) downwind of large sources. The effects on people range from irritation of mucous linings in the eyes and lungs to death from respiratory and heart failure. Death can occur in just 30 seconds at concentrations as low as 3 parts per million.

Sulfur dioxide is also the main cause of **acid rain** (also known as acid precipitation), which is rainfall that is more acidic than normal. Acidity is measured on the **pH scale**, where 7 is neutral and 1 is very acidic (**FIGURE 16-6**). This is a logarithmic scale, so a decrease of 1 in pH indicates a 10-fold increase in acidity. Even unpolluted rainfall is slightly acidic, with an average pH of about 5.6, because water combines with naturally occurring carbon dioxide in the air to form the mild carbonic acid. When sulfur oxides are added to the air, water combines to form the much stronger sulfuric acid, causing acid rain. Technically, it becomes "acid rain" when its pH is below 5.0. In extremely industrialized urban areas, local "acid fogs" can be produced; likewise, in colder regions, "acid snow" may form. In the past, there have been fogs in the Los Angeles basin with a pH of 1.7, more acidic than lemon juice.

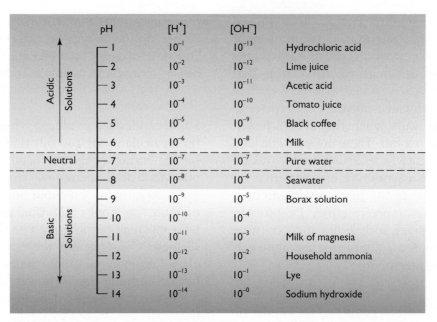

pH	[H⁺]	[OH⁻]	
1	10^{-1}	10^{-13}	Hydrochloric acid
2	10^{-2}	10^{-12}	Lime juice
3	10^{-3}	10^{-11}	Acetic acid
4	10^{-4}	10^{-10}	Tomato juice
5	10^{-5}	10^{-9}	Black coffee
6	10^{-6}	10^{-8}	Milk
7	10^{-7}	10^{-7}	Pure water
8	10^{-8}	10^{-6}	Seawater
9	10^{-9}	10^{-5}	Borax solution
10	10^{-10}	10^{-4}	
11	10^{-11}	10^{-3}	Milk of magnesia
12	10^{-12}	10^{-2}	Household ammonia
13	10^{-13}	10^{-1}	Lye
14	10^{-14}	10^{-0}	Sodium hydroxide

FIGURE 16-6 On the pH scale, 7 is neutral. Values above 7 are increasingly alkaline, and values under 7 are increasingly acidic. A pH of 7 indicates that the H+ concentration is 10^{-7}, or 1 part in 10 million.

Such extremely high acidity usually occurs in confined areas near the sulfur oxide sources. More often, the sulfur oxides disperse downwind, often traveling 3 or 4 days and many hundreds of miles before falling as acid rain. Dry acid deposition can occur when relatively dry sulfate-containing particles, or sulfuric acid salts, settle out of the air. When combined with moisture, these particles may form a very strong acid. The tall smokestacks that many coal-burning power plant companies have built to prevent local particulate pollution have aggravated this problem (**FIGURE 16-7**). As a result, acid rain tends to be concentrated in areas downwind from the large industrial and urban centers that produce the pollution. In North America, acid rain is concentrated in the Great Lakes region, New England, and southern Canada, where the pH has dropped below 4.5 (**FIGURE 16-8**).

FIGURE 16-7 Tall smokestacks that eject emissions higher into the atmosphere contribute to acid rain.

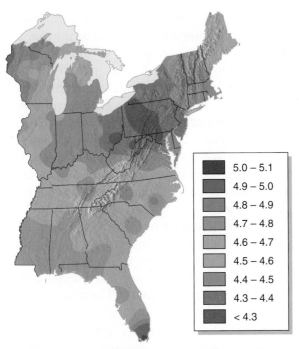

FIGURE 16-8 Average pH values of rain and snow in the eastern United States as of 1999, the latest reliable and complete data available.

■	5.0 – 5.1
■	4.9 – 5.0
■	4.8 – 4.9
■	4.7 – 4.8
■	4.6 – 4.7
■	4.5 – 4.6
■	4.4 – 4.5
■	4.3 – 4.4
■	< 4.3

These areas are downwind of industrialized areas around the southern Great Lakes, such as Cleveland and Pittsburgh. In Europe, acid rain tends to be concentrated in northern Germany and Scandinavia, which are downwind of the industrial, urban areas of England and southern Germany.

Acid rain and acid deposition damage both the physical and biological environment. Physical damage includes paint discoloration; corrosion of metals; and dissolution of marble, mortar, and other building materials that contain lime, which reacts with acid. This has caused many buildings and statues to become pitted and eroded. In the United States alone, acid rain annually causes tens of billions of dollars in equipment repairs and replacement costs. Archaeologists calculate that acid rain has caused more damage to the marble buildings of ancient Greece in the last 60 years than in the entire preceding 2,400 years. Whenever possible, the most valuable historic monuments, such as statues and parts of temples on the Acropolis in Greece, are being moved into museums, and plastic replicas are left in their place. Biological damage from acid rain is most visible in forests and lakes. Coniferous trees, such as pines, are especially sensitive to changes in soil acidity. Symptoms include yellowing and loss of needles (**FIGURE 16-9**). Many of the coniferous forests in New England, Canada, and northern Europe have shown these symptoms—some drastically. For example, it is estimated that more than half of western Germany's forests have been affected by acid rain.

FIGURE 16-9 Acid rain damage.

Lakes suffer biological damage as their waters become more acidic. Generally speaking, lake ecologists have found that larger organisms are more sensitive to acidity:

Game fishes	most sensitive
Invertebrates	
Rough fishes	
Algae	
Microbes	least sensitive

However, much variation exists within this pattern. Invertebrates such as insects and snails vary widely, but many suffer damage below pH 6. Fishes have a lower average pH tolerance because so-called rough fishes, such as garfishes, can tolerate pH values even below 4. On the other hand, many game fishes, especially trout, are affected at pH values as high as 6. Larvae and eggs are usually the most sensitive in all organisms.

Most of the water that fills a lake is runoff that has percolated through the surrounding rocks and soil. In many cases, these rocks and soils are more **alkaline** than the rain and thus neutralize some of the rain's acidity. For example, where limestone and other sedimentary rocks that are extremely alkaline surround lakes, even highly acid rain is often readily neutralized. However, New England, eastern Canada, and northern Europe have surface rocks and soils formed from igneous and other crystalline rocks that are much lower in alkaline chemicals. This is a major reason acid rain affects lakes so strongly in these areas.

Reducing Sulfur Oxides

As with all pollutants, the most effective and cheapest way of reducing sulfur oxides is through input reduction. In this case, this means reducing the amount of sulfur in coal, which produces the vast majority of sulfur oxide problems. The sulfur content of power plant coal typically ranges from 0.2% to 5.5% (by weight). By switching to low-sulfur coal, sulfur emissions can be reduced by 30% to 90%, depending on the original and the new sulfur contents. Unfortunately, approximately 85% of U.S. low-sulfur coal reserves are in the western United States, whereas two thirds of coal consumption is in the east. Coal's large bulk makes it expensive to transport, so the cost of shipping low-sulfur coal is a major deterrent to its use, given the many tons burned per day by each plant (**FIGURE 16-10**). A growing alternative is to clean high-sulfur coal using a variety of methods. For example, sulfur often occurs in heavy minerals, such as pyrite (FeS_2) that readily settle out when the coal is washed in water. Coal cleaning is relatively cheap because it reduces other costs by increasing coal-burning efficiency and reducing waste.

Another option is to use scrubbers. **Scrubbers** are devices that cleanse emissions, usually with water, before they are released into the air. Finely pulverized limestone, mixed with water, is sprayed into the emissions. The gaseous sulfur mixes with the calcium carbonate to form a "sludge" of calcium sulfate (gypsum). Scrubbers can remove approximately 90% of SO_2 from the emissions, but they are expensive: A scrubber adds 10% to 20% to the total cost of a new power plant, and operation costs add about 1 to 2 cents per kilowatt-hour of electricity produced. Scrubbers often corrode and clog, reducing plant efficiency. Finally,

FIGURE 16-10 Some coal-fired power plants in the United States burn as much as 20,000 tons of coal a day, the amount carried by 200 rail cars.

and perhaps most importantly, it is expensive to remove the highly toxic and acidic gypsum sludge that is produced in vast amounts. For example, a 1,000-megawatt plant burning 3% sulfur coal produces enough sludge each year to cover 2.5 square kilometers (1 sq mile) of land to a depth of more than 30 centimeters (1 foot). The cost of disposing of this material is rising rapidly as landfills and other disposal sites fill. Attempts have been made to use the gypsum in construction, but it is not generally economical compared with gypsum from other sources. Scrubbers use as much as 3,785 liters (1,000 gallons) of water per minute in large plants, making them impractical in arid areas.

Positive Trends in Sulfur Oxide Pollution

Thanks to the Clean Air Act Amendments of 1970 and the EPA's Acid Rain Program beginning in 1995, U.S. SO_2 emissions decreased by 38% during the period 1981 to 2000. Even more dramatically, nationwide ambient concentrations of SO_2 dropped by 50% between 1981 and 2000. Thirty-seven percent of that drop came between 1991 and 2000 alone and another 25% decrease from 2000 to 2009, mainly from the use of low sulfur coal. Much of this reduction was achieved from the use of the control methods described and used in industry and stationary combustion sources such as power plants. World Bank data show that sulfur oxide pollution is most common in countries of intermediate wealth. Poor countries do not yet have the industrial and electrical capacity to burn large amounts of coal, whereas the wealthiest countries can afford pollution reduction.

Nitrogen Oxides, Ground-level Ozone, and Volatile Organic Compounds

Nitrogen oxides, ground-level ozone, and volatile organic compounds (VOCs) are discussed together because they are major causes of smog and other photochemical pollutants in urban areas. It is important to distinguish between ground-level ozone and stratospheric ozone. Ground-level ozone is a major pollutant and forms the primary constituent of smog. Ozone also occurs naturally in the stratosphere, where it serves important functions, such as protecting the surface of Earth from high levels of ultraviolet radiation. Both increased concentrations of ground-level ozone and depleted levels of stratospheric ozone pose major environmental and health concerns (see a review of global air pollution for additional information about stratospheric ozone).

Nitrogen oxides are usually referred to as NOX because they are composed of many different oxides, especially NO and NO_2. VOCs include a wide variety of hydrocarbon molecules. A **photochemical pollutant** is produced when sunlight initiates chemical reactions among NOX, VOCs, and other components of air. These reactions are very complex and can produce many different kinds of photochemical pollutants. Nevertheless, we can summarize the basic photochemical reaction as follows:

$$\text{Volatile organics} + \text{NOX} + \text{Sunlight} = \text{Photochemicals}$$

These photochemicals create a variety of environmental problems. Direct effects on people and animals include irritation of eyes, lungs, and other mucous membranes. Plants may experience stunted growth and death. Photochemicals are examples of **secondary pollutants**, meaning that they are produced by reactions among other air pollutants. **Primary pollutants** are those that are directly emitted, such as SOX by coal-burning plants.

Smog originally meant a mixture of polluting smoke and fog, but it now refers to the brown haze of photochemical pollutants in urban areas that irritates membranes and reduces visibility. **Ozone**, or O_3, is the most abundant photochemical pollutant in smog. Oxygen is very reactive, so the O_3 molecule is particularly destructive, leading to much of the lung irritation in people and plant damage. Indeed, ozone alone is thought to be responsible for a large percentage of all air pollution damage to crops, with a total loss of hundreds of millions of dollars to U.S. agricultural productivity annually. Ground-level ozone damages the foliage of plants and is associated with increased susceptibility of plants to diseases, pests, and environmental stresses (such as lack of water or unusually harsh weather). The effects of ozone exposure may not be fully manifested for many years or decades, which means there is the potential for long-term forest and ecosystem decline that may not be readily apparent initially. The reactivity of ozone can also damage materials. For example, it is ozone that causes rubber products to deteriorate and crack. In highly urbanized areas, concentrations of ozone can be more than 10 times higher than the natural near-surface level.

(a)

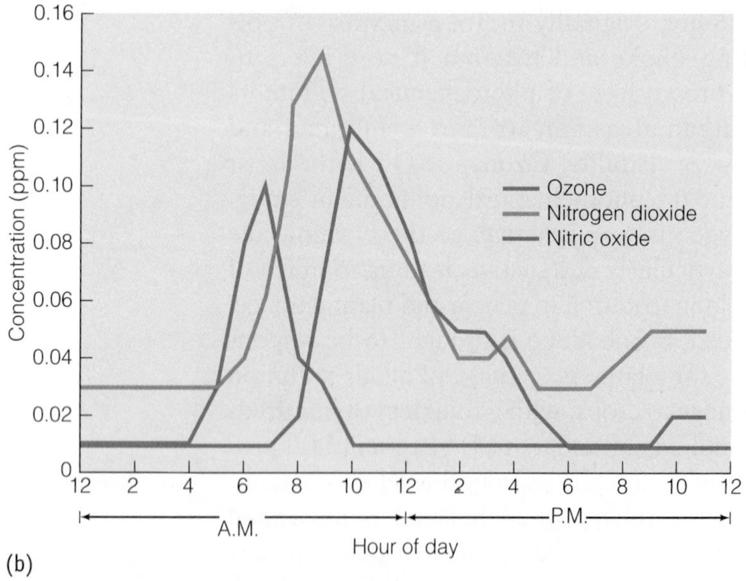

(b)

FIGURE 16-11 (a) Rush hour traffic is a familiar sight to nearly all city dwellers. Such traffic peaks are the main source of many health and other air pollution woes. Seen here is traffic in Buenos Aires, Argentina. (b) Typical concentrations of NOX and ozone in Los Angeles over the course of a day. Morning traffic causes a late-morning peak of ozone (*Source:* Data from U.S. Department of Health, Education, and Welfare).

In addition to their role in smog, nitrogen oxides contribute significantly to acid rain. Although sulfuric acid from SOX contributes about two thirds, nitric acid, formed from NOX and water, contributes about one third of the acid in acid rain.

Sources of Photochemical Pollutants

Nitrogen oxides and VOCs are largely produced by fuel combustion. NOX come mainly from fossil fuels burned in transportation and power generation; VOCs come from transportation and industry. NOX are formed when nitrogen in the air and in fuel combines with oxygen during combustion. VOCs are largely gasoline and other fossil fuel (hydrocarbon) molecules that are not completely burned during combustion.

Because so much NOX and VOC pollution is associated with transportation, urban smog varies considerably with traffic patterns. **FIGURE 16-11** shows how morning "rush hour" traffic in Los Angeles causes a peak of NOX in the early morning. Photochemical reactions on these oxides and VOCs take about 2 hours to produce the ozone peak in late morning.

Reducing Photochemical Pollution

Nitrogen oxides are formed from nitrogen in air and in the fuel during combustion; thus, a major method of reducing NOX has been to alter the combustion process. Because coal-fired power plants emit approximately 25% of the NOX pollution in the United States, emphasis has been placed on designing new technologies that control or reduce NOX production during coal combustion. The so-called low excess air process restricts the amount of air allowed into the combustion chamber. This oxygen-starved environment reduces the oxidation of nitrogen during combustion so that nitrogen in the fuel is released as the harmless N_2 gas. NOX emissions from coal combustion can be reduced by as much as 50% with this method.

Reducing NOX and VOCs from vehicle emissions by changing the combustion process has proved much more difficult than reducing power plant emissions. The problem is that NOX emissions are maximized under the same combustion conditions that minimize VOCs and most other pollutants.

This has been a continuing challenge for vehicle manufacturers, who are required by law to meet certain emission standards. The primary solution has been the development and installation of emission control devices that remove pollutant gases after fuel combustion occurs. These are discussed later in this text. However, a final source of VOC pollution should be mentioned: as much as 20% of an auto's emissions can come from evaporation of gasoline from the gas tank and carburetor. This source has been reduced through increasing use of vapor recovery systems in the tank and carburetor that capture and recycle the fumes through the engine. In addition, gas stations are increasingly using vapor recovery nozzles at the pumps.

Positive Trends in VOC and NOX Pollution

The growing use of vapor recovery systems, emission controls, and other methods has led to a general decline in VOC pollution since about 1970; between 1990 and 2005, VOC emissions decreased by 18%. Most of this decline has resulted from reduced transportation (vehicle) emissions. NOX emissions in the United States decreased by approximately 15% between 1983 and 2002, and by another 30% by 2009.

Carbon Monoxide

Carbon monoxide is a deadly gas in high concentrations, killing more than 300 people every year in North America. Its deadly effects are all the more dangerous because it is so difficult to detect: It is colorless, odorless, and tasteless. Carbon monoxide has little effect on plants or materials, but in people and animals, it interferes with the ability of the red blood cells to carry oxygen to the organs: The CO molecule attaches to the blood cells more readily than do oxygen molecules. Because the brain requires a very large amount of oxygen, dizziness, headaches, and mental impairment are among the first symptoms.

Concentrations as low as 50 ppm will cause headaches and other effects over the course of a few hours. All effects are felt sooner if you are active; the body's cells are depleting oxygen faster and are in greater need of it. This is one reason that jogging during rush hour traffic is often discouraged. Fortunately, the effects of CO are usually reversible, except for extreme exposure. As usual with air pollution, very young people, older people, and people with certain lung and heart ailments are most at risk.

Sources of Carbon Monoxide Pollution

Incomplete combustion when fossil fuels, wood, tobacco, and other organic materials burn under less-than-ideal conditions produces CO. As a result, the carbon is not fully oxidized to CO_2. For example, an engine with a poor air supply or low burning temperature will produce more CO than an engine that has a better air supply or burns at a higher temperature. Thus, the same high temperatures that cause nitrogen to form NOX will minimize CO.

Approximately 60% of CO emissions in the United States are from transportation, mainly vehicle exhausts; in American cities, 95% of the CO can come from automobile exhausts. Similarly, cars and trucks are the main source of CO in nearly all cities of the world. The rest comes primarily from fuel combustion at power plants and industry and solid waste combustion. It is ironic that so much attention is given to these sources while many cigarette smokers deliberately produce much higher levels: Directly inhaled cigarette smoke may contain more than 400 ppm CO, compared with only 5 to 100 ppm in urban air from very congested, stalled traffic. Even second-hand cigarette smoke may increase indoor CO to 20 to 30 ppm.

Reducing Carbon Monoxide Pollution

Control of CO is closely related to control of NOX and VOCs. Vehicle emissions are a major contributor of all three. There are three basic ways to reduce vehicle emissions: precombustion, postcombustion, or changing the combustion process. These methods have been successful in reducing CO emissions (and also VOCs) by 21% since 1993, despite a large increase in the number of vehicles on the road and an increase of vehicle miles driven. National average ambient air levels for CO decreased by 65% since 1983 but leveled in the 1990s. Scientists often correlate high levels of CO emissions to industrialization and lifestyles of overconsumption. In 2009, there was a noticeable drop in CO emissions of just more than 20%; some data suggest this decrease was caused by economic factors which led to less manufacturing and less consumption, not necessarily a permanent shift in consumer and societal behaviors.

Most automobile manufacturers have traditionally favored postcombustion control because it is easiest and most economical. Postcombustion control devices remove the pollutants from the exhaust gas before it is released into the air. Examples include thermal reactors, which are "afterburners," and exhaust recirculators, which divert the exhaust back into the engine. Both of these reduce emissions by further oxidizing the CO and VOCs and thus are less useful for reducing NOX for the reasons noted. In addition, they significantly reduce fuel economy and performance.

For these and other reasons, the auto industry's preferred postcombustion control is the catalytic converter. A "catalyst" is a substance that facilitates and speeds a chemical reaction. A **catalytic converter** is a device that carries out a number of chemical reactions that convert pollutants to less harmful substances. Because CO, NOX, and VOCs each have different chemistries, these reactions are quite complex. CO and VOCs must be oxidized to remove them, whereas further burning (oxidation) will only make NOX pollution worse. The solution has been to make "two-stage" catalytic converters. In the first stage, NOX is converted to harmless nitrogen gas (N_2) by removing oxygen. The second stage acts on CO and VOCs in the opposite way—by oxidizing CO to CO_2 and VOCs to CO_2 and H_2O. Both stages use finely ground particles of substances such as platinum as the catalyst.

Changing the combustion process is another way to control emissions. There are many alternatives to the standard gasoline internal combustion engine, and some of these produce fewer emissions. Diesel engines are currently the most popular alternative. Diesel engines inject fuel directly into a cylinder without a spark plug. By omitting the need for spark plug ignition, diesels are not only more fuel efficient, but also have very low emissions of CO and VOCs. On the negative side, diesel combustion in the cylinder produces high temperatures that create high levels of NOX pollution. Furthermore, high levels of unburned carbon, or "soot," are emitted. Such particulates can be mutagenic and carcinogenic. The NOX and soot problems can be reduced by exhaust recirculation and filters but have created maintenance problems. Another issue with diesel fuels has been the sulfur content, which can be 5,000 parts per million or more. Since 2006, Ultra Low Sulfur Diesel fuels (ULSD), with sulfur contents of 15 ppm or less, have been used in the United States, as required by the EPA under the Clean Diesel program. Mandates follow that railroad fuel will be ULSD by 2012 and marine fuels will need to comply by 2014.

Other alternative engines that have been marketed include the stratified charge engine, gas turbines, and the rotary engine. For economic and other reasons, they have not become popular with consumers, so car companies have tended to meet air quality laws with gasoline internal combustion engines linked to catalytic converters. Instead, the car of the future may be the electric car, which runs on a battery or fuel cell. By eliminating combustion in the car, electric cars can effectively eliminate CO, VOC, and NOX emissions in dense urban traffic. The net environmental impact of electric cars is largely determined by how the electricity to run them is generated. If a coal-fired plant is used, the net impact could be more of many pollutants because coal is not only a very "dirty" fuel, but much usable energy is lost in converting coal to electricity. Urban smog and CO pollution would be reduced, but concentrated air pollution would be shifted to the local area around the power plant. On the other hand, electricity generated by direct solar, wind, or other cleaner fuels would greatly reduce net air pollution everywhere.

Control of CO, NOX, and VOC emissions by precombustion methods is also growing. Examples include "reformulating" gasoline before it is burned and using alternative fuels such as methanol, ethanol, compressed natural gas, propane, or hydrogen. All have various advantages and disadvantages. A promising method of reformulating gasoline is **oxygenation**, which blends oxygen-rich liquids into it. By adding oxygen, CO is converted to CO_2 and VOCs are oxidized better.

Perhaps the most commonly discussed alternative fuels of the future are **methanol** ("wood alcohol") and **ethanol** ("grain alcohol"), which can be produced from natural gas, coal, or biomass. Most commercial methods for producing methanol now use natural gas. Methanol has a much higher octane rating than gasoline, allowing it to burn more completely and reducing CO and VOCs; it also burns at lower temperatures, reducing NOX. Ozone pollution is reduced by half. Methanol also has disadvantages, including higher emissions of formaldehyde; it is also an eye irritant and a possible carcinogen. A practical difficulty is its low volatility, which makes it difficult to start engines on a cold morning. Most important is its low energy content, which is about half that of gasoline; this means gas tanks would have to be twice as large to give cars the same mileage range. A solution is to use a mixture of 85% methanol and 15% gasoline, which eliminates the cold-start problem and improves mileage. Car makers are introducing "flexible fuel" cars that can run on either gasoline or a mix with methanol. In addition to reducing local and regional air pollution, methanol reduces global emissions of CO_2 but only if it is made from natural gas or biomass. If methanol is made from coal, overall CO_2 emissions may more than double using conventional processes, but new clean coal-to-methanol processes are being developed to circumvent these problems, as is methanol extraction from municipal waste. A conversion process from methanol to hydrogen is another alternative fuel source currently being used.

Ethanol is typically produced by the fermentation of agricultural products such as corn, sugar cane, soybeans, or sugar beets (see a discussion of renewable and alternative energy sources). Ethanol is easily used in "flexible fuel" cars that can use alcohol, gasoline, or a mixture of both. A well-tuned car running on ethanol or an ethanol/gasoline mixture (typically 85% ethanol and 15% gasoline) can give better performance and mileage with fewer emissions than a comparable vehicle running solely on gasoline. Flexible fuel vehicles have been used successfully in Brazil for the last 30 years.

Lead

Lead (Pb), like other metals, can accumulate in the body and is especially destructive to the nervous system. Lead poisoning can cause brain damage leading to learning disabilities, seizures, and death at high concentrations. Airborne lead particles are heavy and therefore occur close to the source such as in paint flaking off from a contaminated window frame that is frequently raised and lowered. About one sixth of inhaled lead contributes to poisoning: One third of all inhaled lead particles become trapped in the lungs, and the bloodstream eventually absorbs about half of those.

In 1977, the EPA estimated that 600,000 children in the United States had blood levels of lead that were near, or exceeded, those high enough to cause behavioral changes. Fortunately, human exposure to airborne lead has decreased more than 95% since 1970. The main reason for the reduction in overall lead poisoning cases was the gradual elimination of leaded gasoline during the 1980s and early 1990s. Instead of lead additives to reduce "knocking" in the engine, other anti-knock additives are now used. Nevertheless, unacceptably high atmospheric lead concentrations can still be found around some smelters and battery manufacturing operations. In addition, independent of atmospheric-borne lead, ingestion of lead from water in lead-contaminated pipes and lead-based paint continues to be a major health threat to children, especially in urban areas. Pipes and paint are now required to be relatively lead free, so the threat occurs mainly in older homes.

International Trends for Pollution Levels

In contrast to the United States, where most types of local and regional air pollution are declining, the worldwide total of the six criteria pollutants has generally been increasing. A main reason for these international trends is the increasing industrialization of developing nations and their use of fossil fuel technology. As coal-burning power plants, motor

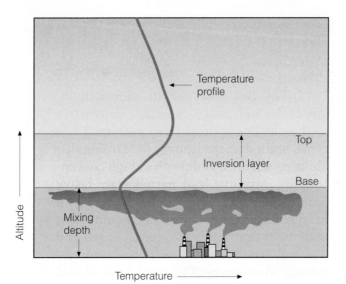

FIGURE 16-12 A temperature inversion. The warm inversion layer prevents pollutants from rising into the air above it.

vehicles, and industries become more common in places such as China, inevitably ever-greater amounts of air pollution are released (see Figure 16-3). The U.S. Department of Energy has estimated that the global fleet of automobiles, trucks, and buses could grow from the current 700 million to 1.1 billion by 2020.

There are many ways to reduce these trends. Most involve promoting industrialization that reduces or even eliminates fossil fuel use, which is the main source of most air pollution. Mass transportation can have a major impact by reducing the use of cars. For example, the average American annually produces more than twice as much carbon dioxide than does the average Japanese, in part because of the very low use of mass transit in the United States. Reformulated gasoline and more fuel-efficient technology will help. However, ultimately the most effective way to reduce fossil fuel pollution is the use of alternative fuels: electric cars, hydrogen fuels, wind power, and so forth. The widespread use of such green technologies in developing nations and their adoption in developed nations can simultaneously improve the environment and the economy.

Weather and Air Pollution

Local weather conditions can strongly influence the impact of air pollution. Tempera-
ture is especially important. For instance, cars release more air pollution when they are first started on cold winter mornings than in the summer. Because incomplete combustion causes much air pollution, a cold engine does not burn fuel as efficiently, and it releases greater amounts of CO and other pollutants. You may have noticed that car exhaust smells different on cold winter mornings. The most important local weather phenomenon that affects air pollution is **thermal inversion**. This occurs when a layer of warm air overlies cooler air, "inverting" the usual condition in which air becomes cooler as altitude increases (**FIGURE 16-12**). The layer of warm air, which can be several yards to several miles thick, acts as a trap for rising air pollution. Such inversions can last as long as several days but are most common at night. Low-lying valleys, such as the Los Angeles area, are most prone to inversions because it is easier for warm air to become trapped. Before air pollution controls were common, inversions could be quite disastrous in urban industrial areas. The most severe urban inversion episode in history occurred from December 5 to December 10, 1952, in London and caused 4,000 deaths. Air pollution controls have reduced the frequency of such disasters, but urban air quality is still generally at its worst during periods of thermal inversions.

In addition to weather affecting air pollution, the reverse can occur: Air pollution can affect weather. For example, industrialized urban areas tend to experience more rain and snow than surrounding areas because of the pollution they produce. Cities release great amounts of thermal pollution. The concentrated burning of fossil fuels and other energy sources, along with the concrete and metal that retain the sun's energy, turns cities into "heat islands" (**FIGURE 16-13**). Rising hot, moist air causes rain when it cools, and this effect is enhanced by the high amounts of particulate air pollution from cities. Particulates such as soot form tiny nuclei on which the cooling moisture can condense into droplets. In addition to affecting local weather conditions, air pollution can affect global climate. This obviously has a much greater potential effect.

16.2 Legal and Social Solutions

We have discussed many technological ways of reducing air pollution. Because most air pollution originates as fossil fuel combustion, these solutions have focused on those fuels. We have seen three ways to control combustion: (1) precombustion, such as higher fuel efficiency or cleaning coal before use; (2) post-combustion, such as scrubbers; and (3) changing the combustion process, such as in electric cars using fuel cells. We have also discussed how input reduction by using less fossil fuel, especially by using alternative energy technology, is often the most effective solution. However, technological reductions will not become widespread unless people have some incentive to use them. In the United States, legal and economic incentives have been used most often.

Legal Solutions

Legal incentives have been the main reason for the decline in many of the local and regional air pollutants since about 1970. Federal legislation actually began with the **Air Pollution Control Act of 1955**. This act mainly provided funds for research, but its enactment was a key event because it initiated federal participation in air pollution control. Until 1955, air pollution was considered a local problem and thus was treated by a hodgepodge of local and state laws that varied widely. The Clean Air Act of 1963 was the first of a series of acts and amendments that exerted increasing federal pressure on air polluters. However, not until the Clean Air Act Amendments of 1970 were widespread legal standards established that were actually enforced. These standards included the National Ambient Air Quality Standards, which established standards of allowable concentrations in surrounding air, and New Source Performance Standards, which established standards of allowable emissions or rates at which polluters could emit. The newly established Environmental Protection Agency was charged with the task of enforcing these standards.

Air Quality Standards

In applying National Ambient Air Quality Standards, the EPA established air quality standards for various major pollutants, such

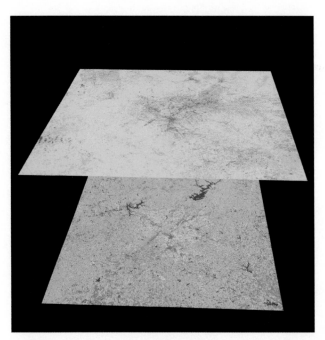

FIGURE 16-13 These satellite images show a temperature map over a land use map of Atlanta, Georgia, derived from the Landsat Thematic Mapper instrument. On the land use map, urban areas are dark gray, suburban areas are light gray, water is blue, and green or orange is vegetation. The strong correlation between urban and suburban areas and higher ground temperatures (oranges and reds) is easily seen. The city's temperature is often as much as 10° higher than that of the surrounding environment. This difference causes more thunderstorms and ground-level ozone events.

as CO, NO_2, O_3, SO_2, PM_{10} (particulate matter), and lead. These are easily measured, occur in significant amounts, and are health threats. They indicate the general "health" of the air. If one or more of these are present in large amounts, we can infer that other, associated air pollutants are also common. Allowable concentrations in California are often lower, illustrating how federal law may allow states to set stricter standards for air quality and many other environmental regulations.

For many years, the Pollutant Standards Index was used, particularly in cities, to monitor air quality. This index ranged from 0 (best air quality) to 500 (worst air quality). The Pollutant Standards Index was based on the major National Ambient Air Quality Standards (NAAQS) pollutants listed in the last paragraph. In 1999, the EPA revised and updated its air quality index, renaming it the **Air Quality Index** based on levels of ground-level ozone, particulate matter, carbon monoxide, sulfur dioxide, and nitrogen dioxide. Like the earlier Pollutant Standards Index scale, the Air

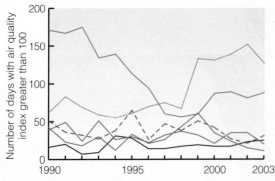

FIGURE 16-14 Air quality has been improving in a number of U.S. cities. (*Source:* Data are from the Environmental Protection Agency.)

Quality Index ranges from 0 (best quality) to 500 (worst). A value of 100 for a particular pollutant is equivalent to the level set for the pollutant as the national air quality standard, so levels below 100 are considered acceptable or safe (0 to 50 is generally considered "good"), whereas levels above 100 indicate various degrees of health concern. **FIGURE 16-14** shows that the air quality has been improving in a number of U.S. cities. An overall Air Quality Index value for a particular day at a given location corresponds to the highest value for any of the monitored pollutants. The current Air Quality Index includes a color-coding scheme (**TABLE 16-1**).

TABLE 16-1	The Air Quality Index	
Air Quality Index (AQI) Values	**Levels of Health Concern**	**Colors**
When the AQI is in this range:	**... air quality conditions are**	**...as symbolized by this color:**
0 to 50	Good	Green
51 to 100	Moderate	Yellow
101 to 150	Unhealthy for sensitive groups	Orange
151 to 200	Unhealthy	Red
201 to 300	Very unhealthy	Purple
301 to 500	Hazardous	Maroon

Emission Standards

The EPA sets New Source Performance Standards to limit certain industries' air emissions. The affected industries include fossil fuel power plants, incinerators, nitric acid plants, petroleum refineries, sewage treatment plants, and ore smelters. The standards are called "New Source" because they are directed at facilities that are to be constructed, not those that existed before the standards were set. This reflects the economic reality that it is usually much cheaper to design a new plant to produce lower emissions than to modify an existing plant that uses older technology. Of course, this begs the question of what to do with older facilities. For instance, by 1977 only about one third of the air quality control regions in the United States were meeting the set standards. This deficiency in the 1970 Clean Air Amendments led to the 1977 Clean Air Amendments, which attempted to deal with these so-called nonattainment regions.

Perhaps the most successful policy to result from these amendments was **emission offsets:** To receive a construction permit, a major new pollution source in a nonattainment area must first find ways to reduce emissions from existing sources. Emission offsets have been very efficient in economic terms because they use the free market to find optimal solutions. For example, someone seeking a construction license may install emission controls on a heavy polluter, even if someone else owns it. Alternatively, the license seeker may buy out the old plant and shut it down.

Clean Air Act Amendments of 1990

In 1990, the U.S. Congress passed a new set of Clean Air Act Amendments. These toughened air pollution controls for the first time since 1977. Some of the amendments' major provisions are listed in **TABLE 16-2**. These provisions focus on three types of local and regional air pollution that were considered inadequately addressed in earlier amendments:

1. Smog is a focus of provisions 1–6, as listed in Table 16-2. They were needed because nearly 100 major cities had not complied with existing regulations. The new amendments required lower tail-

TABLE 16-2	Three Main Air Pollutants, Their Sources, and Major Provisions of the 1990 Clean Air Act Amendments that Seek to Control These Sources*

Source	Provision
Smog	
1. Gasoline	"Reformulated" fuels required in major cities
2. Tailpipe emissions	Reduce NOX and VOC emissions by about half
3. Tailpipe emissions	Emission controls must last 100,000 miles
4. Tailpipe emissions	Low-emission vehicles to be developed
5. Gas stations	Vapors must be captured during refueling
6. Cities	Reduce smog until federal standards are met
Acid Rain	
7. Power plants	Reduce nitrogen emissions by one-third
8. Power plants	Reduce sulfur emissions by one-half
9. Power plants	Can exchange "pollution credits"
Airborne toxics	
10. Businesses	Reduce toxic air pollutants by 90%

*Federal law requires these provisions unless later congressional actions repeal them.

pipe emissions of NOX and VOCs by 1994 and even applied to small industrial sources of smog-producing chemicals, such as bakeries. Car makers were required to redesign pollution controls and encouraged to build electric and other so-called zero- or low-emission cars. Fuel manufacturers were required to produce methanol and the other "reformulated" gasolines. The EPA estimated the total cost of complying with these six provisions at as much as $12 billion yearly.

2. Acid rain is a focus of provisions 7–9. Their goal is to reduce SO_2 and NOX emissions from coal-burning power plants.

3. Airborne toxics (air pollutants that are dangerous in small amounts, such as airborne mercury, arsenic, and many carcinogens) are a focus of provision 10. In 1988, more than 1.2 billion kilograms (2.7 billion pounds) of toxic air pollutants were released in the United States. Only seven such pollutants were regulated until 1990; the new amendments regulated 189 kinds of toxics and more have been added since.

These amendments have been very controversial. They are costly to many businesses, local governments, and citizens. Car prices, fuel prices, local taxes, and business operating costs increased as the provisions were implemented. Some employers who had too many employees driving to work alone rather than car pooling or taking mass transit were fined. As a result, the amendments have encountered considerable resistance from the public and local and state governments. On the other hand, some employers used the amendments in a positive manner, offering parking vouchers or other incentives, such as prime parking spaces, to car poolers.

Economic Solutions

The atmosphere is a classic example of an environmental "commons." It is a shared resource that people will exploit or pollute as long as they are free to do so: Polluters gain many benefits while paying few of the costs of damage to the commons. Legal restrictions are one method of reducing pollution of a commons. Another method is to make polluters pay more of the true costs of damage to the commons. This can be done in many ways, such as taxation, fees, and other government charges on emissions or amount of fuel burned.

Many economists favor economic solutions because they impose costs on pollution and then allow the market to find a cost-effective way of paying for them. We have already discussed how emission offsets do this. The 1990 Clean Air Act Amendments also created a free-market trading system for emission offsets. Using what is known as a cap-and-trade system, Congress identified 110 utilities and assigned each a cap. The cap was based on the utility's history of fuel use. The utility was then granted a limited number of SO_2 emission permits. **Emission permits** allowed the utility to emit a certain quantity of SO_2 into the atmosphere in a particular year. The companies were free to buy, sell, or trade these allowances. Thus, a plant might pay for more efficient emission controls and sell its allowances to a plant that would rather buy the "right to pollute." The idea of a "right to pollute" may seem unattractive, but this method will keep the net regional air pollution below any predetermined limit in an efficient way. The first use of the

1990 Clean Air Act Amendments allowances occurred in May 1992, when the Tennessee Valley Authority bought the pollution rights to 9,100 metric tons (10,000 tons) of SO_2 emissions from Wisconsin Power and Light. In the mid-1990s, the EPA's Acid Rain Program was implemented, setting a cap of 8.1 million metric tons (8.95 million tons) on the total amount of SO_2 that may be emitted by power plants across the country (this is about half the amount emitted in 1980) combined with an emissions trading program.

In 2002, the Bush Administration proposed the **Clear Skies Initiative**, which led to the Clear Skies Act of 2003, which sought to expand the use of sellable pollution permits among industries. Industries strongly favored this initiative, but many environmental groups were strongly opposed to it. They argued that by weakening regulations, the initiative would allow the production of many more tons of smog-producing gases (such as nitrogen oxides) and create large legal loopholes that would allow highly polluting factories to delay implementing pollution controls. These groups also voiced concerns about increases in mercury and other pollutants.

In contrast to tradable emission offsets, which make polluters pay for their outputs, are gasoline, coal, British Thermal Units, and other "green" taxes that make polluters pay for the amount of polluting fuel burned. These "input reduction" solutions are the most efficient means of all because they not only reduce pollution but also conserve resources. A polluter who has to pay higher prices for fuel will use it more efficiently and consume less.

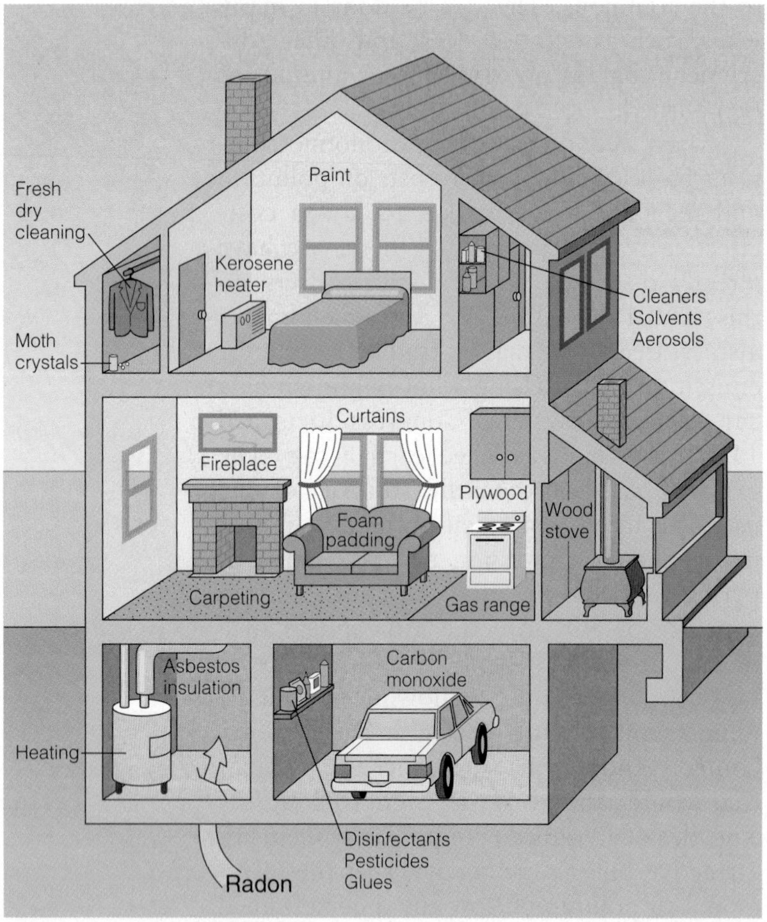

16.3 Indoor Air Pollution

Indoor air pollution is usually a greater direct threat to human health than outdoor air pollution. The cost of health problems from indoor air pollution in the United States is estimated in the tens of billions of dollars. Although there are many kinds of indoor air pollutants, just two, radon and smoking, cause the vast majority of harm.

Indoor air pollution is a major threat to human health for two main reasons. First, the indoor environment tends to concentrate pollutants. Some toxic and cancer-causing pollutants can reach air concentrations that are 100 times greater than outside air. Second, on average, people in industrialized societies spend more than 80% of their lives indoors, including time spent in offices on the job. **Sick building syndrome** is a popular medical term that refers to chronic ailments such as headaches, nausea, allergic reactions, and other symptoms that are caused by indoor air pollutants where we work or live.

Sources and Types of Indoor Pollutants

The sources of indoor air pollution can be divided into three categories: underground diffusion, combustion, and chemical emissions (**FIGURE 16-15**). In nearly all cases, the harmful effects of indoor pollution can be greatly reduced by increasing ventilation in a building. If the ventilation is done properly, it will not significantly reduce the heating (energy) efficiency of many buildings.

FIGURE 16-15 Sources of indoor air pollution in the home. (*Source:* The Environmental Protection Agency.)

Radon: Death by Diffusion

Radon gas is among the most harmful indoor pollutants when found in high concentrations. It generally enters the home through underground diffusion. Radon is a radioactive decay product of uranium, so it is most concentrated in houses built on soil and/or rocks that are rich in naturally occurring uranium minerals and their products.

According to a National Research Council study, radon and related radioactive gases are linked to an estimated 15,000 to 22,000 deaths per year in the United States. These deaths occur mainly from lung cancer that is caused by inhaled radioactive particles that become lodged in the lungs. Radon testing of homes is now common, but efforts to establish the true danger of radon have led to considerable controversy. Most U.S. homes are built on soils that emit relatively harmless amounts of radon, and proving the cause of lung disease is often very difficult.

When underground diffusion of radon is found to be a danger, there are a number of relatively simple and inexpensive ways to reduce indoor concentrations greatly. The precise methods depend on the type of building construction, but most involve increasing ventilation air flow through the building. Increasing air flow through basements, crawl spaces, and other lower parts of buildings is especially effective. For example, large window fans can be used to blow air through basement windows. Another common method is to insert pipes into the basement floor. The pipes draw the radon from the soil and ventilate it to the outside.

Combustion: Especially Smoking

Stoves, fireplaces, and heaters are common appliances that emit air pollution from combustion. Carbon monoxide is a poison caused by incomplete combustion; it can cause headaches and even be lethal in closed areas where stoves or heaters are not functioning properly. Certain cancer-causing chemicals, such as hydrocarbons, are also released by combustion. These problems are relatively minor when appliances are functioning efficiently and the building is well ventilated. However,

in developing countries, where people must often cook and heat with wood or coal, particulate air pollution from ash particles in the home is a major problem. A study by the World Health Organization estimated that indoor particulate levels were more than 20 times higher than safe levels in many developing countries where wood is burned in the home.

The most harmful combustion source of indoor air pollution is tobacco smoke (**FIGURE 16-16**). Many medical studies have found that smoking causes 443,000 to 500,000 deaths per year in the United States, especially from heart disease and cancer. Furthermore, there is strong evidence that nonsmokers are also affected: Second-hand smoke causes an estimated 3,000 to more than 50,000 deaths per year in the United States (numbers vary

FIGURE 16-16 Cigarette smoke is one of the most hazardous indoor air pollutants.

widely, depending on the study cited—second-hand smoke is not only implicated in lung cancer, but also heart disease, asthma, bronchitis, and other diseases, especially in children). In addition to carcinogens, tobacco smoke contains CO, particulates, and other harmful substances. Finally, smoking can enhance the deadly effects of radon. Of 15,000 to 22,000 radon-related deaths per year, only approximately 1,200 to 2,900 are among non-smokers. Smoke particles trapped in the lungs act as attachment sites for radioactive radon particles, increasing the chances for lung cancer. This is a classic, if deadly, example of a synergism where two effects enhance one another.

Chemical Emissions: Source of Many Pollutants

Chemical emissions can come from dozens of sources indoors. Many of these chemical emissions are carcinogenic but only in concentrations that are generally much higher than those that occur in most homes. However, they do often cause headaches, dizziness, and nausea, especially in people with a high sensitivity to certain chemicals.

Mothballs, bleach, shoe polish, cleaning solvents, air fresheners, and many other chemical products emit gases that can be toxic if sufficiently concentrated by poor air circulation. Many materials release carcinogenic fumes, such as benzene and formaldehyde. Plastics, synthetic fibers, and cleaners emit benzene; formaldehyde is emitted from foam insulation, wood products, such as particle board and plywood, and some glues. Asbestos has received much notoriety as a cause of lung cancer and the lung ailment asbestosis. Once used for fireproofing, insulation, tile, and cement, this fibrous material is now being removed from old buildings at great expense. The need for much of this great cost is hotly debated, in part because only some types of asbestos are harmful (see **CASE STUDY 16-2**).

16.4 Noise Pollution

Noise is often defined simply as "unwanted sound." Usually, it is unwanted because the sound is either too loud for comfort or is an annoying mixture of sounds that distracts us, such as a distant conversation. However, noise is partly subjective and depends on one's state of mind and hearing sensitivity: This subjective aspect is important because it adds to the difficulty of formulating policies to control noise. Some people are not as concerned about certain noises as others and are unwilling to pay for controls. For example, spending money for airport noise reduction is often controversial. Still, noise can be harmful to other life forms and often in ways that we are not able to measure. It may disrupt resting patterns or interfere with feeding.

Noise is a kind of air pollution because it is usually transmitted to our ears through the air that surrounds us. Indeed, when we hear any kind of sound, noise or otherwise, our eardrums are receiving vibrations being transmitted to us by colliding air molecules. Air molecules thus transmit sound waves in much the same way that water molecules transmit ocean waves. Whereas a large storm at sea provides the energy that sets the water waves in motion, a large explosion, machinery, or some other source provides the energy that sends the sound waves toward you.

Measuring Noise

Loudness increases with intensity, meaning the amount of energy carried by sound waves. Loudness is measured using the **decibel (dB) scale**. (*Deci* refers to the units of 10 used, and *bel* refers to Alexander Graham Bell, who was a teacher of the deaf as well as the inventor of the telephone.) The scale is exponential in that each unit of 10 increase in sound level (dB) is a 10-fold increase in sound intensity, or energy. The scale ranges from 0 to more than 180, for which 10 is the sound made by a leaf rustling and 180 is a rocket engine. Most daily noises in a busy building or on a city street average around 50 to 60 dB; in a quiet room, they are about 30 to 40 dB.

Health Damage from Noise

The most obvious health damage from loud noises is hearing loss. This occurs from physical damage to the fragile mechanisms in the ear. The damage can occur very rapidly from very loud noises or slowly from long-term exposure to moderately loud noises. Hearing damage begins around 70 dB for long

Is Asbestos a Classic Example of "Environmental Hypochondria"?

VISIT

http://environment.jbpub.com
/mckinney/5e/
for more information

Despite the enormous scope and variety of environmental issues, a few have generated much more intense controversy than many others. One that has inflamed heated passions is the exposure of people to asbestos (**FIGURE 1**). Some forms of asbestos can be quite dangerous to people; for instance, exposure to asbestos can cause mesothelioma, a cancer of the membranes around the lungs and internal organs that can lead to death. If asbestos is so dangerous, why not simply ban it? Indeed, a number of countries have done just that, including Argentina, Chile, Croatia, Iceland, Latvia, and Saudi Arabia.

However, asbestos has many uses, and some people argue that the use of asbestos actually protects and saves more lives than it harms. Another major reason for the debate over asbestos is that some people think the incredibly high economic cost of eliminating all asbestos currently in use—more than $100 billion in the United States—does not clearly yield benefits that are worth the cost. In addition, if asbestos is banned, substitutes will have to be found for many current asbestos uses, and such substitutes may be more expensive than asbestos. Indeed, critics argue that most types of asbestos pose a much smaller hazard than many daily activities, such as walking across a street. They argue that asbestos is a classic example of "environmental hypochondria": an illusory environmental hazard produced by anxiety.

"Asbestos" was a term originally used mainly by geologists to refer to fibrous minerals found in certain metamorphic rocks. Because many of these fibers are heat resistant, asbestos became widely used, especially during the 1950s and early 1960s, for "fireproofing" in building materials. Mines, shipyards, schools, houses, offices, and many structures used asbestos to reduce the chance of fire. Asbestos is also an excellent material for such high-temperature uses as brake pads on vehicles and engine gaskets. In the early 20th century, it was realized that people working directly with asbestos on a regular basis experienced health problems, and by the early 1970s, there was significant scientific evidence that simply breathing asbestos was harmful. The primary problem is that some kinds of asbestos are very crumbly and readily disperse into the air. When inhaled, the needle-like asbestos fibers cannot be broken down in the lungs, so they remain as a constant irritant. In these cases, the body's reaction to this irritation over 10 to 40 years can produce stiffening of lung tissue, which is called "asbestosis." This makes breathing difficult and can lead to lung cancer.

These findings resulted in many lawsuits against makers of asbestos products, costing many billions of dollars in legal fees. Billions more dollars have been spent removing asbestos from buildings. The costs are especially burdensome for the many public school districts with limited funds that must remove asbestos from school buildings. In 1989, the U.S. EPA attempted to ban all uses of asbestos but was unsuccessful in large part because of pressure from the asbestos industry and asbestos users. New uses of asbestos were banned, but older uses were allowed to continue with regulation. Today, asbestos remains in use in the United States in various roofing materials, gaskets, brake pads, and miscellaneous other uses. Another problem is that various levels of asbestos have been found as a contaminant in other (supposedly asbestos-free) products, such

FIGURE 1 A worker wearing a protective suit and mask prepares for asbestos abatement.

Is Asbestos a Classic Example of "Environmental Hypochondria"?

as shingles and insulation, cardboard, crushed rock used in landscaping, fertilizers, and many mineral products in which asbestos may naturally occur in very low concentrations.

Is the energy, time, and money spent so far addressing the asbestos issue worthwhile? For decades this issue has been hotly debated. As the 1990s progressed, there were a growing number of critics of the "antiasbestos" campaign. In 1995, *U.S. News & World Report* stated that there is "now a broad consensus among scientists and physicians that asbestos in public buildings is not much of a threat to health" (February 20, p. 61). A 1995 study by the U.S. Office of Technology Assessment said much the same thing. A 1990 EPA report (the "Green Book") stated that asbestos removal is often not a building owner's best course of action—it is better in some cases to stabilize and cover the asbestos so that it does not enter the air or cause other damage, rather than attempt to remove it.

A primary reason for this rethinking of the asbestos threat was that more detailed research indicated that most of the asbestos in buildings is not the very dangerous variety. There are many kinds of asbestos, including two very different mineral groups. Most of the proof for toxicity centers on the "amphibole" type of asbestos. For example, one study showed that of 33 factory workers who breathed large doses of this type of asbestos in 1953, 19 workers had died of asbestos-related illness by 1990.

However, approximately 95% of the asbestos used commercially is not of this variety. Instead, it belongs to a second mineral group, the "chrysotile" type of asbestos, which evidently is not nearly as toxic. It is very difficult to gather toxicity data because no one can measure exactly how much exposure each worker received, but the only clear proof that chrysotile asbestos is harmful is when it is breathed in industrial settings with exceptionally high levels for long periods of time. The result is that many voluntary risks that people want to engage in are apparently much more harmful than this commonly used asbestos. The Harvard University Energy and Environmental Policy Center estimated that cigarette smoking or walking across the street is more than 200 times more dangerous than asbestos in school buildings. A person is three times more likely to be struck by lightning than to be harmed by asbestos in school buildings.

In recent years, concern about asbestos exposure and the fight to ban all asbestos in the United States have been renewed. In June 2002, Senator Patty Murray (Democrat, Washington State), with the endorsement of the Sierra Club, introduced the Ban Asbestos in America Act into the United States Senate. If it is passed in its proposed form, Senator Murray's bill will not only ban asbestos, but will require EPA studies of asbestos-containing products (including asbestos-contaminated products), additional research into diseases caused by asbestos exposure, a public education campaign focusing on the dangers of asbestos exposure, and a review of the health effects of not only asbestos, but also "unregulated minerals," along with a review of relevant current laws and regulations established for the protection of workers and consumers. In 2003, the bill was referred to the Committee on Environment and Public Works and unfortunately died in committee in 2007. However, internationally over 60 countries have completely banned the use of asbestos. From 1968 to 2009 approximately 9,000 people in the United States lost their lives to asbestosis. Worldwide, approximately 107,000 people die annually from asbestos-related illness.

Questions

1. How serious do you think the asbestos risk is? Is it something you worry about? How have perceptions of the "asbestos risk" changed with time? Do you think asbestos should be banned in the United States? Justify your answers.

2. People often perceive more risk in an activity when they have little control over it. Does this explain why asbestos may be perceived as being riskier than it may really be? Or do you agree with the most severe assessments of the risks of asbestos? Explain.

3. People also tend to perceive more risk when they cannot observe the threat. Is the asbestos threat observable? Explain.

4. List and discuss some other environmental problems that might be criticized as examples of "environmental hypochondria." How do you view the opinion that there is no such thing as "environmental hypochondria" because all environmental and health issues are important and we should not "put a value on human life"?

exposure to a sound such as a loud vacuum cleaner. Outside noises at this level result in widespread complaints by citizens. At about 130 dB, irreversible hearing loss can occur almost instantaneously. Noise can also lead to a variety of ailments that arise from stress and anxiety induced by loud or distracting noises. Such stress-related ailments can be psychological, such as emotional trauma, or physical, such as headaches, nausea, and high blood pressure. Many studies have shown that such stress-related effects reduce productivity in the workplace.

Control of Noise

Noise can be reduced in three ways: (1) at the source, (2) as it travels to the person, (3) by protecting the person. Reduction at the source is often the easiest and most cost-effective means, reflecting the old engineer's adage that "noise means inefficiency." This is because vibrations, friction, and other signs that equipment is not functioning efficiently often cause noise. The solution is to design machinery that minimizes vibrations and friction, such as reducing the number of moving parts. Similarly, the machinery needs to be well maintained, lubricated, and so forth. If equipment still produces loud noises after these steps are taken, as often occurs with industrial machines, the other two methods must be used.

Sound-absorbing or **acoustical materials** achieve a reduction of noise as it travels through the air to the person. Examples include tiles and baffles made of wool or many kinds of synthetic materials that do not transmit sound vibrations efficiently. Car mufflers and jet engine noise deflectors are examples of devices made to reduce noise after it has left the engine source.

The third method of noise reduction, protecting the person, usually employs earplugs or earmuffs. These are made of materials that absorb sound waves before they reach the eardrum—not cotton, which provides little protection when inserted in the ear. Earplugs, and especially earmuffs, can greatly reduce noise to 50 dB. Their efficiency depends heavily on an "acoustical seal," meaning that they must be inserted tightly, with as little exchange of air into the ear as possible.

16.5 Electromagnetic Fields

The flow of electricity through wires produces an **electromagnetic field** (EMF) that can extend through the air for many feet (**FIGURE 16-17**). Concern over the health effects of such fields, in particular the possibility that they may cause cancer, has been growing since the late 1960s. However, numerous studies have yielded conflicting results, so the perceived threat is a matter of controversy. The Electric Power Research Institute and the Congressional Office of Technology Assessment's extensive studies carried out in the late 1980s concluded that health effects of electromagnetic fields could not be proved or ruled out. More recent studies have come to similar, nonconclusive results.

U.S. utility companies have spent more than $1 billion on such studies, with no conclusive evidence of harm yet shown. These and similar high costs for other controversial threats, such as radon and asbestos, are a main reason that the 1995 Congress showed strong interest in benefit–cost analyses. Such analyses try to determine true risks objectively and to estimate appropriate benefits for money spent on them, but many environmentalists note that benefit—cost analyses are often flawed because of the difficulty of accurately measuring benefits and risks. They argue that unforeseen harm could result if

FIGURE 16-17 Power transmission lines produce enormous EMFs as electricity flows through them.

spending on EMF and other potential threats is reduced too soon.

Statistical studies, which test for a correlation between cancer and people exposed to high EMF, usually show a weak correlation or none at all. Biological studies, which directly examine the effects of EMF on living cells and tissues in the laboratory, produce similar results. In some cases, biochemical regulation of cell multiplication does seem to be affected, possibly from changes in the flow of electrically charged calcium ions through cell membranes. The dose–response relationship is clearly complex and affected by many variables:

- Frequency and wavelength of the EMF
- Duration of exposure
- Kind of cells exposed and type of organisms
- Orientation of the EMF with the Earth's electromagnetic field

Considering that individuals of nearly all species vary in their susceptibility to environmental stresses, it will likely be very difficult to precisely determine EMF effects, if any.

One of the most interesting and useful findings so far is that fields generated by toasters, electric blankets, and other indoor electrical equipment may be a greater threat than the high-voltage transmission lines that receive so much publicity. We spend much more time in close proximity to these indoor appliances, and they can sometimes generate a surprisingly strong electromagnetic field.

■ study guide

SUMMARY

- Air pollution is both a health issue and an economic issue—air pollution costs tens of billions of dollars a year in the United States; in the United States, more than 100 million people live in areas where the air is considered unhealthful for some portion of the year.
- Currently, the U.S. EPA recognizes six basic criteria pollutants: particulate matter, sulfur dioxide, nitrogen dioxide, ozone, lead, and carbon monoxide.
- Volatile organic compounds (VOCs) are another important class of pollutants.
- Most air pollutants in the United States have declined during the last 30 years (exceptions are ground-level ozone and nitrogen oxides in some areas).
- Transportation is the largest source of carbon monoxide, nitrogen oxides, and VOCs.
- Coal-burning power plants in particular are a major source of sulfur oxides.
- Sulfur dioxide is the major cause of acid rain, which can cause damage to both biological (such as natural ecosystems) and physical materials (for instance, degradation of building materials).
- Photochemical pollutants, commonly known as smog, are produced by chemical reactions among nitrogen oxides, VOCs, and other components of the air.

- Ozone, the most abundant photochemical pollutant in smog, is very reactive and destructive, causing damage to human health, crops, natural ecosystems, and physical entities (such as causing rubber products to crack).
- Carbon monoxide, a gas dangerous to human health, can be produced by incomplete combustion, such as by fossil fuel-powered vehicles.
- Vehicle emissions can be addressed through such means as catalytic converters, alternative engine designs, and alternative fuels that burn more cleanly.
- Lead poisoning can cause brain damage, especially in children, and other disabilities; however, with the elimination of leaded gasoline more than a decade ago, airborne lead levels have decreased by more than 95% in the United States.
- In the United States, the Clean Air Act and its amendments have gone a long way toward addressing outdoor air pollution.
- Indoor air pollution, caused by sick building syndrome, radon, or tobacco smoke, can be a threat to human health in specific instances and is usually addressed on an individual basis.
- Noise pollution, "unwanted sound," can also be an issue in specific cases.
- A controversy remains about whether electromagnetic fields produced by electricity flowing through wires pose a health hazard.

KEY TERMS

acid rain
acoustical materials
Air Pollution Control Act of 1955
Air Quality Index
alkaline
baghouses
catalytic converter
Clean Air Act
Clear Skies Initiative
criteria pollutants

cyclone collector
decibel (dB) scale
electromagnetic field (EMF)
electrostatic precipitators
ethanol
emission offsets
emission permits
methanol
noise
oxygenation
ozone

particulates
photochemical pollutant
pH scale
primary pollutants
scrubbers
secondary pollutants
sick building syndrome
smog
thermal inversion

STUDY QUESTIONS

1. Distinguish among global air pollution, local and regional air pollution, and indoor air pollution, giving examples of each.
2. Name the six basic criteria pollutants as currently recognized by the U.S. EPA. What are the two main sources of these pollutants?
3. What are some key patterns and historical trends related to criteria pollutants in the United States since about 1970?
4. What are particulates? List some of the terms used to describe them. Why are they the most damaging pollutant to lungs?
5. Where does most acid rain occur in North America? Why does this pattern exist?
6. Describe the methods by which sulfur oxides can be reduced.
7. What does 1 on the pH scale mean? What is the average pH of rainfall? Which fish are most sensitive to acid waters?
8. What is a photochemical reaction? How does smog form?
9. Distinguish secondary pollutants from primary pollutants.
10. What are VOCs, and why are they important?
11. Distinguish ground-level ozone from stratospheric ozone. Which is "good" and which is "bad"?

12. Why can carbon monoxide be a major health hazard to people?
13. What are some basic ways to address carbon monoxide pollution?
14. What is a thermal inversion? Why, when combined with local air pollution, can it have deadly results?
15. What does a value of 100 on the Air Quality Index (AQI) mean?
16. What did the National Ambient Air Quality Standards (NAAQS) and the New Source Performance Standards (NSPS) establish?
17. What was the focus of the 1990 Clean Air Act Amendments? Why are the 1990 Clean Air Act Amendments so controversial?
18. Why is indoor air pollution a generally greater direct threat to human life than smog and many other forms of air pollution?
19. What are the three main categories of sources of indoor air pollution? How can the harmful effects of these sources be reduced?
20. What is noise? Why can it be considered an air pollutant? Discuss ways that noise can be reduced.
21. Describe the controversy over electromagnetic fields (EMFs).

WHAT'S THE EVIDENCE?

1. The authors contend that the most harmful combustion source of indoor air pollution is tobacco smoke. What is the evidence for this statement? Do you agree with this assessment? Why or why not? Are you a smoker? Do your personal habits relative to smoking or not smoking influence your position on the issue?
2. Suppose you have a friend who lives in a house directly adjacent to high-voltage transmission lines. One day you are in a conversation with your friend, and she expresses concern that the electromagnetic field (EMF) from the power lines might adversely affect her health. What would you say to her? Is there definitive evidence that EMFs from power lines are harmless? Is there strong evidence that such an EMF can be detrimental to human health? Should she worry? Would you worry if you were in a similar situation?

CALCULATIONS

1. On the decibel scale, how much greater is the sound intensity (energy) in a noise of 50 dB compared with a noise of 10 dB?
2. On the decibel scale, how much greater is the sound intensity (energy) in a noise of 60 dB compared with a noise of 40 dB?
3. On the pH scale, a substance such as lye (pH 13) is approximately how much more alkaline (basic) than ammonia (pH 11)?
4. If the global fleet of automobiles, trucks, and buses increases from a current 700 million to 1.1 billion by the year 2020, what percentage increase will this represent over this period?

ILLUSTRATION AND TABLE REVIEW

1. Referring to Table 16-1, what level of health concern is indicated by an Air Quality Index (AQI) score of 85? Of 175?
2. Based on Figure 16-11 (Los Angeles traffic), what time of day do nitrogen dioxide levels in Los Angeles peak? How about ozone levels? When is not a good time to go jogging in Los Angeles?

http://environment.jbpub.com/mckinney/5e/

http://environment.jbpub.com /mckinney/5e/

Connect to this text's website at http://environment.jbpub.com/mckinney/5e/. This site features eLearning, an online review area that provides quizzes, chapter outlines, and other tools to help you study for your class. You can also follow useful links for more in-depth information.

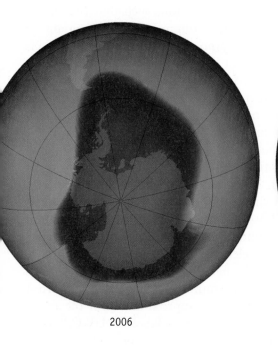

2006

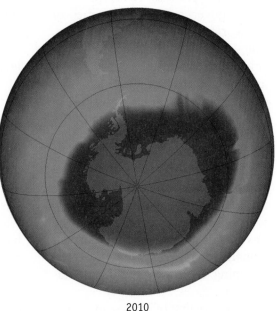

2010

In 2008, the Earth's ozone hole was the fifth largest on record. Since 2000, the overall amount of ozone depleting substances decreased nearly 3.8%. The largest ozone hole ever recorded was 10.6 million square miles, recorded in 2006. The ozone hole in the stratosphere over the Antarctic occurs at the beginning of Southern Hemisphere spring (August). It typically reaches its maximum size in late September or early October. In the following months, the ozone hole begins to break up, dissipating in smaller regions throughout the southern mid-latitudes (including parts of South America, Australia, New Zealand, and South Africa). Human health can be negatively affected by a thinning ozone layer. Plants and other animal species are also affected. Since the 1970s, NASA has been monitoring the status of the ozone layer through satellite observations, see: http://ozonewatch.gsfc .nasa.gov.

Global Air Pollution: Destruction of the Ozone Layer and Global Climate Change

17

Some environmental problems, by their very nature, are global in scope. Perhaps the two most immediate environmental threats that we face as a global community are (1) climatic change, caused by the artificial introduction of large amounts of greenhouse gases into the atmosphere, and (2) abnormally high incidences of ultraviolet (UV) **radiation** on the surface of Earth caused by the destruction of the ozone layer. These problems are interrelated: Both are caused by the introduction of large quantities of human-produced gases into the atmosphere, and in many cases, the same gases are implicated in both problems. For instance, chlorofluorocarbons (CFCs) are not only a major culprit in the destruction of the ozone layer but also a very effective greenhouse gas.

The story of the human damage to the ozone layer now appears to be drawing to a close (although the story is not yet quite over—ozone-depleting compounds continue to be detectable in the atmosphere for years to come). The problem was discovered

Chapter Objectives

After reading this chapter, you should be able to explain or describe the following:

- The importance of stratospheric ozone
- How ozone is created and destroyed in nature and how it can be damaged by human activities
- A brief history of the Montreal Protocol and subsequent international agreements

501

Chapter Objectives
(continued)

- The effects of increased ultraviolet radiation on the surface of Earth
- The greenhouse effect and the gases that cause it
- Evidence that Earth is warming
- Potential consequences of global climate change
- Strategies for dealing with global climate change
- The concept of light pollution

about 1974, and by 1990, all of the major governmental authorities of the world had agreed on a solution. For the first time in history, political leaders, scientists, corporations, and consumers worked together to find an international solution to a global problem that potentially threatened all life on the planet. The "ozone story" sets a precedent as the world tackles even more complex environmental issues. We now know that nations can work together if they must. Perhaps the next major global problem that will be systematically and successfully addressed is global warming caused by the accumulation of greenhouse gases.

In this chapter, we first tell the ozone story. Then we address the issue of global warming.

17.1 Ozone Depletion

Ozone in Nature

Ozone (O_3) is an important natural component of the **stratosphere**. (Ozone is also an important pollutant at ground level; review discussion of local and regional air pollution.) Ozone occurs in scant amounts between about 10 and 50 kilometers (6 and 31 miles) above sea level. The **ozone layer**, where the ozone is most strongly concentrated, is at an altitude of 20 to 25 kilometers (12 to 16 miles)—ozone is formed in the stratosphere (**FIGURE 17-1**) when high-energy **ultraviolet** (UV) **radiation** splits normal oxygen molecules (O_2) into atomic oxygen (O). The atomic oxygen may then combine with a standard diatomic oxygen molecule (O_2) to form triatomic ozone (O_3). Under natural conditions, ozone is also removed from the atmosphere by various reactions. An ozone molecule can absorb UV radiation and split into O_2 and O. In nature, excluding human interference, a dynamic equilibrium exists between ozone production

and ozone destruction, such that the stratosphere always contains a small amount of ozone. The amount of stratospheric ozone is so small that if it were all brought down to sea level, it would form a blanket over the surface of Earth only 3 millimeters (0.118 inches) thick, or slightly more than the thickness of two dimes.

Nevertheless, this stratospheric ozone is essential to the preservation of current forms of life on Earth's surface. The ozone layer acts as a shield that absorbs biologically dangerous UV-B radiation (**FIGURE 17-2**). UV radiation is commonly divided into two bands: UV-A has wavelengths of 320 to 400 nanometers, and the higher energy UV-B has wavelengths less than 320 nm. Sometimes the very shortest wavelength UV, approximately in the range of 200 to 280 nm, is labeled "UV-C," as in Figure 17-2. When an ozone molecule is hit by UV-B wavelengths, it absorbs the radiant energy and photodissociates into O_2 and O while giving off heat. This keeps the UV-B and UV-C radiation from reaching Earth's surface and injuring or killing nearly every living thing on the planet. The layer also creates a temperature inversion in the stratosphere that helps to maintain relatively stable climatic conditions on and near the ground.

Human Damage to the Ozone Layer

The stratospheric ozone layer has remained in a dynamic equilibrium for much of geological time, but by the early 1970s, scientists had discovered evidence suggesting a decline in the ozone concentration because of human interference. In particular, people had been releasing enormous quantities of ozone-destroying substances into the atmosphere, especially a class of chemicals known as **chlorofluorocarbons** (CFCs) that found important uses as solvents, aerosols, fire retardants, and refrigerants.

Other human-produced substances that can attack stratospheric ozone include

High-energy UV radiation — Oxygen molecule — 2 Oxygen atoms   Oxygen atom + Oxygen molecule ⇌ UV radiation → Ozone molecule

FIGURE 17-1 Schematic representation of the formation of stratospheric ozone. Ozone forms when a free oxygen atom, released by the splitting of an oxygen molecule by high-energy UV radiation, combines chemically with another oxygen molecule.

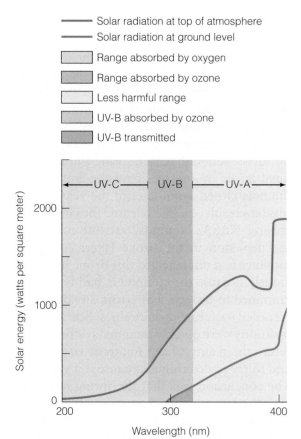

FIGURE 17-2 Absorption of UV light by the atmosphere. (Wavelength is in nanometers; one nanometer equals one billionth of a meter.) (*Source:* Based on data from D. H. Meadows, D. L. Meadows, J. Randers. Beyond the Limits. Post Mills, VT: Chelsea Green, 1992, p. 146.)

Legend:
- Solar radiation at top of atmosphere
- Solar radiation at ground level
- Range absorbed by oxygen
- Range absorbed by ozone
- Less harmful range
- UV-B absorbed by ozone
- UV-B transmitted

the hydrochlorofluorocarbons (HCFCs—substances with only 2–5% the ozone-destroying potential of CFCs, so they are often used as substitutes for CFCs) and the halons, a class of fluorocarbons that contain bromine atoms. The halons release their bromine atoms, which then destroy stratospheric ozone. Halons have been commonly used in fire extinguishers. Methyl chloroform, methyl bromide, and carbon tetrachloride are some other important and widely used chemicals that have ozone layer-destroying properties. Methyl bromide is widely used as a pesticide and is also produced naturally by oceanic plankton; because of its ozone-destroying properties, bans on the production and use of methyl bromide have been proposed.

By the mid-1970s, the possible destruction of the ozone layer by artificial chemicals had become a topic of heated controversy. Large chemical companies and many manufacturers downplayed or denied the possibil-

ity, whereas many environmentally oriented citizens called for immediate, even if economically expensive, action to save the ozone layer. The controversy abated somewhat in the late 1970s after some action was taken to protect the ozone layer (for instance, in 1978, the United States banned the use of CFCs as **aerosol spray** propellants), but it erupted again in the 1980s, when an ozone hole was detected over Antarctica.

Historical Background

The ozone layer's modern troubles began innocuously enough in 1928, when chemists discovered a new class of chemicals that could replace sulfur dioxide and ammonia as the basic fluids in refrigerators. This new class of chemicals, which was produced by DuPont under the trade name of "Freon," became known more generally as the CFCs. CFCs were considered a triumph of modern chemistry; they were safe, nonflammable, stable, unreactive chemicals that proved ideal for many purposes. They were put to use in car air conditioners and refrigerators. They were used as a nontoxic propellant for aerosol cans and as a blowing agent in producing Styrofoam and other plastics. They had insulating properties and could be used as solvents and cleaning agents in the electronics industry.

Applications of CFCs multiplied after their initial discovery, and after World War II, the market for these substances expanded rapidly. By the early 1970s, CFCs constituted an $8 billion per year industry in the United States alone. Since their invention, approximately 20 million metric tons of CFCs have been released into the atmosphere. Most of these chemicals were released directly into the atmosphere during, or immediately after, their use, but some were released once the equipment in which the chemicals were used (such as a refrigerator) was discarded.

However, in 1973 and 1974, two chemists at the University of California at Irvine, Mario Molina and Sherwood Rowland, discovered the effect that CFCs accumulating in the stratosphere have on the ozone layer (for their research they shared the 1995 Nobel prize in chemistry with Paul Crutzen, another researcher of stratospheric ozone destruction).

Molina and Rowland found that CFCs are almost inert in the **troposphere** (the atmospheric layer underlying the stratosphere and covering the immediate surface of Earth), but the chemical stability that makes CFCs so ideal for use on Earth's surface is the very characteristic that allows them to reach the stratospheric ozone layer. CFCs are not soluble in water, so they are not washed out by rain. They drift on air currents up into the stratosphere. In the stratosphere, ultraviolet radiation breaks down the CFC molecules, releasing **atomic chlorine**, a powerful catalyst of ozone destruction. A free chlorine atom reacts with an ozone molecule, removing one oxygen atom and thus converting the O_3 to O_2. However, the chlorine monoxide molecule (ClO) is not stable; it readily reacts with a free oxygen atom to form a molecule of diatomic oxygen (O_2) and a free chlorine atom once again. Thus, once released from the CFC, the single chlorine atom acts as a catalyst that can destroy thousands (perhaps 10,000 to 100,000) of ozone molecules. Eventually, precipitation will wash the chlorine atom out of the atmosphere.

According to Molina and Rowland's calculations, ozone depletion caused by CFCs threatened the very existence of life on our planet. They and other environmentally oriented people quickly lobbied for a ban on CFCs. But the large chemical companies and a few scientists argued that the theory of CFC-induced ozone depletion was speculative and did not justify controlling CFC releases. Part of the problem was that initially there was little empirical evidence to support the Molina–Rowland calculations, largely because of the difficulty of measuring ozone concentrations in the stratosphere. Furthermore, this was not the first time that concerns about ozone depletion had been raised; earlier some researchers had worried about the effects of nuclear bomb detonations on the ozone layer. Still, as news of possible ozone depletion hit the streets, the public responded. By the first half of 1975, the use of CFC-based aerosol spray cans had dropped by 25%, and by 1978, the United States had effectively banned CFCs as an aerosol propellant. Yet CFCs continued to play a major role in other industries, both in North America and around the world.

Despite other scientists' confirmation, including a committee of the U.S. National Academy of Sciences, that the threat to the ozone layer was real, little more action was taken for several years. Then in 1983–1985, a discovery was made that stunned many previously unconcerned observers: Based on 30 years of measurements, the British Antarctic Survey found that each spring a hole developed in the ozone layer over the South Pole; the hole closed again later in the year. Could this be a result of CFCs released into the atmosphere? NASA's **Nimbus-7** satellite confirmed the depletion in the ozone layer; curiously, satellites had not detected this ozone hole previously because the computers had been programmed to disregard as errors any data that recorded such low ozone values. Scientific expeditions were quickly mounted to confirm or refute the reality of the Antarctic ozone hole and to try to determine its cause if it was real. The conclusion was that the spring hole was real (the loss of ozone was about 50% in general, but rose as high as 100% in some spots) and that it was caused by chlorine carried up to the stratosphere in CFC molecules. Molina and Rowland had been correct all along.

International action was soon taken. At a meeting hosted by the Canadian government in Montreal in September 1987, representatives from more than two dozen countries agreed to abide by the **Montreal Protocol**. The countries would freeze CFC production (at 1986 levels) and then gradually decrease CFC and halon production to 50% of 1986 levels by the year 1999. However, it very quickly became clear that the protocol did not go far enough. Evidence of a thinning of the ozone layer over the Northern Hemisphere and continued enlargement of the Antarctic ozone hole indicated that more drastic measures were needed. In addition, the Montreal Protocol did not cover many ozone-destroying chemicals, such as methyl chloroform and carbon tetrachloride. In March 1988, DuPont officially acknowledged the damage CFCs were causing to the ozone layer and agreed to cease CFC production after substitutes were found.

By the late 1980s and early 1990s, stratospheric ozone loss was well documented.

The largest losses occurred over Antarctica. In the Arctic, ozone losses have amounted to about a 10% reduction. Significant losses have also been recorded over the midlatitudes both north and south of the equator. Ozone losses of 3% to 10% or more were recorded over parts of Australia, New Zealand, South Africa, and South America. According to recent data collected by the Nimbus-7 satellite, some of the largest decreases in ozone occurred over the Northern Hemisphere, including North America, Europe, and large portions of Asia, where ozone values had fallen below "normal" (pre-early 1970s) values by as much as 13% to 14%. Lesser ozone losses were documented over the tropics. Averaged over the planet as a whole, it appears that from 1979 to 1991, the Earth lost 3% of its stratospheric ozone.

Given the urgency of the ozone crisis, the Montreal Protocol continued to be amended throughout the 1990s during a series of meetings held in London (1990), Copenhagen (1992), Montreal (1997), and Beijing (1999). As a result, CFC production was banned in industrial countries as of 1996, with the exception of a small amount of production solely for export to developing countries or for use in specialized and essential situations, such as in asthma inhalers. Furthermore, even this production is being phased out. Reportedly all CFC production ceased in Russia as of December 2000, and one of the single largest CFC manufacturing plants in the world, located in the Netherlands, closed at the end of 2005. At the beginning of the 21st century, among the leading producers of CFCs were China and India (both developing nations). As of this writing, both have significantly decreased their overall production and use of ozone-depleting chemicals. As of 2010, the United Nations (UN) reported that a 97% success rate had been achieved for the 195 countries actively working together. During the past 2 decades, it has become apparent that it is not only CFCs that can damage the ozone layer; many other substances have the potential to damage stratospheric ozone, including chemicals that originally were seen as replacements for CFCs. Ozone-damaging substances include hydrochlorofluorocarbons (HCFCs), various halons (such as Halon-1202), methyl bromide (see Case Study 17-1), n-propylbromide, hexachlorobutadiene, 6-bromo-2-methoxynaphthalene, and many other substances. In recent years, there have been, in conjunction with the phase-out and banning of CFCs, restrictions and phase-outs of these and other ozone-harming substances. China reported to the UN that it produced 6,000 refrigerators for the Beijing Olympics using CO_2 as the refrigerant, saving 30% of the energy, totaling an estimated 80 billion kilowatts. The UN is working with India and pharmaceutical companies to develop an alternative delivery mechanism for medicine propellants such as those needed in asthma inhalers. As more chemicals are found to damage the ozone layer, we can expect additional restrictions. Current challenges for the Montreal Protocol beyond 2010 are finding pharmaceutical substitutes, such as chemicals needed to deliver medication from inhalers, eliminating black market trade of ozone-depleting substances through aggressive cooperation with world customs agents, ensuring that replacements for ozone-destroying substances are not environmentally harmful, and continuing the education and economic assistance necessary to keep the current uses of these harmful chemicals on the decline.

The Current Situation

It may appear that the ozone story has come to a happy closure; however, despite the Montreal Protocol and its subsequent amendments, the story is not over. Even with the complete phase-out of CFCs and other chlorine-bearing chemicals, stratospheric chlorine concentrations are expected to remain high well into this century, as previously released CFC molecules rise from the troposphere into the stratosphere (once released, a CFC molecule can take 15 years to make its way from Earth's surface to the stratospheric ozone layer).

Even with the accelerated phase-out mandated by amendments to the Montreal Protocol, the problem will persist. Some of the more common CFCs have life expectancies of between 75 and 110 years, and the chlorine catalyst in the stratosphere can also be quite long lived. The estimated background level of chlorine in the atmosphere, derived from natural sources such as volcanoes, is about 0.6 parts per billion (ppb). Currently, because of human releases of chlorine-bearing chemicals such as CFCs, the concentration of chlorine is about 3.5 to 3.9 ppb; fortunately, this may be a plateau, and it is expected that levels will decline in the future. Another perhaps more direct way to gauge the amount of stratospheric ozone depletion is to monitor the annual spring ozone hole over Antarctica (**FIGURE 17-3**). In September 2006, the U.N. World Meteorological Organization reported that the spring Antarctica ozone hole was 29.5 million square kilometers in size. Unusually cold temperatures can account for the increased size. However, the degree of ozone depletion within the hole can also be measured using Dobson units (essentially a measure of the thickness of the ozone layer above a certain location); 145 Dobson units were recorded in September 2005 (springtime in Antarctica), compared with 98 in 2000 or the record low of 85 in 2006. "Normal" levels of ozone for the Antarctic are about 274 Dobson units. Scientists estimate that the ozone layer will remain depleted to some degree until at least 2040 or 2050, with ozone losses as high as 10% to

30% at times over the northern latitudes, where most of the world's population resides. Although stratospheric ozone should build up again during the next 50 years, measurable amounts of CFCs will continue to reside in the atmosphere well into the 24th century. Thus, even if all nations uphold the Montreal Protocol and its amendments (perhaps an overly optimistic assumption), Earth will be subjected to an abnormally thin ozone layer for decades to come. However, there is some good news in the picture. NOAA scientists reported that in 2010, there had been a 17% improvement in halogen (chlorine, bromine, and other ozone-destroying chemicals) levels over Antarctica and a 31% improvement in the stratospheric halogen levels, which means that if this trend continues, Earth's ozone layer over Antarctica could be restored to the pre-1980s level by 2080.

Natural Causes of Ozone Depletion

A number of naturally occurring substances also have the ability to destroy stratospheric ozone. Most important of these are the hydrogen oxides (HOX) derived from water vapor, methane (CH_4), hydrogen gas (H_2), and nitrogen oxides (NOX). The nitrogen oxides, which are very effective destroyers of stratospheric ozone, not only occur naturally, but also are being supplemented by anthropogenic sources. In particular, nitrogen-containing fertilizers release N_2O into the air; the N_2O can diffuse into the stratosphere and form nitric acid (NO), which has the ability to react with and destroy ozone molecules. There is evidence, however, that in the presence of methane and chlorine (perhaps released from a CFC) NO_2 will form from NO; additionally, solar radiation can break the NO_2 into NO and O. The atomic oxygen (O) can combine with O_2 to form ozone (O_3). Nitrogen compounds will also bind with chlorine atoms, preventing the chlorine from destroying the ozone. Thus, under certain circumstances, nitrogen compounds may actually enhance the ozone layer rather than deteriorate it. Similarly, it has been suggested that stratospheric methane may benefit the ozone layer by helping to remove chlorine atoms. Another substance that breaks down stratospheric ozone

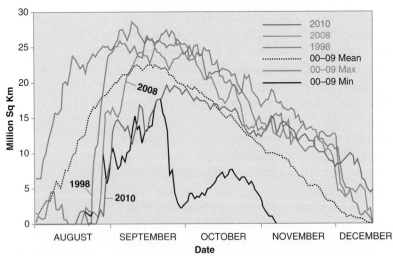

Southern Hemisphere Ozone Hole Area

FIGURE 17-3 Measurements of the spring ozone hole over Antarctica each year show the size the hole is increasing. (*Source:* NASA/Goddard Space Flight Center, Ozone Processing Team.)

is methyl bromide. In the United States and throughout the world, the use of methyl bromide is being phased out; as of 2005, it has virtually disappeared for common uses in the United States, and many hope that within the next ten years it will be a chemical of the past globally. Methyl bromide that people produce was once widely used as a pesticide, but it is also a substance found in nature (for instance, plankton of the oceans produce enormous quantities of methyl bromide—see **CASE STUDY 17-1**).

There are some suggestions that naturally occurring chlorine oxides, such as chlorine monoxide (ClO), may promote the deterioration of stratospheric ozone. During major volcanic eruptions, significant amounts of chlorine may be injected into the stratosphere, but this has never been conclusively demonstrated to be a significant agent of global stratospheric ozone destruction. More troubling, however, is the recent discovery that tiny particulate matter in the stratosphere, known as stratospheric aerosols, may help to promote ozone destruction by providing convenient surface areas where the ozone-destroying reactions can readily take place. In 1982, the eruption of the El Chichùn volcano in Mexico launched 6 or 7 million metric tons of sulfate aerosols into the stratosphere, which may have caused ozone levels in the Northern Hemisphere to temporarily drop by as much as 10%. In June 1991, **Mt. Pinatubo** in the Philippines erupted, injecting an estimated 20 million metric tons of sulfate aerosols into the stratosphere (**FIGURE 17-4**). Some atmospheric scientists believe that this further exacerbated the ozone depletion problem for several years. Additionally, sulfate aerosols released during the burning of fossil fuels are accumulating in the stratosphere.

Effects of Increased Ultraviolet Radiation Reaching Earth's Surface

Why have people been so concerned about the depletion of the ozone layer? Is a little bit more exposure to ultraviolet radiation really all that bad? A small amount of UV radiation is necessary for the well-being of people and other organisms. UV radiation promotes the synthesis of Vitamin D in people and, in small

FIGURE 17-4 The eruption of Mt. Pinatubo in June 1991 sent solid material 30 to 40 kilometers into the atmosphere. Within days, the aerosol cloud from the eruption formed a nearly continuous band that stretched 11,000 kilometers from Indonesia to Central Africa.

amounts, acts as a germicide to control populations of microorganisms. Yet an increased incidence of UV-B radiation at Earth's surface can have numerous adverse effects on the health and well-being of people and other organisms and can cause the degradation of nonliving materials. For every 1% decrease in the ozone layer, the intensity of UV-B radiation on the Earth's surface increases by 2%.

The most widely publicized problems associated with increasing amounts of UV-B are the projected increases in skin cancer and cataracts among people. DNA (the genetic material found in all living organisms) is highly sensitive to and can be damaged by UV-B radiation. A solid body of evidence indicates that skin cancer in people can be caused by UV-induced damage to DNA. The United Nations Environment Programme has estimated that the extra UV-B exposure caused by a 10% loss of global ozone could result in a 26% increase in the incidence of nonmelanoma skin cancers. They are called nonmelanoma because this group of cancers includes all skin cancers except

Methyl Bromide and the Ozone Layer

VISIT

http://environment.jbpub.com
/mckinney/5e/
for more information

Besides CFCs, many other substances can harm the stratospheric ozone layer to various extents, including HCFCs, methyl chloroform, carbon tetrachloride, methyl chloride, and halons (to name just a few). Halons are a class of fluorocarbons that contain bromine atoms; when released in the stratosphere, bromine (like chlorine) has the ability to destroy ozone molecules. Another substance that has been implicated in the destruction of stratospheric ozone is methyl bromide. Indeed, now that the production and use of CFCs and certain other chemicals have been greatly curtailed, according to some researchers, at present the biggest threat posed to the ozone layer may be methyl bromide.

Methyl bromide, a widely used pesticide and fumigant, is applied to everything from alfalfa and tomatoes, strawberries, grapes, walnuts, and wheat—all told, it is used for over a hundred different crops, as well as in forests, nurseries, wood protection, harvest protection, and in quarantine procedures. It has the advantage of killing insects, weeds, and other pests in the soil even before the crop is planted. It is also commonly used to exterminate termites. In the late 1990s, the United States was using over 18 million kilograms (40 million pounds) of methyl bromide each year, three quarters of which went to fumigate soil before planting crops. Worldwide, an estimated 65.3 million kilograms (144 million pounds) a year were being used. The problem was that methyl bromide might also be a potent destroyer of the ozone layer. Some studies indicate that a molecule of methyl bromide can break down ozone about 40 to 50 times as quickly as a CFC molecule. It has been suggested that methyl bromide may currently account for 10% of all stratospheric ozone damage.

Amendments to the Montreal Protocol recognized methyl bromide as an ozone-depleting substance and mandated a phase-out of its use gradually between 1999 and 2005 in developed countries. In developing countries, the phase-out has been extended to 2015. Certain exemptions in both developed and developing countries have been allowed, including the use of methyl bromide for certain quarantine procedures, preshipment fumigation of some commodities, and other so-called critical uses. In 2009, Africa reported to the UN that it had devised a number of substitute agricultural practices drastically limiting the need for methyl bromide.

A great deal of controversy exists about whether methyl bromide should in fact be banned. Unlike the CFCs, methyl bromide is found in nature—it is not simply an artificial, human-made substance. Many natural processes produce methyl bromide, as well as methyl chloride (another ozone-destroying compound). Indeed, the plankton of the oceans produce prodigious quantities of both methyl chloride and methyl bromide—perhaps 30% to 60% of all methyl bromide released into the atmosphere according to various United Nations estimates. One study estimated that salt marshes around the world may produce 10% of all methyl chloride and methyl bromide released into the atmosphere; these same substances are also produced by tropical and subtropical coastal forests and vegetated islands and by both biological and nonbiological degradation of organic matter (such as dead vegetation) in soils. It has been estimated that 10% of all methyl bromide injected into the atmosphere is from the burning of vegetation, and in recent years (before the Montreal Protocol restrictions on methyl bromide use took effect), 20% resulted from fumigation uses.

So why didn't naturally occurring methyl bromide and methyl chloride completely destroy the stratospheric ozone layer long ago? Appar-

malignant melanoma. Of the many types of nonmelanoma skin cancers, basal cell carcinoma and squamous cell carcinoma are the most common. Basal cell carcinoma begins in the lowest layer of the epidermis (the basal cell layer). Approximately 70% to 80% of all skin cancers in men and 80% to 90% in women are basal cell carcinomas. The cancer usually develops on sun-exposed areas, especially the head and neck. Squamous cell carcinomas account for 10% to 30% of skin cancers.

ently there has been an equilibrium in nature, with inputs of these gases being offset by natural processes that destroy them. For instance, in the lower atmosphere, there is an abundance of hydroxyl free radicals (–OH) that can react with and remove these gases from the atmosphere. The fear on the part of many experts is that the anthropogenic release of massive amounts of methyl bromide will upset the delicate balance of nature and lead to significant stratospheric ozone depletion.

Is it really necessary to ban the manufacture and use of methyl bromide? At least some studies indicate that the human contribution of methyl bromide to the atmosphere is far outweighed by natural contributions. As for the methyl bromide used in agriculture, some researchers (although admittedly funded by the methyl bromide industry) contend that most of the methyl bromide used as a pesticide never enters the stratosphere where it could harm the ozone layer. Rather, it is absorbed by the soil or the oceans where it may be broken down into harmless substances. Banning methyl bromide, it is contended, is only harming the agricultural industry and entailing enormous costs (the cost of finding substitutes, and losses in food production and damage to stored harvests) without significantly benefiting the ozone layer.

Unfortunately, no one substitute has been found that can serve all the purposes to which methyl bromide has been applied, and thus, a range of new options and techniques must be developed as methyl bromide is phased out. Alternatives to applications of methyl bromide include employing Integrated Pest Management (see Chapter 14), using beneficial microbes that attack pests, rotating crops, and treating soils with heated water to kill pests before planting of crops. To fumigate foodstuffs and other organic matter (such as wood products) before shipping, hot steam or dry air applications, hot water dips, cold treatments, and subjection to a pure nitrogen atmosphere (therefore suffocating aerobic organisms) have been suggested. Still, critics contend, this variety of processes may be more cumbersome and less effective than treatment with methyl bromide.

Questions

1. Can you draw any parallels between the current controversy over methyl bromide and the controversy in the 1970s over CFCs? (Remember that in the 1970s many CFC industry advocates denied that CFCs could be damaging the ozone layer. At that time, it was often argued that ceasing CFC production would hurt the economy and lead to food shortages as refrigeration units, dependent on CFCs, lapsed and allowed food to spoil.)

2. Given that CFCs and methyl bromide are two different classes of chemicals, is it possible that CFCs really do deserve to be banned, whereas some researchers have exaggerated the dangers of methyl bromide to the ozone layer? How important is it to point out that methyl bromide is produced in large quantities by natural organisms? Can you trust research that is funded by the methyl bromide industry? How about research funded by the EPA or United Nations agencies that may be biased against any chemical suspected of contributing to ozone layer deterioration?

3. If you had to decide, would you allow methyl bromide to be used unchecked? Would you limit but not eliminate the use of methyl bromide, or would you totally ban the production and use of methyl bromide? Justify your position.

VISIT

http://environment.jbpub.com /mckinney/5e/ for more information

An increase in malignant **melanoma** skin cancers, which are much more dangerous than nonmelanoma cancers, would also be expected. In one of the world's southernmost cities, Punta Arena, Chile (near Antarctica), incidents of skin cancer rose by 66% between 1994 and 2001. The U.S. Environmental Protection Agency (EPA) has estimated that continued depletion of the ozone layer could cause an additional 800,000 cancer deaths in the United States during the next century. Recent medical reports indicate that an increase in temperature of 1°C (1.8°F) over an extended period of time could magnify the cancer-causing effectiveness of UV radiation by 10%, especially for those living close to the equator.

Some researchers have downplayed the increased risks of skin cancer by pointing out that the incidence of skin cancer among the general population increases as one moves from higher latitudes toward the equator; presumably, this phenomenon is independent of any loss of ozone in the stratosphere (perhaps in warmer areas, more people spend more time in the sun). The National Academy of Sciences, in a 1975 study, found that in the United States the doubling distance for skin cancer was about 965 kilometers (600 miles) south. From this, one can extrapolate that the risk of skin cancer increases about 1% for every 10 kilometers (6 miles) closer to the equator one lives. Thus, a 26% increase in skin cancer resulting from a 10% loss in ozone would be the same as if everyone simply moved about 250 kilometers (156 miles) closer to the equator (of course, an impossibility for those already living along the equator). Such reasoning seeks to trivialize the effects of ozone depletion. It does not consider the effects of ozone depletion and increased UV-B radiation exposure on delicately balanced ecosystems, and it is of little comfort to people who contract cancer from increased UV-B exposure.

Increased exposures to UV-B have also been linked to increased incidences of **cataracts** (where the lens of the eye becomes opaque), damage to corneas, and retinal disease in people. It has been estimated that a 10% decrease in the ozone layer could cause more than 1.5 million new cases of cataracts each year (currently there are 300,000–400,000 new cataract cases each year). Evidence also suggests that excess dosages of UV light can suppress the human immune system, allowing the spread of infectious diseases. Furthermore, it has been hypothesized that ultraviolet light may help activate the AIDS virus.

Ultimately, the widespread ecosystem damage caused by increased dosages of UV-B may be more important than its immediate effects on people. Abnormally high levels of UV radiation inhibit photosynthesis, metabolism, and growth in a number of plants, including food crops, such as soybeans, potatoes, sugar beets, beans, tomatoes, lettuce, wheat, sorghum, and peas. The UV-B radiation can destroy cells and also cause mutations. Many tree species are particularly sensitive to UV levels, and increasing amounts of UV may result in a major decline in forest productivity. Elevated levels of UV radiation disrupt insect activity, and of course, the other organisms in a typical terrestrial ecosystem depend on the insects.

Many of the plants and animals in freshwater and marine ecosystems are sensitive to UV levels, especially because UV-B can penetrate several meters of water. Phytoplankton (plants and algae that form the bases of many food chains) and fish larvae may be especially susceptible. Reportedly in 1988, when the ozone levels over Antarctica declined by 15% overall, surface phytoplankton levels also decreased by 15% to 20%. Experimental evidence demonstrates that elevated levels of UV-B radiation will damage fish, shrimp, and crab larvae (and certainly many other species that have not yet been tested).

Ultraviolet B exposure can also damage nonliving material objects. For instance, UV exposure will cause or accelerate the breakdown and degradation of various types of paints and plastics, such as polyvinyl chloride (PVC), an important construction material that is used for vinyl siding, garden hoses, and other products that are regularly exposed to sunlight. Such material damage could cost billions of dollars per year.

Although poorly understood at present, major fluctuations in the ozone layer could have significant climatological effects on the Earth's surface. The stratospheric ozone naturally absorbs UV radiation and re-emits it as infrared energy (essentially heat) into the troposphere; this contributes to the Earth's energy budget. At present, the absorption of UV radiation warms the stratosphere, creating a temperature inversion at the boundary between the troposphere and the stratosphere. Some scientists suggest that a decline in the ozone layer could cause a cooling of the stratosphere, which in turn might cause a cooling of the troposphere and ultimately a slight cooling of the Earth's surface (even though more UV radiation would directly reach the ground). However, such a cooling effect could be very small. According to one calculation, a 20% reduction in ozone concentrations in the stratosphere would cause only a 0.258°C (0.458°F) decline in surface temperatures. Before jumping to the conclusion that ozone loss will have one good effect—countering global warming caused by the greenhouse effect—one must consider that many of the chemicals that are contributing to stratospheric ozone declines are also potent greenhouse gases. The increased greenhouse effects of CFCs and other ozone-destroying gases may more than offset any surface cooling that would otherwise have been brought about by destruction of the ozone layer.

Changes in the concentrations of ozone at different levels in the atmosphere could also affect the temperature gradient of the atmosphere, which in turn could affect atmospheric circulation patterns and climatic patterns on the Earth's surface in unpredictable, but potentially adverse, ways. The relative temperature differences among the various layers of the atmosphere are important in determining convective processes that produce clouds and precipitation patterns. Interfering with this system by depleting the ozone layer could have disastrous effects.

17.2 Global Climate Change

The Greenhouse Effect

The basic idea behind the **greenhouse effect** is as follows. Short-wavelength, high-energy, solar radiation shines from the sun onto Earth. The atmosphere reflects some of this incoming solar radiation back into space. Some passes through the atmosphere and is absorbed as it heats the air, and about half reaches Earth's surface (**FIGURE 17-5A**). The surface heats up and in the process gives off longer wavelength, lower energy (infrared, or **heat**) radiation. This **infrared radiation** passes into the atmosphere, but instead of being radiated 100% back into space, much of it is absorbed by the atmosphere and reradiated to the surface (**FIGURE 17-5B**). This phenomenon occurs because many trace gases (the **greenhouse gases**) in the atmosphere are relatively transparent to the higher energy sunlight but trap or reflect the lower energy infrared radiation. Thus, the greenhouse gases act as a one-way filter, letting energy in the form of sunlight in but not allowing the infrared heat to escape at the same rate. This process is crudely analogous to the way glass in a greenhouse allows sunlight to shine in but stops much of the longer wavelength heat from escaping. Even on a cold winter day, the inside of a greenhouse can become quite warm if the sun is shining.

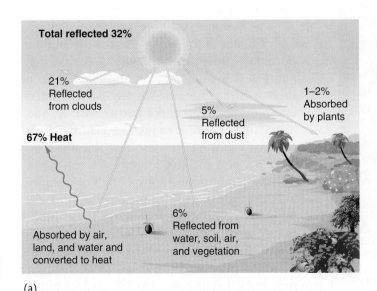

(a)

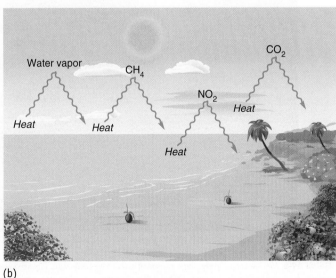

(b)

FIGURE 17-5 (a) Greenhouse gases in the atmosphere allow energy from the sun in but (b) trap or reflect much of the lower-energy infrared radiation (heat) from the Earth's surface.

TABLE 17-1	Greenhouse Gases: Sources and Estimated Relative Contributions to Global Warming	
Gas	**Major Sources**	**Percentage of Contribution**
CO_2	Fossil fuels, deforestation	50
Chlorofluorocarbons (CFCs), other halocarbons	Refrigeration, solvents, insulation, foams, aerosol propellants, other industrial and commercial uses	20
Methane	Rice paddies, swamps, bogs, cattle and other livestock, termites, fossil fuels, wood burning, landfills	16
Tropospheric ozone	Fossil fuels	8
Nitrogen oxides	Fossil fuels, fertilizers, soils, burning of wood and crop residues	6

The estimates of relative greenhouse gas contributions involve significant uncertainties. *Source:* Based primarily on estimates made by World Resources Institute.

Likewise, the Earth's surface would be a frozen mass if it were not for the natural greenhouse effect of the atmosphere. Without this phenomenon, average global temperatures might be on the order of $-18°C$ ($-.049°$). However, with current levels of greenhouse gases, some infrared heat continues to escape. In geologically recent Earth history, a relative steady-state balance has been achieved that maintains the average global surface temperature at about $15°C$ ($59°F$). If no heat escaped,

the surface would continue to heat to unbearable temperatures. The perceived problem is that because of inadvertent human intervention, greenhouse gases are accumulating very quickly in the atmosphere, and some predict this will lead to catastrophic global warming.

Greenhouse Gases

The best-known greenhouse gas (**TABLE 17-1**) is **carbon dioxide** (CO_2). Carbon dioxide is breathed out or otherwise given off by living organisms (including plants) as they undergo aerobic respiration, and is taken in by green plants during photosynthesis. A small amount of carbon dioxide in the atmosphere is an absolute necessity for life on Earth; without it, the planet would be too cold to support living organisms and plants would lack an essential raw ingredient. On the other hand, an excess of carbon dioxide will cause Earth's surface to warm abnormally.

Clearly, people are quickly adding to the amount of carbon dioxide in the atmosphere (**FIGURE 17-6**). In 1850, the average annual concentration of carbon dioxide was about 280 ppm (also referred to as parts per million volume [ppmv]); this increased to 316 ppm by 1959, 379 ppm by 2005, and 387 ppm in 2009. The CO_2 content of the atmosphere continues to increase. Some scientists believe that increasing the parts per million to more than 400 to 500 will do irreparable damage to the climate and is considered the threshold for a stable climate.

The primary way we are increasing the CO_2 content of the atmosphere is through the burning of fossil fuels (**FIGURE 17-7**): coal, oil, and gas. For millions of years, the carbon in these fuels has been out of atmospheric circulation, buried deep under Earth's surface. Suddenly, over a period of just 2.5 centuries (and especially during the last few decades), we have released massive amounts of this fossilized carbon into the atmosphere. For instance, in 2004, more than 7 billion metric tons of carbon was released worldwide by the combustion of fossil fuels—slightly more than 1 metric ton of carbon per person on Earth. Each metric ton of pure carbon released forms approximately 3.66 metric tons of CO_2, so 25.6 billion metric tons of CO_2 was released into the atmo-

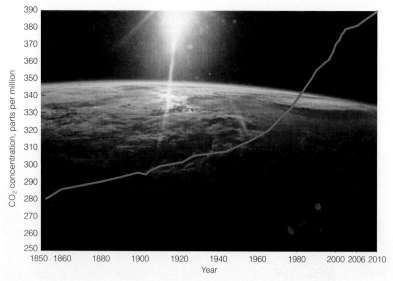

FIGURE 17-6 Average annual atmospheric concentrations of CO_2, 1850–2010. (*Source:* Compiled and updated from various sources, including United Nations Environment Programme, and Worldwatch Institute. Vital Signs 2005. New York: W. W. Norton, 2005, p. 53.)

sphere simply by the burning of fossil fuels in 2004, and 30 billion metric tons in 2009. The people of the industrialized countries disproportionately produce carbon emissions from the burning of fossil fuels. Indeed, the wealthiest 25% of the world's population burn nearly 70% of all fossil fuels. The United States is the biggest single contributor, producing nearly 24% of the world's CO_2 emissions (primarily from burning fossil fuels), while containing less than 5% of the world's population.

An additional 1 to 2 billion metric tons of carbon is emitted into the atmosphere each year from deforestation. Like the fossil fuels, which were once living organisms, the extant forests hold vast stores of carbon. When the trees and other plants die and are burned or allowed to decay, this carbon is converted to CO_2. Furthermore, trees serve the vital function of removing CO_2 from the atmosphere as they grow. Unless deforested areas are quickly replanted, the CO_2 that is emitted directly into the atmosphere remains there, and a vital mechanism for removing excess CO_2 from the atmosphere is destroyed.

Although carbon dioxide is blamed for 50% to 70% of the current abnormal global warming (depending on the authority consulted and how "global warming" is calculated), it is not the only major greenhouse gas. The other major culprits are chlorofluorocarbons (CFCs), **methane** (natural gas, CH_4), tropospheric ozone, and **nitrogen oxides** (NOX).

The chlorofluorocarbons that promote global warming are the same CFCs that are destroying stratospheric ozone. Indeed, they are up to thousands of times more efficient at absorbing heat and promoting global warming than CO_2. Currently, it is estimated that CFCs may account for 15% to 25% of the human contribution to global warming, but this number continues to decline as countries continue to move into compliance with Montreal Protocol standards. This number would have been even higher if steps had not been taken to reduce the CFCs released into the atmosphere. Thus, there are two good reasons to reduce our reliance on CFCs: to save the ozone layer and to reduce global warming.

Methane accounts for an estimated 15% to 20% of current global warming. Currently,

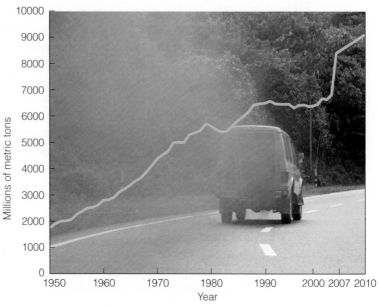

FIGURE 17-7 World annual carbon emissions from fossil fuel burning, 1950–2010. (*Source:* Worldwatch Institute. Vital Signs 2005. New York: W. W. Norton, 2005.)

methane does not occur in anywhere near the concentrations of CO_2 in the atmosphere, but it is as much as 30 times more effective than CO_2 in trapping heat. Like CO_2, methane is a naturally occurring gas that has been present on Earth's surface since its formation. Bacteria in swamps and other areas produce methane as organic matter is decomposed. Termites give off methane as they break down wood, and many animals produce large amounts of methane in their digestive tracts. Ruminants, such as cows in particular, release enormous quantities of methane into the atmosphere. Human habits have intensified these natural sources of methane. We raise billions of cows and other livestock, which serve as natural methane factories. Unfortunately, virtually all of this methane is lost to the atmosphere, although some have suggested that we should try to capture it and burn it as a fuel. Rice farming is another major source of methane; in typical rice paddies, large quantities of organic matter rot under a shallow layer of water to produce methane.

The increased use of fossil fuels has also added significant quantities of "fossil" methane to the atmosphere. During coal mining, underground stored reserves of methane are often inadvertently released into the atmosphere. Likewise, methane occurs above oil in many oil wells. Although this methane is now generally seen as a valuable resource, in the past (and, in some cases, to this day), it was considered not worth bothering with and simply released into the air. The current increased use of methane as the fossil fuel of choice (it burns more cleanly than coal or gasoline) also opens up the possibility that more stored reserves of methane will make their way into the atmosphere, escaping from leaking tanks or pipelines. As much as 15% of all methane released annually into the atmosphere may be leaking from poorly designed and maintained natural gas lines in Eastern Europe and the former Soviet Union.

Another aspect of natural methane could have serious repercussions. An estimated 14% of all the organic carbon of the world is buried (mostly as partially degraded vegetation) and frozen in the permafrost of the tundra. With increased global warming, which many scientists predict will affect the poles to a greater extent than Earth as a whole, vast tracts of tundra could begin to melt. Bacteria would then decompose the organic matter, releasing prodigious amounts of methane into the atmosphere. The added methane would accelerate global warming, which in turn would accelerate the production of more methane, setting off what some fear could be a runaway cycle of increased warming and methane production. Some people even fear that if global temperatures rise high enough, much organic matter found in the soil of temperate regions (where the ground is not frozen) may be converted to methane or CO_2.

A similar effect may occur in the oceans. The oceans, in particular the cold regions, hold huge stores of organic matter, containing more carbon than all the coal reserves on land. Approximately 36,000 billion metric tons of carbon is stored in the oceans, compared with about 4,000 billion metric tons in fossil fuels. As the oceans slowly warm, this organic carbon could decompose, forming methane that will bubble to the surface.

Compared with CO_2, CFCs, and methane, nitrogen oxides (NOX) and tropospheric ozone are relatively minor greenhouse gases, together accounting for perhaps 10% to 15% of global warming. Nitrogen oxides and tropospheric ozone are important air pollutants in other contexts, contributing to petrochemical fog and acid rain (for details see a review of local and regional air pollution); thus, it is important from several standpoints that emissions of these substances be decreased. NOX are formed when chemical fertilizers break down, when coal is burned in power plants, and in general when any fuel is burned at high temperatures (nitrogen and oxygen, the primary components of our atmosphere, combine at high temperatures to form NOX). Large quantities of NOX can also be released during the production of certain synthetic substances, such as nylon stockings and pantyhose.

Other gases also contribute to global warming. Water vapor (H_2O), which can act as a greenhouse gas, is a potentially important factor, but virtually nothing can be done to control the amount of water vapor in the atmosphere. The sun naturally heats the surfaces of the oceans and lakes, and plants naturally transpire, all of which inject large quantities of water vapor into the atmosphere. Of course, this water vapor collects and precipitates out again as rain or snow.

Is Earth Really Warming?

Many researchers would respond to this question with a definitive "yes." Fairly accurate records of global temperatures have been kept since 1880, and they show a fairly steady increase in temperature from about 1880 to 1940. From 1940 to the mid-1960s, temperatures underwent a slight cooling and stabilization, but since the 1960s, there has been a dramatic (although somewhat erratic from year to year) increase (**FIGURE 17-8**), with record highs in the last decade. Based on data through 2010, the warmest year on record was 2010, which tied with 2005. The global average temperature that year was 14.77°C (58.586°F). The next warmest year was 1998, with a global average temperature of 14.71°C (58.478°F). Global temperatures experienced a drop in the early 1990s, but this drop has been attributed to the 1991 eruption of Mt. Pinatubo in the Philippines; the smoke and ash that spewed into the atmosphere had a temporary cooling effect.

The eruption of Mt. Pinatubo and its effect on global temperatures raise an interesting point. The cooling effect of atmospheric pollutants, such as soot and acid particles, may be masking much of the global greenhouse warming. The eruption of Mt. Pinatubo shot a fine mist of sulfuric acid and sulfate aerosols into the atmosphere; this material reflects sunlight back into space and so cools the Earth. Soot, acid particles, and other aerosols are also injected into the atmosphere through human activities—burning fossil fuels, clearing forests, smelting metals, and so forth. For a long time, many climatologists were a bit puzzled when their theoretical models of the effects of increasing concentrations of greenhouse gases predicted much greater temperature increases than have been observed. Only recently has it been widely recognized just how powerful can be the cooling effects of many anthropogenic airborne pollutants. These atmospheric particles not only reflect sunlight back into space, but also form the nuclei of water droplets, promoting the formation of clouds that reflect still more sunlight. According to some recent calculations, atmospheric pollution and particulate matter may be counteracting, or masking, much of the effect of atmospheric greenhouse gases. This is sometimes called **global dimming**. As progress is made toward reducing global air pollution, greenhouse warming may greatly accelerate.

Other evidence for global warming exists in addition to the direct measurements of atmospheric temperatures. Satellite data of ocean-surface temperatures indicate that the oceans have been warming at a rate of about 0.1°C (0.18°F) per decade since the early 1980s. The use of geophysical borehole data is another way to measure temperature changes over time. Heat diffuses from Earth's surface into the soil and rocks, so by plotting a profile of temperatures at increasing depths, it is possible to reconstruct broad temperature trends at the surface. Borehole temperature profiles taken at localities ranging from Alaska to Africa confirm that there has been a distinct warming trend in recent decades.

The retreat of nonpolar glaciers provides another line of evidence for global warming over the last few decades. According to data compiled by the United Nations Environment

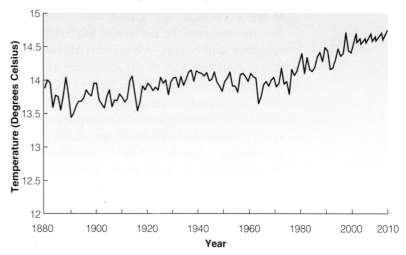

Average Global Temperature, 1880-2005

FIGURE 17-8 Global average temperatures, 1880–2010. (*Source:* Data from Goddard Institute for Space Studies, NASA Goddard Space Flight Center, Earth Sciences Directorate, "Global Temperature Anomalies in 0.01°C, March 2006" and http://www.ncdc.noaa.gov/sotc/global, 2010).

Programme's World Glacier Inventory, most nonpolar glaciers and ice masses (particularly in Asia, Africa, and South America) have been retreating since the turn of the century; in some cases, they have been experiencing particularly rapid melting and decreases in snow cover during the last 40 years. Many glaciologists believe that tropical glaciers (found on high mountain tops) may be especially sensitive to slight increases in global temperature.

The increase in global temperature over the last few decades (Figure 17-8) may not seem dramatic, but from a broader perspective, it is quite rapid indeed. Temperature and climate may be changing more quickly than the ability of many natural species to evolve to the new conditions, with the result that ecosystems are damaged. In 1950, the global average temperature was 13.83°C (56.894°F). In 1955, it was 13.91°C (57.038°F). In 2000, it was 14.41°C (57.938°F), and in 2005, it was 14.77°C (58.586°F; preliminary calculation as this book goes to press); 2002, 2007, and 2009 have been the second, third, and fourth warmest years on record, respectively (2010 and 2005 tied for the warmest year on record). The global average temperature for 2010 was 0.79°C warmer than the 1951 to 1980 average. Based on these data, the current global warming rate is slightly more than 0.5°C (0.9°F) every 50 years.

How Much Will Earth Warm?

If we keep dumping greenhouse gases into the atmosphere, the surface of Earth unquestionably will warm. About this there is no doubt—all one has to do is analyze the surface temperatures of Venus and Mars, which are determined in large part by the carbon dioxide in these planets' atmospheres. The real question is how much the surface temperatures on Earth at different locations will be affected by the equivalent of, for example, a doubling of CO_2 compared with that of preindustrial levels. This topic has been the subject of much debate and speculation, but slowly a consensus seems to be emerging. Many different researchers have come to the conclusion that a doubling of CO_2 levels in the atmosphere will lead to an average global warming of 1.3°C to 4.5°C (2.3°F to 8.1°F) above the 14°C (57.2°F) average global surface temperature over the last 10,000 years. It is important to remember that these numbers refer to average global temperature changes; in many specific areas, the increases could be much greater, especially during the winter. In 1990, a group of about 200 scientists working under the auspices of the United Nations Environment Programme and World Meteorological Organization, the **Intergovernmental Panel on Climate Change (IPCC)**, concluded that given the current situation of increasing greenhouse gas emissions into the atmosphere, there is at least a 50% chance that an average global increase in temperature of 1.6°C to 5.5°C (3°F to 10°F) will take place by about 2050. Their most likely estimate was a 2.7°C to 2.8°C (5°F) increase by 2050. Since 1990, the IPCC has issued revised assessment reports on global climate change. According to the latest report of the IPCC (2010), it is expected that during the next century **carbon emissions** from the burning of fossil fuels will dominate changes in atmospheric CO_2 levels, and by 2100, CO_2 concentrations could be in the range of 540 to 970 ppm. This could lead to global average temperature increases of 3.5°C to 7.1°C (6.3°–12.78°F) between 1990 and 2100 and a projected rise in sea level of anywhere from 9 to 88 centimeters (3.54–35 inches), or more depending on projections. There is a broad range in these figures because of many uncertainties in the raw data, the interpretations of the data, and the models used to make future projections.

Questioning Global Warming

Although they are in the minority, numerous scientists question the concept of global warming both in theory and in practice.

Ice Ages

Over the last few million years, the Earth's surface temperature has undergone natural fluctuations, expressed as the **Ice Ages** (**FIGURE 17-9**). Even in historical times, there have been warmer and colder periods, such as the warmer Medieval Little Optimum around A.D. 900–1100. At this time, Earth's surface was warm enough that ice melted back in the North Atlantic Ocean and the Norsemen colonized Greenland. A couple of hundred years later, the **Little Ice Age** began, and the surface became noticeably colder; the colonies in Greenland failed, and mountain glaciers advanced. Only in the last few hundred years has Earth begun to warm again. No one understands the causes of preindustrial temperature fluctuations (perhaps they are correlated with sunspot activity or other natural phenomena, such as slight variations in the tilt and direction of Earth's axis and the shape of its orbit), but they certainly had nothing to do with industrially produced greenhouse gases.

Over the course of billions of years, there has been a general decline in atmospheric CO_2 concentrations. During the Mesozoic Era (the "time of the dinosaurs," about 245 to 65 million years ago), average global temperatures may have been 5°C (9°F), or more, warmer than at present, and CO_2 concentrations may have been appreciably higher than today, perhaps several times higher. In contrast, CO_2 concentrations may have been only 200 ppm about 40,000 years ago during the last great glaciation. Since then, they rose naturally to the preindustrial levels of about 250 to 280 ppm. Despite the recent rise in CO_2 concentrations, a few researchers suggest that the natural long-term decline of atmospheric CO_2 concentrations over tens of millions of years, if not reversed, eventually could spell the end of life on our planet in the distant future.

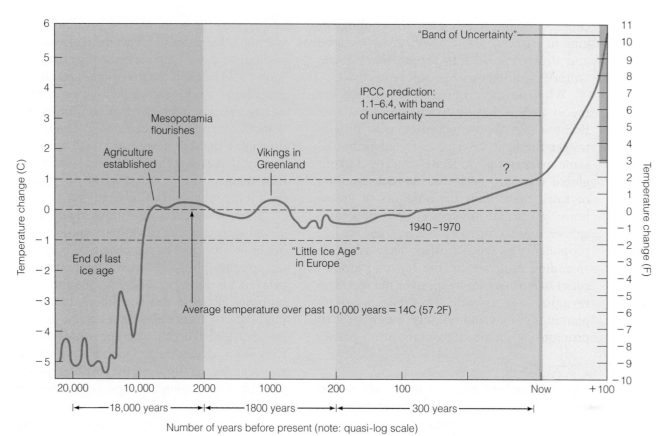

FIGURE 17-9 Variations in average global temperature during the past 20,000 years. Also shown is the IPCC's prediction for the next century. (*Source:* Adapted from A. J. McMichael. Planetary Overload. Cambridge: Cambridge University Press, 1993, p. 84.)

The Pleistocene Ice Ages of the last few million years occurred in cycles of approximately 100,000 to 120,000 years, with each ice age lasting about 100,000 years, followed by an approximately 10,000- to 20,000-year interglacial period. The last ice age ended just more than 10,000 years ago, so if the pattern continues, we could soon enter another ice age. Indeed, some scientists (although a minority) strongly believe that we are on the verge of entering the next ice age.

These scientists do not deny that atmospheric CO_2 levels are on the rise; in fact, they believe this is evidence in support of their hypothesis. According to the theories of John D. Hamaker and Larry Ephron, during a glacial period, ground-up mineral matter fertilizes the soils of the Earth. At the beginning of an interglacial period, plant life thrives, and as a result, CO_2 levels in the atmosphere are relatively low, but toward the end of an interglacial epoch, trees and other plants die off as the mineral nutrients in the soil are exhausted. As forests in particular recede, less CO_2 is removed from the air, and atmospheric concentrations increase. The immediate result of the increased CO_2 levels is greenhouse warming.

But what about the ice age? Ice age proponents believe that most of the warming will occur in tropical and temperate regions, whereas the high latitudes will remain cold (or perhaps become even colder: many scientists disagree with this analysis and believe that the higher latitudes will warm up relative to the lower latitudes). According to certain ice age proponents, greenhouse warming will reinforce the temperature differences between the tropics and the poles. With increased warming, large quantities of water will evaporate from the low latitudes and move toward the higher latitudes, where it will precipitate out as rain and snow. Increased cloud cover, along with increased snow and ice cover, in polar and temperate latitudes will reflect sunlight and eventually cause a cooling trend. After the cooling begins in earnest, glaciers will grow on land and slowly spread into progressively lower latitudes. Another ice age will be on us.

What do the proponents of this ice age theory suggest that we do about the situation? Because greenhouse warming will initially trigger the next ice age, many advocate the same measures as do those concerned about global warming—decrease the consumption of fossil fuels, limit the amounts of greenhouse gases that are spewed into the atmosphere, and so forth. However, some ice age proponents go much further. They advocate spreading minerals (in the form of ground rock) over forest floors all over the world to revitalize tree growth. They also advocate planting billions and billions of new trees to promote CO_2 removal from the atmosphere.

Imprecise Data

Some researchers question the whole notion of global warming, not because they believe that another ice age is imminent, but because they see no clear trend in global temperature changes or they question the whole theory of global warming per se. They point to such data as the slight cooling and stabilization of mean global temperatures during the 1940s to 1960s as indicative of a lack of the predicted global warming. They point out that many of the global warming data are imprecise and that we have a mere century's worth of accurate data. With such data, it is very difficult to separate the signal from the noise, especially because daily temperatures can fluctuate by tens of degrees and seasonal temperatures can fluctuate much more widely, whereas the global warming trend that is being sought is only on the order of 0.5°C (0.9°F) over several decades. Local temperature readings during the past century may also be skewed, depending on where they were taken; large urban areas have progressively become "heat islands" so that temperature readings taken in cities may not reflect real Earth surface conditions. One study found what appears to be a relative increase in the night temperatures throughout the United States since about 1930 without a concurrent increase in daytime temperatures; such data have been used to argue both for and against the seriousness of possible global warming. Increases in nighttime temperatures without an initial increase in daytime temperatures could be expected to occur as greenhouse gases build up, for they would let sunlight in during the day but not let heat (infrared radiation) escape into space at night. Those who downplay the significance of global warming suggest that even if this phenomenon of nighttime temperature increases is real, in and of itself it is probably relatively benign.

Some researchers also argue that even if global warming is occurring, its initial effects might be absorbed by the oceans as they heat up, so atmospheric global warming might be delayed for years, decades, or possibly even centuries. Even if this hypothesis is correct (and data to accurately test it do not exist at present), global ocean warming could have effects just as devastating as global atmospheric warming (especially given the role that oceans play in determining climatic conditions on Earth's surface).

Another aspect of the oceans is that they store a tremendous amount of carbon. The oceans appear to have the capacity to continue to absorb and store much more carbon, but they can do so only at a certain maximum rate. We have already far exceeded the capacity of the oceans to absorb our excess CO_2 as fast as we emit it. The oceans now absorb approximately half of the excess CO_2 emitted by people; the remainder collects in the atmosphere. Nevertheless, without the oceans, CO_2 would collect even faster in the atmosphere than it does currently.

General Circulation Models

Among the primary research tools of climatologists are **general circulation models (GCMs)** of the atmosphere. Using GCMs, some research groups have predicted that a doubling of the atmosphere's CO_2 levels (or the effective greenhouse equivalent of CO_2, such as smaller amounts of CFCs and methane) from about 300 to 600 ppm will increase the average surface air temperature by about 4°C (7.2°F). Proponents of global warming have placed much emphasis on these predictions, but skeptics have pointed out the limitations of the current generation of GCMs.

GCMs are complex mathematical models that, with the help of supercomputers, simulate the Earth's climatic patterns. In typical GCMs, the Earth's surface is conceptually covered with a number of huge boxes (each perhaps a few hundred miles [several hundred kilometers] long on a side and a few miles [several kilometers] high) laid next to each other and stacked one atop another to a height of several tens of miles (tens of kilometers). Each model uses hundreds of thousands of such boxes. For each box, a set of initial conditions is specified, such as air temperature and pressure; H_2O, CO_2, and other gas concentrations; and wind speed and direction. The model also includes expected inflows and outflows from one box to another and from and to the Earth's surface (e.g., CO_2, dust, and other substances spewed into the atmosphere, water evaporated from the oceans, rain and snow precipitated out) and outer space (e.g., incoming solar radiation); equations are used to calculate how matter and energy flow in, through, and out of each box. After the mathematical model is devised and the initial data are inserted, the model is allowed to run reiteratively, generating predictions as to future climatic conditions on different parts of the Earth's surface.

The actual workings of the atmosphere are so complex and so many variables and uncertainties are involved that even the best GCMs are crude in comparison. As a model is run for longer periods of time (to make predictions farther into the future), any errors inherent in the system will accumulate and multiply. Current GCMs are incapable of perfectly predicting present or, retrospectively (also known as retrodiction), past climates with any great accuracy. Critics conclude that predictions of future climatic conditions that these GCMs make are not to be believed. Some adherents of GCM methodology contend that the purpose of these models is not so much to prognosticate the future precisely, but to simply improve our understanding of the atmospheric phenomena being simulated. Even if we must take the actual predictions of the GCMs with a large grain of salt, they can help us make rational choices when it comes to human endeavors that affect the atmosphere and climate.

The use of GCMs is not the only way to model predicted global warming from the greenhouse effect. Based on the greenhouse effects experienced on Venus and Mars, taking into account the atmospheric CO_2 pressures on these planets and their distances from the sun, it has been predicted that increasing the effective CO_2 content of Earth's atmosphere to about 600 ppm will cause an average surface warming of only about 0.4°C (0.72°F), or about one-tenth of that predicted by many GCMs. Of course, many GCM theorists in turn dispute this prediction.

The Consensus View

Despite the questions raised concerning the reality of global warming, it should be firmly emphasized here that the vast majority of scientists who have studied the problem are convinced that global warming is real and will change the distribution of resources across the Earth. There is no doubt that the concentrations of greenhouse gases in our atmosphere are increasing dramatically because of human activities. It is known that these greenhouse gases trap heat that will increase the temperature of the Earth's surface. It is also thought that as the Earth warms, the warming will be unequally distributed, with higher latitudes generally warming more relative to the lower latitudes; the least change will occur in the equatorial regions. If the increase in temperature is great enough, sea level will surely rise because of thermal expansion of the oceans and the melting of ice.

What is unclear to many scientists is exactly what effects global warming will have. Exactly how much will the surface of Earth warm? How high will sea levels rise as glaciers melt and the oceans expand? What effect will global warming have on climatic patterns, including wind directions, rainfall patterns, and ocean currents? Has global warming actually begun, or is it still too early to detect it relative to the "background noise" of daily and seasonal temperature fluctuations? These questions may be partially to completely answered in the next half century as we begin to directly observe more and more of the effects of global warming. Even if we are not certain of exactly how much damage will be done, we should still be practicing the precautionary principle wherever possible because we know that people are having a drastic effect on the environment.

Consequences of Global Warming

Sometimes it is suggested that a little bit of global greenhouse warming might not be such a bad thing, especially for people who live in regions visited by harsh winters (see **CASE STUDY 17-2**). In fact, global warming will not cause overall weather patterns to become consistently warmer and more equitable; indeed, just the opposite may occur in certain regions. As Earth warms, climatic patterns will shift, and in some places, local weather conditions will become much more violent. Cold air currents may be displaced such that, ironically, regions that currently are relatively warm may experience cold snaps and abnormal winter storms. In recent decades, abnormal frosts that have destroyed Florida tomato crops and record low temperatures experienced in Chicago have been attributed to global warming. Because of shifting air currents caused by increased heating of the Earth's surface, cold air masses were displaced from the Arctic.

Climate and Weather

Atmospheric circulation, which ultimately causes what we see as weather (in the short term) and climate (in the longer term), is caused by the differential heating of air masses on the Earth's surface. Although establishing the exact effects global warming will have on these patterns is extremely difficult, some educated predictions can be ventured. As more heat is retained in the system, more air will move across the Earth's surface, producing winds, clashing warm and cold fronts, and generally causing more violent weather conditions. Hurricanes, tornadoes, and other dangerous storms will increase in intensity and frequency. This has been seen recently with Hurricane Katrina, a major hurricane that hit New Orleans in August 2005, causing the entire city to be evacuated and costing hundreds of millions to billions of dollars in damage. This storm was closely followed by Hurricane Rita, which was not as severe, but also was

devastating. Some researchers attribute the record number of very damaging storms in the last decade to global warming. Changes in El Niño weather patterns, including more frequent and more severe El Niño incidents, have been correlated with global warming by some scientists, and certain computer simulations have come to similar conclusions. Other researchers have found that there may not be a simple correlation between El Niño incidents and global climate change. El Niño is a complex weather phenomenon that is still not fully understood.

Global warming will also dramatically change overall climatic patterns. With increased warming, evaporation from the oceans and other large water masses may increase, which will lead to higher levels of precipitation. But the increased precipitation will not necessarily occur where it falls now; with the changing air currents, the areas of rainfall will be displaced. The American Midwest, often referred to as the "breadbasket" of America (and the world), may experience such intense droughts that it will become a desert. The rain that would have fallen in this area may well be pushed north into Canada. Rainfall may also shift from one season to another; some agricultural regions may receive more rain on average than at present, but the bulk of it will come during the winter months, when it is of little use for growing crops. Changing rainfall patterns, coupled with the generally more violent weather, will cause increasing incidences of flash floods. Ironically, some areas will experience droughts and floods simultaneously; during the height of a drought, a violent cloudburst will cause rivers to swell and flood, but it will not replenish reservoirs, which require gentle, protracted rains to recharge.

Because of changing atmospheric circulation patterns, increased heat, and altered rainfall patterns, large portions of temperate regions could experience major declines in soil moisture. The agricultural productivity of these areas could decline, resulting in severe food shortages. An increase of only 2°C (3.6°F) might cause a decline in grain yields of as much as 17% in North America and Europe. Unfortunately, the global food situation is already precarious (as seen by a review of food and soil resources); any disruptions caused by greenhouse warming will not be easily rectified. Some crops will have a hard time growing under even slightly warmer conditions. Many fruit trees and berries need to be subjected to chilling before they will produce fruit.

It is sometimes naively suggested that global warming will simply push many major agricultural belts in the Northern Hemisphere further north; prime cropland will simply be redistributed. To a certain extent, this may be true; however, the difficulty in relocating these crops and the displacement of families and the effects on local and global economies would truly be disastrous in some areas. For instance, Alaska, Finland, and Denmark may experience major increases in crop production. In contrast, the classic breadbaskets of central North America, Europe, the former Soviet Union, and portions of China will become much less productive. However, one must remember that established farmers in the central United States are not likely to pack their bags and move north into Canada as global warming sets in. In addition, many nonclimatic factors must be considered. Even if warmer weather conditions might theoretically favor the cultivation of the northern Canadian prairies and boreal forests, the soils in these areas are not suitable for growing most food crops.

Another consideration is that global warming will not take place all at once. New climatic zones will not simply replace the old. Decades to centuries of very violent, wandering weather patterns may occur before the climate readjusts to a more or less stable equilibrium. It has been suggested that this is already taking place, based on data from insurance companies. For instance, insurance claims from violent weather in 1998 alone were greater than for all of the 1980s combined. Severe storms and unexpected cold snaps or heat spells are very damaging to any agricultural enterprise. The loss of food production could be one of the most serious immediate consequences of global warming.

Not only will atmospheric circulation patterns change with global warming, but the paths of ocean currents also will be modified. For instance, currently, cold water sinks in the Arctic Ocean and travels southward along the bottom of the seas toward the equator. In the tropics, the water is warmed and then moves north once again as a surface current, forming what can be thought of as a giant convection cell. With global warming, the northern polar regions may heat up enough to disrupt the convection cell; adequate quantities of cold water will no longer sink to the bottom to drive the surface currents back to the pole. The Gulf Stream in particular may slow, stop flowing, or even change its direction. This could dramatically change local weather and climate patterns. For instance, the warm waters of the Gulf Stream passing by its shores heavily influence England's weather. If global warming disrupts the Gulf Stream, British winters may become much colder and harsher. In addition, disrupting deep-water currents may affect the circulation of nutrients in the oceans, exacting a heavy toll on delicate ocean ecosystems.

Sea Levels

One of the most discussed effects of global warming is the predicted **sea level rise**. One study by the EPA predicted that sea levels could rise by as much as 2.2 meters (7 feet) by the year 2100, although some more recent studies, based on IPCC data, predict only a 48-centimeter (19-inch) increase by that time. However, any significant rise in sea level, even a mere 48 centimeters, could have devastating effects on the human population and global ecosystems.

Approximately one-third of all people live within 60 kilometers (37 miles) of the sea, often at elevations close to sea level, so a rise in sea level would affect them directly, driving them to higher ground. On the east coast of America, such cities as Boston, New York, and Washington, D.C. (to name just a few) would be prone to flooding. Of course, productive agricultural areas could be destroyed as well. Freshwater supplies could be contaminated with seawater. Coastal wetlands could be wiped out, and other unpredicted changes could occur. The IPCC estimated that a 1-meter (3-foot) rise in global sea levels (a distinct possibility within the next

The Benefits of CO$_2$

VISIT

http://environment.jbpub.com
/mckinney/5e/
for more information

Some researchers argue that increasing the concentration of CO$_2$ in the atmosphere will have benefits that will far outweigh any deleterious "greenhouse effects." As is well known, CO$_2$ is an essential raw ingredient on which all photosynthetic green plants rely. There is evidence that throughout much of Earth's history the atmospheric CO$_2$ content was higher than it has been during humankind's existence; thus, many

plants may have evolved to grow optimally in somewhat higher concentrations of CO$_2$. Experiments have shown that doubling the CO$_2$ content of the air (e.g., from 330 ppm [the level in the middle 1970s] to 660 ppm CO$_2$) will raise the productivity of many plants by about one-third on average. That is, the plants produce more organic matter more efficiently, but such productivity ultimately will peak; however, this depends on the particular crop and where the productivity of the plant increases (for example, leaves versus fruit). In addition, in ecosystems, different plants respond differently, and this can have drastic effects on ecosystem stability. Some authorities contend that even if the rate of photosynthesis for many plants increases in the presence of elevated levels of CO$_2$, not all of the cultivated crops that people currently depend upon may respond in this way (**FIGURE 1**).

Increasing CO$_2$ concentrations also lead to a decrease in the transpiration (water loss by evaporation from the leaves) rate in many plants. Experiments have demonstrated that on average, the amount of water lost by plants in this manner is decreased by one-third if the CO$_2$ content is doubled from 330 to 660 ppm. Overall, this means many plants become much more efficient producers of organic matter as CO$_2$ concentrations increase; they produce more with less loss of water. Because of these benefits, many commercial nurseries routinely use a CO$_2$-enriched environment to grow plants.

FIGURE 1 To determine how wheat might respond to elevated levels of atmospheric CO$_2$, scientists near Phoenix, Arizona, are raising the levels artificially around the plants in a study called Free Air Carbon Dioxide Enrichment.

few centuries given current trends in global warming) would flood almost 402,250 kilometers (250,000 miles) of coastlines around the world.

Global warming can promote rises in sea level through several mechanisms. It is often suggested that increases in mean global temperature will cause mountain glaciers and polar ice caps to melt (**FIGURE 17-10**). As the ice melts, the water eventually will empty into the oceans, raising their level. The gravest

concern is that the Antarctic ice cap, which is 3.2 kilometers (2 miles) thick in places and contains an estimated 80% of all the ice in the world, may undergo significant melting. If the entire Antarctic ice cap were to melt, the mean average sea level would be raised by almost 90 meters (300 feet). (In fact, the average level of the oceans is about 60 meters [200 ft] higher today than at the end of the last ice age about 10,000 years ago; much of this rise is attributed to the melting

Even if increased concentrations of atmospheric CO_2 accelerate plant growth (although there currently is no unambiguous evidence that the increase in CO_2 in the atmosphere has produced this result), the ultimate outcome may not be beneficial from a human perspective. Weeds may proliferate more rapidly at the expense of other types of plants. Insects and other pests that feed on plants may increase in numbers. Some studies demonstrate that certain insects feed more rapidly on plants in a high CO_2 environment. Apparently, the fast-growing plant leaves contain lower concentrations of essential nutrients (particularly nitrogen), so the insects must eat more in the same period of time. Other insects are incapable of using this strategy (perhaps they already feed as quickly as they can) and thus are not as healthy under elevated CO_2 conditions. It might just be that the latter groups of insects are necessary components of the local ecosystem, perhaps the pollinators that ensure plant reproduction.

Human crops may also grow faster when subjected to higher concentrations of CO_2, but the growth may be concentrated in parts of the plant that are not consumed or used otherwise. Thus, the rapid growth may yield little net gain for people.

More critical than any potential increase in plant production are the disruptions that will certainly occur from increasing surface temperatures on Earth. Insect–plant relationships will be assaulted. In higher latitude temperate regions,

the freezing conditions of the annual winter season are important for containing the populations of insects that feed on human crops and other plants. If global warming occurs to the extent that it prevents or mitigates freezes and frosts in certain regions, insect pest populations will proliferate. Of course, the increases in violent storms and abnormal weather patterns that are expected from global warming will not benefit plant production either, nor will encroaching oceans from rising sea levels along currently productive coastal areas. These latter disruptions, which might actually decrease the overall net growth of many types of plants, are simply a function of global warming per se, whether this global warming is caused by CO_2 or other greenhouse gases (which do nothing to enhance the growth of plants).

Questions

1. Can the results of greenhouse experiments necessarily be extrapolated to the scale of the entire Earth?

2. Even if increased concentrations of CO_2 may accelerate the growth of some types of plants, is this always beneficial? What types of consequences might result from increased atmospheric concentrations of CO_2?

3. In your opinion, do the predicted "benefits" of increased atmospheric CO_2 concentrations justify (as some people argue) not taking action to counter the buildup of CO_2 in the atmosphere?

VISIT

http://environment.jbpub.com
/mckinney/5e/
for more information

of great continental glaciers.) Realistically, not many scientists believe that the Antarctic ice cap as a whole will melt significantly or very quickly. This ice cap has lasted for about 10 million years, withstanding many warm and cold spells. If it did begin to melt on a large scale, the melting probably would take place very slowly; the core of the ice cap is extremely cold (much colder than ice in many Northern Hemisphere glaciers, for instance), and ice forms a very good insulator. Even

with extreme global warming, melting all of Antarctica's ice would take many thousands of years.

Of course, only a very small percentage of Antarctica's ice needs to melt to raise sea level by a few feet, causing global devastation. In recent decades, ice shelves have been calving and breaking up, apparently because of warmer temperatures (**FIGURE 17-11**). The breakup and melting of floating ice shelves and icebergs have only a minor effect on global

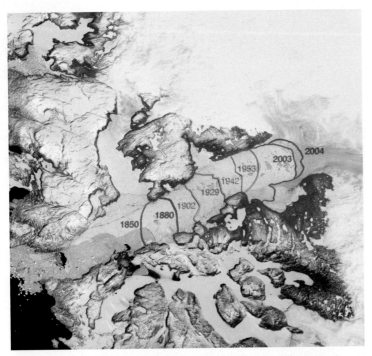

FIGURE 17-10 The Jakobshavn Isbrae glacier in Greenland has been gradually receding since 1850, but from 1997 to 2003, the speed of the glacier's retreat has almost doubled. It is Greenland's largest outlet glacier, draining 6.5% of Greenland's ice sheet area.

sea levels per se (because the ice is already in the water). But when Antarctic coastal shelf ice melts and breaks away, land-based ice creeps or slides down into the sea, replacing the ice that broke away; consequently, more water is added to the oceans, and the sea levels rise.

Global warming could also cause a rise in sea levels through thermal expansion of water.

FIGURE 17-11 Global warming is threatening the Antarctic home of these emperor penguins.

Liquid water expands and contracts slightly with changes in temperature. Given the tremendous volume of water in the world's oceans, even a slight expansion of the water from global warming will cause a measurable rise in sea level.

When we speak of changes in sea level, we usually refer to global averages—but to the people or other organisms living near the shore, the height that the sea reaches during exceptional high tides or major storms may be even more important. On a local or regional scale, global warming will contribute to this aspect of "sea level" also. Increasingly violent storms such as typhoons and hurricanes, accompanied by high winds and changes in barometric pressure, will drive water high onto land. Changes in atmospheric pressure can actually pull the water up, locally raising the sea level. Such storm surges will only increase as global warming continues.

Climate Zones and Biodiversity

In another major effect of greenhouse warming, the climatic zones on the Earth's surface will tend to shift from the equator toward the poles. The effects will be devastating for many ecosystems, but forests will be especially hard hit. As climatic zones migrate away from the equator, trees will find themselves living in environments for which they are not adapted. Very small changes in average temperatures could spell death for large tracts of forest. The decline of red spruce populations in the eastern United States since 1800 has been attributed to global warming. One analysis suggests that if the effective concentration of CO_2 in the atmosphere doubles, the ranges of such trees as the eastern hemlock, yellow birch, beech, and sugar maple would have to migrate 480 to 965 kilometers (300 to 600 miles) north. If the warming occurs rapidly, as is expected, many of the trees will simply die.

Rapid global warming could further reduce biodiversity, already under assault from humankind. Many species of plants and animals require very narrow ranges of temperature and moisture. As greenhouse warming modifies their habitats, they will not be able to adapt or migrate quickly enough and so will succumb; however, other species may be able to expand their ranges, sometimes with devastating results from a human perspective (see **CASE STUDY 17-3**). Species that live on mountaintops or in the high-latitude Arctic and Antarctic regions will have nowhere to relocate. Human barriers (such as roads, farms, and urban areas) will impede other animals that could possibly migrate to higher latitudes. Mangrove swamps and other coastal wetlands will be flooded by rising sea levels. Some wildlife will find themselves trapped in artificial wildlife preserves that originally were designed to protect them. Already, global warming may be exacting its toll on some relatively immobile species. In Canada, long-term studies have documented decreased amounts of ice and longer ice-free seasons on some lakes, and this is affecting the relative populations of various aquatic species. In the Caribbean and around the world, corals have been dying; some biologists believe this is the result of rising sea temperatures. Overly warm water appears to stress the corals (corals are a type of sedentary animals) to the point where they expel the algae, which live symbiotically within them, become "bleached" (they lose their color without the algae), and ultimately die (**FIGURE 17-12**). In addition, such factors as pollution, mining, overfishing, and overharvesting of reefal organisms are also damaging coral reefs. It has been estimated that more than 25% of the world's coral reefs may already be damaged beyond repair. Coral reefs contain a wealth of biodiversity and are the breeding grounds for many ocean organisms, so their loss is particularly significant.

General Strategies for Dealing with Global Warming

There are three basic attitudes toward and ways of dealing with the potential of massive global warming. (1) The Waiting Strategy holds that because all of the data are not yet in, we should wait before taking substantive measures to combat global warming (perhaps the predictions are simply wrong). (2) The Worst Case Scenario Strategy says we should assume that the most pessimistic predictions are valid and act accordingly. (3) The

FIGURE 17-12 Pollution, mining, overfishing, overharvesting of reefal organisms, and global change all contribute to coral reef damage.

Compromise Strategy says we should follow a path that is environmentally safe and promotes the general well-being of the planet, regardless of whether global warming turns out to be a valid phenomenon.

Each of these strategies has proponents and opponents. The Reagan and first Bush administrations strongly espoused the Waiting Strategy. Their position was that the reality and magnitude of global warming and its causes had not yet been definitively "proven" scientifically, and without such proof, taking serious steps to reduce CO_2 emissions would be premature and inappropriate. National policies, it was argued, must be based only on hard facts. Any actions that would significantly reduce emissions of greenhouse gases would impose a heavy financial burden on the country; enormous amounts of money would be spent. Economic output could decrease, and jobs might be lost. Such pain might be for naught if the predicted global warming turned out to be either false or grossly overestimated.

The problem with the Waiting Strategy is that if global warming is real, dealing with the eventual problems (both economic and otherwise) will be more painful and costly than taking measures at an earlier stage. This line of thinking leads some people to argue that even though all the data are not yet in,

Global Warming and Disease

http://environment.jbpub.com
/mckinney/5e/
for more information

A major theme of this book is that the complex connectedness of environmental systems leads to unpredictability of human impacts. Our individual actions can accumulate and have unseen consequences that are not evident for many years.

A good example is a 1995 landmark study released by the World Health Organization on the effects of global climate change on human health. Until the mid-1990s, the vast majority of research on global climate change had focused on physical impacts, such as a rise in sea level, but there were growing signs that the potential effects on human health are no less serious. This possibility came to public attention in 1993, when a controversial article published in the medical journal *The Lancet* concluded that the 1991 cholera outbreak in South America was related to localized warming of Pacific Ocean waters from global climate changes. The article argued that the warming had caused the rapid growth of plankton that harbors the cholera bacterium, leading to thousands of deaths. During the last decade, numerous studies have linked global climate change with disease threats and adverse health effects, not just for people but for other organisms as well.

Some of the projected adverse effects on health and well-being are physical in nature: strong, hot, and prolonged heat waves in some areas and warmer nights in particular. Prolonged heat can enhance the production of smog in populated areas, and one predicted impact of continued global warming is that many cities will experience "killer heat waves" that will increase deaths from bronchitis, asthma, and many other ailments. Indeed, in some areas, it has been projected that the number of deaths that heat waves directly cause could double between 2000 and 2020. Global climate change is also increasing the number and severity of weather fluctuations and extremes, such as severe droughts and massive rainstorms. These erratic weather phenomena can cause harm to people outright, such as by drowning, dehydration, or starving (and, of course, also severely affect ecosystems upon which people are dependent), but they can also promote the spread of infectious diseases.

The promotion and spread of diseases are perhaps the most obvious, and pernicious, biological effects that global warming causes. Since the time an outbreak of cholera was first linked to global warming a decade ago, increased incidences and new outbreaks of numerous diseases have been correlated with global warming, including West Nile virus, schistosomiasis, malaria, dengue fever, yellow fever, various forms of encephalitis, Rift Valley fever, filariasis, onchocerciasis (river blindness), trypanosomiasis (sleeping sickness), tuberculosis, and hantavirus.

Many of the infectious pathogens or their carriers are opportunists that thrive on the sequential extremes in environmental conditions brought about by global climate change. Some diseases, such as tuberculosis, increase in situations of overcrowding, as when people are displaced by flooding and cramped into a smaller area with perhaps inadequate sanitary facilities. Microorganisms or viruses cause many diseases, but animal carriers, or "vectors," transmit the pathogens. For example, various species of mosquito (vectors for malaria, dengue fever, yellow fever, various forms of encephalitis, filariasis, Rift Valley fever, and West Nile virus), black flies (river blindness), tsetse flies (sleeping sickness), ticks (Lyme disease and some forms of encephalitis), water snails (schistosomiasis), and rodents (hantavirus) can all spread disease among people. Mosquitoes and other vectors are very sensitive to temperatures; cold, such as during the winter season or at higher (and cooler) latitudes and altitudes, will kill mosquito adults, larvae, and eggs. As global warming progresses, temperatures are rising and remaining at much warmer levels longer and over larger geographic areas, allowing mosquitoes and other disease-carrying populations to expand. For example, *Anopheles* mosquitoes that spread malaria only do so effectively in a sustained manner in regions where temperatures are maintained above about 15.56°C (60°F) (**FIGURE 1**). Currently, malaria kills several thousand people daily, and more than 45% of the world's population lives in areas where there is a risk of malaria; based on some models of global warming, by the end of the 21st century, 60% of the world's population may live in areas with a risk of malaria. Similarly, *Aedes aegypti* mosquitoes, those that carry yellow fever and dengue fever, must be in regions where the temperatures seldom fall below 10°C (50°F) in order to spread diseases; as global warming occurs, their range will expand.

In the tropics, the altitude on mountains where it remains below freezing year-round has climbed nearly 150 meters (500 feet) since about 1970; these temperature changes correlate with the melting of mountain glaciers around the world and also allow insect populations to spread to higher upland elevations that were once "safe" from many diseases. In the past, many people, particularly 19th century European colonists, lived in the uplands and mountains of the tropics to escape both the heat and the insects, along with the diseases they brought. Now insect-borne diseases are spreading to elevations above 1,600 meters (1 mile), unheard of 50 to 100 years ago, in the mountains of east and central Africa, Asia, and South and Central America. In higher latitudes, such as temperate regions, many diseases and their vectors were controlled by the cold weather of the winter months that killed off large numbers of insects that could otherwise rapidly multiply and spread disease. Now, with milder winters in New England and elsewhere over the last decade, this natural mechanism of insect (and other disease-carrying organisms) population control has been failing.

To make matters worse, mosquitoes and other vectors may become more active in warmer environments (as long as it is not too warm to kill them). For instance, mosquitoes tend to breed more quickly and bite more victims more often in warmer temperatures, increasing the chance of picking up a pathogen and transmitting it to a healthy victim. Furthermore, the disease-causing microorganisms within the mosquitoes also reproduce more quickly in warmer temperatures. The malaria-causing parasitic microorganism, *Plasmodium falciparum*, takes 26 days to develop inside the mosquito at 20°C (68°F) but only half that time at 25°C (77°F). This further increases the effectiveness of the carrier mosquito to spread the disease. Increasing winter temperatures and nighttime temperatures will also effectively increase the mosquito season, meaning that diseases can be spread for a longer period.

Global climatic disruption, resulting in sustained droughts and intense rainstorms with flooding, can cause increases in mosquito and other insect populations and the diseases they carry. Abnormal weather may negatively affect species such as birds, fishes, amphibians, and insects that are predators on certain insects and thus help to control their population numbers. For instance, ladybugs prey on mosquitoes. Amphibians (such as frogs and salamanders) have been shown to be one of the most sensitive groups to pollution and other environmental effects because they literally "breathe" through their skin. Both floods and, somewhat ironically, droughts can increase the spread of insects such as mosquitoes that breed in water. Flooding may allow dormant, but viable, eggs to hatch. Drought often results in stagnant pools of what water remains (including artificial pools, perhaps even a barrel of water created by people as a storage mechanism) that can be perfect incubating grounds for insect larvae. Even a drop of standing water is sometimes enough for some mosquito species to breed. Droughts can also lead to the congregation of people and animals near the few remaining sources of water, and the overused water itself may become contaminated and spread disease. The crowding of people or other organisms allows disease to spread more readily.

Flooding can also promote the spread of waterborne (non–insect-carried) diseases, and so can

VISIT

http://environment.jbpub.com /mckinney/5e/
for more information

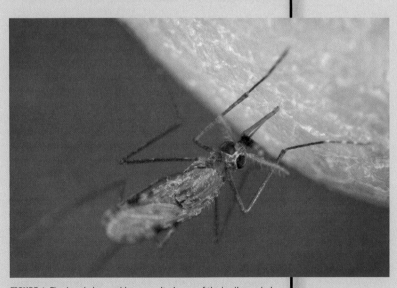

FIGURE 1 The Anopheles gambiae mosquito is one of the leading malaria vectors in the world.

VISIT

http://environment.jbpub.com
/mckinney/5e/
for more information

drought conditions. Flooding can flush sewage and pathogens, as well as fertilizers from farms and even yards, into drinking water supplies. Algal blooms can produce toxins directly harmful to people and contaminate fishes and shellfishes. Algal blooms can also encourage the growth of pathogens, such as *Vibrio cholerae,* the cause of cholera. Droughts can promote waterborne diseases as supplies of safe drinking water are depleted and contaminated supplies are used. In addition, a lack of water can lead to a lack of proper hygiene, further promoting the spread of diseases.

Nobody really knows how much increase in human disease can be expected from global warming. Of course, the answer will depend in part on how much global warming actually takes place, which in turn depends on what measures we take in the coming years and decades to stop greenhouse gases from accumulating in the atmosphere. Already, billions of people are at risk, and the numbers will surely increase substantially. One study projected that an increase of global temperatures by just a few degrees could result in 50 million or more new malaria cases a year. It is not only human populations that are being affected by increases in diseases caused by global climate change; nonhuman populations and ecosystems are also feeling the effects. With slight increases in temperature, slime molds have

been found growing more quickly on eelgrasses (a group of important aquatic plants that grow in and help support productive coastal areas). Eastern oyster disease has been found in the waters of Maine, where previously it was too cold for the disease to survive. It seems certain that more diseases linked to warming temperatures will be found in nonhuman populations if global climate change continues.

Questions

1. How many people do you think are aware or ponder that a trip to the supermarket in their gasoline-powered automobile (which spews greenhouse gases) can contribute to an increase in worldwide malaria and other diseases? Does this knowledge have any impact on your personal choices for transportation?

2. If major wide-scale problems, such as global warming, result from the accumulation of millions of personal choices each day, how can such problems be solved?

3. Does a nation such as the United States, which contributes more than its "fair share" of greenhouse gases to the atmosphere, bear any responsibility to help treat increased incidences of malaria and other diseases in developing countries?

we should not take the chance that the more pessimistic scenarios will turn out to be true. Those advocating this Worst Case Scenario Strategy believe that as a form of insurance, we should do everything we can to prevent potential global warming. Thus, we should significantly decrease the burning of fossil fuels, halt deforestation, reduce cattle production, reforest areas that have been clear-cut, and so forth. If all of the appropriate actions are taken and extreme global warming never occurs, we may never know whether human actions prevented a disaster or whether the predictions were simply incorrect; however,

taking no action at all would be taking an unacceptable risk.

The Compromise Strategy, sometimes referred to as the "no regrets strategy" or the "tie-in strategy," takes a middle-of-the-road approach. Its advocates argue that we should develop policies and take actions that will help prevent the worst of global warming while benefiting the environment and society otherwise. For instance, increasing conservation and raising energy efficiencies, curbing the use of fossil fuels, and placing more reliance on renewable energy sources will benefit humankind in the long run whether global warming is real or not. Independent of the greenhouse effect, acid rain and general air pollution will be reduced. Decreased coal mining and oil drilling will help preserve the landscape, and reduced dependence on oil in particular will lessen dependence on foreign energy supplies. The main problem with the Compromise Strategy is that if the most dire predictions turn out to be true, the moderate response may be inadequate—too little, too late.

International Efforts to Combat Global Climate Change

Through the 1990s, many governments pursued a weak compromise strategy and took relatively little action to combat global warming. As a result of the 1992 Earth Summit sponsored by the United Nations in Rio de Janeiro, Brazil, more than 160 countries signed the **United Nations Framework Convention on Climate Change (UN FCCC)**. Although the convention established no legal obligations or specific target dates, it required signing countries to use their best efforts to control emissions of greenhouse gases. The convention suggested that emissions of greenhouse gases be stabilized at 1990 levels by the year 2000. By the beginning of 1994, enough nations had ratified the convention to make it a legal document, and by 2005, fully 189 nations and the European Union were parties to the convention. According to its provisions, certain nations were required to produce a national action plan explaining the policies they were following to reduce greenhouse emissions and projecting the levels of emissions they intend to allow in the future. The national action plans would then be submitted to a conference of parties (established by the convention) for review. The hope was that the participants would then be able to establish specific, legally binding emissions levels.

According to many scientists at the time, the Climate Convention did not go nearly far enough; for instance, the IPCC estimated that greenhouse gas emissions should be reduced to 60% of 1990 levels just to stabilize the climate. In 1995, delegates of the countries that ratified the 1992 Rio agreement met in Berlin and agreed to negotiate a new set of targets for reduced emissions of greenhouse gases

within the next few years. These talks resulted in the **Kyoto Protocol** of 1997 (named after the Japanese city where meetings were held—the basic protocol was developed in 1997 but not ratified), which among other actions, collectively called for industrial and former Eastern communist bloc countries to reduce their greenhouse gas emissions by 5.2% below 1990 levels by and during the period 2008–2012 (with different targets of emissions reductions for different countries). The 1997 Kyoto Protocol includes provisions for the trading of emissions permits, the use of carbon sinks such as forests, and credits based on carbon-reducing and saving initiatives as ways of garnering credits for reducing greenhouse gas emissions. The Kyoto Protocol also commits developing countries to the monitoring and ultimate reduction of their greenhouse gas emissions.

Between 1998 and 2000, governments met for negotiations to attempt to finalize the rules and provisions of the Kyoto Protocol so that it could be finally ratified and implemented. William (Bill) Jefferson Clinton, a Democrat, was president of the United States from 1993 to 2001 (with the environmentally oriented Al Gore as vice president) during the initial development of the Kyoto Protocol. In January 2001, the Republican George Walker Bush (son of the former president George Herbert Walker Bush, 1989–1993) was inaugurated president, and one of his first major international acts was to announce in March 2001 that he would not sign the Kyoto Protocol, and the United States effectively withdrew from the negotiating process. Some commentators believe that Bush's announcement may have spurred other countries to reach some form of international agreement on the issues all the more quickly. In July 2001, a meeting held in Bonn, Germany, resulted in 178 nations reaching agreements on various elements of the Kyoto Protocol, and further meetings (with various agreements and compromises reached) were held in Morocco later that year. Japan and all 15 of the European Union nations had ratified the Kyoto Protocol in 2002, and Russia, which is responsible for 17% of global emissions, signed on in 2004. However, it was not until 2005 that enough countries had ratified the treaty for it to come into force. The only two industrialized countries that have not are the United States and Australia. Because the United States is responsible for approximately 25% of global greenhouse gas emissions, it will be difficult to make steady progress without the nation's compliance.

Why does the United States refuse to sign the Kyoto Protocol? A main reason given is that doing so would harm the U.S. economy by increasing the price of energy. President Bush cited a study suggesting that implementing the Kyoto Protocol would decrease U.S. economic productivity by $300 billion. Ironically, this is approximately the amount of damage that Hurricane Katrina caused in 2005. Proponents of the Kyoto Protocol note that the United States would be much better off economically by signing the Protocol and taking steps to reduce the number and intensity of hurricanes that many climate models suggest are caused by global warming. Other proponents argue that turning to more sustainable energy sources, such as wind and solar, will benefit the U.S. economy in the long term (for additional details see an overview of environmental science, a review of people and natural resources, and a discussion of the fundamentals of energy, fossil fuels, and nuclear energy).

Technological, Social, and Economic Measures to Combat Global Warming

The principal ways to control the greenhouse effect are to cut down on emissions of CO_2 and other greenhouse gases into the atmosphere and to restore or provide additional sinks for CO_2 (including artificial means of sequestering carbon dioxide). Interestingly, virtually all measures that curb greenhouse gas emissions have additional benefits as well. Thus, reducing emissions of CFCs and related gases will help protect the ozone layer and help mitigate the greenhouse effect. Increasing energy efficiency and decreasing global dependence on fossil fuels can readily control CO_2 emissions (for instance, switching to more environmentally friendly fuels, such as clean-burning hydrogen gas). Of course, decreasing our use of fossil fuels offers other benefits as well, including reducing air pollution and acid rain,

protecting the landscape, and extending the time that our finite supplies of these fuels will last. Fossil fuels and petroleum products are also important for many other products and services in society, such as plastics, computer discs, and much more. Of particular importance to the United States and other developed countries is that a decrease in dependence on fossil fuels (and oil in particular) will mean a decrease in dependence on foreign supplies from volatile regions of the world, such as the Middle East. Switching to other fuel sources and increasing energy efficiency not only will benefit the environment but also will save billions of dollars in military expenses and actually prevent open conflicts and save lives in a direct sense. Alternative, nonpolluting, cleaner energy sources, such as solar and wind technologies, need to be further developed and implemented, although great strides have been made along these lines in recent years. Some researchers suggest that at least on a global scale, nuclear power may play a much larger role in the future because this form of energy emits very little in the way of greenhouse gases (although typically greenhouse gases are emitted when fossil fuels are used to power the mining operations that obtain the uranium fuel). However, there are other drawbacks to increasing our dependence on nuclear power (see a discussion of the fundamentals of energy, fossil fuels, and nuclear energy).

One way to discourage the burning of fossil fuels is to impose a **"carbon tax"** on their consumption. A carbon tax is levied in proportion to the amount of carbon emitted during combustion of a particular fossil fuel; accordingly, coal generally would be taxed at a higher rate than natural gas. Such a tax would encourage more efficient energy use and the development of non–carbon-emitting energy technologies; it would also bring the price of burning fossil fuels more in line with the environmental costs they entail. The amount of money that could be raised with carbon taxes is truly staggering: The Congressional Budget Office has estimated that a tax of $28 per ton (0.907 metric ton) of carbon content in fossil fuels would raise more than $30 billion a year. Although not strictly a carbon tax, every penny added to the federal taxes on 1 gallon (3.785 L) of gasoline contributes about $1 billion annually in revenue to the U.S. government.

An argument often heard against carbon taxes is that they are regressive. That is, relatively poorer people pay a much higher proportion of their income for energy than do richer people; therefore, any carbon tax will extract a relatively higher proportion of the income of the poor than of the rich (even if the rich pay a higher amount of carbon tax in absolute terms). A solution might be to couple a stiff carbon tax with lower Social Security and income taxes for the poor.

Another suggestion is to institute a standardized system of emissions permits on both a national and a global level that would allow a company or organization to emit only a certain amount of CO_2 and other pollutants into the atmosphere. These permits could be sold or traded, so a firm that did not emit its allotted amount of pollutants could sell the "right" to emit the pollutants to another firm. Such a system has been used to control air pollution in California for a number of years and was established at a national level by the 1990 Clean Air Act Amendments.

In addition to the burning of fossil fuels, another important source of greenhouse gases is the CO_2 released during deforestation (**FIGURE 17-13**). A "simple" solution to this problem is to stop cutting and burning forests, but of course, real life is not so simple. Deforestation occurs because people or governments find it to be economically profitable. Furthermore, sometimes forests catch on fire and burn despite people's best attempts to put out the fire. This, of course, releases enormous amounts of carbon dioxide into the atmosphere, and it is far from clear how much carbon in the long-run (given that forest fires and various decay processes that release carbon back into the atmosphere occur naturally) can be permanently or even semipermanently sequestered and stored in forests. Various environmental groups recently have de-emphasized the role of forests as a way of mitigating global warming in the long run, suggesting that planting and preserving forests should not be seen as a substitute for increasing energy efficiency measures and developing alternative fuels. Nevertheless, saving the Earth's forests is a

FIGURE 17-13 Deforestation can be seen clearly in this Landsat satellite image of part of the Amazon Basin, Brazil. The original natural forest is dark green; the pale greens, yellows, and browns mark leveled forest.

worthy pursuit in its own right, and not only may help abate the greenhouse effect, but will also contribute to the preservation of biodiversity, help protect soils, ameliorate climatic extremes, and serve many other useful and necessary functions.

In addition to reducing emissions of greenhouse gases, we can develop sinks that will absorb CO_2. This has given rise to a new field, sometimes referred to as "carbon sequestration," which aims to "geoengineer" the Earth and its atmosphere by using various natural and artificial means to remove atmospheric carbon dioxide. In recent years, the United States Department of Energy and the EPA have increased their interest in and promotion of various carbon sequestration techniques.

Perhaps one of the easiest ways that has been widely promoted is to plant trees where none currently exist. Trees, like all photosynthetic plants, convert gaseous CO_2 into a solid form such as cellulose (essentially wood fiber), where it is stored until the tree is burned or allowed to decompose. It is suggested that planting billions and billions of trees not only will help reduce CO_2 levels in the atmosphere but also will restore many of our devastated forests, add scenic beauty, help stabilize eroding soils, provide habitats for wildlife, and help ameliorate climatic extremes. A couple of well-placed shade trees can help keep a house cool in the summer, thus significantly cutting down on the amount of energy needed to air condition or otherwise cool the house. But, as noted, it is doubtful that simply planting trees (or other plants, for that matter (even plants that have been bioengineered to absorb abnormally large quantities of CO_2) will solve the problem of global climate change. To keep the carbon dioxide out of the atmosphere after a tree (or even a part of a tree, such as a leaf or limb) or other plant dies, it must be further sequestered and preserved. If it burns or decomposes, the CO_2 will once again be released. Thus, some researchers have advocated the permanent landfilling of suburban yard waste, rather than allowing it to be composted and otherwise decomposed to retrieve the valuable nutrients stored in the organic material.

Another method that has been proposed for removing carbon dioxide from the atmosphere is to fertilize the oceans with iron, nitrogen, phosphorus, and other limiting nutrients so that the growth of photosynthetic plankton will be promoted, thus absorbing carbon dioxide. The theory is that after the plankton die, they will fall to the ocean floor and thus store the carbon. In the 1990s, an experiment was carried out in the waters of the equatorial Pacific where 500 kilograms of iron (in the form of iron sulfate) was dumped into the ocean. This caused a temporary and localized phytoplankton bloom that consumed hundreds of tons of carbon dioxide; however, most or possibly all of the CO_2 used by the phytoplankton may have already been dissolved in the ocean water, and it is not clear that any significant amount of carbon dioxide was transferred from the atmosphere to the water. It is also unclear what the environmental and ecological repercussions of the massive fertilization of phytoplankton would be, and some researchers suggest that it could reduce oxygen levels (with possible detrimental effects) in the deep oceans.

Rather than fertilizing phytoplankton, another suggested approach is to inject carbon dioxide directly into deep ocean areas 1,000 or more meters below the surface. The water can absorb the CO_2; however, it will make the water more acidic, and this could have detrimental effects on deepwater organisms. Another unknown is how CO_2-enriched water at that depth will interact with nonenriched seawater. Furthermore, the technology to pump carbon dioxide deep below the ocean surface, perhaps using long pipelines from tankers, may end up being quite expensive.

Carbon dioxide can also be pumped into geologic formations, either below the ocean floor or below the surface of the dry land. For instance, in 1996, 4 million metric tons of carbon dioxide removed from natural gas deposits in the North Sea was pumped into a thick sandstone layer nearly 1 kilometer below the ocean floor, rather than releasing the CO_2 into the atmosphere. Similarly, it has been suggested that carbon dioxide could be pumped into other porous rock formations around the world, or even into depleted underground coal, oil, and gas deposits. However, cost factors could be prohibitive, and more importantly, there are concerns over how much such carbon sinks may "leak" in both the short and long term. There also may be a much more limited capacity for sequestering carbon artificially than is released, particularly in certain regions, and the costs of transporting carbon dioxide to other regions with more storage capacity, at least in the short term, could be extremely expensive. Furthermore, sequestration of carbon may at best be a finite solution—ultimately, the global capacity to store excess carbon will be reached. In the very long term, our best bet may be to pursue energy efficiency measures and cleaner fuel sources, rather than continuing to rely on fossil fuels and then attempting to find ways to safely dispose of the carbon dioxide.

17.3 Global Light Pollution: A Developing Problem?

Besides stratospheric ozone depletion and the potential of global climate change, there is another issue that is gaining increasing attention and appears to have the potential to affect the atmosphere around the world. Many people in urban, suburban, and even relatively rural areas have never seen a truly dark night sky because of light pollution. **Light pollution** is generally considered to be the excess "waste" light given off by outside light sources at night, such as street lights, illumination around buildings, automobile headlights, and so forth, plus such sources as light emanating outdoors from the interiors of illuminated buildings (**FIGURE 17-14**). In many communities, the sky at night is never fully dark but rather glows (perhaps in shades of pink, orange, white, yellow, or gray) because of all of the excess light in the local atmosphere. The light further reflects off of clouds, dust, and atmospheric pollution. This phenomenon of light pollution is often referred to as **sky glow**, and many people now grow up thinking that this is how the sky naturally appears!

For many years, astronomers in particular, both amateur and professional, have bemoaned light pollution. Light pollution can make it very difficult to see celestial objects at night, even with sophisticated, high-powered telescopes. It has been estimated that under natural conditions (without light pollution), on a clear, cloudless night, one can observe with the unaided eye the Milky Way and more than 10,000 stars. Today in many urban areas, the residents can no longer identify the Milky Way, and even on the clearest night, a big-city dweller might see only around 100 stars. It is reported that during a power outage in Los Angeles during the 1990s, many residents were actually afraid of the strange illuminated "cloud" in the night sky that they had never seen before. They were observing the Milky Way for the first time.

Does it really matter if we can see the stars at night? Is this just a minor aesthetic nuisance for some people? Isn't it better to have well-illuminated streets and buildings so that people can see, traffic accidents can be avoided, and burglaries and violent crime re-

duced? For millennia, people have "feared the dark" and attempted to stave off the night; it is not without reason that the invention of the incandescent light bulb was heralded with such fanfare. However, some people contend that more light is not necessarily better. Unnecessary lights waste energy (estimated at a value of $2–$8 billion a year in the United States, depending on how it is calculated). Much of the electricity comes from fossil fuels ultimately and thus adds to the greenhouse effect. Excessively bright lights in automobile drivers' eyes and unwanted glare may actually cause some accidents, and artificial over-illumination at night may otherwise impair visual acuity and performance. Another concern is the right to privacy: Bright lights from elsewhere shining on one's property may be seen as violating privacy at night.

There may be many more subtle negative ramifications of light pollution, effects that we are only beginning to become aware of. Many people feel that they cannot get a good night's sleep in some cities. Artificial illumination shining into the bedroom may be so intense that normal blinds and shades do not block it completely. There is research indicating that artificial light (even at very low levels over extended periods of time) can disrupt natural hormone production and circadian rhythms. Circadian rhythms are physiological cycles with an approximately 24-hour periodicity in people. This may lead to a poor night's sleep and ultimately to severe sleep deprivation over time, which can reduce waking-hour mental and physical performance and quality of life. The disruptions can also induce illness and disease (including, possibly, the risk of certain types of cancer).

Effects on Animals and Plants

The effects of light pollution on flora, fauna, and entire ecosystems may be profound, but we are just beginning to understand such issues. Artificially produced sky glow can affect areas that are far away from obviously visible artificial lights, and the sky glow may in turn disrupt wild animals and plants. The migrations of birds may be affected, and birds may be attracted to artificially lit areas (much to their detriment, in some cases being killed by

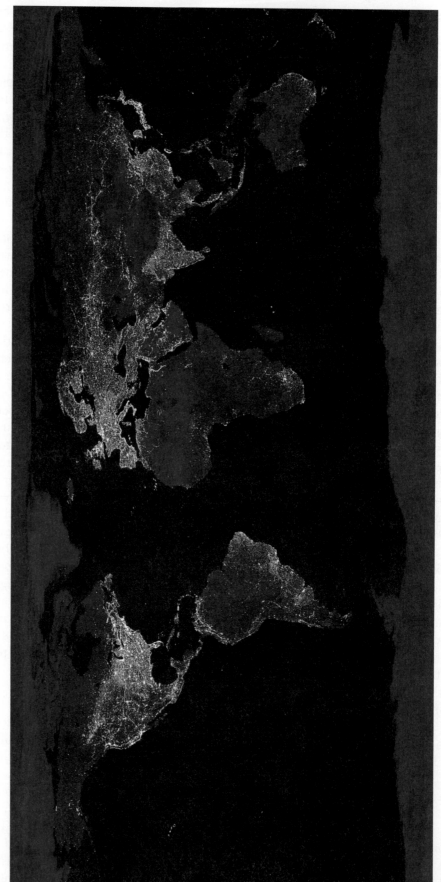

FIGURE 17-14 Earth at night, dramatically illustrating the extent of artificial lighting in populated areas around the globe. This image is a composite of hundreds of photographs taken by orbiting satellites.

CHAPTER 17 Global Air Pollution: Destruction of the Ozone Layer and Global Climate Change

electrocution on electrified towers). The cycles of many nocturnal species, such as various mammals, frogs and other amphibians, and even aquatic worms, may be disrupted. In Florida, sea turtles lay their eggs on the beaches, and when the baby turtles hatch, they instinctually move toward the open ocean at night, apparently drawn by the light of the moon and stars reflecting off the ocean. Now, with the development of hotels and other well-lit structures along the coast, the hatchlings become disoriented and may migrate inland and die. In another example of the disruptive effects of light pollution, it was found that tiny aquatic creatures known as phantom midges migrate to the surfaces of lakes at night to feed on algae. Now there is evidence that sky glow can prevent the phantom midges from coming as close to the surface at night as they would under natural conditions, and there is fear that this could disrupt a delicate ecological balance in some lakes. If the midges do not eat enough of the algae at night, an excess of algae may grow, disrupting other organisms in the lake and ultimately the entire local ecosystem. Light pollution may also affect the health and life cycles of various plants, which of course are the producers of our terrestrial ecosystems upon which all other life is dependent. At this point, we have too little information to fully comprehend the long-term effects of light pollution.

Curbing Light Pollution

Light pollution can be addressed in many ways. The obvious "solution" is to simply turn off unnecessary lights, but in many urban settings, night illumination is considered necessary for safety, security, and economic reasons (people work, shop, or engage in leisure activities after dark—especially in the winter). However, in many situations, current light illumination is poorly directed, and as a result, a good percentage shines into the sky, causing light pollution without benefiting anyone. Shielded lights are available that shine down in a directed manner so as to illuminate only the areas meant to be lit. Appropriate lighting, not too bright or harsh for the purpose at hand, can be used. Ordinances can be passed against one property owner shining unwanted light on another property (violations of which are sometimes known as "light trespass").

■ study guide

SUMMARY

- Two major environmental issues, by their nature, are global in scope: stratospheric ozone depletion and global climate change (global warming).
- Ozone naturally occurs in the stratosphere and shields Earth's surface from dangerous levels of ultraviolet (UV) radiation.
- Declining stratospheric ozone concentrations were first documented in the 1970s and are linked with increased levels of human-produced gases, such as chlorofluorocarbons (CFCs), hydrochlorofluorocarbons (HCFCs), and halons.
- Initial international action to protect the ozone layer culminated in the 1987 Montreal Protocol, which was amended several times during the 1990s.
- As a result of international agreements, CFCs and other ozone-depleting chemical production and use have been greatly curtailed and, it is hoped, will be almost completely eliminated within the next decade or two.
- It is estimated that ozone depletion has peaked and the ozone layer may return to "natural" conditions by the middle or end of the century.
- Increased UV radiation on Earth's surface can adversely affect human health (causing skin cancer and cataracts), cause widespread damage among ecosystems, and affect climates by modifying temperatures at different levels in the atmosphere.
- Greenhouse gases, such as carbon dioxide, methane, and nitrogen oxides, trap infrared radiation (heat) and thus increase the temperature near the surface of Earth.
- During the last two centuries or so, people have measurably increased the atmospheric concentrations of carbon dioxide and other greenhouse gases, such as by burning fossil fuels.

- Although the reality of human-induced global warming continues to be questioned by some researchers (and Earth has experienced natural fluctuations in climate over geologic time), the consensus view among climate scientists is that global warming is real.
- Apparent and projected consequences of global warming include changing weather patterns (including more violent weather and altered rainfall patterns), shifting of climatic zones (which will affect ecosystems and the biodiversity they contain), melting of glaciers covering continents (e.g., Antarctica) and in the mountains, and increasing sea level.
- The best ways to address human-induced global climate change are to reduce the release of carbon dioxide and other greenhouse gases into the atmosphere (e.g., by reducing the use of fossil fuels) and remove carbon dioxide from the atmosphere (e.g., encourage reforestation of deforested areas).
- On an international level, negotiations over reducing emissions of greenhouse gases have been slow; the Kyoto Protocol (1997), aimed at reducing greenhouse gas emissions, took until 2005 to be ratified by enough nations to come into force.
- Light pollution that artificial illumination causes may be a growing problem that has the potential not only to affect adversely human health and well-being but also to disrupt natural ecosystems.

KEY TERMS

aerosol spray
atomic chlorine
carbon dioxide (CO_2)
carbon emissions
carbon tax
cataracts
chlorine
chlorofluorocarbons (CFCs)
general circulation models (GCMs)
global dimming
greenhouse effect
greenhouse gases

heat
Ice Ages
infrared radiation
Intergovernmental Panel on Climate Change (IPCC)
Kyoto Protocol
light pollution
Little Ice Age
melanoma
methane
Montreal Protocol
Mt. Pinatubo
Nimbus-7

nitrogen oxides (NOX)
ozone
ozone layer
radiation
sky glow
stratosphere
troposphere
ultraviolet radiation (UV)
United Nations Framework Convention on Climate Change (UN FCCC)

STUDY QUESTIONS

1. What is the history of ozone layer deterioration? When was the problem first discovered? How long did it take for the world community to be convinced that action should be taken?
2. What are CFCs?
3. Why were CFCs invented? What applications did they have?
4. How do CFCs damage the ozone layer?
5. Describe how people will be able to manage without CFCs and other ozone-destroying chemicals.
6. Name some other chemicals that are destructive to the ozone layer. Are they naturally occurring or artificial?
7. What is the importance of the ozone layer? What effects are caused by significant thinning of the ozone layer?
8. Discuss the Montreal Protocol and its subsequent amendments.
9. How do the problems of ozone destruction and global warming differ from one another? In what ways are they similar?
10. What is the nature of the controversy surrounding potential global warming? Is the evidence for substantial global warming convincing to you? Why or why not? What is the evidence supporting global climate change?
11. What are some of the major greenhouse gases?
12. Describe how the greenhouse effect promotes global warming.
13. Which countries are currently the biggest contributors to potential global warming?
14. How fast have greenhouse gases been accumulating in the atmosphere during this cen-

tury? Based on the work and projections of various climatologists, how much might the Earth warm? Would the warming be evenly distributed over the Earth's surface?

15. List some of the potential consequences of significant global warming.

16. What can, is being, and should (in your opinion) be done to combat global warming? How serious do you think the problem really is? (Be sure to fully justify your answer.)

17. If global warming is real, do you think we should "learn to live with it," try to reduce global warming by reducing greenhouse gas emissions, or address the issue by developing more sinks for carbon dioxide and other greenhouse gases? Or should we use some combination of strategies to deal with global climate change?

18. What is light pollution? Do you think this is a serious issue in the area where you live? Why or why not? What, if anything, should be done to address light pollution? Justify your answer.

WHAT'S THE EVIDENCE?

1. The authors state that the consensus view of scientists who have studied the issue is that Earth's atmosphere is warming and at least a component of this warming is the result of human activities. Is consensus the way issues are ultimately decided in science, or should evidence and "facts" be the ultimate foundation upon which scientific opinions are based? In this case, what is the evidence for the assertion that global warming is real and at least in part caused by people? Do you find the evidence convincing?

2. The authors suggest that one consequence of global climate change will be the spread of disease among human populations. What evidence and rationale is given in support of this contention? Do you find it persuasive?

CALCULATIONS

1. Given that the concentration of CO_2 in the atmosphere increased from 317 ppm in 1960 to 377 ppm in 2004, what percentage increase occurred over this period?

2. Given that the concentration of CO_2 in the atmosphere increased from 280 ppm in 1850 to 377 ppm in 2004, what percentage increase occurred over this period?

ILLUSTRATION AND TABLE REVIEW

1. Referring to Figure 17-8 (global average temperatures, 1880–2004), what is the lowest average global temperature recorded during this period, and in what year did it occur? What is the highest average global temperature recorded during this period, and in what year did it occur? What is the difference between these two extremes?

2. Referring to Figure 17-9 (variations in average global temperature over the past 20,000 years), approximately how much have average global temperatures increased since the end of the last ice age? When did the last ice age end?

3. Referring to Figure 17-9, how much have average global temperatures fluctuated between approximately 3,000 years ago and 150 years ago? Is this variation, from the warmest to the coldest temperatures, more than, less than, or comparable with the variation that has occurred during the last 150 years?

4. Referring to Figure 17-7 (world annual carbon emissions from fossil fuel burning, 1950–2004), global annual carbon emissions in 2000 were approximately how many times greater than those in 1950?

http://environment.jbpub.com/mckinney/5e/

http://environment.jbpub.com /mckinney/5e/

Connect to this text's website at http://environment.jbpub.com/mckinney/5e/. This site features eLearning, an online review area that provides quizzes, chapter outlines, and other tools to help you study for your class. You can also follow useful links for more in-depth information.

Locally, regionally, nationally, and globally, solid and hazardous wastes continue to be critical concerns. With a global population of over 7 billion people, these concerns are further intensified. These are not problems with easy solutions since developing countries often follow in the footsteps of industrialized countries that are just recently making significant strides to replace "cradle to grave" industrial design models with more sustainable "cradle to cradle" practices. At the same time, eco-minded consumers are also just beginning to collectively exercise the power they have to organize through social media and other means to positively affect the manufacturing processes of international corporations vying for consumer dollars and investments in a tight global economy. Athouh the future holds great promise for more sustainable practices, small local communities often still pay the price of irresponsible corporate practices. Shown here, this sign warns residents of Calexico, California, to stay out of the New River, which has been the recipient of long-term contaminates from industry and sewage. See, http://www.calexiconewriver.com/media/photographs and http://www.mcdonough.com/writings/cradle_to_cradle-alt.htm.

18 Municipal Solid Waste and Hazardous Waste

Chapter Objectives

After reading this chapter, you should be able to explain or describe the following:

- What constitutes solid waste
- Basic approaches to solid waste management
- Problems with incineration
- Issues surrounding landfills
- Ways to mitigate solid waste problems
- Types of reuse and recycling, including their benefits
- Hazardous waste
- The Superfund Program
- Toxics Release Inventory

The amount of waste generated each year is staggering. No one really knows how much waste people generate, but much of it originates from the developed countries (impoverished people tend to generate less waste). Just as the United States is a leader in energy consumption and pollution, this country also produces the most waste per capita. Estimates of the amount of solid waste the United States generates range from around 5.4 to 9 billion metric tons a year or more (depending on how "solid waste" is defined and how the estimates are calculated). Using the former figure, this amounts to about 18 metric tons of waste a year for every American or approximately 49 kilograms (108 pounds) of waste a day. The average person does not literally throw away 49 kilograms of trash each day. The trash that individuals put out for the garbage collector, the **municipal solid waste** (old newspapers, packaging materials, empty bottles, and so forth), makes up a small percentage of America's waste (**FIGURE 18-1**). Most waste in the United States comes from mining, agricultural, and industrial operations, but ultimately, these operations exist to feed and provide for the necessities and desires of the consuming public. The average American directly generates about 4.3 pounds of trash and garbage each day. After subtracting the amount composted or recycled, the final discards (including combustion with energy recovery) are about 2.4 pounds per person per day, about the same in 2010 as they were in

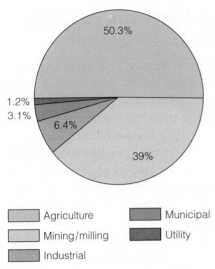

FIGURE 18-1 A breakdown, by percentage, of the approximately 5.4 to 9.0 billion metric tons of solid waste generated annually in the United States. (*Source:* Environmental Protection Agency.)

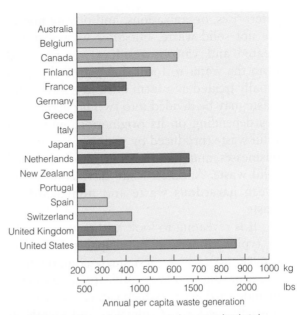

Annual per capita waste generation

FIGURE 18-2 Annual municipal waste generated per person in selected developed nations. (The data in this figure are for comparative purposes only; more recent data across all countries were not available. Because of differences in methods of estimating and accounting, the annual waste generated per person in the United States as shown here does not precisely agree with the data given in Table 18-1.) (*Source:* Data compiled from World Resources Institute, World Resources 1992–1993. New York: Oxford University Press, 1992, p. 319.)

1970, even as waste generation increased by more than a pound per person per day during the same period (**TABLE 18-1**); however, this amount is still quite a bit more than is generated per person in various other countries (**FIGURE 18-2**).

18.1 Defining Solid Waste

Solid waste, broadly defined, includes a number of items that are not generally thought of as "solid." Certainly, household **garbage**, **trash**, **refuse**, and **rubbish** are all solid waste, but so too are solids, various semisolids, liquids, and even gases that result from mining, agricultural, commercial, and industrial activities. Often substances such as liquids and gases are confined in solid containers and disposed of with more conventional solid wastes. Sewage, effluent, and wastewater from commercial

TABLE 18-1	Generation, Materials Recovery, Composting, and Discards of Municipal Solid Waste (MSW), 1960–2010											
Pounds per Person per Day												
Activity	**1960**	**1970**	**1980**	**1990**	**2000**	**2003**	**2004**	**2005**	**2007**	**2008**	**2009**	**2010**
Generation	2.68	3.25	3.65	4.50	4.63	4.53	4.61	4.54	4.63	4.52	4.34	4.43
Recovery for recycling	0.17	0.22	0.35	0.64	1.03	1.05	1.07	1.08	1.15	1.11	1.09	1.15
Recovery for composting*	Neg.	Neg.	Neg.	0.09	0.32	0.36	0.38	0.38	0.39	0.40	0.37	0.36
Total materials recovery	0.17	0.22	0.35	0.73	1.35	1.41	1.45	1.46	1.54	1.51	1.46	1.51
Combustion with energy recovery†	0.00	0.01	0.07	0.65	0.60	0.63	0.64	0.62	0.58	0.57	0.52	0.52
Discards to landfill, other disposal‡	2.51	3.02	3.24	3.12	2.62	2.49	2.52	2.46	2.51	2.44	2.36	2.40
Population (thousands)	179,979	203,584	227,255	249,907	281,422	290,850	293,660	296,410	301,621	304,060	307,007	309,051

*Composting of yard trimmings and food scraps and other MSW organic material.
†Includes combustion of MSW in mass burn or refuse-derived fuel form and combustion with energy recovery of source separated materials in MSW (e.g., wood pallets and tire-derived fuel).
‡Discards after recovery minus combustion with energy recovery. Discards include combustion without energy recovery. Details may not add to totals because of rounding. (*Source:* Environmental Protection Agency.)

enterprises, organizations, and private homes are not solid waste, but after wastewater is treated and various residues are removed from the water to form sludge, the sludge is usually treated as a form of solid waste. Solid waste may be divided into two broad categories depending on its origination: municipal solid waste (produced by various institutions, businesses, and private homes) and **industrial solid waste.** Another useful distinction is between **hazardous waste** and **nonhazardous waste.**

It is revealing to look at the composition of typical American municipal solid waste (**FIGURE 18-3**). The largest single component is paper products; other significant components include yard waste (which can easily be composted), food waste, plastics, and metals. If we analyze the waste in terms of products discarded (Figure 18-3b), the single largest component by weight is containers and packaging (including paper, plastic, and metal packaging), followed by nondurable goods (such as newspapers, telephone directories, clothing and textiles, and various consumer electronic products that have a short life span), durable goods, yard trimmings, and food scraps. Americans are composting more, but we are still throwing out approximately 40% of our yard trimmings and virtually all of our food scraps; together they account for almost 25% of U.S. waste.

18.2 Alternative Paradigms for Waste Management

For most of human history, wastes were usually disposed of by a **"dilute and disperse"** strategy. Early gathering and hunting cultures simply left their trash where it fell and moved on. In the pre-Industrial and early Industrial Ages, settlements, communities, and factories were often located near waterways. Streams and rivers not only supplied fresh water from upstream, but also provided a convenient way to get rid of wastes. Waste materials were dumped into the river and washed away. As the water flowed downstream, the wastes and pollutants became less concentrated and were also naturally filtered out or absorbed by organisms and sediments. Wastes were often simply dumped into the seemingly infinite reservoirs of the atmosphere and the oceans. Most rivers eventually run to the sea, and coastal cities could dump raw sewage and waste directly into the ocean. Tall smokestacks could inject waste, in the form of gases and fine particulate matter (smoke), into the skies.

At the beginning of the Industrial Age, 220 years ago, many philosophers and social thinkers could not conceive of a time when, on a global scale, we would begin to run out of clean air and fresh water—but this is in fact just what is happening (as we saw

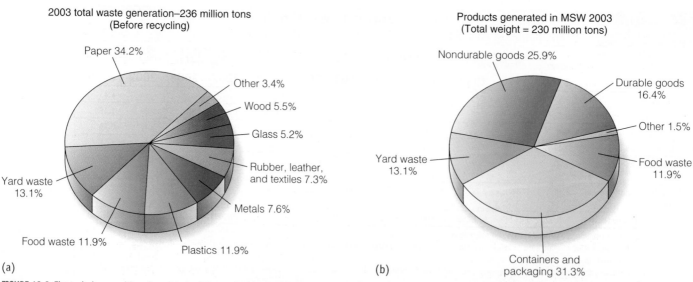

FIGURE 18-3 The typical composition of municipal solid waste in the United States according to the weight of various types of materials. (*Source:* Environmental Protection Agency.)

in reviews of water pollution, local and regional air pollution, and global air pollution). Today, the pollution of industrial humanity is found throughout the globe, even in the most remote and isolated areas. It has become clear that the dilute and disperse approach to waste management is no longer valid as a general paradigm. The capacities of Earth's natural ecosystems and reservoirs (sinks) for absorbing human-produced wastes are being reached and surpassed. In many cases, the natural systems have been so overloaded that not only are they incapable of dealing with all the human-produced waste hurled into them, but they also are actually damaged by the excess waste, with the result that they can no longer function effectively and efficiently.

People have produced innumerable new types of waste that nature was never equipped to deal with, such as new elements (for instance, plutonium and other artificially produced elements and isotopes), synthetic compounds (chlorofluorocarbons, assorted types of plastics, synthetic pesticides, fertilizers, and so forth), and highly concentrated quantities of substances that are found in nature only in very dilute form, such as heavy metals given off by factory operations. Even under the best of conditions, with certain types of waste, the dilute and disperse strategy will do only that—dilute and disperse—because natural systems lack the capability to absorb and nullify these wastes. Long-lived radioactive isotopes or synthetic compounds simply accumulate throughout the globe.

By the early 20th century, it was clear that the dilute and disperse strategy had reached its practical limits, and a new concept was introduced—"**concentrate and contain.**" Concentration of waste has been used to a certain extent for many millennia. North American natives concentrated oyster shells in huge middens, and the ancient cities of the Old World developed large urban trash heaps (which are now much sought after by archaeologists). In many cases, concentration was only the first step toward ultimate dispersal. Trash might be collected and brought to a dump to be burned and the ashes then thrown into a river or harbor. The relatively modern concept of concentration and containment is to collect waste

and then permanently isolate it from the rest of the environment. **Sanitary landfills** and hazardous and radioactive waste disposal sites are tangible expressions of the concentration and containment strategy of waste management. Often the dilute and disperse and concentrate and contain strategies are used in tandem, as in some incinerator operations. Rubbish may be burned in an incinerator, dispersing into the atmosphere a good percentage of its original volume (60% or more), but the remaining bottom ash and collected fly ash constitute a highly concentrated, and often toxic, waste product that must be isolated from the environment, usually by being permanently contained in a specially designed **landfill**. (Bottom ash collects in the bottom of the incinerator, and fly ash is removed from the smoke in the incinerator's chimney.)

There are numerous problems with the concentrate and contain strategy, especially as it is applied to modern industrial wastes. Many modern wastes are synthetic or artificial in nature and are extremely hazardous, especially in concentrated form. Therefore, they must be contained securely for very long periods of time, essentially forever. Yet we have learned from experience that perfect containment is very difficult to achieve; even the best-designed landfills may leak, containers can break, accidents and natural disasters happen, and there is always the risk that toxic substances in containment will be purposefully released, such as by a terrorist group. In addition to these practical problems, containment is also subject to deeper criticisms. Is it fair and right for one generation to use resources for its own advantage and then remove them forever from use at the expense of all future generations? It can be strongly argued that ethically we should not be depriving future people of valuable material resources by converting those materials into hazardous or toxic waste that must be taken out of circulation permanently.

As a result of such considerations, waste management analysts have suggested a third strategy, variously referred to as "resource recovery," "industrial ecosystems," "sustainable waste management," and most optimistically "resource management." The basic philosophy underlying this strategy is that there should be no such thing as material waste, although the energy powering the system may produce waste heat. Under sustainable waste management, unnecessary waste (that is, unnecessary use of materials and energy) is first reduced at the source. After materials (goods) are used for the designated purpose, all remains are re-used or recycled. However, we currently are a long way from achieving this goal. Even if the best reuse and recycling programs currently available were put into place globally, human-produced waste would still exist, and we would still be throwing out valuable material. Max Spendlove, with the U.S. Bureau of Mines, popularized the term "urban ore" to describe the raw resources now buried in our dumps and landfills. In 1971, he said, "All the tin cans, wire coat hangers, bottle caps, pots, alarm clocks, electric knife sharpeners—almost everything the housewife chucks out is rich in iron, aluminum, copper, zinc, tin, lead, and brass. We can mine all that stuff. Our refuse is richer than some of our natural ores." We may yet resort to unearthing our garbage dumps and landfills, but if we want to stop throwing out what we can reuse, a major change in the technology of manufacturing will be necessary. In the future, there may be no such thing as waste—only potential resources. Recent developments in zero waste and zero emission production plants and design initia-

tives are a step in the right direction but still a fledgling worldwide undertaking. We will have to continue to address the question of waste management for the near future.

Traditional Means of Waste Management

For the past century, people have dealt with solid wastes in three basic ways: (1) by burning the waste, thus essentially injecting much of it, converted to gases and smoke, into the atmosphere; (2) by storing wastes, including the leftover ash from burning, in dumps, impoundments, and most recently sanitary landfills; and (3) by injecting or burying wastes in rock cavities deep underground (a method proposed for the disposal of industrial and conventional toxic or hazardous waste, as described later, and also for radioactive waste [see a review of the fundamentals of energy, fossil fuels, and nuclear energy]). Each of these waste disposal methods has its proponents and its critics. In the next few sections, we consider each briefly.

Incineration

In the industrial technique of **incineration** (**FIGURE 18-4**), trash and garbage are burned in a large furnace at high temperatures to get rid of as much of the refuse as possible. Of course, burning trash is a time-honored procedure, but the use of large incinerators dates back only to the late 19th century.

During their first 50 years of existence, incinerators were in and out of fashion. Many early incinerators were relatively inefficient, caused massive pollution, and left large quantities of ash and other nonburnables. But by World War II, some 700 new and improved incinerators were operating throughout the United States, on both a large scale and a small scale. Some apartment buildings even had small incinerators to burn the residents' trash. Nevertheless, incinerators continued to cause problems. Aesthetically, incinerators were an offensive intrusion on the skyline, and people who lived near them complained of the odors and said the smoke and gases caused respiratory problems. As early as the 1950s, many incinerators were closed. The increasing environmental awareness of the late 1960s and early 1970s continued to

FIGURE 18-4 A trash incineration plant.

erode people's confidence in and tolerance of incinerators. The Air Quality Act of 1967 and the Clean Air Act Amendments of 1970 established new emission standards that many existing incinerators did not meet; most operators simply closed their incinerators rather than adding costly emission control devices.

Shutting down the incinerators meant that the trash and garbage had to be disposed of some other way. The preferred alternative was sanitary landfills. By the late 1970s and early 1980s, many cities and municipalities were finding that their landfills were running out of space, and there were fewer and fewer sites available, at least politically. A new breed of incinerators known as resource recovery plants presented an apparent solution to this predicament.

The basic idea behind **resource recovery facilities** (or "plants") was that trash and garbage would be burned, and the heat generated could be recovered and applied to some useful end, such as generating electricity. Resource recovery facilities, also commonly referred to as *waste-to-energy facilities*, currently take two basic forms: refuse-derived fuel facilities (RDFs) and mass-burn incinerators. In an RDF, the solid waste is fed into the plant on a conveyer belt. Then, through a series of mechanical operations such as pulverization, shredding, sieving, density separation, and magnetic separation, the waste is sorted into various components—paper, wood, and other combustibles; iron, steel, and other magnetic metals; aluminum; glass; and so on. In this manner, recyclable materials can be recovered from the waste, whereas the shredded combustibles can be burned onsite to generate electricity or formed into a fuel (often in the form of pellets) that can be sold to conventional coal-fired plants to be burned in place of coal. RDFs are often expensive to operate because of the complicated machinery involved and also because there are often large fluctuations in the markets for the recycled materials and fuel produced by such a plant. In addition, the refuse-derived fuel cannot always be sold easily because it may not be of consistent quality.

Mass-burn incinerators take a more direct approach to the waste (**FIGURE 18-5**). The un-

FIGURE 18-5 The interior of the RESCO (Refuse Energy System Company) refuse-to-energy plant in Bridgeport, Connecticut. The crane is transporting waste from the refuse pit to feed the high-temperature furnace.

sorted trash and garbage are simply fed into a furnace that burns the refuse at very high temperatures (around 1800°F to 2000°F, or 980°C to 1100°C). The heat from the burning refuse is used to produce steam that drives a turbine to generate electricity. Whatever is not burned in the incinerator is removed and simply disposed of (for instance, in a landfill).

Problems With Incineration

Incinerators not only burn trash, reducing it to ash and gases, but they also give off (indeed, manufacture) huge amounts of toxic pollution. As an example, a typical 1990s state-of-the-art incinerator (one that may be in use today) that meets all regulatory requirements burning 2,042 metric tons of refuse a day could emit 4.5 metric tons of lead and 15.4 metric tons of mercury per year, as well as 263 kg of cadmium, 263 kg of nickel, 2,040 metric tons of nitrogen oxides, 774 metric tons of sulfur dioxide, 705 metric tons of hydrogen chloride, 79 metric tons of sulfuric acid, 16 metric tons of fluorides, and 89 metric tons of small dust particles. Perhaps most dangerous of all, incinerators produce and expel dioxin.

Dioxin (actually not a single substance as the term is commonly used but a group of more than 75 related compounds) is considered by some experts to be one of the most toxic chemicals that people have ever manufactured.

Exposure to dioxin can cause skin eruptions (which resemble acne), headaches, dizziness, digestive disorders, birth defects, and various forms of cancer. Dioxin was an unintended contaminant of the defoliant Agent Orange that was used during the Vietnam War. It is formed as a by-product in the production of some herbicides, certain bleached papers, and other products. Dioxin can form inadvertently in many situations. In particular, dioxin can be formed when organic wastes containing chlorine are burned. Plastic containers, paper, and other trash typically found in municipal solid wastes can often give rise to dioxin when burned in incinerators.

Not until the late 1970s was it even realized that dioxin production from incinerators might be a problem. At first, many experts believed that the high temperatures of incineration would destroy any dioxin in trash. In fact, this was essentially true: Dioxin is destroyed in the furnace of the incinerator, but dioxin is then synthesized from the chemicals present in the cooler parts of the incinerator beyond the furnace and rises up the smokestack.

Modern incinerators are equipped with all sorts of mechanisms to capture the dioxins, heavy metals, and other toxic substances that would otherwise pour out of their stacks. The gases are subjected to various scrubbers, electrostatic precipitators, and fabric filters that catch the fly ash and toxic components of the gases. However, even in the best cases a commercial incinerator that meets all regulations can release substantial quantities of dangerous substances through its stacks. Furthermore, after incineration, a substantial bulk of both bottom ash and fly ash remains to be disposed of, usually in a landfill or the equivalent. Some bottom ash may contain toxic levels of contaminants, but in other cases, it may be safely deposited in an ordinary landfill or even used to make building materials such as cinder blocks. The fly ash, which contains the dioxin, heavy metals, and other toxic substances that were removed before the gases left the stacks, typically is much more hazardous than the bottom ash.

This hazardous material must be carefully isolated from the environment. In some cases, it is chemically stabilized and hardened to a rock-like consistency before being deposited in a specially designed landfill. Unfortunately, abuses have occurred, to everyone's detriment. For instance, incinerator ash containing toxic substances has been improperly dumped in older landfills from which the toxics escaped into the general environment.

Because of the regulations involved, incineration of waste is generally the most expensive form of waste disposal, costing much more per ton of waste than landfilling (landfilling generally costs between $60 and $270 per metric ton, whereas land-based incineration costs range from about $400 to $550 per metric ton). Currently, the construction of a typical mass-burn incinerator that can dispose of 1,815 metric tons of garbage a day will cost about $250 million. Assuming a rate of garbage production of 1.8 kilograms (4 pounds) per person per day, such an incinerator would service about 1 million people. To avoid many of the regulations that are applied to land-based incinerators, large amounts of material (especially waste liquids such as oils and solvents) are regularly burned on ship-based incinerators in the open oceans (ocean-based incineration can cost half the price of land-based incineration). This may cut costs by skirting some regulations that do not apply, or cannot be enforced, on the high seas, but of course, the hazardous materials released by ship-based incinerators still enter the biosphere.

Dumps and Landfills

One of the most ancient and most common ways of disposing of solid waste is to simply pile it up in a convenient but out-of-the-way place. Accordingly, most cities, towns, and other populated areas have generally had open **dumps** on their outskirts. There, trash and garbage would simply be left to sit or to be picked through by garbage pickers and sifters. In other cases, liquid or semiliquid wastes were contained in artificial or natural lakes or impoundments. But open dumps and impoundments tend to be associated with many problems, especially as they grow larger. Aesthetically, they can be smelly and unsightly. They can present serious health hazards, serving as breeding grounds for

FIGURE 18-6 Smoke rises from an open garbage dump in El Salvador.

disease vectors and attracting pests. Dumps can significantly contribute to local air pollution, especially if the trash is periodically ignited and allowed to burn—as is done at some open dumps to reduce the volume of waste (**FIGURE 18-6**). Chemicals can also leak from an impoundment or leach from a dump, perhaps as it is rained upon, and the resulting runoff can seriously pollute the local surface and groundwater.

Despite the problems associated with open dumps, they remain a common form of solid waste disposal, especially in developing countries. Even in the United States, open dumps are not hard to find. A few of these dumps are officially sanctioned, but most open dumps in this country are relatively small and illegal—essentially, accumulations of rubbish in relatively deserted areas. The percentage of solid waste that actually ends up in American open dumps is quite small (**CASE STUDY 18-1** describes a type of solid waste that often ends up in large open piles).

Replacing the ancient dumps have been modern techniques of solid waste disposal, such as incineration and the lineal descendant of the open dump, the modern sanitary landfill. In the simplest sense, a modern sanitary landfill is essentially a closed dump (**FIGURE 18-7**). The solid and semisolid wastes are confined to a specific area, usually a giant hole in the ground; after being compacted, they are covered over daily with a layer of soil. Thus, the hole in the ground is filled ("landfill"), and the waste is isolated from the general environment by the periodic dirt covering, making the whole complex relatively "sanitary" as compared with open dumps.

The earliest recorded predecessors of sanitary landfills date to the early 1900s in the United States and to the 1920s in Britain. By the 1930s, sanitary landfills were catching on throughout the United States. In 1945, approximately 100 American cities had sanitary landfills; by 1960, approximately 1,400 cities had adopted this method of solid waste disposal. Today, more than half of U.S. municipal solid waste is disposed of in landfills.

Through their hundred years of existence, the philosophy and design of sanitary landfills have undergone major evolutionary shifts. From the beginning, the central idea was that after a landfill was completely filled, it could be covered with a final layer of clay

FIGURE 18-7 A sanitary landfill in the process of being capped.

18.2 Alternative Paradigms for Waste Management   545

Tires

Every year, Americans dispose of an estimated 270 million car, truck, bus, and tractor tires. These tires, although only about 1.8% of the American municipal solid waste stream by weight, pose a major disposal problem (**FIGURE 1**). An estimated 300 to 500 million (numbers vary, depending on the source) used tires are stockpiled in America today, many of them in large, unsightly outdoor heaps. Many communities no longer allow tires to be discarded with other rubbish; tires are classified as "special waste," and disposing of them may cost $2 or more a piece. One of the problems is that tires contain pockets of air and are not easily flattened completely, so in older landfills where the tires were not properly disposed of (perhaps by weighting them), they can "float" up to the top of the landfill. In floating to the top, they can disrupt and destabilize the other components of the landfill, causing serious damage. They can even float up and break the clay caps that have been placed over landfills. At some older landfills that once accepted tires, the tires that have surfaced must periodically be skimmed off. Today, more than 30 states ban whole tires from landfills.

Piles of tires are not just eyesores; they contribute to other environmental and health problems. The tires can catch on fire, producing billows of noxious black smoke and fumes. A tire pile is mostly air, which promotes the fire, and the rubber in the tires burns at very high temperatures. As they burn, tires give off dangerous oily liquids. Putting water on the fire in an attempt to extinguish it can cause these liquids to spread, causing more environmental damage. Extinguishing the fire can be extremely difficult. Some tire fires have burned out of control for weeks or even months. A major tire fire in Hagersville, Ontario (Canada), in 1990 consumed 14 million tires and temporarily drove 4,000 people from their homes. In 1999, a tire fire near Westley, California, burned for 30 days and released an estimated 8 million pounds of pollutants into the atmosphere.

Piles of tires also provide a breeding ground for disease-spreading pests. Water can collect in tires, providing the perfect incubation site for mosquitoes. Rats and other rodents make their homes in tire piles, especially if the piles are located near a garbage dump (as is often the case) that provides a ready source of food.

Fortunately, not all scrapped tires are simply sent to landfills or stored in unsightly heaps. Of the estimated 290 million scrap tires generated in the United States in 2003, an estimated 233 million went to some kind of market for reuse, although that still left more than 50 million tires that were simply "trashed." Here are the numbers for the tires that went to market in 2003:

- Used as fuel: 130 million (44.7%)
- Used in civil engineering projects, such as building erosion barriers or retaining walls: 56 million (19.4%)
- Tires processed into ground rubber: 30 million (12.1%)
- Tires exported out of the country: 9 million (3.1%)
- Tires cut up and made into stamped or punched products: 6.5 million (2%)
- Tires used for agricultural or other miscellaneous uses: 3 million (1.7%)

Unfortunately, only a limited number of practical uses have been devised for used tires. Perhaps the most obvious way to deal with the tire problem is to cut down on the number of tires entering the

FIGURE 1 A typical tire dump.

waste stream by making each tire last longer. Proper inflation, balancing, and periodic rotation can all extend the life of tires. Some brands and types of tires last significantly longer than other types, and retreading can extend the life of a used tire. Because of their initial expense, many bus and truck tires are commonly retreaded, but most individuals prefer to buy new tires for their passenger cars because new tires tend to cost little more than retreads. In 2003, it was estimated that only 24 million retread tires were sold in the United States and Canada combined, but even doubling, tripling, or quadrupling the life span of the average tire will not solve the tire disposal problem. With so many cars on the road, tens of millions of tires will continue to be added to America's waste stream each year.

Reclaiming and recycling the rubber from tires is difficult and expensive. The various synthetic rubber, glass, and steel layers in radial tires (1.13 kilograms [2.5 pounds] of steel is embedded in a typical steel belted radial passenger tire) make reclamation all the more difficult and generally impractical. Whether whole or cut up, tires have only limited uses. Some playground equipment is composed largely of used tires (**FIGURE 2**). Tires are used on boat docks and piers as bumpers; sandals, floor mats, and other products can be made from cut-up tires, and rubber chips produced from chopped-up tires can sometimes replace traditional mulch in gardens. Whole, shredded, and chipped tires have been successfully used in the construction and maintenance of landfills. Tires, weighted with concrete, have been used to start artificial reefs on coastlines (although not always with satisfactory results); likewise, tires have been used as a primary building material for breakwaters and erosion barriers on beaches. There has been some use of tires packed with dirt or cement as a structural material inside the walls of buildings. A promising large-scale secondary use for tires is in "asphalt rubber." Simply put, tires are ground up and mixed with asphalt to produce a material, asphalt rubber, that can be used to surface, repair, and restore roads. Million of tires have been used in asphalt rubber

in both the United States (especially on federal highways) and Europe. However, most secondary uses of tires are rather limited (e.g., only so many playgrounds or breakwaters are needed) and do not solve the problem of what to do with the hundreds of millions of stockpiled tires and the tens of millions more entering the waste stream annually.

Tires can be burned as fuel in specially designed power plants; indeed, as you will have noted from the previously cited statistics, this is the single largest market for used tires. Currently, approximately 70 facilities in the United States use approximately 130 million recycled tires each year as fuel across the following industries:

- Cement industry–41%
- Pulp and paper mills–20%
- Electric utilities–18%
- Industrial/institutional boilers–13%
- Dedicated tire-to-energy facilities–8%

FIGURE 2 This playground and park used 3,000 recycled tires.

VISIT

http://environment.jbpub.com
/mckinney/5e/
for more information

Tires can also be burned with other materials in more conventional power-generating incinerators or as a source of heat and energy for industrial operations, such as paper production. Critics of tire burning are concerned about the toxic air pollution and ash that it may produce. In many areas, tire burning has not found favor with the public.

Questions

1. Make a list of possible uses for used tires; include both current uses that you are aware of as well as possible new uses.

2. Referring to your list of possible uses for old tires, which uses are the most environmentally friendly? Which are the least? Are some uses of tires worse than simply stockpiling them?

3. Do you advocate the burning of tires as a source of fuel? Is it better to keep the carbon (and various other substances found in tires) sequestered rather than release it into the atmosphere?

4. How can we reduce the number of tires that are disposed of annually?

and dirt and the "reclaimed" land devoted to some other purpose. This is the concept of sequential land use: What was "wasted" land in the past, perhaps a natural swamp or gully, becomes a place to deposit waste, and after the land has been completely filled, the reclaimed land can serve as the site for homes, a park, a factory, or any of numerous other uses. However, over the decades, there has been a dramatic shift in thinking as to what land is suitable for landfills. In the early decades of landfill development, the most suitable and desirable locations were thought to be swamps, marshes, and other low-lying wetlands. Wetlands were viewed as classic "waste land" that served no useful purpose: They were too wet to build on or grow crops and were believed to breed only disease, insects, and other pests. By filling the wetlands with garbage interbedded with layers of soil, the land could be raised in elevation and made dry and usable. Of course, we now understand the importance of wetlands in the overall global scheme; far from filling them in, efforts are under way to preserve remaining wetlands. To make matters worse, building sanitary landfills on wetlands certainly does not isolate the waste from the rest of the environment; in fact, the opposite is true. The

natural wetlands provide an easy conduit for liquids, including very hazardous and toxic substances, to drain from the landfill and enter surface and groundwaters. A swamp or other wetland is just about the worst possible place to site a sanitary landfill.

Today, landfills are sited in areas where contamination of surface and groundwaters will not occur. Some of the best sites for sanitary landfills are in arid regions with little rainfall to produce leachates from the landfill. **Leachate** is formed as water percolates through the refuse, either from the top, such as from rain falling on the landfill, or laterally because of groundwater flow intercepting the landfill. Leachate is essentially an aqueous solution containing any chemicals and particles that can be dissolved, leached, or removed from the trash, including in some instances live disease-producing microorganisms. Leachate can be a hazardous and concentrated brew of substances.

Modern sanitary landfills usually are begun by digging a huge hole at a suitable site. In some cases, a pre-existing hole, such as an abandoned coal or copper mine, can be used but usually not because the surrounding rock will be too permeable. After the hole (which may be 15 meters [50 feet] deep) is excavated, the next step is to line the sides of the cavity with thick layers of dense clay, a thick, sealed plastic liner, and sand and gravel. Many older sanitary landfills are not lined or have inadequate linings, but newer landfills are invariably lined. The liner serves as a barrier to the uncontrolled migration of the leachate out of the sanitary landfill. In the newest landfills, the leachate is not allowed to collect in pools at the bottom of the landfill; if it did, the pool eventually would overflow the sides of the landfill. Instead, pipes and drains in the bottom of the landfill collect the leachate so that it can be removed.

Landfill operators want their landfill to stay as dry as possible, so the leachate is removed, collected, and treated. The leachate may be treated like sewage, either being treated at a special plant on site at the landfill or sent to the local municipal sewage facility. Typically, the water is separated from the leachate, purified, and then released into a local river or the ocean. The remaining solid sludge may be dumped into the landfill once again, or it may be burned and then the ashes placed in the landfill. In some cases, the sludge can be used as a fertilizer, or it may be safely dumped into the ocean. In other cases, the leachate and sludge contain enough toxic substances that they are considered to be a hazardous waste and therefore must be disposed of in a suitable hazardous waste facility.

Besides leachate, sanitary landfills produce **methane gas** as bacteria decompose certain organic components in the landfill. This methane must be removed from the landfill on a regular basis, or it could accumulate and form a hazard if it ignited. In some cases, pipes perforated with holes are inserted into wells drilled down into the landfill; the gas can then be collected from these pipes. Another method is that a lattice of perforated pipes may be included within the landfill as the refuse is accumulating. At some landfills, the methane is simply burned or released directly into the atmosphere (which, of course, contributes to the greenhouse effect; see a review of global air pollution), but at other facilities, the methane is collected, purified, and sold as fuel.

Once a sanitary landfill is full, it is capped with a thick layer of compacted clay and soil. It can then be seeded, landscaped, and developed to serve any number of purposes. Parks, golf courses, and similar facilities are ideally suited for areas underlain by old landfills, but even houses or industrial buildings can be situated on these sites. However, the materials in the landfill may continue to compact for a number of years after closing, causing surface and subsurface subsidence; it is unwise to build major structures on a newly capped landfill. In addition, methane gas will continue to form in the landfill for many years, as may leachate. The methane must continue to be removed long after the landfill has been closed. Likewise, the former landfill and the area adjacent to it should be monitored closely to ensure that no pollutants begin to escape.

Currently (statistics for 2009), about 50% of U.S. municipal solid waste is disposed of in landfills; 33% is recycled or composted, and the remainder is incinerated or otherwise burned (including in waste-to-energy plants). Despite this continued high reliance on landfills, during the past two decades, many landfills have closed (**FIGURE 18-8**).

According to the Environmental Protection Agency (EPA), in 1978, there were about 20,000 operating landfills in the United States; by 2005, there were only 1,754 landfills, and that number remains steady (2009). However, during the same period, the average size of landfills has increased. On a national level, according to the EPA, landfill capacity (ability to receive more wastes) has remained relatively constant and does not appear to be a problem, although locally a particular

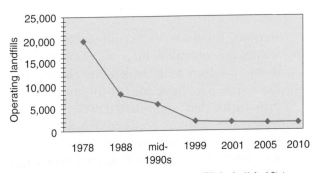

FIGURE 18-8 The declining number of operating landfills in the United States.

community may fill a landfill to capacity and then need to seek alternative waste disposal means (perhaps opening a new landfill or shipping wastes to another landfill combined with more intensive recycling). It is increasingly difficult to open new landfills (both because of their expense and the not-in-my-back-yard [NIMBY] syndrome).

New EPA regulations are increasing the pressure on the nation's landfill capacity. The requirements include stricter guidelines for landfill liners, leachate collection systems, and groundwater and methane monitoring. This will increase the cost of opening and operating a landfill to an estimated average of $125 million and may force closure of many older landfills that cannot economically be upgraded to meet the new requirements. As of the mid-1990s, the estimated average remaining life expectancy of an American landfill was about 14 years.

Some researchers contend that the whole question of landfills running out of capacity is a political issue, rather than a scientific problem, for the United States. They argue that the only reason it is harder to site new landfills is the NIMBY syndrome; people do not like the thought of "disgusting" landfills in their communities. One economist has estimated that all of America's projected solid waste for the next 500 years could be easily fit within a single landfill 32 kilometers (20 miles) on each side and about 91 meters (100 yards) deep. This would certainly be a huge landfill, but it would represent only a small dot on a map of the continental United States. Another researcher has suggested that the ideal location for such a national landfill (or landfills) would be in the thick shales that cover most of western South Dakota.

Deep-Well Injection

Various types of industrial or hazardous wastes are sometimes disposed of by injecting them into deep wells drilled into the crust of the Earth. The basic idea is to pump wastes into rocks that are situated well below all freshwater aquifers so as to avoid contaminating any groundwater supplies. A typical well might be hundreds to several thousands of meters deep. For decades, the oil industry has used deep-well injection to dispose of liquid wastes, such as salty brines, that are often pumped out of the ground in association with oil production—the brines are simply injected back into the rock. Using **deep-well injection** to get rid of other types of wastes is associated with many potential problems and as a result is opposed by many environmentalists.

Natural earthquakes may divert groundwater flow and allow previously isolated injected waste to contaminate other bodies of groundwater. This could ultimately lead to the spread of the waste and contamination of drinking water. Furthermore, the increased fluid pressure that results from injecting liquid waste deep underground may actually initiate earthquakes. Apparently, the increased fluid pressure allows rocks to move or slide along pre-existing joints and fractures. Earthquakes caused by the injection of wastes into deep wells have been reported in Colorado, Texas, Utah, and California. In another case, the chemical wastes that had been pumped down a well blew back up and spilled into Lake Erie. Despite such accidents, proponents of deep-well injection point out that taken as a group, most operations have no such problems.

Another concern over deep-well injection is that what constitutes a "deep well" at present may not seem so deep in the future. As we deplete the readily available freshwater aquifers, we need to drill deeper and deeper for usable water. In the future, we may need to drill as deep or deeper than the injected waste, but in the process, the injected waste may contaminate our efforts. In addition, not all wastes should be injected into all rocks. Certain wastes may react adversely with some types of rocks or with the natural pore fluids that already occur in the particular rock.

In areas where deep-well injection of waste is used, after the waste is injected into the rocks and the well plugged, the job is not over. The injection field must be permanently monitored to ensure that the waste does not migrate out of the confining rocks into which it was injected, perhaps through the natural pores of the rock or along natural or artificial cracks, holes, or fractures. After many years, even a properly sealed well may begin to leak.

Garbage as a Source of Revenue for Poorer Communities

Some communities have found that an easy way to attract money and jobs is to take the garbage from other areas. A classic recent example is Kimball County, Nebraska, which allowed a hazardous-waste incinerator to be built in its jurisdiction. The incinerator represented a $60 million investment in the financially depressed region. Likewise, New York City shipped treated sludge to a ranch in Hudspeth County, Texas, which created 35 badly needed jobs for local residents.

On a global scale, waste is shipped from richer to poorer nations (**FIGURE 18-9**). Hazardous wastes in particular are shipped from industrialized nations, where tight regulations make the materials expensive to dispose of, to developing nations with looser regulations or lax enforcement of existing regulations. Such overseas shipments are hard to monitor, especially because the material may not be properly identified as hazardous waste. The United States and Western Europe are shipping millions of tons of hazardous waste to Asia, Africa, Latin America, and Eastern Europe every year. Sometimes host countries are happy to take the trash; before German unification, East Germany willingly accepted wastes from West Germany and Denmark for a fee. In other instances, the disposal is totally surreptitious. For instance, in 1988, the Nigerian government discovered that hazardous wastes from Italy were being dumped in a Nigerian port. To stem the illegal movement of waste, the United Nations Environment Programme oversaw the development of a treaty, adopted by many nations in 1989, to regulate the international movement of hazardous wastes.

Although such waste exchange arrangements can be mutually advantageous, as already pointed out, critics argue that the short-term economic gains are more than offset by the long-term costs and hazards of accepting other people's waste. Ultimately, the waste may pollute and otherwise degrade the environment, causing a self-perpetuating cycle of decreasing property values and the movement of wealth from the area. In the case of a typical landfill or incinerator/landfill

FIGURE 18-9 Containers of Australian computer waste being impounded by Filipino authorities in Manila.

operation, eventually the space will be filled; at that point, what happens in the accepting community? When the waste no longer flows in, neither will money and jobs; the community may be left with a pile of garbage that still needs to be monitored.

Dealing with Rubbish and Other Waste in the Near Future

Many nations, experts, and organizations (including the EPA) agree on a common approach, at least in principle, to the problem of waste management. Often known as the "waste management hierarchy," this approach involves a list of options ordered from the most desirable to the least desirable along the following lines:

1. **Source reduction.** Reduce the generation of waste in the first place.
2. **Reuse** of products. For instance, washing and reusing beverage containers directly.
3. Recovery and **recycling.** Using "waste" as the raw materials of secondary industrial processes, such as collecting old aluminum cans, melting them down, and using the recovered aluminum to manufacture new products (see "Tires" Case Study 18-1).

4. Waste treatment and incineration. This may include recovering energy as the trash is burned.

5. Storage and disposal. The residual ash and solids of incineration or other waste treatment ultimately must go somewhere. In some cases, the material can be used to manufacture items, such as cinder blocks from certain types of incineration ash, but more typically what should be the last resort is used first. The waste may be permanently disposed of in a landfill. This last method does not force people to deal with the issue of waste nor does it change their behavior.

Reducing Consumption

Certainly, the most immediate, and ultimately most beneficial, solution to the solid waste problem is source reduction—using less raw material in the first place. Source reduction can be accomplished in many different ways, including the use of alternative raw materials (such as substituting lighter and stronger aluminum for other materials in some goods), altering the product being produced, and changing manufacturing procedures. Ultimately, simply consuming less can accomplish source reduction. Many waste management experts suggest that the best way for the average citizen to help solve the waste management problem is to not purchase or use unnecessary goods. But source reduction alone can never solve all of our waste problems. People require a minimum number of material products to survive, so appropriate source reduction must be combined with other objectives farther down the waste management hierarchy.

Reuse is next on most people's lists of ways to deal with the garbage crisis. In some ways, reuse is really just a form of source reduction. Instead of making 50 soft-drink bottles, use one bottle 50 times. Yet, reuse has not captured the popular imagination to the same degree as recycling.

Lately, recycling has become increasingly popular, and most people view this as a positive development (recycling is discussed in more detail later in the chapter). However, some environmentalists argue that as a solution to the refuse problem, recycling is deceptive and misleading and perhaps even contributes to the problem by diverting energy and resources that could be used for more beneficial endeavors—namely, actual waste reduction and reuse of goods. Instead, it is argued, we should change our basic consumption habits. We can use and reuse while consuming at a lower rate. For instance, virtually all food-carrying containers could be manufactured so as to be reusable—from the bottles that your beverages come in to the containers for your cereal (although they may have to be manufactured from something other than cardboard). Reusing, if properly organized, should be no more inconvenient than a recycling or ordinary garbage collection program. Massive reuse would save energy and resources.

Waste treatment, such as incineration, and permanent disposal in landfills are on the bottom of many people's waste management hierarchy. Because of the air pollution and other dangers inherent in some forms of incineration, some environmentalists contend that landfilling and other forms of permanent storage should be placed before incineration in the hierarchy. In contrast, many proponents of the "waste-to-energy" industry argue that incineration should be much higher on the hierarchy than it already is. Waste-to-energy incineration plants are touted as a form of "recycling"; heat from burning trash is converted to usable energy.

Finally, it is possible to view the whole notion of a "waste management hierarchy" as counterproductive. People who hold this view argue that individual communities should be allowed to pursue an "integrated approach" to waste management, choosing the technologies that best suit local needs. A strong argument against this line of thinking is that waste management is no longer simply a local problem. The world must soon begin addressing this issue in concert if we are not to be overwhelmed by trash and pollution on the land, in our waters, and in the world's atmosphere.

Waste Disposal and Recycling in the United States

According to EPA statistics (2009), in the United States, 54.3% of municipal solid waste (by weight) is discarded (for instance, in landfills or by incineration without energy recovery), 11.9% is combusted with energy recovery, and the remaining 33.8% is recycled. These numbers represent a significant change in municipal solid waste disposal approaches (**FIGURE 18-10**). In addition to reducing the waste stream, using recycled secondary materials decreases our energy use, water use, mining wastes, and air and water pollution (**TABLE 18-2**).

In the United States, four prominent categories of materials are commonly recycled: glass, metals (especially aluminum, steel, and iron), paper, and plastics. But not all recycling is equal. In just these four categories, we have both **closed-loop recycling** (or nearly closed-loop), represented by the glasses and metals (these substances can be recycled completely and indefinitely), and various types of **open-loop recycling**, with paper and to an even greater extent plastics (these substances can be recycled only to a limited extent). As cases in point, we now take a brief look at the current status of recycling efforts for these four categories of materials.

Glass

In many ways, pure glass is the ideal substance to recycle for it is virtually 100% recyclable. Glass bottles and jars can be melted down to make new glass bottles and jars, and this process can be repeated in a virtually endless cycle without damaging the raw constituents of the glass—primarily silica-based sand (quartz) and a few other ingredients. Thus, under ideal conditions, glass recycling can form a closed-loop system: Only energy needs to be added to make new glass products from old. Recycling glass, as compared with producing new glass from virgin materials, entails enormous energy savings, decreases the amount of mining that needs to be carried out, and significantly reduces the amount of air and water pollution generated.

In the United States, approximately 26% (based on 2009 data) of glass containers in municipal solid waste are recycled. Billions of glass jars and bottles are recycled every year, but more could be done. Theoretically, virtually all glass containers could easily be

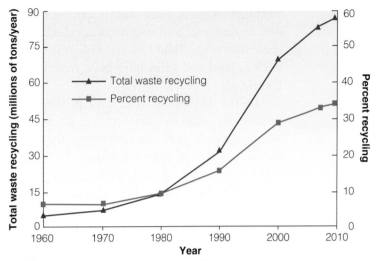

Waste recycling rates from 1960 to 2005

FIGURE 18-10 Waste Recycling Rates of Municipal Solid Waste from 1960 to 2010 in the United States. (*Source:* Environmental Protection Agency.)

recycled. However, not all glass products can be easily and directly recycled, however. Objects that contain glass and other substances (for instance, metals), such as light bulbs, television tubes, and mirrors, are difficult to recycle.

Even when glass cannot be recycled directly into similar glass products, it can be used for construction materials, such as fiberglass (although this usually requires stringent limits on the chemical composition of the glass). Crushed glass can also be substituted for crushed stone in asphalt used in paving roads and other surfaces (sometimes known as "glasphalt").

TABLE 18-2	Environmental Benefits Derived From Substituting Secondary Materials for Virgin Resources			
Environmental Benefits	**Aluminum**	**Steel**	**Paper**	**Glass**
Reduction of energy use	90% to 97%	47% to 74%	23% to 74%	4% to 32%
Air pollution	95	85	74	20
Water pollution	97	76	35	–
Mining wastes	–	97	–	80
Water use	–	40	58	50

Source: Based primarily on data found in R. C. Letcher, M. T. Scheil. Source separation and citizen recycling. In W. D. Robinson, ed. *The Solid Waste Handbook.* New York: John Wiley and Sons, 1986.

Metals

Many metals offer the same recycling advantages as glass. Pure metals can be melted and reused over and over without damage to their substance (**FIGURE 18-11**). Precious metals such as gold and silver have been recycled for thousands of years.

In the United States, steel is the most commonly recycled material; more steel is recycled every year than all other materials combined. More than 45 billion kilograms (100 billion pounds) of steel is recycled in the United States annually (much of this, of course, is from sources other than municipal waste), and more than half of all steel cans from the municipal waste stream are recycled; overall, an estimated two-thirds of all steel products are recycled. Steel is relatively easy to separate for recycling because it is magnetic, and producing new steel from scrap steel instead of virgin ore entails tremendous savings in energy and mining expenses and considerable reductions in the pollution produced.

Aluminum is also easily and 100% recyclable, forming the ideal closed-loop system. Beverage cans are the best-known example, but aluminum foil, pie plates, other disposable containers, and even parts of cars and appliances can be recycled. Producing aluminum from ore (bauxite) is a very energy-intensive process, so recycling can result in energy savings as great as 90% or more (and also reduces pollution given off during manufacturing by comparable amounts; see Table 18-2). Currently, about 50.7% of all aluminum cans in the United States are recycled. Although this statistic is admirable, you may wonder why we cannot recycle as close to 100% as possible. One challenge to recycling is regional differences in attitudes and access to convenient recycling programs. The EPA reports that although more than 80% of people in the Northeast have access to curbside recycling, only 30% of all southern states have curbside recycling programs. There are more than 700 landfill facilities in the South and only 134 in the Northeast. Data indicate in locations where curbside recycling exists, people recycle more.

Many other metals are also recycled in the United States, although mostly by specialized scrap metal dealers. Some of the more common metals include copper (from copper pipes, wiring, and other products), lead (especially from automobile batteries), zinc, nickel, titanium, and chromium. More than 93% of lead-acid automobile batteries are recycled. An emerging industry is the recycling and processing of old computers, including the circuit boards and chips, for valuable metals such as copper, nickel, cobalt, silver, gold, palladium, and platinum.

Paper

About 30% of America's municipal solid waste is composed of paper and paperboard products. Americans are profligate users of paper, using more than 272 kilograms (600 pounds) of paper a year per person. For comparison, annual per capita paper use in parts of the former Soviet Union is 11.3 kilograms (25 pounds), and in China, it is less that a kilogram (2 pounds); of course, not everyone can get a daily newspaper in China. Fortunately, a very high percentage of all paper (more than 40%) is recycled in some form.

FIGURE 18-11 A goldsmith pours molten recycled gold into water to create granulated gold.

Paper, especially cardboard and other corrugated material, is one of the more commonly recycled products in the United States. Indeed, about half of all corrugated paper materials are recycled. Many businesses that depend heavily on cardboard boxes for shipping and packing goods have been recycling the used cardboard for years, so networks for recycling cardboard are well established.

Printing and writing paper, computer paper, glossy paper, and newsprint are all recycled to various degrees in different parts of the country. Currently, about 60% of all U.S. newspapers are recycled (**FIGURE 18-12**), but newspapers are still a major component of landfills. More than half of all high-grade office paper and a quarter of magazines are recycled.

Recycling paper is not exactly comparable to recycling materials such as glass or metals because paper recycling is ultimately not a closed-loop system. Paper cannot be used over and over endlessly because the fibers eventually degrade and become unusable for most purposes (see **CASE STUDY 18-2**). Realistically, many paper fibers can be recycled no more than six to eight times. To offset this effect, virgin material is often mixed with recycled fibers such that even "recycled" paper may contain a significant percentage of virgin fiber. When purchasing "recycled" paper, it is important to distinguish between "preconsumer" and "postconsumer" recycled content. Preconsumer waste is primarily waste produced during the manufacturing of the paper, such as scraps from cutting sheets of paper; paper companies traditionally have reused this preconsumer waste. Postconsumer content is made from other paper that was actually used for some purpose by the public and then returned to be recycled.

The paper recycling business historically has been a roller coaster. Sometimes old newspapers can be sold for a profit, but at other times, a hauler must be paid to take bundled newspapers away. In the late 1980s and early 1990s, some warehouses were full of paper waiting to be recycled. In the past, because of the public's distrust of recycled paper, paper companies often were unwill-

FIGURE 18-12 Currently, about 60% of all U.S. newspapers are recycled.

ing to produce it. However, today the public generally is more accepting of paper containing recycled content. Indeed, domestically and globally there is a continuing strong demand for paper and paper products, and recycled paper is helping to fill the need. During the 5 years from 1990 to 1994, more than 85 paper mills that could recycle paper products were built in the United States. In the early 1990s, old newspaper was of negligible value, but 10 to 15 years later, recyclable newspaper averaged around $55 to $64 per metric ton, and in some areas of the country, prices reached $160 a metric ton. Old newspapers and other scrap paper have become such a valuable commodity across the country that in some areas theft of papers from curbside recycling bins has become a significant problem.

Using scrap paper to manufacture other paper is not the only way that paper can be recycled. For instance, recycled paper is already used to manufacture construction materials, such as wallboard, roofing paper, padding, and insulation. Some farmers successfully use shredded newspaper in place of straw for bedding material for animals. Although not exactly recycling in the traditional sense, paper can be either burned directly as fuel or converted into fuel pellets.

Paper Versus Plastic

Which is less damaging to the environment, paper or plastic? Which fills up our land-fills? Which should you use over the other? Many grocery stores give the shopper the option of paper or plastic grocery bags. Many people instinctively answer that paper is less degrading to the environment; after all, paper is made from natural plant fibers, is biodegradable, and is recyclable. In contrast, most plastic is manufactured using scarce petrochemicals (although there are a few plant-based plastics being developed). Plastic is not always easily or economically recyclable, and once manufactured, plastic may last virtually indefinitely. Almost no common plastic products are truly biodegradable.

However, in the United States, paper and cardboard products are the single largest component of municipal waste. Although paper and wood products theoretically may be biodegradable, in most landfills, they do not biodegrade; the anaerobic conditions within a landfill generally are not amenable to microorganismal growth. In addition, with continuing concerns about global warming, biodegradability is not necessarily a good thing. When an organic substance biodegrades, it is ultimately broken down into primarily water and carbon dioxide, and in many cases, significant methane is released. Carbon dioxide and methane are powerful greenhouse gases, and some researchers think it is better to leave the carbon sequestered in solid form (for example, in the form of nonbiodegraded paper) rather than release it into the atmosphere. Thus, for the environmentally aware consumer, the choice between disposable plastic and paper products is not always clear-cut. Here, we review some of the pros and cons of using paper versus plastic.

Tremendous numbers of trees are chopped down every year to be pulped and turned into paper products. However, as everyone knows, paper is "recyclable," so why do we have to cut down more trees? Why can't we just recycle a larger proportion of the massive amounts of paper that end up either deposited in landfills or burned in incinerators? A major hurdle to paper recycling is that the paper fibers are modified as they are used and reused. To be recycled, discarded paper must be de-inked and converted to pulp before it can be manufactured into new paper. In the process of pulping, the paper fibers are broken. Therefore,

paper made from recycled fibers will be composed of shorter fibers than the original paper, resulting in a lower quality paper of inferior strength. As paper is recycled over and over again, the fibers become progressively shorter, and the paper's quality and strength diminish. The value of paper made from progressively recycled fibers and its potential uses drop sharply. At some point, the fibers become too short to produce serviceable paper for most purposes. For instance, newsprint or other paper that is too weak cannot be run through typical high-speed presses without tearing or other problems. Even under ideal conditions, paper fibers cannot be recycled indefinitely, and a percentage of virgin wood pulp must be injected from time to time. For this reason, newspapers cannot always be printed on 100% recycled paper; rather, paper containing about 80% recycled material and 20% virgin fibers might be used. Yet newsprint made of 80% recycled fiber is certainly preferable to newsprint made of 100% virgin fiber. In addition, there are many important uses for very low-quality, short-fiber paper, including toilet tissue.

Of course, paper not can only be recycled, but many paper products also can be reused. Paper bags, cardboard boxes, even paper from photocopy machines can be reused. For instance, grocery bags typically need not be discarded after one use. Manila envelopes and boxes often can be used to package and send things through the mail several times. Photocopies often are only on one side of a sheet of paper. Why not photocopy on both sides of the paper (this not only saves paper, but also saves space) or use the backs of old one-sided photocopies as note paper? Rather than being insulted that someone writes to you on the back of "scrap paper," feel proud that the writer is environmentally conscientious enough to reuse the paper.

Compared with paper, plastics generally have fewer uses and accordingly make up a lesser volume of our solid waste. However, many of the common uses of plastic are identical to some of the uses of paper. Both plastics and paper are used heavily in packaging, indeed often in tandem (for example, the plastic bottle with a paper label, the cardboard box or drink container that is plastic lined, the cardboard backing to an otherwise plastic container). Plastic bags can be used over and over again, often more times than a comparable paper bag. However, after the plastic bag is torn or broken, it typically cannot be recycled as easily as paper. Over the years, plastic bags, containers, and other items have become lighter and less bulky while retaining their strength, so after they end up in the landfill, they occupy even less room. On the other hand, when plastics (and paper in some cases) are burned in an incinerator, they may give off many toxic substances, including deadly dioxins. Recently, much research has been devoted to developing reuse and recycling options for standard plastics, and truly **biodegradable plastics** may play a significant role in the future.

Fully biodegradable plastics made from non-petroleum products, such as wheat starch, maize (corn), potatoes, and other plants (sometimes referred to as **bioplastics**), are potentially the wave of the future. One important class of bioplastics is the family of polylactic acid polymer-based plastics (also referred to as polylactide polymers, or PLA). Such plastics can be truly and fully biodegradable; potentially they can be composted in the same manner as yard trimmings and food scraps. The use of bioplastics is still in its infancy; currently, such materials are used primarily for certain medical and hygiene products, packaging, and agricultural and horticultural materials. In some cases, bioplastics are mixed with petroleum-based plastics, detracting from their environmentally friendly nature. Various laboratories and companies (including the car manufacturer Toyota) are aggressively pursuing the development of bioplastics, and we can expect bioplastics to make up a progressively larger share of the plastics market in coming decades. However, petroleum-based nonbiodegradable plastics dominate the current market. The bioplastics industry is growing rapidly, with some companies experiencing a 100% growth rate in 2008–2009. Despite this, bioplastics still hold just a little more than 1% of the total worldwide biodegradables market.

Many modern environmentalists suggest that when it comes to the question of paper versus nonbiodegradable plastics, there is no clear-cut answer. In fact, this may be the wrong question to ask. What is really much more important is to cut down on unnecessary usage and reuse whenever possible. When you go to the grocery store, take your own stash of used grocery bags and use them again. Better yet, invest in a couple of durable cloth bags that you can carry with you and use repeatedly for many different purposes, whether it is buying groceries or carrying books. Ultimately, for many people, there may be bigger and more important issues that they should consider instead of fretting over whether to ask for paper or plastic bags at the grocery store. How did you get to the store in the first place? Did you walk or ride a bicycle? If so, excellent. Did you take public transportation? Very good. Did you drive a super-low emissions and super high-mileage hybrid car? Well, good. Or did you drive your gas-guzzling sport utility vehicle or out-of-tune clunker? If the latter is the case, there are more important things for you to worry about, from an environmental point of view, than your use of paper versus plastic. Put your priorities into order and find a more fuel-efficient and more environmentally friendly mode of transportation when you take that trip to the store. You need to look at the entire picture and keep various concerns in perspective.

Questions

1. The obvious question is: You are in a grocery store and offered a choice of paper or plastic bags—which do you choose? Justify your answer. How much does your individual answer to this question depend on specific factors in your personal life at the moment (e.g., whether you can more easily reuse a paper or a plastic bag)?

2. Overall, from a proenvironmental point of view, do you think more emphasis should be placed in the future on paper or plastics, or should the current balance be maintained?

3. Perhaps you would like to drive a state-of-the-art, fuel-efficient, and super–low-emissions hybrid vehicle, but you just cannot afford it. You are grateful to have your gas-guzzling, exhaust-spewing clunker. What else can you do to help preserve the environment? Is there reliable and safe public transportation offered in your area? If so, do you use it? If not, perhaps you're stuck driving the clunker. Frankly and honestly assess your own situation and lifestyle. Is there more that you could be doing to help the local and global environment? Is there more that you are willing to do?

VISIT

http://environment.jbpub.com/mckinney/5e/
for more information

Plastics

Traditionally, plastics have been the bane of environmentalists (**FIGURE 18-13**). Plastics make up a little more than 12% of U.S. municipal solid waste by weight but pose many problems. Most plastics are synthetic compounds composed of polymers containing hydrogen, carbon, and oxygen (usually manufactured from petroleum and its derivatives). Typical plastics in use today are not biodegradable or otherwise readily broken down in nature; they not only clog our landfills, but they also produce unsightly litter across the landscape. When burned or incinerated, otherwise inert, and thus nontoxic, plastics can give off many toxic substances, including carcinogens such as dioxin.

The large-scale recycling of plastics faces many practical obstacles. For one, the many different types of plastic, which look very similar, must be sorted according to specific plastic resins. Some common products, such as squeeze bottles for food, are composed of several layers of different types of plastics, making it very difficult to separate them for recycling. Furthermore, even under the best conditions, most plastics cannot be recycled to their original use. Common plastics can be recycled only in an open-loop system. Unlike a glass bottle, a plastic soda bottle cannot simply be melted down and made into another

plastic soda bottle. Not only does the quality of the plastic diminish, but the temperatures at which plastics are melted and remolded are not always high enough to sanitize the plastic for use as a food or beverage container. In order to recycle plastics, secondary uses generally must be found for them.

To make it easier to identify different types of plastics for potential recycling, many plastic manufacturers have adopted a voluntary coding system that appears on containers, bags, and other products. This code takes the form of a number within a recycling triangle and generally an acronym below the triangle (**TABLE 18-3**). Of the categories of plastic listed in Table 18-3, PETE and HDPE are currently the most commonly recycled plastics; however, the amounts of all plastic types that are recycled continues to increase, and demand for good recyclable plastic is strong. As of 2003, about 25% of plastic soft drink containers were recycled in the United States, but when it comes to other plastics, the rate was less impressive— a 10% rate of plastic recovery and recycling for plastic containers and packaging overall and less than a 5% rate for plastics in discarded durable goods.

Despite the significant progress that has been made in recycling plastics, it should be emphasized that recycling plastics once is quantitatively and qualitatively different from the reuse of paper fibers half a dozen times (after which the fibers theoretically can be composted, as described later) or the virtually endless recycling without degradation of some types of glass and metal. Eventually, second-generation plastic products, such as lawn furniture made from recycled milk jugs, must be disposed of. In addition, today's typical plastics are manufactured from scarce fossil fuel resources in an energy-intensive fashion, often producing significant amounts of pollution as a by-product. For these reasons, many environmentally concerned citizens continue to look askance at plastics.

Composting

Nearly half of all yard wastes (leaves, grass clippings, and so forth) are composted. Still, many experts agree that **composting** gener-

FIGURE 18-13 The ready markets for cheap plastic trinkets from China and other Asian countries contribute to the solid waste problem.

	TABLE 18-3 — Coding System for Plastics
	Polyethylene terephthalate (also known as PET)—this is a transparent plastic (although it may be colored) commonly used to make 2-liter soda bottles and other containers, such as peanut butter jars. PETE is used in approximately 25% of all plastic bottles. It can be recycled into such items as strapping (for packaging), fiberfill for winter clothing, carpets, surfboards, sailboat hulls, and the like.
	High-density polyethylene—this plastic is commonly used to manufacture plastic milk jugs, bleach and detergent bottles, motor-oil bottles, plastic bags, and other containers. HDPE is used in more than 50% of all plastic bottles. It can be recycled into trash cans, detergent bottles, drainage pipes, base cups for soda bottles, and other similar items.
	Vinyl or polyvinyl chloride (also known as PVC)—this is used in the manufacture of vinyl siding, plastic pipes and hoses, shower curtains, and so forth; it is also used to make some cooking-oil and shampoo bottles, as well as bottles for some household chemicals. It can be recycled into fencing, house siding, handrails, pipes, and similar items.
	Low-density polyethylene—a plastic commonly used to make cellophane wrap, it is also used to manufacture bread bags, trash bags, and other types of containers. It can be recycled into grocery and garbage bags.
	Polypropylene—a lightweight plastic commonly used in packaging some foods (e.g., some margarine and yogurt containers) and for certain types of lids and caps. It can be recycled into car-battery cases, bird feeders, and water pails.
	Polystyrene—this is the substance that is commonly known as "Styrofoam," used to make coffee cups, plastic peanuts for packing, egg cartons, meat trays, plastic utensils, videocassettes, and so forth. It has been recycled into tape dispensers and reusable cafeteria trays.
	Other—plastic resins other than the six basic categories listed above, as well as objects produced from several different types of plastics mixed together. Mixtures of plastics can sometimes be recycled into "plastic lumber" and used to manufacture benches, lawn furniture, picnic tables, marine pilings, and other types of outdoor equipment.

ally is underused in the United States (**FIGURE 18-14**). About 14% of our solid municipal waste, by weight, consists of yard wastes and food wastes, which are easily composted to make fertile humus or topsoil (60% of yard wastes were recycled in 2009). It is ironic that such organics are often incinerated or buried in landfills, where they are permanently removed from the ecosystem and generally do not biodegrade. Yard and food wastes could be used to produce natural fertilizer or topsoil, substances that are sorely needed for our agricultural lands (see a review of food and soil resources). In addition, if not contaminated with heavy metals or other toxic substances, sewage sludge (consisting in large part of dead bacteria—sewage sludge is not the same as raw sewage) can be composted to make a natural fertilizer. In addition, the paper, wood, leather, textiles, and other biodegradable substances that make up about 50% of our municipal waste stream could

FIGURE 18-14 A community compost center.

18.2 Alternative Paradigms for Waste Management

provide low-grade compost and be used for nonagricultural purposes, such as landfill cover, strip-mine reclamation, reforestation projects, or clean fill in areas along roads or construction sites.

A fully implemented program of reuse, recycling, and composting, combined with separation and removal of hazardous waste for special disposal, would leave very little material to be placed in a landfill. Under such a system, landfills would be receptacles primarily for synthetic, nonrecyclable, and nonbiodegradable substances such as certain types of plastics and synthetic textiles.

Composting has occurred in nature for hundreds of millions of years. It is a free service provided by the microorganisms of the soil. When leaves and other organic material fall to the ground in a natural ecosystem, they are "eaten" by microscopic decomposers that attack the organic matter and break it down into smaller molecules, which can then be recycled in the ecosystem. This is the natural process by which humus and soil form.

Virtually anyone can facilitate composting; it can be carried out on a small or large scale, in a low-tech or high-tech fashion. Many individuals and families have compost piles or bins in their backyards. Grass clippings and other yard refuse, perhaps along with kitchen scraps, animal manure, and some dirt, are piled together. As long as the pile is properly aerated (it helps to stir the compost or turn it over occasionally) and has sufficient moisture (in temperate climates, the rain provides all of the moisture needed), the microorganisms will do their job. The organic matter will decompose to form good compost. During the composting process, the aerobic microorganisms generate heat (the interiors of backyard compost piles can reach temperatures as high as 70°C [160°F]) that naturally kills many pathogenic bacteria and weeds, thus "sterilizing" the resultant compost. There is also a process of composting that allows people to compost their food waste in their homes using newspaper, dirt, and earthworms.

Depending on the quality, compost has numerous uses. High-quality compost can be used to help replace lost topsoil on farms.

The amounts of compost that could be used for this purpose are enormous; it takes an estimated 59 metric tons of compost to add just 2.5 centimeters (1 inch) of topsoil to about half a hectare (acre) of land, but then the potential for producing good compost is also large. Composting the organic wastes of a community of 1 million people could produce an estimated 544 metric tons of compost daily. Compost is also widely used in the nursery and landscaping industries; it is necessary for many types of land reclamation projects, and of course, sanitary landfills need compost or soil between the trash layers and as a cap on top once the landfill is closed.

Recycling—Closing the Loop

Millions of Americans carefully sort their trash, then dutifully deliver the various categories of recyclables to a recycling center or place the separated trash outside for curbside recyclable collection (**FIGURE 18-15**). Certainly, recycling will help address the solid waste problem in the United States, but sorting trash and taking it to a recycling center are only the first steps. The materials must actually be recycled. Unbeknownst to many citizen recyclers, not all the material that is collected for recycling is actually recycled. The problem is not that the material could not be recycled, but that markets for recyclables do not always exist.

Although communities and the general public have been eager to sort and collect recyclables, the capacity for processing recyclable materials has not always expanded accordingly. As discussed, the market for used paper was once so poor that millions of tons of it were accumulating in warehouses. Similarly, mountains of plastic bottles and sorted glass sometimes await recycling. In September 1992, the company Waste Management of Seattle, Inc., had a reported 5,445 metric tons of glass for which it could not find a ready market. In other cases, segregated paper and other trash intended for recycling have been incinerated or landfilled because there was no market for the material and storing it was uneconomical. However, it looks like the market for recyclables is now, in the early 21st century, gaining strength and momentum. If this trend continues, these problems will be a thing of the past.

Still, such situations have presented a dilemma for some recycling advocates who have had to decide whether to tell the public that their "recycling" efforts were for naught. Some leaders of the recycling movement have been known to practice deception, believing that it is important to encourage the recycling habit, even when the material is not recycled. They hoped that if the beginning of the recycling loop (sorting and collection) remained intact, the remainder of the loop (processing plants and markets) eventually would develop. Based on the current strength of the recyclable market, early indications are that this strategy has worked.

Strengthening the Market for Recyclables

Collecting cannot lead to recycling if there are no markets for recyclables. Various ways of strengthening the market for recyclables have been suggested, not all of which have proven equally effective. Well-meaning legislators passed laws requiring the collection of paper for recycling without requiring the use of recycled paper, which would have developed a market for the product. In some instances, such short-sighted laws have actually driven trash collectors into bankruptcy when they were forced to spend extra time and money picking up recyclables but could not sell them. However, the federal government and many state governments now require all paper used in government offices to have a recycled component.

Indeed, there is a popular misconception that the collection, sorting, and selling of recyclable trash should be a profit-earning venture for a community or at worst a break-even situation. In fact, recycling may be costly, at least until markets develop. Even if markets must be "artificially" created through tax incentives and other means, recycling can still pay off if the ultimate cost is less than or equal to the cost of waste disposal. Indeed, disposal of waste by any means is expensive. According to one estimate, the United States spends approximately $30 billion on municipal waste disposal annually.

If we are to increase the rate of recycling, increased research and development and eco-

FIGURE 18-15 Providing separate containers and curbside pickup makes it easier for people to recycle.

nomic incentives will be needed. Although reuse and recycling have been carried out on a limited scale for ages, recycling of 21st century goods on a 21st century scale requires new technologies. New ways to sort, clean, and reuse materials will be welcome. Perhaps even more importantly, technical innovations may allow products manufactured from recycled materials to be of equal or better quality than products produced from virgin materials. To close the recycling loop, consumers must purchase products manufactured from recycled materials. This means that the recycled goods must be competitive on the open market, from both a quality and a cost perspective. Currently, some products manufactured from virgin materials cost less than they would were it not for various tax breaks and subsidies that apply to the virgin materials; tax incentives and subsidies instead could favor products manufactured from recycled materials. Local, state, and national governments can give an economic boost to recycling efforts by mandating the use of certain minimum levels of recycled materials in various new products, thus creating a ready market for recycled materials.

The concept of redistributing unwanted goods through the Internet is becoming popular with the help of the nonprofit Freecycle Network's free Internet bulletin boards. The Freecycle Network was started in 2003 to promote waste reduction in Tucson, Arizona. The bulletin boards are organized at a local community level and provide individuals with an electronic forum to give away just about anything they no longer need or want. In this way, items are reused rather than thrown into the waste stream.

Industrial Ecosystems

Natural ecosystems do not produce substantial quantities of material waste; the "waste" of one organism is the lifeblood of other organisms. Likewise, an **industrial ecosystem** does not produce waste; the effluents of one industrial process form the raw materials for another industrial process. Already, some companies are finding that they can sell their effluents for a profit, whereas previously they had to pay to dispose of the same effluents as "waste." For example, a steel-processing company may "recycle" the sulfuric acid it uses in the steel mills and sell the resulting iron sulfate compounds to magnetic tape companies. Otherwise, the "used" sulfuric acid might simply be discarded. A key component to the success of industrial ecosystems is communication. The company with a particular effluent must be able to locate the potential users

of the material. Accordingly, information networks are being established in many countries (for example, France, Germany, and Belgium) and in various regions of the United States to put potential consumers of effluents in touch with the appropriate producers.

If the industrial ecosystem is to mimic natural ecosystems properly, virtually all materials must be reused/recycled. This means that not only manufacturing effluents but also goods discarded by society at large must re-enter the system as raw materials.

However, because of the human penchant for treating symptoms rather than root problems, in some instances, we are backsliding from the ideals of the industrial ecosystem. Take the typical automobile as an example: in general, it is much more difficult to recycle all of the components found in a car today than it was in the 1920s. A modern automobile is manufactured from numerous different alloys of metal, including iron, steel, aluminum, nickel, lead, and copper, various polymers of plastic, rubber, cardboard, and so forth. The typical car has a very short life span and produces enormous amounts of waste during its operation; it not only exhausts pollution into the atmosphere, but also runs through tires that typically end up in landfills (refer to Case Study 18-1).

Eventually, the car is junked, and although many of its materials could be recycled, the sad truth is that often they are not. It requires a lot of time and effort to strip a car and separate its components for reuse or recycling (**FIGURE 18-16**). Furthermore, because of impurities, goods manufactured from the recycled materials may be inferior to goods manufactured from virgin materials. For decades, in some countries, such as Germany and Japan, many parts of the car have been labeled in order to make the process of recycling them easier and more efficient. Each type of metal and plastic is labeled, put into a category, and then re-used after the car is discarded. UNEP estimates that by 2020 in European Union (EU) countries, more than 40% of all the discarded end-of-life vehicles could be recycled if current trends there continue. In 2000, Ford teamed up with automotive industry partners in Japan and Europe to create the

FIGURE 18-16 Much of the material in junked cars could be recycled.

International Materials Data System to facilitate a more streamlined approach to documenting materials used in car manufacturing and documenting recycling efforts. As a result the United States, as well as a growing number of other countries, are now properly labeling their car parts in a standardized way to make world-wide distribution of recyclable parts easier. By 2009, 95% of all the materials used in new Ford vehicles were recyclable. This move to more recyclable supply chain input will keep an estimated 13.6 million kg (30 million pounds) of plastic from entering landfills each year.

Despite the efforts to design recyclable materials for cars in recent year, the fact remains that an increasing numbers of vehicles have been put into service, they have produced escalating levels of pollution. A typical response to this problem has been to design ever more sophisticated cars that produce less pollution; the symptom, pollution from car exhausts, is treated rather than the problem. A much simpler, although surely less popular, alternative would be to attack the root problem—namely, too many cars.

In the 1970s, catalytic converters began to be widely used on the exhaust systems of automobiles to reduce various pollutants in the exhaust before they enter the atmosphere. But catalytic converters also have drawbacks; while helping to solve one environmental problem, they exacerbate another. Standard catalytic converters rely on the use of scarce platinum group metals (iridium, osmium, palladium, platinum, rhodium, and ruthenium). Before the extensive use of catalytic converters, these rare and valuable metals typically were recycled at efficiencies of 85%. Early catalytic converters were not designed with recycling in mind, so their introduction substantially reduced the rate of recycling among platinum group metals. The inability to recycle the metals is counter to the ideal of the industrial ecosystem. Fortunately, systems are now being developed to remove catalytic converters from old cars and recover the platinum group metals. Likewise, the metals from discarded computers and other electronic equipment increasingly are being recycled (see **CASE STUDY 18-3**).

18.3 Hazardous Waste

The Comprehensive Environmental Response, Compensation, and Liability Act (CERCLA), the federal statute passed in 1980 that established Superfund, defines a hazardous substance as "any substance that, when released into the environment, may present substantial danger to public health, welfare, or the environment." CERCLA goes on to define extremely hazardous substances as substances that "could cause serious, irreversible health effects from a single exposure." Hazardous waste is essentially waste composed of such hazardous substances. Hazardous waste can originate from the home (such as household cleaners or pesticides), local or national government operations, agricultural use, industry, or other sources (**TABLE 18-4**). Hazardous waste often includes substances that are chemically reactive, corrosive, flammable, explosive, or toxic to living organisms (toxic materials are harmful or fatal when consumed by organisms in relatively small amounts) (**FIGURE 18-17**).

There are three main categories of hazardous waste: *organic compounds, inorganic compounds and elements,* and *radioactive waste.* Organic compounds are carbon-based substances that also contain a substantial percentage of hydrogen and oxygen. Some organic compounds are naturally formed molecules. Some are derived from once-living material, such as petroleum and other fossil fuels, and some are totally artificial substances synthesized in the laboratory. Many fertilizers, pesticides, organic dyes, plastics, and other substances are organic compounds. Theoretically, at least, organic compounds should be degradable into simple, nontoxic substances, such as carbon dioxide and water (plus some residue). However, in reality, many are not easily decomposed in nature and persist for very long periods of time. Sometimes incineration can be applied successfully to break down and detoxify hazardous organic compounds; another approach is to use microbes that can chemically attack and degrade some such substances. Remediation of hazardous waste is discussed more fully in this text.

Inorganic hazardous compounds have little or no carbon, but they commonly contain

TABLE 18-4	Health Effects of Selected Hazardous Substances	
Chemical	**Source**	**Health Effects**
Pesticides		
DDT	Insecticides	Cancer; damages liver, embryo, bird eggs
BHC	Insecticides	Cancer, embryo damage
Petrochemicals		
Benzene	Solvents, pharmaceuticals, and detergent production	Headaches, nausea, loss of muscle coordination, leukemia, linked to damage of bone marrow
Vinyl chloride	Plastics production	Lung and liver cancer, depresses central nervous system, suspected embryo toxin
Other organic chemicals		
Dioxin	Herbicides, waste incineration	Cancer, birth defects, skin disease
PCBs	Electronics, hydraulic fluid, fluorescent lights	Skin damage, possible gastrointestinal damage, possibly cancer causing
Heavy metals		
Lead	Paint	Neurotoxic; causes headaches, irritability, mental impairment in children; damages brain, liver, and kidneys
Cadmium	Zinc processing, batteries, processing fertilizer	Cancer in animals, damage to liver and kidneys

Source: Based primarily on data compiled by World Resources Institute and published in *World Resources 1987*. New York: Basic Books, 1987.

heavy metals, such as lead, mercury, cadmium, copper, arsenic, iron, aluminum, manganese, chromium, beryllium, nickel, selenium, zinc, silver, and others. Some heavy metals are necessary to life in small, or trace, amounts, but all are toxic to organisms in larger doses. All of these elements occur naturally in the environment, usually in low concentrations, and are cycled through the Earth's chemical and biological systems. The problem is that modern industry has mined and released much higher concentrations of heavy metals than the natural geochemical and biological systems are capable of readily handling. Therefore, once released into the environment, these substances tend to slowly accumulate, often to toxic levels, and like all hazardous substances, they are very difficult to remove. For safe disposal of toxic inorganic compounds, they sometimes are combined with various chemicals that transform the material into a stable, cement-like block that can be buried or otherwise disposed of. The solid block resists leaching and thus does not allow the inorganic compounds to escape into the environment; however, the best way to deal with such inorganic compounds is to not allow them to mingle with the general environment. Rather than being discarded, waste substances containing heavy metals should be recycled whenever possible (see Case Study 18-3).

Radioactive waste is the third generally recognized category of hazardous waste. Radioactive waste is generated by nuclear power plants, nuclear weapons manufacturing facilities, medical facilities (radioactive equipment and treatments), some research labs, and many common commercial, industrial, and home functions (for instance, the

FIGURE 18-17 Sampling water from a cyanide-poisoned lagoon.

E-Waste and Technotrash

VISIT

http://environment.jbpub.com
/mckinney/5e/
for more information

The electronic age of computers, the Internet, e-commerce, instant global communication, mobile phones, and even MP3 players has brought about a revolution in the way we (at least those of us connected to the system) interact, work, and relax.

Mobile (cellular) telephone use has seen a meteoric rise in the last 15 years. In the late 1980s, there were only a few million cellular phone subscribers worldwide; now there are more than 4 billion, and the number continues to climb. For many people, especially in developing countries, mobile phones are their only phones. Electronic communications and media hold the promise of reducing waste and energy consumption per amount of information that can be stored and transferred. For instance, the equivalent of tens of thousands of pages of text can easily be fit onto a single CD-ROM. Millions of pages can be stored on a USB drive that can be held in the palm of a hand. Photographic film processing is becoming a thing of the past. The savings in paper and other raw materials can be enormous—or so some contend, but while modern electronics may reduce certain types of wastes, the Electronic Revolution is producing entirely new forms of hazardous wastes, variously referred to as e-waste (electronic waste) or technotrash (technological waste) at a prodigious rate. To manufacture the semiconductors and other components used in various electronic devices, massive amounts of water and chemicals (many highly toxic) are required, resulting in equally massive amounts of wastewater and many hazardous wastes. It is difficult to get accurate statistics, but manufacturing the components of a typical personal computer generates on the order of thousands of liters of wastewater and several kilograms of hazardous waste by-products, and requires hundreds of kilowatt-hours of electricity. The wastes of the electronics industry do not just disappear; they must go somewhere. In fact, they now contaminate many sites. The place credited with the birth of the U.S. semiconductor industry, Santa Clara County, California, now contains more U.S. EPA Superfund sites than any other county in the country. IBM (International Business Machines) and National Semiconductor have faced lawsuits involving allegations of cancer and birth defects caused by the exposure of workers to toxic chemicals, and in at least one case in the 1980s, IBM settled out of court for an undisclosed sum over an instance of a child being born with birth defects to parents who had worked at an IBM plant in New York.

Computers, cell phones, and other electronic equipment typically have a very short half-life—often about 18 months or less. Such devices are not discarded because they are broken and cannot be repaired but perhaps more typically because they have become obsolete as more speed and power are packaged into smaller and less-expensive computer components and other electronic devices. Indeed, some electronic gadgets, such as cell phones, are for all practical purposes "disposable." Yet these electronic devices are difficult and dangerous to recycle. It can be very difficult to extract minute quantities of valuable rare metals from integrated circuit boards and chips; modern electronic devices are not designed to be disassembled. Furthermore, they contain many harmful substances. A typical cathode ray tube computer monitor contains 1.8 to 3.6 kilograms of lead (a dangerous heavy metal linked to nervous system disorders); cadmium (linked to increased risks of cancer and damage to reproductive systems) is found in many batteries used in electronic devices, and flat panel screens contain mercury (a highly toxic substance that damages the nervous system, especially in children). It is estimated that some 40% of all lead in U.S. landfills is from computer monitors; accordingly, over 15 states including Massachusetts, Maryland, Wisconsin, and California, have banned cathode ray tube monitors from landfills and incinerators; 50% of all states have set restrictions on other technotrash such as cell phones and video game components. Nevertheless, the National Safety Council estimates that 315 to 600 million desktop and laptop computers in the United States will soon be obsolete and in need of disposal.

So where do obsolete computers, discarded cell phones, and other trashed electronic devices end up? Millions of tons of electronic waste are dumped in U.S. landfills every year, but this is not

E-Waste and Technotrash

the only place that electronic wastes end up. Estimates suggest that from 50% to 80% of electronic waste collected in the United States to be recycled is actually exported to poor Asian countries. Such recycling may, superficially, seem like a good thing, but the poorly paid rural villagers dismantle the computers by hand and are routinely exposed to hazardous and toxic chemicals while their villages, rivers, and groundwater supplies are poisoned. As documented in a 2002 report by the Basel Action Network and the Silicon Valley Toxics Coalition, in villages such as Guiyu, China, poorly or totally unprotected adults and children dismantle the computers for about $1.20 a day. The laborers smash cathode ray tube monitors to extract the copper and other metals; they strip wires and sort wires and plastics from computers by hand. They burn unusable plastics and circuit boards in open pits (releasing carcinogens into the environment). They de-solder circuits using crude burners and open fires, and they use strong acids in uncovered vats to extract gold, silver, and other precious metals from electronics components (**FIGURE 1**). This activity has polluted the local air, soil, river, and groundwater. The local river water was found to contain 190 times the levels of some pollutants allowed under World Health Organization guidelines. In fact, drinking water must be shipped in from 18 miles away. Guiyu is not an isolated instance; similar situations occur elsewhere in Asia, including in India and Pakistan.

In 1989, an international treaty commonly known as the Basel Convention (Convention on the Control of Transboundary Movements of Hazardous Wastes and Their Disposal) sought to restrict the transfer of hazardous wastes, including electronic wastes, across national borders. The treaty encouraged disposal of wastes close to the points of origin. In 1994 and 1995, the Basel Convention was amended to prohibit the export of wastes from developed industrial (rich) nations to nondeveloped (poor) nations, even for supposed "recycling" purposes. Although not yet fully legally binding (all of the amendments have yet to be ratified by a sufficient number of countries, including the United States), many countries abide by the Basel Convention and its amendments. China has banned the importation of electronic waste, yet it still flows in a steady stream to Guiyu and other areas of the country. This, critics suggest, is in large part because the United States—the world's single largest producer of hazardous waste—does not fully honor the Basel Convention and its amendments, and the U.S. government has effectively made export part of its overall strategy for dealing with electronic wastes.

However, as the crude "recycling" methods and dangers of e-waste are being exposed and brought to the attention of consumers, there is increasing pressure from environmental and human rights organizations (such as the Basel Action Network, Greenpeace China, Silicon Valley Toxics Coalition, Society for Conservation and Protection of the Environment, and Toxics Link India) to implement genuine recycling programs in the country of origin rather than simply shipping such waste to developing countries. Japan has instituted an Appliance Recycling Law that requires manufacturers (after a fee is paid by the consumer) to take back and recycle various appliances, such as air conditioners and refrigerators. As of this writing, it is anticipated that computers and other electronic devices will be added to the list covered by the Japanese law. Likewise, various European nations have instituted "take-back programs" for electronic equipment, and the cost is

FIGURE 1 A rural Chinese woman melts the lead-based solder from a circuit board and exposes herself to the toxic fumes.

either borne by the original manufacturer or paid by the consumer.

However, the solution to e-waste and technotrash ultimately may not be "after-the-fact" recycling, but instead the design of electronic components and devices that can be manufactured in an environmentally friendly manner. The products should be easily disassembled so that the components can be reused and recycled and be free of as much toxic and hazardous substances as possible. At the individual level, reducing consumption by resisting the urge to upgrade to the latest model of cell phone, MP3 player, computer, or other electronic gadget would help alleviate the problem.

Questions

1. Do you think manufacturers should be held responsible for the entire life cycle of their products, from cradle to grave (sometimes known as the principle of "extended producer responsibility" or EPR)? In other words, should their responsibility for the product continue after the purchase of the product by the consumer? Should a manufacturer be required to "take back" a product at the end of its life cycle? Should this be the case just with electronic products, only certain types of products, or for all consumer items? Justify your answer.

2. Should a manufacturer be held legally responsible and liable if the products it sells are found to be causing human and environmental damage, whether during routine use or after being discarded?

3. Should manufacturers be required to design electronic equipment with ease of disassembly and recycling of components in mind and produce electronic components using less toxic and poisonous substances? What if such procedures significantly increase the cost, perhaps by a factor of two or three times or more, of the end product? Or do you think such procedures could be encouraged on a voluntary basis? Would you purchase a more expensive, but more environmentally friendly, computer or cell phone?

4. Should manufacturers be encouraged (or required) to design longer lasting, more easily repairable and upgradeable products? Based on what you know about computers and other electronic devices, do you think this is technically feasible, economical, or practical? Why or why not?

5. The principle of "**environmental justice**" is that no group of people, on the basis of economic status, social status, or race, should be forced to bear the environmental risks of others. Is the shipping of wastes, such as e-waste and technotrash, from rich, developed nations such as the United States to poor developing nations a form of "environmental injustice" (the concept that a group of people, community, or country is the recipient of the wastes, pollution, and environmental problems created by another group that is generally better off materially)? Put simply, do you believe it is morally wrong for rich nations to ship their wastes to poor nations? Why or why not?

VISIT

http://environment.jbpub.com /mckinney/5e/ for more information

radioactive elements of home smoke detectors). High-level radioactive waste, from nuclear power plants and weapons facilities, is disposed of according to special regulations and handling procedures. Much low-level radioactive waste is considered to be below regulatory concern, and no special precautions are taken in disposing of it. In some cases, radioactive material has been incinerated along with other rubbish or disposed of in municipal dumps and landfills.

Hazardous waste is a major problem because of its inherent dangers and the amount that continues to be produced each year. No one really knows how much hazardous waste is produced in the world every year, but one estimate is that the United States alone produces more than 260 million metric tons annually. World production must be at least twice this amount. Much of this waste comes from the chemical, paper, and petroleum industries.

Like other waste, the best way to control the influx of hazardous substances into the waste stream is to reduce the amounts used in the first place and then reuse and recycle such substances whenever possible. Because many manufacturing processes use or generate hazardous substances, making the processes more efficient can reduce the amounts of some of these substances. Better information networks would enable hazardous material generated by one manufacturer to be passed on to another company that could use it (see the earlier discussion of industrial ecosystems). Sometimes biodegradable and environmentally benign substances can be substituted for artificial, synthetic, or less environmentally friendly chemical substances; an example would be using natural, biologically based fertilizers, pesticides, and herbicides in place of synthetic chemicals. Even around the house, the average consumer can buy less toxic, more environmentally friendly paints, sprays, cleansers, and other chemicals.

Hazardous waste disposal is now recognized as a special problem that requires particular techniques and solutions. However, all too often in the past, hazardous wastes were disposed of improperly with little concern for long-term consequences, even though their disposal was technically legal at the time. Items now recognized as hazardous substances (for example, lead-acid batteries) were often thrown away along with ordinary wastes. Now we are haunted by hazardous waste dump sites across the nation and around the world. One of the first to come to the attention of the nation was **Love Canal** (**FIGURE 18-18**).

From 1947 to 1953, Hooker Chemical and Plastics Corporation used Love Canal, located in the town of Niagara Falls, New York, as a disposal site. An estimated 19,873 metric tons of chemical wastes were deposited in the abandoned canal, mostly contained in 55-gallon (208-liter) steel drums. Numerous hazardous wastes, including powerful carcinogens, were buried in the canal, substances such as dioxin, benzene, chloroform, and dichloroethylene. Later, an elementary school and numerous homes were built in the area, only to be abandoned in the late 1970s when toxic chemicals began to seep from the ground. Ultimately, it cost hundreds of millions of dollars to clean up the mess.

Superfund

After the chemical contamination of Love Canal was exposed in the 1970s, the United States realized that something had to be done about hazardous waste sites. In 1980, Congress passed the Comprehensive Environmental Response, Compensation, and Liability Act (CERCLA), authorizing the EPA to identify all hazardous waste sites in the United States, determine what health dangers each posed, and list what measures would be required to clean up each site. The agency then compiled a National Priorities List (NPL) of high-priority sites that merited action. The cleanup and related operations were to be funded by taxes on the chemical and petroleum industries, as well as assessments on various corporate polluters. Initially, $1.6 billion was allocated to a Hazardous Substance Response Fund, commonly referred to as "**Superfund**," to be used for cleanup activities.

FIGURE 18-18 An evacuated home in the Love Canal community, Niagara Falls, New York.

After more than 30 years, the Superfund program has grown and evolved. Tens of thousands of hazardous waste sites have been identified, and more continue to be located. In the first 20 years of the program, more than 6,400 actions were taken to reduce health and environmental threats from such sites, and cleanup was completed at more than 750 sites (the work being carried out by the responsible parties at 70% of the sites). Through the Superfund program, private parties paid or funded more than $18 billion in settlements. No one really knows how much it will cost to clean up all of the hazardous waste sites identified by the EPA (**FIGURE 18-19**). As of July 2006, 1,244 sites were listed on the National Priorities List (sites are added to and removed from the list periodically). According to one estimate, the average cleanup cost for a single site ranges from $21 to $30 million; on that basis, it could cost $26 to $37 billion to clean up all the sites on the National Priorities List. Some analysts have estimated that the cost of cleaning up all hazardous waste nationwide could range from $750 billion to $1 trillion and require more than 50 years of sustained effort. To solve some of the continuing financial problems involved with cleaning up hazardous waste, "no-fault" programs have been proposed. Under no-fault policies, all companies in an industry would be required to help finance cleanup efforts, regardless of whether they were directly involved in creating the problem. Of course, many "clean" companies and organizations regard such policies as inherently unfair: Why should they help pay for damage caused by other companies? Should the federal government simply pick up the entire tab, which means that the public ultimately pays for the cleanups? There are no easy answers as to how to fund the enormous cleanup costs that we face in the future.

The necessity and effectiveness of Superfund cleanup operations also have been questioned. Some researchers suggest that the health risks of many hazardous waste sites have been overstated, and in some cases, cleanup efforts may disturb toxic materials and create more of a hazard than if the site

FIGURE 18-19 The clean up of the Bunker Hill, Idaho, Superfund site included bringing down this stack.

were simply left alone. Various EPA studies have suggested that problems such as indoor pollution and outdoor air pollution pose a greater health risk than does hazardous waste—the logical conclusion being that limited resources might better be expended on cleaning up other environmental problems rather than focusing on hazardous waste sites.

The United States is not the only country that faces huge price tags to clean up its hazardous waste problems; similar situations exist around the globe. Some of the worst toxic dump sites and environmental hazards are in Eastern Europe and the countries of the former Soviet Union. Toxic dumps and rampant pollution also plague developing nations; in many cases, the dumping of hazardous waste is inadequately regulated. For instance, in Mexico City, wastewater contaminated with heavy metals and toxic organic chemicals is discharged into the municipal sewer system. From there, with very little treatment, it is transported to agricultural areas for use in irrigation, with the result that toxic substances have been found in vegetables and other crops. Likewise, in China, some 400 million metric tons of industrial wastes and mining tailings are dumped annually on the outskirts of cities or into lakes, streams, and rivers with devastating results—60,000 hectares (148,260 acres) of land in China may be covered with hazardous material.

The Bhopal Disaster

One of the best-known industrial chemical disasters occurred at the Union Carbide pesticide-producing plant in Bhopal, India, on December 3, 1984. Apparently, a faulty valve allowed water to leak into a tank containing approximately 15 metric tons of liquid methyl isocyanate, an extremely toxic substance. The water and methyl isocyanate reacted chemically, producing high temperatures and pressures that vaporized the liquid; ultimately, the material escaped as a deadly cloud above the city of 800,000 people. Unfortunately, there were no emergency plans or procedures in place, and no one knew how to deal with the situation. The results were devastating; several thousand people died immediately, and over the next few years, thousands more (estimates run as high as 10,000) died as a result of methyl isocyanate poisoning. In addition to the deaths, an estimated 50,000 to perhaps 500,000 were injured to some degree as a result of the disaster.

The Bhopal disaster should not have occurred. A number of safety devices at the Union Carbide plant were not operational at the time of the disaster; indeed, some had been turned off for months. The Bhopal disaster suddenly made people aware of the inherent dangers of many industrial processes, especially if plans are not in place to deal with extraordinary disaster. In the United States, the Bhopal incident caused Americans to look with increasing suspicion at the chemical industry and helped inspire Congress to pass the Emergency Planning and Community Right-to-Know Act, which established the important Toxics Release Inventory.

Toxics Release Inventory

In October 1986, Congress passed the Emergency Planning and Community Right-to-Know Act as Title III of the Superfund Amendments and Reauthorization Act. Under this law, larger companies and organizations must report their production, use, and release of hundreds of potentially hazardous chemicals listed by the EPA. Based on this information, the EPA produces an annual report, known as the **Toxics Release Inventory (TRI)**. With the release of the 1987 TRI, for the first time Americans could get an overall picture of how much toxic material the country's industries were emitting. Based on the material gathered for the TRI, in 2004, 23,675 facilities reported to EPA's TRI Program. These facilities reported 4.24 billion pounds of onsite and offsite disposal or other releases of almost 650 toxic chemicals. More than 87% of the total was disposed of or otherwise released onsite; almost 13% was sent offsite for disposal or other releases (see a review of the principles of pollution control, toxicology, and risk).

During the past 15 years, the TRIs have had some very positive benefits. They have enabled citizens and communities, armed with data, to put pressure on companies to cut emissions of hazardous substances. Likewise, people in high-level management positions were themselves appalled by the amounts of toxic chemicals their own factories were releasing. Many industries are now taking voluntary actions to reduce the amount of toxic material released into the environment.

Despite the good that the TRI has done, it is not without criticism. Perhaps most critically, the TRI is merely a compilation of certain toxic chemicals released directly by large facilities. Many different types of businesses, especially smaller businesses such as dry cleaners and photographic processors, are not required to submit information to the EPA. Originally, companies were required to report releases only if they manufactured or processed 25,000 pounds (11,340 kg) or more, or otherwise used 10,000 pounds (4,536 kg) or more, of listed chemicals. Eventually, the reporting thresholds for certain substances were lowered significantly, and the reporting system was revamped. However, the TRI still does not include all of the relatively small releases by individuals, small businesses, municipalities, and other organizations, and these can add up. Furthermore, the list of nearly 650 toxic chemicals compiled by the EPA is just a small percentage of the thousands of chemicals in use. In addition, many hazardous and toxic chemicals are incorporated into end products (paints, stains, coolants, and so forth) that are used by a consumer, but eventually released into the environment; again, these are not accounted for in the TRI. While certainly an important step toward accounting for our chemical waste, the TRI may represent merely the proverbial "tip of the iceberg."

Technologies for Dealing with Hazardous Waste

Engineers and scientists have long grappled with the problem of what to do with hazardous wastes. Common solutions have involved isolating such wastes from the environment in permanent waste disposal sites, such as a landfill or deep in the ground through deep-well injection. Another approach is to treat hazardous and toxic wastes so as to eliminate or reduce their harmful properties. For decades, this was attempted by simply burning the waste, but in many cases burning at temperatures that are not sufficiently high actually compounds the problem by creating and releasing molecules that are more harmful than the initial substance (for example, dioxin and heavy metals may be released during standard incineration, as discussed earlier).

In recent years, several laboratories have concentrated on developing more efficient ways of destroying and neutralizing toxic chemicals. Westinghouse Plasma Corporation operates several plasma torches that can generate temperatures from $5000°C$ to $10,000°C$. Toxic organic chemicals are forced through the plasma, which breaks down the hazardous materials into relatively simple and harmless gases. Westinghouse has a portable plasma torch that can be brought to hazardous waste dump sites. In this way, extra transportation of hazardous waste can be avoided, lessening the chance that an accident might happen on the way to a disposal facility. A potential drawback of plasma torch technology is that it can consume large amounts of energy to destroy a significant amount of hazardous waste.

The Solar Energy Research Institute in Colorado and Sandia National Laboratory in New Mexico, both U.S. government organizations, are researching ways to destroy toxic chemicals using solar energy. One method resembles in principle some solar energy collectors described elsewhere in this text. Water contaminated with toxic organic chemicals is mixed with catalysts and then pumped through tubes while the sun's light is focused on the mixture. The ultraviolet light of the sun has the ability to break down and destroy 90% or more of the contaminants. In another method, the sun's light is focused on a special vessel containing toxic organic chemicals. The solar energy can increase the temperature inside the vessel to very high levels, converting the more complex chemicals into simple substances, such as carbon monoxide and hydrogen gas. It has even been suggested that these products could be used as a fuel; either the CO and H_2 can be used to form methanol, or the hydrogen gas can be burned directly. In a related process, magnified sunlight focused on a container full of dioxin broke down 99.999% of the dioxin into simpler, relatively harmless substances.

Bioremediation

The biological action of microorganisms has been a necessary part of nature since life originated nearly 4 billion years ago. Microorganisms convert kitchen waste into compost, and bacteria have been used commercially for years to treat sewage and even to concentrate metals such as copper and nickel from low-grade ores.

Bioremediation, or biotreatment, is the use of organisms, ranging from bacteria and other small organisms (such as single-celled and multicellular microbes and fungi) to trees and other plants, to clean up or reduce unwanted concentrations of certain substances. Microorganisms have been used in a variety of hazardous materials applications, at toxic and hazardous waste sites, and as biological "scrubbers" to remove dangerous pollutants from factory emissions before they leave the facility. Essentially, the microorganisms eat or digest unwanted chemicals, transforming them into simpler, less harmful, or useful forms. Certain strains of bacteria and fungi can digest even highly dangerous chemicals, such as dichlorodiphenyl trichloroethane (the insecticide DDT), **trinitrotoluene** (the explosive TNT), **polychlorinated biphenyls** (PCBs), dioxins, toluene, naphthalene, xylene, carbon tetrachloride, toxic nitrates, asphalt products, creosote (a wood preservative), and many other substances.

If a bacterium cannot be found in nature to perform a certain job, researchers may be able to genetically engineer one in the laboratory. Of course, the substances produced as organisms to digest dangerous compounds may also be dangerous. For example, DDT may be degraded to an even more toxic substance than the original pesticide. One must be careful in applying bioremediation techniques in specific cases.

Especially encouraging is the discovery that certain bacteria can effectively attack oil and related organic substances, converting the material into primarily water and carbon dioxide. After the Exxon Valdez oil spill of March 1989 (**FIGURE 18-20**), simple bioremediation techniques were used on a very small scale with quite encouraging results. Liquid farm fertilizers were sprayed on a portion of an oily beach, and after approximately 2 weeks, the oil on the beach was significantly reduced compared with a test area that had been left untreated. No microorganisms per se had been added to the beach; essential nutrients had simply been spread to encourage the growth of already present bacteria that would digest the oil. The success of this simple experiment has had wide ramifications for the prospects of bioremediation. Before this time, the EPA generally did not sanction the use of microorganisms for cleaning up oil spills or hazardous waste sites. Since the initial success of the Exxon Valdez experiment,

bioremediation has been put to work cleaning up hundreds of sites overseen by the EPA.

Today, dozens of companies are developing and marketing biotreatment products—everything from bacteria and nutrients to help clean out a home septic system to mixtures that are specifically designed to deal with gasoline spills and other hazardous materials. Bioremediation tends to be cheaper than traditional treatment methods such as incineration of contaminated soil, and it has the distinct advantage that it often can be applied on site. For example, imagine that some toxic substance, such as a petroleum product, has leaked into the ground. Instead of digging up hundreds of tons of contaminated soil and transporting it to an incinerator, large augers can drill into the soil and mix in bacteria and nutrients. The bacteria then attack the substance, breaking it down into harmless chemicals. Such techniques already have been used to treat soil contaminated with creosote at a lumber-treatment plant and at a 16-hectare (40-acre) site near Los Angeles where the soil had been contaminated with marine fuel.

Biotreatment is also finding many useful applications to avoid releasing pollutants into the environment. Microbes are of particular use in treating volatile organic compounds (VOCs) that are often released as gaseous fumes from factories and small businesses, such as dry cleaners, print shops, auto body shops, and even bakeries. In one technique, the fumes are passed through layers of well-humidified and aerated peat moss and compost that contain bacteria. The gases are caught on wet films covering the compost and peat moss and can then be broken down into harmless gases (mostly water and carbon dioxide) by the bacteria. This is a relatively low-tech, inexpensive, but very effective method of substantially reducing the release of harmful volatile organic compounds. Bacteria simply do the work of chemical degradation for free. Such systems are finding wide use in Europe.

Bioremediation is not limited to the use of microbes. Various trees and other plants can also be used to clean up pollutants. For instance, hybrid poplar trees have been used successfully to help with the remediation and

FIGURE 18-20 A cleanup crew works on a beach shortly after the Exxon Valdez oil spill of March 1989.

cleaning of soil and groundwater contaminated with chlorinated hydrocarbons. It has been found that vegetation cover used in conjunction with microbes can help promote the microbe-based biotreatment as microbes congregate on and around the plant roots and the plants bring contaminants to the surface by evapotranspiration. The microbes then attack the contaminants, and simply being exposed to the oxygen environment near the surface of the soil layer may chemically break down some contaminants. The vegetation itself may also absorb some contaminants, transforming them into less dangerous or noxious substances (in some cases, trees and other plants are being bioengineered specifically for this purpose). Furthermore, the plants will help contain any remaining contamination on site. The planting of trees and other vegetation for bioremediation purposes can save millions and millions of dollars per site as compared with the pumping of groundwater and removal of soil for treatment or disposal, and the vegetation used in bioremediation also controls erosion, attracts wildlife, and can be aesthetically pleasing.

■ study guide
SUMMARY

- Solid waste consists of municipal solid wastes (garbage and trash, refuse, and rubbish from homes and various businesses) and industrial solid waste.
- Alternate traditional paradigms for solid waste disposal are the "dilute and disperse" (e.g., burning and dispersing the smoke) versus "concentrate and contain" (e.g., landfilling) strategies.
- A third paradigm is "resource recovery," including reuse and recycling.
- Incineration is the burning of trash and garbage; the heat may be used to generate electricity or otherwise be put to use in a resource recovery facility.
- Problems with incineration include the dioxins and other toxic substances that may be emitted.
- Landfills isolate and contain solid waste, but the isolation may not be complete; for instance, hazardous leachate may drain from the landfill or methane gas may be produced.
- There is increasing emphasis among waste management experts to address current solid waste issues through source reduction, reuse

of products (rather than disposal of them) and recycling of materials.
- Some materials, such as glass and metals, can be recycled virtually indefinitely in closed-loop systems.
- Other materials, notably paper and plastics, can be recycled only a limited number of times (open-loop systems) before the material is too degraded for additional use.
- Hazardous wastes are substances that, when released, may present a substantial danger to public health, welfare, or the environment and include chemically reactive, explosive, and toxic substances; such materials originate from industries, businesses, governments, and households.
- Hazardous waste accumulations can be a major issue locally; in 1980, the U.S. Congress passed legislation that set up a Superfund to be used for cleanup activities at hazardous waste sites.
- In the United States, the Toxics Release Inventory (TRI) provides data on the billions of kilograms of toxic chemicals released or disposed of by industry each year.

KEY TERMS

biodegradable plastics
bioplastics
bioremediation
closed-loop recycling

composting
concentrate and contain
deep-well injection
dilute and disperse

dioxin
dumps
environmental justice
garbage

hazardous waste
incineration
industrial ecosystem
industrial solid waste
landfill
leachate
Love Canal
methane gas

municipal solid waste
nonhazardous waste
open-loop recycling
recycling
refuse
resource recovery facilities
reuse
rubbish

sanitary landfill
solid waste
source reduction
Superfund
Toxics Release Inventory
trash

STUDY QUESTIONS

1. Define solid waste and its subcategories.

2. Discuss the alternative paradigms that historically have been used for waste management. What are the pros and cons of each? Which are still in common use?

3. Discuss both the benefits and the drawbacks of the following techniques of waste disposal: incineration, resource recovery facilities, dumps and landfills, and deep-well injection.

4. Define hazardous waste. What different types of hazardous waste are there?

5. What are some of the new technologies that are being developed to deal with hazardous waste?

6. How can garbage be a source of revenue for some communities and nations? What are the positive and negative aspects of such an enterprise?

7. Describe the "waste management hierarchy." Is there universal agreement on this hierarchy?

8. Describe the advantages and disadvantages of source reduction, reuse, and recycling. What can and should be done to encourage recycling?

9. Why do some environmentalists object to a heavy emphasis on recycling?

10. How do glass, metal, paper, and plastic recycling differ from one another?

11. How are most municipal wastes disposed of in the United States? How much is recycled?

12. Describe the extent of the solid and hazardous waste problem in the United States. How serious are these problems compared with other environmental issues discussed in this text?

13. Briefly describe the process and benefits of composting. The process of composting can produce greenhouse gases, so is composting always a good thing? Justify your answer.

14. Comment on the controversies surrounding the paper versus plastics debate.

15. Briefly describe the concept of an industrial ecosystem.

16. Briefly describe the Love Canal and Bhopal incidents. What practical effects, especially legislatively, did they have?

17. What is Superfund? What has it accomplished? Why has it been controversial?

18. What is the Toxics Release Inventory? What purpose does it serve?

WHAT'S THE EVIDENCE?

1. Today, more than half of all municipal solid waste in the United States is disposed of in landfills, and many people are concerned that we are quickly running out of landfill capacity. Does the evidence support this assertion? Name some of the factors that pertain to, and affect, landfill capacity.

2. Some people argue that it is wrong to ship wastes, especially wastes that contain hazardous or toxic components, from developed nations to developing nations. Part of the rationale behind this point of view is that even if the developing country gains short-term economic benefits from accepting the waste, in the long term, it will suffer. Is there hard evidence to support this assertion? What is your position on this issue? Justify your answer.

CALCULATIONS

1. Based on the data presented in Table 18-1, the per capita daily garbage output of the average American increased by what percentage from 1960 to 1980?

2. Based on the data presented in Table 18-1, the per capita daily garbage output of the average American increased by what percentage from 1980 to 2005?

3. Based on the data presented in Table 18-1, in 1960 in the United States, what percentage of municipal solid waste was recovered for recycling?

4. Based on the data presented in Table 18-1, in 2005 in the United States, what percentage of municipal solid waste was recovered for recycling?

ILLUSTRATION AND TABLE REVIEW

1. Referring to Figure 18-3 (the typical composition of municipal solid waste in the United States according to the weight of various types of materials), what are the categories of products that make up the first and second largest components of municipal solid waste?

2. Referring to Figure 18-3, what are the categories of materials that make up the first and second largest components of municipal solid waste?

3. Referring to Figure 18-10 (waste recycling rates from 1960 to 2003 in the United States), how many more million tons per year of municipal solid waste were recycled in the United States in 2003 than in 1960?

http://environment.jbpub.com/mckinney/5e/

http://environment.jbpub.com /mckinney/5e/

Connect to this text's website at http://environment.jbpub.com/mckinney/5e/. This site features eLearning, an online review area that provides quizzes, chapter outlines, and other tools to help you study for your class. You can also follow useful links for more in-depth information.

Social Solutions to Environmental Concerns

"The economy is like a huge digestive tract, with a flow of inputs and outputs. Sustainable economies will mimic more efficient, not bigger digestive tracts."

Herman Daly, environmental economist

"A culture of performance will not come quickly. We can expect no instant revolutions in social values. All we can realistically hope for is painfully slow progress . . . punctuated by rapid advances. When most people see a large automobile and think first of the air pollution it causes rather than the social status it conveys, environmental ethics will have arrived."

Alan Durning, How Much Is Enough? (1992)

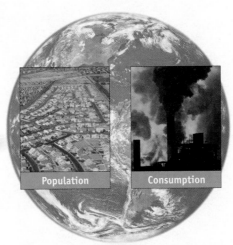

Population Consumption

**1. The Environment
and Humans
(Chapters 1–2)**

Lithosphere Biosphere

Hydrosphere Atmosphere

**2. The Environment
of Life on Planet
Earth (Chapters 3–5)**

Lithosphere Biosphere

Hydrosphere Atmosphere

**3. Resource Use and
Management
(Chapters 6–13)**

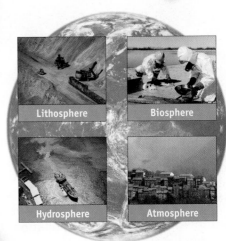

Lithosphere Biosphere

Hydrosphere Atmosphere

**4. Dealing with
Environmental
Degradation
(Chapters 14–18)**

Lithosphere Biosphere

Hydrosphere Atmosphere

**5. Environmental Issues: Social Aspects and
Solutions
(Chapters 19–20)**

19 Environmental Economics

Chapter Objectives

After reading this chapter, you should be able to explain or describe the following:

- The importance of economics to environmental problems

- How sustainability produces more jobs

- What sustainable growth is and how to achieve it

- Why the free market needs adjusting to solve environmental problems

- The importance of including true environmental costs in consumer pricing

- Why global poverty is a growing problem for the environment

- Why economic aid and investments from wealthy nations are essential

Economics can be defined as the study of methods of allocating finite resources, especially when human wants and desires outstrip the resource base. Economics is thus concerned with environmental issues in two basic ways. The most obvious is resource scarcity. In a review of resources, we see how people are depleting a variety of resources, or inputs, to society. Economics is also concerned with outputs, or pollution. It addresses fresh air, clean water, and other basic needs when they become scarce from pollution and other forms of environmental degradation. Economic realities are an inevitable part of solving all environmental problems, so we have discussed them throughout this text. Here, we provide an overview that covers many basic principles of economics and the environment.

19.1 Economics Versus Environmental Science?

Economics and environmental science are frequently depicted as mutually exclusive and natural enemies. Indeed, economists and environmental scientists are sometimes critical of each other. A main reason for the conflict is that the two disciplines have different ways of looking at the world, especially in three areas (**TABLE 19-1**):

- *Time span of outlook.* Environmental science historically has been concerned with species interactions, natural cycles, and other aspects of nature that reflect

TABLE 19-1	Three Areas in Which Economics and Environmental Science Differ	
	Economics	**Environmental Science**
Time span	Short term	Long term
Priorities	Society	Environment
Social solutions	Incentives	Laws, persuasion

the long-term impacts of resource depletion and pollution on the biosphere. Such a long-term view is useful when considering such problems as global warming or radioactive waste, whose impacts last hundreds to millions of years. In contrast, economics is a social science that deals with human timescales and focuses on problems, such as unemployment and taxes, that are measured in yearly, quarterly, or even monthly terms.

- *Priorities.* Traditional economists have been concerned mainly with society, especially how society produces, distributes, and consumes goods and services. They tend to view nature as a resource to be used for these purposes. Environmental scientists have tended to place higher priority on preserving the natural environment than producing more goods and services.

- *Social solutions.* Economists are social scientists and know that incentives are often the most effective way to regulate human behavior. Environmental scientists have tended to emphasize persuasion and laws such as the Endangered Species Act as solutions to environmental problems.

The popular news media tend to reinforce the oversimplified idea that economics and environmental science are fundamentally opposed by presenting such stories as "development versus wildlife preserves." One of the best-known examples is the spotted owl controversy, which was widely discussed in terms of jobs (logging the forest) versus environment (preserving the forest habitat of the spotted owl).

In a limited sense, these apparent conflicts are real. A logger who loses his job to an endangered species would certainly see a conflict. However, in a broader sense, the conflict between economics and the environment is a false dichotomy. A **false dichotomy** is when a complex question becomes polarized into an "either–or" issue. The dichotomy of jobs or environment is false because usually the ultimate choice is much more complex than that. In the long run, environmental preservation typically produces more jobs. For instance, in the spotted owl debate, if the old-growth forest were harvested, many loggers would be out of a job in a few years as the remaining forest disappeared. On a larger scale, in many countries where the tropical rain forests have been destroyed, the local inhabitants have lost their way of life and live in abject poverty. The same could be said for fishermen in the overharvested fishing grounds of offshore New England, farmers who have exhausted the soils in Africa or Haiti, and countless other examples.

Sustainability Produces More Jobs

These examples show that controlled use of resources often provides more employment in the long term. Fishermen, loggers, farmers, and many others can remain employed only if their resource base is not depleted. In other cases, the loss of environmentally damaging occupations usually will be more than compensated by new occupations that promote environmental preservation. Throughout this text, we have discussed many examples in which activities that help preserve the environment create more jobs than they eliminate.

Take, for instance, the energy industry. In 2008, the average electricity consumption for a U.S. residential utility customer was 11,040 kWh, which is about 920 kilowatt hours every month. About 15% of this total amount was used in lighting, 16% for cooling, and 9% for water heating. The remainder of electricity use is divided among other household uses. There are about 5,400 power plants in the United States with multiple generators on-site. Some of the plants have generators that derive their power from more than one source.

By converting generators or replacing old ones with ones compatible with renewable technology, the company not only diversifies its options, but also creates opportunity for workers to gain new training for the transition and creates new jobs. Consider the fast-growing wind energy market. From 2004 to 2009, wind generation grew by 39%. In 2000, wind energy generation was less than 3,000 megawatts. Currently, wind energy can power more than 9.7 million homes, which amounts to 35,000 megawatts. In 2009, just over 10,000 megawatts of new wind energy capacity was installed. This was the most ever in the United States. Still, only 1.8% of all power comes from wind; however, wind comprises 50% of all the renewable resources of power, which include wind, solar, hydro-electric, geothermal, and bioenergy, used in U.S. generators.

When considering a growing job market such as wind, one must consider the whole picture. Renewable energy such as wind provides many jobs beyond actual power plant work. Some of these include the manufacturing of the turbine parts, quality control, safety, and shipping, as well as on-site jobs such as land acquisition, inspection, community education, and a whole range of other jobs related to engineering, marketing, advertising, and business investment.

In some states, such as Ohio, where the unemployment rate hovers at close to 10%, jobs in the renewable energy field allow workers to take advantage of reduced education tuition for training and allow growth not only at the much-needed factory worker level, but also for highly trained workers in management, marketing, and development. Despite the current economic challenges one such company, Cardinal Fasteners in Bedford Heights, Ohio, is experiencing growth because of its competition in a world market made possible by wind energy.

Approximately 85,000 people in the United States are employed in wind energy-related fields, but this does not reflect the additional numbers of people currently in training internships or those who are working crossover jobs. Although statistics vary about the growth in total green energy job markets, conservative government figures indicate that 2.5 million people will be employed in green energy jobs by 2025.

This figure translates to a job potential that could employ 8% of the 30 million unemployed people in the United States. Current government-sponsored loans, training programs, and industry internships provide additional help to level the playing field while also employing support personnel not traditionally thought of when considering the impact of green energy on the U.S. job market. Additional subsidies, incentive plans, and policies to support and expand green energy will need to evolve to keep the jobs in the United States and to help fuel the current momentum.

While considering the positive trend in wind energy, for instance, that offers jobs growth, career mobility, and long-term employment prospects, the other growth sector in environmental jobs traditionally has been in environmental remediation to clean up ill-planned use of resources. Since the early 1970s, the U.S. environmental cleanup business, including pollution control and toxic waste remediation, has grown dramatically. This growth has created millions of new jobs in more than 75,000 new firms. With better power plant engineering, cleaner energy, and increasing consumer demand, clean energy offers the potential to make significant positive changes in the way we get our energy and manage natural resources in the first place to diminish the need for future generations to have the remediation challenges we continue to face.

Even more jobs are created by indirect environmental activities, such as increased efficiency, recycling, and renewable resource use. For example, the Worldwatch Institute reports that to produce 1,000 gigawatt-hours of electricity, the following numbers of workers are needed:

Unsustainable fuels:

- Nuclear plant—100 workers
- Coal-fired plant—116 workers

Sustainable fuels:

- Solar thermal plant—248 workers
- Wind farm—542 workers

In another example, data from the New York City Department of Sanitation show that as far back as the early 1990's before "green" technologies were as popular as they are now, recycling produced at least 10 times more jobs per 1 million metric tons of waste than did landfills (**FIGURE 19-1**). Environmental laws, increased business interest in efficiency, and increasing public demand for thousands of "green" consumer items such as organic foods and recycled clothing, likely will lead to additional growth in environmental jobs which are currently employing over 2.3 million workers, worldwide.

Environmentally sustainable activities produce so many jobs because they often are labor intensive. They rely on people instead of large amounts of energy and machinery. In contrast, many highly polluting, resource-depleting industries are just the opposite, using higher amounts of energy and fewer people.

The Commission for Environmental Cooperation reports that approximately 90% of all the toxic pollutants released or transferred in North America originate from 15 industrial sectors and derive from just 30 substances. Of this 5.5 billion kilograms (12.1 billion pounds) of toxins released, the top polluting industries are metal mining, activities related to oil and gas extraction, fossil fuel power plants, chemicals manufacturing, and primary metals manufacturing. Some regional variations occur because of reporting requirements and data collection methods. Consider the comparison between a fossil fuel plant that ranges in efficiency from 28% to 45% at best and is in the top four of toxic polluters, and a hydroelectric plant that operates at a 90% efficiency rate and generates virtually no toxic pollutants by comparison. The average 300-MW coal-fired power plant employs about 53 full-time workers, with numbers decreasing each year for the past 5 years, primarily because of worker retirement. The average rate of compensation was about $22.00/hour. Meanwhile, the average salary of a worker in a hydroelectric plant is almost $10/hour more. With the growth in other green energy such as wind (70% from 2007 to 2008) and solar (100% from 2009 to 2010), the case for investment in renewables, especially by young professionals looking for a sustainable employment future, makes good

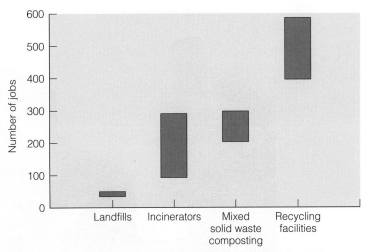

FIGURE 19-1 The number of jobs created for each 1 million metric tons of waste processed in New York City. A range is shown for the different processes because some facilities (such as different incinerators or different recycling facilities) are more efficient than others. Recycling creates the most jobs, even when the variation is included. (*Source:* Adapted from M. Renner Worldwatch Paper 104, Jobs in a sustainable economy. Washington, DC: Worldwatch Institute. Copyright © 1991.)

sense from an economic, and not just environmental, perspective.

Despite the greater job-generating capacity of many environmentally related activities, some older industries promote the belief that environmental preservation causes job losses. Companies threaten to close or relocate (often to another country) if they are forced to comply with environmental laws. This is sometimes called **greenmail**, or job blackmail. Such industries are often inefficient, rely on depleted resources, or have declining employment from increased automation or international competition (**FIGURE 19-2**). In fact, annual reports consistently show that only about 0.1% of all U.S. job layoffs are related to environmental regulations.

19.2 Environmental Economics: Blending the Dichotomy

The increasing integration of jobs and environmental preservation has led to the development of a new discipline, called **environmental economics**. (This discipline and its subdisciplines are also sometimes called *ecological* or *green economics*.) Environmental economics studies society and the natural environment as a single system. It treats both the short-term need for jobs and the long-term need to

FIGURE 19-2 Older, obsolete industries sometimes blame environmental regulations for their closing or layoffs, but such regulations actually cause only a tiny fraction of total job losses. Seen here is an abandoned limestone mill in Bainbridge, Pennsylvania.

protect the environment as goals. Similarly, producing goods and services and protecting the environment have equal priority.

Traditional economics focused on the production and consumption of goods in society as if they were isolated from the environment. Traditional economics largely ignored the environmental costs of extracting source materials from the natural environment, such as the costs of depletion to future generations. Similarly, traditional economic theory largely ignored the environment as a sink for waste and pollution. Old economic texts sometimes described the air, water, and other environmental resources as "free goods." In contrast, environmental economics incorporates environmental considerations into its theory and practice. This is accomplished through the valuation of ecosystem services, such as water purification and carbon sequestration. It is also accomplished by incorporating the environmental costs of mining, pollution, and other production activities in economic calculations. These

positive and negative benefits are called **externalities**. A reason for this different approach is that when traditional economics arose, human populations and technological impacts were much smaller than they are now (**FIGURE 19-3**). Society has greatly increased in size, whereas the environment has remained finite, even with technological advancements. The human impacts are now so great that they can no longer be ignored.

What Is Sustainable Growth?

Traditional economics is based on the principle that pure economic growth is extremely desirable. This view arose because economic growth, by increasing the quantity and quality of goods and services, has led to a general improvement in the human condition. As long as human populations and the technological capacity to consume and produce were relatively low, environmental costs of growth were not of much concern. Environmental economics agrees that economic growth can be desirable,

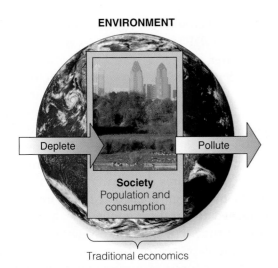

ENVIRONMENT

Deplete

Pollute

Society
Population and
consumption

Traditional economics

(a) Small population and little technology:
Society has low impact on environment

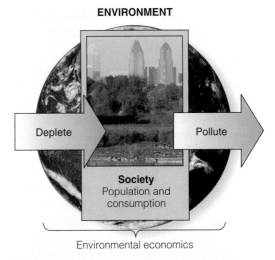

ENVIRONMENT

Deplete

Pollute

Society
Population and
consumption

Environmental economics

(b) Larger populations and increased technology:
Society has great impact on environment

FIGURE 19-3 (a) Traditional economics originated when society (human population) was smaller and had less impact on the environment. Economic theory could effectively ignore depletion and pollution impacts. (b) Society is much larger now, with greater impacts. Environmental economics acknowledges this and includes the environment in its theories.

such as where people are poor, but argues that the goal should be growth that can be sustained for a long time, not growth at any cost. **Sustainable growth** both preserves the environment and permits a nation's economy to increase in size, creating jobs and improving the human condition.

Sustainable growth differs from unsustainable growth in four basic ways (**TABLE 19-2**):

1. *Resource use.* As we have seen, traditional economics was based in part on the idea of limitless resources. Accordingly, unsustainable growth maximizes resource use and thus the rate of resource depletion. Economist Kenneth Boulding has called this approach **cowboy economics** because it views the environment as a source of endless space and riches just as the pioneers viewed the Old West. In contrast, sustainable

growth recognizes that resources are finite and that the rate of resource depletion must be slowed if we are to have growth in the long term. Accordingly, sustainable growth emphasizes minimizing resource use by reducing inputs to society through increased efficiency, reuse/recycling, and substitution of renewable resources (as seen in a review of people and natural resources).

2. *Pollution control.* Reducing inputs not only reduces the rate of resource depletion but also reduces the pollution associated with mining, processing, manufacturing, using, and disposing of goods. In contrast, unsustainable growth, with its emphasis on maximizing input, concentrates on output management, that is, controlling pollution after the resource is used. Consequently, unsustainable growth relies on abatement devices, such as wastewater treatment and smokestack scrubbers, that are much more expensive than reducing the waste stream on the input side.

3. *Resource fate.* Sustainable growth emphasizes reuse and recycling of matter instead of discarding matter in the one-way flow that is characteristic of the unsustainable **throwaway society**. Reuse

TABLE 19-2	Unsustainable Growth Versus Sustainable Growth	
	Unsustainable	**Sustainable**
Resource use	Inefficient	Efficient
Pollution control	Output reduction	Input reduction
Resource fate	Matter discarded	Matter recycled
Resource type	Nonrenewable	Renewable

and recycling not only reduce pollution from solid waste but also reduce depletion of virgin resources.

4. *Resource type.* Sustainable growth substitutes renewable resources where possible. For example, paper containers are preferable to plastic made from petroleum. Future generations will be able to manufacture more paper from trees, but the petroleum from which plastic is made likely will be gone well within the next century. In addition, renewable resources tend to be biological in origin and therefore are often more biodegradable. Renewable energy is preferred over nonrenewable forms.

Efficiency improvements, reuse/recycling/upcycling, and substitution, the hallmarks of sustainable growth, are among the major kinds of **green technologies** that currently provide many new jobs. Green technologies tend to reduce, rather than increase, the flow of goods through society. The improvement in living standards that occurs with sustainable growth comes mainly from increasing the quality of items produced, not the quantity. The well-known environmental economist Herman Daly has compared the economy with a digestive tract. A sustainable economy does not increase the size of the tract, but it does improve the digestion process. Thus, the change is qualitative, not quantitative. The United States may be falling behind in developing these critical technologies because of its greater emphasis on resource consumption compared with other modernized nations, especially in Western Europe.

Measuring Sustainable Growth

The traditional economic measurement of **Gross National Product** (GNP) can be defined as the market value of all final goods and services during a certain period of time, such as a year, for a particular nation. Officially since 1991, the **Gross Domestic Product** (GDP) has replaced the GNP in the United States for most purposes. The U.S. GDP covers goods and services within the borders of the country, whereas the U.S. GNP includes all United States residents, even if some or all of their economic activity is outside of the nation's borders. Economic measurement in terms of GNP or GDP (for the purposes of the discussion later, the terms can be used somewhat interchangeably) is not very satisfactory because GDP does not subtract environmental costs that may occur when wealth is created. For example, when trees are cut and sold for timber, they add to the GDP. However, in counting only immediate, short-term dollars, GDP ignores the greater net wealth that might have been created if that forest were managed to produce goods over a long period of time. For example, Nigeria was once a major exporter of timber, but it overcut most of its forest resources and now imports many more forest products than it exports.

Other examples also illustrate how misleading GDP can be, as follows:

1. The *Exxon Valdez* oil spill in March 1989 actually caused the U.S. GDP to rise because much of the $2.2 billion spent on labor and equipment for cleanup was added to income.

2. The tens of billions of dollars spent on health care for illnesses that air pollution caused in the United States generally add to GDP.

3. The timber and increased agricultural production obtained through rain forest destruction add to the GDP of those countries even though the destruction leads to the widespread extinction of many tropical species. The permanently lost genetic variations in those species might have produced lifesaving drugs, foods, and many other economic benefits that would have been much greater than the immediate income from the timber and agriculture.

4. Poverty, disease, and human misery increased globally during the 1980s and 1990s, although the Gross World Product (GWP or the global equivalent of GDP) actually increased by more than 50% between 1980 and 2001. World output of goods and services increased by a factor of eight between 1950 and 2004, from $7 trillion to $56 trillion.

Examples such as these have caused some economists to seek indicators of economic well-being that account for environmental costs of the loss of ecosystem services. One of the best-known alternatives to the GDP is the **Index of Sustainable Economic Welfare** (ISEW), which economists Herman Daly and John Cobb developed. The ISEW helps us measure sustainable growth that creates income and jobs without exacting a high environmental cost. It includes several environmental measures such as depletion of nonrenewable resources, loss of farmland from erosion and urbanization, wetlands loss, damage from air and water pollution, and even estimates of long-term damage from global warming and ozone depletion. In effect, these are a loss of ecosystem services. A later refinement of the ISEW is the **Genuine Progress Indicator** (GPI), which factors in additional measures, for instance aspects of the cost of crime and the breakdown of families.

Because it includes so many variables, the ISEW, or GPI, is much more difficult to compute than GDP, especially for developing nations where environmental monitoring is poor. However, enough U.S. data are available, and when the U.S. GPI is computed, it presents a very different picture of the nation's well-being than does GDP: Instead of steadily increasing, GPI has remained relatively static since the late 1960s (**FIGURE 19-4**). It seems likely that after data are available for other countries, their GPIs will show declines in many

cases. Developing countries with high populations and rapid environmental destruction are especially likely to show declines.

Dozens of other new approaches have been developed, including the **Green Gross Domestic Product** (Green GDP). Green GDP is an index of economic growth that factors in the environmental consequences of that growth. The Green GDP monetizes the loss of biodiversity and accounts for costs caused by climate change.

Economics of Sustainable Growth

Two basic economic systems have been used in the modern world, and neither has fostered sustainable growth. In **command economies**, the government tells individuals what to produce and sets prices for those products. There is no real incentive for efficiency because the rates and types of products being produced are dictated. The collapse of the Soviet Union demonstrated the major problem with command economies: The system develops widespread inefficiencies because product costs, which are artificially set, do not reflect real costs. These massive inefficiencies are also the reason more environmental damage has occurred in the countries of the former Soviet Union and other command economies than in countries with the other type of economy. These **market economies**, as they are called,

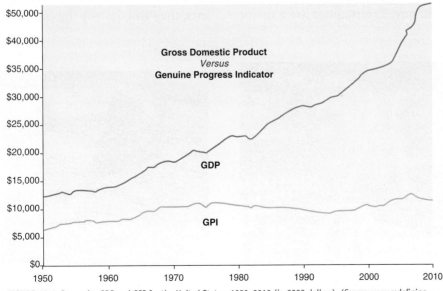

FIGURE 19-4 Per capita GDP and GPI for the United States, 1950–2010 (in 2000 dollars). (*Source:* www.redefining progress.org.)

are based on the "invisible hand," as Adam Smith first discussed more than 200 years ago. In a pure market economy, competition among individuals determines production and costs. Smith argued that self-interested individuals promote optimum allocation of resources in society, as if driven by an invisible hand. This system is more efficient than command economies because the laws of supply and demand set prices that generally reflect real costs to society.

19.3 Externalities: Market Failures and the Environment

Despite their general efficiency in allocating resources, market economies are not perfect. Economists have long noted that market failures occur when the individual pursuit of self-interest does not always benefit society. For example, until the enactment of laws regulating child employment, children were used as a cheap source of labor in the sweatshops of the late 1800s (**FIGURE 19-5**). Child labor exacted a toll on society in the form of the psychological and physical injury to the children's development that was not included in the price of the cheap clothing or other products they made. Economists call such market failures externalities: Some costs of production are not included in (are external to) the price of the product. With externalities, society as a whole pays a cost that the producer and consumer of the product should pay. Externalities are a major

reason no pure market economy exists. Regulations and other forms of intervention, which we discuss shortly, are necessary. Almost all real economies in the world are **mixed economies**, which combine varying degrees of market and command traits. In the United States, the market economy traits generally are more prominent than in Western Europe, although both have mixed economies.

Pollution, resource depletion, and the many other ways that markets degrade the environment are environmental externalities, which society pays because they are not paid by the industry or other activity that produces the degradation. How can markets generate such damaging externalities? Markets generally are assumed to be efficient because they are based on individuals making rational decisions to maximize benefits and minimize costs. Thus, in theory, markets should act to reduce environmental degradation before the environmental costs exceed the benefits of production. The problem is that markets work only if the commodity is privately owned.

In an overview of environmental science, we examined Garrett Hardin's 1968 essay *The Tragedy of the Commons*. Resources that are public goods, or commons, eventually will become depleted or polluted if the free market is left unchecked. Unlike a privately owned good, no one person or organization has a vested interest in taking care of public goods. If individuals are left to pursue their own interests, they will destroy the commons, rendering

(a)

(b)

FIGURE 19-5 During the Industrial Revolution, (a) children often worked long hours in U.S. and European factories. (b) Child labor continues to be used in many developing nations today.

it useless to everyone, including themselves. This behavior is actually quite logical: Because no one owns the commons, someone else will deplete or pollute the resource if you do not.

Tragedy of the Commons: Depletion and Pollution

The economic law of supply and demand stimulates depletion of a public resource: The price of goods increases as the supply diminishes. Thus, the scarcer the resource, the more expensive it will become. For example, the Organization of the Petroleum Exporting Countries (OPEC) embargo of the early 1970s caused petroleum prices to skyrocket but also sparked an incentive for more renewable technology. A key implication for public goods is that as the resource becomes scarcer, there will be increasing pressure to exploit the resource even more. Scarcity promotes even greater scarcity because rising prices reward people who exploit what resource remains. Among the many examples that could be cited are the current overexploitation of fishing grounds in many parts of oceans, overexploitation of game animals such as rhinos, soil exhaustion and erosion from overuse, and depletion of the highest-grade ore deposits in the United States.

Unsustainable growth is also characterized by pollution of the environmental commons for the same reason: Because no one has a vested interest in the commonly held property, people will freely pollute it. Thus, air pollution occurs because the atmosphere is a commonly held good that no one owns. The same is true for water pollution, for virtually no one owns an entire lake, much less an entire river, or ocean. Litter is usually worst in parks, along highways, and in other public places.

19.4 Solution: Internalizing Environmental Externalities

The solution to negative environmental externalities is to internalize the environmental costs of the loss of ecosystem services by requiring the depleters and polluters to pay for the damage incurred to public environmental property. In the case of depletion of a commons, the greater price that accompanies scarcity must be counteracted. One way

to do this is to impose a **severance tax**, which is levied on minerals as they are extracted. Many states levy severance taxes on coal that is mined. To offset the high price of scarce supplies, such a tax ideally would increase as the resource became depleted. Miners, drillers, loggers, fishermen, and other resource harvesters would then be discouraged from harvesting further. Who decides whether the resource is depleted, however? For renewable resources such as fishes or lobster, the tax would decrease once the supply (populations) had rebounded. For miners and other harvesters of nonrenewable resources, the tax would stay high, encouraging the industry to invest in finding alternative resources to use, conservation, and recycling. Similar taxes and other methods discussed later could be used in the case of pollution of a commons to force polluters to pay the costs of pollution.

What's the Environment Worth? Cost Problems

A major obstacle to internalizing, or paying for, environmental costs has always been determining what those costs are. Four main costs influence and thus complicate the calculation of overall environmental costs of an activity:

- Intangible costs
- Hidden costs
- Future costs
- Unequally distributed costs

Intangible costs include destruction of scenery and other subjective costs that are very difficult to measure because people place different values on them. To some people, a view of wilderness landscape has no value; to others, it is priceless. **Hidden costs** are environmental degradations that we are unaware of; for example, the full effects of pesticides and other pollutants with chronic toxicity may not be evident for many years. **Future costs** are costs that are passed on to subsequent generations. In theory, most environmental damage, especially to nonrenewable resources, includes at least some future costs. Keep in mind that the market economy and people's mentality usually only extend into the short term. However, some impacts may have only future costs: They have no discernible cost now, but the cost will rise in the future. For example, most plants now becoming extinct have not been studied for their food and medicinal potential. A plant species that has gone extinct might have been the best way to fight a disease that arises in the future. A major problem with all three of these costs is that society tends to ignore them. Intangible costs are not real costs to people for whom the value of a wilderness landscape is in its utilization. Beautiful scenery can mean little to someone who is struggling to feed his or her family every day. Hidden costs are not visible. Future costs are ignored by current generations who "discount the future" because it is uncertain and their needs are in the present.

Unequally distributed costs occur because environmental costs are almost never distributed evenly throughout society. Poorer people tend to pay relatively more of many environmental costs (**FIGURE 19-6**). In a sense, the problems are shifted onto those who cannot afford to stave off environmental problems. The factory that has to pay higher costs for pollution control simply passes that cost on to the consumer by raising the price of its product. Whether the product is a car, tires, food, or clothing, the poorer person feels the impact of the higher cost more strongly. Similarly, poorer people tend to be the ones who become unemployed when a factory is closed because it pollutes too much or because the

costs of controlling pollution are too high for it to operate profitably. Unequal distribution of cost creates much of the environmental controversy you hear about in the news media. The spotted owl controversy was in large part attributable to loggers in the area bearing much more of the cost of preserving the owl's habitat than, for example, avid bird watchers living in Texas. The same is true when nuclear power plants are shut down, factories are closed, and many other activities are slowed for environmental reasons.

Another way to deal with negative environmental externalities is to place a dollar value on ecosystem goods and services. For example, putting a price on the value of clean water, carbon sequestration, or recreation can help more effectively estimate the economic benefits and costs of degrading ecosystem goods and services. However, this requires the privatization of these ecosystem services.

Pay Up! Persuasion, Regulation, Motivation

One can attempt to overcome these problems and determine true environmental costs. The **Contingent Valuation Method** uses surveys and other techniques to find out how much people are willing to pay for things such as visiting national parks. This helps measure the intangible cost if a park is degraded. However, the science of making such measurements is just beginning, and estimates are often controversial. Despite such problems, attempts at estimating environmental costs must be made because society can no longer afford to ignore them. Using such cost estimates, industrial societies have begun to discourage depletion and pollution by requiring resource depleters, and especially polluters, to pay the true costs of environmental degradation. As we will see, the current payments generally are far below the level needed for a sustainable global economy, but social pressures are growing to attain that goal. Polls regularly show that most U.S. citizens are willing to pay extra for a cleaner environment.

Society can extract payment for formerly free environmental externalities in three basic ways: persuasion, government regulation, and economic motivation or incentives (**TABLE 19-3**). Persuasion usually influences only a few peo-

FIGURE 19-6 Poor families in many nations tend to bear a greater share of the burden of environmental problems. This lower income family lives next to a waste site.

TABLE 19-3	Methods of Promoting Payment for Externalities		
	Effectiveness	**Cost of Implementing**	**Best Suited For**
Persuasion	Least	Lowest	As a supplement to other means
Regulation	Moderate	Highest	Situations with a few major violators
Economic motivation	Most	Moderate	Situations with many violators

ple. It is particularly ineffective for industrial activities because cheaters are rewarded by having lower costs and will soon drive out of business conscientious industries that pay more environmental costs. However, persuasion's low cost makes it a useful supplement in conjunction with regulation or economic incentives.

Government regulation restricts depletion and pollution by passing laws. Regulation is deceptively appealing. It is the most direct way to prevent environmentally damaging activities: ban plastic, require pollution controls, fine people for littering, close polluting factories, and so forth (**FIGURE 19-7**). However, many economists, including prominent environmental economists such as Lester Brown, agree that regulation is often more inefficient and costly than using economic incentives. One reason is the high cost of enforcing regulations where many people are involved. Costs of enforcement are increased further because people often try to evade the laws. Thus, enforcement entails costs of monitoring plus costs of catching lawbreakers, prosecuting them, and sometimes keeping them in jail. For example, imagine the costs of trying to enforce a law requiring people to take public transportation to work once a week. In contrast, an economic incentive such as higher gasoline prices will encourage smaller cars, fuel economy, carpooling, and use of public transportation. Similarly, more people may recycle items if they are paid for doing so than if a law is passed requiring them to recycle.

Another criticism of regulations is that they tend to be relatively inflexible. For example, a law that requires all factories to meet the same rigid air quality standards for emissions does not take into account the age of

the factories or what they produce. Such laws can lead to high prices for the consumer and force factories to close. The 1990 Clean Air Act Amendments, which allow trading and buying of **pollution permits,** achieve the same level of control at less social cost by recognizing that some factories can eliminate many emissions very cheaply whereas others cannot. Older factories that would otherwise have to close can stay open and save jobs by buying pollution permits from newer factories that can more easily cut emissions because of their more advanced technologies.

FIGURE 19-7 Laws against environmental damage are easy to make but sometimes difficult and costly to enforce. This photograph of illegal dumping was taken in Alaska.

19.4 Solution: Internalizing Environmental Externalities 589

In the United States, pollution permits have been used mainly for reducing sulfur emissions that produce acid. In this regard, they have been very successful, and many economists have argued that pollution permits should be extended to other kinds of air pollution. In 2003, the Bush administration promoted its "Clear Skies Initiative" (see a review of local and regional air pollution) that sought to extend the use of pollution permits to other kinds of factory pollution. Many environmental groups opposed this initiative on the basis that it created too many loopholes for factories and increased overall air pollution levels above those that the Clean Air Act allowed. In addition, there is an important problem that sometimes arises with pollution permits, that of pollution "hot spots," which are areas with a high concentration of badly polluting factories. If many polluting factories buy pollution permits, very high levels of air pollution may be created in areas near those factories.

Despite its drawbacks, regulation has been used successfully during the last few decades by many industrial nations, especially the United States. Regulation is most useful where there are few violators and each violator can do great harm to the environment. Nuclear waste disposal fits both of these criteria and thus probably is best controlled by regulation. Manufacturing chlorofluorocarbons (CFCs), hunting whales, and trading in endangered species are other examples of easily regulated activities. The great improvement in air and water quality in the United States has been due largely to the Clean Air Act, Safe Drinking Water Act, and other laws passed in the early 1970s that set uniform standards. In recent years, rising costs of regulation, changing economic conditions, increasing emphasis on input reduction, and many other factors have led to increasing interest in economic incentives as a way of sustaining the environment.

Economic Incentives to Sustain the Environment

By rewarding individuals monetarily, economic incentives help meet many goals effectively. Incentives are less expensive than regulation because they save on enforcement costs. Indeed, incentives are the only practical way to discourage environmentally harmful behavior where thousands of individual citizens are involved. For example, imagine the difficulty of trying to catch every person who litters. In addition, incentives are more flexible: Producers and consumers of pollution both decide how to pay for the change to meet the goal. As each chooses the least costly way to achieve the goal, society as a whole has lower indirect costs. Environmental economists most often promote two basic types of economic incentives: government incentives and privatization.

Government incentives include taxes, subsidies, licenses, fees, vouchers, and many other ways that local, state, and national government can reward or discourage behaviors. These government incentives are often called green fees. The government incentives most often suggested are various types of taxes, or **green taxes**. For example, sulfur dioxide emissions can be discouraged directly by taxing emissions or indirectly by taxing high-sulfur coal, electricity consumption, or electricity generation (**FIGURE 19-8**). Many economists advocate such taxes as a relatively simple way to improve the functioning of the market to reflect an activity's true environmental cost, meaning the cost of externalities. Green taxes can do this by (1) serving as an incentive to slow depletion and pollution and (2) raising money that can go toward paying other environmental costs.

Green taxes are often an effective incentive because they promote all of the major aspects of a sustainable economy: less depletion, recycling and use of renewable resources, and less pollution at the source. For example, a carbon tax on fossil fuels promotes conservation, encourages alternative energy research, and reduces greenhouse gas pollution. Similar taxes can be levied on water to promote conservation and desalination, on metals to promote conservation and recycling, and so on. Raising the price encourages people to stop discounting the future, conserve resources, and slow pollution and solid waste disposal, as well as save money. Green taxes also raise money for environmental uses. Taxing polluting substances and scarce resources, especially from a carbon tax on fossil fuels, could raise many billions of dollars.

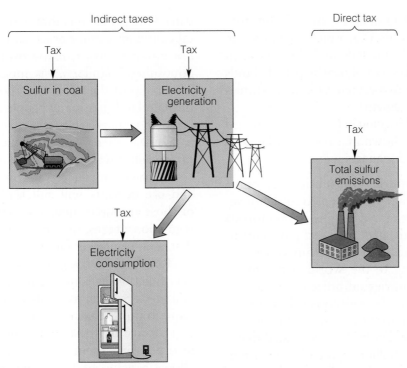

FIGURE 19-8 Green taxes can be indirect, such as a tax on sulfur in coal or on electricity generation or consumption, or direct, such as a tax on sulfur emitted.

Many estimates suggest that the overall burden of green taxes need not be great. The United States already has among the lowest taxes of all industrialized countries and imposes relatively few green taxes compared with Western Europe. More important, green taxes need not add new taxes; instead, taxes can be shifted from one activity to another. For instance, the revenue that the green taxes generate could reduce income taxes (**TABLE 19-4**). Taxes tend to discourage the activity being taxed. Instead of discouraging work, as income taxes do, green taxes discourage depletion and pollution.

Stopping subsidies on environmentally damaging activities could also reduce the burden of green taxes. Many nations have a "reverse green tax," sometimes called **dirty subsidies**. Instead of discouraging environmental damage with taxes, these actually encourage damage with government subsidies (payments). Examples in the United States include the following:

1. Subsidized mining of virgin ore deposits, which discourages recycling of metals (as shown in a review of mineral resources).
2. Subsidized irrigation water for farmers, as in California (as shown in a review of water resources).

				Revenue
TABLE 19-4		**Examples of "Tax Shifting" from Income to Environmentally Harmful Activities**		
Nation	**Year Implemented**	**Taxes Cut On**	**Taxes Raised On**	**Revenue Shifted (%)**
Sweden	1991	Income	Carbon, sulfur emissions	1.9
Denmark	1994	Income	Energy, water, landfills	2.5
Spain	1995	Wages	Gasoline	0.2
Netherlands	1999	Income	Fossil fuels, landfills, water	0.9
Germany	1999	Wages	Energy use	2.1
Source: Modified from L. Brown. Eco-Economy, Table 11-1. New York: Norton, 2001.				

19.4 Solution: Internalizing Environmental Externalities   591

3. Subsidized grazing of animals on government land. U.S. ranchers often can graze cattle on federal land for much less than the cost of nearby private lands (as shown in a review of land resources and management).
4. Subsidized logging in government-owned forests (as shown in a review of land resources and management).

Many other countries also subsidize resource use and encourage rapid depletion. Until recently, Brazil promoted logging of rain forests by building roads, making direct payments to timber companies, and using many other subsidies. According to the World Bank, many agricultural countries subsidize pesticides.

Such subsidies decrease costs below market price, resulting in massive waste and pollution. For example, the World Bank estimated that a staggering 5,445 million metric tons of carbon pollution could have been eliminated between 1991 and 2000 if developing countries and the former Soviet countries had stopped subsidizing fossil fuels. Most consumption subsidies are artifacts of outdated economic policies and protectionist trade policies. For instance, the United States once wanted to encourage mining of ores to provide a growing manufacturing economy with cheap raw materials. Environmental economics argues that such subsidies do not promote sustainable economies. By lowering costs well below the true cost, they discourage conservation, recycling, and pollution reduction (see an overview of environmental science).

The second basic economic incentive is **privatization**, in which environmental resources become the property of individuals or businesses. By transferring resources from the commons to individuals or companies that have a vested interest in them, the basic cause of the tragedy of the commons is removed. Examples include commercial forests that raise trees for timber, commercial parks that raise endangered species, and individual ownership of water rights in the western United States.

Despite some success with privatization, many environmental economists believe that it is often less useful than green taxes (and other green fees) in solving environmental prob-

lems. One problem is that many environmental commons cannot practically be privatized. For instance, how can you own a part of the atmosphere? Similarly, it is impractical to have ownership of the oceans or rivers that travel through land that many people own. Another problem is that even where private ownership is practical, there is no guarantee that market prices will always reflect true environmental costs. For example, private ownership of oil and ore deposits will still lead to depletion of those resources unless society implements regulations, taxes, or some other mechanisms that prevent market failure.

Recently emerging ecosystem service markets function as a means to value the goods and services, such as clean water and carbon sequestration, that ecosystems provide. Putting a price on these goods and services and incorporating them into the market system allows a value to be placed on them, thus more effectively allowing privatization. However, these markets, unless purely voluntary, require environmental standards to protect natural resources.

The Low Cost of Sustainability

Many environmental economists believe that the transition to a sustainable economy can be made with surprisingly little sacrifice, if it is begun soon. They say that the savings that will be realized when currently wasted resources are used more efficiently can pay for the transition. Three types of change will help to pay for existing externalities:

1. Cost-free changes to eliminate policies that cause environmental damage.
2. Investments that are economically and environmentally profitable.
3. Changes that cost money but are environmentally profitable.

The first two types of changes require no economic sacrifices. First, we can simply stop subsidies and other government policies that actively promote environmental damage, such as subsidized mining and rain forest destruction. Second, public and private investments in water supplies, conservation, and other activities both pay for themselves and make a monetary profit. For example, a review of renewable and alternative energy sources shows that many investments in energy efficiency not only reduce fossil fuel pollution and conserve resources, but also increase net profits by reducing waste.

In some cases, private investments in businesses that develop such green technology can benefit both the environment and the investor. **Green investing** is a type of socially responsible investing that has become increasingly popular as an alternative to investments made on the basis of profits alone (see **CASE STUDY 19-1**). Ending current taxpayer subsidies that promote unsustainable businesses, such as the fossil fuel and nuclear industries, would encourage green investing by giving sustainable industries a level playing field. For example, businesses would have a greater incentive to develop more fuel-efficient cars and alternative fuels.

Only the third group of changes requires actual monetary sacrifice. Examples would be green taxes on smokestack emissions and landfill wastes, which would increase the cost of products and garbage removal; however, even these measures have no net cost when the benefits of a cleaner environment are included in the calculation.

The basic effect of these changes would be to restructure the economy and rechannel industry and jobs into producing goods and services that the environment can sustain for many centuries. To repeat a key point, this change will not lead to unemployment but will shift employment to sustainable jobs. For example, as internal combustion engines become less important, employment making solar batteries will increase. As noted, such sustainable activities tend to provide more jobs than unsustainable ones. Of course, the transition to sustainability will not always be easy, but economic transitions historically have required social adaptations (**FIGURE 19-9**). The difficulties can be minimized if the transition is gradually carried out. A phase-in period of 5 to 10 years is often suggested. In his book *Eco-Economy*, Lester Brown discusses a number of examples of fast-growing industries that are not only profitable but also more sustainable and cleaner than traditional industries: fish farming, bicycle manufacturing, wind farm construction, wind turbine and fuel cell

FIGURE 19-9 Economic transitions often involve the demise of some careers, forcing people to adjust by seeking new careers. The village blacksmith, as depicted in this late 18th-century American engraving, is now a thing of the past.

manufacturing, hydrogen generation, light rail construction, and tree planting.

However, the current transition to sustainability, especially in the United States, is too gradual for most who are concerned about the environment. In the United States, there has been very little support for green taxes. In 1990, Congress approved a 4 cent per gallon (3.785 liters) gasoline tax, but this is a small fraction of the tax required to significantly discourage auto use, lead to more efficient cars, and reduce emissions. Studies show that current prices must be doubled or tripled to shift consumption patterns significantly. In the case of gasoline, a tax of at least $1 per gallon (and probably more) will be required. Such a tax would put U.S. gasoline costs at about the level of those in Western Europe and Japan, where the high costs have greatly improved conservation and efficiency. Sadly, it may have taken a natural disaster to reawaken

Would You Be a Green Investor?

http://environment.jbpub.com
/mckinney/5e/
for more information

Economists often say that people vote with their dollars, especially in a market economy. This means that our spending reflects what we truly value. A good example is the growing popularity of socially responsible investing, also known as SRI to investment brokers.

Socially responsible investing refers to financial investments that are based on more than merely the desire to maximize your profit return. For instance, you tell your investment broker that you want to avoid investing in certain companies because you disapprove of some of their activities and do not want to support them. Popular examples are socially screened mutual funds that seek to invest only in companies that have a clean record on the environment or other social issues. The first socially screened fund was the Pax World Fund, created in 1970 to avoid investing in suppliers to the Vietnam War.

In the early 1990s, the number of socially responsible investors grew dramatically. According to the Investor Responsibility Research Center (IRRC), which promotes SRI by supplying subscribers with information about corporate behavior on social issues, the number of socially screened mutual funds grew at four times the rate of traditional funds. Total assets of these funds topped $3 trillion by the end of 2011. The specific social issues promoted vary among the mutual funds. Some funds consider a wide range of

issues. Although these issues usually include the environment, they also include a company's hiring practices on women and minorities and its record on animal testing. Investment in tobacco and alcohol companies may also be avoided, but some funds are strictly environmental; in fact, a 1993 survey of money managers concluded that the issue that concerned most investors is the environment. Green Century Capital Management, the Calvert Fund, and Merrill Lynch's Ecological Trust are examples of funds that focus on a company's environmental record.

As always, one must be wary of the "greenwashing" problem. Greenwashing refers to false environmental claims that are sometimes made by companies to use the popularity of environmental issues to attract consumers or, in this case, investors. For instance, a number of waste management companies are known as environmental sector stocks because they are in the pollution control and trash collection business. Thus, some brokers have included them as desirable green investments, but many of these companies have been fined millions of dollars for flagrant environmental violations, such as illegal dumping and disposal of waste. Most environmental funds avoid them, preferring instead to invest in companies with no environmental violations, especially if the company actively promotes sustainability by making recycled prod-

interest in energy conservation. Only when gasoline prices skyrocketed in late 2005 after Hurricane Katrina disrupted Gulf Coast refineries did the federal government and the American public begin programs to examine energy use and ways to reduce energy waste.

The speed and power of green taxes to improve the environment were demonstrated in 1989, when a high tax on leaded gasoline went into effect in the United Kingdom. Just 1 year later, in 1990, the use of unleaded gasoline had climbed from 4% to 30%.

19.5 Poverty and the Global Environment

Global environmental problems are tightly interwoven with poor economic conditions. Poor nations find themselves in a vicious cycle. Poverty often correlates with overpopulation, which then leads to ecological decline as greater numbers of people overuse resources and pollute. As natural resources dwindle and are degraded, poverty increases, often leading to more population growth. Many nations,

ucts, helping rebuild wetlands on its property, and so forth.

Another point to consider is how important profitability is to you. Some money managers caution investors that environmental (and all SRI) funds and stocks may, on average, produce lower rates of return. This is not always true; the Parnassus Fund is often cited as a consistently high-performing investment, but brokers often note that because SRI funds are constrained in where they can invest, they may not be as productive. On the other hand, many investors are willing to risk lower profits because they are concerned about how a company makes its money. Does this potential sacrifice have any effect? Does SRI work? Although SRI is still relatively new, it appears to be having a significant effect. One sign of this is the rapid growth of organizations such as the IRRC that provide subscribers with information about a company's record on social issues. Perhaps even more significant is that the major pension funds, the real titans of investment, are beginning to consider SRI.

With the vast amounts of money they have at their disposal, these funds can strongly influence a company by threatening to withdraw their investments. In 1990, the world's largest pension funds, the teachers' and professors' retirement funds, with assets of more than $100 billion, began allowing members to put money into so-cially screened accounts. These members often send resolutions to companies requesting that they adhere to specific environmental principles. As one official noted, "Companies who receive a resolution from a public pension fund take it very seriously."

So does the investment future look green? It is already becoming standard practice before company mergers and buyouts to do an analysis on a company's environmental record. According to a veteran Wall Street broker, "The companies that are . . . getting themselves into cleaner businesses are the ones that look good for the future. We are in between a long period when it paid to be dirty and when it will be seen as profitable to be clean."

Questions

1. If you had $10,000 to invest right now, would you be a "green" investor? Why or why not? Which companies would you specifically prefer to invest in?
2. If you had $10,000 to invest right now, would socially responsible investing (of any kind) be an option for you? Why or why not? Which companies would you specifically prefer to invest in?
3. Some critics have called SRI a mere ploy to gain money from "bleeding hearts." Is this unjustified cynicism? Do they have a valid point?

VISIT

http://environment.jbpub.com /mckinney/5e/ for more information

even those as different as Poland and Haiti, illustrate how economic problems are closely related to environmental problems.

In 2002, approximately one-fifth of the world's population lived in abject poverty, with incomes of $1 a day or less. Billions more have standards of living well below that of U.S. and Canadian citizens. Increasing the standard of living in these countries involves far more than simply "developing" their economies by transferring current technologies to make their agricultural economies more industrial. Most current technologies use far more resources than the world could afford if everyone used them. Environmental economist Herman Daly has estimated that for all developing nations to attain the income levels of the leading industrial nations such as Japan and the United States, the consumption of the Earth's resources would have to increase by a factor of 36, or 3,600%, if current technologies are used to industrialize.

This would require the natural resources provided by three or more planets like Earth. Many geologists estimate that the world's

VISIT

http://environment.jbpub.com /mckinney/5e/ for more information

In his classic 1973 book *Small Is Beautiful,* E. F. Schumacher argued eloquently for the importance of "appropriate technology." These are sustainable technologies that are affordable and can be produced and maintained locally. Such technologies not only have less environmental impact than traditional industrial technologies, but also promote sustainable economic growth and increase the standard of living in poor nations.

The immense power of such small-scale local technologies occurs from the cumulative effects when many people use them. Thus, as often happens, large-scale change is a product of seemingly mundane trivial changes multiplied many times. One of the most dramatic examples is the cookstove. Half the people in the world, mainly in developing nations, prepare food and heat their homes with fires that burn dung, wood, or other combustible materials.

Although seemingly rather harmless, these cookstoves are major causes of massive environmental destruction. Many thousands of acres of forests and other ecosystems are denuded and degraded as people seek firewood and other biomass fuels. In addition to the obvious harm to the ecosystem, deforestation and plant denudation are a major cause of soil erosion. Human health also suffers. Smoke from indoor fires often exceeds 20 times the limit of ash and other pollutant levels recommended by the World Health Organization.

Traditional cookstoves can be improved in a number of ways. The most basic is to increase the heating efficiency. Traditional open fires with a pot above are only about 10% efficient; 90% of the heat goes into the air and is wasted. Traditional metal stoves are not much better. Lining the stove with ceramic, brick, or clay increases the heating efficiency to at least 20% and often as much as 40%.

Increasing efficiency to just 20% can reduce the amount of firewood (ecological damage) and smoke (human health damage) by half, a good example of the power of a simple change. Using less firewood or other fuel also saves money. In Kenya, switching to more efficient cookstoves typically saves about $65 per year, as much as one-fifth the annual income of many families. Because the cost of such lined cookstoves usually is less than $5, this is a good investment. China has more than 120 million insulated cookstoves because of a government program promoting them. Millions of similar stoves are becoming popular throughout Asia and Africa thanks to dozens of government and private programs.

A second, usually better, type of improvement is the solar cookstove or oven (**FIGURE 1**). When the sun shines, the food cooks in a glass-covered box that traps the heat. The walls and floor of the box are lined with reflective metal or foil. On a sunny day, meat stews and rice dishes can be fully cooked in 2 to 5 hours. No biomass fuel is used, and no unhealthy smoke is produced.

---

entire oil supply would be gone in just a few years if all countries used as much per person as the United States. One answer to this dilemma, as noted in many parts of this book, is to lift developing nations out of poverty by transferring sustainable, or green, technologies that produce sustainable growth (see **CASE STUDY 19-2**).

Another solution to the problem of dwindling resources is somewhat easier to implement: reduce our own consumption of resources. The wealthiest 25% of the world's population now consume about 75% of the world's resources. In addition to consuming more resources, the technologies that the wealthiest nations use also produce much more solid waste, greenhouse gases, and other pollution per person (**FIGURE 19-10**).

Origins of Global Poverty

Economic theories of the 1940s through 1970s generally maintained that the poverty of the developing (then called Third World) countries

FIGURE 1 A solar oven being used in Kenya.

VISIT

http://environment.jbpub.com
/mckinney/5e/
for more information

The solar cookstove is still relatively new and is not yet as widely used as insulated biomass cookstoves. Cost is a factor, too, as solar cookstoves, at $20 to $40 each, are considerably more expensive, but mass production of solar cookstoves, often by local craftspeople, is expected to greatly reduce prices, just as it has with the insulated biomass cookstoves over the last few years. Considering that solar energy greatly reduces the need to search for or buy firewood or other fuels and thus saves even more time and money, they could become quite popular.

The experience with cookstove improvements has been educational in many ways. Early efforts to introduce the new technologies were unsuccessful because many social factors, such as community needs and customs, were not considered. In many areas, people will adopt newer cookstoves only when community groups are allowed to modify the designs and local artisans become stove makers. Improved cookstoves thus not only reduce harm to the environment, they also produce jobs for the community.

Questions

1. What do you think is meant by the phrase "small is beautiful" in the context used here?
2. Think of examples of other kinds of "appropriate technology" that can benefit the environment and the economy of a country.

was a temporary situation. Their agricultural economies, which produced little wealth, were largely a product of 19th-century colonial policies that had made little effort to modernize the native societies. Supposedly, trade with richer nations and some monetary aid would in time provide the technology and other necessities for the poor economies to reach **economic takeoff**. This was the point at which poor countries could produce enough wealth to support themselves financially, ex-

port manufactured goods, and eventually attain the living standards of the United States, Canada, Europe, and other industrialized (First World) countries. As early as 1948, President Harry Truman, in his inaugural address, urged not only the rebuilding of Europe with the Marshall Plan, but also the development of the world's poor countries. President John F. Kennedy declared the 1960s to be the "decade of development" and pledged 1% of the U.S. GNP toward this goal.

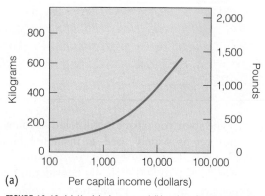

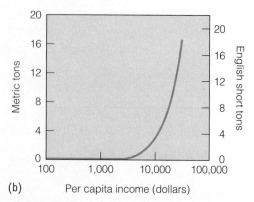

FIGURE 19-10 (a) Municipal waste and (b) carbon dioxide are just two of many pollutants that increase rapidly with increasing wealth. (*Source:* Adapted from World Development Report 1992. The International Bank for Reconstruction and Development/The World Bank. Oxford: Oxford University Press, Inc., 1992.)

However, economic takeoff has not occurred in most poor countries. Instead of becoming richer, poor nations have become poorer relative to the richer nations (**FIGURE 19-11**). In 1950, the average income in the rich nations was about seven times as much as the average income in the poorest nations. By the early 2000s, the differential had risen to more than 20.

Why haven't the poorer countries experienced economic takeoff? Economists disagree on the answer, but a number of factors seem important. First, much of the money that developing nations have acquired has been badly mismanaged. Instead of investing in roads, education, and other necessities for building a prosperous economy, they have often spent the money on military equipment to help keep dictators and elite social classes in power. Recent trends toward democracy in Africa, Asia, and Latin America may alleviate this problem. A second and more important reason is that the world is much more competitive than when the United States and other developed countries were industrializing. It is very difficult for less developed countries to make products that can compete with advanced technologies.

As a result, poorer countries must sell whatever products they can, which usually means their natural resources. Often these are nonrenewable: oil, high-grade ores, native species being driven toward extinction, and so forth. Rather than investing in equipment and other capital that will create more wealth, these countries are spending their resources in ways that will not provide them with long-term sustainable growth. As seen by a review of the principles of pollution control, toxicology, and risk, the loss of these irreplaceable natural resources is accelerating because they are cheap. The price paid for these resources, mainly by consumers in the industrialized nations, does not reflect their true environmental costs. The consumers who buy such resources discount (and often omit) the value of them to future generations, to local peoples, and to local ecosystems.

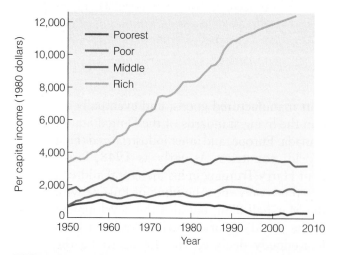

FIGURE 19-11 Per capita income has steadily increased in the rich nations but has remained relatively stable or even decreased in the poorer nations. (*Source:* Adapted from P. Ekins. Green Economics. London: Gaia Books, 1992, p. 32.)

Eliminating Global Poverty

The growing disparity between poor and rich countries has led to concern that a new economic war between North and South will

FIGURE 19-12 This polluted slum in Brazil is an all-too-common example of the economic disparity between poor tropical countries and the rich countries to the north.

replace the now-defunct military Cold War between East and West. Most of the wealthier nations are in the higher latitudes, mainly in the Northern Hemisphere (an exception is Australia). Poor countries are found to the south, around the equator, and in most of the Southern Hemisphere (**FIGURE 19-12**). These poor tropical nations share many common interests, as is readily apparent at the United Nations, where poor countries often form voting blocs on international issues. One solution to this growing poor–rich division is political: National governments must meet and reach agreements on international issues that affect them all. International meetings on economic and military issues have been held for years, but not until 1992 did international environmental issues came to the fore. In that year, the **Earth Summit** in Rio de Janeiro, which was attended by 178 nations, addressed global concerns such as global warming, biodiversity, and deforestation. Although the summit was a great success in demonstrating global concern for these issues and some important agreements were signed, many participants fear that many of the agreements will be ineffective because they either lack means of enforcement or take steps that are too little or too late to rectify major environmental problems.

Economic Aid and Debt Cancellation

We noted previously that economic aid has not led to economic takeoff because of money mismanagement and the competitive global economy, among other factors. For these and other reasons, many wealthier countries have reduced their aid to poor countries in recent years. Most industrialized countries have never come close to giving the 1% GNP pledged in the early 1960s. The United States annually gives among the lowest proportionate aid levels of any nation. Each year, the United States donates less than 0.20% of its GNP to foreign aid. Norway and the Netherlands rank among the highest in the world, donating about 1% of their GNPs. Furthermore, much U.S. aid has always gone to just a few nations, such as Israel and Egypt, for military reasons.

The lack of donated money has meant that poor countries have had to borrow from richer countries, causing a debt crisis, also called a "debt bomb" (as seen in an overview of environmental science). In the early 2000s, the debt of the developing nations exceeded $2 trillion, amounting to about half of those countries' collective GNP. Because the interest on this debt is higher than the foreign aid that rich countries provided, since 1982, there has been a net flow of money from the poor nations to the rich nations.

As a first step in resolving this debt crisis, the United Nations estimates that the environmental and developmental programs proposed at the Earth Summit would require rich countries to contribute at least 0.7% of their GNP as aid to the poor countries. This sum is less than 10% of the annual military spending of the United States. In most other wealthy nations, it would be less than 33% of the military budget.

In addition to direct payments, the rich nations can provide many other types of economic aid to ease the debt crisis. One is cancellation of debt. The United States, Germany, Canada, Great Britain, and other developed countries have canceled tens of billions of dollars in loans to developing nations, largely for humanitarian reasons. However, this is a tiny fraction compared with the more than $2 trillion the poor countries owe. The Worldwatch Institute estimates that about 60% of this debt will have to be canceled if the poor countries are to save enough money (instead of making loan payments) to begin sustainable development; however, outright cancellation of such a large debt is very unpopular among the taxpayers of richer nations, especially the United States.

To make the cancellation more appealing, some observers have suggested that the debt be traded for something valuable. Among the most notable trades are the **debt-for-nature-swaps** that biologist Thomas Lovejoy suggested. With these, a conservation organization pays about 15% to 30% of the debt to the lender, and the rest of the debt is canceled. In exchange, the poor country carries out specified environmental programs. In a review of land resources and management, we saw that land was set aside for preservation of tropical forests. Other examples include Costa Rica, where the Dutch government in 1989 swapped $33 million in debt for $10 million worth of programs in reforestation, water resource management, and soil conservation. Similarly, the United States has agreed to allow Poland to cancel 10% of its debt if the money was spent instead on toxic clean-up and other badly needed environmental restoration.

Private Investments and Business Loans

Debt cancellation, debt swaps, and even direct donations of money are largely short-term solutions. The fundamental problem is that the developing nations are not being paid the true costs for their resources when such values as environmental uses are considered. If the value of reducing global warming, conserving rare species and rare genes for agriculture and pharmaceuticals, and preserving uses for future generations were included in the payments, money would flow into these poor nations. Such a flow is not aid; it is simply the proper payment for the true value of their resources.

The best way to obtain the true value of resources is to produce a healthy local economy so that profits are channeled to the local inhabitants of a poor nation. This usually requires investments and business loans from wealthy nations. An important kind of business loan is the **microloan**, which is a small loan made to a small business that promotes sustainable activities. This is another example of the grassroots or bottom-up approach

to environmental problems. In this case, such loans improve both the economy and the natural environment of a poor nation. For example, microloans have been made to family businesses in Brazil that practice sustainable harvesting of the rain forest and to small businesses in India that build solar ovens (and thereby reduce deforestation; see Case Study 19-2). These loans empower communities and promote women's roles in communities when the businesses are owned or operated by women. Government agencies of wealthy nations, the World Bank, and other international lending agencies sometimes give microloans. Nongovernmental organizations, often called NGOs, also give them.

After enough money is made available to developing countries through a mixture of loans and gifts, most economists believe that further funding will be added by economic growth. Private investment from richer countries will occur as the poorer countries develop stable, growing economies. This economic growth will itself produce profits that can be reinvested and help pay debts; however, the environmental goal is for this growth to be sustainable with minimal ecological harm.

19.6 Developing a Sustainable Economy

Most environmental economists agree that a sustainable world economy can still be built, but only if major steps are taken within the next 2 decades. These steps can be summarized as follows: (1) increase the flow of money into developing countries, reversing the present flow away from them; (2) channel the money flow into activities that not only produce wealth but are sustainable; and (3) establish some social order that promotes fair distribution of that wealth or access to it.

Many experts believe that these steps can be most effectively carried out if a central global financial institution oversees the money flow. For instance, if richer nations were to contribute 0.7% of their GNP, as discussed previously, the money would be given to this central institution, which would then distribute it. A possible candidate is the World Bank, which is already the largest international lender in the world. Despite the many advantages of having a single, central institution, political differences among countries probably render it impractical. In addition, there is no guarantee that the money would be distributed in the best way. In the early 1980s, the World Bank and similar institutions came under intense criticism for disregarding the environmental impacts of their loans. They funded dam building and irrigation projects that destroyed huge tracts of land and polluted rivers; they also funded massive construction and deforestation projects in many countries (**FIGURE 19-13**). This criticism led to the creation of an Environment Department in the bank whose staff investigates the environmental impacts of funded projects. The department has been widely criticized as being too small and ineffective; environmentally damaging loans still occur.

Aside from the World Bank, an increasingly important institution influencing developing nations is the **World Trade Organization (WTO)**. The WTO was founded in 1995 to reduce trade barriers and settle trade disputes between nations. Because of its focus on economic growth, the WTO has been increasingly criticized by environmentalists as well as advocates for social justice (see **CASE STUDY 19-3**).

In addition to governmental organizations, private groups can promote sustainable economies. The importance of nongovernmental organizations is considered in a review of historical and cultural aspects of environmental concerns. Even private individuals can make a difference. Ted Turner, founder of CNN, donated $1 billion to the United Nations in 1997 to support work on population stabilization and environmental protection. He created the UN Foundation through which other billionaires could "give back" to the global community. A similar effort has since been made by Bill Gates, founder of Microsoft, in starting a large foundation for coping with health and population issues in poor nations. Given the enormous wealth in the private sector of wealthy nations, these types of initiatives can potentially promote even greater changes than are possible with governmental initiatives.

Is a sustainable global economy just a lofty ideal that is unrealistic? Not in basic economic calculations. **TABLE 19-5** outlines the estimated costs and long-term benefits of many environmental programs that can be undertaken in developing countries. The costs are estimated as a percentage of the GDP of developing countries in the year 2000. (Remember, GDP is the amount of wealth produced within a country, as opposed to GNP, which includes exports and wealth produced from outside the country.) The key point is that the cost of all these environmental programs is only a small fraction of the wealth produced by a healthy, growing economy. The total estimated cost of the programs listed in the table is about $75 billion, or only about 1.4% of the combined GDPs of developing countries in the year 2000. Of course, Table 19-5 is not comprehensive, but the World Bank estimates that even if rain forest protection, pollution cleanup, alternative energy development, and other comprehensive items were included, the cost would still not exceed 3% of the GDP of the developing countries. These estimates are obviously very approximate, but they provide an essential view of the true economic costs of a global, sustainable society. They show that a sustainable global society is a realistic, attainable goal.

FIGURE 19-13 Financing the building of dams in developing nations was once a very common, and very ecologically destructive, practice of the World Bank. Persistent pressures by many environmental groups have helped reduce, but not eliminate, such financing of unnecessary environmental damage. Shown here is the Itaipu Dam, a Brazilian hydroelectric project.

The World Trade Organization: Wealth Versus Environment and Justice?

Today, global environmental issues are inexorably associated with the broader issues of the relationships between the North and the South—the industrialized nations and the developing nations—as well as the movement to liberalize world trade. Furthermore, the concept of sustainable development includes the notion of equity between different peoples and nations of the world, so it is only fair that the developed countries help the developing countries address environmental issues of global impact.

Clearly, unilateral action by a single nation will do little to correct global environmental problems that are collectively caused by many countries. For instance, curbs on greenhouse or CFC emissions by a single country could not have solved the global problem of atmospheric deterioration. In addition, we now realize that environmental deterioration and the depletion of resources, particularly in developing countries, often promote cycles of poverty that lead to further environmental deterioration. The North is increasingly pressuring the South to end environmental deterioration and protect its remaining natural resources (such as tropical rain forests) for the good of the planet as a whole, yet the North has not set a very good example of resource conservation. Furthermore, many developing countries are heavily in debt to developed countries and must exploit their natural resources to service that debt.

International trade agreements, such as the North American Free Trade Agreement (NAFTA), can also have direct environmental ramifications. Major issues include whether free trade in goods and products will also entail free trade in pollutants or environmentally destructive practices. A country that requires recyclable, and thus more expensive, packaging places an added burden on foreign manufacturers who do not have to comply with the same requirements at home. The result can be trade disputes and international discord. NAFTA acknowledges that individual nations may enact environmental, health, and safety laws as long as they are "necessary" and based on solid scientific data, but defining "necessary" and solid science in this context can be difficult. As some commentators have noted, many U.S. environmental laws are based more on political

compromise than science and thus may not meet the standards of NAFTA or other international trade agreements.

The most visible place where these issues have recently come to public attention is with the World Trade Organization (WTO). Founded in 1995 to replace GATT (General Agreement on Tariffs and Trade), the WTO has more than 100 nations in its membership who attend meetings intended to promote global trade and economic growth. The WTO has gained a reputation among many critics who argue that it emphasizes economic growth at the expense of environmental regulations and social justice. For example, critics note that whenever the WTO has negotiated a case in which there has been a conflict between trade and environmental regulations or social injustice

FIGURE 1 Protesters at the 1999 World Trade Organization meeting in Seattle.

(such as the use of human labor under poor living conditions), the WTO generally has ruled in favor of trade.

As a result, WTO meetings typically attract many protesters. In late 1999, a meeting in Seattle, attended by 5,000 delegates from more than 150 nations, also attracted 50,000 protesters who used civil disobedience and other means to voice their concern over the direction that the WTO was taking (**FIGURE 1**). WTO meetings in the early 2000s have attracted similar numbers of protesters.

Questions

1. Do you think civil disobedience at the WTO meetings is justified? Is it the right course of action? Would you do it?
2. Should the WTO be eliminated? Why or why not?

VISIT

http://environment.jbpub.com /mckinney/5e/ for more information

TABLE 19-5	Estimated Costs and Long-Term Benefits of Selected Environmental Programs in Developing Countries			
Program	**Billions of Dollars/Year**	**As a Percentage of GDP**	**As a Percentage of Growth**	**Long-Term Benefits**
Increased investment in water and sanitation	10.0	0.2	0.5	More than 2 billion more people provided with service. Major labor savings and health benefits. Child mortality reduced by more than 3 million a year.
Controlling particulate matter (PM)	2.0	0.04	0.1	PM emissions virtually eliminated. Large reductions in emissions from coal-fired power stations, respiratory illnesses, and acid deposition.
Reducing acid deposition from new coal-fired stations	5.0	0.1	0.25	
Changing to unleaded fuels; controls on the main pollutants from vehicles	10.0	0.2	0.5	Elimination of pollution from lead; more than 90% reductions in other pollutants.
Reducing emissions, effluents, and waste from industry	10.0–15.0	0.2–0.3	0.5–0.7	Appreciable reductions in levels of ambient pollution despite rapid industrial growth. Low-waste processes often a source of cost savings from industry.
Soil conservation and reforestation, including extension and training	15.0–20.0	0.3–0.4	0.7–1.0	Improvements in yields and productivity of agriculture and forests. Lower pressures on natural forests. All areas eventually brought under sustainable forms of cultivation.
Additional resources for agricultural and forestry research, in relation to projected levels, and for resource surveys	5.0	0.1	0.2	
Family planning (incremental costs of an expanded program)	7.0	0.1	0.3	Long-term world population stabilizes at less than 10 billion instead of 12.5 billion.

Source: World Development Report 1992. New York: World Bank, 1992, p. 174.

SUMMARY

- Environmental economics blends the short-term need for jobs and the long-term need to protect the environment.
- Environmental economics recognizes the need for sustainable growth, which uses minimal resources, controls pollution by input control, and relies on renewable resources.
- The Gross National Product does not subtract the environmental costs incurred when wealth is created. A well-known alternative is the Genuine Progress Indicator, which includes several environmental measures.
- Market failures occur in a market economy when society is not benefited by this individual pursuit of self-interest. These market failures, known as externalities, do not include all costs of production in the price of a product.
- When dealing with the environment, externalities include resource depletion and pollution and the negative impacts on ecosystem services.
- The solution to environmental externalities is to internalize the cost by forcing consumers to pay for the true environmental costs of ecosystem goods and services. Environmental costs

are complicated by intangible costs, hidden costs, future costs, and unequal distributions of cost.
- Society can extract payment for environmental externalities by persuasion, government regulation, and economic motivation, including green taxes and privatization.
- The solution to global poverty and many global environment problems is sustainable development.
- Developing countries have not experienced economic takeoff for various reasons, including their mismanagement of money and the competitive world in which they are becoming industrialized.
- The developing nations' debt crisis may be at least partly resolved by direct economic aid and cancellation of debts.
- Environmental economists believe that a sustainable world economy may be built by increasing the flow of money to developing nations and channeling money into wealth-producing, sustainable activities within a social structure that fairly distributes the wealth.

KEY TERMS

command economies
Contingent Valuation Method
cowboy economics
debt-for-nature swaps
dirty subsidies
Earth Summit
economics
economic takeoff
environmental economics
externalities
false dichotomy
future costs

Genuine Progress Indicator (GPI)
Green Gross Domestic Product
   (Green GDP)
green investing
greenmail
green taxes
green technologies
Gross Domestic Product (GDP)
Gross National Product (GNP)
hidden costs
Index of Sustainable Economic
   Welfare (ISEW)

intangible costs
market economies
mixed economies
microloans
pollution permits
privatization
severance tax
sustainable growth
throwaway society
unequally distributed costs
World Trade Organization
   (WTO)

STUDY QUESTIONS

1. What are two ways in which economics is concerned with environmental issues? Name three ways that the economic approach differs from that of environmental science.

2. What is a false dichotomy? Why is the issue of jobs versus environment a false dichotomy?

3. Why do environmentally friendly activities produce so many jobs? What is greenmail?

4. Define environmental economics, sustainable growth, and cowboy economics.

5. How does traditional economics measure the total output of goods and services of a country? What environmental measures does the genuine progress indicator include?

6. What are the two basic economic systems of the modern world? Which is more efficient? Why?

7. What are externalities? What is the tragedy of the commons?

8. Why does scarcity promote even greater scarcity? Give an example.

9. What are some criticisms of government regulation to solve pollution and other problems?

10. List and define the four main costs that complicate the calculation of the overall environmental costs of an activity.

11. What are three basic ways that society can extract payment for a formerly free environ-

mental externality? Define them. Which is least effective?

12. What is economic takeoff? What are some factors that explain why it has not occurred in many poor nations?

13. In what year did the net flow of money begin to flow from poor to rich nations? What is the fundamental problem causing this net flow in the wrong direction? Give examples.

14. Explain the concept of green taxes. Provide examples and explain the advantages and disadvantages of their use.

WHAT'S THE EVIDENCE?

1. The authors state that the GNP/GDP is misleading. What evidence is presented to support this?

2. This chapter implies that green taxes are an effective way of letting the free market solve environmental problems. Is the evidence presented for this convincing? Why or why not?

CALCULATIONS

1. If the U.S. GDP was about $11.7 trillion per year (2004 figure) and the United States donated 1% of that to poor nations, how much money would be donated?

2. If a U.S. GDP of $12 trillion grew by 10% over 5 years, how much would be donated to poor nations 5 years from now if 1% were given?

ILLUSTRATION AND TABLE REVIEW

1. Based on Table 19-5, in your judgment, which environmental program gives the most long-term benefits?

2. Based on Table 19-5, which environmental program or programs will be the most expensive to implement? The least expensive?

http://environment.jbpub.com/mckinney/5e/

http://environment.jbpub.com /mckinney/5e/

Connect to this text's website at http://environment.jbpub.com/mckinney/5e/. This site features eLearning, an online review area that provides quizzes, chapter outlines, and other tools to help you study for your class. You can also follow useful links for more in-depth information.

On February 9, 1880, the Atchison, Topeka and Santa Fe Railway Company celebrated the arrival of its first train into the Santa Fe, NM, depot. The train's arrival marked an end of an era of covered wagon travel and horseback as the main way to transport goods to the area. Though the time period is steeped in cultural clashes, the arrival of the train to Santa Fe began the accelerated development of the area. This development significantly impacted land use, local economy, and subsequent cultural exchange. Native Americans who were displaced to reservations began to create arts and crafts to market to railway travelers, as did local Hispanic artisans. Santa Fe's railyard became a bustling destination as a more permanent city began to take shape. The area's diversity evolved into the crux of its appeal. Many of the descendants of settlers and Native Americans passed down stories of the train-town's famous travelers and prominent events. In the mid to late 20th century, the Santa Fe depot fell into extreme disregard, by the early 2000s, however, a diversity of citizens came together to restore the depot and make it again a great destination spot. Today's depot complex houses a teen art center, several cultural arts venues, a variety of multicultural restaurants, and a large farmer's market. More than 130 years from its first arrival, the train depot of Santa Fe is still making its mark on the town's economy, but today, an inclusive community group monitors its sustainable expansion. For more information about the current use of the Santa Fe depot complex, see: http://railyardsantafe.com.

20 | Historical and Cultural Aspects of Environmental Concerns

Chapter Objectives

After reading this chapter, you should be able to explain or describe the following:

- Various social, historical, and legal aspects of current environmental concerns
- The history of American environmentalism
- How Western societal values may affect environmental issues
- Sustainable development/ sustainable growth
- The fundamental concepts underlying environmental laws and legal instruments
- Methods of decision making in the public arena relative to environmental issues

E ver since the origin of our species, humans have interacted with the natural environment. Our current relationship with the environment is the culmination of definitive historical trends and specific human actions carried out across sequential generations. Although we may not have intended to bring about global environmental changes, humankind clearly and unquestionably has had a global impact on natural systems. The depletion of the stratospheric ozone layer by artificial chlorofluorocarbons (CFCs), to cite just one example, cannot be attributed to natural causes.

People are social creatures, and a human society's most characteristic attributes are its **social institutions**—its familial, political, educational, religious, and economic systems. The social institutions of any society reflect and express its underlying beliefs and values. Some critics argue that the values, beliefs, and institutions of modern, technologically oriented Western society (as it originated in Western Europe and then spread to North America and the rest of the world) are directly responsible for fomenting much of the environmental degradation that we currently face. More optimistically, it has been suggested that although Western society may have brought on current global environmental concerns, it also contains the necessary technology, infrastructure, and institutions to address those concerns. Other observers disagree; they believe that the world's environmental problems will be resolved only by a radical remodeling of Western values and social institutions.

Here, we briefly explore social, legal, and historical aspects of current environmental concerns. Because the focus of this book is environmental science and not environmental studies, we do not deal with these issues in as much depth as we have dealt with some of the more scientific aspects. Nevertheless, science cannot be divorced from society. Science does not exist in a vacuum, and of all of the sciences, environmental science is perhaps the most intimately bound up with social, political, and historical factors.

Here we focus primarily on Western social institutions and values, particularly as manifested in the United States, for several reasons:

1. As we have already noted, many critics blame much of the current global environmental degradation on Western values and social institutions.

2. In many regards, these Western values and institutions find their fullest expression in the United States.

3. Much of the world is apparently striving to emulate the United States in many respects, such as in industrial development and material affluence.

4. Because the United States arguably leads the world in pollution emissions, resource exploitation, energy use, and general environmental degradation, it can provide a particularly valuable case study.

5. The environmental history of the United States is in some ways more accessible than that of other major countries, at least since the time of European contact. Since European explorers first reached the "virgin" continent about five centuries ago (of course, the Native Americans had already modified the "natural" environment to some extent), there has been a fairly continuous recorded history of the environmental changes that have occurred. In comparison, major human-induced environmental changes have been taking place in Europe, Africa, and Asia since before recorded history (although great strides have been made toward reconstructing the environmental history of Europe and, indeed the whole world).

20.1 A Historical Perspective on North American Environmentalism

When Europeans first colonized North America, it was certainly not an "uninhabited wilderness." The Americas had been inhabited for more than 15,000 years by peoples who had crossed over the Bering Strait from Asia and possibly from Europe, following the edges of the glaciers, at the end of the last Ice Age. These indigenous peoples, whom the Europeans called "Indians" (and are now commonly referred to as "Native Americans" or "Native Peoples"), had spread throughout the two continents. North America may have had on the order of 5 million human inhabitants at the time of European contact (**FIGURE 20-1**).

It is a myth that the native North Americans did not modify the environment, for they certainly did. They cleared fields, planted crops, hunted game, built dwellings, and so forth, but they did not exploit the natural environment in the same way that the European settlers did. On the whole, research looking

FIGURE 20-1 This hand-colored woodcut depicts Dutch merchants trading with Native Americans on Manhattan Island in the 1600s.

at artifacts, oral histories, written journals, and cultural practices support the understanding that early Native Americans lived in a relatively sustainable manner for thousands of years. (However, some researchers argue that the first early humans may have been at least partially responsible for massive extinctions of large mammals, such as the mastodon, that occurred at the end of the last Ice Age, about 10,000 years ago.) In part, this may have been due to their small populations and relatively simple material culture, but their complex belief systems also fostered sustainable living. Historically and currently, Native American tribes believe in the communal stewardship of land and nature rather than individual ownership. Out of necessity, cultural belief systems, and circumstances, they live lives more as a part of nature than apart from nature.

The earliest European colonists certainly viewed nature and human society as a dichotomy. People were separate and above nature, not a part of it. Some scholars have argued that an interpretation of Judeo-Christian theology, which saw the world and nature as having been created by God for humankind, encouraged human domination and exploitation of the land. The European settlers considered it their right to subdue, tame, and exploit the "uninhabited wilderness" of North America for humankind's benefit.

The natural resources of the vast continent were ripe for the taking. Sustainable use was not an issue because the resources seemed limitless; there was always more to the west, on the frontier. In the 1600s, the English colonies in New England survived in large part by shipping lumber, beaver skins, and other natural resources back to Europe in exchange for manufactured goods. When the 16th- and 17th-century Europeans decimated the forests and animal populations in coastal areas, they knew they could simply move farther inland. This attitude of the earliest colonists, now sometimes referred to as a **frontier mentality**, continued to exert a major influence on American development for the next 400 years.

The forests in particular were seen not just as a resource to be exploited, but also as a nuisance and fearful place to be eliminated. Because the new Europeans were largely ignorant of the complexities found in the forests, the wilderness was often viewed as a dark unknown, sheltering savages and dangerous wild beasts. Accordingly, the forests were cleared to make agricultural fields and gardens. The once-wild land was cultivated, as most of the land of Europe had been for many centuries (or even millennia).

This mindset of exploiting the land and its natural resources and then moving on dominated American environmental thinking for about 250 years (approximately from the 1620s through the 1870s) as the country continued to expand westward. Wildlife, such as the American bison (see **CASE STUDY 20-1**), was decimated in many areas. Forests were destroyed, and soils eroded as they were put under the plow.

In the 1840s, a related mindset took hold that heavily influenced American expansion. John O'Sullivan, an influential magazine editor, coined the phrase **"Manifest Destiny"** to describe the belief that Americans had a divinely given right to annex all of the land to the west of the Mississippi. "Our manifest destiny," he wrote, "(is) to overspread the continent allotted by Providence for the free development of our yearly multiplying millions."

The U.S. government encouraged the expansion of the frontier by transferring government-owned land and natural resources to private hands. The Homestead Act of 1862 gave each qualified settler in the Great Plains area and elsewhere 160 acres free of charge. Likewise, the Railroad Acts of the 1850s and 1860s gave away large tracts of land to railroad companies to encourage them to build railroad lines that could move people and goods across the country. Vestiges of the frontier mentality and Manifest Destiny remain an integral part of the American psyche to this day.

The First Century of American Environmentalism

However, even as the West was being "won," certain thinkers were questioning the rampant exploitation of natural resources. The New England **transcendentalists**, such as

The North American Bison

The story of the near extinction of the North American bison, *Bison bison* (also known as the American buffalo) is a complex tale of exploitation, sad mismanagement, and, in some cases, deliberate destruction of a species. Only a few centuries ago, an estimated 30 to 60 million bison roamed North America. Two subspecies inhabited the continent. The more numerous plains bison lived mainly east of the Sierra Nevada and inhabited most of what is now the Midwest and Eastern United States, except the Great Lakes area, New England, and parts of the southeast coast. The plains bison extended north into Manitoba, Saskatchewan, and eastern Alberta (Canada). The woodland, wood, forest, or mountain bison (generally considered a distinct subspecies) inhabited the Rocky Mountain region from Colorado to Alberta and even farther north.

Europeans and their descendants were not the only people to hunt the bison en masse. The animals were often hunted in small groups by indigenous Paleo hunters sneaking up on the animals with spears. Another method involved collaborative hunting aimed at separating a small part of the herd and directing them over a bluff. Stampeding bison are unpredictable and, at times, more bison than intended met with death over a cliff. At one site in southeastern Colorado, dating from about 6500 B.C., indigenous Paleo hunters drove a herd of close to 200 bison to their death over the edge of a gorge. Certain Native American tribes survived almost exclusively by hunting the bison, often using the entire carcass for everything from food, fuel, and shelter to tools, dyes, and clothing. Historical writings indicate that there are more than 100 different uses for all parts of the bison. More often than not, the men hunted a manageable number of bison that could then be processed by women to meet nearly all the material needs of various tribes.

In historical times, tribal use of the bison sustainably supported the needs of Native Americans until the arrival of horses, firearms, and the conflicts brought about by European settlement of the west. After the arrival of firearms, the North American Indians sometimes hunted buffalo in larger numbers, aided by the horse and firearms, which they adopted from the Europeans. This often resulted in conflict among the tribes about how to manage the bison.

Another conflicting factor was railroad expansion, which was essentially a death sentence for the bison in several ways. European hunters who traveled by train for the sport of killing a bison became a more common occurrence, as did the fact that trains made it easier for European trappers to transport a large number of hides at one time without regard or regulation for the indigenous use of the animal. In addition, segments of the various bison herds were separated from one another due to the sprawling railroads' expansion in the west. In addition, as bison herds slowed the train when using the passes cut through rough terrain, they were often shot. As time went on, larger guns were employed by the settlers. This made killing higher numbers of buffalo, even up to 150 bison a day, possible from anywhere from 200 to 600 yards. As settlement progressed, culling the bison served as a means for settlers to control and remove a major resource base of the Native American tribes who had depended on the bison to support their needs for thousands of years.

As the massacre of Native American tribes and removal to reservation lands continued throughout the West, the bison populations continued to decline. As a means of defense, native tribes would often burn grasses in an attempt to herd the bison away from European hunters, some of which were commissioned to kill the bison as a way of convincing remaining Native Americans to concede land. Historically, tribes would camp by the killed bison for months at a time and process the meat and prepare other items from the kill.

Conflicts between the settlers and native tribes escalated as time went on. In 1883, for example, the Lakota and settlers were involved in direct combat over the bison. The Lakota tribe was being driven from their land and was suffering from starvation as the bison population dwindled. The Lakota tribe worked to fight back by burning the grasses to try to keep the herd close and away from the European settlers.

Between hunting by the Native Americans and settler attacks, huge numbers of bison were decimated. Ten thousand bison were killed in 3 days from one herd alone. Most were killed by the hired

VISIT

http://environment.jbpub.com/mckinney/5e/ for more information

The North American Bison

MAP
ILLUSTRATING
THE EXTERMINATION OF
THE AMERICAN BISON
PREPARED BY
W.T. HORNADAY.

FIGURE 1 Map of historical bison range.

abundant elsewhere in North America as late as the 1860s (**FIGURE 1**). On the plains, herds of bison 8 kilometers (5 mi) in breadth and some 80 kilometers (50 mi) long could be seen, with the animals so closely spaced that "the whole country was covered with what appeared to be a monstrous moving brown blanket" (letter from Colonel C. Goodnight in the early 1860s).

The further expansion of the railroad made it easy for professional hunters to reach the rest of the herds and for bison products to be shipped back East. As amateurs joined professional hunters, hundreds of thousands and then millions of bison were slaughtered. So many carcasses were left to rot on the plains that settlers complained of the stench. By 1875, so few members of the southern herd were left that both Kansas and Colorado passed laws attempting to protect the buffalo. Still, 100,000 more buffalo were killed during the winter of 1877–1878. A few hundred survivors made their way to Texas, but by 1889, they too had been shot—and thus ended the southern herd. By 1884, the northern herd was virtually wiped out in the United States, and by 1885, the wood or forest buffalo of Canada was almost extinct in the wild.

Why were the buffalo hunted so heavily? Mainly the European hunters wanted the hides, which could be made into leather goods, and fur traders sought "buffalo robes" to be used as overcoats and wraps. The growing market for bison meat was a significant impetus. Buffalo fat could be used in the manufacture of soap and candles, and buffalo horns were used in hat racks and other accessories.

Many buffalo were simply killed by Europeans for pure sport, similar to decimation of herds of elephants and other large mammals in Africa and India during the same time. In America, expeditions were arranged where interested parties could test their shooting skill by slaughtering wild buffalo from a railway car. Finally, the Native Americans' reliance on the bison was a prime incentive for the European settlers to decimate the bison herds. Once a tribe was cut off from its food source and suffered irreversible damage to its culture, then they were more easily moved to

hunters with larger guns who took only the skin, some of the meat, and little else. Despite these abuses, the major blame for the near extinction of the bison must be placed on the ignorance of the European settlers who had no understanding of the permanence of their actions. These settlers were by no means immune to the consequences of their actions. As the bison neared extinction, trappers and entrepreneurs that led hunting expeditions actually went out of business.

East of the Mississippi River, bison were extinct by about 1880. They remained extremely

reservations with less fighting since there were no option left but death.

Even after the buffalo were gone, the ground in many areas was literally covered with their bones (**FIGURE 2**). Homesteaders collected the bones and burned them as fuel. The bones were shipped back East by the trainload where they were ground up for fertilizer and used in sugar refining (to neutralize acids). The hooves and horns were used in the manufacture of glue. Well into the 20th century, buffalo bones were being shipped to the East for commercial use.

Only after the fact did the Canadian and U.S. governments intercede to try to halt the destruction. Canada passed legislation to protect the buffalo in 1885, but the U.S. government waited until 1899. By the 1890s, only two groups of bison remained in North America: a mixed lot of plains and forest or wood buffalo in Canada and a small herd of plains buffalo in Yellowstone Park. William T. Hornaday, Director of the New York Zoological Park, made a census of all known living North American buffalo and found just under 1,100. In 1905, Hornaday founded the American Bison Society, with the primary goal of saving the American buffalo from extinction.

The American Bison Society, with support from President Theodore Roosevelt and others, established reservations for the bison, where they were provided with shelter and fodder as necessary. Results were achieved quickly: In 1910, there were more than 2,100 bison in North America, and by the 1930s, there were over 20,000. Today, ranchers keep herds throughout the continental United States, and there are sizable herds in government reservations such as Yellowstone Park and the Wood Buffalo Park in Alberta, Canada.

A sad footnote to the otherwise happy resurrection of the North American buffalo is that the wood or forest bison no longer exists as a distinct subspecies. The last remaining herd of pure forest bison inhabited the Wood Buffalo Park, but in the 1920s, plains bison were introduced to the same park. Naturally, the two subspecies mixed and hybridized, and the rarer forest bison ceased to exist as a pure-breed form. Today the "wild" herd in Yellowstone is made up of about 3,000 individuals. Nationally only about 400,000 individuals exist in smaller privately owned herds, farms, and zoos.

Despite the hard work of a range of people and enhancements to selective breeding programs promoting genetic diversity the plight of the bison is not over. Legislation passed in the late 1990s allows for a culling of the herd when individual bison wander onto designated ranch land, or

VISIT

http://environment.jbpub.com /mckinney/5e/ for more information

FIGURE 2 Piles of North American bison (buffalo) bones were gathered throughout the West in the late 19th century to be used for fertilizer.

The North American Bison

VISIT

http://environment.jbpub.com
/mckinney/5e/
for more information

otherwise grow in numbers within Yellowstone Park, or wander outside the management of Yellowstone. Organized hunts for the bison still exist, although they are seasonally managed. Even as Native American tribes on reservations still seek to establish and maintain their own herds of bison, instead of relocating these wandering bison to groups that offer to take them in, they are often killed instead. Even since 1985, more than 6,000 buffalo have been killed in Yellowstone, 1,094 in 2008, and 219 in 2009–2010. This continued killing is particularly problematic when considering the most genetically pure herd exists only in Yellowstone. In addition, sometimes management practices, including penning, increase the chance of disease among the buffalo.

The future of the bison is dependent on individuals organizing for improved management strategies and enhanced communication among stakeholders involved in the management of bison populations. Like so many species, the future of the bison remains uncertain.

Questions

1. Why do you think it took so long for the U.S. government to take action in an attempt to save the North American bison?
2. The near extinction of the North American bison is only one case of people exploiting a species that was once extremely abundant. Can you name some other examples (you might refer to previous chapters in this book)?
3. Organizations like the Buffalo Field Campaign (http://www.buffalofieldcampaign.org) serve to promote education about topics that challenge a particular species, in this case the bison. They are largely dependent on volunteers who spend time on site documenting the herd. Would this type of activity interest you why or why not? What are some of the pros and cons of such organizations?

Ralph Waldo Emerson (1803–1882) and Henry David Thoreau (1817–1862), decried the human destruction of the environment. Thoreau mourned the decline and loss of numerous species from his native eastern Massachusetts, such as bear, moose, deer, porcupines, wolves, and beavers, as well as the great reduction in the size and diversity of the forests. To gain a better appreciation for nature, he built a cabin in the woods on Walden Pond near Concord, Massachusetts. From his two years of living alone there came *Walden, or Life in the Woods* (1854). *Walden* has become an American classic, inspiring generations of naturalists, ecologists, and environmentalists to this day.

In 1864, George Perkins Marsh published the exhaustive study *Man and Nature; or Physical Geography as Modified by Human Action*. Marsh had grown up in Vermont but traveled widely. His compilation documented how human intervention had resulted in the destruction of forests, soils, and waters and greatly modified flora and fauna. Marsh's work, using clear scientific evidence and case studies, challenged the notion of an inexhaustible Earth. Marsh did not argue against humankind's intervention in and transformation of nature, but he did suggest it must be done with knowledge and foresight, if irreparable damage and negative consequences were to be avoided. His work laid the groundwork for the late 19th-century conservation and preservation movements.

Two figures dominate American environmentalism in the late 19th and early 20th centuries: Gifford Pinchot (1865–1946) and John Muir (1838–1914) (**FIGURE 20-2**). Although both supported environmentalism, they took very different approaches to the problem. Pinchot, although born in Connecticut, was professionally trained in Europe

(a)

(b)

FIGURE 20-2 Working across ideological differences has always been a part of American environmentalism. Pictured here, (a) Gifford Pinchot and (b) John Muir devoted much of their life work to environmental causes. Gifford Pinchot, first chief of the forest service (1905 to 1910) supported responsible commercial use of forest products and promoted the need for personal involvement in forest conservation stating, "Unless we practice conservation, those who come after us will have to pay the price of misery, degradation, and failure for the progress and prosperity of our day . . . The vast possibilities of our great future will become reality only if we make ourselves responsible for that future." John Muir, founder of the Sierra Club in 1892, took a preservationist position against commercial exploitation of the environment with quotes, such as, "The battle we have fought, and are still fighting for the forests is a part of the eternal conflict between right and wrong, and we cannot expect to see the end of it.... So we must count on watching and striving for these trees, and should always be glad to find anything so surely good and noble to strive for." Their ground-breaking work is still being carried on today, see: http://www.pinchot.org and http://www.sierraclub.org.

as a forester and believed in using the latest scientific knowledge to manage the land. His aim was to produce the maximum sustained yield in forestry. Taking a utilitarian approach, Pinchot basically thought that the forests and other natural resources should be managed so as to obtain the most benefit for the greatest number of people. As such, he was an advocate of **conservationism**. From 1898 to 1910, Pinchot was chief forester for the U.S. Division of Forestry (re-established as the U.S. Forest Service in 1905, with Pinchot at the helm), so he was in a position to implement his philosophy.

Muir was born in Scotland, but came to America as a child. Largely on his own, he explored much of the U.S. and Canadian wilderness, becoming an accomplished naturalist. He also visited Asia, North Africa, Australia, and New Zealand, writing about his travels for newspapers and magazines. Through his experiences, Muir came to believe that nature has an **inherent value** and right to exist in and of itself independent of any value it may have

for humankind. Thus, Muir took a strict **preservationist** position. He had a particular interest in the Sierra Nevada Mountains of western North America and founded the environmentalist group the Sierra Club in 1892. Muir advocated the creation of national parks that would be protected from any type of human intervention. In particular, he was instrumental in convincing the government to establish Yosemite as a National Park in 1890 (**FIGURE 20-3**).

Although strong allies in the general environmental movement, Muir and Pinchot clashed over many philosophical and practical issues. Their greatest battle was over the damming of the Hetch Hetchy Valley (adjacent to Yosemite Valley) through which the Tuolumne River flows. This valley was of great scenic beauty, and Muir, with his preservationist stance, advocated preserving and protecting it. Pinchot, taking the more utilitarian approach of the classic conservationist, supported damming the river to develop a supply of freshwater for the San Francisco region. After protracted debate, Pinchot's views won,

FIGURE 20-3 Theodore Roosevelt (left) and John Muir, founder of the Sierra Club, in Yosemite. John Muir often spoke of the restorative aspects of preserving the natural world as evidenced in the following quote: "Thousands of tired, nerve-shaken, over-civilized people are beginning to find out that going to the mountains is going home; that wilderness is necessity; that mountain parks and reservations are useful not only as fountains of timber and irrigating rivers, but as fountains of life."

role in protecting forests and other natural resources for the public. In 1872, Yellowstone National Park was established, and in 1891, the Forest Reserve Act permitted the president to establish forest reserves (which later became national forests). The 1890 U.S. census declared that there was no longer a definable American frontier, conceptually closing an era in American environmental history.

Theodore Roosevelt (1858–1919), who served as president from 1901 to 1909, contributed greatly to the environmental cause. Roosevelt loved the outdoors and was an established naturalist and conservationist in his own right. Pinchot served as a key adviser to Roosevelt, and under his administration, large tracts of land were added to the national forests. Roosevelt also protected the Grand Canyon and other areas that would later become national parks, and he sponsored a White House Conference of Governors to discuss conservation issues. In 1908, the National Conservation Commission was charged with making an inventory of the country's natural resources.

The Great Depression of the late 1920s and 1930s and the administration of Franklin D. Roosevelt (1882–1945; president 1933–1945) affected attitudes toward natural resource management and conservation. The economic depression of this time can be attributed, at least in part, to the exhaustion and degradation of the land in parts of the United States. For instance, in the Appalachian region of the Southeast, the forests had been decimated and farming had exhausted the soil, causing widespread erosion and flooding. In the Midwest and Great Plains, the once-fertile soils had been over cultivated, and severe soil erosion set in. With just a few dry years, the "Dust Bowl" was created. Windstorms stripped the land of the soil, creating huge dust clouds and causing havoc for the population and the ecosystem (**FIGURE 20-4**).

During the depression, the federal government instituted massive programs aimed at creating employment and restoring the environment. The Civilian Conservation Corps put otherwise unemployed citizens to work planting trees, developing and maintaining park and recreation areas, restoring waterways, building flood control devices (such as

and in 1913, the Hetch Hetchy Valley was dammed. Some have speculated that this defeat contributed to Muir's death the following year. Muir's reputation and message have been enhanced with time. The Sierra Club remains one of the foremost environmental groups in North America today. Meanwhile, modern environmentalists continue to debate the merits of conservationism versus preservationism.

Found in both men's life work was a strong environmental ethic that brought the environment to the forefront as an entity on which we are dependent and to which we must be responsible stewards.

In the late 19th and early 20th centuries, the most widely held environmental stance was **progressive conservationism**. Influenced by Pinchot and like-minded individuals, the federal government took an increasingly active

(a)   (b)

FIGURE 20-4 The Dust Bowl. (a) A photo album montage of a dust storm approaching Stratford, Texas, 1935. (b) This famous photograph, by Dorothea Lange, of a California migrant worker captured the destitution and despair caused by the Depression. The woman in the photograph was just 32 years old and had seven children.

levees and dams), controlling soil erosion, and protecting wildlife. During this period, the Tennessee Valley Authority was established to address economic and resource management issues in the depressed Tennessee Valley. As part of this program, forests were replanted and dams (for hydroelectric power and flood control) were built. Likewise, dams were constructed in many arid regions of the western United States. One of the most famous is Hoover Dam on the Colorado River (near the Nevada–Arizona border); built in 1935, it created the water reservoir known as Lake Mead. Dam construction promoted agriculture and provided hydroelectric energy to fuel industry. Thus, these government programs were viewed as not only conserving natural resources, but also promoting economic growth.

Other developments included the founding of the Soil Erosion Service (later renamed the Soil Conservation Service) in 1933 and the creation of the Agricultural Adjustment Administration (later to become the Agricultural Stabilization and Conservation Service). Researchers studied the causes and prevention of soil erosion and shared their findings with farmers. In addition, the government began to pay some farmers to reduce their crop produc-

tion, thereby helping to stabilize prices and reduce soil erosion. To reduce soil erosion caused by overgrazing, some limits were placed on the grazing of animals on federal lands.

The onset of World War II (America's direct involvement on the battlefield lasted from 1941 to 1945) saw conservation and resource management issues take a backseat to other concerns. The war forced the United States to mobilize; industrial and technological efforts concentrated on military production and winning the war. The experience of this war generally reinforced the American public's belief in social progress through economic growth and technological achievements. For instance, the development of the atomic bomb led to the "atoms for peace" program and the construction of commercial nuclear power plants that presumably would generate clean and inexpensive energy for all. Science and technology could finally make the American Dream come true.

Environmentalism Since World War II

Unfortunately, the technological advances of World War II and the immediate postwar period had the potential to unleash devastating effects on the environment. Production methods, using newly developed technologies, shifted from labor-intensive processes to energy-intensive processes that consumed enormous quantities of energy, often in the form of fossil fuels. Synthetic chemicals increasingly substituted for "natural" resources. This was the age of plastics and the automobile. Individualism, perhaps best expressed in the American ideal of at least one car per family and the freedom to drive on the open road, led inevitably to an unprecedented per capita consumption of natural resources. The Green Revolution, which fed the burgeoning population, was based on energy-intensive, as well as fertilizer-, herbicide-, and pesticide-intensive "factory farming." Economic growth reached unprecedented levels.

All was not perfect, however. With the new technologies and consumption habits came new environmental concerns. Greater and greater levels of toxic chemicals and other dangerous wastes were generated and, in many cases, released into the environment. Nuclear power, in some people's opinions, was a catastrophe waiting to happen. Conservationists began to suggest that we were quickly depleting our resources and degrading the quality of our air, water, and land. Concern over the quality of life, not just the quantity of material goods, became the focus of much environmentalism in the 1960s and 1970s.

In 1962, Rachel Carson's book *Silent Spring* described the adverse effects of pesticides and warned of the environmental disaster that might befall us if we continued to pollute. The book's title refers to one of the consequences—a springtime without songbirds. Carson's book made a strong impact, and some historians consider its publication the beginning of the modern era of American environmentalism. Other books also heralded this new awareness of our limited resources, including *The Population Bomb* (1968) by Paul Ehrlich; *The Limits to Growth* (1972) by Donella H. Meadows, Dennis L. Meadows, Jorgen Randers, and William W. Behrens III; and *Ecology and the Politics of Scarcity* (1977) by William Ophuls. The 1960s and early 1970s were a time of change in American society. The civil rights movement was challenging long-

accepted forms of discrimination (and from this, in part, arose the concept of "discrimination" against animals—see **CASE STUDY 20-2**). The hippie counterculture was questioning traditional values, and as U.S. involvement in the Vietnam War escalated, more and more people were protesting the establishment. Many of the initial environmental concerns were people oriented: We needed to protect the quality of the air, water, and land and preserve wildlife and wilderness for the benefit of people. With the Apollo missions to the Moon (culminating with the first human landing on the Moon in 1969), photographs of the Earth taken from outer space were widely circulated. The concept of "Spaceship Earth" gained wide currency: We are all on one tiny planet drifting through space; Earth is all we have, so we had better take care of it.

On April 22, 1970, much of the country enthusiastically celebrated the first Earth Day. In many ways, this was a time of idealism. Even as people finally acknowledged the existence of environmental problems, they assumed that the problems could be solved. In the 1970s, rising public activism and litigation led to the passing of a number of federal laws that still form the basis for environmental protection in the United States.

- 1970 National Environmental Policy Act (NEPA) requires environmental impact studies before land development projects.
- Environmental Protection Agency (EPA) created.
- Clean Air Act Amendments (CAAA).
- 1972 Clean Water Act (CWA) reduces pollution of lakes and rivers.
- Coastal Zone Management Act (CZMA) begins cleanup of coastal ocean waters.
- 1973 Endangered Species Act (ESA) enacted to preserve endangered species.
- 1974 Safe Drinking Water Act (SDWA) requires EPA to set and enforce drinking water standards.
- 1976 Toxic Substances Control Act (TSCA) helps limit the amount of poisonous chemicals made and sold in the United States.

Animal Rights

VISIT

http://environment.jbpub.com
/mckinney/5e/
for more information

Do animals have rights? Do animals have moral status? Or do only people have moral status and rights? For that matter, should all humans have rights? After all, 150 years ago, some people in some parts of the United States still kept human slaves, and a century ago, American women were not allowed to vote (**FIGURE 1**). Today, there is certainly agreement that all human beings have **moral status** and certain basic rights. What about nonhuman animals, however?

Western thought has had a long tradition of treating all nonhuman animals and plants as objects. Classical Kantian ethics treated animals and plants as merely a means for humans to achieve their goals and ends. The influential philosopher René Descartes (1596–1650) treated animals and plants as simply thoughtless machines that might respond to stimuli but did not possess feelings or consciousness; thus, they did not rightfully deserve moral status. One could legitimately hunt, kill, eat, pen, or even "torture" animals (of course, by Cartesian standards, it would not really be torture because animals have no feelings) with no moral qualms.

In recent times, an increasing number of people have begun to wonder whether animals, particularly the "higher" animals such as domesticated mammals and pets, should have moral status—that is, inherent rights. In practical terms, this issue is often linked to the question of whether various types of animals can experience pleasure and pain and to what degree they experience these sensations. A related issue is how much animals understand, how (if at all) they perceive the past, and how much they can anticipate the future. In other words, how sentient are animals? Some of these questions can be addressed scientifically and clearly demonstrate the interplay between scientific information and ethical theories.

People who are in intimate contact with animals on a regular basis, such as with dog or cat companions, have long intuitively realized that at least some animals are sentient beings. For instance, higher mammals seem to possess a keen awareness of pleasure and pain and can sense and express a gauntlet of emotions, as well as to some degree remember the past and anticipate the future. Evolutionary theory indicates that it is only to be expected that animals and humans should share similar experiences, given our collective common origins and kinships. Anatomical and physiological studies demonstrate that at least the higher nonhuman animals have the appropriate neural structures and networks to transmit pleasure and pain impulses. Indeed, the neural anatomy of higher mammals is remarkably similar to that of humans. It seems clear that at least the higher animals are fully sentient as evidenced in recent studies indicating that elephants, some whales, and several dolphin species have a clearly distinct understanding of self and others as individuals—the psychological caveat needed for higher emotional intelligences such as empathy.

If nonhuman organisms are fully sentient, then many theorists reason that they are due some form of moral consideration. In response to this idea, a strong animal rights movement has developed over the past few decades. Animal rights activists object to the use of live animals in testing new drugs and cosmetics: The animals suffer pain and death simply for the benefit of humans. They may object to hunting animals, wearing furs

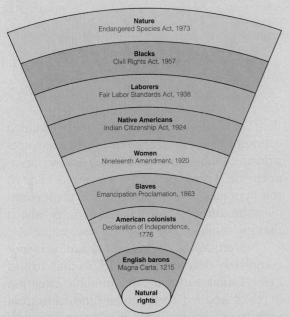

FIGURE 1 The expanding concept of rights. (*Source:* Adapted from R. F. Nash. The Rights of Nature. Madison, Wisconsin: University of Wisconsin Press, 1989, p. 7.)

Animal Rights

and leather, exploiting animals on farms for food, locking up animals in cages or zoological parks, or even keeping animals as pets where the animal takes a position subservient to a human "master." In a larger context, animal rights activists argue that what is good for humans might not be good for animals, and human interests and concerns should not always outweigh animal interests. The benefit to humans of developing a piece of pristine forest may not be sufficient to offset the suffering of the animals that currently live in the forest.

Animal rights theorists view their ideas as part of the historical development of morality (**FIGURES 1** and **2**). Just as humans (and in particular human males) have learned to extend moral consideration successively beyond the confines of gender (acknowledging the moral status of women) and the local tribe, nation, and race, so humans must learn to extend moral consideration beyond the confines of our own species to other sentient organisms. Although some people may think it absurd to give moral consideration to animals, animal rightists point out that at one time some white American males thought it absurd to disrupt the basic economy of the southern United States simply to relieve the alleged suffering of a few black slaves. It is just as unjustifiable for members of one particular species (humans, *Homo sapiens*) to regard the feelings and concerns of other

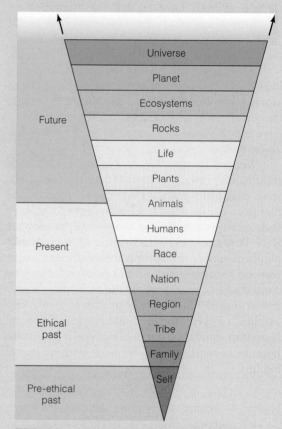

FIGURE 2 The evolution of ethics. (*Source:* Adapted from R. F. Nash. The Rights of Nature. Madison, Wisconsin: University of Wisconsin Press, 1989, p. 5.)

However, by the mid-1970s, public concern with the environment had declined. Oil shortages were occurring (although they were artificially created by the Organization of the Petroleum Exporting Countries [OPEC] cartel), and the country faced economic recession. All too often, it seems, when the public has to choose between the economy and the environment, the economy wins. By the late 1970s and early 1980s, an active backlash against environmentalism was under way. Many of the dire predictions made by environmentalists in the 1960s did not seem to have come true. The administration of President Ronald Reagan (1981–1989) valued economic growth over environmental protection and long-term responsible resource management. Then, the 1980s was the decade of the "yuppies"—the young urban professionals who often seemed to care more about making money than anything else.

Nevertheless, environmentalism was not completely dead in the late 1970s and 1980s. Love Canal came to the public's attention in 1978 and led to the Superfund legislation with its efforts to clean up toxic and hazardous wastes. The Three Mile Island nuclear power plant accident in Pennsylvania occurred

organisms as of less moral significance as it is for one race of humans to dismiss another race as of less moral significance.

Some practical criticisms of the typical animal rights stance have been advanced. If an animal rights theorist stresses the concept of sentience as the basis for moral standing, where do we draw the line? Certainly, cats, dogs, rhinos, zebras, and hippos are sentient and therefore bear moral consideration by this criterion. What about snakes, frogs, lobsters, shrimp, clams, and bacteria? What is the ultimate basis for sentience as a criterion for moral standing in the first place? Do animals have a degree of moral standing only insofar as they resemble humans in their abilities to feel pleasure and pain and perhaps to process cognitive information? Do life forms that differ significantly from humans receive no moral credit? What about vegetable rights? After all, plants are living organisms too, so should not plants also be accorded some degree of moral consideration?

Some theorists also ask whether all sentient animals should be accorded equal moral status. For instance, is a stray domestic cat just as important from a moral point of view as a member of a rare, endangered species of rhino? If we had to choose between saving 200 stray cats or a dozen rhinos, should we assume that each individual has the same moral status and therefore on utilitarian

grounds maximize the good by saving the 200 cats while letting the dozen rhinos (which might be the last members of their species) expire? To many people, such reasoning seems intuitively wrong—the rare rhinos should be saved first (although in actuality, many people expend much time, energy, and money supporting animal shelters for stray domestic cats and dogs even as many rare and endangered species go extinct in the wild). Yet what justification would we have for favoring rhinos over stray cats? Do rhinos bear a higher moral status than stray cats, or would we save the rhinos simply so that we humans would have the pleasure of knowing that the rhinos exist as a species?

Questions

1. In your opinion, do animals have rights? If so, to what degree or extent? If not, why not? Should all animals have the same rights, or do some animals deserve more rights that others? If so, on what basis?
2. Are there laws in your local community against, for instance, torturing cats and dogs? If there are, do cats and dogs have a legal status and rights?
3. Are some animals, either particular individuals or entire species, more important than others? If so, how does one judge relative importance? And who does the judging?

in 1979, effectively putting a damper on the further development of nuclear power in the United States. Then the Chernobyl nuclear power plant disaster in Russia in 1986 confirmed some people's worst fears. Even as the Reagan administration downplayed environmental issues, environmental groups such as the Sierra Club counted hundreds of thousands of new members. The radical environmental organization Earth First! was organized in 1980 and quickly grew in membership. By the end of the decade, the United States had joined other nations in signing the Montreal Protocol to save the ozone layer (1987), and the

1989 oil spill in Prince William Sound, Alaska, helped bring environmental issues to the forefront of the public's attention once again. In 1990, the passage of the Oil Pollution Act (OPA) strengthened the EPA's ability to prevent and respond to catastrophic oil spills, and the Pollution Prevention Act focused industry, government, and public attention on reducing the amount of pollution through cost-effective changes in production, operation, and raw materials use. The 1990–1991 Gulf War fought during the George H. W. Bush administration (1989–1993; Bush had been Ronald Reagan's vice-president) led many people to once more reassess the wisdom of America's heavy reliance on foreign oil. Record heat waves and unexpected severe storms at the end of the decade caused many to take seriously the notion of global warming, and the global population continued to increase with no end in sight even as per capita food production declined. All in all, by the end of the 1980s and early 1990s, environmental issues were coming once again to center stage. Earth Day 1990 received more attention than any Earth Day since the first one in 1970. The end of the Cold War, the breakup of the former Soviet Union, and the reunification of East and West Germany in 1990 allowed (indeed, required) us to focus attention on cleaning up the world and developing societies that can survive the 21st century and beyond.

In 1993, a Democratic administration took the reins from the Republicans, who had held the White House for 12 years. President Bill Clinton (1993–2001) expressed concern and interest in environmental issues, but perhaps more significantly, Vice President Al Gore, author of the widely read book *Earth in the Balance: Ecology and the Human Spirit* (1992), was an avowed environmentalist. Before joining the Clinton ticket, Gore had launched a presidential campaign based in part on an explicit pro-environmental platform.

However, in these same years, a strong backlash against the environmental movement arose once again. Politicians who were unsympathetic to environmental concerns continued to be elected to Congress. In Washington and in state houses across the country, environmental legislation was stalled, and environmental lobbying groups lost some of their political clout. Polls showed that while the public was still generally interested in environmental issues, the economy, crime, health, and welfare were more immediate concerns. Many of the older, established environmental groups faced budget shortfalls (as generous contributions became less common) and declining memberships. For instance, Greenpeace and the Wilderness Society witnessed dramatic drops in membership. The Sierra Club faced its worst financial situation in decades with multimillion-dollar deficits. A *USA Today* poll found that the 10 largest environmental groups in America collectively lost 6.5% of their membership between 1990 and 1994.

Part of the problem with the largest environmental groups and agencies may be that they grew into massive bureaucracies, and in the process, they lost the trust of many grassroots environmentalists. For instance, the Sierra Club operated with an annual budget in the tens of millions of dollars, including a Washington lobby group, a book business, and a worldwide ecotourism operation. Environmental agencies in the state and federal governments became notorious for their regulations, bureaucratic hoops, and red tape. Washington pollster Celinda Lake quipped that "today, if you ask a typical American to name a bureaucrat, he'll point to someone in some environmental agency."

However, even as larger, established environmental organizations were declining, many newer, smaller, local and grassroots organizations were springing up, often in response to environmental threats in their own communities. Grassroots organizing seems to be the wave of the environmental future. Estimates are that more than 6,000 grassroots environmental organizations are active in the United States. On college campuses, interest in the environment is on the upswing; environmental studies programs are expanding and evolving. Environmental awareness has become part of the worldview of the generation that is now entering adulthood, and their concern is not just with local recycling programs or nuclear power plants. They are focusing on the broader issues that must be faced in the 21st century— the achievement of long-term sustainability and the fundamental changes in society that this may entail. With wider use of the internet and the pooling of resource groups such as the Natural Resources Defense Council (founded in 1970), which works for a broad range of issues, including understanding climate change, promoting green energy choices, reviving oceans, promoting conservation and endangered animal survival, diminishing pollution, increasing drinking water safety, and fostering sustainable communities. Through branching into interest groups, more than 1.3 million citizens and more than 350 professionals such as scientists, policy makers, and lawyers pool their talents and demonstrate strength in numbers across a wide variety of issues that affect everyone's community.

In 2001, George W. Bush (son of former President George H. W. Bush) was inaugurated president of the United States after a closely contested battle with Al Gore. President Bush quickly made it clear that he was not the best friend of many staunch environmentalists. He advocated increased development of fossil fuels, and he was accused of weakening environmental and health standards. For example, in 2003 the Clear Skies Act (CSA) was passed to reduce and cap emissions of sulfur dioxide (SO_2), nitrogen oxides (NOX), and mercury from electric power generation to approximately 70% from 2000 levels. The act implemented a market-based cap-and-trade program to control power plant emissions caps without specifying the specific methods used to reach those caps. Furthermore, the law actually repealed or reduced environmental protections of the Clean Air Act and cut the budget for enforcement.

In many people's lives, environmental issues took a backseat to other affairs after the Al Qaeda (also spelled Al-Qaida) attacks (masterminded by Osama bin Laden) on the United States on September 11, 2001. The U.S. military quickly attacked the Taliban regime in Afghanistan, which was sheltering Al Qaeda elements, and in March 2003, the United States initiated a war against Iraq, deposing the regime of Saddam Hussein. In the second term of his administration (2005–2009), Bush continued to promote an economy based on fossil fuels, doing relatively little to implement subsidies or legislation to promote renewable energy. Barack Obama's presidency (2011 to the time of writing) has witnessed major environmental disasters both domestically (the Deepwater Horizon blowout and resulting oil spill in the Gulf of Mexico, April through July 2010) and internationally (the March 2011 earthquake off the coast of Japan and the resulting tsunami that had major ramifications, including severe damage to the Fukushima Nuclear Power Plants and associated structures). It is unclear at this time what the full ramifications of these disasters will be environmentally, politically, and culturally. Will the warning call be heeded?

20.2 Are Western Societal Values to Be Blamed for Current Environmental Problems?

Every human society has social institutions—based ultimately on fundamental beliefs, behaviors, and cultural norms—that govern the interactions of the vast majority of its people with each other and with the surrounding physical and biological environment. The most basic social institutions of a society are those governing (1) family organization and relationships, which provide most importantly for the rearing of children; (2) economic concerns, which provide for the production and distribution of material goods and services; (3) political concerns, which provide for the distribution of power and prestige and the protection of members of society from each other and from external enemies; (4) educational concerns, which provide for the transmission of knowledge and the cultural heritage from one person and generation to another; and (5) religious concerns, which provide for the establishment of a moral and ethical code and an explanation of the meaning and purpose of human life within the society.

Fundamental values and norms, and thus the social institutions through which these values and norms are expressed, may vary from one society to another. The social institutions of any particular society constrain the manner in which members of the society interact with each other and their environment. For instance, a particular society may value the concept of private ownership and the right of landowners to do anything they please with their land. A landowner who clear-cuts a large forest and replaces it with an open dump may cause severe problems for the other members of the society as leachate from the dump flows onto their property. Yet the norms and social institutions of the society may limit the ability of the other members to stop the landowner from pursuing such activities. The situation may effectively be irresolvable until the social institutions are modified, perhaps by incorporating the concept that private landowners can do whatever they want on their land as long as it does not affect anyone else detrimentally. Many thinkers contend that social institutions (and the underlying norms and values) must change and evolve over time as a human society expands and develops technologically. Thus, for instance, not all of the norms and values of 2,000 or 3,000 years ago are necessarily appropriate to the 21st century.

Western Values and Social Institutions

According to many observers, the dominant social institutions in the modern world are those that incorporate the values, norms, beliefs, and ideologies found in Western European and American society. Indeed, this is arguably true despite that much of the world is not, on the face of it, "Western." For instance, one fifth of the world's population is Muslim, a "non-Western" culture and religion, yet even as some radical Muslims reject Western beliefs and ideals, many Muslims are influenced by and adapt aspects of Western thinking, institutions, and technologies to their own cultures. The same can be said for other non-Western religions and traditions, such as Hinduism and Buddhism. Around the world, indigenous cultures are going extinct, only to be replaced by the homogenizing Western culture. The entire world seems to be industrializing and adopting Western technology and simultaneously adopting Western values, beliefs, and social institutions, either explicitly or in very subtle ways. Here we use Western beliefs and social institutions in a broad sense to include the entire spectrum from classic capitalism (with its emphasis on private property and private ownership of the means of production) to various forms of socialism and communism (which de-emphasize private property and stress varying degrees of public property and government intervention in the means of production and the distribution of goods and services). Non-Western beliefs and social institutions include those represented by traditional Islamic, Hindu, Buddhist, and Confucian societies (to give a few major examples) and those found among many indigenous peoples, such as the Native Americans or Australian aborigines before European contact.

Classic Western institutions (whether capitalist, socialist, or communist) are based on the concept of increasing rates of economic growth, increasing development, increasing the provision of material goods and services, and increasing the standard of living for everyone. In simple terms, these institutions rely on the "pie" increasing in size so that everyone's slice of the pie gets larger. For instance, in an ideal capitalist system, the rich and elite may get richer, but the poor don't get poorer. As the pie gets bigger, even the small slices that the lower classes receive will, in absolute terms, get bigger. As long as the pie continues to grow, everyone makes material progress over time, and general standards of living rise. But it appears we will soon reach, or perhaps have already reached, the absolute limits to the size of the pie. It can be argued that we can no longer continue to expand—to pursue "business as usual"—because the physical environment will not support continued human expansion. The strategy that worked so successfully for hundreds of years, as long as there was a relative abundance of untapped natural resources, must now be modified.

It can be argued that current environmental concerns can ultimately be traced to a few fundamental beliefs inherent to Western culture:

1. Western society and culture are dominated by an anthropocentric and humanistic worldview. Humans are superior to all other living things, and human interests come before those of other species or inanimate objects. There is a dichotomy between nature and human society. Humans have a right to conquer, tame, and subdue nature.

2. Western culture emphasizes the individual human and often promotes individual achievements, even to the detriment of the collective good. An individual who exploits a natural resource, such as an oil field, for his or her own gain and profit is regarded as an industrious, ingenious person to be admired even though his or her actions may cause damage to society in the long run.

3. Western culture is focused on materialism, the idea that the production and consumption of material goods are necessary

for a good life. To satisfy this materialistic urge, Western society must draw on the physical environment.

4. According to traditional Western thinking, the natural world is virtually unlimited, containing a wide variety of free (except for the human labor needed to procure them) and inexhaustible resources. In this view, not only can we take from the natural world with impunity, but we can dump our wastes back into nature. As a sink to absorb wastes, the atmosphere, rivers, oceans, and even the surface of the land itself are, for all practical purposes, infinite. Nature is the ultimate provider of resources and the ultimate sink for wastes.

5. Western society and culture are based on a growth ethic. Progress is measured in terms of growth; bigger and more are perceived as better. People who produce and consume large quantities of goods are better off. Social institutions depend on an expanding economy. Sustainable living, which does not involve continually exploiting new resources and new frontiers, is antithetical to this basic Western worldview.

6. Western society and cultural institutions emphasize technological and scientific knowledge and achievements over other types of knowledge, such as moral or aesthetic knowledge and achievements. This is not to say that the arts are not valued in the West, for they certainly are, but technology that can be used to manipulate the physical environment to achieve the material goals of society is even more important. Indeed, Western society seems almost to have a blind faith that technology can solve all problems and achieve all goals. We sometimes hear comments such as this: "If we can put a man on the moon, we should be able to eliminate greed and social strife." The fallacy here is that putting a person on the moon is a relatively straightforward technological problem (**FIGURE 20-5**), but eliminating greed or social strife is a complex sociocultural problem that is not directly amenable to technological solu-

FIGURE 20-5 Edwin E. Aldrin, Jr., stands on the Moon in a photograph taken by Neil A. Armstrong, July 1969.

tion, although of course, technology may contribute to the solution by providing an adequate supply of food and material goods for everyone.

Some critics, such as the protesters at the 1999 World Trade Organization meeting in Seattle, maintain that Western economic and political institutions in particular promote beliefs that lead to abuse of the environment. For instance, in capitalistic systems, private property owners make the decisions about the means of production of goods and services. These property owners attempt to maximize their own profits while often ignoring or discounting the damage that might ensue to the general environment that is shared with the rest of the population. Even in socialist and communist systems, where the government makes economic decisions about the means of production and the modes of distribution of goods and services, the traditional emphasis is on increasing economic growth and supplying ever more goods and services to the populace. In practice, environmental considerations take a backseat to sociopolitical concerns. The net result is that in traditional Western systems, the environmental costs of goods and services are not factored into the price of a product. The market fails to take account of "externalities," the social and environmental costs of goods and services that are not reflected in the price that consumers pay. For instance, the varying rate that a consumer may pay for a gallon of gasoline in the United States primarily reflects the cost of pumping, refining, and transporting the petroleum; it does not reflect the cost of the health and environmental damage that burning the gallon of gasoline will cause, nor does it necessarily reflect the government expenses of protecting and securing the supply of oil—think of the wars that have been fought over oil supplies. In the case of gasoline, even exploration, refining, and marketing are partially subsidized by the U.S. government. Globally, the rapid industrialization of the developing world, especially China and India, has placed even more stress on oil fields and refining capacity. For example, such factors, coupled with the devastating effects of Hurricane Katrina (September 2005) and armed conflicts in the Middle East, caused a surge in gasoline prices in the United States.

Political decisions in traditional Western systems often promote an economic system that leads to environmental abuse. In capitalistic systems, business groups and business-funded lobbyists are extremely influential, and such groups typically wish to keep to a minimum legislation that protects the environment at the expense of business, such as legislation that would factor externalities into the cost of products. Lobbyists claim such legislation will reduce profits—reduce economic growth (on which the entire system is based)—causing unemployment and social disruption. In socialist and communist systems, much the same line of thinking takes place except that the government and business are one and the same. Indeed, recent history has demonstrated that repressive communist systems (many of which have now collapsed) may cause even greater damage to the environment than do capitalistic systems; in communist systems, any dissenters who might speak on behalf of the environment may be crushed so that the state can successfully conceal environmental degradation from the populace and the world at large.

Alternatively, one can argue that the fundamental beliefs and social institutions of Western society are not the source of environmental problems, but rather may be our only hope for a solution. In this view, the concept of special Western guilt for global environmental destruction is a myth. Primordial innocence never existed; neither did the noble savage who lived in true harmony with nature. Aboriginal peoples in such areas as Madagascar, Hawaii, and New Zealand caused numerous species to go extinct and generally wrought ecological devastation on their lands. The extinction of many large mammal species at the end of the last Ice Age, about 10,000 years ago, may have been due in part to human overhunting. For thousands of years, the Chinese have been involved in the wholesale extermination of many wildlife species, including elephants, tigers, and rhinoceroses (**FIGURE 20-6**). Indigenous, traditional East Asian pharmacology

FIGURE 20-6 Traditional Chinese medicines often include ingredients derived from rare species. In recent years the World Wildlife Federation and other groups have made progress working with practitioners to reduce the use of rare plants and animals in the traditional remedies. See, http://www.worldwildlife.org/what/globalmarkets/wildlifetrade/traditionalchinesemedicine.html.

advocated the use of preparations made from rhinoceros horn or tiger penis as remedies for all sorts of afflictions; Western medical science is much more effective and less damaging to the environment. Only through Western technology and human ingenuity, promoted by Western concepts of individualism, resource expansion, and development, can we ever satisfy the needs and cravings of all the world's population—or so some thinkers argue.

According to one school of thought, there are no resource limits as long as we do not limit human creativity to develop and exploit new resources. For example, in the 19th century, some pessimists believed that modern civilization eventually would run out of high-quality energy sources as coal and other fossil fuels were depleted. In the 20th century, Western scientists learned to harness the hitherto unimagined energy contained within the atom. We now have hundreds of fission nuclear power plants in operation around the world, and the development of fusion as a controllable energy source is just a matter of time. Likewise, aluminum was once an unknown metal. For several decades after it was first discovered (around 1825), it was extremely rare and costly. Now aluminum is very common and is put to the most mundane uses, such as packaging food and beverages. To cite just one more example, the Green Revolution after World War II allowed increased harvests that would have been unthinkable just 50 years earlier. Even the physical limits of Earth, according to extreme advocates of this point of view, represent no absolute limits. We have already proved that we are capable of space travel, and the universe is unimaginably large. Similarly, Europe may have seemed very small and crowded in 1450, but a century later, Europeans were colonizing an entire New World (North and South America) that was unimagined by the pessimists of the mid-15th century.

It is not our purpose here to decide whether Western values and social institutions are the cause of or the solution to current environmental concerns. Ultimately, it may be that Westernization was both. Only time will tell.

Sustainable Development/Sustainable Growth

Classically, it has been argued that a strong correlation exists between economic development and growth and environmental degradation. As people increase their rates of growth and expand their production of goods and services, more and more resources are depleted. More wastes are dumped into the environment, and the environment suffers. However, does economic growth have to lead to environmental degradation? Perhaps not. Many environmentalists believe that we must adopt a new paradigm—a new way of thinking and acting, a new ideology; this ideology has been labeled **sustainable development** or "sustainable growth." Sustainable development focuses on making social, economic, and political progress to satisfy global human needs, desires, aspirations, and potential without damaging the environment and the ability of future generations to meet their needs. In this context, development does not refer to old notions of simply increasing industrial output and increasing consumption of goods and services. Rather, "Development involves a progressive transformation of economy and society" (World Commission on Environment and Development, 1987). Sustainable development emphasizes equity in managing the world's resources—equity among different peoples and nations of the world and equity between one generation and another. Nonsustainable growth means leaving a degraded environment to future generations simply to live a certain lifestyle in the present. One country should not disproportionately use global resources or cause irrevocable environmental degradation that affects other peoples or the entire Earth. The key word is sustainability, meaning to prolong indefinitely into the future the ability of Earth to support and provide for not only people, but also all life on the planet.

Addressing Environmental Concerns by Refocusing Western Political–Economic Structures

As we have already discussed, some people believe that one of the major flaws of traditional Western economics is the failure to include the environmental consequences of goods and services in market prices. The environmental costs are externalized and not paid directly by the producer or the consumer; instead, the populace as a whole absorbs the price. For too long, it is contended, companies and individuals have been allowed to freely exploit raw resources and dump their wastes back into the commons (such as a factory dumping waste into a public river or the global atmosphere). This allows a few to profit at the expense of the majority. By fostering economic growth, the traditional Western political–economic structure actually encourages resource depletion and pollution, at least as long as there continue to be resources to deplete and some place to dump the resultant wastes. Economic growth and the increasing profits and employment that it brings have been the traditional goal of both Western business and Western governmental/political institutions. This goal is very short sighted if the methods used to achieve short-term economic growth discount long-term environmental degradation. In this section, we briefly discuss a few of the many ways that these concerns might be addressed. We hope that you, the reader, will continue thinking about such issues.

In the United States, an environmentally conscientious company often may feel trapped. If it practices business as usual, externalizing environmental costs of production, damage to the environment results. However, if the company internalizes the environmental costs, perhaps by using a more expensive manufacturing system that eliminates harmful emissions, the product will cost more; because the public generally will purchase the cheaper product, the company will not be able to compete and eventually may go bankrupt. Likewise, the politician who is perceived as putting the environment before jobs (although this is often a false dichotomy) may soon be voted out of office.

This analysis is too simplistic, however. In at least some cases, the public does not automatically purchase the cheapest product or always vote for the politician who will lower taxes and increase income. Education has gone a long way toward changing public perceptions and actions. The fact that a product is environmentally sound can be used to help market it even if the product does cost a bit more than its competitors. Likewise,

throughout the world, "green" politicians are gaining influence and power.

However, playing on the environmental sympathies of the public can carry some businesses only so far—governments must also become involved. Environmentally progressive businesspersons may favor government intervention that allows for the internalization of externalities across an industry. That way, different businesses and corporations can compete fairly while also reducing or eliminating the environmental damage they cause. Government intervention can take the form of regulations that protect the environment. Government subsidies that promote the exploitation of virgin resources can be eliminated. Government-imposed limitations on corporate liability for environmental damage could be eliminated; likewise, government-subsidized insurance programs that cover environmental hazards and damages could be curtailed. Why, for instance, do we need the Price-Anderson Act of 1957, which limits the liability of the nuclear industry in case of an accident, other than to promote the development of nuclear power? Laws can be (and are) passed that set limits on permissible amounts of pollution that can be pumped into the environment. Certain types of toxic or hazardous materials can be banned. Particular resources, such as a forest or wilderness area, can be preserved through government intervention. Government can also help internalize environmental costs by instituting stiff "green" taxes on waste emissions, resource depletion, energy use (such as carbon taxes or taxes on nuclear energy production to reduce the accumulation of radioactive wastes), and so forth. Such environmental taxes can go a long way toward correcting the market by internalizing the indirect costs of products.

The government could also actively subsidize environmentally compatible technologies and activities. For instance, it could give tax breaks to companies that install environmentally friendly technologies (such as solar heating systems or pollution-reducing equipment) or pay bonuses for devising more efficient means of production. Some analysts argue that we should also create markets for the "right" or "privilege" to pollute. For instance, the government sells permits that allow the holder to dump wastes, up to a certain level, into a particular river or the air. The sum total of the pollution permitted would not be more than the environment can safely and sustainably handle. Pollution permits could be bought, sold, and traded on the open market. A manufacturer that wanted to discharge pollutants into a river would have to pay for the appropriate permits and thus would be forced to internalize the cost of the pollution. A competing manufacturer might be able to produce the same product by a different (perhaps more expensive) non-polluting technology and therefore would save the expense of purchasing pollution permits. Concerned citizens, environmental organizations, or even the government might buy and hold pollution permits to curb waste emissions. Savvy investors might put their money into pollution permits just as today they invest in real estate. The core of such a "cap-and-permit" program was established in the United States with the Clear Skies Act of 2003 for fossil fuel electric power generation plants of greater than 25 megawatts, capping nitrogen oxides, sulfur dioxide, and mercury at roughly 70% below 2000 emissions levels (reductions and caps on emissions are being implemented in phases through the year 2018) and allowing the trading and selling of permits to emit pollutants.

A conceptual problem with pollution permits concerns who has the right to establish and sell them in the first place. Perhaps a global governmental authority would be needed because some types of pollution are global in nature, but should the "privilege" to pollute be legitimized and privatized in such a manner? On the face of it, one can argue that such a system will increase inequity between peoples—those who hold pollution permits will be a rung above those who do not. Those who are allowed to pollute can cause damage and injury to those who cannot, whereas the latter have no way to reciprocate. As the global population expands (and every indication is that it will continue to grow for many years), pollution inevitably will increase, and those holding the limited number of pollution permits will gain in power and wealth at the expense of everyone else. Thus, pollution permits may be at odds with the concept of sustainable development and the goal of global equity among all peoples. Alternatively, we could simply insist that individuals and corporations not pollute and instead carry on their businesses in a sustainable manner that does not threaten the environment or human health.

In the United States and many other countries, corporations can be forced to internalize external costs to a certain extent through the court system. An individual or community that is hurt or damaged by a company's negligence or illegal acts can file a lawsuit against the corporation. The courts may force the company to comply with existing regulations, change its activities, or pay a large monetary settlement. In this manner, the company is made to internalize certain factors that previously were mere externalities. For example, a company may be compelled to take responsibility for the injuries that ensue when it dumps toxic waste into a river. Using the judiciary system in this way leads to numerous problems, however. On a practical level, considerable knowledge and a substantial initial outlay of money generally are needed to attack a large corporation that can afford to hire good lawyers. In most areas, the court systems are already crowded, and a case may take years to come to trial or arbitration. In many situations, such as damage caused by air pollution, it may be difficult or impossible to identify the particular corporation that is directly responsible for the damage in order to take it to court. Lawsuits are perhaps most effective when used by individuals or small groups of victims to remedy damages that can be clearly attributed to the actions of a single or possibly a few corporations.

Political solutions to environmental problems are becoming increasingly prominent as more and more citizens are becoming environmentally aware. Citizens can form environmental interest groups that lobby the government at all levels, or an individual can write directly to governmental officials about environmental concerns. Citizens can support (with donations, time, and votes) candidates who are pro-environment. Individuals and groups can also influence public opinion and governmental decisions through the educational system and the media. Sitting on a local city council or writing letters to the local newspaper can be very effective.

Beyond working within the confines of the current political system by voting, lobbying, testifying before committees, and so forth, one can engage in political action outside the established route. Some environmentalists contend that because the established political system encourages the continued degradation of the environment, we have a moral obligation to challenge the legitimacy of the current system and the status quo. Increasingly, some people are calling for an environmental revolution that will radically change our values and the way we do business. Grassroots movements of concerned individuals on a local, national, and ultimately international level may be needed to challenge and overturn the status quo. Tactics that can be used include marches, boycotts, protests, and mass meetings demanding change (**FIGURE 20-7**). If nothing else, such activities can focus awareness and concern, forcing politicians to make

what were once peripheral issues central concerns. Just as the U.S. civil rights movement used such tactics with great success, so too might the current environmental movement (**CASE STUDY 20-3**).

20.3 Environmental Law

Law and the Environment—A General Overview

Over the past century, and especially over the last 60 years, society has turned increasingly to legal avenues to protect the environment. A number of laws dealing with environmental issues have been enacted on the local, national, and international levels. The efforts to clean up pollution have been highly successful in some areas, including lakes and rivers, toxic waste dumps, and ocean dumping. It is not only the United States and Canada that are making great strides in cleaning up wastes and preserving the environment; many other countries, such as those of Western Europe, are as well. At the international level, a number of important pieces of environmental legislation have been established, such as the 1973–1975 Convention on International Trade in Endangered Species of Wild Fauna and Flora (CITES) and the 1987 Montreal Protocol on Substances that Deplete the Ozone Layer.

Yet there are many gaps in these laws, for this field is still developing. Initially, the prime motivating force behind most environmental legislation was to protect human safety and welfare, but now certain aspects of the environment (for instance, endangered organisms) are coming to be viewed as having legal standing and legal rights unto themselves. Increasingly, environmental advocates and their attorneys are defending the environment from assault and injury by human beings.

In the broadest sense, the field of **environmental law** encompasses all of the laws, statutes, regulations, agreements, treaties, declarations, resolutions, and the like that have bearing on environmental issues. Environmental laws range in scale from local community ordinances prohibiting litter on the streets to international treaties regulating trade in endangered species or the release of stratospheric ozone-destroying substances.

FIGURE 20-7 Antinuclear protesters block a railway track in Germany to protest the shipment of spent nuclear fuel from France to a storage facility in northern Germany.

Ecofeminism—Overview of a Theory in Practice

The Ecofeminism movement unites many of the underlying tenets of feminism with a connection to environmental sustainability. Just like other philosophies, there are many branches of ecofeminism. At the core of the ecofeminism in all its forms is a conversation that places humans in relationship with nature (as opposed to hierarchical models of domination and exploitation). In their relationship with nature, women are often most directly connected to the actions that strive toward a more amicable and sustainable relationship with the resources nature provides. Through their association as primary-care givers, household managers, and educators, women are often in the forefront of grassroots activism that brings about a more rich understanding of sustainability. Since the term *ecofeminism* was coined in 1974 by Françoise d'Eaubonne, the philosophy and inclusive nature of the movement has provided a framework for various interpretations and is still evolving today. Like many of the social and environmental movements that can be traced back to the late 1960s and early 1970s, ecofeminism addresses the implications of power and policies that affect human interaction with the natural world. One of the most widely publicized examples of ecofeminist action was the Chipko Movement in the 1970s where the phrase "tree hugger" was coined. In Hindi "chipko" means to cling to (as in an embrace).

As an action of nonviolent protest and active resistance to contractors who were coming to fell a few thousand trees, a group of nearly 30 women arrived and held hands around the base of the tree, preventing the contractors from cutting the trees. Though often used as an insult to degrade environmental activists, the "tree huggers" were ingeniously and nonviolently saving their way of life. Without the trees, there would be no fuel wood, no shade for livestock, flooding, and increased desertification over time. This hilly already overforested Himalayan area provided the sustenance of a way of life for this community. The women's action drew international attention to their cause.

This type of grassroots action exemplifies the history of ecofeminist activity rooted in social responsibility and environmental sustainability.

One of the important leaders in ecofeminism is Dr. Vandana Shiva, winner of the 1993 Right Livelihood Award (Alternative Nobel Prize) and the 2010 Sydney Peace Prize. Through her activism, she gained international support for the Chipko demonstration. Her work also sheds light on the connection women have to the environment and the integral role that seed-sharing and planting have on the relationship among farmers and their communities. She argues that monoculture crops, such as those promoted by large multinational companies, such as Monsanto, change the societal landscape of farming and the practice of community agriculture in ways that can be harmful to indigenous cultures, crop biodiversity, and the sustainability of agriculture as a whole.

Meeting women in their everyday lives in places such as India and Africa, where women's work is still largely inextricably linked to the health of the environment around them, is a goal of much of global ecofeminist work. Expanding grassroots activism and local political involvement, spreading education, promoting stewardship of the environment and advocating for long-term sustainable practice is at the root of historical and current ecofeminism.

Dr. Wangari Maathai, founder of the Greenbelt Movement and 2004 Nobel Peace prize recipient, spent much of her life promoting sustainable community practices, advocating for the education of women, and demonstrating an activism rooted in the an appreciation for the interdependence between humans and the environment. With women as her focus group, Maathai promoted education, sustainable farming practices, and developed partnerships for stronger local economies. Her work is an example of how ecofeminists bring together men, women, and children in the building of communities that practice sustainable management of Earth's resources. Currently (2012) the Green Belt Movement has more than 600 community networks spread across Kenya that together care for 6,000 tree nurseries. The people within these networks have participated in planting more than 40 million trees in various locations, including private and public land, reserves, culturally significant sites, and urban centers. Trees prevent

VISIT

http://environment.jbpub.com
/mckinney/5e/
for more information

Ecofeminism—Overview of a Theory in Practice

flooding, help mitigate erosion, provide shade, purify the air, provide wildlife habitat, and address myriad other human needs.

Current work by researchers in the United States, such as philosopher Dr. Carolyn Merchant (2010), promotes a "Partnership Ethic" when examining humanity's relationship with the environment. In this approach, the natural world and humans hold equal moral standing, which she suggests allows for a more equitable framework for sustainability. (For instance, what a river needs to be healthy is viewed as equal to what a community needs to be healthy.) In this framework, the environment becomes a stakeholder in the decision-making process. She suggests that this philosophical perspective can be useful when communities make decisions about how to use natural resources and can aid in sustainable decisions that do not exploit natural resources. As the challenges facing the global citizens of the 21st century continue to grow in complexity, Merchant's partnership ethic advocates a movement away from human domination of the natural world. She promotes a more focused ethic that examines equity between human and nonhuman community, which includes all other organisms and all shared resources, such as minerals and water. This ethic gives moral consideration for other species, respect for cultural diversity and biodiversity, and incorporates women, minorities, and nonhuman nature in the code of accountability. It also promotes ecologically sound management that is consistent with the health of human and nonhuman communities.

Such ecofeminist philosophy can be employed by corporations and other institutions. For instance, by examining corporations that operate under the "cradle to cradle" design model (see the discussion on people and natural resources) and view "waste as food" and strive toward zero waste, we can see the ecofeminist "partnership ethic" in practice. The bottom line is minimal impact to the earth while producing the goods and services that humans need to support a certain standards of living. The cradle-to-cradle industry model, congruous with ecofeminist ideals, views nature as a partner and is both profitable and sustainable. Currently, there are more than 150 companies producing over 200 products that hold a silver or gold ecolabel through the Cradle to Cradle certification process. These ecologically sustainable factories and design models are in direct contrast to corporate design models that view nature as a closed system that can be controlled and exploited with the bottom dollar profits often given a higher priority over absolute safety and zero waste practices. Outdated mechanistic models do not promote long-term sustainability and can result in damaging consequences for humans and the environment. Recent examples of damaging mechanistic models would include the Deepwater Horizon oil spill on April 20, 2010, in the Gulf of Mexico, where investigators concluded that the short-sighted drive for bottom dollar profit and lack of disaster contingency plans as contributing factors of the magnitude of the event. Equally disturbing was the Upper Big Branch mine explosion that occurred on April 5, 2010, killing 29 miners, making it the worst U.S. mine disaster since 1970.

Many authorities divide environmental **legal instruments** (laws, treaties, regulations, conventions, and so on) into two categories: **hard laws**, which are legally binding and mandatory, and **soft laws**, which are not legally binding, but act more as a guide to policy. This distinction is particularly useful in analyzing international treaties, conventions, and agreements. Many international legal instruments are not binding on the signing nations, but tremendous moral and public pressure may be brought to ensure that the signatories more or less conform to the agreement. In contrast, hard laws force the parties to comply or else. Or else what? Enforcing hard laws at the international level

In the 5 years up to the explosion, the Massey Energy company that operates the mine had received more than 1,000 violations, including several in the weeks just before the disaster citing numerous issues, including problems with ventilation and with escapes routes, which were cited as implicating factors in the incident. Ecofeminists support an alternative ethic that considers possible consequences to life and the environment and promotes long-term sustainability of resource management, business practice, and community development. A catastrophe often stirs people into action for change; such has been the case in many of the examples of ecofeminism in the United States.

Many of the women most active in the environmental movement, and in ecofeminism, were and continue to be everyday ordinary citizens who could no longer look the other way and began organizing for change. When Rachel Carson addressed Congress in 1963 with her documentation about the damage of the pesticide DDT, she was already a grandmother. When Lois Gibbs began organizing neighbors in the Love Canal area of New York in the late 1970s to seek answers regarding why her children and those around their town were becoming seriously ill, she was a just a concerned mother advocating for the health needs of her children. Now Gibbs' Center for Health, Environment & Justice helps other citizens address environmentally based concerns in communities around the world.

The philosophy of ecofeminism can be applied in many different cultural and socioeconomic contexts. For instance, until her death in 2008, Val Plumwood's work explored a nondominating interaction with the natural world that allows the Earth to flourish while at the same time "acknowledge[ing] our own animality and ecological vulnerability" (1993, 1985, 1999). Karen Warren's recent work on a "care sensitive ethic" also supports the role of humans as equal to other forms of life and that to live healthy sustainable lives, humans need to explore the assumptions inherent in the beliefs they practice and the way in which they express ethics related to resources. Some ecofeminists apply the philosophy to meat consumption and choose to promote vegetarian and vegan lifestyles as part of their philosophical interpretations of ecofeminism. Others interpret ecofeminism through a more spiritual lens and write about the healing and restorative aspects of nature.

As with any philosophy, ecofeminism provides a framework by which to consider one's place in context of our interactions with others. Ecofeminism promotes community action and urges humans to examine the power inherent in the relationships with one another, in their communities, and in the decisions made with regard to using resources from the natural world. Furthermore, ecofeminist thought urges people to examine how those relationships extend beyond the self and the immediate community and into the relationships and interactions that involve the environment. Ecofeminist thought and action continue to evolve and provide an opportunity for people to work together to better understand our interdependence with nature and to secure systemic practices that provide the generations that follow with a framework for socially responsible and sustainable environmental practices.

may be difficult. Countries wishing to see an environmentally friendly treaty honored may impose trade or other sanctions against an offender, but to this date, nations rarely go to war over environmental protection per se, although certainly countries have gone to war over the control of scarce natural resources (such as metals, oil, or water). Many international environmental treaties are initiated by, or fall under the aegis of, the United Nations (UN), but the UN has only limited powers of enforcement.

The legal and philosophical basis for environmental laws can be approached from several different directions. In a narrow sense, one can argue that pollution is a nuisance or

CASE STUDY 20-3

Ecofeminism—Overview of a Theory in Practice

VISIT

http://environment.jbpub.com
/mckinney/5e/
for more information

Questions

1. Many of the people who become environmental leaders are responding to an issue in a local community and decide to do something about it. Are you currently involved in any environmental issues in your community? If so which ones, and, if not, what sort of issues would make you feel like getting involved?

2. Many of the ideas of ecofeminism have become mainstream in recent years, such as the increase in sustainable business practices, increased awareness of earth stewardship, and increased interest in supporting local economies. When a movement goes mainstream, do you think it loses or gains influence? Explain your position.

3. In recent years, social media has played an important role in social movements throughout the world. What untapped potential do you see for the use of social media in the advancement of sustainable community initiatives around the globe?

References

http://grad.berkeley.edu/lectures/event.php?id=735&
    lecturer=473
http://muse.jhu.edu/login?uri=/journals/ethics_and_the_
    environment/v007/7.2cuomo.pdf
http://cnr.berkeley.edu/departments/espm/env-hist/
http://nature.berkeley.edu/departments/espm/env-hist/
    Moses.pdf
http://www.luvei.com/?p=1757
http://www.rightlivelihood.org/
http://clinton2.nara.gov/WH/EOP/OVP/24hours/carson.html
http://www.rachelcarson.org/
http://media.pfeiffer.edu/lridener/courses/ecowarrn.html
http://www.bu.edu/wcp/Papers/Gend/GendWarr.htm
http://www.vandanashiva.org/?p=536
http://www1.american.edu/TED/chipko.htm
http://www.vandanashiva.org/?p=536
http://kurungabaa.net/2011/01/18/being-prey-by-val-
    plumwood/
http://www.bu.edu/wcp/Papers/Gend/GendWarr.htm
http://muse.jhu.edu/login?uri=/journals/ethics_and_the_
    environment/v007/7.2warren.html
http://www.caroljadams.com/interviews3.html
http://nobelprizes.com/nobel/peace/2004a.html
http://www.greenbeltmovement.org/w.php?id=59
http://mbdc.com/detail.aspx?linkid=2&sublink=8
http://www.sourcewatch.org/index.php?title=Upper_
    Big_Branch_Mine_Disaster
http://www.sourcewatch.org/index.php?title=Gulf_
    of_Mexico_Oil_Spill
http://www.sourcewatch.org/index.php?search=deepwater+h
    orizon&ns0=1&title=Special%3ASearch&fulltext=Sear
    ch&fulltext=Search
http://www.register-herald.com/disaster/x993489321/
    Drilling-efforts-to-vent-deadly-mine-could-take-up-to-
    48-hours
http://www.washingtonpost.com/wp-dyn/content/article/
    2010/04/05/AR2010040503877.html?hpid=topnews&s
    id=ST2010040505519
Merchant, Carolyn. *The Death of Nature.* HarperSan
    Francisco, 1980.
Merchant, Carolyn. *Reinventing Eden the Fate of Nature in
    Western Culture.* Routledge, 2003.
Warren, Karen J. *Ecofeminist Philosophy: A Western Per-
    spective on What It Is and Why It Matters.* Rowman &
    Littlefield Publishers, Inc., 2000.
Zell, Stacy. *Science & Education,* "Ecofeminism and the
    Science Classroom: A Practical Approach" Volume 7,
    Number 2, 143–158, 1998. http://www.springerlink
    .com/content/u82461289777870u/

negligent behavior that detracts from others' property and personal rights. Thus, pollution in particular and environmental degradation in general can be viewed as **common law torts** (wrongs) that are within the jurisdiction of the law. More broadly, one can contend that fundamental human rights include the right to a safe, healthy environment. Any government should use its authority and power to promote the general welfare of its citizens and subjects, and this includes enacting laws that address environmental concerns. Using the U.S. legal system as an example, some scholars have argued that the Ninth Amendment to the U.S. Constitution (which reads, "The enumeration in the Constitution, of certain rights, shall not be construed to deny or disparage others retained by the people.") can be interpreted as guaranteeing the peoples' right to a healthy, clean, and safe environment. Nowhere in the Constitution are environmental issues mentioned explicitly (of course, such topics were perhaps not a major concern in the late 18th century), but one can argue that the right to a healthy and healthful environment is so fundamental that it did not need to be explicitly enumerated in the Constitution.

In the broadest context, one can justify environmental legislation as simply protecting the fundamental rights of not only people, but also all organisms and even inanimate objects and Earth itself (of course, some people deny that nonhumans have fundamental rights, but then at one time human slaves were also denied fundamental rights). Our mandated environmental laws are necessary only to the degree that people fail to be considerate of the larger nature of which they are necessarily a part. If people did not exploit and abuse the environment, there would be no need for environmental legislation.

International Environmental Law

At the international level, a number of environmental treaties, conventions, agreements, and protocols (more than 200 at last count) have been established. These legal instruments are binding and enforceable to varying degrees, and all are dependent on independent and autonomous governments becoming signatories and pledging their support. Unfortunately, many "global" treaties and agreements have received less than global support, and even when technically binding on participating countries, they are very difficult to enforce. Nevertheless, in the latter half of the 20th century and early 21st century, we have witnessed great strides in international cooperation to protect and preserve the environment. Conventions and agreements have been established that provide for, among other things, the protection of areas of outstanding natural or cultural value; restrictions on trade in endangered species; protection of migratory species; a legal framework for the use of the oceans and their resources, including which areas fall under national jurisdictions; and the prevention of marine pollution.

Some of the most important soft legal instruments of recent decades trace their origins to the United Nations Conference on Environment and Development (UNCED; popularly known at the time as the "Earth Summit"), which was held in Rio de Janeiro in June 1992 (**FIGURE 20-8**). It is generally considered to be one of the largest summits ever held on any topic—delegates from more than 175 countries, including more than 100 heads of state, attended. This conference can be viewed as the final ending of the Cold War, marking a new beginning where money that once would have been spent on military expenditures could now be funneled into environmental protection and sustainable global development (although a decade later, some were arguing that now money must be spent fighting terrorism, detracting from environmental spending). As a result of the Earth Summit, a number of soft treaties and declarations were formulated that lay the groundwork for future discussions and agreements.

- *Climate Treaty (United Nations Framework Convention on Climate Change)*. This treaty, formulated at the Rio Conference, established broad principles and outlined the potential moral and legal obligations of nations to curb the release of greenhouse gases that could cause global warming. Overall, the treaty is very weak and effectively nonbinding, but it set a precedent and lay the groundwork for further international discussions on the topic (see a review of global air pollution, destruction of the ozone layer, and global climate change).
- *Biodiversity Treaty (United Nations Convention on Biological Diversity)*. This treaty is aimed at promoting

FIGURE 20-8 Delegates meet during the Earth Summit held in Rio de Janeiro, June 1992.

the preservation and careful management (including sustainable use) of biological diversity and begins to address the issue of genetic engineering using genes found in rare species. However, it imposes no binding legal obligations on the signing nations and does not totally clarify the relationship between nations that exploit the genes of species and the nations that are the repositories of the species bearing the genes. For instance, if a representative of a developed country (such as a scientist employed by a large pharmaceutical corporation) finds that a gene of a newly discovered indigenous species in a developing country has great medicinal and therefore economic value, who shares in the fruits of this discovery? Does the developing nation have a right to profit from the discovery? It is generally hoped that the preliminary Biodiversity Treaty formulated at the Earth Summit in 1992 will lay the groundwork for hard agreements addressing these types of issues.

- *Forest Agreement (Statement of Agreement on Forest Principles).* Environmentalists hoped that the Earth Summit would provide a forum to develop an international treaty aimed at protecting the world's remaining forests; however, the developing nations (especially India and Malaysia) generally took the stance that forests within their boundaries were subject only to their authority and thus could not be the subject of an international agreement. Part of the problem is that the developed nations (the North) long ago cut down much of their forest areas en route to achieving their wealthy, developed status. The developing nations (the South) demand the right to follow a similar path in promoting their own internal development, even though much of their natural resources are still used to pay their international debt to industrialized countries. Essentially, the countries of the South do not want to remain in relative poverty while protecting their forests for, as they see it, the relative benefit of the North. The final general statement on world forests that the Earth Summit produced contained few new ideas and accomplished little in the way of promoting the preservation of the Earth's forests.

- *Rio Declaration.* This statement, which came out of the Earth Summit, was signed by many nations, generally committing participants to pursue sustainable development and work to rid the world of poverty. The statement declares that developed countries have a special responsibility to help in global restoration efforts because they have the financial resources to afford such efforts, and historically, the developed countries have been the largest contributors to global environmental degradation. The Rio Declaration, like other Earth Summit documents, is nonbinding, but it does carry some moral and political obligations.

- *Agenda 21.* This 800-page document from the Earth Summit contains recommendations as to how countries can pursue sustainable development and protect the environment. It covers such topics as decentralization of decision making relative to the management of local natural resources, land reform to increase the land rights of rural and indigenous peoples, ways to improve the status and participation of women in the context of sustainable development, and the development and adoption of taxes and other economic incentives that promote sustainability with the concomitant removal of subsidies that are at odds with the preservation of natural resources.

The negotiation of international environmental treaties and other legal instruments, such as those just outlined, is an important step. Nevertheless, it is generally extremely difficult to monitor and enforce international treaties. Indeed, some critics argue that many environmental treaties currently in existence are virtually meaningless because they are not enforced. Many treaties require signatory countries to self-report on their progress in fulfilling the requirements of the treaty, but this self-monitoring generally is not sufficient. In some cases, only a minority of nations submits the required reports, and even then the data may be incomplete.

Some treaties, such as the Montreal Protocol, include voting mechanisms that allow a majority of participating nations to vote stricter measures into effect after the fact; these measures then become binding on all original signatories. Thus, unanimous consent is not required to modify the agreement (some argue that consensus may actually be reached more quickly because dissenting countries know they may be outvoted). Still, after the agreement is in place, ensuring that all participating countries live up to it can be difficult.

Short of military action, which very few people would advocate, peer pressure, moral persuasion, and public embarrassment have been found to be surprisingly effective in forcing nations to comply with the provisions of environmental treaties. For most nations and their leaders, "public face," or how the world community views the nation and its government, is very important. To maintain or improve their image, governments may comply with environmental treaties rather than risk being publicly embarrassed. In recent years, nongovernmental organizations (NGOs), such as citizens' groups, business organizations, and environmental coalitions, have been extremely helpful in forcing compliance of international environmental treaties. NGOs are often quite willing to gather and publicize incriminating information on nations that are unwilling to live up to the letter, or even spirit, of an environmental treaty. Acknowledging the value and contributions of NGOs, the UN and its affiliated agencies have been encouraging the increased involvement of NGOs not just in monitoring environmental treaties but also in helping to initiate and formulate them.

Money and trade can also be useful in implementing and enforcing international environmental agreements. Understandably, many developing nations are much more willing to implement a treaty if they receive financial assistance. Thus, under the 1990 amendments to the Montreal Protocol, a fund containing several hundred million dollars was established to help developing countries finance the phase out of ozone-depleting chemicals.

Trade incentives and disincentives are ways to force compliance with environmental treaties. For instance, Montreal Protocol members are forbidden to purchase CFCs from countries that have not signed the convention. A country or group of countries can impose trade embargoes on nations that do not fulfill the requirements of an environmental treaty. However, some environmentalists fear that international trade agreements may interfere with the use of trade sanctions to force nations to behave in an environmentally sound manner. Depending on the interpretation, certain provisions of the World Trade Organization (WTO), which sets the basic rules for global trade, may make it illegal to restrict trade based on certain environmental concerns.

Environmental Law and Regulation in the United States

In the United States, the national government is composed of three branches: the legislative, the executive, and the judicial. The **legislative branch** makes the basic laws and also allocates funds; the federal legislative body is the Congress, composed of the House of Representatives and the Senate. The **executive branch**, consisting of the president, vice president, and the president's appointees, is charged with administering and enforcing the laws of the country. The **judicial branch**, or the court system (at the highest level, the U.S. Supreme Court), interprets the laws of the land, including the U.S. Constitution, and renders judgments in trials and disputes. Each state has a somewhat similar government. At the local level, a city may have a mayor (the executive) and a city council (the legislative branch).

At all levels, all three branches of government have a role to play in environmental issues. Legislatures may pass laws that protect, or harm, the environment or at least regulate human affairs that affect the environment. The actual influence that a particular law has depends on how it is administered and enforced. Finally, disputes over the implementation of laws are often decided in the courts. In some cases, the judiciary may define and clarify what originally may have been a fairly vague law.

Extremely important on a national level are the **regulatory agencies** that deal with environmental concerns. After Congress enacts a law, it is typically administered by a regulatory agency in the executive branch of government. From an environmental perspective, some of the more important national regulatory agencies are the Environmental Protection Agency, the Department of Energy, the Department of the Interior, the Department of Agriculture, and the Bureau of Land Management. The effectiveness of a particular law often depends on the agency administering it. Furthermore, the regulatory agencies often have considerable latitude in interpreting and implementing a law. For instance, Congress may pass a law requiring certain industries to maintain emissions of specific pollutants at acceptable levels; the appropriate regulatory agency must then determine exactly what constitutes "acceptable" levels. The influence that regulatory agencies can have was demonstrated during the Reagan administration. The Reagan era was marked by the weakening and dismantling of much of the federal bureaucracy charged with administering environmental laws. As a result, between 1981 and 1988, enforcement of laws governing pollution and hazardous waste was extremely lax, and many government-owned natural resources were made available to individuals and companies for private profit. For instance, the Reagan administration increased the leasing of continental shelf areas and national forests for oil and mineral exploration, exploitation, and timbering. These actions were not simply the result of negligence when it came to environmental concerns, but were pursued as an integral part of the administration's pro-business, free-market approach to government and economic affairs.

Regulatory agencies are subject to the influences of political opinion and professional lobbyists. As the agencies set standards and guidelines for the enforcement of environmental laws, they often expend considerable energy gathering information and conducting studies. Advisory committees may make recommendations, and interested parties may testify at public and private hearings. For instance, the levels set for "acceptable" emissions of pollutants may be the result of negotiation and compromise among different interests. Ultimately, the existence of any regulatory agency is dependent on the legislature, for it supplies the agency's funds. If members of Congress perceive strong public opposition to the policies of a certain agency, they may simply cut the agency's funding.

Decision Making in the Public Arena

As administrators implement environmental laws by developing specific rules, regulations, and guidelines, many decisions must be made at many different levels. All decisions must include some political, social, and ethical values; there is no such thing as a politically neutral decision, although certainly some decisions come closer than others. Many "rational" decisions are based on evidence and "proof." At a fundamental level, the standards of proof that are accepted can vary greatly depending on a particular person's outlook and inclinations. For instance, given the question of whether small concentrations of a certain chemical are harmful to human or environmental health, an identical study can be interpreted in more than one way. The chemical's manufacturer may presume the chemical to be harmless until "proven" otherwise, so a scientific study that simply suggests (perhaps based on a weak statistical correlation) that the chemical has ill effects will be rejected. On the other hand, an environmentalist or health advocate may, in the name of safety, presume the chemical is dangerous if there is any evidence, no matter how weak, to that effect. Thus, both the manufacturer and the environmentalist will use the same study to bolster their arguments. In this case, the disagreement boils down to the use of different standards of proof for the chemical's harmfulness or harmlessness. How stringent a standard of proof should be is not a scientific question but a political, moral, and philosophical one. In such a case, there is no neutral middle ground. Accepting a middle-of-the-road standard of proof for our hypothetical chemical is not a neutral decision but a decision that essentially weighs environmental and health issues equally with the economic concerns of the manufacturer and thus arrives at a practical compromise between the two camps.

Making policy decisions concerning environmental affairs usually relies on three basic techniques: (1) soliciting the best professional judgments of experts in the relevant field or fields, (2) basing new policies on the precedents of previous ones (sometimes known as **bootstrapping**), and (3) using **benefit–cost analyses** (also referred to as risk–benefit analyses). These techniques are not mutually exclusive; in practice, they often are used in conjunction with one another. Here we briefly discuss each of these techniques.

Judgments of Experts

Relying on the judgments of experts would seem to make good sense—for who is more knowledgeable in a particular field than a trained expert? Such a strategy quickly runs into problems, however. Equally competent and well-credentialed experts within even a narrowly circumscribed field may disagree. On the other hand, if the professionals agree, it may simply be because they have all been subjected to similar training and backgrounds and therefore carry similar underlying assumptions and biases—they represent a single point of view, but not necessarily the only point of view. For any given situation, which experts should be consulted? If a decision has to be made as to whether, how, and to what extent to log an old-growth forest, who should be consulted? Professional foresters, zoologists, logging company executives, professional employees of environmental organizations, economists, and others might all be considered relevant experts, but they could have very different views on the situation. Ultimately, the relevant knowledge that professionals can offer should be taken

into consideration, but as we have already pointed out, the ultimate decision-making process must include more than simply objective knowledge. A decision can be reached only by combining the knowledge and judgments of the experts with political, social, and ethical values.

Precedents

In many situations, new or revised policies and decisions are based on the precedents that older decisions and policies set. Previously established standards are applied to new situations. As an example, in attempting to establish health standards for the use of a new synthetic chemical, one might allow risks to the health of the public and the environment that are similar to the risks posed by other chemicals that are already in use. This method of decision making and policy setting puts great emphasis on the status quo; essentially, the decision maker assumes that the current standard is the best until trouble ensues. Then the policy or standard is revised just enough to solve the immediate problem.

This mode of policy making is sometimes referred to as "bootstrapping" (after the old adage of pulling oneself up by one's bootstraps—that is, without anyone else's help) because current policies and decisions are the result of the slow accumulation and refinement of past policies and decisions. This method of policy making is also referred to as **muddling along** or following the political path of least resistance. Because the decision makers are tied to precedent, they have little opportunity to make major policy revisions or consider new types of data and radical alternatives to current policies. Such muddling along tends to promote business as usual and can be an effective way of dealing with slight variations on old themes. However, it may fail completely when novel and unprecedented situations arise. For instance, scientific and technological innovations such as bioengineered organisms, new medical techniques, or sophisticated electronic wizardry initially may fall outside of the boundaries of the issues that established policies addressed. New policies, not just refinements of the old, may be needed.

Another popular approach to decision making and policy formulation is benefit–cost (or risk–benefit) analysis. The task of a benefit–cost analysis is to establish what potential decisions could be made and then determine for each the costs, or risks, versus the benefits. For any particular decision, there may be a statistical probability of good or bad effects, and these probabilities must be taken into account. Benefit–cost analyses usually quantify in one way or another the benefits and costs of each potential decision, and the rational course to pursue is considered to be the decision that bears the least cost but most benefit. Benefits and costs are often quantified in monetary terms, but other measures can also be used, such as human lives lost or saved, infant mortality rates, loss of productive work time at a factory, and so forth. Classic Western benefit–cost analysis is often closely associated with **utilitarianism** and its efforts to maximize "the greatest good for the greatest number." Of course, it is often difficult, if not impossible, to maximize two different variables simultaneously.

Benefit–cost analyses often have a broad appeal because, at least superficially, they use numbers and appear to be scientific and rigorous. Naively, people may believe that such analyses can lead to purely rational, objective decisions and policies. This is not to deny the importance of benefit–cost analysis, which can lead to refined and critical thinking about a problem, but as we have already stressed, no decision can be made in a political, social, or ethical vacuum. Unfortunately, benefit–cost analyses may contain built-in, and perhaps unacknowledged, political, social, and ethical variables that become incorporated into the ultimate decision.

Benefit–cost analyses often consider very different types of entities and need to equate them to quantify them. For instance, in designing a new car, an automobile manufacturer may do a benefit–cost analysis. The more the company must spend to make a better car, the higher the price it has to charge for the car. All other things being equal, higher priced cars tend to sell more slowly, but better quality cars tend to sell more quickly. A benefit–cost analysis could help the manufacturer optimize profits by determining how well a car should be built; the benefits of better sales are weighed against the costs of building the better car, and the increase in sales of cheaper cars is weighed against the decline in sales if the car is more expensive. What if the more expensive, but less profitable, car is safer and better for the environment? The typical capitalistic business perspective does not take such considerations into account as long as they do not affect the bottom line—the company's profit. However, from an ethical perspective, such considerations might be more important than the bottom line.

If the car manufacturer is forced to internalize the costs/risks the cars pose to human health and the environment, these considerations will enter the analysis. For instance, assume that if the car crashes, passengers are likely to die because the car was poorly designed and inexpensively built. The manufacturer is sued and has to pay, on average, $1 million per life lost in such accidents. In terms of the manufacturer's benefit–cost analysis, a human life is now worth $1 million. Assume further that such a death occurs in 1 in every 100,000 cars and that the manufacturer would have to pay $100 per car to avoid the problem that leads to the deaths. A simple benefit–cost analysis indicates the manufacturer is much better off paying $1 million to the family of a dead victim once in every 100,000 cars than paying $10 million ($100 × 100,000) to make 100,000 cars safer. Of course, this overly simplistic analysis ignores such issues as the loss in sales that might ensue from bad publicity. However, many people would find the decision to build a cheaper car at a greater profit, even though an occasional life is lost, to be morally reprehensible. This example shows one of the major limitations of benefit–cost analysis: All factors must be quantified in terms of a common unit (in this case, dollars).

Many benefit–cost analyses do not take all considerations into account or at least do not weight them equally. Typically, considerations that directly affect the decision or policy maker are given the most weight. Returning to our hypothetical car manufacturer, perhaps a more expensive car that yields less profit also produces less pollution and waste and thus is considerably less costly (that is, less damaging) to the global environment. This will be of no direct concern to the car manufacturer who is only trying to maximize profits, as long as the company does not need to pay the costs of environmental damage caused by the cheaper and thus more profitable cars. However, a strong governmental authority charged with regulating the automobile industry may factor environmental costs into its benefit–cost analysis when determining what policies and regulations to set for the industry. But even if we take environmental damages into account, how do we place a dollar figure on them so they can be incorporated into our quantitative analysis? What is the value of a forest undamaged by acid rain? Is the value simply the value of the lumber that resides in the trees, or is it something more? How much is it worth, in dollars, to have a city free of photochemical smog? Do we simply add the medical costs of pollution-induced diseases and the costs of pollution-induced damage to buildings and other structures? Or is there a value in having a clean, healthy environment beyond the immediate expenses avoided by not living in a polluted environment? What is the value of a scenic vista, a wildlife preserve, or a healthy forest with no immediate utilitarian use?

These examples are only a few of the issues that can be very difficult to factor into a benefit–cost analysis. Another criticism is that very often benefit–cost analyses consider only a subset of contemporary people (in our hypothetical example, perhaps only the owners of the automobile manufacturing company). However, it can be cogently argued that the environment belongs not just to living people, but also to future generations and to nonhuman beings as well. In a benefit–cost analysis of, for example, open-pit coal mining or a major hydroelectric power dam, to what extent should the benefits and costs for nonhuman organisms and future generations be taken into account?

These criticisms of benefit–cost analyses do not mean that this technique is without legitimate uses. Such analyses often can be extremely useful, even if they are not used as the sole criterion for arriving at a decision. Especially when opposing groups apply independent benefit–cost analyses to the same problem, these techniques can help clarify differences in underlying assumptions and values that lead to disagreements. Perhaps the greatest benefit of the technique is that when used correctly, it can promote thoughtful, rational, critical thinking on important issues.

Ethical Considerations

Another way to arrive at decisions and establish policy is to formulate opinions solely on the basis of an underlying well-considered philosophy and ethics (see, for example, **CASE STUDY 20-4**). This means making decisions and establishing policies on the basis of what is considered to be inherently right and good, not by simply balancing immediate, or even long-term, costs and benefits. Of course, not everyone agrees on what is ultimately "right" and "good" or shares the same values, and some persons consider making decisions on the basis of a benefit–cost analysis to be inherently "right." Others think decisions based on ethical considerations are nonrational and simply political or ideological. However, as we have tried to show, any particular benefit–cost analysis is inherently value laden simply by virtue of the way it is formulated and the potential alternatives, costs, and benefits it incorporates. As long as people must make decisions, there will be room for discussion and argument as to which paths should be followed.

Mobilization Bias: Why Do Small Groups Have So Much Influence?

A perplexing aspect of many environmental problems is that so many people are in favor of certain solutions but they have so little influence on public policy. For example, opinion polls consistently show that the majority of Americans are in favor of cleaner air, preserving more land, reducing urban sprawl,

Global Population and Ethical Considerations

VISIT

http://environment.jbpub.com
/mckinney/5e/
for more information

With more than 7 billion people on Earth, it can be argued that virtually every major environmental problem is exacerbated by overpopulation. Many environmental activists regard overpopulation as the root cause of most current global environmental problems, yet in many circles, discussion of population control is taboo. Out of religious or ethical considerations, many people object to any control on human reproduction. To have children is a fundamental desire of many persons. Some have suggested that the family is the fundamental and natural unit of society and that it is a fundamental human right for the family itself to make all choices and decisions about family size, that is, how many children a couple chooses to have. Those who advocate population control have been accused of racism because they tend to be wealthy whites from the industrialized nations urging the use of birth control in nonwhite developing nations. Abortion, which is often used where contraception fails, is thought to be ethically wrong by many who see it as the murder of the unborn.

Out of equally strong ethical convictions, many people regard population control as a primary duty. Given the current state of affairs, limiting human population is the only way the globe can be saved. Neglecting population control in developing countries results in increased poverty, starvation, and death among the already poor peoples of the world, and increasing the wealthy population of the developed world places further burdens on the already overstressed global environmental support system. In the end, if the global population continues to grow, we may all be doomed.

The "right" of a couple to have as many children as they please when they please ends when this "right" affects the "rights" of other humans (and possibly nonhuman organisms also) to live in a healthy, safe, functioning environment.

Taking a somewhat utilitarian view, we might wish to maximize the "good" in the world. If we focus only on human good, then we might maximize the total amount of human good by increasing the number of people in the world. As long as a human life, on average, contains more good than bad, then the more people there are, the more good there is. According to this way of thinking, population control will be called for only when adding more lives begins to bring more bad into the world than good. But how do we determine when this point is reached? Who can judge for the world as a whole? Even in a region plagued by starvation and disease, not everyone will necessarily agree that on average another human life will create more bad than good, and in an affluent region, not everyone will necessarily agree that an additional human life will bring more good than bad.

Even within this framework, other considerations must be taken into account. What are the future implications of increasing (or decreasing) the human population now? Perhaps increasing the human population will increase the amount of human good in this generation but bring about untold misery in the long run. Do we factor potential future generations into our analysis?

Maximizing the total human good in the world may not even be an appropriate strategy. If we maximize the total human good by increasing the population to the point where adding more

promoting mass transit, recycling, and achieving many other environmental goals. Yet, as we have seen throughout this text, in most cases these goals are rarely met. Why is this so? Is not a democratic government supposed to reflect public opinion?

In their book, *One with Ninevah: Politics, Consumption, and the Human Future* (Island Press, 2004), Paul and Anne Ehrlich discuss the importance of **mobilization bias** in affecting government policy. Mobilization bias is a term used by political scientists to describe the disproportionate influence that small but powerful groups may have on public policy. Of course, most small groups are not powerful in terms of government influ-

people will start to decrease the total amount of good, we will end up with many people whose lives are only marginally good. This consideration has led some people to argue that instead of maximizing the total human good in the world, we should maximize the average amount of good per human life. Essentially, this approach emphasizes quality over quantity. By this way of thinking, a small population where individual lives are characterized by a great deal of "good" (however good is defined) would be preferable to a larger population where the average human life contains even slightly less good. In extreme cases, one could even contend that it is ultimately better to allow an overly large population to reduce its size by such natural (if somewhat cruel) means as starvation and disease so that those who survive can have a higher average amount of good in their lives. The alternative might be to barely maintain, perhaps through relief efforts, those on the brink of starvation—but such actions can result in a general lowering of the average amount of good per individual. Certainly, many people find it morally more acceptable to control population growth by the use of birth control so that more lives are not conceived in the first place. Actual population reductions will come about as people die of old age.

Those who espouse the maximization of the average human good generally believe the current world needs a strong policy of population reduction. In both developed and developing nations, most people probably would be better off, on average, if there were fewer people around. Fewer people would mean more resources per person and less poverty, starvation, and crowding throughout the world.

So far in this discussion, we have focused only on human good, but what about the rights, interests, and therefore the good of the nonhuman organisms, species, and ecosystems with whom we share the planet? If we take these other creatures into account, as the ecocentrists (those taking a holistic and nonanthropocentric, nonhuman-oriented perspective that focuses on larger entities, such as entire ecosystems) insist we must, clearly the people already inhabiting the globe are having detrimental effects on other life forms and the Earth as a whole (as is discussed at length elsewhere in this text). From the ecocentric perspective, it is clear that drastic reductions in the human population are needed.

Questions

1. Do you believe that each person has a fundamental human right to decide how many children to have? Or do people have a fundamental duty to control their own breeding? Justify your answer.

2. Do you think that it is appropriate to attempt to maximize the "good" in the world? If so, what exactly is "good" and how should it be maximized? If you do not believe that the "good" should be maximized, explain why not.

3. Do nonhuman species have "rights" or "interests"? Justify your answer. If they do have rights and interests, how do those rights and interests affect the issue of global human population?

VISIT

http://environment.jbpub.com/mckinney/5e/ for more information

ence, so mobilization bias is often associated with money and the small groups that are very wealthy.

Consider an example that the Ehrlichs cite in their book. The real estate industry in the United States makes billions of dollars each year by converting ecologically important coastal marshes into housing subdivisions or boat marinas. So this powerful group spends huge amounts of money to influence local, state, and federal politicians; however, aside from the few people who live in these houses or uses the marinas, most of the rest of society are "losers" in this transaction. The thousands of acres of coastal swampland that are lost each year provides essential habitat for many economically important ocean species. Thus, the loss of these areas has a strong negative impact on the commercial and sport marine fishing industries and the shrimp industry. Dramatic increases in water pollution occur because coastal vegetation is reduced while sources of pollution increase. Aside from sport fishing, other recreational uses of these marshes are also lost, such as bird watching or just enjoying the natural experience.

Mobilization bias gives us great insight into why governmental policy so often promotes environmental degradation, despite that the majority of people do not want it. In the case of coastal real estate development, there are relatively few people who benefit involved (developers, a few home buyers, and boaters), but they each derive very high benefits (money, personal needs). The vast majority of local citizens, who are not interested in buying a new house on the coast, suffer an environmental cost, but that cost *per person* is relatively small. In other words, there are many more social "losers" in this transaction, but the loss to each one person is so small that very few of them are generally ever motivated to do anything about it.

On the other hand, the social "winners," such as the land developers, each have very high benefits, so each one is strongly motivated to try to persuade policy makers to promote their view. For example, real estate developers often lobby politicians to weaken regulations that limit land development of ecologically important areas. The ultimate outcome is often that the small group of powerfully motivated and wealthy people exerts more influence on policy makers at many levels of government than do the much larger but weakly motivated general public. This outcome is, of course, not limited to real estate development. Many powerful industries, such as the fossil fuel (coal, petroleum), mining, logging, chemical, and road-building industries, have huge impacts on the environment and exert enormous influence on the political process. Many environmental groups, notably the Sierra Club, have argued that the oil industry has exerted far too much influence on federal energy policy in the last few years, producing less interest in energy conservation and alternative energy sources.

One way to measure mobilization bias is by the increase in lobbyists, who represent specific interest groups to influence politicians. In the 1950s, only about 5,000 lobbyists were registered in Washington, D.C. In 2004, there were more than 20,000 registered lobbyists in Washington, D.C., which translates to more than 38 lobbyists for each member of Congress. Each year, these lobbyists spend more than $1.5 billion to influence federal policy in Congress.

There is no simple solution to mobilization bias. In almost any democracy, wealthy and passionate citizens will always have a disproportionate impact. The best hope for the environment is for the vast but silent majority to become more active, more outspoken, and more influential in public policy. However, it usually takes an extreme environmental crisis to motivate this silent majority. Human activities greatly enhanced the devastating costs of Hurricane Katrina in 2005; the costs ran into the hundreds of billions of dollars in New Orleans and nearby areas. Poorly planned coastal development, devegetation of thousands of square miles of barrier islands and wetlands, poor engineering strategies of the Mississippi River, and many other short-sighted activities clearly increased property loss. Rising sea level and increased hurricane strength from global warming from human activity may also have contributed to losses. We need more of the general public to be educated and motivated to speak out against policies that promote environmental harm before such crises occur.

SUMMARY

- Social institutions reflect the beliefs and values of human societies.
- Early Europeans coming to North America initially tended to tame, use, and "exploit" the "uninhabited wilderness."
- By the middle 1800s, writers and thinkers, such as George Perkins Marsh, documented how human intervention was modifying the natural environment.
- In the late 19th century, the beginning of modern environmentalism took root with the conservationism of Gifford Pinchot and the preservationism of John Muir.
- "Modern" U.S. environmentalism came to the fore in the 1960s and 1970s with such events as the publication of Rachel Carson's *Silent Spring* (1962), the establishment of the Environmental Protection Agency (1970), and the first Earth Day (1970).
- In some people's opinions, Western societal values have promoted current environmental problems, but others argue that it is Western values and technologies that are best suited to address current environmental concerns.
- Sustainable development focuses on social, economic, and political progress without damaging the environment.
- Environmental law, at both national and international levels, can be viewed as protecting the fundamental right of all people (and perhaps all organisms) to live in a healthy environment.
- At a public level, decisions affecting the environment may be made using any of the following techniques/bases: (1) professional judgments of experts in relevant fields, (2) basing new policies on precedents, (3) benefit–cost analyses, and (4) philosophical, ethical, and moral considerations.
- It is important for more of the public to speak out about environmental issues because all too often mobilization bias occurs, in which small groups of people get the most say.

KEY TERMS

benefit–cost analyses
bootstrapping
common law torts
conservationism
environmental law
executive branch
frontier mentality
hard laws

inherent value
judicial branch
legal instruments
legislative branch
Manifest Destiny
mobilization bias
moral status
muddling along

preservationist
progressive conservationism
regulatory agencies
social institutions
soft laws
sustainable development
transcendentalists
utilitarianism

STUDY QUESTIONS

1. Give a brief historical overview of American environmentalism. What are the major chapters or periods in this history? What was accomplished in each?

2. Discuss the historical trends in federal legislation and interest in and public sympathy toward environmental issues. What is the mood of the country relative to these matters today?

3. List some of the basic social institutions of Western society. In your opinion, are they to be blamed for the current environmental concerns?

4. How might it be argued that Western social institutions and technology are the keys to solving our environmental problems?

5. Compare and contrast the approaches of conservationists and preservationists with environmental issues. Who was the outstanding early advocate for each?

6. Describe the environmental stance that was known as progressive conservationism.

7. How did the Great Depression and World War II affect environmental attitudes in the United States?

8. Describe and discuss the concept of sustainable development. Do you believe we should be striving toward this goal? If your answer is yes, how might it be accomplished? If your answer is no, what alternatives do you propose?

9. What are the philosophical and historical bases for environmental laws?

10. Distinguish between hard and soft laws.
11. Describe some of the major international environmental laws, treaties, conventions, declarations, and protocols that are currently on the books.
12. What was accomplished at the 1992 Earth Summit?
13. Why is it difficult to enforce international environmental laws? How can compliance be encouraged?
14. Explain how environmental laws and regulations are arrived at and instituted in the United States. Who makes the laws? Who makes the regulations? Who enforces them?
15. List the various ways that environmental decisions, especially public policy decisions, may be arrived at.
16. What are the pros and cons of benefit–cost analyses? Why is this method of decision making often strongly favored as well as sometimes strongly criticized?

WHAT'S THE EVIDENCE?

1. Cite evidence for and against the thesis that "Western societal values are to be blamed for current environmental problems." Is this an overstatement? Is it simply true? Or is it false?
2. When it comes to decision making, some people put most of their faith in benefit–cost analyses. Others, in contrast, believe that such analyses can be dehumanizing and may overlook important factors that are not readily quantified (perhaps the value of a human life, the concept of "quality of life," or aesthetic considerations). Develop a compelling argument for or against the use of benefit–cost analyses, citing evidence that supports your position.

http://environment.jbpub.com/mckinney/5e/

http://environment.jbpub.com /mckinney/5e/

Connect to this text's website at http://environment.jbpub.com/mckinney/5e/. This site features eLearning, an online review area that provides quizzes, chapter outlines, and other tools to help you study for your class. You can also follow useful links for more in-depth information.

The technology of the Information Age can now reach all over the globe. A nomadic woman in a remote area of southwest China's Tibet Autonomous Region talks on a mobile phone in front of her tent.

Epilogue: Environmental Literacy

"The ecologically literate person has the knowledge necessary to comprehend interrelatedness, and an attitude of care or stewardship. Such a person would also have the practical competence required to act on the basis of knowledge and feeling. . . . Ecological literacy presumes that we understand our place in the story of evolution. It is to know that our health, well-being, and ultimately our survival depend on working with, not against, natural forces."

David Orr

Our global society and the global environment continue to change at an astounding pace. With the beginning of the third millennium, we are, indeed, entering a new age in which we increasingly need to cultivate environmental awareness and literacy across the global community. Humankind can no longer afford to be insensitive toward environmental issues, for if we do not take concerted action to address environmental problems, they will quickly overwhelm us. Environmental issues are no longer a "special interest" but are taking center stage in the political, economic, and social arenas. Every citizen of the planet must have a working knowledge of environmental issues; everyone must be environmentally literate.

In the past, environmental science often was considered a marginal discipline that held little hope for the future. Throughout this book, we have tried to demonstrate the various ways that environmental science can, and indeed must, become the "science of reality." This means that environmentalists must do more than criticize the status quo. It requires that we move beyond simply reacting to problems and relying on emotional persuasion and abstract formulations. Instead, we must proactively seek long-term solutions that are practical, sustainable, and realistic, solutions grounded in factual knowledge. Doing so will not always be easy, for in some cases it will require more than simple technological fixes: the very fabric and mindset of modern society may have to be modified. This can come about only through increasing environmental literacy.

In his book *Ecological Literacy,* the philosopher David Orr discusses the importance of moving beyond heated dialogues about various issues toward more comprehensive cooperation in building a sustainable society for all living things. This means that we should approach environmental problems by first becoming environmentally literate, by taking the time learn the basic facts. Without this step, political, social, and personal biases will determine the outcome of environmental problem solving, often to the long-term detriment of both the environment and society. Admittedly, environmental literacy is not easy in today's "information age," which immerses all of us in a sea of facts. Sorting out the relevant from the irrelevant, the important from the trivial, can be extremely difficult. But this is what textbooks are for, to provide a distillation and overview of the fundamentals. Here we have provided a foundation upon which you, the reader, can build as you learn more about environmental issues and how they apply to your local community.

But we hope you will not be satisfied to simply learn about environmental issues. Become actively involved! You should never think that you, as an environmentally literate individual, cannot make a difference. Indeed, we have attempted to demonstrate to you many ways throughout the text that individuals can make important changes through coordinated efforts. With the innovative use of social media, global citizens can work together more efficiently than ever before to bring about lasting change. Further, you can have an impact through the way you think, the way you vote, the way you spend your money, and the way you live your life. Nearly all so-called global environmental problems are aggregations of local problems. Global warming, species extinctions, ozone depletion, and certainly many regional problems result from the nearly countless actions of numerous individuals. Because the problems arise from individuals, the solutions also must arise from that level. Global change will occur; the global society of the future will not be the society we know today. The question is: How will society change? Will the change be environmentally sustainable? The collective actions of all of us will determine the future. When global

society consists entirely of people who care enough about the environment and future generations to act accordingly, we will finally be able to solve our environmental problems. Don't be overwhelmed by the issues; you can do a lot to be helpful. Carpool or bike and walk more. If you have to drive, choose a car that gets better gas mileage. Take shorter showers, purchase green cleaning products and other household supplies, recycle, purchase less stuff, install solar panels or other technology to green your home, compost! The important thing is to be always mindful of "sustainability" in all aspects of your life and those around you and work to implement changes in your day-to-day life and convince others around you to do the same. The key is committing to the little things and doing them consistently. As 2004 Nobel peace prize winner and environmental leader Wangari Maathai stated: "It's the little things citizens do. That's what will make the difference. My little thing is planting trees." When she started planting trees in the 1970's she would have no way of knowing that by the time of her death this past September (2011), the organization she founded would have planted over 40 million trees impacting thousands of people's lives in many positive ways! Her life's work inspires action:

> *"In the course of history, there comes a time when humanity is called to shift to a new level of consciousness, to reach a higher moral ground. You cannot protect the environment unless you empower people; you inform them; and you help them understand that these resources [trees, farm land, wells] are their own, that they must protect them . . . we have to shed our fear and give hope to each other . . . We cannot tire or give up. We owe it to the present and future generations of all species to rise up and walk! . . Recognizing that sustainable development, democracy and peace are indivisible is an idea whose time has come."*

In an increasingly complex global community supported by shared natural resources, it is in all our best interests to work together toward a more sustainable world. We look forward to a time in human consciousness where the very notion of global environmental problems will exist only as a historical relic; and, we look forward to a society that is at peace with a restored and sustainable environment.

Appendix: English/Metric Conversion Tables

These tables are meant to be handy reference guide to some of the basic English and metric units involving length, mass (weight), area, volume, and temperature. They are not intended to be comprehensive. For common measures of energy and power, see Fundamentals 7-2, page 230.

Conversion values (from English to metric, or vice versa) are rounded unless designated as exact equivalents.

English to Metric		
	English	**Metric**
Length:	1 inch (in)	2.54 centimeters (cm) (exactly)
	1 foot (ft)	30.48 cm (exactly) 0.3048 meters (m) (exactly)
	1 yard (yd)	91.44 cm (exactly) 0.9144 m (exactly)
	1 mile (5280 ft)	1609.34 m (exactly) 1.60934 kilometers (km) (exactly)
Mass (weight):	1 ounce, avoirdupois	28.3495231 grams (gr)
	1 pound (lb), avoirdupois	453.59237 gr (exactly) 0.45359237 kilograms (kg) (exactly)
	1 ton, net or short (2000 lb)	907.18474 kg 0.90718474 metric ton
Area:	1 square inch (in²)	6.4516 cm² (exactly)
	1 square foot (ft²)	0.092903 m² (exactly)
	1 square yard (yd²)	0.836127 m²
	1 acre (43,560 ft²)	4046.856 m²
		0.4046856 hectares (ha)
	1 square mile	258.9975 ha 2.589975 km²
Volume:	1 cubic inch (in³)	16.38706 cm³
	1 cubic foot (ft³)	0.028317 m³
	1 cubic yard (yd³)	0.764555 m³
	1 gallon (U.S.)	3.785 liters (L)
	1 liquid quart (U.S.)	0.946 L
	1 liquid ounce (U.S.)	29.573 milliliters (ml)

Temperature: 1 degree Fahrenheit equals 5/9 (.555555) degree Celsius (centigrade). To convert a temperature in degrees Fahrenheit to the equivalent temperature in Celsius, subtract 32 and divide by 1.8: degrees C = (degrees F − 32)/1.8

	Metric	English
Length:	1 centimeter (cm)	0.393701 in
	1 meter (m)	39.3701 in 3.28084 ft 1.09361 yd
	1 kilometer (km)	3280.84 ft 0.62137 mile
Mass (weight):	1 gram (gr)	0.0352739619 ounce, avoirdupois
	1 kilogram (kg)	2.204623 lb
	1 metric ton	1.10231 short tons
Area:	1 cm^2	0.155 in^2
	1 m^2	10.7639 ft^2 1.19599 yd^2
	1 hectare	2.47105 acres
	1 km^2	247.105 acres 0.3861 mile^2
Volume:	1 cm^3	0.06102 in^3
	1 m^3	35.3145 ft^3 1.3079 yd^3
	1 liter (L)	1.057 quarts 0.2642 gallon (U.S.)
	1 milliliter (ml)	0.0338 liquid ounce (U.S.)

Temperature: 1 degree Celsius (centigrade) equals 9/5 (1.8 exactly) degrees Fahrenheit. To convert a temperature in degrees Celsius to the equivalent temperature in Fahrenheit, multiply by 1.8 and then add 32: degrees F = (degrees C × 1.8) + 32

Glossary

abatement devices "End of the pipe" pollution control equipment.

abortion The termination of a pregnancy before the normal term is up (before the child is scheduled to be born).

absorptive ability The ability of a natural sink (such as a river, lake, or the atmosphere) to absorb pollutants or potential pollutants.

acid rain Precipitation (rain, snow, sleet, and so forth) that is more acidic than normal (generally due to human-produced air pollutants); also known as acid precipitation.

acoustical materials Sound-absorbing materials that can be used to reduce noise.

active solar techniques Mechanisms, such as flat-plate collectors, that are designed to actively collect the energy of sunlight and use it; for example, to heat a building or to heat water.

acute toxicity When toxic substances are harmful shortly after exposure.

additivity With reference to toxicology, when the toxic effects of two substances are added together.

ADMSE Acronym for absorption, distribution, metabolism, storage, and excretion—the five basic steps in the path of a toxic substance through the body.

aeration Exposing water to the air; often results in the release into the atmosphere of gaseous impurities found in polluted water.

aeration, zone of The zone above the water table where the voids in the soil or rock may contain water but are not fully saturated.

aerosol spray Products that are sprayed as a fine mist during use, such as canned spray paints, deodorants, and so forth. For many years such products used ozone-depleting CFCs as propellants.

age structure profile The population age structure profile is a graphic representation of the age structure of a population at a given time.

age structure The relative proportion of individuals in each age group in a population.

agricultural (Neolithic) revolution The advent of domestication and agriculture beginning around 10,000 years ago.

agriculture The cultivation or raising of plant crops and livestock.

Air Pollution Control Act of 1955 An early piece of federal legislation where air pollution was recognized as a national problem that needed to be addressed on a federal level; it was acknowledged that the nation needed to be made more aware of the ramifications of poor air quality and pollution.

Air Quality Index A daily measure of air quality; how clean or polluted the air is and the possible health effects of the air quality on a daily basis. There are five basic parameters measured; ground-level ozone, particle pollution, carbon monoxide, sulfur dioxide, and nitrogen dioxide (five major air pollutants regulated by the Clean Air Act). Higher AQI number means worse air quality.

albedo Albedo refers to the amount of radiation reflected by a surface; it is the ratio of scattered to unscattered surface reflectivity of electromagnetic radiation.

alkaline The opposite of acidic; basic. Alkaline soil or rock may neutralize acid rain.

alpha particles A type of radiation essentially composed of energetic helium nuclei.

alternative energy sources Energy sources, such as solar power, wind power, and so forth, that are alternatives to the fossil fuels, nuclear power, and large-scale hydroelectric power.

amensalism A form of symbiosis where one species inhibits another while being relatively unaffected itself.

amino acids The complex molecules that are the "building blocks" of proteins.

ammonification The conversion (by decomposers in nature, for instance) of organic nitrogen and hydrogen to ammonium ions.

animal rights The concept that nonhuman animals have inherent ethical moral status and rights.

annual growth rate How much a population grows in one year. *See also* percent annual growth.

antagonism With reference to toxicology, when substances work against each other and cancel each other's effects.

anthracite Hard coal.

anthropocentric Human-centered.

anthropogenic Formed, produced, or caused by humans.

appropriation law A law under which owners of land may be denied the right to withdraw water from a lake or stream if a more beneficial use for the water is found (government can appropriate the use of the water).

aquaculture Refers to aquatic organism farming in general and sometimes to freshwater organism farming in particular.

aquiclude A relatively impermeable rock layer that obstructs the flow of water.

aquifer A relatively permeable rock layer below the water table that contains a significant amount of water.

area curve *See* species area curve.

artesian well A well from which water flows freely without pumping due to water pressure built up in the recharge area.

artificial ore An artificial mixture or conglomeration of different metal-bearing materials, such as might be the result of a scrap yard where scrapped metal products are crushed together.

asthenosphere A layer in the mantle that is relatively weak and viscous; lies directly underneath the solid lithosphere.

atmosphere The sphere or "layer" of gases that surrounds the Earth.

atmospheric cycles The large-scale movements in the Earth's atmosphere that give rise to the major wind belts and so forth.

atom The smallest particle of an element.

atomic chlorine Chlorine atoms influence cycles that are linked to ozone destruction; this greenhouse gas is potentially toxic to plant and animal life.

background extinction Background extinction refers to the relatively constant or general rate of disappearance of organisms and species over geological time. Background extinction is the opposite of mass extinctions, which occur very rapidly and decimate many organisms and species over a short period of time.

baghouses A series of fabric bags that act as filters to remove particulate matter from polluted air.

battery A device that when charged with electricity stores the energy in the form of chemical energy. When the battery is discharged, the energy is converted back into electrical energy.

behavioral toxicology The study of chemicals that cause disorders of behavior or learning.

benefit-cost analysis (BCA) A method of comparing the benefits of an activity to its cost; also known as risk-benefit analysis.

benthic Refers to the bottom-dwelling zone of the aquatic (specifically marine) biome.

beta particles A type of radiation, essentially high-speed electrons.

Bhopal The city in India where, in December 1984, a major leak occurred at the Union Carbide pesticide-producing plant, ultimately killing thousands and injuring tens of thousands more. The Bhopal incident helped inspire Congress to pass the Emergency Planning and Community Right-to-Know Act that established the Toxics Release Inventory. *See also* Toxics Release Inventory.

big five energy sources Coal, oil, natural gas, large-scale hydroelectric, and nuclear power.

bioassay An hierarchical sequence of procedures used to test the toxicity of a chemical substance. Initially, a relatively short, inexpensive test is administered; if harmful effects are indicated, then progressively more thorough tests are carried out.

biocentric Life-centered.

biochemical conversion The harnessing of microorganisms to convert biomass into certain fuels.

bioconcentration Bioaccumulation or biomagnification; the tendency for a substance to accumulate in living tissue. An organism may take in a substance, but not excrete or metabolize it; then, a predator eats the organism and ingests the substance already accumulated in the prey.

bioconcentration factor (BCF) A measure of the tendency for bioconcentration by a given substance.

biodegradable plastics Biodegradable generally refers to a substance that can be degraded, decomposed, or broken down by microorganisms into simple compounds such as water and carbon dioxide. A biodegradable plastic is a plastic that can be broken down in such a manner; traditionally, most plastics have been synthetic compounds that are not biodegradable.

biodiversity Biological diversity, or the variety of living things in a given area. Biodiversity can be defined in many different ways, such as in terms of genetic diversity or ecosystem diversity, but it is most often defined in terms of species diversity.

bioengineering Genetic manipulations and engineering to produce new varieties and types of organisms.

biogas digester A special chamber or reactor used to promote biochemical conversion of biomass.

biogeochemical cycles The cycles of elements and compounds through the atmosphere, lithosphere, hydrosphere, and biosphere.

biological control *See* Integrated Pest Management.

biological environment The living world or biosphere.

biological impoverishment The loss of variety in the biosphere (even when species have not gone completely extinct).

biological oxygen demand (BOD) The amount of oxygen used by organisms and chemical processes in a particular stream, lake, or other body of water to carry out decomposition.

biomass The total weight of living tissue in a community.

biomass energy Energy produced by the burning of such biomass as organic wastes, standing forests, and energy crops.

biomass pyramid A graphic depiction of the amount of biomass that occurs at each trophic level in a particular community or ecosystem.

biome A large-scale category that includes many communities of a similar nature.

bioplastics Plastics produced from organic and generally biodegradable compounds (such as vegetable polymers), and generally containing little to no petroleum products. Many bioplastics, when in the proper compost setting, will break down.

bioprospecting The developing field wherein biologists, chemists, and other researchers are compiling a database of the commercial potential of many species; bioprospecting can, for example, draw from the knowledge of native healers or shamans in tropical countries about the medicinal properties of local plants and animals, but it ultimately seeks information about all species, including even deepwater marine organisms.

bioremediation The use of bacteria and other small organisms (such as single-celled and multicellular microbes and fungi) to clean up or reduce unwanted concentrations of certain substances; also known as biotreatment.

biosphere The sphere or "layer" of living organisms on Earth.

biotechnology The artificial use and manipulation of organisms toward human ends.

biotoxin A poisonous substance produced by an organism, often of high potency such that a small amount can kill an adult human.

birth control The artificial control, or prevention, of unwanted births.

birth rate How often members of a population give birth, or number of births per unit time per unit number of members of a population. *See* crude birth rate.

bituminous coal Soft coal.

blackout When a region served by an electric power plant is left without power for an extended period of time, perhaps due to a major breakdown.

bootstrapping Basing new policies on the precedents of previous ones; sometimes referred to as "muddling along" or following the path of least resistance.

bottom-up approach The development and encouragement of sustainable uses of biodiversity that provides incentives to save species while also respecting the right of all people to support their families and have a decent quality of life.

breeder reactor A nuclear reactor that is especially designed to actively convert nonfissionable isotopes into fissionable isotopes that can then be used as fuel.

brownfield development The improvement of run-down or abandoned lands, usually contaminated with hazardous substances or pollutants, by renovating them for new industries, stores, businesses, or multifamily housing.

brownout When the capacity of a power plant is exceeded by a few percent and the voltage to consumers is inadequate such that lights often dim.

BTU British thermal unit; the amount of energy that when converted completely to heat will warm one pound of water by one degree Fahrenheit.

buffer zone In a preserve, an area of moderately utilized land that provides a transition into the unmodified natural habitat in the core preserve where no human disturbance is allowed.

Cambrian explosion The period when many types of fossil organisms first appear in the fossil record, dated in real time to approximately 570 to 540 million years ago. *See* explosion of life.

calorie The amount of energy that when converted completely to heat will warm one gram of water by one degree Celsius.

carbamates A group of synthetic organic pesticides that are less persistent but more toxic to humans than chlorinated hydrocarbons; includes sevin.

carbon cycle The biogeochemical cycle of carbon.

carbon dioxide CO_2, the primary greenhouse gas.

carbon efficiency The amount of economic output per unit of carbon released.

carbon emission The emission of carbon, primarily as carbon dioxide, into the atmosphere during the burning of fossil fuels and other organic matter and similar activities.

carbon tax A tax levied on fossil fuels (or any fuels) in proportion to the amount of carbon emitted during combustion.

carcinogen A cancer-producing substance.

carrying capacity The maximum population size that can be sustained by a certain environment for a long period of time (potentially indefinitely); often represented by the symbol K.

carryover grain stocks Stocks of grain that are saved from one harvest year and remain at the beginning of the next harvest year.

cassandras People who argue that humans have always altered the environment and managed things poorly; they maintain that ultimately exponential growth of populations and technologies will totally degrade the environment.

catalytic converter A device that carries out a number of chemical reactions that convert air pollutants to less harmful substances.

cataract A condition where the lens of the eye becomes opaque.

chain reaction In a nuclear reactor, when the fissioning of one atom releases neutrons that induce the fissioning of other atoms, and so forth.

channelization The artificial straightening of a river or stream.

chaparral Savanna-like environments with dense, scrubby vegetation, found, for instance, in the southwestern United States and Mexico.

charismatic species A high-profile endangered or threatened species that attracts broad public concern.

chemical prospecting The search for data on the chemical properties and potential of various species (such as for use in medicines, cosmetics, genetic engineering, and so forth).

chlorinated hydrocarbons A group of synthetic organic pesticides that includes chlordane and DDT.

chlorine A reactive element, found in chlorofluorocarbons and other substances, that has been implicated as a prime factor in the deterioration of the ozone layer.

chlorofluorocarbons (CFCs) Artificially produced compounds composed primarily of carbon, fluorine, and chlorine. CFCs have been implicated in the deterioration of the ozone layer.

chronic toxicity Long-term effects of toxic substances, especially relative to long exposures of low doses.

circle of poison Due to the fact that much produce consumed in the United States is grown in developing countries, U.S. citizens continue to be exposed to various pesticides that have been banned in the United States but are still used in various developing countries; this is sometimes referred to as the circle of poison.

Clean Air Act A federal statute enacted in 1963 that was the first of a series of acts and amendments that exerted increasing federal pressure on air polluters to clean up their emissions.

Clean Water Act (CWA) A federal statute enacted in 1972 that has been very successful in improving the water quality of lakes and rivers.

Clear Skies Initiative A market-based approach of emission caps and trading enacted in 2003 by President George W. Bush establishing federally mandated caps on sulfur dioxide, mercury, and nitrogen oxide pollutants.

clear-cutting The harvesting of trees such that an entire stand of trees is completely removed; also known as even-aged management.

climate The average weather in a certain area over time ranges of decades to millennia to hundreds of millions of years.

climax community The last, relatively stable and diverse, community in the sequence of community succession.

closed system An isolated system that exchanges nothing with other systems.

closed-loop reclamation (water) Treating wastewater to the level needed before direct reuse.

closed-loop recycling The indefinite recycling of a material or substance without degradation or deterioration, such as the recycling of many metals and glasses.

clustered housing A way to increase the efficiency of land use by clustering houses fairly close together while leaving relatively large amounts of open or undeveloped land that can be used for social commons, parks, wildlife habitats, or other non-residential use.

coal A general term used to refer to various solid fossil fuels.

cogeneration A power plant that produces several types of energy simultaneously, such as electricity and heat, that can be used locally.

cold fusion Promotion of fusion reactions at relatively low ("room temperatures") temperatures.

coliform bacteria Bacteria, such as *Escherichia coli*, that live in the human intestine in huge numbers. The abundance of coliform bacteria in water is a good indicator of how much human feces had recently entered the water.

combined cycle turbine plant A power plant that not only uses a gas turbine, but utilizes excess heat to heat water and power a steam turbine.

command economy A national economy where the central government tells individuals and companies what to produce and sets prices for the resultant products (for example, the economy of the former Soviet Union before its collapse). *See also* market economy for comparison.

commensalism A form of symbiosis where one species benefits and the other is not affected.

commercial extinction When a species becomes so rare that harvesting it is no longer economically viable.

common law A body of law based primarily on judicial rulings, custom, and precedent.

common law torts Basically "wrongs" or wrongful acts against a person, people, or institution.

community All of the populations of different species that inhabit a certain area.

community succession The sequential replacement of species in a community by immigration of new species and the local extinction of old species. The first stage is the pioneer community, and the last stage is the climax community.

compact development A way to minimize the amount of geographic area occupied by a city, for instance, by setting boundaries on the city's growth; compact development can encourage development that builds "up" rather than "out."

competition Organisms competing ("fighting") for the same limited resource.

competitive exclusion A situation where niche overlap is very great and competition is so intense that one species eliminates another from a particular area.

complexity How many kinds of parts a system has.

composting The decomposition of organic materials by microorganisms; produces various forms of "soils."

compound Two or more atoms chemically bonded together. For instance, water is a compound of hydrogen and oxygen.

concentrate and contain A method of waste disposal that concentrates waste solids and liquids and then contains them, such as in a storage facility, so as to reduce exposure to humans and the environment.

compressed air energy storage (CAES) Electricity is used to pump air under pressure into a storage reservoir; when the energy is needed, the pressurized air is released.

cone of depression The localized lowering of the water table around a well from which water is being withdrawn faster than it is replenished.

confined aquifer An aquifer bounded above and below by aquicludes.

conservation Refers to attempts to minimize the use of a natural resource.

conservation biology A subdiscipline of biology that draws on genetics, ecology, and other fields to find practical ways to save species from extinction and preserve natural habitats.

conservation easements A preservation tool that may be used by a land trust or conservation group to limit development; under a conservation easement a landowner may give, lease, or sell the development rights to the land to a preservation group.

conservation of matter and energy, law of The concept that matter and energy cannot be created or destroyed; they can only be transformed.

conservationism The classical view that forests and other natural resources should be managed to provide the most benefit for the greatest number of people.

conservationists A branch of the larger environmental movement that aims to politically and socially conserve all species within an ecosystem for the future. George Perkins Marsh and U.S. President Theodore Roosevelt are often touted as famous early conservationists. In the early 1900s the environmental movement was split into conservationists like Gifford Pinchot, an American forester who wanted to use resources for the greatest good for the longest period of time, and preservationists. *See also* preservationists.

consumed water Water that is withdrawn and not returned to its original source.

consumption The use of goods and services, materials and energy, by humans.

containment structure In a nuclear reactor, a housing around the reactor vessel designed to protect the outside environment from major radioactive contamination if an accident should occur.

contaminated water Water that is rendered unusable for drinking.

continental crust The layer of rock that underlies the continents, generally containing a higher percentage of silicon dioxide than oceanic crust. *See* crust.

Contingent Valuation Method (CVM) A method that attempts to "objectively" measure the dollar values of changes in environmental quality; often uses questionnaires and other surveys that ask people what they would pay for various environmental improvements.

contraception The prevention of conception or impregnation.

control rod Made of some substance with the ability to absorb neutrons, such as cadmium or boron, the control rods are used in a nuclear reactor to control or even halt the nuclear chain reaction.

convection cell A circulation pattern set up in a hot fluid, such as in the Earth's mantle or in the atmosphere. A hot liquid rises, flows laterally, and sinks as it cools.

Convention on Biological Diversity (CBD) A document signed at the 1992 Earth Summit that acknowledges that developing nations deserve a share of the profits generated from their genetic resources by agricultural activities and genetic technology.

Convention on International Trade in Endangered Species (CITES) An international convention that outlaws trade in endangered species and products made from endangered species.

core (a) The innermost portion of the Earth; thought to be composed primarily of an iron-nickel alloy. (b) The interior of a nuclear reactor containing the fuel, moderator, and control rods.

Coriolis effect The Coriolis effect is due to the Earth's rotation deflecting air (and other liquids also) that has already been set in motion by the pressure gradient force. In the northern hemisphere, the objects deflect to the right and in the southern hemisphere, to the left. The more the air is moving, the more it will be deflected.

cornucopians People who argue that human ingenuity always has, and always will, overcome environmental limitations.

corrosive When used in reference to hazardous wastes, generally referring to liquids that are highly acidic, very alkaline, or otherwise chemically very reactive.

cowboy economics A perspective that views the environment as a source of endless space and riches (as the frontier in the Old West was once viewed).

"cradle to grave" Another name for Life Cycle Analysis (LCA), a methodology that analyzes each step of a product's life cycle, from the mining or harvesting of raw materials through ultimate disposal by the consumer and what happens to the product after it has been disposed of, with the goal of making the process more efficient and environmentally friendly.

criteria pollutants With reference to air pollution, the six basic criteria pollutants are particulate matter, sulfur dioxide, nitrogen dioxide, ozone, lead, and carbon monoxide.

critical mineral A mineral that is necessary for the production of essential goods.

crop rotation Planting different crops on a particular field in different years—the same crop is not planted on the same field year after year.

crude birth rate The number of births per year per 1000 members of a population.

crude death rate The number of deaths per year per 1000 members of a population.

crust The outermost layer of rock that forms the solid surface of our planet Earth; divided into continental crust and oceanic crust.

cultural extinction The extinction of a tribal people or indigenous culture.

cyclone An intense storm that typically develops over a warm tropical sea.

cyclone collector A method of removing particulate matter from the air, where emissions are spun in a mechanically produced vortex, causing heavier particles to collect at the bottom to be removed.

dams Structures that obstruct river or stream flow to form artificial lakes or reservoirs.

daughter products The atoms resulting from the splitting, or fission, of a large atom such as uranium or plutonium.

dead zone A polluted area around a mine in which no vegetation or animal life can survive.

debt bomb The more than $1.5 trillion that the poor, developing countries (generally found in the Southern Hemisphere) owe the developed countries; sometimes referred to as the Third World debt crisis.

debt-for-nature swaps In a typical debt-for-nature swap, an environmental organization pays a percentage of a debt to a lender on behalf of a poor country and in exchange, the country carries out a specified environmental program, such as a reforestation program or setting aside land for a preserve.

decentralization The movement away from large, centralized sources of power and production.

decibel (dB) scale A scale used to measure the loudness of sounds.

decommission To take out of service, dismantle, and dispose of a nuclear power plant.

deep ecology A philosophical outlook that takes a holistic and nonanthropocentric perspective, and rejects the human versus environment/nature dichotomy.

deep-well injection A method of disposing of liquid wastes, such as industrial or hazardous wastes, wherein they are pumped or injected down wells deep below the Earth's surface.

deficit areas (water) Areas that receive less precipitation than is needed by well-established vegetation.

deforestation The removal of forest cover from an area.

Delaney Clause A controversial provision, passed in 1958, of the Food, Drug and Cosmetics Act that explicitly prohibits even minuscule amounts of pesticides and other chemicals in processed foods if the chemicals are found to harm laboratory animals. Critics of the clause argue that the costs of regulating and removing such minuscule amounts of chemicals (which may cause little, if any, harm in humans) far outweigh any potential benefits.

delisted species Species that are removed from the endangered species list; a species can be delisted either because it is no longer endangered or threatened, or because it has gone extinct.

dematerialization The reduction of the size of products, particularly as a way to conserve mineral resources.

demographic transition Essentially, the concept or theory that as a nation undergoes technological and economic development, its population growth rate will decrease.

demography The study of the size, growth, density, distribution, and other characteristics of human populations.

denitrification The chemical reduction of nitrate and nitrite (such as by certain types of bacteria) into gaseous nitrogen; in many respects the opposite of nitrogen fixation.

density-dependent regulation Biological interactions, such as competition and predation, acting to control the abundance of a population of organisms.

density-independent regulation Physical processes, such as droughts or volcanic eruptions, acting to control the abundance of a population of organisms.

deontology The study of moral obligations and duties.

depth diversity gradient The concept that among aquatic communities species richness generally increases with water depth down to about 6560 feet (2000 m) and then declines with farther depth.

desert pavement The resulting soil composition that exists after wind and other erosional factors have removed smaller soil matter such as sand and silt, and what is left are rocks and pebbles that pack together tightly to form a pavement-like surface.

descriptive ethics The cataloging and description of the various ethical approaches that have been used by humans.

desertification The spread of desertlike conditions due to human exploitation and misuse of the land.

developer pays An approach where a city requires that a developer of a residential subdivision pay for the required infrastructure of roads, sewer lines, utility lines, and connections to other services that are often covered by the local government and thus ultimately the taxpayers; this is a method to discourage the growth of urban and suburban sprawl and instead use land, such as brownfields, that is already available more efficiently.

dilute and disperse A method of handling waste where the substance is diluted and then dispersed into the environment. This is a difficult method to justify in more modern times because of the sheer amount and toxicity level of much of the waste generated by industry and commercial sectors.

dilution The reduction in concentration of a pollutant when it is discharged into water.

diminishing returns, law of As the supply of a resource declines, increasing efforts to extract the resource produce progressively smaller relative amounts of the resource.

dioxin A group of more than seventy-five related compounds that are extremely toxic, artificially produced chemicals. Dioxins can be inadvertently synthesized in incinerators when trash and garbage are burned.

direct value The value of utilizing a particular resource in such a way that it may be depleted or destroyed—for instance, the value of logging a forest.

discharge The volume of water carried by a channel.

dirty subsidies A payment or other incentive by government or a large corporation that promotes waste, over-use of natural resources for the benefit of the government or corporate entity, or causes other harmful effects such as pollution or health problems.

discounting by distance Ignoring or not fully paying for the environmental costs of our actions on people living in another area.

discounting the future Focusing on the present and ignoring or discounting future costs of resource depletion and environmental degradation.

distillation A method of desalination whereby salt water is evaporated so as to remove the dissolved salts.

DNA Deoxyribonucleic acid, the material of which genes are composed.

dosage The amount of a substance, such as a poison, medicine, or vitamin, administered to an individual organism. Dosage is generally measured as the amount of a substance administered to an organism divided by the body weight of the organism.

doubling time The amount of time that a population of a given size at time zero, increasing at a fixed rate, will take to double in size.

drainage basin The region drained by a particular network of rivers and streams.

drought A period of abnormally low rainfall for an extended period of time in a particular area.

dump A place where trash, garbage, and other waste are piled—often in a relatively out-of-the-way area such as on the outskirts of a town.

durable goods Material goods or products that are designed to last a relatively long period of time.

duty A moral or required action or obligation.

Earth Summit The United Nations Conference on Environment and Development (UNCED) held an international meeting in Rio de Janeiro in June 1992 to discuss environmental and development issues; delegations from over 175 countries, including more than a hundred heads of state, attended.

earthquake Shock waves that originate when large masses of rocks, generally located below the surface of the Earth, suddenly move relative to each other.

ecocentric Ecological-centered.

ecofeminism Analyzes societal and cultural traits that may have led to the degradation of the environment; specifically concentrates on the oppression of women by a traditional patriarchal society as a major form of social domination that has precipitated other forms of exploitation and domination, including exploitation of the environment.

ecological extinction Occurs when a species, although not totally extinct in an area, has become so rare that it has essentially no role or impact on its ecosystem.

ecological footprint The amount of ecological impact caused by an individual, group, corporation, town, nation, and so forth (both micro- and macro-). Measuring footprints is a tool to determine resource use with regard to environmental sustainability.

ecological release The population of a particular species increases greatly in size when a competitor is removed.

ecological succession The successive groups of plants and animals that will colonize a newly cleared patch of land or uncolonized body of water.

ecology The study of how organisms interact with each other and their environment.

ecology, first law of Garrett Hardin's concept that "we can never do merely one thing"; sometimes referred to as the "law of unintended consequences."

economic take-off The point at which a poor country can presumably produce enough wealth to financially support itself, export manufactured goods, and eventually attain the living standards of a wealthy, industrialized country.

economic value Directly tangible and monetary value of a resource, such as when a resource is bought or sold, perhaps for food, materials, or energy.

economics The study of the production, distribution, and consumption of goods and services; in an environmental context, economics is often defined as the study of scarcity and how humans can cope with scarcity.

ecoscam *See* greenwashing.

ecosystem A biological community plus the surrounding physical environment.

ecosystem services The processes by which Earth produces resources often taken for granted by humans; these include clean water, timber, and habitat for game animals and fish, as well as pollination of native and agricultural plants. Ecosystem services data are used to determine the true cost of utilizing natural resources.

ecosystem simplification Occurs when the number of species in an ecosystem declines.

ecoterrorism Committing terrorist-type acts, such as sabotage against property or people, in the defense of the environment and environmental issues and ideals.

ecotone A sharp boundary between adjacent biological communities.

ecotourism Responsible travel to natural areas, often to see wild flora and fauna, that conserves the local environment and supports the local people.

ecotoxicology The study of how chemicals affect entire ecosystems.

edge effects Disturbances from the surrounding area (perhaps dogs, housecats, humans, wind, temperature changes, pollution, and so forth) penetrate along the edges of a preserved area, resulting in habitat loss.

efficiency The useful work that is performed relative to the total energy input to a system.

efficiency improvements Innovations that reduce the flow of throughput by decreasing the per capita resource use.

effluent charges (water) Taxes or fees levied on the discharge of industrial wastewater.

effluent taxes Taxes charged on the basis of how much pollution (effluent) is discharged by a company, industry, or operation.

electricity An electric current or flow of electrons through a conductor.

electromagnetic field The physical influence, reach, strength, or "field" of various types of radiation, including visible, gamma, microwave, and so forth. Electromagnetic fields are given off, for instance, by modern power lines.

electrostatic precipitator A device that uses a high voltage to remove charged particulates from polluted air (such as the smoke coming out of a smokestack).

element A fundamental substance that cannot be broken down further into other elements by standard chemical means. Gold, iron, hydrogen, and oxygen are examples of elements.

embodied energy The energy used in producing a product.

emissions offsets The concept that in order to receive a construction permit to build a factory or other industrial operation that may be a new source of air pollution in a particular area, ways must be found to reduce emissions from existing sources in that area.

emissions permit An administrative method of providing economic incentives to reduce pollution emissions. A certain amount of pollutants are allowed into the environment, and industry trades the "rights" to pollute in the form of permits. This type of incentive leaves allocation up to the free market after the initial regulation.

emotional value Emotional bonds by a human or humans to a resource; the value of a resource beyond "practical value" or simple sensory enjoyment.

Endangered Species Act of 1973 An act that directs the U.S. Fish and Wildlife Service to maintain a list of species that are endangered (in immediate danger of extinction) or threatened (likely to be endangered soon).

endemic species A very localized species that inhabits only a relatively small area.

energy The ability to do work.

energy budget The Earth's energy budget is, collectively, all of the various flow pathways of all energy on Earth.

energy conservation Decreasing the demand for energy.

energy efficiency The usable output per unit of energy.

energy farm A farm that produces biomass to be used as an energy source.

energy intensity A measure of the amount of energy required to create a good or service (e.g., a car, an acre of agricultural land, etc.).

energy intensity index The energy consumption per gross national product of a nation.

energy minerals The fossil fuels (oil, coal, and natural gas) and uranium ore.

energy storage Storing energy in a form that is readily accessible to humans.

enriched Referring to uranium, see enrichment.

enrichment Increasing the percentage of fissionable uranium above that found in natural uranium as it is mined and processed.

entropy The amount of low-quality energy, or the amount of disorder and randomness, in a system.

environment In the broadest sense, all aspects of the natural environment plus human manipulations and additions to the natural environment.

Environmental Impact Statement (EIS) A key part of the National Environmental Policy Act, anyone proposing action on public lands, such as logging or construction of roads or buildings, must first perform a study that documents the project, reasonable alternatives to the proposed project, an overview of the natural environment that will be affected, and the environmental consequences of the project. Most states also require environmental impact statements at the state level.

environmental economics Economics with an emphasis on the study of the integration of jobs and environmental preservation.

environmental equity Treating all persons, regardless of color, creed, or social status, equally when developing environmental policies and enforcing environmental laws and regulations.

environmental externalities The actual cost of many goods and services does not factor in natural capital. Resources are usually considered free, although they provide many benefits to humans and the economy such as healthy air, clean water, and so forth. The use of resources has unintended (external, often indirect) costs that are not factored into the price of a product; for instance, the price of a gallon of gasoline does not necessarily include the cost of the damage, such as in terms of air pollution and global climate change, that burning the gasoline entails.

environmental hormones Pollutants that send false signals to the complex hormonal systems that regulate reproduction, immunity, behavior, and growth of organisms.

environmental justice The concept of implementing environmental equity and also reducing the environmental risks to all people.

environmental law Laws, statutes, regulations, treaties, agreements, declarations, resolutions, and the like that have bearing on environmental issues.

environmental racism Racial discrimination in environmental policy making and the racially unequal enforcement of environmental laws and regulations.

environmental science The systematic study of all aspects of the environment and their interactions.

environmental service Values of a resource in providing services that allow humans to exist on Earth, such as the production of atmospheric oxygen by photosynthetic plants.

environmental wisdom The ability to sort through facts and information about the environment and make correct decisions and plan long-term strategies.

epidemiological statistics Statistics used to examine correlations between environmental factors and the health of humans and other organisms.

erosion The deterioration and weathering away of soil or rock.

esthetic (aesthetic) value Value of a resource in making the world more beautiful, more appealing to the senses, and generally more pleasant.

estuaries Transitional ecosystems between ocean and freshwater biomes.

ethanol Ethyl alcohol, commonly used biomass-derived fuel.

ethical value Value of a resource unto itself, regardless of its value to humans.

ethics The philosophical study of moral values.

eukaryote A cell with a true nucleus, true chromosomes, and various specialized cellular organelles.

eutrophication A rapid increase in algae or plant growth in an aquatic system due to the influx of a limiting nutrient that was in short supply previously.

evaporite A mineral deposit formed when salt ions precipitate out of a natural body of water.

evapotranspiration The transfer of water into the atmosphere by evaporation and transpiration (the release of water vapor by plants).

executive branch The branch of the government charged with administering and enforcing the laws; the federal executive branch consists of the president, the vice president, and the president's appointees.

exotic species A nonnative species that is artificially introduced to an area.

explosion of life A phrase sometimes used to refer to the rapid diversification of life on Earth about 570 to 540 million years ago. *See* Cambrian explosion.

exponential growth The relatively rapid ("geometric") growth phase experienced by many populations during their history. Such growth can be expressed, at least approximately, using an exponential equation or curve.

exponential phase Period or phase of exponential growth in a population. *See* exponential growth.

exponential reserve How long a resource reserve, such as a mineral reserve, will last if demand changes (specifically, if it increases) over time.

externality Environmental, social, and other costs of production that are not included in the price of the product that causes the costs. *See also* market failure.

extinction The loss or death of a group of organisms.

extinction vortex Even if some individuals of a species or population survive disturbances, the population may never fully recover if it becomes too small. The species is said to fall into an extinction vortex and is doomed to eventual extinction.

extirpation The extinction of a species or other group of organisms in a particular local area.

extrinsic values Values that are external to a resource's own right to exist; values based on the ability of a resource to provide something valued by humans.

fallacy of enlightenment The idea or notion that education of people alone, without any other action, will solve overpopulation and other associated environmental problems of the world.

false dichotomy When a complex question inappropriately becomes polarized into an "either-or" issue (for instance, jobs or the environment).

Federal Insecticide, Fungicide, and Rodenticide Act (FIFRA) A law that requires anyone wishing to manufacture a pesticide to register that product with the EPA.

ferrous Referring to iron and related metals that are commonly alloyed with iron, such as nickel and chromium.

fertile isotope An isotope, such as U-238, that can absorb a neutron and form a new element/isotope (such as Pu-239) that is readily fissionable.

fertility rate The number of births in a population. The general fertility rate is the total number of births in a population in any given year as a function of the number of women in their reproductive years. The total fertility rate is the number of children a woman in a given population will have, on average, during her childbearing years.

fertilizer A substance, often an artificial chemical mixture, that is spread on or through the soil to make it more fertile.

filtration The percolation of water through sand and other settled sediment to remove suspended particles.

first law of thermodynamics Often referred to as the Law of Conservation of Energy because energy can move through systems but can never be created nor destroyed.

fission The splitting of an atom, such as uranium or plutonium, to release energy.

fissionable atom An atom that is easily split by neutron penetration, such as U-235.

five e's The five potential values of environmental resources: esthetic (aesthetic), emotional, economic, environmental services, and ethical.

flat-plate collector A device that usually consists of a black metal plate that absorbs heat from the Sun; the heat can be transferred to a liquid and then used as desired (for instance, to heat a building).

flood A high flow of water that overruns its normal confinement area and covers land that is usually dry.

fluidized bed combustion A way to reduce air pollution by burning very small coal particles at very high temperatures in the presence of limestone particles (the limestone helps to capture sulfur and other pollutants).

flywheel energy storage system Uses a rapidly rotating flywheel to store energy.

food web A graphic depiction of the interrelationships by which organisms consume other organisms.

Food, Drug and Cosmetics Act A U.S. federal act (1938) that authorizes the Food and Drug Administration (FDA) to test foods, drugs, and cosmetics to see if they are safe to market for human use.

fossil fuels Coal, oil, natural gas, and related organic materials that have formed over geologic time.

frontier mentality The general concept that humans are separate from and above nature, that humans need to subdue the uninhabited wilderness, and that there are always plenty of fresh natural resources to exploit on the edge of the frontier.

fuel cell In a fuel cell electrons are removed from hydrogen atoms to form an electric current; the hydrogen ions combine with oxygen to form water.

fusion The combining or fusing of isotopes of light elements to form a heavier element—in the process, energy may be released.

future costs Environmental costs of a product or service that are not paid now, but rather are passed on to future generations.

Gaia hypothesis The hypothesis that the Earth is similar to an organism and its component parts are integrated analogously to the cells and organs in a living body.

gamma rays A type of radiation consisting of short-wavelength, high-energy electromagnetic radiation.

garbage "Wet" and generally edible (perhaps by pigs or other animals) discarded matter, such as old food remains, yard clippings, dead animals, leftovers from meat packing operations and butcher shops, and so forth.

General Circulation Models (GCMs) Complex mathematical models that, with the help of supercomputers, simulate the Earth's climatic patterns.

generator A machine that converts mechanical energy (rotational energy) into electrical energy.

gene The basic unit of heredity.

genetic patent rights The patenting of genetic resources, often advocated as a way to allow native peoples to receive profits derived from pharmaceuticals, agriculture, and other uses of genetic resources found in their territories.

genetically modified (GM) A specific type of technology that is used to alter the genetic makeup of living organisms. This process involves combining genes from various organisms to gather the best "traits" and create a more commercially desirable organism. This technology is subject to debate, although often used on agricultural products.

Genuine Progress Indicator (GPI) An alternative to Gross Domestic Product as a way to measure the well-being of a country. Beyond pure economic productivity and growth, the GPI factors in environmental externalities and various other factors affecting human happiness. The GPI can be viewed as a long-term measure of the health of the economy.

geothermal energy Energy (heat) originating from deep within the Earth.

global dimming A gradual reduction in the amount of solar radiation that penetrates Earth's surface due to the increased reflectivity of clouds and pollution that results from fossil fuel emissions. Global dimming may create a global cooling effect that has partially offset the effects of greenhouse gases on global warming.

grain Cereals and similar plants that are used for human or animal consumption, such as wheat, corn (maize), rice, barley, and so forth.

grain production The growing, or production, of grains such as rice, wheat, maize (corn), sorghum, barley, oats, rye, millet, and so on.

grassroots activism Local participation in environmental issues.

gray water Untreated or partially treated wastewater that is used for such purposes as watering golf courses and lawns or flushing toilets (rather than using cleaner water of drinkable quality).

Green Movement A political and social movement of the late twentieth century and twenty-first century that addresses the root causes of environmental degradation and generally advocates a revolutionary break from the past and the establishment of new, Earth-amenable social institutions.

Green Revolution Modern, chemically based, usually mechanized agriculture that was first used on a large scale in the industrialized countries after World War II.

green fees Fees or taxes (such as fees that increase the price of a resource) that are applied to generally reduce throughput (reduce the use of a resource and/or to reduce the production of pollutants).

green investing Investing in environmentally friendly and sustainable businesses and industries.

green taxes *See* green fees.

green technologies Environmentally friendly technologies, including technologies that promote sustainability via efficiency improvements, reuse/recycling, and substitution.

greenbelt zones Zones or areas in or around a city where the removal of native vegetation is prohibited and/or parks and other open, undeveloped, and vegetated space is protected.

greenhouse effect The warming up of the lower atmosphere due to the accumulation of greenhouse gases that trap heat near the surface of the Earth.

greenhouse gases Gases, such as carbon dioxide, methane, and CFCs, that are relatively transparent to the higher-energy sunlight, but trap lower-energy infrared radiation. Greenhouse gases that accumulate in the atmosphere promote global warming.

greenmail When companies threaten to close or relocate (often to another country) if they are forced to comply with environmental laws; also known as job blackmail.

greenwashing Marketing that unscrupulously seeks to profit from environmental concerns; also known as ecoscams.

Gross Domestic Product (GDP) One of several means of measuring the economic activity of a country, which includes all the goods and services produced within a given year. This index does not factor in environmental externalities or true measures of human benefits beyond dollar amounts.

Gross National Product (GNP) A standard index that measures a nation's output of goods and services; GNP differs from GDP (see above) in that it includes economic activity by all nationals, including economic activity that takes place abroad.

groundwater A general term for the water beneath the Earth's surface.

growth rate The rate at which a population is increasing (or decreasing, in the case of a negative growth rate) in size. *See also* percent annual growth.

habitat The general place or physical environment in which a population lives.

habitat conservation plan A plan where habitat of an endangered species is preserved (in some cases, a portion of the habitat may be destroyed but the remainder is protected).

habitat fragmentation Habitat disruption where natural habitat is broken into small, relatively isolated fragments.

hard laws Statutes or legal instruments that are legally binding and mandatory.

hard technologies Energy technologies that depend on large scale, centralized, complex, and expensive plants and infrastructures.

hardness The amount of calcium, magnesium, and certain other ions in water; "hard water" contains high amounts of these ions.

hazardous waste Wastes that are particularly dangerous or destructive; specifically characterized by one or more of the following properties: ignitable, corrosive, reactive, or toxic.

Healthy Forests Initiative A 2003 federal initiative that calls for the thinning of forests to prevent wildfires and maintain the health of forest ecosystems.

heat Infrared radiation. *See* infrared radiation.

heavy metals Elements, such as lead, mercury, zinc, copper, cadmium, and so forth, that may be required in trace amounts by organisms, but can cause damage when ingested in larger quantities (such as binding with enzymes and thus impairing their functions).

herbicide A chemical substance used to kill plant weeds.

hidden costs Environmental costs of a product or service that we are initially unaware of, such as future environmental degradation from the full effects of pesticides or other pollutants.

holistic Seeking connections among all aspects of a problem, concern, or issue.

hot dry rock technology A technology used to literally "mine" the heat coming off of rocks buried beneath the surface of the Earth. Through this process, water is injected and pumped through rock and then circulated back up. The water hitting the rock creates steam as it flows through the rock, which can then be used for energy.

hot spot An area of exceptionally high species richness, especially of concentrations of localized rare species that occur nowhere else.

Hubbert's bubble In the 1950s the geologist M. King Hubbert accurately predicted that U.S. oil production would peak around 1970 and subsequently decline.

Human Suffering Index (HSI) An index developed by the Population Crisis Committee in the 1980s. This index is a summation of a number of different ratings of a country, such as GNP per capita, food sufficiency, inflation, accessibility of clean drinking water, literacy, energy consumption, growth of the labor force, urbanization, and political freedom.

hunger Not having enough food to sustain oneself adequately; a weakened condition that is brought on by the lack of food.

hybrid engine This type of engine is typically smaller than the traditional internal combustion engine and uses a rechargeable energy storage system (RESS) to propel the vehicle. These vehicles are lower users of fossil fuels and are much more efficient. The term usually refers to a hybrid of electric and fossil fuel power.

hydroelectric power *See* hydropower.

hydrologic cycle Movement of water about the surface of the Earth, driven by energy from the Sun.

hydropower The use of artificial or natural waterfalls to generate electricity.

hydrosphere The liquid water sphere or "layer" on Earth; it includes the oceans, rivers, lakes, and so on.

hydrothermal fluid reservoir An area where hot rock occurs at relatively shallow depth and natural groundwater is heated, sometimes to extremely high temperatures.

hydrothermal processes Hot water dissolves, transports, and subsequently reprecipitates and concentrates elements and minerals into deposits.

ice ages Intervals in the history of the Earth, especially during the last two million years, when average global surface temperatures were lower than they are currently and continental ice sheets were much more extensive than they are today.

igneous rock A rock formed or crystallized from molten rock (magma).

ignitable A substance that will easily ignite and burn rapidly; a characteristic of some types of hazardous waste.

immigration The movement of people into a country or, more generally, the movement of individuals into a particular population.

immunotoxicology The examination of how chemicals affect the immune system.

incineration The burning of trash and garbage at high temperatures in a large furnace so as to get rid of as much as possible.

independent power producers (IPPs) IPPs construct electricity-generating plants and then sell the electricity to the large utilities.

Index of Sustainable Economic Welfare (ISEW) An alternative to the traditional Gross National Product, the ISEW measures sustainable growth. The ISEW includes environmental measures such as depletion of nonrenewable resources, loss of farmland from erosion and urbanization, loss of wetlands, damage from air and water pollution, and so forth.

indicator species A species in a community or ecosystem that is more susceptible to disturbances than most other species.

indirect value The way that a resource may be valued other than its direct value, such as the emotional and aesthetic values of a forest.

Industrial Revolution A series of industrial and technological inventions, beginning in England in the late eighteenth and early nineteenth centuries, that resulted in the cheap and efficient mass production of many commodities.

industrial ecosystem An industrial situation that mimics the principles of ecosystems in nature. No "waste" is produced; rather, the effluents of one industrial process form the raw materials for another industrial process.

industrial materials Nonmetallic materials/minerals necessary to industry, such as salts, fertilizer components, sulfur, asbestos, abrasive minerals, and so forth.

industrial solid waste Solid wastes produced by industries, including, for instance, wastes from large-scale manufacturing, mining, resource processing, and so forth.

infant mortality rate The number of babies that die before their first birthday, given that they are born alive.

infrared radiation Low-energy, long-wavelength electromagnetic radiation that humans perceive as heat.

infrastructure Roads, bridges, buildings, electric power lines, water and sewer pipes, and other relatively durable material objects upon which modern societies depend.

inherent value An objective value or "worth" that does not require a subjective aspect; that is, it is independent of a person's subjective opinion. In a certain context the inherent value of the metal gold is independent of what the gold has been manufactured into; the gold in a raw lump weighing four grams is worth the same as the gold per se in an ancient coin weighing four grams, although the ancient coin may have added value over the inherent value of the gold it is made of. In environmental ethics, inherent value usually refers to features of nature, or nature as a whole, which are said to have inherent value for their own sake independent of human opinion. In economics, inherent values are a component of a theory of value where a good or service contains value within or unto itself.

input The materials and energy that enter the societal system and flow through society as throughput and exit the societal system as output (deposited into sinks or possibly reused or recycled in the cases of some materials, in which case the output becomes input and throughput once again); such input ultimately comes from natural resources (also known as sources). *See* output and throughput.

input reduction Reducing the flow of materials and energy through society.

intangible costs Environmental costs of a product or service that include destruction of scenery or other subjective costs that are very difficult to measure because different people place different values on them.

Integrated Pest Management (IPM) Integrated Pest Management and the concept of biological control follow the basic philosophy that the farmer should not try to totally eliminate pests, but should simply attempt to control them so that they do not cause serious damage. IPM and biological control often use "natural" controls, such as the pests' natural biological predators. IPM seeks to reduce the use of artificial chemical pesticides.

integration The strength of the interactions among the parts of a system.

Intergovernmental Panel on Climate Change (IPCC) A large, international group of officials, scientists, and other researchers who, under the auspices of the United Nations, have been investigating the issue of global climate change, particularly potential future global warming.

intermittent power source A power source, such as the Sun, that can only be used periodically (the Sun can only be used directly during daylight hours).

intrinsic factors The subtle ways that the complex biochemistries of even closely related species can differ such that a chemical that has no effect on one species may have acute or long-term effects on another species.

intrinsic rate of increase The potential for increase in a given population, often symbolized by r; in large part it is determined by the birth rate and death rate of the population.

intrinsic value The value of a resource unto itself, regardless of its value to humans; often considered the ethical value of a resource, or the right of the resource to exist.

introduction The release of species into new areas that they did not formerly (naturally) occupy, such as the introduction of African savanna species onto game preserves in Florida and Texas.

ionizing radiation Radiation that upon hitting electrically neutral atoms may cause the atoms to lose electrons and thus gain an electrical charge—that is, become ionized.

irrigation The artificial watering of land.

isotopes Atoms of the same element that differ from each other in weight because they have differing numbers of neutrons.

joule A basic unit of energy and work, equal to a force of one newton times one meter.

judicial branch The branch of a government that interprets the laws of the land; in the United States, this is the court system, and the highest federal level is the U.S. Supreme Court.

keystone species A certain species that one or more other species are dependent upon for food, reproduction, or some other basic need.

Kyoto Protocol An amendment to the United Nations Framework Convention on Climate Change, which brings together industrial and developing countries to lower their greenhouse gas emissions in a unified fashion. The United States and Australia are the only developed countries that have refused to sign the agreement. The protocol opened for signatories in 1997 and was implemented in 2005.

lag phase The relatively slow growth exhibited by many populations at the beginning of their history.

Land Trust Alliance An association of hundreds of local land trusts, dedicated to preserving open space and natural habitat, in the United States.

landfill In the simplest sense, a hole in the ground where solid waste is deposited. In a modern sanitary landfill, the hole is lined so that materials will not escape, and it is covered with layers of dirt as it is progressively filled. When completely filled, it is capped and sealed with more dirt and topsoil.

latent heat storage Heat storage based on phase changes in material objects, such as the melting of ice.

latitudinal diversity gradient The steady decrease in species richness in most groups as one moves away from the equator.

law of diminishing returns The concept that eventually there comes a time when the return on an investment in a good or service begins to diminish. For example, a diamond mine will eventually produce fewer and fewer high-quality diamonds, or the satisfaction received by consuming an item will diminish after a certain point.

law of the minimum Holds that the growth of a population is limited by the resource in shortest supply.

leachate A liquid solution that forms as water percolates through waste, such as refuse in a landfill or old mining tailings. Leachate may contain any chemicals that can be dissolved, particles, and even live microorganisms.

legal instruments Laws, treaties, regulations, conventions, and the like.

legislative branch The branch of a government that is responsible for making basic laws; in the United States, the federal legislative body is the Congress, composed of the House of Representatives and the Senate.

less developed countries (LDCs) The less rich and less industrialized countries of the world; basically, the countries that are not considered to be more developed countries. *See also* more developed countries.

Lethal Dose-50 (LD-50) The dose of a toxic substance that will kill 50% of a certain population. LD-50 has become the standard reference for summarizing the toxicity of substances.

Life Cycle Analysis (LCA) Sometimes known as cradle to grave analysis, LCA analyzes the entire "life cycle" of a product, from procurement of the raw materials, through its use, to its eventual disposal and the possible reuse or recycling of its components.

life expectancy The average number of years that a typical person at a certain age can expect to live.

life history Age of sexual maturation, age of reproduction, age of death, and other important events in an individual's lifetime, particularly as they influence reproductive traits.

light pollution Excess "waste" light given off by outside sources (or sources visible from the outside) at night.

light water reactor (LWR) A common type of commercial reactor that uses ordinary (light) water as the moderator.

lignite A soft, dark brown, coal-like material.

limiting nutrient The nutrient in shortest supply in a particular ecosystem.

lithosphere The rock sphere or layer that forms the surface of the Earth; composed of the crust and uppermost portion of the mantle.

Little Ice Age A period in Earth history, beginning in late medieval/early renaissance times (ca. A.D. 1300) and ending only two or three hundred years ago, during which average global temperatures were slightly lower than immediately before or after.

loss of life expectancy (LLE) The average amount that a life will be shortened by a particular risk under consideration.

Love Canal A site in the town of Niagara Falls, New York, that gained national attention in the late 1970s when hazardous chemicals that were buried in the area began to adversely affect the residents. The disaster of Love Canal helped to spur Congress to pass the Comprehensive Environmental Response, Compensation, and Liability Act of 1980.

magmatic processes Processes during which certain minerals may selectively crystallize out of a hot, molten body of rock.

malnutrition The condition of a human or other organism where the proper amount of energy and nutrients is not maintained through intake of calories, protein, and various vitamins and minerals. The amount of nutrients required for each person depends on age, sex, body build, and other characteristics.

manifest destiny The 19th century nationalistic belief, not a specific policy, that the United States had a mission to expand, spreading democracy and freedom and also conquering territory and resources from the Atlantic Ocean to the Pacific Ocean.

mantle The thick, at least partially molten, layer of rock found between the Earth's core and crust.

mariculture Saltwater seafood farming. *See also* aquaculture.

marine protected areas Areas of ocean that are set aside as preserves for marine life.

market economy A national economy in which production and costs are determined by competition among individuals. *See also* command economy for comparison.

market failure When market prices do not reflect all the true costs of a product or service (often referred to as an "externality").

mass extinction A catastrophic event in Earth history that kills large numbers of species. Some mass extinctions killed more than 60% of all living species on Earth at the time of the extinction.

maximum contaminant levels (MCLs) The highest concentrations allowed by the EPA in water designated for certain uses.

maximum sustainable yield (MSY) The concept that the optimum way to exploit a renewable resource is to harvest as much as possible up to the point where the harvest rate equals the renewal rate.

megacity A large city, generally with a population of more than 10 million.

megalopolis A single vast urban area formed by the expansion and merging of adjacent cities and their suburbs.

megawatt (MW) A million watts.

melanoma A condition of malignant skin cancer.

meltdown A major accident at a nuclear power plant where the fuel assembly (core) is heated beyond its melting point.

membrane method A way of producing freshwater from saltwater by forcing the saltwater through a fine membrane; also known as the filter or reverse-osmosis method.

Mercalli Scale The Mercalli Intensity Scale is a rating scale used by seismologists to measure earthquake intensity and effects on a scale from I on the low end to XII on the high end. Much of the rating information is taken from subjective data, usually by people who experienced the effects of the earthquake.

metallic minerals Minerals containing significant (and often minable) amounts of iron, aluminum, copper, zinc, lead, gold, silver, and so forth.

metamorphic rock A rock formed by modifying a preexisting rock, usually through high temperatures and/or pressures.

methane Natural gas, CH_4, a fossil fuel and potent greenhouse gas.

methane gas The principal component of natural gas, methane is a widely distributed chemical in nature that is odorless and colorless. Methane is usually mixed with other stronger smelling compounds (sometimes simply as a safety feature, so that gas leaks can be easily detected) to produce commercial fuels, and methane is a greenhouse gas that contributes to global warming, some of which comes from the flatulence and belching discharge of factory farm animals.

methanol Methyl alcohol; a fuel that can replace gasoline in many situations and can be produced from natural gas, coal, or biomass.

microirrigation Method of irrigation in which water is transported to crops through pipes and then dripped onto the plants through tiny holes in the pipes, which are installed on or below the surface of the soil; sometimes called drip irrigation.

microloan A small loan made to a small business that promotes sustainable activities.

milling process The process, including crushing, grinding, and leaching, during which a mineral, such as copper or uranium oxide (yellowcake), is removed from the raw ore and concentrated.

mineral A naturally occurring inorganic solid that has a regular crystalline internal structure and composition—for instance, the mineral quartz.

mineral deposit An area where a certain mineral has been concentrated by natural processes.

mineral resources Minerals and earth materials (sometimes including the energy minerals, such as fossil fuels and uranium ore) that form natural resources from which humans draw.

minimum viable population The smallest population size (for a certain population or species) that can stay above the extinction vortex.

mitigation When humans replace a natural ecosystem that was destroyed with a new ecosystem.

mixed economies A term used to identify economies which stray from the more conventional notions of free market or planned economies, and instead include a mixture of both. This model includes some central planning, public and private freedom, and some mandates and intervention by environmentalists and social justice sectors.

mobilization bias Organizations, institutions, groups, and individuals mobilize according to what will bring certain rewards that are valued by the culture, group, or institution, such as money, success, power, and so forth, and reflect the current value structure of the society.

moderator A substance, such as water, graphite, or beryllium, used in a nuclear reactor to slow down fast neutrons.

molecular toxicology The examination of the interaction of toxic chemicals with cellular enzymes.

monoculture A form of agriculture where only a single species is grown in a particular field, such as a field devoted entirely to wheat.

Montreal Protocol An agreement reached in 1987 at a meeting in Montreal, Canada, whereby a number of industrialized countries pledged to freeze CFC production at 1986 levels and then gradually decrease CFC production to 50% of 1986 levels by 1999.

moral antirealists Persons who maintain that moral values are created by humans, either consciously or unconsciously.

moral realists Persons who believe that moral values are objective and real and can be discovered by humans.

moral status The question of who, or what, is to be considered worthy of having rights (or what level of rights) in an ethical system.

moral/moral values Relating to the principles of right or wrong, good or bad behavior. Moral values involve distinguishing or discovering the principles of right and wrong and determining their relative importance and relations.

more developed countries (MDCs) The richer, industrialized countries of the world; basically, the countries of North America, Europe, Japan, Australia, New Zealand, and the former Soviet Union.

Mount Pinatubo The site of a 1991 volcanic eruption in the Philippines that spewed so much smoke and ash into the atmosphere that it temporarily depressed global temperatures (had a minor cooling effect that was experienced worldwide for several years).

muddling along *See* bootstrapping.

multiple-use development The use of a piece of land for different purposes simultaneously, such as the use of riverside land for water filtration by plants, recreation, flood buffers, and wildlife habitat.

multiple-use philosophy/principle When land is put to many uses at the same time, such as logging, mining, grazing, farming, oil exploration, hunting, fishing, and so forth.

multistage flash distillation (MSF) A method of distillation used to desalinate seawater; cold seawater is run through a series of coils in chambers that become progressively hotter.

municipal solid waste The solid waste produced by the residents and businesses of a city, town, or other municipality; includes old newspapers, packaging materials, empty bottles, leftover foods, leaves and grass clippings, and so forth.

mutagen A substance that causes genetic mutations in sperm or egg cells.

mutation A change in a gene, ultimately caused by a change in the DNA sequence.

mutualism A form of symbiosis that benefits both species.

National Environmental Policy Act (NEPA) A federal act (1969) requiring that all major federal actions, including activities on federal public lands, be reviewed for their environmental impacts before such actions take place. *See* Environmental Impact Statement.

National Marine Sanctuary A marine area that is protected under the auspices of the U.S. government.

natural environment The physical and biological environments independent of human technological intervention.

natural gas A term used for fossil fuels in the gaseous state, particularly methane.

natural hazard An "unpredictable" natural event, such as an earthquake, volcanic eruption, flood, avalanche, drought, fire, tornado, hurricane, and so forth.

natural selection The basic mechanism of evolutionary change, first formulated by Charles Darwin. Biological populations of organisms exhibit variations among the individuals, and the individuals with more advantageous traits tend to contribute more offspring (bearing the advantageous traits) to the next generation. Thus, natural selection determines which individuals will contribute most to the next generation.

natural siting criteria An approach where the locations (sites or sitings) of roads, buildings, and other structures are selected to be where the geological and biological factors are most favorable; essentially, "working with nature" when selecting locations for human-made structures.

Nature Conservancy Founded in 1951, the Nature Conservancy is an organization that preserves natural areas and wildlife sanctuaries around the world.

neo-Malthusian A term often used to refer to persons who believe that the modern rapid increase in the human population is extremely detrimental.

net primary productivity (NPP) The rate at which producer or primary, usually plant, biomass is created.

net secondary productivity (NSP) The rate at which consumer and decomposer, or secondary, biomass is created.

net yield The concept of net yield for nonrenewable resources holds that a resource can continue to be extracted as long as the resources used in extraction do not exceed the resources gained.

neutron A subatomic particle that has approximately the same mass as a proton, but does not bear an electric charge.

niche An organism's "occupation," or how it lives.

Nimbus-7 A NASA satellite that confirmed the depletion of the ozone layer over the South Pole in the 1980s.

nitrogen cycle The biogeochemical cycle of nitrogen.

nitrogen fixing The conversion of atmospheric nitrogen, usually by nitrogen-fixing bacteria, into forms such as ammonia that are more chemically reactive than atmospheric nitrogen and can also take a nongaseous form under various conditions found on the surface of Earth.

nitrogen oxides NO_x, important components of both lower atmospheric pollution and the upper atmospheric greenhouse gases that promote global warming.

noise Unwanted sound.

nonferrous Metals such as gold, copper, silver, and so forth that are not commonly alloyed with iron. *See also* ferrous.

nonhazardous waste Waste that is not classified as hazardous waste. *See* hazardous waste.

nonmetallic minerals Structural materials, such as sand, gravel, and building stone, and nonmetallic industrial minerals, such as salts, sulfur, fertilizer components, abrasives, gemstones, and so forth.

nonpoint source Pollution discharge that is not easily traced to a single source or specific location, but rather discharges from many locations. Nonpoint source pollution is more dispersed and therefore more difficult to regulate than point source pollution. *See also* point source.

nonrenewable resource A resource, such as fossil fuels, that does not significantly regenerate itself on a human time scale.

normative ethics Drawing upon a certain ethical system, normative ethics renders ethical opinions and judgments of actions and prescribes appropriate behavior.

nuclear power The use of nuclear fission reactions to generate electricity.

nuclear proliferation The spread of atomic weapons.

ocean A large body of saltwater on the Earth's surface, such as the Atlantic or Pacific Ocean; also refers to the entire body of saltwater that covers much of the surface of our planet.

ocean energy Waves, tides, differential heat layers, and other sources of energy directly related to the world's oceans.

ocean thermal energy conversion (OTEC) A proposed way of extracting usable energy from the oceans based on the fact that in tropical areas the surface waters are warmer than the waters at depth.

oceanic crust The layer of rock that underlies the ocean basins, generally containing a lower percentage of silicon dioxide than continental crust. *See* crust.

oil Petroleum and related liquid fossil fuels.

oil shale A rock that contains the organic precursors of oil known as kerogen. Kerogen can be converted to oil by heating.

open system A system that is not isolated in that it exchanges matter and/or energy with other systems.

open-loop recycling A situation where a material or substance can be recycled once or a few times, but not indefinitely because the material is damaged or degraded each time it is recycled (for instance, paper fibers or some plastics).

openness Refers to whether a system is isolated from other systems.

optimum sustainable yield (OSY) The concept that the optimum harvestable rate for a renewable resource, such as a wild population of fish, must consider factors beyond simple maximum yield, for instance, effects on other species in the ecosystem and various human uses of the ecosystem. *See* maximum sustainable yield.

ore deposit A mineral deposit that can be economically mined at a certain time and place with a certain technology.

organic farming Avoiding the use of synthetic chemicals, such as synthetic or artificial fertilizers, pesticides, and herbicides, when farming; many organic farmers also avoid the use of genetically modified or bioengineered organisms.

organophosphates A group of synthetic organic pesticides that are less persistent, but more toxic to humans, than chlorinated hydrocarbons; includes malathion and parathion.

output Materials and energy that flow out of the societal system. *See* input and throughput.

overshoot When a population exceeds its carrying capacity.

oxygenation A method of reformulating gasoline by blending oxygen-rich liquids into it such that carbon monoxide will be converted to carbon dioxide and volatile organic compounds will be better oxidized.

ozonation The use of the reactive chemical ozone (O_3) to disinfect water.

ozone An O_3 molecule. Ozone contributes to air pollution in the troposphere, but is an important natural component of the stratosphere. The stratospheric ozone layer protects the Earth's surface from excessive levels of ultraviolet radiation.

ozone layer A layer of ozone in the stratosphere, most concentrated at an altitude between about 12 and 16 miles (20–25 km).

Pangaea The name given to the landmass formed when all of the continental landmasses were joined into a single supercontinent around 240 million years ago. *See* plate tectonics.

parasitism Occurs where one species (the parasite) lives off another species (the host) and may actively harm the host; often considered a form of symbiosis.

particulates Small particles of dispersed matter, in either a solid or a liquid state, that are larger than individual molecules and are one of the categories of air pollution.

passive solar design A type of architecture that uses the inherent characteristics of a building to capture heat and light from the Sun.

peak load The amount of electricity needed at the time of highest demand.

peat A thick accumulation of partially decayed plant material.

pegmatite A rock type formed from residual magma that is often characterized by very large crystals that may contain high concentrations of otherwise relatively rare elements.

pelagic Refers to the water column zone of the aquatic (specifically, marine) biome.

percent annual growth The rate of natural increase expressed as a percentage of the given population.

persistence How long a pollutant stays in the environment in unmodified form.

pesticide A chemical substance used to destroy animal pests, such as insects, that might attack a crop. More generally, a pesticide can be any chemical manufactured to kill any organisms that humans consider undesirable.

pesticide treadmill The concept that to keep up with the rapid pace of evolving pesticide immunity, new pesticides must constantly be developed.

petajoule 10^{15} joules.

Petkau effect The general concept that sustained low doses of radiation may be more damaging to organisms than single large doses. More specifically, the Petkau effect documents how cell membranes are damaged by low but sustained doses of radiation.

petroleum Essentially, oil or liquid fossil fuel.

pH scale A scale that is used to measure acidity; 1 is very acidic, 7 is neutral, and 14 is very basic (alkaline).

pheromones Biochemical scents used by males or females to attract the opposite sex.

philosophical ethics The consideration of ethical controversies and underlying values, principles, and rules at an abstract level.

photic zone The upper part of the aquatic biome where light can penetrate the water.

photochemical pollutant A pollutant produced when sunlight initiates chemical reactions among NO_x, volatile organic compounds, and other substances found in the air.

photosynthesis The process by which organisms such as green plants convert light energy to chemical energy and synthesize organic compounds from water and carbon dioxide.

photovoltaics The use of semiconductor technology to generate electricity directly from sunlight.

Physical Quality of Life Index (PQLI) An index developed by the Overseas Development Council in the late 1970s that rates a country on the basis of such factors as the average life expectancy, infant mortality, and literacy rates of its citizens.

physical environment The natural physical nonliving world, including the lithosphere, hydrosphere, and atmosphere.

pioneer community The initial community of colonizing species in a particular area.

pioneer stage of succession A natural process that follows a set of steps leading to a climax in ecosystems. The pioneer stage represents the initial set of organisms that colonize the area and begin the process of succession. The pioneer organisms are often well evolved to deal with harsher climates, such as greater exposure to elements like wind and solar radiation.

pioneer stage *See* pioneer community.

placer deposit Heavy mineral grains that have weathered from rocks, been transported, sorted, and finally settled in an ore deposit, such as gold nuggets in the bottom of a stream.

plate tectonics The concept or theory that the Earth's lithosphere is divided into numerous plates that are in motion relative to each other; the continents ride on the backs of the plates and thus move ("drift") over geologic time. Plate tectonics accounts for the distribution of most earthquakes, volcanoes, and other important geological phenomena.

plutonium A heavy element that contains 94 protons. Fissionable isotopes of plutonium can be used as fuel in nuclear reactors and can also be manufactured into bombs.

point-of-use treatment A method that treats water pollution/contamination at the point of use, such as through individual consumer filters on faucets or similar methods.

point source Discharge of pollutants from easily identifiable single locations, such as a pipe dumping waste into a river or a smokestack emitting waste into the atmosphere. Industrial and domestic wastes are often released from point sources.

Pollutant Standards Index (PSI) An index that measures air quality, ranging from 0 (best air quality) to over 400 (worst air quality).

polluted water Water that is rendered unusable for its intended purpose.

pollution A term often used to refer to excess outputs by society into the environment. Depending on how the term is used, pollution can also refer to excess outputs by natural processes (such as "pollution" caused by a volcanic eruption).

pollution permits A program suggested in 1990 and implemented by the first Bush administration in 1994 as a market-based solution to trade "permits" based on a certain quota of emissions. Essentially, the government will set a cap on a certain pollutant and then allow a certain number of permits based on that number, which companies are allowed to trade according to supply and demand.

population A group of individuals of the same species living in the same area.

population bomb A term used to refer to the fact that the vast majority of the world's population growth is currently taking place in the Southern Hemisphere.

population momentum A population may decrease its fertility rate to, or below, replacement level (generally considered 2.1 children per woman), but if the initial population is composed of many young people, as the population ages it will continue to increase in size as women reach their reproductive years and bear offspring.

positive feedback The process in which part of a system responds to change in a way that magnifies the initial change.

potable water Water that is safe to drink.

power Work (requiring energy) divided by the time period over which the work is done.

power plant A plant that converts some form of energy, for instance, chemical energy found in coal, into electrical energy.

precautionary principle The principle that advises that, in the face of uncertainty, the best course of action is to assume that a potential problem is real and should be addressed ("better safe than sorry").

precipitation Removal of water from the atmosphere as rain, snow, sleet, or hail.

precycling The reduction of packaging materials by manufacturers.

predation The process in which certain organisms kill and consume other organisms.

preservation Refers to nonuse, such as a "preserve" that is set aside and protected in its pristine natural state.

preservationism The classical view that nature should be preserved for its own sake.

preservationists A branch of the environmental movement that is committed to preservation of the natural world; preservationists believe that some land and resources should be set aside and not used by humans. The beginning of this school of thought is primarily associated with John Muir, who established the Sierra Club. The extreme preservationists, often referred to as Deep Ecologists, believe that ecosystems and biodiversity should be preserved no matter what the cost to humans.

primary pollutants Pollutants that are directly emitted, such as SO_x by coal-burning power plants and other industries.

primary standards Under the drinking water standards of the Safe Drinking Water Act, primary standards specify contaminant levels based on health-related criteria.

primary treatment The use of physical processes, especially screening and settling, to remove materials from water.

privatization When a resource, such as an environmental resource, is transferred from the commons to an individual or company that has a vested interest in the particular resource.

progressive conservationism The view that the federal government in particular should take an active role in protecting forests and other natural resources for the public; popular during the late nineteenth and early twentieth centuries.

property rights Rights associated with ownership or use of material property, including land.

protocells Cell-like structures artificially produced by heating amino acids and other organic molecules.

pumped hydroelectric storage (PHS) In PHS, electricity is used to drive pumps that transfer water from a lower reservoir to a higher one. Electricity can then be regenerated by allowing water to flow through a turbine on its way back to the lower reservoir.

quad A quadrillion BTUs.

radiation Electromagnetic radiation; includes visible light, heat, ultraviolet radiation, gamma rays, X rays, and so forth. The term radiation is also sometimes used to refer to the emission of particles (such as alpha and beta particles) from a radioactive atom.

radioactivity The emission of particles (such as alpha and beta particles) and rays (energy, such as gamma rays) from a nucleus as it disintegrates.

radionuclides Unstable atoms that undergo spontaneous disintegration and give off radiation in the process.

rate of natural increase The crude birth rate minus the crude death rate of a population.

reactive With reference to hazardous waste, materials that are very active chemically and can easily cause explosions and/or release harmful fumes.

reactor vessel In a nuclear power plant, a thick steel tank that usually contains the reactor core and primary water loop.

recharge area An area where rainfall can infiltrate into an aquifer.

recovery rate The amount of the original material that can actually be recovered and recycled.

recovery ratio The recovery rate expressed as a ratio. *See* recovery rate.

recycling Using the same resource over and over, but in modified form.

recycling loop The use of a resource, followed by the discarding and reprocessing of the resource, and then the reuse of the resource.

refuse Refers to both trash and garbage. *See also* garbage; trash.

Refuse Act Federal legislation passed in 1899 that prohibited the dumping of waste into major waterways. This act laid the foundation for future water pollution control laws.

regulatory agencies Agencies that are part of the executive branch of the government and are responsible for administering and enforcing laws enacted by Congress.

reintroduction Release of plants and animals back into habitat that they formerly occupied, but in which they may have become locally scarce or extinct.

relativists Persons who believe that there is no objective or rational way to determine right or wrong, good or bad, in an ultimate (versus relative) sense.

remediation Efforts to counteract some or all of the effects of pollution after it has been released into the environment.

renewable energy An energy source that, from an Earth perspective, is continually renewed (for example, solar energy).

renewable resource A resource that will regenerate within a human time scale; for example, crops and energy received from the Sun.

replacement level fertility The number of children needed to keep a population at a stable size; generally considered to be a total fertility rate of about 2.1 children per woman.

reprocessing facility A facility designed to reprocess spent nuclear fuel in order to recover fissionable materials.

reproductive rights Essentially, a woman's right to determine for herself if, and when, she will bear children.

reserve A resource that has been located and can be profitably extracted at the current market price. For minerals in particular, a reserve is an identified ore deposit that has yet to be exploited.

reserves This term refers to a resource that has potential to be used, but is not currently serving the economy; it can be put to future use.

residence time (a) The amount of time that a certain atom or molecule spends, on average, in a certain portion of its biogeochemical cycle. (b) The time it takes for a pollutant to move through the environment.

resource A source of raw materials used by society.

resource recovery facility A plant where trash and garbage are burned and the heat generated is recovered and applied to some useful end, such as generating electricity.

respiration Biological combustion or the "burning" of food molecules in an organism. During respiration large organic molecules are broken down into simpler organic molecules, and energy is released.

restoration The process of returning a degraded resource to its natural state.

reuse Using the same resource over and over in the same form.

Richter scale Seismologists use the logarithmic Richter as a measure of the amount of seismic energy released by an earthquake.

right Something to which one has a just, inherent, or natural claim—such as the right to life.

riparian law A law under which the owner of land has the right to withdraw water that is adjacent to the land, such as from a river or lake.

riparian zone Area along a river bank or stream bank.

risk A dangerous situation, or the chance of loss or peril to human life or other valuable entities; generally refers to any potential danger.

rock cycle The cycling of solid earth materials—rocks—from one form to another. Sedimentary rocks may be subjected to great pressures and/or temperatures and be transformed into metamorphic rocks. Metamorphic rocks may be heated to the point where they melt, forming a magma that later crystallizes as igneous rocks. Igneous rocks may be weathered and broken down into grains that form the basis of a sedimentary rock.

Roadless Area Conservation Rule Put into effect under President Bill Clinton in 2001 to prevent tens of million of acres of national forest land from road building, mining, logging, and other prospecting. The public submitted over 1.6 million comments, mostly in favor of the act. In 2005, President George W. Bush repealed the act and instead replaced it with a state management plan.

rubbish A very general term that includes trash, garbage, and other items such as construction and demolition debris.

Safe Drinking Water Act (SDWA) A federal statute enacted in 1974 that aimed explicitly at improving the quality of drinking water by establishing primary and secondary standards for contaminant levels in water.

salinization An increase in soil salt content that sometimes occurs due to prolonged irrigation, especially in poorly drained arid regions.

sanitary landfill *See* landfill.

saturation, zone of The region below the water table where all voids in the soil and rock are fully filled with water.

scaling factors In the analysis of dose-response assessment and toxic risk assessment of substances, a way of extending animal results to humans by taking differences in body sizes (masses) into account.

scarcity With reference to natural resources, such as minerals, a lack of such resources as compared to demand for the resources.

scoping When local government agencies and the general public are involved in producing an Environmental Impact Statement for a proposed project.

scrubbers Devices that cleanse emissions, usually with water, before they are released into the air.

sea level rise Worldwide rises in sea level (sea level height increasing relative to the elevations of the continents), such as has been predicted as a result of global warming.

secondary pollutants Pollutants produced by reactions among other air pollutants, such as photochemical pollutants.

Second law of thermodynamics In an isolated system that is not in equilibrium, entropy ("useless energy" or disorder) will increase over time, until reaching a maximum value.

secondary standards Under the drinking water standards of the Safe Drinking Water Act, secondary standards specify contaminant levels based on nonhealth-related criteria (such as color or odor, which are not strictly health issues).

secondary treatment The use of biological (microbial) processes to remove materials from polluted water.

sedimentary processes Natural processes that may concentrate minerals through precipitation from a solution or by differential settling of grains in moving or still water.

sedimentary rock A rock formed on or near the Earth's surface by the settling or precipitation of materials. A sedimentary rock may be composed of grains or clasts of particulate matter (usually older rocks or mineral grains), or it may be formed from the chemical or organic precipitation of minerals from an aqueous solution (such as certain limestones formed in an ocean).

sedimentation The settling out of suspended particles from a body of water (or in some cases, very fine particles settled from the air or blown by the wind).

selective cutting The harvesting of trees such that only certain trees are cut down and the land is not stripped bare; also known as uneven-aged management.

sensible heat storage Allowing a material substance to heat up and then release its heat again.

septic tank A large concrete tank buried underground to receive household sewage. Septic tanks are used in areas without a sewer system.

sequential use development When land that is no longer needed in one capacity is used in another way, as for instance when a capped landfill is converted to a park.

severance tax A tax levied on minerals or other resources as they are extracted.

shore zone Transitional zone of a body of water with land.

sick building syndrome Refers to ailments such as headaches, nausea, allergic reactions, and other symptoms that are caused by indoor air pollutants in a building where people work or live.

sinkhole A type of land subsidence, taking the form of a large depression in the ground, caused by water withdrawal. A sinkhole occurs when a thin layer of rock overlying an underground cavern collapses.

sinks Environmental reservoirs that receive the throughput of society.

six e's The six potential values of wild biological resources: esthetic (aesthetic), emotional, economic, environmental services, ethical, and evolutionary.

sky glow The phenomenon when the sky at night in a certain area is not dark as it would be naturally, but glows (perhaps in shades of pink, orange, white, yellow, or gray) due to excess light in the local atmosphere. *See* light pollution.

slash and burn agriculture A form of agriculture where trees and other vegetation are cut down and burned in order to clear the land and release nutrients into the soil.

smelting A process in which concentrated ore, such as copper one, is roasted and subjected to high temperatures in a smelting furnace to produce crude metal (for instance, crude copper).

smog Originally referred to a mixture of smoke and fog, but now generally refers to the brown haze of photochemical pollutants found in some urban areas.

social ecology An approach to environmental concerns that attempts to identify social factors that are the underlying causes of current environmental degradation.

social institutions Generally, the institutions or constructs that are important in regulating society, such as familial, political, educational, religious, and economic systems and institutions.

soft laws Legal instruments that are not legally binding, but act more as guides to policy.

soft technology Energy technologies that are generally small scale, relatively inexpensive, and localized.

soil A mixture of weathered rocks and minerals, decayed organic matter, living organisms, air, and water.

soil degradation The damaging or destruction of natural soils; often due to overuse, abuse, and neglect by humans.

soil fertility The ability of the soil to support plant life and associated fauna.

soil horizons Layers of soil that form approximately parallel to the surface of the land; may include the topsoil, subsoil, a layer of partially disintegrated rock, and finally the underlying bedrock.

solar energy Energy derived from the Sun.

solar thermal technology The use of the Sun's energy to heat substances such as water to produce steam that drives a turbine and generates electricity.

solar-hydrogen economy An economy based primarily on solar power (in all its forms, including wind power and hydroelectric power) and using hydrogen as a convenient way to store and transport energy.

solid waste Broadly defined, soil waste includes such items as household garbage, trash, refuse, and rubbish, as well as various solids, semisolids, liquids (such as sludge or liquids in solid containers), and gases (often contained in solid containers, such as gas canisters), that result from mining, agricultural, commercial, and industrial activities.

source reduction As applied to solid waste, reducing the generation of waste in the first place (as opposed to later reusing or recycling waste).

sources The environmental resources (matter and energy) that are taken from nature and used by society.

species Often defined, at least for sexual organisms, as all of the organisms that can interbreed (or potentially interbreed) to produce fertile offspring. Different species are reproductively isolated from one another.

species area curve A curve or graph that shows the number of different species found in a gradually enlarged area of sampling.

species recovery Increasing the population size of a threatened or endangered species so as to ensure that it will not go extinct.

species richness The number of different species that occur in at given area.

species triage The concept that conservation efforts should be directed at those species that are at risk of extinction and can still be saved.

static reserve How long a resource reserve, such as a mineral reserve, will last if demand does not change with time.

stock pollutant Refers to synthetic materials that are not found in nature and tend to be among the most persistent pollutants.

stormwater runoff Runoff of rain and snowmelt from impervious surfaces such as roads, highways, and parking lots that floods and pollutes rivers and streams and carries surface pollutants downstream.

strategic mineral A critical mineral that a particular country must import from areas that are potentially unstable politically, militarily, or socially.

stratosphere The thermal layer of the atmosphere above the troposphere in which temperature increases with altitude. The ozone layer occurs within the stratosphere.

strip-mining A form of surface mining, especially for coal, that is very destructive to the landscape.

structural materials Nonmetallic materials/minerals used in building, such as building stone, sand, gravel, and other components of cement and concrete.

substitutability Finding substitutes for various materials. *See* substitution.

substitution The use of one resource in place of another.

suburban sprawl The spreading of a city's population out into the surrounding countryside, forming suburbs.

superconducting magnetic energy storage (SMES) A system in which superconducting loops or coils would be used to store electrical energy by allowing the current to circulate around a closed loop of nearly zero resistance.

superconductivity Superconductors are substances through which electrons can pass with virtually no friction or resistance, thus allowing almost 100% energy transmission.

Superfund The common name for the federal Hazardous Substance Response Fund that is used for cleanup and related expenses associated with hazardous waste sites on the EPA's National Priorities List.

supertoxin A poisonous substance of such high potency that less than a drop can kill an adult human.

surplus areas (water) Areas that receive more precipitation than is needed by

well-established vegetation, including crops.

sustainability Meeting the needs of today without reducing the quality of life for future generations.

sustainable development Development that focuses on making social, economic, and political progress to satisfy global human needs, desires, aspirations, and potential without damaging the environment; sometimes known as sustainable growth.

sustainable economy An economy that produces wealth and provides jobs for many human generations without degrading the environment.

sustainable growth *See* sustainable development.

sustainable harvesting The sustainable use/harvesting of nuts, fruits, and other products that can be extracted from an ecosystem without causing damage; sometimes known as extractive forestry.

sustainable technology Technology that permits humans to meet their needs with minimum impact on the environment.

swidden techniques A shifting method of agriculture where vegetation is cut down, allowed to dry, and burnt as a way to enrich the soil. Often used as a synonym of slash and burn agriculture (see above). After a few years the land must be allowed to lay fallow for a period before being cultivated again; the purpose is to allow the soil to regenerate between cultivations.

symbiosis Organisms of different species living together; includes such relationships as amensalism, commensalism, mutualism, and parasitism.

synergism With reference to toxicology, when the toxic effects of different substances are multiplied through interaction.

syngas A mixture of hydrogen gas and carbon monoxide; also known as coal gas or town gas.

system A set of components functioning together as a whole.

systems approach The study of phenomena as systems and subsystems, allowing the isolation of various parts or portions of the world so as to focus on those aspects that interact with one another more closely than others. *See* system.

tailings In mining, the residue after high-grade ore is extracted.

taxonomists Practitioners of taxonomy. *See* taxonomy.

taxonomy In biology, the description, classification, and naming of groups of organisms.

teleological tradition According to this way of thinking, every being and object in nature (as well as among human artifacts) has a purpose, function, end, final cause, or utility in the overall natural design and order of the world and universe.

teratogen A substance that adversely affects fetal development, such as when a pregnant woman ingests harmful chemicals.

terawatt (TW) 10^{12} watts.

thermal convection cells The process of thermal convection can occur in fluids (air, water, molten rock, etc) that are exposed to different temperatures. The warmer fluid rises to the surface while the denser, cooler fluid falls to the bottom creating a convection system that can propel processes like plate tectonics and atmospheric currents.

thermal inversion Occurs when a layer of warm air overlies cooler air in the trophosphere (lower atmosphere), thus inverting the usual condition in which air becomes cooler as altitude increases.

thermochemical conversion The heating of biomass in an oxygen-deficient environment to produce substances that can be used as fuels.

thermochemical heat storage The use of reversible chemical reactions to store heat.

thermocline The boundary between warmer surface waters and colder deeper waters in a body of water.

thermodynamics The study of energy and energy conversions.

thermodynamics, first law of States that energy can be neither created nor destroyed, but only transformed.

thermodynamics, second law of States that when energy is transformed from one form to another, it is degraded. This is sometimes known as the law of entropy because energy transformations increase the entropy of a closed system.

Third World debt crisis *See* debt bomb.

thorium An element that contains 90 protons; thorium-232 can be used in a breeder reactor to produce fissionable U-233.

three R's Return, repurify, and reuse (with reference to water).

threshold of exposure The smallest amount of a poison or other toxic substance that is necessary to cause harm.

throughput The movement of materials and energy through society, or the materials and energy so moved. There can be continuous flow from input to throughput to output. *See* input and output.

throwaway society A society based on the one-way flow of matter; products are simply discarded after being used.

tidal power The harnessing of the tides to produce energy in a form that humans can readily utilize.

tokamak A large machine that uses magnetic fields to confine and promote controlled fusion reactions.

topsoil An upper layer of the soil, composed primarily of a mixture of organic matter and mineral matter; it is alive with microscopic and small macroscopic organisms.

tornado Typically consists of a rapidly rotating vortex of air that forms a funnel.

torts *See* common law torts and toxic tort.

total fertility rate *See* fertility rate.

Toxic Substances Control Act (TSCA) A federal statute enacted in 1976 to regulate toxic substances not covered under other laws. The basic goals of the TSCA are to require industry to produce data on environmental effects of chemicals and to prevent harm to humans and the environment, while not creating unnecessary barriers to technology.

toxic Toxic materials or substances are harmful or fatal when consumed by organisms in relatively small amounts.

toxic threshold *See* threshold of exposure.

toxic tort A lawsuit against a manufacturer of a toxic substance for harm caused by that substance.

toxicology The study of the effects of chemicals that are harmful or fatal when consumed by organisms in relatively small amounts.

Toxics Release Inventory (TRI) A report compiled annually by the Environmental Protection Agency on toxics released by U.S. industries based on data reported to the EPA by those industries under the Emergency Planning and Community Right-to-Know Act of 1986.

toxin A poisonous substance produced by an organism; the term is sometimes used to refer to any toxic substance in general.

Tragedy of the Commons The concept that property held in common by many people will generally be overused until it deteriorates or is even destroyed.

transcendentalism A general philosophy espoused in the nineteenth century by such writers as Ralph Waldo Emerson and Henry David Thoreau. In many ways, transcendentalism emphasized the spiritual over the material, especially the materialism of society. This resulted in a concern for the natural environment and a mourning and decrying of the destruction of the natural environment by humans.

transcendentalist An adherent of the philosophy of transcendentalism. *See* transcendentalism.

transgenic crops Genetically transformed crops; crops that have been artificially engineered using bioengineering.

trash Waste that is "dry" (as opposed to liquid or gas) and nonedible, such as newspapers, boxes, cans, containers, and so forth.

troposphere The lowermost thermal layer of the atmosphere, wherein temperatures normally decline with increasing altitude; the layer of the atmosphere in which most weather phenomena take place.

true environmental costs The concept that the cost or value of a resource should include all indirect, as well as direct, values and costs of the resource.

turbine A machine that converts the lateral motion of a liquid or gas into rotational motion.

tsunami A series of water surface waves generated when a large amount of water is displaced, such as by a landslide into the ocean, a meteoritic impact, or an underwater earthquake.

ultraviolet (UV) radiation Relatively high-energy, short-wavelength, electromagnetic radiation (light). UV has wavelengths in the range of 200 to 400 nanometers (one nanometer equals a billionth of a meter).

umbrella species A large, charismatic species (e.g., the Florida panther). When the habitat for such a species is protected, many other species will be protected as well.

unequally distributed costs Environmental costs of a product or service that are not distributed evenly among socioeconomic groups of people, such as when poorer people either pay relatively more or absorb relatively more of the environmental costs (for instance, being forced by economics to live in more polluted areas).

unique species Species that are not closely related to any other living species.

United Nations Framework Convention on Climate Change (UN FCCC) A convention agreed to by many nations at the 1992 Earth Summit. Although the convention does not establish legal obligations or specific target dates, it requires signing countries to use their best efforts to control emissions of greenhouse gases.

uranium A heavy element that contains 92 protons. Fissionable isotopes of uranium can be used as fuel in nuclear reactors and can also be manufactured into bombs.

urbanization The trend toward increasing numbers of people living in cities.

utilitarianism A philosophy whose basic principle is that the overall good should be maximized ("the greatest good for the greatest number").

value The relative worth, utility, or importance of an object or idea. *See also* moral values.

virgin ores or resources Original natural ores or resources that are extracted from the Earth or from "Nature."

volcano A place on the Earth's crust where hot, molten rock (magma) wells up to the surface.

waste-to-energy The burning of municipal solid waste to produce energy.

water table The boundary between the zones of aeration and saturation.

waterlogging The rising of the water table over time, and the soaking of soils, in areas where irrigated land is poorly drained. Water-logging is often associated with salinization.

watt A common unit of power defined as one joule of work or energy per second.

weather Short-term, daily perturbations in the atmospheric/hydrologic cycles.

wetland An area on land that is dominated by water covering the surface and/or waterlogging of the soil for all or much of the year, such as a marsh or a swamp. Wetlands play vital roles in ecosystems by acting as buffer zones and natural filters, removing pollutants and refreshing the water supply. The water saturation influences the soil as well as the flora and fauna that can exist in a particular wetland. Wetlands can be found on every continent except Antarctica.

Wilderness Act of 1964 A federal statute that allowed for the designation of wilderness areas on federal lands that are to be managed so as to retain their primeval character.

Wildlands Project An ambitious plan to designate a network of interconnected wildlife preserves and habitat corridors throughout the United States.

wildscaping Retention of native soil, vegetation, and other natural features when building on the land, rather than the removal of soil, vegetation, and natural features followed by artificial landscaping once building is completed.

wind farm A vast trace of land covered with wind-powered turbines that are used to drive generators that produce electricity.

wind power The harnessing of the wind's energy for human applications.

Wise Use Movement A movement with the avowed goal of nullifying or eliminating most environmentally based laws and regulations. Wise Users generally see such laws as overly expensive, disruptive of free enterprise, and infringing on property rights.

withdrawn water Water that is taken from its source (such as a river, lake, or aquifer); it may be returned to its source after use.

work A force applied to a material object times the distance that the object is moved.

World Trade Organization An international organization founded to reduce trade barriers and settle trade disputes between nations.

xeriscaping Landscaping designed to save water.

xerophytes Plants, such a plants found in deserts, with adaptations that allow them to survive where there is a scarce and unpredictable water supply.

yellowcake Uranium oxide (U_3O_8) or "natural uranium."

zone of aeration A zone where there are pores and cavities in rock, soil, or other materials that are either dry or only partially or temporarily filled with water.

zone of saturation A zone where the pores and cavities within rock, soil, or other materials are completely saturated with water. The upper boundary of the zone is the water table and can shift during wet or dry periods.

Photo Credits

Study 10-1 Courtesy of Dr. David Hyndman, University of New South Wales; 10-6 © haveseen/ShutterStock, Inc.; 10-7 © Photodisc.

Chapter 11. Chapter opener © Eric Gevaert/ShutterStock, Inc.; 11-1 © Steven Blandin/ShutterStock, Inc.; 11-5 Courtesy of Peggy Greb/USDA Agricultural Research Service; Case Study 11-1 © Peter Yates/Science Photo Library/Photo Researchers, Inc.; 11-6 © New Zealand Conservation Department, HO/AP Photos; 11-7 Courtesy of the John Hay Library, Brown University; 11-8 Courtesy of NOAA; 11-9 Courtesy of Mary DeDecker/US Fish and Wildlife Service; 11-10 Courtesy of Lynn Betts/USDA /Natural Resources Conservation Service; 11-12 Courtesy of Robert K. Boone/U.S. Fish and Wildlife Service; Case Study 11-2-1 Courtesy of Claire Dobert/U.S. Fish and Wildlife Service; Case Study 11-2-2 © Anthony Mercieca/Photo Researchers, Inc.; 11-15a Courtesy of John and Karen Hollingsworth/U.S. Fish and Wildlife Service; 11-15b Courtesy of Steve Hillebrand/U.S. Fish and Wildlife Service; 11-16 © PMLD/ShutterStock, Inc.

Chapter 12. Chapter opener Courtesy of NASA/USGS; 12-2 Courtesy of NPS; Case Study 12-1-1 Courtesy of NPS; Case Study 12-1-2 Courtesy of NPS ; 12-4 © Photodisc; 12-5 © Photodisc; 12-6 © Lee Prince/ShutterStock, Inc.; 12-9 © Mighty Sequoia Studio/ShutterStock, Inc.; 12-10a © Terranova International/Science Source/Photo Researchers, Inc.; 12-11 © Photodisc; 12-12 © Photodisc; 12-13a Courtesy of Florida Keys National Marine Sanctuary/NOAA; 12-15 Courtesy of Lynn Betts/USDA Natural Resources Conservation Service; 12-17a Photo provided by the Environmental Protection Agency; 12-17b Photo provided by the Environmental Protection Agency; Case Study 12-2 Courtesy of Ijams Nature Center; Case study 12-3 © Paul Matthew Photography/ShutterStock, Inc.; 12-18 Courtesy of the Natural Resources Council/USDA.

Chapter 13. Chapter opener © Iryna Rasko/ShutterStock, Inc.; 13-1b © Peter Menzel/Photo Researchers, Inc.; 13-2 © REUTERS/Mohammed Salem/Landov; 13-4a Courtesy of USDA; 13.04B_back Courtesy of USDA; 13.05 Courtesy of Image Science and Analysis Laboratory, NASA-Johnson Space Center. "The Gateway to Astronaut Photography of Earth."; 13-6a © Photos.com; 13-6b Courtesy of Lynn Betts/USDA ARS; 13-7 © William Bachman/Photo Researchers, Inc. ; 13-8 Courtesy of Jeff Vanuga/USDA Natural Resources Conservation Service; Case Study 13-1 Photo by Robert Schoch, August 2005; 13-11 Courtesy of Tim McCabe/USDA Natural Resources Conservation Service; 13-12 © Djordje Zoric/ShutterStock, Inc.; 13-13 Courtesy of P. Alejandro Díaz; Case Study 13-2 Courtesy of Scott Bauer/Agricultural Research Service/USDA; 13-14 Courtesy of Tim McCabe/USDA Natural Resources

Conservation Service; 13-17a Courtesy of Gene Alexander/USDA Natural Resources Conservation Service; 13-17b Courtesy of Lynn Betts/USDA ARS; (Page 407) © tommaso delpiano/ShutterStock, Inc.

Chapter 14. Chapter opener © M. Cornelius/ShutterStock, Inc.; 14-2 © John D. Cunningham/Visuals Unlimited; 14-3 © AbleStock; 14-4 © Ishbukar Yallifatar/ShutterStock, Inc.; 14-7a © TAOLMOR/ShutterStock, Inc.; 14-7b © Photodisc; 14-8 Inserts © Photodisc; 14-10 © Al Goldis/AP Photos; Case Study 14-2 © Jeff Greenberg/age fotostock; 14-14a Courtesy of Scott Bauer/USDA ARS; 14-14b Courtesy of Scott Bauer/USDA ARS; 14-14c Courtesy of Scott Bauer/USDA ARS; 14-14d Courtesy of Scott Bauer/USDA ARS; 14-15 © Roy David Farris/Visuals Unlimited; 14-16a © Dinodia Photos/Alamy; 14-16b Courtesy of Tim McCabe/USDA; 14-18a Courtesy of Scott Bauer/USDA ARS; 14-18b Courtesy of Scott Bauer/USDA ARS; 14-19 © Sondeep Shankar/AP Photos.

Chapter 15. Chapter opener © David R. Frazier Photolibrary, Inc./Photo Researchers, Inc.; 15-1 © David S. Addison/Visuals Unlimited; 15-4 © Photodisc; 15-5 Courtesy of Tim McCabe/USDA Natural Resources Conservation Service; 15-6 Courtesy of Tim McCabe/USDA Natural Resources Conservation Service; 15-8 © Alon Brik/ShutterStock, Inc.; 15-9 Courtesy of Ken Winters/U.S. Army Corps of Engineers; 15-12 © Geoffrey Whiting/ShutterStock, Inc.; 15-15 © Liviu Toader/ShutterStock, Inc.; Case Study 15-2 Courtesy of NASA/GSFC/JPL, MISR Team; 15-16 Courtesy of Eric Erbe, Colorization by Christopher Pooley/USDA ARS; 15-17 Courtesy of Jeff Vanuga/USDA Natural Resources Conservation Service.

Chapter 16. Chapter opener © Andrew F. Kazmierski/ShutterStock, Inc.; Case Study 16-1 © AP Photos; 16-3a Courtesy of Edwin P. Ewing, Jr./CDC; 16-3b Courtesy Jacques Descloitres/MODIS Land Rapid Response Team at NASA GSFC.; 16-7 © Dominik Dabrowski/ShutterStock, Inc.; 16-9 © Kenneth Keifer/ShutterStock, Inc.; 16-10 © Ivars Zolnerovics/ShutterStock, Inc.; 16-11a © steve estvanik/ShutterStock, Inc.; 16-13 Courtesy of Landsat 5 Thematic Mapper; C.P. Lo, University of Georgia/NASA; 16-16 © Leonid Nishko/ShutterStock, Inc.; Case Study 16-2 Courtesy of Adrien Lamarre/US Army Corps of Engineers; 16-17 © Photodisc.

Chapter 17. Chapter opener Courtesy of NASA/NASA Ozone Hole Watch; 17-4 © Photodisc; 17-6 Courtesy of NASA; 17-7 © Photodisc; Case Study 17-2 Courtesy of Jack Dykinga/USDA Arrgricultural Research Service; 17-10 Courtesy of NASA/Goddard Space Flight Center - Scientific Visualization Studio, Goddard TV; 17-11 Courtesy of Guiseppe Zibordi/Michael Van Woert, NOAA NESDIS, ORA; 17-12 © JonMilnes/ShutterStock, Inc.; Case Study 17-3 Courtesy of James D.

Gathany/CDC; 17-13 Courtesy of Jacques Descloitres/MODIS Land Rapid Response Team, NASA/GSFC; 17-14 Courtesy of C. Mayhew & R. Simmon (NASA/GSFC), NOAA/NGDC, DMSP Digital Archive.

Chapter 18. Chapter opener © ZUMA Wire Service/Alamy; 18-4 © Ana Gram/ShutterStock, Inc.; 18-5 © Hank Morgan/Science Source/Photo Researchers, Inc.; 18-6 © Peter von Bucher/ShutterStock, Inc.; 18-7 Courtesy of Robert Etzel/US Army Corp of Engineers; Case Study 18-1-1 © Tonis Valing/ShutterStock, Inc.; Case Study 18-1-2 © fotomine/ShutterStock, Inc.; 18-9 © Mark Warford/Greenpeace; 18-11 © Maximilian Stock Ltd./Photo Researchers, Inc.; 18-12 © Anne Gro Bergersen/ShutterStock, Inc.; 18-13 © Jeff Davies/ShutterStock, Inc.; 18-14 © Bernd Wittich/Visuals Unlimited; 18-15 © Sorin Alb/ShutterStock, Inc.; 18-16 © Ljupco Smokovski/ShutterStock, Inc.; 18-17 Courtesy of Harry Weddington/US Army Corp of Engineers; Case Study 18-3 © Behring-Chisholm/Greenpeace; 18-18 © Lisa Bunin/Greenpeace; 18-19 Courtesy of HTRW Cleanup Seattle/US Army Corps of Engineers; 18-20 Courtesy of EXXON VALDEZ Oil Spill Trustee Council/NOAA.

Chapter 19. Chapter opener © Angelo Giampiccolo/ShutterStock, Inc.; 19-2 © Jeffrey Howe/Visuals Unlimited; 19-3a Inserts © Photodisc; 19-3b Courtesy of NASA; 19-5a Courtesy of Lewis W. Hine/U.S. National Archive and Record Administration; 19-5b © REUTERS/Kamal Kishore /Landov; 19-6 © Gary Kazanjian/AP Photos; 19-7 © Ken Graham/Greenpeace; 19-9 © North Wind Picture Archives; Case Study 19-2 Courtesy of Daniel M. Kammen, University of California, Berkeley; 19-12 © guentermanaus/ShutterStock, Inc.; 19-13 © Andrew Davies/Greenpeace; Case Study 19-3 © Beth A. Keiser/AP Photos.

Chapter 20. Chapter opener © SuperStock/Alamy; 20-1 © North Wind Picture Archives; Case Study 20-1-1 Courtesy of J. Schmidt/Yellowstone National Park/NPS; Case Study 20-1-2 Courtesy of Hugh Lumsden/Yellowstone National Park/NPS; 20-2a Courtesy of Library of Congress, Prints & Photographs Division, [reproduction number LC-USZ62-3906]; 20-2b Courtesy of Library of Congress, Prints & Photographs Division [reproduction number LC-DIG-ggbain-06861]; 20-3 © Theodore Roosevelt Collection, Harvard College Library (560.51 1903-1918); 20-4a Courtesy of George E. Marsh Album/NOAA; 20-4b Courtesy of Library of Congress, Prints & Photographs Division [reproduction number LC-USZ62-95653]; 20-5 Courtesy of NASA; 20-6 © Doug Scott/age fotostock; 20-7 © REUTERS/Christian Charisius /Landov; 20-8 © Eduardo DiBaia/AP Photos.

Epilogue. Chapter opener © GAISANG DAWA/Xinhua/Landov.

Index